AF323163

Springer Series on
ATOMIC, OPTICAL, AND PLASMA PHYSICS 29

Springer Series on
# ATOMIC, OPTICAL, AND PLASMA PHYSICS

The Springer Series on Atomic, Optical, and Plasma Physics covers in a comprehensive manner theory and experiment in the entire field of atoms and molecules and their interaction with electromagnetic radiation. Books in the series provide a rich source of new ideas and techniques with wide applications in fields such as chemistry, materials science, astrophysics, surface science, plasma technology, advanced optics, aeronomy, and engineering. Laser physics is a particular connecting theme that has provided much of the continuing impetus for new developments in the field. The purpose of the series is to cover the gap between standard undergraduate textbooks and the research literature with emphasis on the fundamental ideas, methods, techniques, and results in the field.



Series homepage—http://www.springer.de/phys/books/ssaop/

Nils Andersen
Klaus Bartschat

# Polarization, Alignment, and Orientation in Atomic Collisions

Foreword by Joachim Kessler

With 183 Illustrations

Nils Andersen
Niels Bohr Institute
Oersted Laboratory
Universitetsparken 5
DK-2100 Copenhagen
Denmark

Klaus Bartschat
Physics Department
Drake University
Des Moines, IA 50311
USA

Library of Congress Cataloging-in-Publication Data
Polarization, alignment, and orientation in atomic collisions / Nils Andersen, Klaus Bartschat.
     p. cm. — (Atomic, optical, and Plasma Physics ; 29)
    Includes bibliographical references and index.
    ISBN 0-387-98989-7 (hardcover)
    1. Collisions (Nuclear physics)   2. Polarization (Nuclear physics)   3. Nuclear orientation.
I. Andersen, Nils, 1947–  II. Bartschat, Klaus, 1956–  III. Atomic, molecular, and optical
physics (New York, N.Y.) ; 29.
QC794.6.C6 P65 2000
539.7′57—dc21                                                00-027555

Printed on acid-free paper.



Production managed by Jenny Wolkowicki; manufacturing supervised by Erica Bresler.
Camera-ready copy provided by the authors.
Printed and bound by Hamilton Printing Co., Rensselaer, NY.
Printed in the United States of America.

9 8 7 6 5 4 3 2 1

ISBN 0-387-98989-7           SPIN 10750932

Springer-Verlag New York Berlin Heidelberg
*A member of BertelsmannSpringer Science +Business Media GmbH*

# Preface

Since the early work of Étienne Malus, polarization studies have played an increasingly important role in science as a tool for understanding natural phenomena. In the field of atomic collision physics, the polarization of photons and electrons emerging from the collision zone is in many cases intimately related to the dynamical evolution of the collision complex. Therefore, a quantitative understanding of the origin of the observed polarization and the development of appropriate concepts for describing the emission process have been a central goal of theoretical and experimental studies since the early days of collision studies in the 1920s. The idea of a "perfect" or "complete scattering experiment," introduced in the late 1960s, as well as the concept of "alignment" and "orientation" were of crucial importance for the further advancement of the field. The latter allowed for a proper disentanglement of dynamical and geometrical features, and aimed at describing the shape of the collisionally excited atomic electron cloud and its intrinsic currents.

Since then, systematic and coordinated theoretical and experimental investigations have gradually led to a very satisfactory situation concerning the understanding and accurate description of many fundamental electronic and atomic collision processes in nature. Thus, at the end of the 1990s, atomic collision theory has developed into a mature state in which collision-induced polarization phenomena for a few benchmark systems have been measured and predicted with an unprecedented accuracy and degree of detail over a wide range of impact energies. At the same time, these studies often reveal a very detailed and direct picture of the collision dynamics. Hence, the experimental methods, the general framework, and the numerical approaches have now been developed to a level that provides an excellent starting point for the study of even more complex collision problems. Apart from being of high intrinsic interest for the community of atomic physicists, this gain in understanding is also expected to be useful for applications in neighboring fields of science, such as surface, solid state, and nuclear physics, astronomy, and the physics of lasers, plasmas, and planetary atmospheres.

Based on current knowledge, the present book aims at introducing the nonexpert reader to this field without having to go through the individual papers or the comprehensive, technical reviews in the literature. With this in mind, Part I of the book describes polarized light (Chapter 2) and polarized electrons (Chapter 3), experimental and theoretical methods for scattering studies (Chapter 4), the density matrix formalism as a convenient tool for the theory of measurement (Chapter 5), and basic numerical methods (Chapter 6) that are widely used in calculations supporting ongoing experiments. This part of the book is aimed primarily toward graduate students in physics, and has been used as a basis for graduate courses in atomic physics. Therefore, some chapters are supplemented with suggested problems.

Part II presents in a systematic way a series of case studies selected with the aim of illustrating applications of the framework developed in Part I to fundamental atomic collision processes for electronic (Chapter 7) and heavy-particle (Chapter 8) impact. No attempt is being made toward complete coverage, but rather the goal is to "tell a story" concentrating on a few benchmark systems. Chapters 9 and 10 on propensity rules and particle-impact ionization, respectively, address further topics of intense current research. These sections, together with Chapter 11 on selected applications of polarization studies in other areas of physics, indicate the potential of the formalism and its expected impact on future research. Hence, in addition to the graduate student reader mentioned above, this part is intended for active researchers in atomic collision physics who would like to place their own work in a broader perspective, and to researchers in neighboring fields who want to evaluate the potential of applying polarization studies in their own field.

With the form of presentation chosen for this book, an uninitiated reader may be carried to the frontier of the field in a short time and with a relatively modest effort. On the other hand, the approach hides the long, fascinating, and sometimes tortuous development that, with numerous detours, has led to the present situation. Furthermore, it does not always give sufficient credit to the early pioneers on whose joint efforts this presentation rests. In particular, it hides the interesting and often very educational paths along which the critical concepts were developed. In an attempt to remedy this situation, we have selected for Part III some of the early, groundbreaking publications in this field from the period 1925–1976. These papers are marked by an asterisk in the reference sections to in the individual chapters. Owing to space limitations, a few sections from the longer papers have been omitted, and sometimes short initial notes of the research results have been chosen instead of the longer follow-up publications that give more results and sometimes important details. Consequently, the book is accompanied by a CD-ROM that contains in PDF format all the papers of Part III, as well as a few related publications. We have used these papers successfully as reading material in graduate student courses to give a flavor of what really happened when the principal ideas were born. Furthermore, the CD-ROM contains QuickTime movies of theoretically predicted electron charge clouds and currents for some of the cases discussed in Part II. In order to read the PDF files and to play the movies, it is necessary for readers to have the freely available software installed on their computer.

Many people have helped us during the preparation of this book. We are particularly indebted to Gordon Drake for suggesting the book project as a follow-up to our contributions to the "Handbook on Atomic, Molecular, and Optical Physics" and to Maria Taylor and Jenny Wolkowicki for administering the project with Springer-Verlag, to Bill Baylis, Albert Crowe, Markus Drescher, Friedrich Hanne, Ulrich Heinzmann, Joachim Kessler, Mette Machholm, Bill McConkey, Andy Mikosza, Jim Williams, Dehong Yu, and Copenhagen graduate students for comments on early drafts of the manuscript; to Igor Bray, Dmitry Fursa, and Don Madison for providing some of their results in numerical form; to David Loveall for producing the movies; and to Duane Jaecks, Joe Macek, Don Madison, Friedrich Hanne, and Joachim Kessler for letting us borrow precious (sometimes the very last!) reprints of their publications for reproduction. Excellent technical support was provided by Lisbeth Dilling and David Loveall, and the project would not have been possible without financial support from the National Institute of Standards and Technology for the review series on "Alignment and Orientation in Atomic Outer Shells" initiated by Jean W. Gallagher at JILA, as well as continued funding from the Danish Natural Science Research Council and the United States National Science Foundation. Last but not least, we thank our many colleagues who have indirectly contributed to this book through extensive discussions over the years, and our families for their support over the course of the production.

Copenhagen and Des Moines, October 1999

*Nils Andersen*
*Klaus Bartschat*

# Foreword

The field covered by this book has been the focus of broad activities over the past decades. Individual researchers, small groups, and centers of excellence (thus or similarly named) contributed to the enormous progress of our knowledge of polarization, alignment, and orientation in atomic collisions.

A comprehensive overview of these developments has been given in a series of reviews published in Physics Reports by groups of researchers headed by the authors of the present monograph. Since these articles give a detailed and (at the time of their publication) complete survey of the field, I was somewhat surprised when I learned that Andersen and Bartschat were preparing a book on almost the same subjects covered by their earlier reviews. I must admit that I was first a little bit skeptical, wondering what the new message of such an enterprise should be. However, when I went through the manuscript of the book it did not take long to convince me that here a new and extremely useful approach has been developed.

In contrast to the aforementioned reviews, it is the purpose of the present publication to give an *introduction* to this area of modern atomic physics. The time of such a project is well chosen since, from the wealth of separate experimental and theoretical results, a coherent picture has been put together in the past decade. It is therefore quite appropriate to speak of a "relatively mature field" as the authors do. This does not mean, however, that its development will soon level off. I rather believe that the field will continuously grow, thereby shifting from simple atomic systems to molecular and even more complex systems.

A requirement for this growth is, of course, that a number of bright researchers can be enthused for such topics, a task which the present monograph is perfectly suitable to fulfill. Its concept differs from the technical reviews written so far that strive for completeness while giving a well-structured compilation and careful analysis of the existing data, thus aiming at the researcher already familiar with the field. But for the curious scientist trying to get an idea of the matter the material accumulated there may be overwhelming, if not discouraging. The much more compact form of the present introduction is, however, appropriate for becoming, in a reasonable time, familiar with the basic concepts as well as with the latest developments. A feature of the book adding much to its appeal are the links between a variety of processes that seem to be unconnected, since a general framework was established that allows for a superior treatment of different phenomena.

Owing to the wide scope of the field the authors are frequently compelled to present and comment on the facts without being able to give detailed derivations such as those found in textbooks. The reader who wants to get deeper insight into one topic or the other has to resort to other sources. This turns out not to be a problem since the authors assist by providing plenty of references not only to the primary literature, but also to introductory books.

Working through the book, the reader soon will notice a few features that give the subject its particular flavor. First of all it is a fine blend of experimental and theoretical work without rigid borderlines between these two sections of physics. The principal reason for this is that most of the research has been done in coordination between experimentalists and theorists, where the partners of the two sections defined and solved tasks for each other. Personally, I found this to be the most rewarding aspect when working in this field. As a result of this close cooperation, intuitive models were developed so that the reader does not have to put up with results in the form of abstract parameters that cannot be easily visualized. In fact, the "charge cloud movies" provided on the accompanying CD present a delightful example of such a visualization, having become possible through recent advances in data and image processing.

A special feature of the book is that it really goes back to the roots. Part III gives a selection of fundamental papers published between 1925 and 1976 that were crucial for the development of the field. I welcome this idea very much and not only because it is often necessary for gaining a proper perspective to go back to the primary sources. One also gets an impression on how the style of publications has changed over the years. Finding that in this respect the early papers compare favorably with most modern ones, one might be encouraged to return to the old standards. Besides, as a consequence of the information explosion, easy access to early literature is no longer granted in most libraries.

Needless to say, the monograph benefits considerably from the fact that the authors have participated actively in the field over the years with a great amount of fundamental and instrumental contributions. This, together with the effort they had to go through in writing their comprehensive reviews, enabled them to give a complete picture of polarization, alignment, and orientation in atomic collisions. Graduate students and researchers will derive abundant profit from it when studying this subject of central importance.

Münster, December 1999                                                  *Joachim Kessler*

# Contents

## Part III: Selection of Historical Papers (1925–1976)

PART I

# BASIC CONCEPTS

# 1. Introduction

The motivation for the study of polarization, orientation, and alignment phenomena in atomic collisions is summarized. Following a brief historical overview of the pioneering experiments, a classification of the studies according to the level of detail and the corresponding amount of available information is introduced.

## 1.1 Motivation

The study of electronic and atomic collision processes continues to be a very active and rapidly developing field of physics. Many areas of this field of research have now reached a high level of maturity since the early pioneering investigations by Ramsauer's generation. The constant development of new experimental techniques together with the ever increasing power of modern computers has propelled the level of comparison between experiment and theory to a very high degree of sophistication [1.1]. A special role here is played by so-called *complete* or *perfect scattering experiments* [1.2–4], since, under favorable conditions, they probe our understanding of fundamental collision processes at the deepest possible level, the quantum-mechanical scattering amplitudes. Despite the highly complicated techniques employed in such studies, they frequently reward the researcher by displaying in a very simple way the underlying collision dynamics, often allowing for interpretation in terms of intuitive models. A key element in this understanding is the careful analysis of the polarization of the emitted photons, owing to the fact that the polarization properties of the light reveal the orbital motion of the electrons participating in the reaction in a direct way, and thus provide information on the shape and dynamics of the excited electron charge cloud. This area of study has therefore served as a crucial testing ground for atomic scattering theories; methods that have passed this test successfully are now being widely used for a series of practical applications as well. These include the production of atomic data for the modeling of fusion plasmas, where the atomic and molecular processes in the edge region serve as diagnostic tools [1.5–7], as well as data needed for astrophysics, laser physics, and the physics of planetary atmospheres.

## 1.2 Historical Perspective

"If a unidirectional beam of electrons ... is projected through a vapor at very low pressure ... there is reason to suspect that the radiation emitted perpendicular to the electron beam should be partially polarized ..." This is the opening sentence of one of the very first papers on light polarization in atomic collisions [1.8]. During the years 1925–1935, many experimental [1.8–23] and theoretical [1.24–29] papers on the polarization of optical radiation from electron–atom collisions appeared for targets such as He, Ne, Na, and particularly Hg. An early summary was given in the 1933 review by Bethe [1.29].

The quantum theory of electron-impact excitation was pioneered by Oppenheimer [1.24–26] and Penney [1.27,28], the latter summarizing his views in 1932 as follows: "It seems quite certain that first-order cross sections will not do for calculations on polarization ..." [1.28]. The classic 1958 paper of Percival and Seaton [1.30] (see also [1.31]) marks the next major step in understanding polarization phenomena. But it took until the late 1960s before progress in experimental techniques (laser sources, polarized electron beams), theoretical developments (now greatly helped by the access to modern computers), and — last but not least — conceptual understanding made time ripe for further steps forward.[1] Seminal papers by Bederson [1.2-4], Kleinpoppen [1.32], and Macek and Jaecks [1.33] pointed to the perspectives in and techniques for attempting complete experiments, thereby elucidating not only the magnitude, but also the phase of the quantum mechanical scattering amplitudes. A landmark in this development was the review paper by Fano and Macek [1.34] that created a powerful, general framework for future investigations in this field. Shortly after, it was demonstrated by Hertel and Stoll [1.35] that equivalent information could be extracted from experiments using the time-reverse scheme. The idea was conveniently realized with the then newly available laser light sources, an idea soon put on firm theoretical grounds by Macek and Hertel [1.36]. This period also witnessed the first experiments fully exploiting photon coherence and correlation analysis by the groups in Orsay [1.37] and Lincoln [1.38] for heavy-particle collision processes and in Stirling for electron-impact excitation [1.39–41]. In addition, pioneering work at Münster [1.42,43] and New York [1.44] involving spin-polarized electron and atom beams was performed. Part III of this book contains a selection of some of these groundbreaking papers, written during the period 1925–1976.

Impressed by the perspectives in these pioneering studies, many groups all over the world responded to the challenge and contributed significant results over the past 25 years, which have witnessed an explosion of theoretical and experimental data. Following the wise advice of H.W.S. Massey, who

---

[1] In a visionary "Letter to the Editor" in 1955, Ramsey [Phys. Rev. **98** (1955), 1853] was the first to propose the idea of exploring alignment effects as a tool for studying collision phenomena.

stated that "[These experiments] are very important for deepening our understanding of atomic collisions and it is essential that their very complexity should not be allowed to obscure their importance. If, as soon may be the case, experiments of this magnitude are beyond the range of any individual university laboratory, collaborative arrangements should be made on a national or even international basis" [1.45], this development has been a truly international effort, as will become clear from a study of the literature. The series of biennial international workshops on "Polarization and Correlation Phenomena", started in 1981 in Gaithersburg (Maryland, USA), has been instrumental in this aspect.

The development after 1976 has been so voluminous and multifaceted that an attempt at a historical summary is far beyond the limits of this book. Instead we refer to the partial reviews that have appeared during this period for narratives of this progress. (See the list of "Further Readings".) Finally, we note that our presentation is based on the three volumes of the review series "Collisional Alignment and Orientation of Atomic Outer Shells" [1.46–48] and on the review "Complete Experiments in Electron-Atom Collisions" [1.49].

## 1.3 Modern Approaches

In this section, we establish a hierarchy of approaches to collision studies, each level being determined by the symmetry of the experimental geometry.

A generic experiment in this field addresses collisions of two particles, A and B. Here A is an atom, while B may be an electron, atom, or ion, but we exclude more complex collision partners, such as molecules, clusters, or solids. A and B are initially isolated and in quantum states described by $\alpha$ and $\beta$. They are then allowed to interact for some time, after which they separate again in quantum states $\alpha'$ and $\beta'$, respectively. For example, at the expense of some kinetic energy $\Delta E$, A may have ended end up in an excited state $A(\alpha')$ after the collision, while B may or may not be excited. We write this process symbolically as

$$A(\alpha) + B(\beta) \rightarrow A(\alpha') + B(\beta') - \Delta E. \tag{1.1a}$$

Subsequently, as shown in Figure 1.1(a), the excited state $A(\alpha')$ may decay by photon emission,

$$A(\alpha') \rightarrow A(\alpha) + h\nu. \tag{1.1b}$$

A common experimental technique is to record the total number of photons emitted per unit of incoming beam flux, a number that may be converted into a total, angle-integrated cross section $\sigma_{\mathrm{T}}^{\mathrm{Exp}}(v)$ whose value is generally a function of the collision velocity $v$.

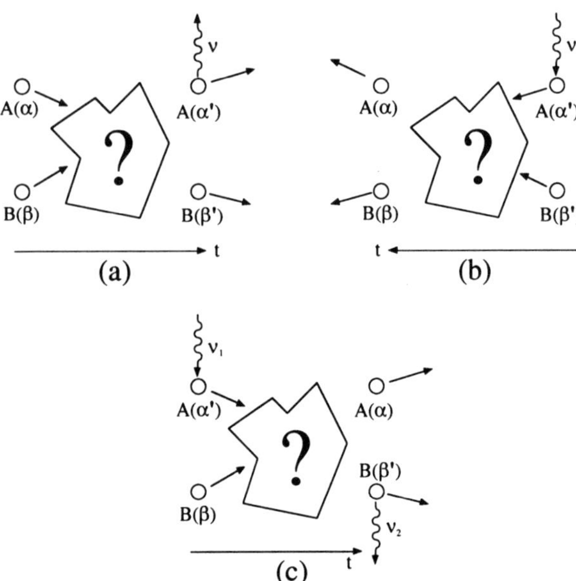



**Fig. 1.1.** (a) A schematic diagram for the collisional excitation of atom $A(\alpha \rightarrow \alpha')$, which subsequently decays by emission of a photon $\nu$. (b) The time-reverse of (a), with initial optical preparation of the atom A into state $\alpha'$ before collisional de-excitation $A(\alpha' \rightarrow \alpha)$ takes place. (c) A combination of (a) and (b), with initial optical preparation of the atom A into state $\alpha'$ before the collision by a photon $\nu_1$ and detection of a photon $\nu_2$ after the collision. In most experiments, one of the collision partners is effectively at rest, so that the incoming momentum vector of the other one defines a preferred axis in the system.

On the theoretical side, several approaches are available, depending on circumstances to be specified in Chapter 6. One method often used for atom–atom and ion–atom collisions involves the selection of an impact parameter $b$ and the solution of the time-dependent Schrödinger equation as particle A moves along its trajectory with velocity $v$. This may be done by selecting a basis in Hilbert space and assuming a model Hamiltonian, based on qualified guesses of the A–B interaction. The final outcome of such a calculation consists of the asymptotic final-state coordinates $(..., a_M, ...)$ in Hilbert space, with the coordinates $a_M = a_M(b, v)$ being complex numbers. These, however, cannot be directly compared to experiment. Further evaluation is necessary to achieve comparison, and the procedure may be summarized as follows:

$$\sigma_{\mathrm{T}}^{\mathrm{Theory}}(v) = \sum_{B^*} \sum_{A^*} \sum_{M} \int_0^\infty 2\pi b \, db \, |a_M(b, v)|^2.$$

$$(5)\,(4)\,(3) \leftarrow (2) \leftarrow (1)$$

(1.2)

The formula indicates the following steps:

(1) The magnitude of the excitation amplitude $a_M$ for the state $\alpha'$ is squared to get the excitation probability.

(2) The probability is integrated over all impact parameters, since this is done in the experiment.[2]

(3) The magnetic sublevels $M$ of $\alpha'$ are summed over.

(4) All channels leading to emission of the photon $\nu$ are summed over, since the level emitting the photon may have been populated directly in the collision process, or by cascade population from a higher lying level.

(5) All channels in which the collision partner B is also excited are summed over.

In addition, it may be necessary to average over various distributions of initial states, depending on the preparation of the beams for A and B before the collision.

Only after these steps, a number $\sigma_{\mathrm{T}}^{\mathrm{Theory}}(v)$ is obtained that can be compared to the experimental $\sigma_{\mathrm{T}}^{\mathrm{Exp}}(v)$. In each of the steps $(1) - (5)$, however, information is lost. The aim is thus to design an experiment, or a set of experiments, that to as large an extent as possible avoids these steps. The ultimate, often unachievable goal is to reach a determination of the full amplitude set $(..., a_M, ...)$ in a so-called *complete* or *perfect scattering experiment* [1.2–4].

The principle in the development of recipes for approaching this goal is to break the symmetry of the experimental setup. We therefore introduce a classification of experimental approaches based on their symmetry:

In a *first-generation* experiment, an integration is performed over the entire solid angle $4\pi$ for the process. The typical parameter determined here is a *total* or *scattering-angle integrated* cross section, which we shall denote by $\sigma_T$.

In a *second-generation* experiment, the *cylindrical* symmetry of the experimental setup is exploited. The photon radiation pattern from the decay of the excited state A($\alpha'$) will generally exhibit a corresponding anisotropy. This is revealed, for example, by the linear polarization of the radiation, as measured in a direction perpendicular to the incident beam axis. This dimensionless quantity is called the *total* or *scattering-angle integrated* polarization, and we shall denote it by $P_T$. From this number information on the distribution of population among the magnetic sublevels may be extracted, i.e., the step (3) is avoided.

Another powerful method in this category is the (angle-)*differential* cross section, which measures the probability for population of the level A($\alpha'$) as function of the scattering angle $\theta$. (This angle is sometimes, but not always, uniquely related to the impact parameter $b$.) For brevity, we shall denote it by $\sigma(\theta)$ in the following, although a more illuminating description would be $d\sigma(\theta)/d\Omega$. If combined with sufficiently powerful angular resolution and

---

[2] The integral $\int b\,db$ assumes cylindrical symmetry of the process; in the most general case, it must be replaced by a surface integral $\iint d^2b$.

energy loss analysis, the integral (2) as well as the summations (3) and (5) above may be decomposed, whereas the operation (1) and the summation (3) remain.

A *third-generation* experiment fully exploits the planar symmetry of the scattering event. In Chapter 4, we shall encounter experiments in which the emitted photon is measured in coincidence with the scattered particle. If the photon detection is combined with a Stokes-vector determination, all operations $(1) - (5)$ can be resolved into their components. Hence, third-generation experiments allow for a very detailed comparison between theory and experiment. In some cases, Quantum Mechanics will actually not allow us to go any further and hence a *complete scattering experiment* has been achieved. If, however, the spin of the collision partners can be prepared before and analyzed after the collision, and the spin plays an active role in the outcome, the level of detail can be pushed even further. At this point, we will not introduce any fourth-order (or higher) classification and defer the discussion to the case studies in Part II of this book.

An alternative experimental approach may be to study the *time-reverse* process, involving optical preparation of atom A before the collision, as shown in Figure 1.1(b):

$$A(\alpha) + h\nu \rightarrow A(\alpha'); \qquad (1.3a)$$

$$A(\alpha') + B(\beta') \rightarrow A(\alpha) + B(\beta) + \Delta E. \qquad (1.3b)$$

In this approach, atom A is prepared optically in an excited state before the collision, and one measures the superelastic cross section for the de-excitation process $A(\alpha' \rightarrow \alpha)$. Although this method yields, in principle, the same information as the technique outlined in Figure 1.1(a) [1.35,36],[3] it may in practice give much better counting statistics. Whenever we shall present results produced by this method, we will use them in discussions of the time-reverse process.

A significant step in further refinement may be achieved in experiments that, in some sense, are their own  time reverse; i.e., they involve optical preparation of atom A before the collision by a photon $\nu_1$ and detection of another photon $\nu_2$, which may come from either A or B (now excited as well), after the collision. The basic scheme is shown in Figure 1.1(c) and can be formulated as follows:

$$A(\alpha) + h\nu_1 \rightarrow A(\alpha'); \qquad (1.4a)$$

$$A(\alpha') + B(\beta) \rightarrow A(\alpha) + B(\beta'); \qquad (1.4b)$$

$$B(\beta') \rightarrow B(\beta) + h\nu_2. \qquad (1.4c)$$

Experiments of this kind are presently emerging, although the background photons from the laser pump beam may present a major obstacle if the two

---

[3] This statement is not exactly true, since, in the optical pumping process, higher order multipoles may be introduced, thereby giving access to additional information.

wavelengths involved are not very well separated. A step in the direction of Figure 1.1(c) is represented by experiments in which, for example, the charge state and the energy loss of the scattered particle is determined for different optical preparations of the target.

# References


1.1 G.W.F. Drake (ed.), *Atomic, Molecular, and Optical Physics Handbook*, AIP Press, Woodbury, New York 1996.
*1.2 B. Bederson, Comm. At. Mol. Phys. **1** (1969) 41.
*1.3 B. Bederson, Comm. At. Mol. Phys. **1** (1969) 65.
*1.4 B. Bederson, Comm. At. Mol. Phys. **2** (1970) 160.
1.5 R.K. Janev (ed.), *Atomic and Molecular Processes in Fusion Edge Plasmas*, Plenum Press, New York 1995.
1.6 C.D. Lin (ed.), *Review of Fundamental Processes and Applications*, World Scientific, Singapore 1993.
1.7 R. Hoekstra, Comments At. Mol. Phys. **30** (1995) 361.
1.8 A. Ellett, P.D. Foote, and F.L. Mohler, Phys. Rev. **27** (1926) 31.
*1.9 W. Kossel and C. Gerthsen, Ann. Phys. (Leipzig) **77** (1925) 273.
*1.10 H.W.B. Skinner, Proc. Roy. Soc. A **112** (1926) 642.
1.11 J.A. Eldridge and H.F. Olson, Phys. Rev. **28** (1926) 1151.
*1.12 H.W.B. Skinner and Appleyard, Proc. Roy. Soc. A **117** (1927) 224.
1.13 R. Quarder, Z. Physik **41** (1927) 674.
*1.14 K. Steiner, Z. Physik **52** (1928) 516.
1.15 A. Ellett and W.A. MacNair, Phys. Rev. **31** (1928) 180.
1.16 W. Elenbaas and M.G. Peteri, Z. Phys. **54** (1929) 236.
1.17 W. Hanle and R. Quarder, Z. Phys. **54** (1929) 819.
1.18 W. Elenbaas, Z. Phys. **59** (1929) 289.
1.19 W. Elenbaas and L.S. Ornstein, Z. Phys. **59** (1929) 306.
1.20 A. Ellett, Phys. Rev. **35** (1930) 588.
1.21 H. Schüler and J.E. Keyston, Z. Phys. **72** (1931) 423.
1.22 L. Larrick and N.P. Heydenburg, Phys. Rev. **39** (1932) 289.
1.23 J.A. Smit, Physica **2** (1935) 104.
*1.24 J.R. Oppenheimer, Proc. Nat. Acad. Sci. **13** (1927) 800.
*1.25 J R. Oppenheimer, Z. Phys. **43** (1927) 27.
*1.26 J.R. Oppenheimer, Phys. Rev. **32** (1928) 361.
1.27 W.G. Penney, Phys. Rev. **39** (1932) 467.
*1.28 W.G. Penney, Proc. Nat. Acad. Sci. **18** (1932) 231.
*1.29 H. Bethe, in *Handbuch der Physik*, Vol. XXIV, Springer, Berlin 1933, p. 508.
*1.30 I.C. Percival and M.J. Seaton, Phil. Trans. Roy. Soc. London A **251** (1958) 113.
1.31 D.R. Flower and M.J. Seaton, Proc. Phys. Soc. London **91** (1967) 59.
1.32 H. Kleinpoppen, in *Physics of One- and Two-Electron Atoms*, North-Holland, Amsterdam 1969, p. 613.
*1.33 J. Macek and D. Jaecks, Phys. Rev. A **4** (1971) 2288.
*1.34 U. Fano and J. Macek, Rev. Mod. Phys. **45** (1973) 553.
1.35 I.V. Hertel and W. Stoll, J. Phys. B **7** (1974) 570; ibid. 583.
*1.36 J. Macek and I.V. Hertel, J. Phys. B **7** (1974) 2173.
*1.37 G. Vassilev, G. Rahmat, J. Slevin, and J. Baudon, Phys. Rev. Lett. **34** (1975) 444.

*1.38 D.H. Jaecks, F.J. Eriksen, W. de Rijk, and J. Macek, Phys. Rev. Lett. **35** (1975) 723.

*1.39 M. Eminyan, K.B. MacAdam, J. Slevin, and H. Kleinpoppen, Phys. Rev. Lett. **31** (1973) 576.

*1.40 M. Eminyan, K.B. MacAdam, J. Slevin, and H. Kleinpoppen, J. Phys. B **12** (1974) 1519.

*1.41 M.C. Standage and H. Kleinpoppen, Phys. Rev. Lett. **36** (1976) 577.

*1.42 K. Jost and J. Kessler, Phys. Rev. Lett. **15** (1965) 575.

*1.43 G.F. Hanne and J. Kessler, Phys. Rev. Lett. **33** (1974) 341.

1.44 K. Rubin, B. Bederson, M. Goldstein, and R.E. Collins, Phys. Rev. **182** (1969) 201.

1.45 H.S.W. Massey in *Fundamental Processes in Energetic Atomic Collisions*, H. Lutz, J.S. Briggs, and H. Kleinpoppen (eds.), Plenum Press, New York 1983, p. 659.

1.46 N. Andersen, J. W. Gallagher, and I.V. Hertel, Phys. Rep. **165** (1988) 1.

1.47 N. Andersen, J.T. Broad, E.E.B. Campbell, J.W. Gallagher, and I.V. Hertel, Phys. Rep. **278** (1997) 107.

1.48 N. Andersen, K. Bartschat, J.T. Broad, and I.V. Hertel, Phys. Rep. **279** (1997) 251.

1.49 N. Andersen and K. Bartschat, Adv. At. Mol. Opt. Phys. **36** (1996) 1.

# 2. Polarized Light

Detailed analysis of light emitted from excited atomic states provides an efficient key to the understanding of many atomic collision processes. The mathematical description of coherent electromagnetic radiation is briefly reviewed, including light polarization, the Stokes-vector parameterization, and the Poincaré sphere. The electric dipole radiation pattern emitted from an excited P state is then discussed and related to the scattering amplitudes for excitation of the atomic state. The connection between light polarization and parameters characterizing the atomic state and the electron charge cloud is discussed in some detail, including experimental geometries for their determination. Finally, effects due to loss of full coherence and nonconservation of atomic reflection symmetry are considered.

## 2.1 Polarization of Coherent Electromagnetic Radiation

### 2.1.1 Maxwell's theory of electromagnetic radiation

According to Maxwell, the following set of equations yields a satisfactory description of electric and magnetic phenomena:

$$\nabla \cdot \boldsymbol{E}(\boldsymbol{r},t) = \rho(\boldsymbol{r},t)/\epsilon \,, \tag{2.1a}$$

$$\nabla \cdot \boldsymbol{B}(\boldsymbol{r},t) = 0 \,, \tag{2.1b}$$

$$\nabla \times \boldsymbol{E}(\boldsymbol{r},t) = -\frac{\partial \boldsymbol{B}(\boldsymbol{r},t)}{\partial t} \,, \tag{2.1c}$$

$$\nabla \times \boldsymbol{B}(\boldsymbol{r},t) = \mu \, \boldsymbol{J}(\boldsymbol{r},t) - \mu\epsilon \frac{\partial \boldsymbol{E}(\boldsymbol{r},t)}{\partial t} \,. \tag{2.1d}$$

Here $\boldsymbol{E}(\boldsymbol{r},t)$ and $\boldsymbol{B}(\boldsymbol{r},t)$ are the time-dependent electric and magnetic fields at the point $\boldsymbol{r}$ in space, $\rho(\boldsymbol{r},t)$ and $\boldsymbol{J}(\boldsymbol{r},t)$ are the charge and current densities, $\epsilon$ is the permittivity, and $\mu$ is the permeability of the material, respectively.

For the purposes of this book, it is sufficient to consider the case where both $\rho$ and $\boldsymbol{J}$ vanish, $\epsilon \approx \epsilon_0$ and $\mu \approx \mu_0$, where $\epsilon_0 = 8.85410^{-12}\mathrm{C}^2/\mathrm{Nm}^2$ and $\mu_0 = 1/(\epsilon_0 c^2)$, with $c$ being the velocity of light in vacuum. These values are sufficiently accurate for air under low and normal pressures. Applying the curl operation to (2.1c), using the identity

$$\nabla \times (\nabla \times \boldsymbol{f}) = \nabla(\nabla \cdot \boldsymbol{f}) - (\nabla \cdot \nabla)\,\boldsymbol{f} \tag{2.2}$$

together with (2.1a), and interchanging the space and time derivatives in (2.1d) yields the well-known wave equation

$$\Box \boldsymbol{E}(\boldsymbol{r},t) = \Big(\frac{\partial^2}{\partial x^2} + \frac{\partial^2}{\partial y^2} + \frac{\partial^2}{\partial z^2} - \frac{1}{c^2}\frac{\partial^2}{\partial t^2}\Big)\boldsymbol{E}(\boldsymbol{r},t) = 0\,. \tag{2.3}$$

Starting with the curl operation on (2.1d), the partial differential equation (2.3) for the electric field $\boldsymbol{E}(\boldsymbol{r},t)$ may also be derived for the magnetic field $\boldsymbol{B}(\boldsymbol{r},t)$. Using standard vector analysis, one can verify that

$$\boldsymbol{E}(\boldsymbol{r},t) = \boldsymbol{E}_0 \exp\left[\mathrm{i}\left(\boldsymbol{k}\cdot\boldsymbol{r} - \omega t\right)\right]\,, \tag{2.4a}$$

$$\boldsymbol{B}(\boldsymbol{r},t) = \boldsymbol{B}_0 \exp\left[\mathrm{i}\left(\boldsymbol{k}\cdot\boldsymbol{r} - \omega t\right)\right]\,, \tag{2.4b}$$

with complex amplitude vectors $\boldsymbol{E}_0$ and $\boldsymbol{B}_0$ are solutions for a "monochromatic plane wave" with a single angular frequency $\omega = 2\pi\nu$, provided that

$$\omega = ck\,. \tag{2.5}$$

Here $k = |\boldsymbol{k}| = 2\pi/\lambda$ is the magnitude of the wavevector $\boldsymbol{k}$, $\lambda = c/\nu$ is the wavelength, and $\nu$ is the frequency.

Throughout this book, we will assume that such a monochromatic plane wave is a sufficiently accurate approximation to the true wavepacket that is obtained by a Fourier integral over different frequencies. Since the wavepacket is generally produced by dipole radiation from excited atoms or molecules, this requires that the light detector be far enough away from the source to neglect the spherical character of the wave. Finally, we note that only the real parts of the solutions represent the physical electric and magnetic fields, respectively.

We are now in a position to show the transversality of the fields as well as the orthogonality of $\boldsymbol{E}$ and $\boldsymbol{B}$. Using the vanishing divergence (see 2.1a,b), we immediately obtain

$$\boldsymbol{E}_0 \cdot \boldsymbol{k} = 0\,, \tag{2.6a}$$

$$\boldsymbol{B}_0 \cdot \boldsymbol{k} = 0\,. \tag{2.6b}$$

Furthermore, (2.1b) yields

$$\boldsymbol{k} \times \boldsymbol{E}_0 = \omega \boldsymbol{B}_0\,. \tag{2.7}$$

Consequently, the triple $(\boldsymbol{E}_0, \boldsymbol{B}_0, \boldsymbol{k})$ forms a mutually orthogonal right-hand system of vectors with $|\boldsymbol{B}_0| = |\boldsymbol{E}_0|/c$.

## 2.1.2 The polarization ellipse

Without loss of generality, we may choose the positive direction of the $z$ axis along the wavevector $\boldsymbol{k}$, i.e.,

$$\boldsymbol{k} = k\,\hat{\boldsymbol{z}};\tag{2.8}$$

see Figure 2.1(a). With a proper choice of zero time, the time dependence of the electric field at a fixed point on the $z$ axis may thus be written in the form

$$E_x(t) = E_{x0}\cos\left(\omega t\right),\tag{2.9a}$$
$$E_y(t) = E_{y0}\cos\left(\omega t - \delta\right),\tag{2.9b}$$

with $E_{x0}, E_{y0} > 0$. Equation (2.9) shows that the electric vector traces out an ellipse in time, as seen in Figure 2.1(b). We notice especially the following values of $\delta = 0, \pi/2, \pi, 3\pi/2$; see Figure 2.1(c). The first and third picture correspond to purely linearly polarized light, while some degree of circular polarization is found in the regions in between, $0 < \delta < \pi$ and $\pi < \delta < 2\pi$. In particular, if $\delta = \pi/2$ or $3\pi/2$ and the amplitudes in the $x$ and $y$ directions are the same, we have fully circularly polarized light. Following the notation of classical optics [2.1], we use the term *right-hand-circular* polarization (RHC) when the electric vector moves *clockwise*, as seen by an observer looking *toward* the light source, as is the case for $\delta = 3\pi/2$. Similarly, the term *left-hand-circular* polarization (LHC) is used when the electric vector moves *counterclockwise*, in the same geometry (e.g., $\delta = \pi/2$).

Since different conventions are used in the literature, we shall briefly elaborate on the connection to concepts such as photon helicity; see Figure 2.1(d). This shows the variation along the $z$ axis of the electric vector at a fixed time, for RHC and LHC polarized light, respectively. Thus RHC corresponds to a corkscrew-shaped surface for the electric vector in space. In this picture, no explicit reference to the direction to the light source is required, since a corkscrew is the same whether one screws in or out. For LHC polarization, the surface turns the opposite way. In the photon picture of light, RHC corresponds to a *negative* photon helicity, i.e., loosely speaking, each photon carries one unit of angular momentum $\hbar$, pointing in the backward direction. Similarly, LHC photons have *positive* helicity, each one carrying one unit of $\hbar$, pointing forward.

## 2.1.3 Parameterization of polarization: Stokes vectors

A complete, quantitative, and operational characterization of the polarization state of a light ray was introduced by Stokes [2.2]. For this purpose we here assume access to perfect linear and circular polarizers and refer to standard textbooks [2.1] for a discussion of their actual realization and procedures for determination of and correction for imperfections. For a given ray of light,

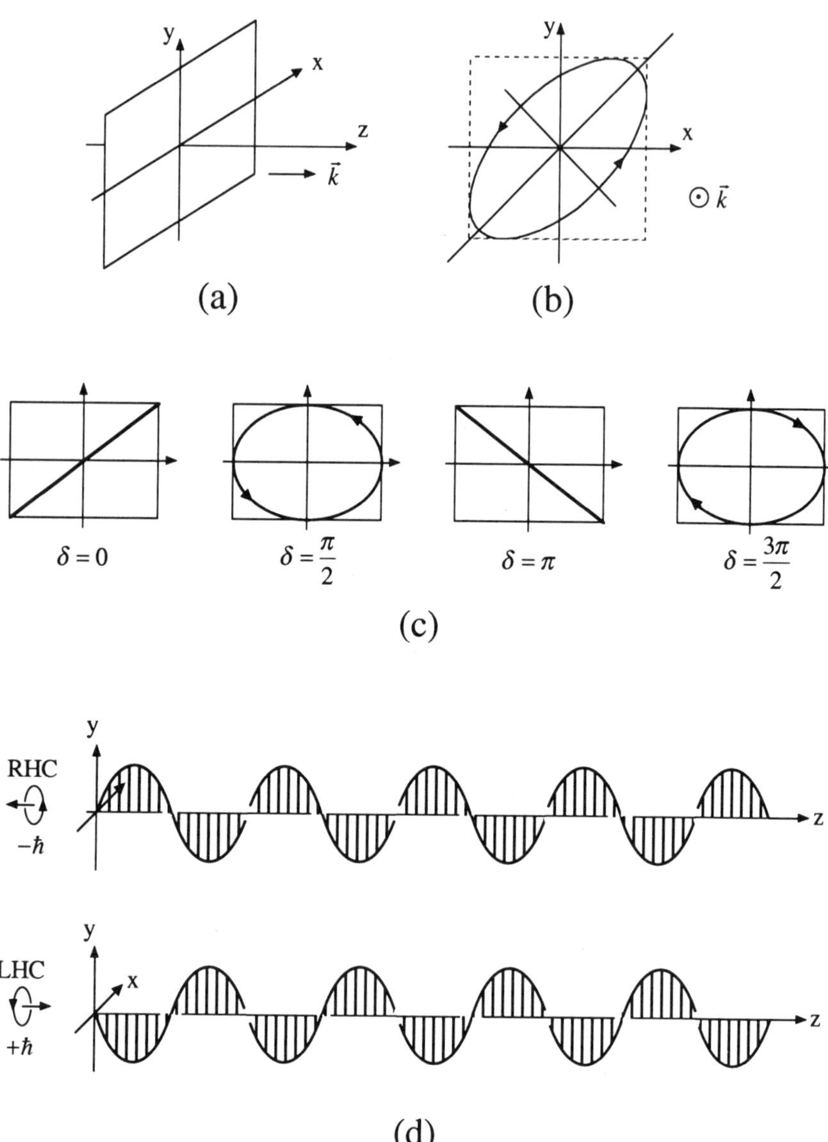



**Fig. 2.1.** (a) Choice of coordinate system, with the positive $z$ axis in the direction of the wavevector $\boldsymbol{k}$ of the light wave. (b) The endpoint of the electric vector $\boldsymbol{E}$ traces out an ellipse in the $xy$ plane. (c) Shape of the polarization ellipse for selected values of the phase angle $\delta$. (d) For fixed time, the electric vector forms a corkscrew surface in space for RHC polarization, while for LHC polarization the surface turns the other way. The corresponding helicities in the photon picture are also indicated.

in our case specified by (2.9), four Stokes parameters $(s_0, s_1, s_2, s_3)$ may be determined from intensity measurements of the light transmitted through polarizers set in a systematic way:

$(i)$   We denote by $s_0$ the total intensity, proportional to the time average of the squared amplitudes of the electric vector components, i.e.,

$$s_0 = \tfrac{1}{2}(E_{xo}^2 + E_{yo}^2), \tag{2.10a}$$

where the factor $\tfrac{1}{2}$ originates from averaging the term $\cos^2(\omega t)$ over time. This intensity is also equal to the sum of the two intensities transmitted through a polarizer set at any two mutually orthogonal transmission directions.

$(ii)$   We denote by $s_1$ the difference in intensities transmitted through a linear polarizer with its transmission direction parallel to the $x$ axis and $y$ axis, respectively, i.e.,

$$s_1 = \tfrac{1}{2}(E_{xo}^2 - E_{yo}^2). \tag{2.10b}$$

$(iii)$ We denote by $s_2$ the difference in intensities transmitted through a linear polarizer with its transmission direction set at angles of 45° and $-45°$, respectively, with respect to the $x$ axis. A little algebra (Exercise 2.1) shows that

$$s_2 = E_{xo}E_{yo}\cos\delta. \tag{2.10c}$$

$(iv)$  Finally, we denote by $s_3$ the difference in intensities transmitted through a circular polarizer with its transmission set for RHC and LHC polarizations, respectively. Again, a little algebra (Exercise 2.1) shows that

$$s_3 = -E_{xo}E_{yo}\sin\delta. \tag{2.10d}$$

We note from (2.10) the relation

$$s_1^2 + s_2^2 + s_3^2 = s_0^2. \tag{2.11}$$

In cases where only the polarization state and not the total intensity is of interest, the dimensionless Stokes vector $\boldsymbol{P} = (P_1, P_2, P_3)$ with components defined by

$$P_i = s_i/s_0; \quad i = 1, 2, 3 \tag{2.12}$$

is useful. According to (2.11), $\boldsymbol{P}$ is a unit vector; i.e.,

$$P_1^2 + P_2^2 + P_3^2 = 1. \tag{2.13}$$

Other notations may be found in the literature, in particular [2.3]

$$I = s_0, \; M = s_1, \; C = s_2, \; S = s_3, \tag{2.14a}$$

and [2.4,5]

$$\eta_1 = P_2, \; \eta_2 = -P_3, \; \eta_3 = P_1. \tag{2.14b}$$

### 2.1.4 The principal frame

It is of interest to discuss how the Stokes parameters come out in a coordinate frame rotated by an angle $\phi$ with respect to the $xy$ frame. Of the four Stokes parameters, $s_0$ and $s_3$ are frame independent, while $s_1$ and $s_2$ are not. Owing to the relation (2.11), the parameter $s_\ell > 0$, defined by

$$s_\ell^2 = s_1^2 + s_2^2 \,, \tag{2.15}$$

is also invariant under a frame rotation. It turns out (see Exercise 2.2) that

$$s_1(\phi) = \phantom{-}s_1 \cos 2\phi + s_2 \sin 2\phi \,, \tag{2.16a}$$
$$s_2(\phi) = -s_1 \sin 2\phi + s_2 \cos 2\phi \,. \tag{2.16b}$$

Thus,

$$\begin{pmatrix} s_0(\phi) \\ s_1(\phi) \\ s_2(\phi) \\ s_3(\phi) \end{pmatrix} = \begin{pmatrix} 1 & 0 & 0 & 0 \\ 0 & \cos 2\phi & \sin 2\phi & 0 \\ 0 & -\sin 2\phi & \cos 2\phi & 0 \\ 0 & 0 & 0 & 1 \end{pmatrix} \begin{pmatrix} s_0 \\ s_1 \\ s_2 \\ s_3 \end{pmatrix} . \tag{2.17}$$

We can now define the principal frame $x'y'$ for the polarization ellipse, i.e., the frame with the $x'$ axis along the major axis of the ellipse, forming an angle $\gamma$ with respect to the $x$ axis; see Figure 2.1(b). In this frame

$$s_1(\gamma) = s_\ell \,, \tag{2.18a}$$
$$s_2(\gamma) = 0 \,. \tag{2.18b}$$

The angle $\gamma$, defining the principal frame for the polarization ellipse, may thus be determined from the two equations

$$s_1 = s_\ell \cos 2\gamma \,, \tag{2.19a}$$
$$s_2 = s_\ell \sin 2\gamma \,, \tag{2.19b}$$

or, in a single equation, from

$$s_1 + i s_2 = s_\ell e^{2i\gamma} \,. \tag{2.20}$$

### 2.1.5 The Poincaré sphere

A useful graphical representation of light polarization was introduced by Poincaré [2.6]. In this picture, the vector $(P_1, P_2, P_3)$ is visualized as a point on the unit sphere, see Figure 2.2. Two orthogonal polarizations correspond to opposite points on the unit sphere. For example, the North (South) Pole corresponds to RHC (LHC) polarized light. Furthermore, the equator describes linearly polarized light with $\gamma$ angle varying with geographical longitude, as shown. Approaching the North Pole from the Equator along a meridian corresponds to increasing the degree of circular polarization while keeping the $\gamma$ angle fixed. For a further discussion of the properties and uses of the Poincaré sphere, see, e.g., [2.7].

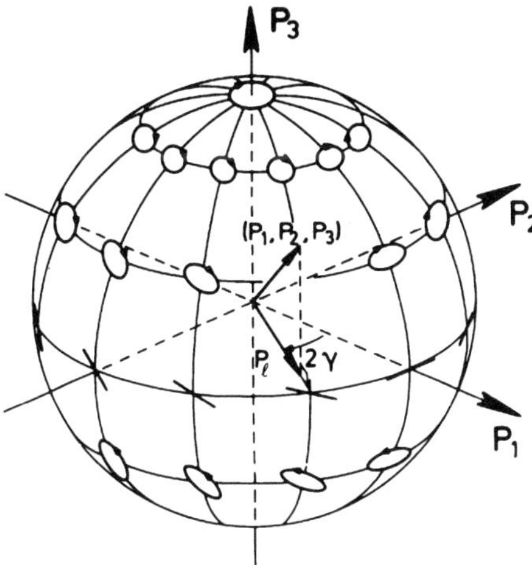


**Fig. 2.2.** The Poincaré sphere. The Stokes vector $(P_1, P_2, P_3)$ characterizing the polarization ellipse is mapped onto a unit sphere.

## 2.2 Electric Dipole Radiation from Atomic Transitions

### 2.2.1 Coordinate frames, scattering amplitudes, and Stokes parameters

In this section we explore the connection between an excited atomic state and the electric dipole radiation pattern, as it is emitted when the state decays to a lower-lying state. For the sake of simplicity, we first neglect possible effects of electron and nuclear spin (fine and hyperfine structure) and thus assume that the *orbital* angular-momentum quantum numbers $(LM_L)$ of the $|nLM_L\rangle$ states involved in the transition completely characterize the radiation pattern.

For symmetry reasons, the radiation pattern from an S state $(L = M_L = 0)$ that decays to a P state $(L = 1, M_L = -1, 0, +1)$ must be isotropic as long as the magnetic quantum number of the final state is not detected. Hence, the first nontrivial case is the radiation pattern from a P state, which for simplicity we assume will decay to an S state. Simplifying the picture even further, we first analyze the case in which the P-state wavefunction has positive reflection symmetry with respect to a plane. This is a very common case in atomic collisions where the P state is produced by collisional excitation from an S state, also exhibiting positive reflection symmetry with respect to the scattering plane, by a force that conserves reflection symmetry, such as the Coulomb interaction. (See Exercise 2.3 for further discussion.)

The electric dipole radiation pattern is determined by the angular parts of the wavefunctions only. Again, since conventions differ in the literature, we here state explicitly our notation for spherical harmonics $Y_{LM_L}(\theta, \phi)$ used

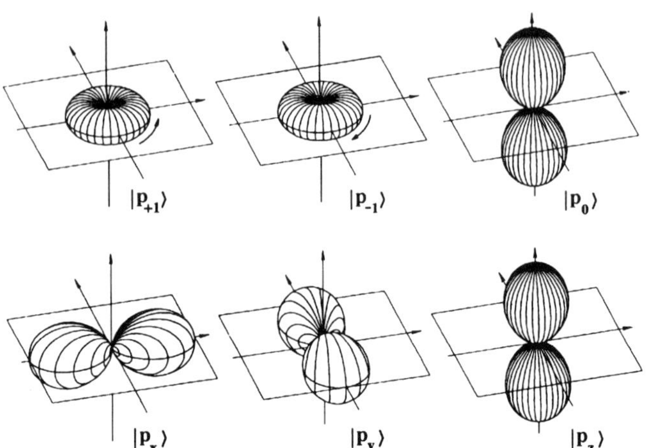



**Fig. 2.3.** Two sets of P state basis functions are commonly used. The upper panel shows the probability distributions for the helicity or "atomic physics" basis $(|p_{+1}\rangle, |p_0\rangle, |p_{-1}\rangle)$ of (2.23). The lower panel shows the real or "molecular physics" basis $(|p_x\rangle, |p_y\rangle, |p_z\rangle)$ of (2.24).

in the helicity or "atomic physics" basis of ket vectors $(|p_{+1}\rangle, |p_0\rangle, |p_{-1}\rangle)$ and the real-valued functions used in the p-orbital or "molecular physics" basis of ket vectors $(|p_x\rangle, |p_y\rangle, |p_z\rangle)$.

The corresponding probability densities, i.e., the absolute squares of the wavefunctions, are shown in Figure 2.3. Omitting the common radial part of the atomic basis functions, we obtain the following dependence on polar angles $(\theta, \phi)$:

$$\langle \hat{\boldsymbol{r}} | p_{+1} \rangle = -\sqrt{\frac{3}{8\pi}} \sin\theta\, e^{i\phi}, \tag{2.21a}$$

$$\langle \hat{\boldsymbol{r}} | p_{0} \rangle = \sqrt{\frac{3}{4\pi}} \cos\theta, \tag{2.21b}$$

$$\langle \hat{\boldsymbol{r}} | p_{-1} \rangle = \sqrt{\frac{3}{8\pi}} \sin\theta\, e^{-i\phi}, \tag{2.21c}$$

and

$$\langle \hat{\boldsymbol{r}} | p_{x} \rangle = \sqrt{\frac{3}{4\pi}} \sin\theta \cos\phi, \tag{2.22a}$$

$$\langle \hat{\boldsymbol{r}} | p_{y} \rangle = \sqrt{\frac{3}{4\pi}} \sin\theta \sin\phi, \tag{2.22b}$$

$$\langle \hat{\boldsymbol{r}} | p_{z} \rangle = \sqrt{\frac{3}{4\pi}} \cos\theta. \tag{2.22c}$$

The transformation formulas from one frame to the other are

$$|p_{+1}\rangle = -\tfrac{1}{\sqrt{2}}(|p_x\rangle + \mathrm{i}\,|p_y\rangle)\,, \tag{2.23a}$$

$$|p_0\rangle = |p_z\rangle\,, \tag{2.23b}$$

$$|p_{-1}\rangle = \tfrac{1}{\sqrt{2}}(|p_x\rangle - \mathrm{i}\,|p_y\rangle)\,, \tag{2.23c}$$

and

$$|p_x\rangle = -\tfrac{1}{\sqrt{2}}(|p_{+1}\rangle - |p_{-1}\rangle)\,, \tag{2.24a}$$

$$|p_y\rangle = \tfrac{\mathrm{i}}{\sqrt{2}}(|p_{+1}\rangle + |p_{-1}\rangle)\,, \tag{2.24b}$$

$$|p_z\rangle = |p_0\rangle\,. \tag{2.24c}$$

An orbital's reflection symmetry with respect to the $xy$ plane is determined by its behavior under the mirror operation $\theta \to \pi - \theta$. From (2.21–24) we thus see that the orbital $|p_0\rangle = |p_z\rangle$ has negative reflection symmetry, whereas the other ones have positive reflection symmetry.

Two coordinate frames are particularly useful for the discussion of atomic scattering events, cf. Figure 2.4(a). One frame, the so-called *collision frame*, labeled with superscript "$c$," has the axis of quantization $z^c$ parallel to the vector $\boldsymbol{k}_{in}$ of the incoming beam and $y^c$ parallel to $\boldsymbol{k}_{in} \times \boldsymbol{k}_{out}$. The other frame, called the *natural frame* and labeled with superscript "$n$," has $x^n$ parallel to $\boldsymbol{k}_{in}$ and the axis of quantization $z^n$ parallel to $\boldsymbol{k}_{in} \times \boldsymbol{k}_{out}$. Hence, $x^n = z^c$, $y^n = x^c$, and $z^n = y^c$. Though the outcomes of a specific collision event in the two coordinate frames must be identical, it turns out that in many cases the algebra becomes simpler and the physics more transparent if the description is performed in the natural frame. (For the *atomic coordinate frame*, see Exercise 2.9.) Since both frames are used in the current literature, we now introduce the fundamental equations in both formulations. The P-state wavefunction $|\psi_{\mathrm{P}}\rangle$ may be expressed as

$$|\psi_{\mathrm{P}}\rangle = f^c_{+1}|p^c_{+1}\rangle + f^c_0|p^c_0\rangle + f^c_{-1}|p^c_{-1}\rangle \tag{2.25a}$$

$$= f^c_0|p^c_z\rangle - \sqrt{2}f^c_{+1}|p^c_x\rangle \tag{2.25b}$$

$$= f^n_{+1}|p^n_{+1}\rangle + f^n_{-1}|p^n_{-1}\rangle \tag{2.25c}$$

$$= -\tfrac{1}{\sqrt{2}}(f^n_{+1} - f^n_{-1})|p^n_x\rangle - \tfrac{\mathrm{i}}{\sqrt{2}}(f^n_{+1} + f^n_{-1})|p^n_y\rangle\,. \tag{2.25d}$$

Here we have used (2.23) and the fact (Exercise 2.3) that the assumption of positive reflection symmetry imposes the conditions

$$f^n_0 = 0\,, \tag{2.26a}$$

$$f^c_{-1} = -f^c_{+1}\,. \tag{2.26b}$$

The coefficients $f_{M_L}$ are the complex probability amplitudes for the individual substates of the three-dimensional subspace spanned by the basis vectors. They are uniquely defined except for a common, arbitrary phasefactor. In this chapter we simply take them for granted. Ways to evaluate them numerically as scattering amplitudes for the collision event of Figure 2.4(a) will be summarized in Chapter 6.

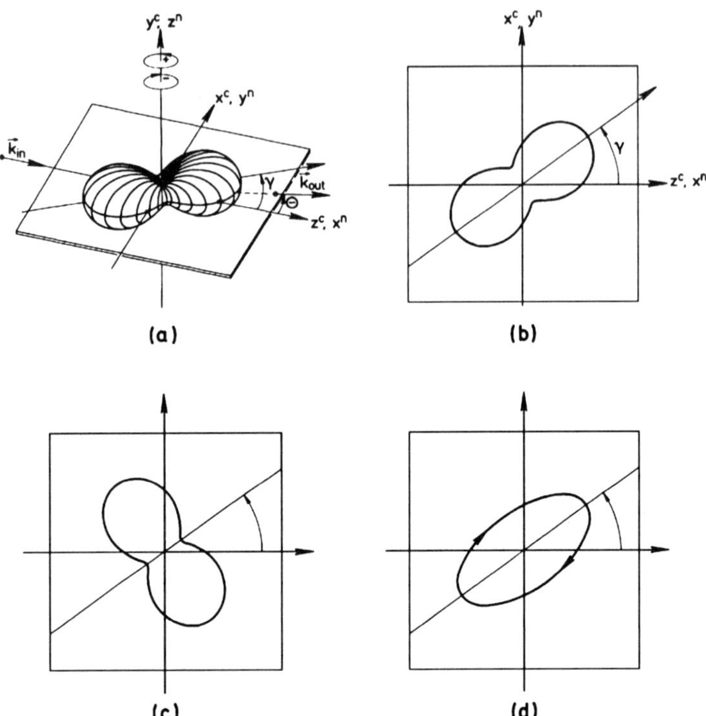



**Fig. 2.4.** (a) A side view of the angular part of an excited P-state electron charge cloud having positive reflection symmetry with respect to the $x^c z^c = x^n y^n$ scattering plane. (b) A cut through the electron charge cloud in the scattering plane. (c) The angular distribution of photons emitted in the scattering plane. (d) The polarization ellipse for light observed along the direction $+z^n = +y^c$. The case shown has parameters $L_\perp = -0.8\hbar$ and $\gamma = 35°$. See text for the definition of the orientation parameter $L_\perp$ and the alignment angle $\gamma$.

Since $|p_z^c\rangle = |p_x^n\rangle$ and $|p_x^c\rangle = |p_y^n\rangle$, we note from (2.25) the following transformation formulas, valid for excitation of P states with positive reflection symmetry:

$$f_0^c = -\tfrac{1}{2}(f_{+1}^n - f_{-1}^n), \tag{2.27a}$$

$$f_{+1}^c = \tfrac{i}{2}(f_{+1}^n + f_{-1}^n), \tag{2.27b}$$

and their inverse

$$f_{+1}^n = -\tfrac{1}{\sqrt{2}}f_0^c - i f_{+1}^c, \tag{2.28a}$$

$$f_{-1}^n = \tfrac{1}{\sqrt{2}}f_0^c - i f_{+1}^c. \tag{2.28c}$$

We also read from (2.25) the transformation formulas from the helicity basis, (2.21), to the real-values basis, (2.22). In the collision coordinate frame one obtains

$$f_x^c = -\sqrt{2}f_{+1}^c \,, \tag{2.29a}$$

$$f_y^c = 0 \,, \tag{2.29b}$$

$$f_z^c = f_0^c \,. \tag{2.29c}$$

Thus, $f_{+1}^c$ and $f_0^c$ are almost the amplitudes in the basis $(|p_x^c\rangle, |p_y^c\rangle, |p_z^c\rangle)$. In the natural coordinate frame one obtains

$$f_x^n = -\tfrac{1}{\sqrt{2}}(f_{+1}^n - if_{-1}^n) \,, \tag{2.30a}$$

$$f_y^n = -\tfrac{i}{\sqrt{2}}(f_{+1}^n + if_{-1}^n) \,, \tag{2.30b}$$

$$f_z^n = 0 \,. \tag{2.30c}$$

More useful transformation formulas and general recipes for their derivation are discussed in Chapter 5.

The electric dipole radiation pattern emitted in the $P \to S$ transition is now readily derived, using the close analogy between the results of the quantum theory of radiation and those from the classical theory of dipole radiation. Equations (2.25b) and (2.25d) correspond to two coherently excited classical dipoles perpendicular to each other with the strengths and phase relations described by the coefficients shown. Defining polarizer angles with reference to the incident beam direction, the Stokes parameters for light emitted in the direction $z^n = y^c$ perpendicular to the scattering plane are thus, except for a common proportionality factor, given by (Exercise 2.4):

$$s_0 = |f_0^c|^2 + 2|f_{+1}^c|^2 \,, \tag{2.31a}$$

$$s_1 = |f_0^c|^2 - 2|f_{+1}^c|^2 \,, \tag{2.31b}$$

$$s_2 = -2\sqrt{2}|f_0^c||f_{+1}^c| \cos \chi \,, \tag{2.31c}$$

$$s_3 = 2\sqrt{2}|f_0^c||f_{+1}^c| \sin \chi \,, \tag{2.31d}$$

in terms of the collision frame parameters. Here $\chi$ is the angle from $f_0^c$ to $f_{+1}^c$ in the complex plane,

$$\chi = \arg(f_{+1}^c f_0^{c\,*}) \,, \tag{2.32}$$

where the asterisk denotes the complex-conjugate quantity. Alternatively, the parameters are given by (Exercise 2.4)

$$s_0 = |f_{+1}^n|^2 + |f_{-1}^n|^2 \,, \tag{2.33a}$$

$$s_1 = -2|f_{+1}^n||f_{-1}^n| \cos \delta \,, \tag{2.33b}$$

$$s_2 = 2|f_{+1}^n||f_{-1}^n| \sin \delta \,, \tag{2.33c}$$

$$s_3 = |f_{-1}^n|^2 - |f_{+1}^n|^2 \,, \tag{2.33d}$$

in terms of the natural frame parameters for the helicity basis, where $\delta$ is the angle from $f_{-1}^n$ to $f_{+1}^n$ in the complex plane; i.e.,

$$\delta = \arg(f_{+1}^n f_{-1}^{n\,*}) \,. \tag{2.34}$$

In several cases, particularly for low-energy collisions between heavy particles, the scattering amplitudes in the real-valued basis in the natural coordinate frame $(|p_x^n\rangle, |p_y^n\rangle, |p_z^n\rangle)$ are of interest (Exercise 2.4). In this context, the alternative basis labels $(|\sigma\rangle, |\pi^+\rangle, |\pi^-\rangle)$ may be found in the literature.

### 2.2.2 Atomic state parameters, electron charge clouds, and their experimental determination

For experimental investigation and comparison with theoretical predictions, convenient ways of parameterizing an atomic P state have been introduced. Since the state is completely determined by two complex scattering amplitudes, three real numbers are required for a full characterization (keeping the common, arbitrary phase factor in mind). It is convenient to select these as one cross section and two dimensionless quantities, one relative size and one relative phase.

A natural choice for a cross section is the quantity

$$\sigma = |f_0^c|^2 + 2|f_{+1}^c|^2 \tag{2.35a}$$
$$= |f_{+1}^n|^2 + |f_{-1}^n|^2 , \tag{2.35b}$$

i.e., proportional to $s_0$, the intensity of light emitted perpendicular to the collision plane.

In the collision frame, the traditional choice of dimensionless parameters has been $(\lambda, \chi)$ [2.8], with the relative size $\lambda$ defined as

$$\lambda = \frac{|f_0^c|^2}{\sigma} \tag{2.36}$$

and the relative phase $\chi$ given by (2.32). With these definitions, the dimensionless Stokes-vector components (2.12) are, according to (2.31) (Exercise 2.5),

$$P_1 = 2\lambda - 1 , \tag{2.37a}$$
$$P_2 = -2\sqrt{\lambda(1-\lambda)} \cos \chi , \tag{2.37b}$$
$$P_3 = 2\sqrt{\lambda(1-\lambda)} \sin \chi . \tag{2.37c}$$

In the natural frame, the traditional choice of dimensionless parameters has been $(L_\perp, \gamma)$ [2.9] . Here the relative size is the orientation parameter $L_\perp$, the expectation value of the $z^n$-component of the transferred orbital angular momentum, i.e.,

$$\frac{\langle \psi_P | \boldsymbol{L} | \psi_P \rangle}{\langle \psi_P | \psi_P \rangle} = (0, 0, L_\perp) , \tag{2.38a}$$

or

$$L_\perp = \frac{|f_{+1}^n|^2 - |f_{-1}^n|^2}{|f_{+1}^n|^2 + |f_{-1}^n|^2} . \tag{2.38b}$$

The alignment angle $\gamma$ of the polarization ellipse is related to the relative phase angle $\delta$ of the scattering amplitudes (2.34) through (Exercise 2.5)

$$2\gamma = \pi - \delta \,. \tag{2.39}$$

The degree of linear polarization $P_\ell$ is defined by

$$P_\ell \equiv s_\ell/s_0 = \sqrt{P_1^2 + P_2^2} = \sqrt{1 - P_3^2} = \sqrt{1 - L_\perp^2} \,. \tag{2.40}$$

and is evidently frame independent.

With these definitions, the dimensionless Stokes-vector components (2.12) are obtained from (2.33) (Exercise 2.5)

$$P_1 = P_\ell \cos 2\gamma \,, \tag{2.41a}$$
$$P_2 = P_\ell \sin 2\gamma \,, \tag{2.41b}$$
$$P_3 = -L_\perp \,. \tag{2.41c}$$

The equation analogous to (2.20) reads

$$P_1 + iP_2 = P_\ell e^{2i\gamma} \,. \tag{2.42}$$

We thus conclude that the full set of parameters characterizing the P-state in the simple case with positive reflection symmetry may be determined by a Stokes-vector analysis in the direction perpendicular to the scattering plane. This technique, involving a full determination of the polarization ellipse, is often called *coherence analysis* and may be realized experimentally by using, e.g., scattered-particle–polarized-photon coincidence techniques. These will be further described in Chapter 4.

The wavefunction (2.25) can be written as (Exercise 2.6)

$$|\psi_{\mathrm{P}}\rangle = \sqrt{\frac{\sigma}{2}} \left[ \sqrt{1 - P_3}\, |p_{+1}^n\rangle - \frac{1}{P_\ell}(P_1 + iP_2)\sqrt{1 + P_3}\, |p_{-1}^n\rangle \right] \tag{2.43a}$$

$$= \sqrt{\frac{\sigma}{2}(1 + L_\perp)}\, |p_{+1}^n\rangle - e^{2i\gamma} \sqrt{\frac{\sigma}{2}(1 - L_\perp)}\, |p_{-1}^n\rangle \,. \tag{2.43b}$$

It is instructive to evaluate the angular part of the P-state electron charge cloud. The result is (Exercise 2.6)

$$|\psi_{\mathrm{P}}|^2(\theta, \phi) = \sigma\tfrac{1}{2}[1 + P_1 \cos 2\phi + P_2 \sin 2\phi] \sin^2 \theta \tag{2.44a}$$

$$= \sigma\tfrac{1}{2}[1 + P_\ell \cos 2(\phi - \gamma)] \sin^2 \theta \,. \tag{2.44b}$$

We may thus associate relative length ($\ell$), width ($w$), and height ($h$) parameters with the electron cloud, the length being related to the major symmetry axis of the cloud in the collision plane, the width to its minor axis, and the height to the density in the perpendicular direction. We see from (2.44) that

$$\ell = \tfrac{1}{2}(1 + P_\ell) \,, \tag{2.45a}$$
$$w = \tfrac{1}{2}(1 - P_\ell) \,, \tag{2.45b}$$
$$h = 0 \,, \tag{2.45c}$$

with the normalization

$$\ell + w + h = 1 \, . \tag{2.45d}$$

Note the relation

$$P_\ell = \frac{\ell - w}{\ell + w} \, , \tag{2.46}$$

which justifies the name *shape parameter* for $P_\ell$.

Equation (2.44) has been used to generate Figure 2.4(a) with $P_\ell = 0.6$ and $\gamma = 35°$. Figure 2.4(b) shows a cut through the electron charge cloud in the $xy$ plane,

$$|\psi_{\mathrm{P}}|^2(\theta = 90°, \phi) = \tfrac{1}{2}\left[1 + P_\ell \cos 2(\phi - \gamma)\right] \tag{2.47a}$$
$$= \tfrac{1}{2}\left[1 + P_1 \cos 2\phi + P_2 \sin 2\phi\right] \, . \tag{2.47b}$$

Owing to the characteristic properties of electric dipole radiation, this shape can, except for a rotation by $90°$, be recorded by a rotatable photon detector tracing the intensity distribution $I(\phi)$ in the direction $\phi$ in the collision plane, Figure 2.4(c). Hence,

$$I(\phi) = \tfrac{1}{2}\left[1 + P_\ell \cos 2(\phi - \gamma - \tfrac{\pi}{2})\right] \tag{2.48a}$$
$$= \tfrac{1}{2}\left[1 - P_\ell \cos 2(\phi - \gamma)\right] \tag{2.48b}$$
$$= \tfrac{1}{2}\left[1 - P_1 \cos 2\phi - P_2 \sin 2\phi\right] \, . \tag{2.48c}$$

This technique, involving a determination of the shape of the dipole radiation pattern, is called *correlation analysis* and may be realized experimentally by using scattered-particle–emitted-photon coincidence techniques. The radiation emitted in the $xy$ plane will be linearly polarized in this plane, but the detection system need not be sensitive to the polarization. The method has been widely used at wavelengths for which (circular) polarizers are only available with low efficiency, such as the resonance transitions for the rare gases. Correlation analysis may thus extract the parameters $(P_1, P_2)$ or $(P_\ell, \gamma)$, but the sign of $L_\perp$ is left undetermined. The sense of rotation of the electric vector $\boldsymbol{E}$ of the polarization ellipse, see Figure 2.4(d), is not accessible to detection by this technique.

So far we have neglected the effects of fine and hyperfine structure. Though this is often an excellent approximation during the collision, these internal forces may have ample time to modify the shape and dynamics of the electron charge cloud after the collision, but before the optical decay leading to photon emission takes place. However, using standard techniques (see, e.g., [2.4]) one can evaluate these effects if the atomic lifetime and the fine and hyperfine coupling constants are known. In this section, we briefly explain the physical effects and postpone the general, quantitative treatment to Chapter 5.

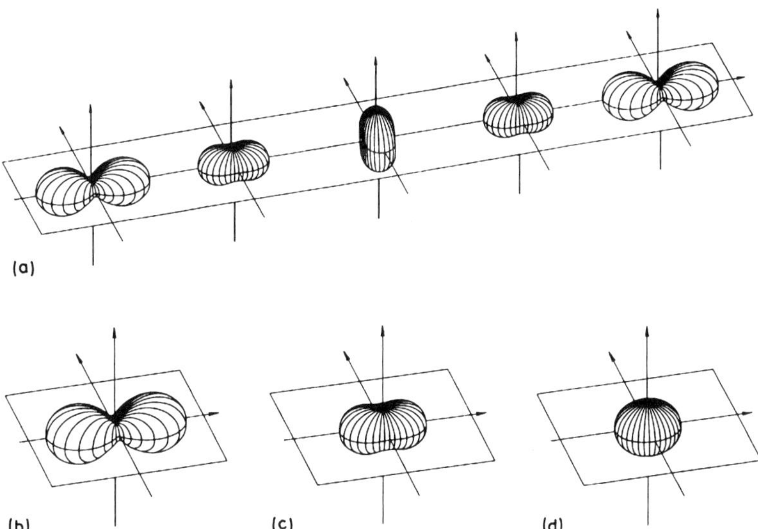

**Fig. 2.5.** (a) The time evolution of the shape of the electron charge cloud for an atomic P state with shape parameter $P_\ell = 0.6$ under the influence of spin-orbit coupling for an electron spin $s = \frac{1}{2}$. The five pictures of the quantum beat sequence cover a full precession period and correspond to $\omega_{\mathrm{fs}} t = 0, \pi/2, \pi, 3\pi/2$, and $2\pi$, respectively. The sequence repeats itself until the optical decay. The lower row shows the depolarizing influence on the time-averaged charge cloud by (b) → (c) an electron spin $s = \frac{1}{2}$, and (c) → (d) a nuclear spin $I = \frac{3}{2}$.

Figure 2.5(a) shows the time evolution of the electron charge cloud depicted in Figure 2.4(a) under the influence of an electron spin of $s = \frac{1}{2}$. The quantum beats  are followed through a full period $T$ of the spin precession time ($T = 2\pi/\omega_{\mathrm{fs}}$, with $\hbar\omega_{\mathrm{fs}} = \Delta E_{\mathrm{fs}}$ equal to the fine structure splitting) and repeat themselves until the optical decay takes place. The depolarizing influence on the time-averaged charge cloud is illustrated by the sequence (b) → (c) of Figure 2.5. The problem can be treated analytically using the tools discussed in Chapter 5. (See also [2.4] and Appendix B of [2.10].) The analysis shows that the corresponding effect on the Stokes vector measured in the $z^n$ direction is a reduction in linear polarization, whereas the circular polarization is unaffected. In detail,

$$P_1 \to \frac{3}{7} P_1 , \tag{2.49a}$$

$$P_2 \to \frac{3}{7} P_2 , \tag{2.49b}$$

$$P_3 \to P_3 . \tag{2.49c}$$

It is thus tempting to introduce a so-called *reduced Stokes vector* $\bar{\boldsymbol{P}}$ with components

$$\bar{P}_1 = \frac{7}{3}P_1 \, , \tag{2.50a}$$

$$\bar{P}_2 = \frac{7}{3}P_2 \, , \tag{2.50b}$$

$$\bar{P}_3 = \phantom{\frac{7}{3}}P_3 \, . \tag{2.50c}$$

to recover the Stokes vector of the *nascent* electron charge cloud. Although the *measured* degree of polarization $P = |\boldsymbol{P}|$ is generally less than unity, the equation $|\bar{\boldsymbol{P}}| = 1$ holds for the *reduced* degree of polarization. The sequence (c) $\rightarrow$ (d) of Figure 2.5 shows the additional depolarizing effect of a nuclear spin of $I = \frac{3}{2}$, as seen, for example, in sodium.

The example above demonstrates the possibility of defining a *reduced* Stokes vector for P $\rightarrow$ S transitions [2.11] of the form

$$\bar{P}_1 = c_2 P_1 \, , \tag{2.51a}$$
$$\bar{P}_2 = c_2 P_2 \, , \tag{2.51b}$$
$$\bar{P}_3 = c_1 P_3 \, , \tag{2.51c}$$

that corrects for post-collisional fine-structure and hyperfine-structure effects and thus recovers the Stokes vector of the nascent charge cloud, with the degree of polarization being unity.

### 2.2.3 The incoherent case with conservation of atomic reflection symmetry

Reduction in the degree of polarization may be caused by the experimental setup being unable to distinguish among several in principle distinguishable channels, which accordingly will be summed over. An example is impact excitation of sodium by an unpolarized electron beam. Without any knowledge or control of the electron spin, the experiment will add singlet and triplet channels for the total (target + projectile) spin incoherently, and the resulting Stokes parameter will be the weighted sum of the two contributions. An adequate description of such effects is elaborate and will be outlined in Chapter 5; examples will be encountered in Part II.

In the present context we restrict ourselves to the observation that the resulting Stokes vector is generally no longer a unit vector; i.e., we now have the inequality $P \equiv |\boldsymbol{P}| \leq 1$ for the degree of polarization. Some authors also introduce the degree of coherence $\mu$ [2.1]. However, since $\mu$ depends on the coordinate frame and does not convey any new independent information compared to $P$, we shall discard it here.

The degree of linear polarization $P_\ell$ may now be selected as an independent (and coordinate-frame independent) parameter. We can thus extract a total of *three* independent, dimensionless parameters from the radiation pattern, namely the set $(L_\perp, \gamma, P_\ell)$.

## 2.2.4 The incoherent case without conservation of atomic reflection symmetry

We end this chapter by discussing the case where the assumption of conservation of reflection symmetry of the atomic charge cloud with respect to the collision plane breaks down. This may occur, for example, in electron-impact excitation of heavy atoms. Here, the $|p_0^n\rangle$ atomic state may be populated if the electron spin flips, for example, due to spin-orbit coupling. The relative population of the $|p_0^n\rangle$ state is a measure of the relative spin-flip cross section [2.12], and it can be significant for atoms such as heavy rare gases, or mercury, that we shall encounter in Part II. For these atoms, $LS$-coupling may no longer be a good description for the excited state. Here we shall consider the case where the total angular momentum of the excited state is $J = 1$ and the fine structure is completely resolved. Since this state radiates like a set of classical oscillators completely analogous to a $^1P_1$ state, we shall maintain our previous description, simply replacing the term electron charge cloud density by oscillator density. Also, we keep the traditional notation $L_\perp$, although $J_\perp$ might seem more appropriate.

We keep the notation from the previous part of this chapter, but modified where necessary with an upper index "+" to indicate parameters related to the part associated with positive reflection symmetry. We thus get

$$L_\perp^+ = -P_3 = \frac{|f_{+1}^n|^2 - |f_{-1}^n|^2}{|f_{+1}^n|^2 + |f_{-1}^n|^2}, \tag{2.52a}$$

$$P_\ell^+ e^{2i\gamma} = P_1 + iP_2, \tag{2.52b}$$

$$P^+ \equiv \sqrt{P_1^2 + P_2^2 + P_3^2} = \sqrt{P_\ell^{+\,2} + L_\perp^{+\,2}}. \tag{2.52c}$$

The population of the state (2.21b) with negative reflection symmetry may be described by a probability amplitude $f_0^n$, and the relative height $h$ of the classical oscillator density will thus be

$$h = \frac{|f_0^n|^2}{\sigma}, \tag{2.53a}$$

now with

$$\sigma = |f_{+1}^n|^2 + |f_0^n|^2 + |f_{-1}^n|^2. \tag{2.53b}$$

In this case, the oscillator density $D$ for the excited atomic state can be written as (Exercise 2.7)

$$D(\theta, \phi) = \sigma\{(1 - h)\tfrac{1}{2}[1 + P_\ell^+ \cos 2(\phi - \gamma)] \sin^2 \theta + h \cos^2 \theta\}. \tag{2.54}$$

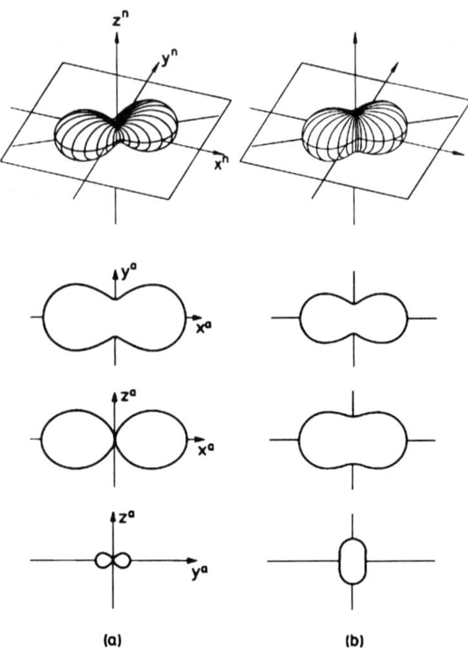

**Fig. 2.6.** The top row shows classical oscillator densities with linear polarization $P_\ell = 0.6$ and alignment angle $\gamma = 35°$. The height parameter is $h = 0$ (a) and 0.25 (b), respectively. Shown below are cuts along the principal planes of the densities.

Figure 2.6 shows an example of the effect of a finite height by comparing density plots for $h = 0$ and 0.25, respectively.

The relations (2.45a–c) are replaced by

$$\ell = (1 - h)\tfrac{1}{2}(1 + P_\ell) , \tag{2.55a}$$

$$w = (1 - h)\tfrac{1}{2}(1 - P_\ell) , \tag{2.55b}$$

$$h > 0 , \tag{2.55c}$$

with the normalization condition (2.45d) being preserved as

$$\ell + w + h = 1 . \tag{2.55d}$$

The relation (2.46) still holds; i.e.,

$$P_\ell^+ = \frac{\ell - w}{\ell + w} . \tag{2.56}$$

In the collision plane, the classical oscillator corresponding to (2.21b) or (2.22c) emits radiation that is polarized perpendicular to this plane. A straightforward experimental approach to use for determination of its relative strength is thus to determine the Stokes vector corresponding to a direction *in* the collision plane perpendicular to the beam, i.e., along the $y^n$ direction. This will have the form $(P_4, 0, 0)$, the two last components being zero for symmetry reasons. The parameter $h$ may then be determined from the equation (Exercise 2.8)

$$ h = \frac{(1 + P_1)(1 - P_4)}{4 - (1 - P_1)(1 - P_4)} . \tag{2.57}$$

In the general case of a $J = 1 \to J_f = 0$ transition, therefore, *four* parameters, conveniently selected as $(L_\perp^+, \gamma, P_\ell^+, h)$ can be extracted from the radiation pattern. Observation from *two* directions are required for their determination. Alternatively, the parameter $h$ may be derived from a correlation analysis by an additional measurement involving a direction out of the collision plane. (See Section 3.3.1 of [2.10] for a quantitative evaluation of this approach.)

### 2.2.5 Summary of parameterization for P-state excitation

Table **2.1** Summary of cases of increasing complexity, and the orientation and alignment parameters necessary for characterization. The table is valid for unpolarized beams only.

| *Example case* | e–He | e–Na | e–Hg |
|---|---|---|---|
| Forces | Coulomb | + exchange | + spin-orbit |
| Representation | wavefunction | density matrix | density matrix |
| Reflection symmetry | + | + | +, − |
| *Parameters* | | | |
| Angular momentum | $L_\perp$ | $L_\perp$ | $L_\perp^+$ |
| Alignment angle | $\gamma$ | $\gamma$ | $\gamma$ |
| Linear polarization | $P_\ell$ | $P_\ell$ | $P_\ell^+$ |
| Degree of polarization | $P = 1$ | $P \leq 1$ | $P^+ \leq 1$ |
| Height | $h = 0$ | $h = 0$ | $h \geq 0$ |
| Independent parameters | 2 | 3 | 4 |
| Observation directions | 1 | 1 | 2 |

The results of the preceding sections are summarized in Table 2.1. The table gives the three cases of increasing complexity, summarizing their reflection symmetry with respect to the scattering plane, the maximum number of independent alignment and orientation parameters, their definitions, and the number of observation directions necessary for a full characterization of the radiation pattern. The top line lists some collision systems that belong to the three categories, respectively, together with the forces active during the excitation process.

The discussion may be generalized further to states with $L > 1$. Already the case of D state excitation becomes considerably more involved, but the theory and notation can be developed as a natural extension of the framework outlined above. A number of theoretical and experimental papers have addressed this problem, which requires triple-coincidence techniques for full elucidation [2.13,14]. At the time of writing, collisional excitation of F states or states with even higher angular momentum has, except for a few pioneering studies [2.15-17], been addressed at the level of total, or scattering-angle integrated, cross sections and polarizations — concepts to be discussed systematically in the following chapter.

## Exercises

1.  Derive the four equations (2.10a–d) from (2.9) and the definitions of the Stokes vectors.
2.  Show Equations (2.16) for the Stokes parameters in a rotated frame.
3.  Consider the conservation of positive reflection symmetry.
    a)  What are the reflection symmetries of the three P-state components in the "atomic physics basis" (2.21) and in the "molecular physics basis" (2.22)?
    b)  Prove that conservation of positive reflection symmetry for $S \rightarrow P$ excitation leads to (2.26).
    c)  Generalize (a) to the basis sets for D states.
4.  Generate expressions for the Stokes parameters of excited P states.
    a)  Derive (2.31) for the collision coordinate frame.
    b)  Derive (2.33) for the atomic physics basis in the natural coordinate frame.
    c)  Derive expressions for the Stokes parameters using the natural coordinate frame and the molecular physics basis $(|p_x^n\rangle, |p_y^n\rangle, |p_z^n\rangle)$, and compare to (2.31).
5.  Express the Stokes parameters in terms of the P-state parameters.
    a)  Derive (2.37) for the collision coordinate frame.
    b)  Derive (2.41) and (2.42) for the natural coordinate frame in the atomic physics basis.
6.  Prove Equation (2.43) for the P-state wavefunction in terms of the Stokes-vector components.
7.  Discuss the loss of positive reflection symmetry for the atomic $J = 1$ state. In (2.54), the two parts of the wavefunction with positive and negative reflection symmetry have been added incoherently. Why is this procedure valid?
8.  Derive the relation (2.57) for the height parameter $h$ by evaluating the linear polarizations $P_1$ and $P_4$ for emission from three orthogonal classical oscillators aligned along the $x, y, z$ axes.

9.  The atomic coordinate frame $(x', y', z')$ is obtained from the natural frame by rotation through the angle $\gamma$ around the $z = z'$ axis. Hence, the $x'y'$ coordinate frame is the principal frame of the polarization ellipse.

a) Translate the formulas (2.43) and (2.44) into atomic-frame parameters using $(i)$ the helicity basis $(|p_{+1}\rangle, |p_0\rangle, |p_{-1}\rangle)$ and $(ii)$ the real-valued basis $(|p_x\rangle, |p_y\rangle, |p_z\rangle)$.

b) Establish transformation formulas that relate scattering amplitudes in the atomic frame to those in the collision frame and the natural frame.

# References


2.1 M. Born and E. Wolf, *Principles of Optics* (4th edition), Pergamon Press, New York 1970.

2.2 G.G. Stokes, Trans. Cambr. Phil. Soc. **9** (1852) 399. Reprinted in: *Mathematical and Physical Papers*, Vol. III, Cambridge University Press 1901, p. 233.

2.3 E.B. Brown, *Modern Optics*, Reinhold Publ. Corp., New York 1965.

2.4 K. Blum, *Density Matrix Theory and Applications* (2nd edition), Plenum, New York 1996.

2.5 J. Kessler, *Polarized Electrons* (2nd edition), Springer, Heidelberg 1985.

2.6 H. Poincaré, *Théorie Mathematique de la Lumière* (Chapter 12), G. Carré, Paris 1889.

2.7 W.A. Shurcliff, *Polarized Light: Production and Use*, Harvard University Press, Harvard 1962.

*2.8 M. Eminyan, K.B. MacAdam, J. Slevin, and H. Kleinpoppen, J. Phys. B **7** (1974) 1519.

2.9 N. Andersen, I.V. Hertel, and H. Kleinpoppen, J. Phys. B **17** (1984) L901.

2.10 N. Andersen, J.W. Gallagher, and I.V. Hertel, Physics Reports **165** (1985) 1.

2.11 N. Andersen, T. Andersen, C.L. Cocke, and E.H. Pedersen, J. Phys. B **12** (1979) 2541.

2.12 N. Andersen, K. Bartschat, and G.F. Hanne, J. Phys. B **28** (1995) L29.

2.13 N. Andersen and K. Bartschat, J. Phys. B **30** (1997) 5071.

2.14 A.G. Mikosza, J.F. Williams, and J.B. Wang, Phys. Rev. Lett. **79** (1997) 3375.

2.15 D. Cvejanovic and A. Crowe, Phys. Rev. Lett. **80** (1998) 3033.

2.16 D.V. Fursa and I. Bray, Phys. Rev. A **59** (1999) 1297.

2.17 J.B. Wang, C.M. Maloney, and D.H. Madison, J. Phys. B **32** (1999) 1067.

# 3. Polarized Electrons

The physics of spin-polarized electrons has been described in detail by
Kessler [3.1]. A brief review is presented in this chapter, including the most
important equations together with a basic experimental setup to measure the
spin polarization of an electron beam.

## 3.1 The Dirac Equation

The Dirac equation for a free particle with linear momentum $\boldsymbol{p}$,

$$i\hbar\frac{\partial\Psi}{\partial t} = (c\boldsymbol{\alpha}\cdot\boldsymbol{p} + \beta mc^2)\,\Psi ,\tag{3.1}$$

allows for four-component spinor solutions

$$\Psi = \begin{pmatrix} \psi_1 \\ \psi_2 \\ \psi_3 \\ \psi_4 \end{pmatrix}.\tag{3.2}$$

In (3.1),

$$\alpha_x = \begin{pmatrix} 0 & \sigma_x \\ \sigma_x & 0 \end{pmatrix},\ \alpha_y = \begin{pmatrix} 0 & \sigma_y \\ \sigma_y & 0 \end{pmatrix},\ \alpha_z = \begin{pmatrix} 0 & \sigma_z \\ \sigma_z & 0 \end{pmatrix},\ \beta = \begin{pmatrix} 1 & 0 \\ 0 & -1 \end{pmatrix}\tag{3.3}$$

are $4 \times 4$ matrices constructed from the Pauli spin matrices

$$\sigma_x = \begin{pmatrix} 0 & 1 \\ 1 & 0 \end{pmatrix},\qquad \sigma_y = \begin{pmatrix} 0 & -i \\ i & 0 \end{pmatrix},\qquad \sigma_z = \begin{pmatrix} 1 & 0 \\ 0 & -1 \end{pmatrix}\tag{3.4}$$

and the $2 \times 2$ unit matrix $\mathbf{1}$.

In the rest frame, the solutions of (3.2) correspond to electrons with spin
"up" ($\psi_1$) or "down" ($\psi_2$) and positrons with spin "up" ($\psi_3$) or "down"
($\psi_4$), respectively, where "up" and "down" are defined with respect to the
quantization ($z$) axis. Note that the spin is contained in the formalism as

an intrinsic property of spin-$\frac{1}{2}$ particles (and their antiparticles). Mathematically, the spin property corresponds to an angular momentum of magnitude $\sqrt{\frac{3}{4}}\,\hbar$ with possible projections $\pm\frac{1}{2}\hbar$ along the quantization axis.

In the following, we are only interested in the electron spin component of the wavefunction, particularly in the description of a spin-polarized electron beam. Before we can define this polarization, we note that the general spin state of an electron can be written as the linear combination

$$|\chi\rangle = a_+|+\tfrac{1}{2}\rangle + a_-|-\tfrac{1}{2}\rangle\,, \tag{3.5}$$

where $a_+$ and $a_-$ are complex numbers and

$$|+\tfrac{1}{2}\rangle \equiv \begin{pmatrix} 1 \\ 0 \end{pmatrix}, \qquad |-\tfrac{1}{2}\rangle \equiv \begin{pmatrix} 0 \\ 1 \end{pmatrix}, \tag{3.6}$$

are orthogonal and normalized two-component spinors describing spin up or down with respect to the quantization axis. The normalization condition,

$$\langle \chi | \chi \rangle \equiv 1\,, \tag{3.7}$$

requires that

$$|a_+|^2 + |a_-|^2 = 1\,. \tag{3.8}$$

## 3.2 Pure Spin States: State-Vector Description

Suppose we have a beam of electrons whose spin states are all described by the same vector (3.5). In this case, the beam is said to be completely polarized with polarization vector $\boldsymbol{P} = (P_x, P_y, P_z)$, whose components are obtained as the expectation values of the Pauli spin matrices, i.e.,

$$P_i = \langle \chi | \sigma_i | \chi \rangle; \qquad i = x, y, z. \tag{3.9}$$

It follows (see Problem 3.1) that the degree of polarization, i.e., the magnitude of the polarization vector, is unity in this case, i.e.,

$$P = \sqrt{P_x^2 + P_y^2 + P_z^2} = 1\,. \tag{3.10}$$

Furthermore, it can be shown (see Problem 3.2) that the spinor with components $a_+ = \cos\theta/2$ and $a_- = e^{i\phi}\sin\theta/2$ corresponds to

$$\begin{pmatrix} P_x \\ P_y \\ P_z \end{pmatrix} = \begin{pmatrix} \sin\theta\cos\phi \\ \sin\theta\sin\phi \\ \cos\theta \end{pmatrix}, \tag{3.11}$$

where $(\theta, \phi)$ are the polar angles of the polarization vector.

## 3.3 Mixed Spin States: Density-Matrix Description

In practice, an electron beam is only partially polarized; i.e., the degree of polarization $P$ is less than unity. This situation is illustrated in Figure 3.1, where, on average, eight out of ten electrons have their spin component aligned along the positive $z$ direction, while the other two have it aligned along the negative $z$ direction. In this case, the polarization component along the $z$ direction is defined as

$$P_z = \frac{N_\uparrow - N_\downarrow}{N_\uparrow + N_\downarrow}, \tag{3.12}$$

where $N_\uparrow$ ($N_\downarrow$) is the number of electrons in the beam with spin up (down) in this direction. Furthermore, we assume that the average spin projections along the $x$ and $y$ directions vanish; i.e., $P_x = P_y = 0$.

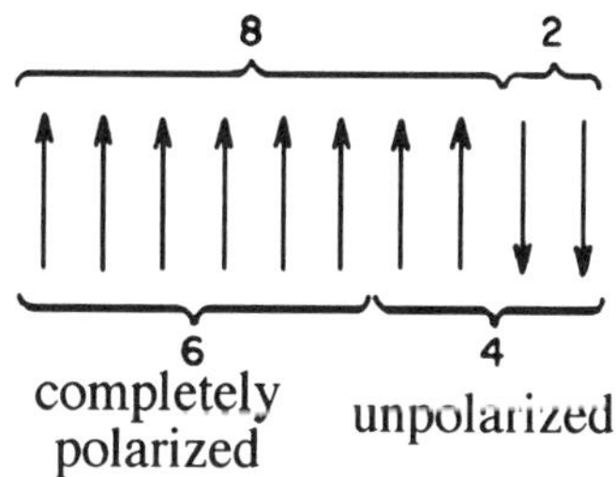


**Fig. 3.1.** Definition of a partially spin-polarized beam (from Kessler [3.1]).

In this case the spin distribution in the electron beam can no longer be described by a single state vector; instead, a density matrix description must be used (see also Chapter 2). In the particular example of Figure 3.1, we can represent the result ($P_z = +60\%$) as an *incoherent* superposition of two beams, each polarized completely along the positive or negative $z$ direction, but with relative weights $w_\pm$, respectively. In other words,

$$P_z = w_+ \langle +\tfrac{1}{2} | \sigma_z | + \tfrac{1}{2} \rangle + w_- \langle -\tfrac{1}{2} | \sigma_z | - \tfrac{1}{2} \rangle, \tag{3.13}$$

where $w_+ = 0.8$ and $w_- = 0.2$ in Figure 3.1.

There are two important points that should be noted here:

($i$)  A *coherent* superposition of the two beams would produce a different result (for example, a completely polarized beam along a different direction).

($ii$)  More generally, the beam could have been considered as a superposition of several beams, each completely polarized along a given direction and entering with a relative weight $w_i = N_i/N$, where $N_i$

is the number of electrons in beam "$i$" and $N = \sum_i N_i$ is the total number of electrons in the beams.

The latter situation can be compactly expressed in terms of the spin density matrix. It describes the statistical mixture of several states through the density operator

$$\rho = \sum_i w_i |\Psi_i\rangle\langle\Psi_i|, \tag{3.14}$$

where the weight $w_i$ is the probability for finding the system in the pure quantum state $|\Psi_i\rangle$. Note that this density operator contains the maximum available information about a physical system [3.2]. Consequently, it can be used to calculate expectation values for any operator $\mathcal{A}$ that represents a physical observable. In general,

$$\langle\mathcal{A}\rangle = \mathrm{tr}\{\mathcal{A}\rho\}/\mathrm{tr}\{\rho\}, \tag{3.15}$$

where $\mathrm{tr}\{\rho\}$ in the denominator of (3.15) ensures the correct result for an arbitrary trace of the density matrix.

For our case of interest, the spin polarization of an electron beam, the density matrix is a hermitian $2 \times 2$ matrix. If we choose the standard normalization of trace unity, it can be parameterized by three independent quantities, namely the three components of the polarization vector [3.2]. One finds [3.1]

$$\rho = \frac{1}{2}\begin{pmatrix} 1 + P_z & P_x - iP_y \\ P_x + iP_y & 1 - P_z \end{pmatrix}. \tag{3.16}$$

Using (3.15) with $\mathrm{tr}\{\rho\} = 1$, the individual components of the density matrix can be obtained according to

$$P_i = \mathrm{tr}\{\sigma_i\rho\}; \quad i = 1, 2, 3. \tag{3.17}$$

Also (see Problem 3.3), a pure spin state corresponding to a completely polarized beam with $P = 1$ can be identified and therefore distinguished from the general case of mixed states through the condition [3.2]

$$\mathrm{tr}\{\rho\} = \mathrm{tr}\{\rho^2\}. \tag{3.18}$$

## 3.4 Experimental Determination of Electron Polarization

Determining the spin polarization of an electron beam is significantly more complicated than the corresponding measurement of a light polarization, mainly because electron spin polarizers and analyzers have a significantly lower efficiency than the nearly 50% that can routinely be achieved with polarization filters for visible light. For many years, the standard process used for spin polarization measurements has been the so-called "Mott scattering." This mechanism is based on the fact that, because of relativistic effects, the differential cross sections for scattering of electrons with spin component "up" and "down" relative to the normal vector $\hat{n}$ of the scattering plane are different [3.3]. Consequently, an unpolarized electron beam may become polarized perpendicularly to the scattering plane after the collision, and an existing spin polarization can be measured through a left-right asymmetry in the differential cross section [3.1]. In detail, the spin polarization of an initially unpolarized electron beam after scattering at an angle $\theta$ is given by

$$\boldsymbol{P} = P_{\hat{n}}(E,\theta)\,\hat{\boldsymbol{n}} \equiv S_P(E,\theta)\,\hat{\boldsymbol{n}}, \tag{3.19}$$

while the left-right asymmetry in the differential cross section is determined by

$$\frac{\sigma_\uparrow(E,\theta) - \sigma_\downarrow(E,\theta)}{\sigma_\uparrow(E,\theta) + \sigma_\downarrow(E,\theta)} = \frac{\sigma_\uparrow(E,\theta) - \sigma_\uparrow(E,-\theta)}{\sigma_\uparrow(E,\theta) + \sigma_\uparrow(E,-\theta)} \equiv S_A(E,\theta). \tag{3.20}$$

Both the spin polarization function $S_P(E,\theta)$ and the asymmetry function $S_A(E,\theta)$ depend on the scattering angle $\theta$ and the energy $E$ of the beam. Also, owing to conservation of parity, the differential cross section for scattering of spin-up electrons to the left $(\theta)$ is the same as that for scattering of spin-down electrons to the right $(-\theta)$ [3.1].

As a consequence of (3.20), the count rates in electron detectors positioned at angles $\pm\theta$ can be combined as follows:

$$\frac{N_l - N_r}{N_l + N_r} = S_A(E,\theta)\,P_{\hat{n}} \tag{3.21}$$

where $N_l$ and $N_r$ are the rates for scattering to the left and right, respectively. Hence, provided $S_A(E,\theta)$ is known, the polarization component $P_{\hat{n}}$ may be determined.

The calibration of such a "Mott detector"  can be performed through a "double scattering experiment" (see Figure 3.2) where an unpolarized electron beam is first polarized through scattering and the polarization is subsequently measured via the left-right asymmetry. For elastic scattering, where $S_A(E,\theta) = S_P(E,\theta) \equiv S(E,\theta)$ due to time-reversal invariance of the interaction [3.1], such an experiment allows for the determination of the absolute value of the "Sherman function" $S(E,\theta)$ through

$$\frac{N_l - N_r}{N_l + N_r} = S^2(E,\theta). \tag{3.22}$$

Figure 3.3 shows a typical setup of such a Mott detector where a low-energy electron beam, whose polarization is to be determined, is accelerated through a voltage of 120 keV and the left-right asymmetry for elastic scattering from a thin gold foil at $\theta = 120°$ is determined. For single scattering from

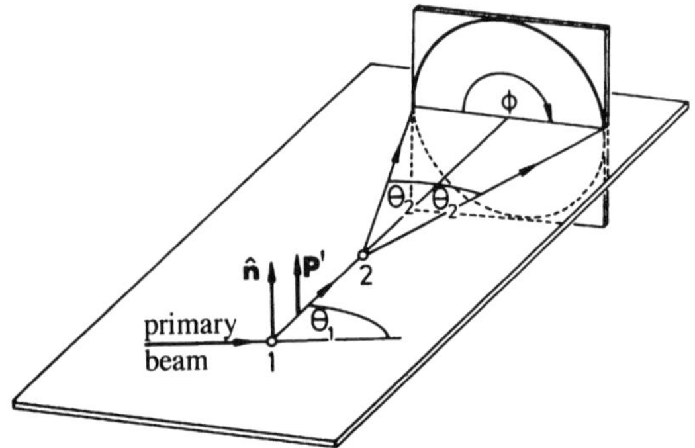



**Fig. 3.2.** Double scattering experiment to produce and analyze a spin-polarized electron beam (from Kessler [3.1]).

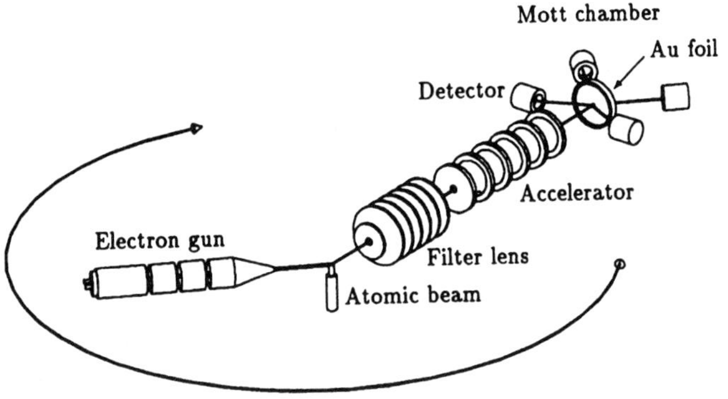



**Fig. 3.3.** Setup of a Mott detector (from Müller and Kessler [3.7]).

a gold atom in such a kinematic arrangement, the Sherman function can be calculated with sufficient accuracy, so that at least the sign of $S(E,\theta)$ may be taken with confidence from theory for this setup.

The production of spin-polarized electrons beams as well as electron polarimetry have been the subject of intensive study over many years. The principal efforts for the sources concentrate on high polarization with sufficient intensity. A frequently used technique, based on the Fano effect [3.8] (see Chapter 11), is the emission of polarized electrons from GaAs crystals irradiated by circularly polarized laser light [3.9]. Detector efficiency and accuracy combined with a convenient operation of the detector and affordable prices are important factors as well. In addition to classical Mott scattering, optical polarimetry has become a promising alternative. The basic idea in this approach is the transfer of an electron polarization to an excited atom via inelastic exchange scattering, followed by a measurement of the circular light polarization in a subsequent optical decay [3.4-6]. An example will be presented in Section 7.1. The status of polarized electron sources and electron polarimetry has been reviewed by Pierce [3.9] and Gay [3.10], respectively, to which we refer for further details.

## Exercises

1. Prove (3.10) for the magnitude of the polarization vector.
2. Prove the spinor representation (3.11) and give explicit expressions for state vectors describing an electron beam that is completely polarized
   a) along the positive $x$ direction,
   b) along the negative $y$ direction,
   c) in the $xy$ plane at an angle of $30°$ relative to the $x$ direction.
3. Show that the condition (3.18) holds for the density matrix of a completely polarized beam.
4. Show that $\mathrm{tr}\{\rho\} > \mathrm{tr}\{\rho^2\}$ holds for a partially polarized beam with $P < 1$.

## References


3.1 J. Kessler, *Polarized Electrons* (2nd edition), Springer, Heidelberg 1985.
3.2 K. Blum, *Density Matrix Theory and Applications* (2nd edition), Plenum, New York 1996.
3.3 N.F. Mott and H.S.W. Massey, *The Theory of Atomic Collisions*, Clarendon, Oxford 1965.
3.4 P.S. Farago and J.S. Wykes, J. Phys. B **2** (1969) 747.
*3.5 J.S. Wykes, J. Phys. B **4** (1971) L91.
3.6 T.J. Gay, J. Phys. B **16** (1983) L553.
3.7 H. Müller and J. Kessler, J. Phys. B **27** (1994) 5893.
3.8 U. Fano, Phys. Rev. **178** (1969) 131.
3.9 D.T. Pierce, in *Experimental Methods in the Physical Sciences — Atomic, Molecular and Optical Physics: Charged Particles*, F.B. Dunning and R.G. Hulet (eds.), Academic Press, New York 1995, p. 1.
3.10 T.J. Gay, in *Experimental Methods in the Physical Sciences — Atomic, Molecular and Optical Physics: Charged Particles*, F.B. Dunning and R.G. Hulet (eds.), Academic Press, New York 1995, p. 231.

# 4. Experimental Geometries and Approaches

An overview of the basic experimental setups that have been used for atomic collision experiments involving electron, atom, and ion beams is given. Starting from angle-integrated measurements with unpolarized beams without observation of the collision partners after the scattering, the level of detail is increased step by step, resulting in angle-differential setups, preparation of polarized beams, coincidence techniques, and polarization analysis after the collision. Special emphasis is given to "time-reversed" studies involving laser-prepared targets.


## 4.1 Integrated Cross Sections and Alignment

In this section, we describe some experiments that were performed without observation of the projectile scattering angle or the recoil angle into which the target atom was propelled. Further case studies will be discussed in Part II.

### 4.1.1 Schematic setup for angle-integrated measurements

The schematic diagram of such experiments is shown in Figure 4.1, where electrons ($e^-$), charged ions ($A^+$), or neutral atoms (A) collide with a target B. An important parameter describing this process is the angle-integrated or total cross section $\sigma_T$. (Throughout this book, we will typically be interested in a particular transition between well-defined initial and final states, so we will not use the term "total cross section" for the sum of angle-integrated cross sections over all possible transitions.) Note that the *absolute* calibration of the total cross section is a major difficulty in practical experiments.

Assuming for the moment that neither the projectile nor the target beams are polarized in any way, the setup shown in Figure 4.1 exhibits cylindrical symmetry around the incident beam axis. Consequently, this axis and any direction orthogonal to it are not necessarily equivalent from the physical point of view. A straightforward verification of this nonequivalence is the measurement of the linear polarization $P_1 = P_T = P$ of light emitted by either the excited projectile or the excited target with a photon detector positioned perpendicular to the incident beam direction. A nonvanishing value for $P_T$

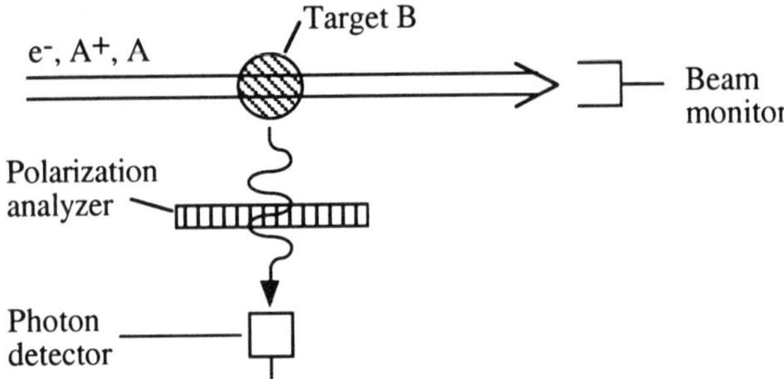


**Fig. 4.1** Schematic diagram of collision experiment with cylindrical symmetry around the incident beam axis.

in such an experiment is determined by an angle-integrated "alignment," i.e., an anisotropic population of the magnetic sublevels during the collision.

For the setups described above, $P_2$ and $P_3$ have to vanish because of the cylindrical symmetry of the problem. If a mirror is put into the plane defined by the incident beam direction and the direction of photon observation, the directions $\pm 45°$ as well as the definitions of RHC and LHC would be reversed in the mirror, although the problem itself is invariant against such operation. However, nonvanishing values of $P_2$ and $P_3$ can occur, for example, in impact excitation by spin-polarized electrons [4.1]. Some examples will be discussed in Part II of this book.

### 4.1.2 Setups with results for electron-impact and atom-impact excitation

As two practical examples of such experimental setups, we show the apparatus used by Ott *et al.* [4.2] for electron-impact excitation of atomic hydrogen in Figure 4.2 and by van Eck *et al.* [4.3] for proton impact of helium in Figure 4.3, together with some selected results for total cross sections and the linear light polarization $P_T$, respectively. In these experiments, the light polarization was measured in order to relate the optical excitation function, i.e., the number of photons observed in the decay of the excited states, to the actual excitation cross section.[4] Although the basic experimental scheme has now been used for many decades, we note that high precision data, as produced in the work of James *et al.* [4.4,5] (see Figure 4.2) or Werner and Schartner [4.6], continue to be of interest in this field.

We also note that the development of a special circular polarization analyzer for the VUV regime, described in detail by Westerveld *et al.* [4.7], has

---

[4] In terms of the magnetic sublevel cross sections $\sigma_{M_L}$, the total excitation cross section is given by $\sigma = \sigma_0 + 2\sigma_1$. After averaging over the azimuthal angle of the collision plane, one finds $I \propto \sigma_0 + \sigma_1$ for the light intensity perpendicular to the beam. This implies that $\sigma \propto I/(1 - P_T)$.

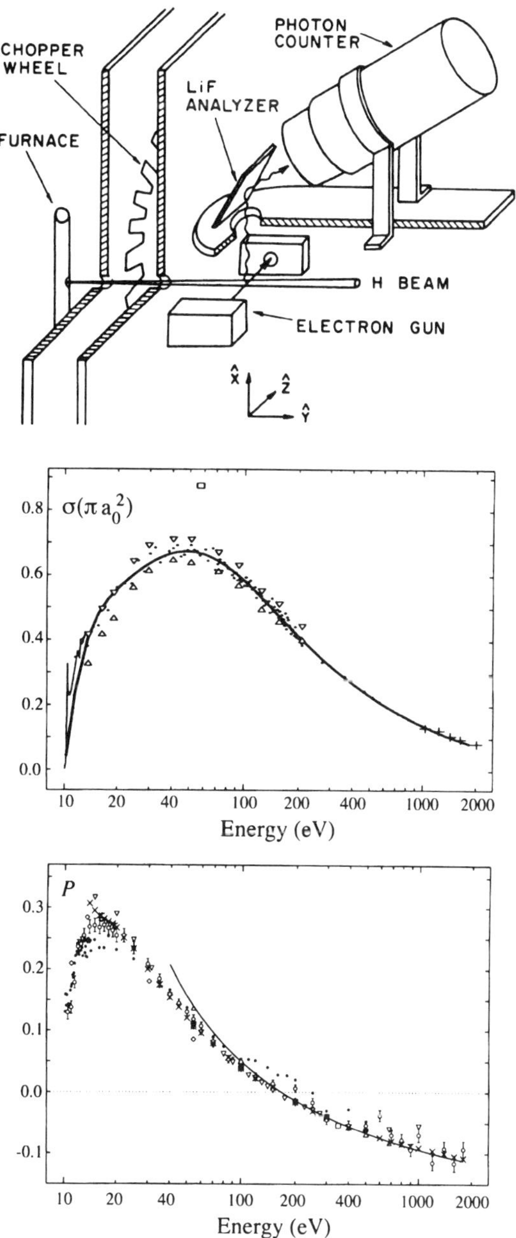


**Fig. 4.2** Experimental setup of Ott *et al.* [4.2] for electron-impact excitation of atomic hydrogen, together with some recent experimental cross section [4.4] and polarization [4.5] results for excitation of the H(2p)$^2$P state. Details of the various experiments and theories can be found in the above references.

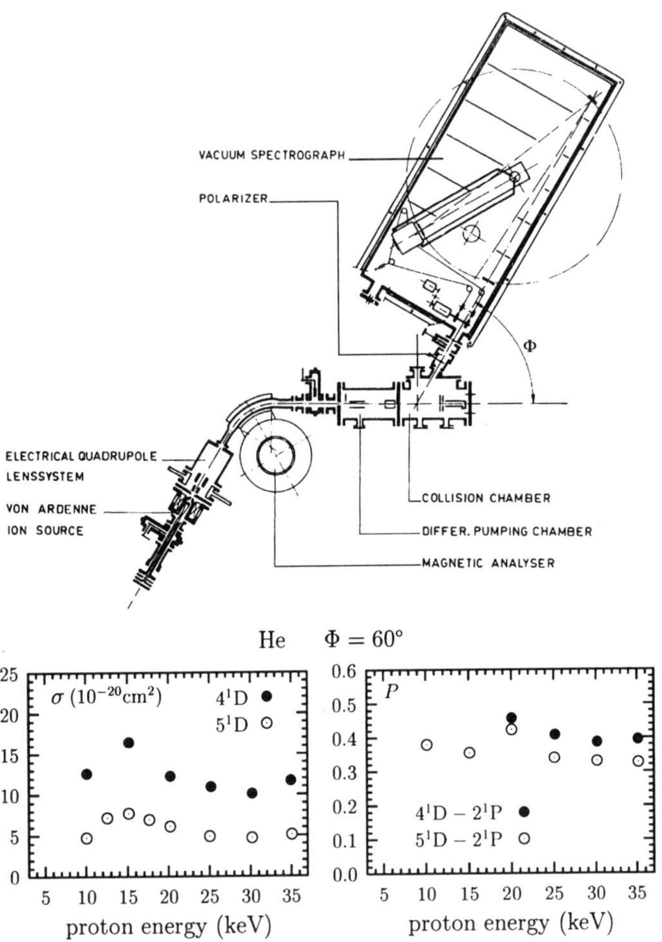


**Fig. 4.3** Experimental setup of van Eck *et al.* [4.3] for proton impact excitation of helium, together with some of their cross section and polarization results.

been of crucial importance for the measurements involving such wavelengths. Furthermore, the use of spin-polarized incident electron beams can lead to nonvanishing values of all three Stokes parameters $(P_1, P_2, P_3)$ [4.1]. Following early work at Münster [4.8], several experiments of this kind have recently been performed in the Perth [4.9] and Lincoln [4.10] groups. Some examples will be discussed in Part II.

## 4.2 Differential Cross Sections

### 4.2.1 Schematic setups for angle-differential measurements

The analysis of the collision process shown in Figure 4.1 becomes significantly more detailed if the scattering angle of the projectile or the recoil angle of the target is observed. This situation is shown schematically in Figure 4.4, where a detector is placed at the angle $\theta$. For the moment, we will not consider any light emission by either the projectile or the target after the collision, and both the beams are still assumed to be unpolarized.

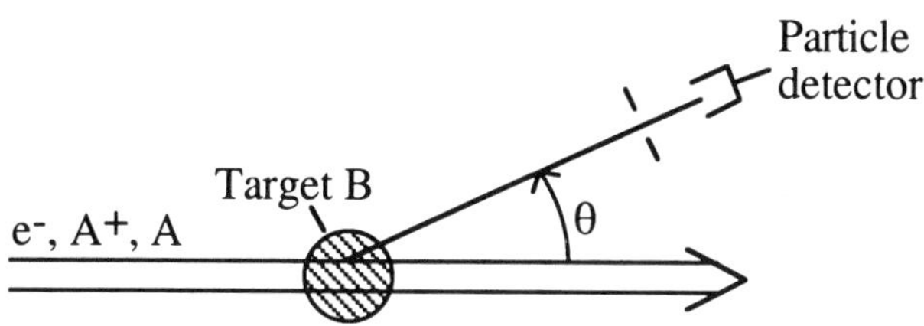


**Fig. 4.4** Schematic diagram of an angle-differential collision experiment.

The standard observable of interest in such experiments is the differential cross section $\sigma$. Some examples of experimental setups and results are shown below. An alternative method to the scheme of Figure 4.4 is to use the recoil-atom technique [4.11].

### 4.2.2 A setup with results for electron–atom collisions

Figure 4.5 shows the apparatus used by Holtkamp *et al.* [4.12] to measure absolute differential cross sections for elastic electron scattering from mercury atoms. An electron beam whose energy spread is reduced by a monochromator passes through a scattering cell, and the scattering angle can be varied by a rotatable electron spectrometer. As in the case of angle-integrated cross sections, a major difficulty in this kind of experiment arises again from the *absolute* normalization of the differential cross section. Some example results, compared with two theoretical predictions by Walker [4.13] and Haberland [4.14], are also shown in the figure. The comparison indicates that the theoretical treatment of even such a relatively simple collision problem is still not perfect, although the overall angular structure is reproduced in agreement with experiment.

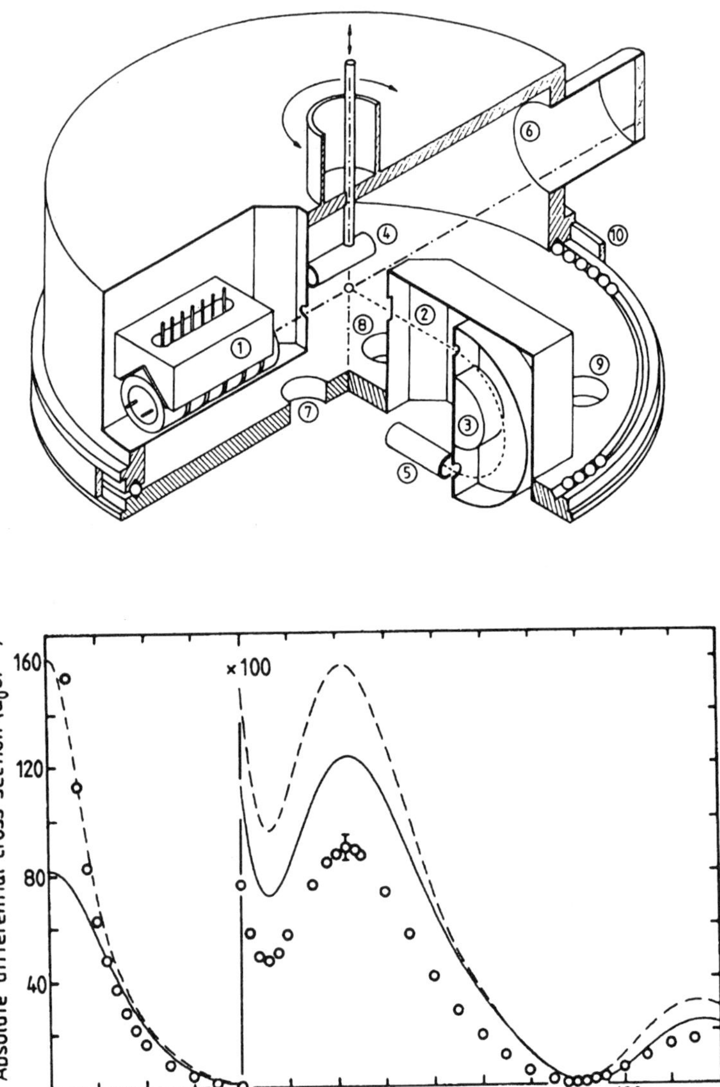


**Fig. 4.5** Experimental setup of Holtkamp *et al.* [4.12] for elastic e–Hg scattering. The individual elements are 1: electron beam; 2: aperture system; 3: analyser; 4–6: Faraday cups; 7: bypass; 8: connection to pressure gauge; 9: connection to mercury oven; 10: Teflon collar. Also shown are results for an electron energy of 150 eV, compared with theoretical predictions from Walker [4.13] (solid line) and Haberland [4.14] (dashed line).

### 4.2.3 A setup with results for atom-impact excitation

Figure 4.6 shows the experimental setup used by Everhart's group [4.15] for their pioneering angle-differential studies of electron transfer in ion-atom collisions in the keV impact energy range. The scattered particles are analyzed with respect to their charge state after the collision. Results for the $He^+$–He interaction [4.16] are shown in Figure 4.6 for a fixed scattering angle of $5°$. They revealed for the first time oscillations, nearly equally spaced in collision time, in the probability $P_0$ for transfer of an electron from the target to the projectile.

## 4.3 Planar Scattering Symmetry: Alignment and Orientation Parameters

### 4.3.1 Schematic setups for coherence and correlation analysis

As the next step of refinement, we consider cases where not only the scattering or recoil angle $\theta$ is resolved, but in addition the light, emitted after collisional excitation of either the projectile or the target, is observed in coincidence with the particle detected under the angle $\theta$. The schematic setup of such an experiment is shown in Figure 4.7. Alternatively, the time reverse experiment can be performed in which the de-excitation rate of a laser-prepared target via superelastic scattering is determined as a function of the laser polarization.

If the projectile and target beams are unpolarized before the collision, the problem is symmetric with respect to the reaction plane defined by the incident beam axis and the direction of $\theta$. When compared to the cylindrical symmetry discussed in Section 4.1, the higher planar symmetry of the present problem allows for additional nonvanishing observables to be determined.

For the purposes of this book, the most important examples of these additional parameters are the shape of the excited-state charge cloud together with the angular-momentum transfer during the collision. As discussed in Chapter 2, the shape of a P-state charge cloud with positive reflection symmetry can be determined by either an angular correlation measurement, where the coincidence rate is mapped in the reaction plane, or by a coherence measurement, where the linear light polarizations $P_1$ and $P_2$ are measured with the photon detector placed perpendicular to the reaction plane. The angular-momentum transfer can only be determined in a coherence measurement, where the circular light polarization $P_3$ is measured. In the case of an excited P state with positive reflection symmetry, however, the magnitude (but not the sign) of $P_3$ and therefore $L_\perp$ can be obtained from either an angular correlation or a coherence experiment with $P_1$ and $P_2$ only by using the condition (2.13) for completely polarized light. Note, however, that such a

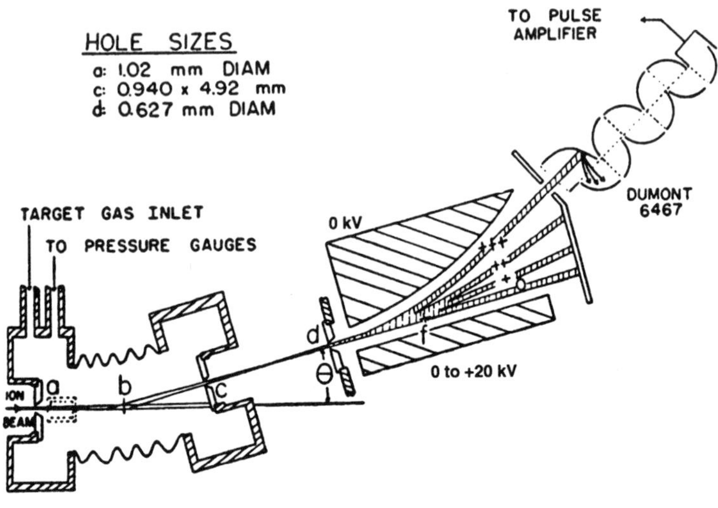


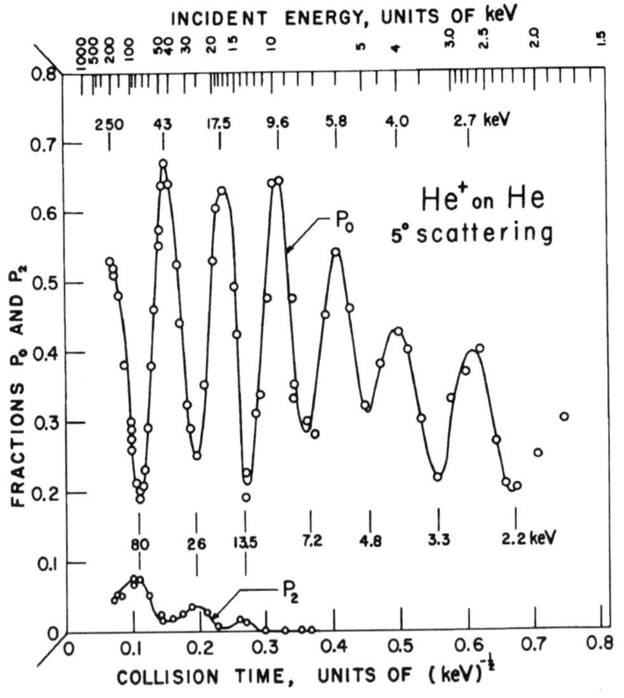


**Fig. 4.6** Experimental setup for a differential cross section measurement of electron transfer in ion-atom collisions (from [4.15]). The results shown are for He$^+$ − He collisions (from [4.16]).

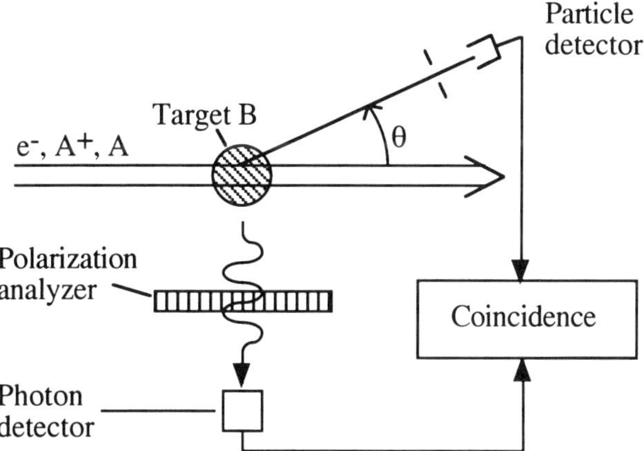


**Fig. 4.7** Schematic diagram of a scattered-particle–polarized-photon coincidence experiment.

procedure removes the possibility of experimentally checking the condition $P = 1$, with a violation indicating a systematic error.

### 4.3.2 Setups with results for electron-impact excitation and de-excitation

The angular correlation setup of the Stirling group [4.17] is shown in Figure 4.8, together with results for electron-impact excitation of the $2^1$P state at an incident electron energy of 60 eV. Note how the intensity pattern observed in the scattering plane is reproduced quite well in the first Born approximation (FBA) at the small scattering angle of 16°, while a dramatic discrepancy is observed at 40°. This demonstrates how such experiments can serve as detailed tests of theoretical predictions.

The apparatus was later modified by Standage and Kleinpoppen [4.18] to allow for polarization measurements of the emitted radiation. Their schematic setup and results for the circular polarization $P_3$ in the $3^1$P $\to$ $2^1$S transition of helium are shown in Figure 4.9.

Finally, we show in Figure 4.10 the corresponding superelastic scheme used by Hertel and Stoll [4.20] for electron-impact de-excitation of sodium. The current status of the e–Na collision system will be the subject of a detailed discussion in Part II.

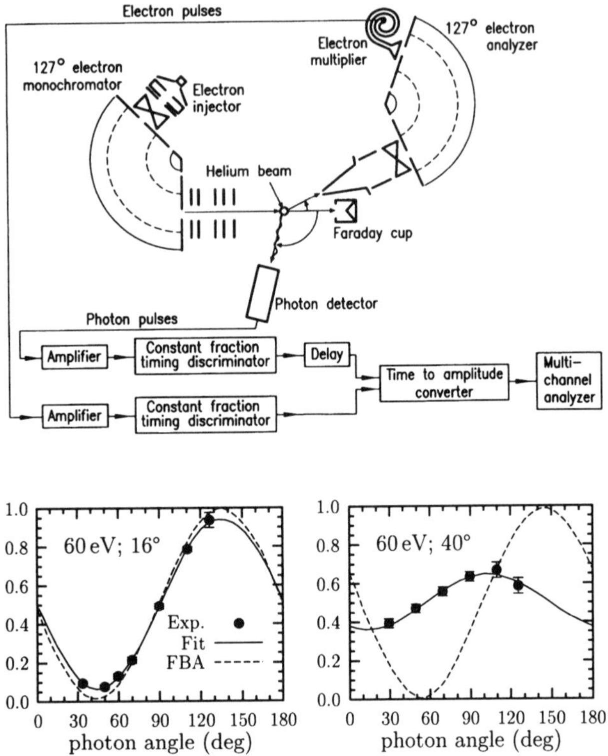


**Fig. 4.8** Angular correlation setup of the Stirling group, together with electron–photon coincidence rates after electron-impact excitation of He $(2^1\mathrm{P})$ at 60 eV incident energy and scattering angles of 16° and 40° [4.17].

### 4.3.3 Setups with results for atom-impact excitation

The first scattered-particle–emitted-photon coincidence studies for heavy particle impact were done in Lincoln and Orsay, but did not employ photon polarization analysis [4.21,22]. In this way, relatively small differential cross sections for specific atomic states could be determined by exploiting the vastly enhanced energy resolution obtained by photon detection.

The same groups, however, later used linear polarization analysis for studies of He $3^3\mathrm{P}$ excitation in $\mathrm{He}^+$–He collisions [4.23,24]. Circular polarization analysis was not attempted. In addition, the Orsay group used the angular correlation technique to study the He $2^1\mathrm{P}$ channel. Figure 4.11 shows the Orsay apparatus used for linear polarization analysis [4.23]. Results for He $3^3\mathrm{P}$ excitation, obtained at an incident energy of 150 eV and a laboratory scattering angle of 13.5° corresponding to a center-of-mass scattering angle of 27° are also presented in Figure 4.11. Each data point typically represents 24 hours of data accumulation with a true coincidence rate of 2 counts/hour.

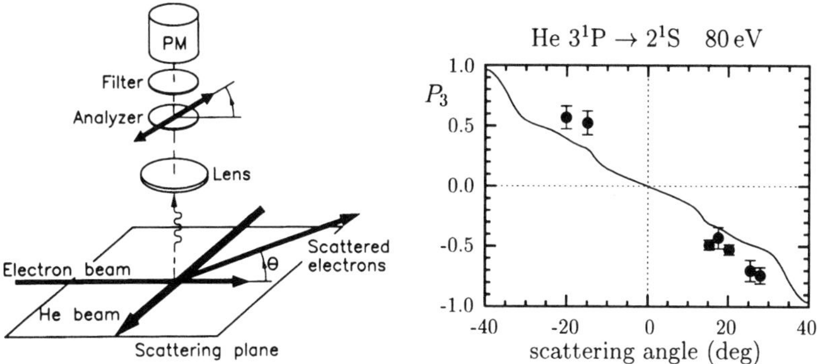



**Fig. 4.9** Electron–photon coincidence setup of Standage and Kleinpoppen [4.18], together with results for the circular light polarization $P_3$ observed in the He $3^1$P → $2^1$S transition after impact excitation by 80 eV incident electrons. The theoretical curve is based on a multichannel eikonal calculation by Flannery and McCann [4.19].

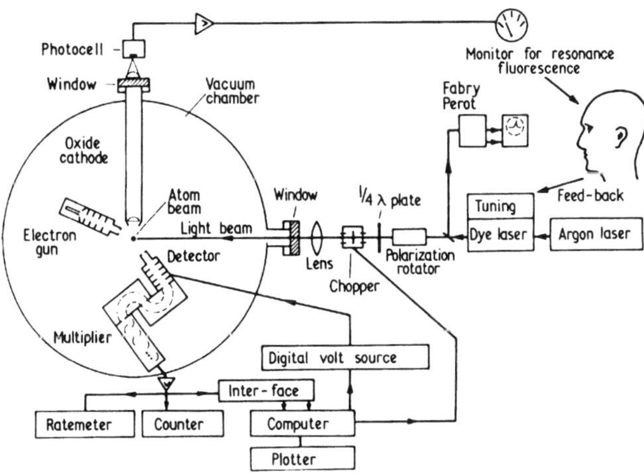



**Fig. 4.10** Schematic diagram of the electron-impact de-excitation setup used by Hertel and Stoll [4.20].

The intensity transmitted through the linear polarizer shows a clear maximum for polarizer direction parallel to the internuclear axis. This contrast is consistent with an interpretation in terms of pure $\Sigma - \Sigma$ coupling in the quasi-molecule being responsible for the excitation.

The first experiment involving determination of all three Stokes parameter components for a heavy particle collision was published by the Freiburg group for the K-Ar system [4.25].

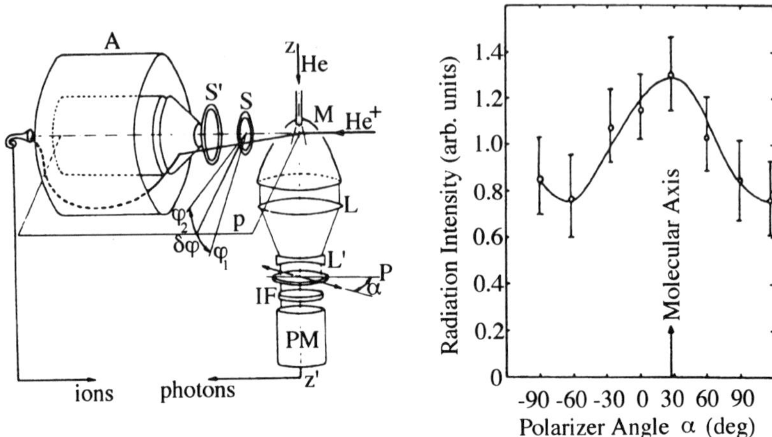



**Fig. 4.11** Experimental setup for the determination of the linear polarization of the coincidence light in $He^+ - He$ collisions [4.23]. Also shown is the ion–photon coincidence rate for 3889 Å photons from the $He(3^3P)$ state as a function of polarizer angle for 150 eV $He^+ - He$ collisions.

## 4.4 Generalized *STU* Parameters for Electron Collisions

The next level of complexity is reached when spin-polarized projectile or target beams are used, and the spin polarization is analyzed after the collision process. For electron scattering, the Münster group developed an experimental setup in which polarized electrons are scattered from unpolarized atoms and the electron polarization after the collision is determined.

This type of experiment allows for the determination of the so-called "generalized" *STU* parameters [4.26] that fully describe the change of an arbitrary initial electron polarization through scattering from any ensemble of unpolarized target atoms for both elastic and inelastic scattering. The basic scheme is shown in Figure 4.12 in the notation of the natural coordinate frame. There are seven *relative* generalized *STU* parameters with the following physical meaning: The polarization function $S_P$ gives the polarization of an initially unpolarized projectile beam after the scattering, while the asymmetry function $S_A$ determines the left-right asymmetry in the differential cross section for scattering of spin-polarized projectiles (see also Chapter 3). Furthermore, the contraction parameters $T_x$, $T_y$, $T_z$ describe the change of an initial polarization component along the three Cartesian axes, while the parameters $U_{xy}$ and $U_{yx}$ determine the rotation of a polarization component in the scattering plane. Together with the *absolute* differential cross section $\sigma_u$ for the scattering of unpolarized electrons, these eight parameters de-

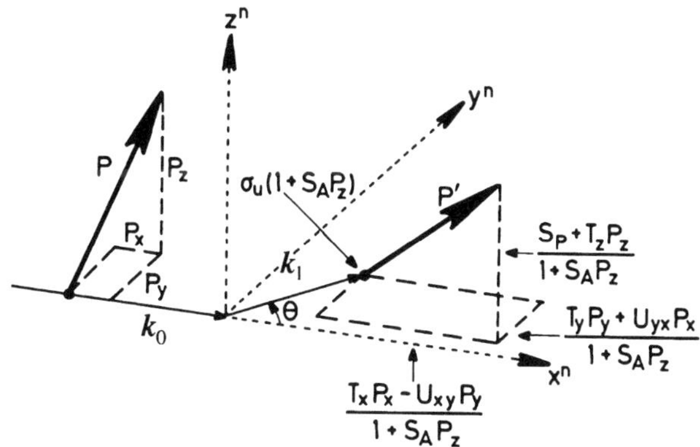



**Fig. 4.12** Physical meaning of the generalized *STU* parameters for an initial spin polarization **P** which is changed to a final spin polarization **P′** through the scattering process.

scribe the maximum information that can be obtained from preparation and analysis of electron polarization alone.

The essential parts of the apparatus used by Berger and Kessler [4.27] to measure the *STU* parameters for elastic electron scattering from mercury and xenon atoms are shown in the top part of Figure 4.13. The spin-polarized electrons were produced by photo-emission from a GaAsP crystal irradiated by circularly polarized light from a He–Ne laser. The development of such highly efficient polarized-electron sources has been crucial for the feasibility of many of the studies discussed in this book. Also note the use of a Wien filter for rotating the electron spin polarization, which was analyzed by a standard Mott detector (see also Chapter 3).

For this particular case of a target without angular momentum in the initial and final states, the generalized *STU* parameters are not independent of each other. In fact, for elastic scattering from such targets, one finds (see Chapter 5 and [4.26] for more details):

$$S_P = S_A = S \,, \tag{4.1a}$$

$$T_x = T_y = T \,, \tag{4.1b}$$

$$T_z = 1 \,, \tag{4.1c}$$

$$U_{xy} = U_{yx} = U \,. \tag{4.1d}$$

Results from the work of Berger and Kessler [4.27] together with the absolute differential cross section $\sigma_u$ of Holtkamp *et al.* [4.12] for elastic e–Hg scattering at 50 eV are shown in the bottom part of Figure 4.13. All the *relative* parameters are in very good agreement with the relativistic calculation of McEachran and Stauffer [4.28], whereas the *absolute* differential cross section is again overestimated by the theory (see also Figure 4.5). The most important reason for this disagreement is the neglect of "absorption",

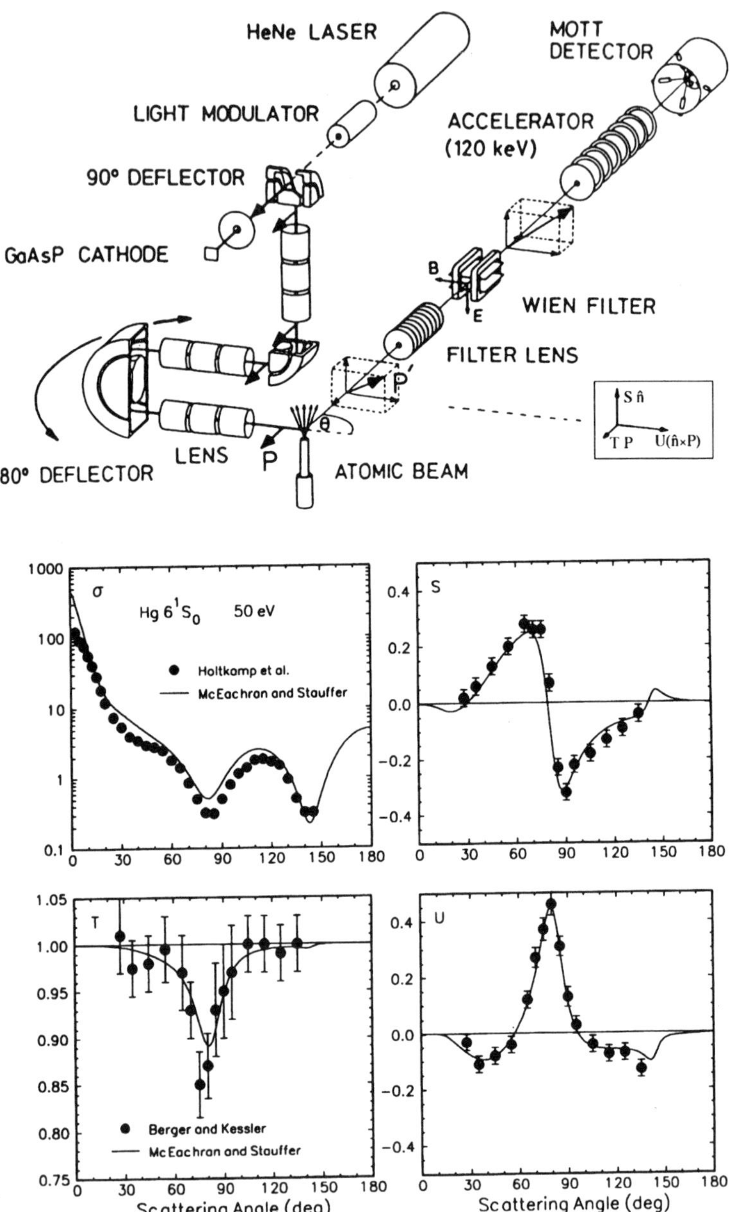


**Fig. 4.13** Experimental setup for measurement of the change of the electron polarization vector in elastic scattering (from [4.27]). Also shown are the differential cross section $\sigma_u$ and the $STU$ parameters for elastic e–Hg scattering at 50 eV. The experimental data of Holtkamp *et al.* [4.12] for $\sigma_u$ and of Berger and Kessler [4.27] for $(S, T, U)$ are compared with results from a relativistic calculation by McEachran and Stauffer [4.28].

i.e., loss of flux into inelastic channels, in the calculation. This problem can be remedied, for example, by including a semiempirical complex absorption potential [4.29].

Such measurements have also been performed for excitation processes and even for polarized-electron scattering from laser-prepared polarized targets, such as in the e–Cs work of the Bielefeld group [4.30]. Some of these studies will be further discussed in Part II.

## 4.5 Generalized Stokes Parameters for Electron–Atom Collisions

A further milestone in the electron–photon coherence and correlation analysis described in Section 4.3 was reached in the use of spin-polarized incident electrons. The first such experiment was performed in the Münster group [4.31], with the most recent refinements achieved by Sohn and Hanne [4.32]. A detailed analysis of the information contained in the experiment will be discussed in Part II. Here we only introduce the crucial concept of the "generalized" Stokes parameters defined for specific values of the incident projectile spin polarization [4.33].

The basic idea is shown in Figure 4.14. With a photon detector placed in the $\hat{n}$ direction, four light intensities are measured for orthogonal positions of the light-polarization analyzers and electron-beam polarizations $\pm P$. These intensities are then combined in a Stokes *matrix*. The elements $\left(Q_{1j}^{\hat{n}}\right)_P$, $j = \{1, 2, 3\}$, in the first row are defined as follows:

$$I_u^{\hat{n}} \left( Q_{11}^{\hat{n}} \right)_P = I_P^{\hat{n}}(0°) + I_{-P}^{\hat{n}}(0°) - I_P^{\hat{n}}(90°) - I_{-P}^{\hat{n}}(90°)\,, \tag{4.2a}$$

$$I_u^{\hat{n}} \left( Q_{12}^{\hat{n}} \right)_P \equiv I_P^{\hat{n}}(0°) - I_{-P}^{\hat{n}}(0°) - I_P^{\hat{n}}(90°) + I_{-P}^{\hat{n}}(90°)\,, \tag{4.2b}$$

$$I_u^{\hat{n}} \left( Q_{13}^{\hat{n}} \right)_P \equiv I_P^{\hat{n}}(0°) - I_{-P}^{\hat{n}}(0°) + I_P^{\hat{n}}(90°) - I_{-P}^{\hat{n}}(90°)\,, \tag{4.2c}$$

where

$$I_u^{\hat{n}} \equiv I_P^{\hat{n}}(0°) + I_{-P}^{\hat{n}}(0°) + I_P^{\hat{n}}(90°) + I_{-P}^{\hat{n}}(90°) \tag{4.3}$$

is proportional to the light intensity measured with unpolarized electrons, independent of the light-analyzer setting. Similarly, one defines the elements $\left(Q_{2j}^{\hat{n}}\right)_P$ and $\left(Q_{3j}^{\hat{n}}\right)_P$, $j = \{1, 2, 3\}$, of the second and third rows by replacing $(0°, 90°)$ by $(45°, 135°)$ and, for circular polarization analysis, by (RHC,LHC) or $(\sigma^-, \sigma^+)$, respectively.

The first column of the generalized Stokes parameter matrix, i.e., the parameters $\left(Q_{i1}^{\hat{n}}\right)_P$, $i = \{1, 2, 3\}$, is the standard Stokes vector $(P_1, P_2, P_3)$ for unpolarized incident electrons. Similarly, the parameters in the third column,

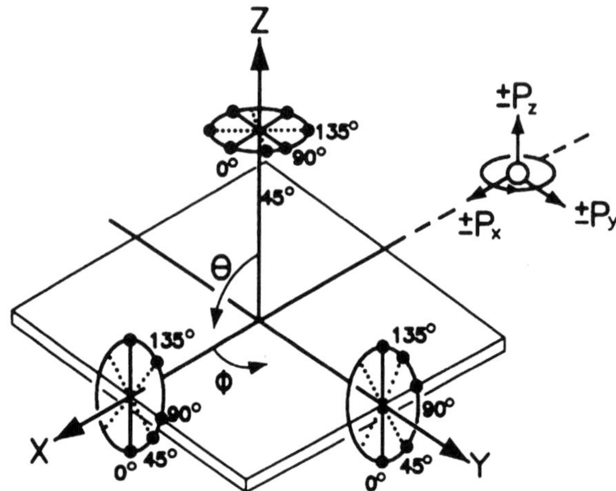


**Fig. 4.14.** (a) Definition of generalized Stokes parameters [4.33]. The linear polarizer settings in the directions $\hat{n} = x, y, z$ are shown for polarizer angles $\beta = 0°$, $45°$, $90°$, and $135°$, following the notation of Blum [4.34]. The incident electron beam is characterized by spin polarization components $\pm P_x, \pm P_y$, or $\pm P_z$, as indicated.

$\left(Q_{i3}^{\hat{n}}\right)_{P}$, $i = \{1, 2, 3\}$, yield an "optical asymmetry" that compares light intensities measured with spin-up and spin-down electrons, independent of the light-analyzer setting. Finally, the elements of the second column, $\left(Q_{i2}^{\hat{n}}\right)_{P}$, $i = \{1, 2, 3\}$, correspond to observables whose measurement requires both spin-polarized incident projectiles and photon-polarization analysis.

The setup of Sohn and Hanne [4.32] to perform such measurements is shown in the top part of Figure 4.15. Once again, the spin-polarized electron beam is produced by a standard GaAs source. The coincidence spectra are taken under almost complete computer control, since a very stable apparatus is necessary for the long data-accumulation times.

The light polarizations $(P_1, P_2, P_3)$ and $(P_4, P_5, P_6)$, with the latter corresponding to measurements with the photon detector placed in the scattering plane, are also shown in Figure 4.15 [4.32] and compared with predictions from a five-state Breit–Pauli R-matrix calculation based on the work of Scott *et al.* [4.36]. This numerical model (see also Chapter 6) is still the present state-of-the-art for such computer simulations. In light of the complexity of the problem and the detail of the comparison, the agreement between experiment and theory is quite satisfactory in this case. The relationship between the results of Sohn and Hanne, the generalized Stokes parameters, and the charge cloud of the excited state depending on the electron spin polarization [4.37], will be further discussed in Section 7.2.4.

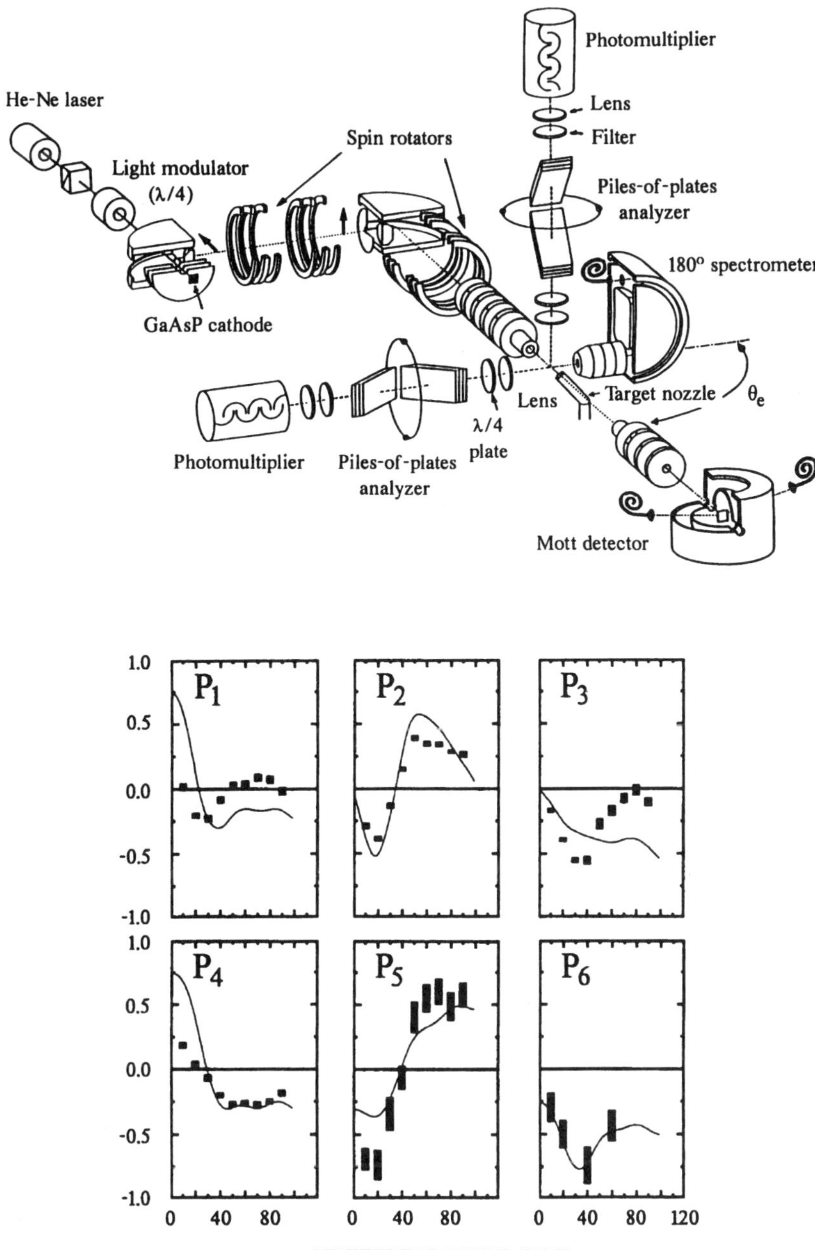


**Fig. 4.15.** Experimental setup of Sohn and Hanne [4.32] and light polarizations $(P_1, P_2, P_3)$ and $(P_4, P_5, P_6)$ in electron–photon coincidence experiments after impact excitation of Hg $6^3P_1$ by spin-polarized electrons. The experimental results [4.32,35] are compared with predictions from a five-state Breit–Pauli R-matrix calculation based on the results of Scott *et al.* [4.36].

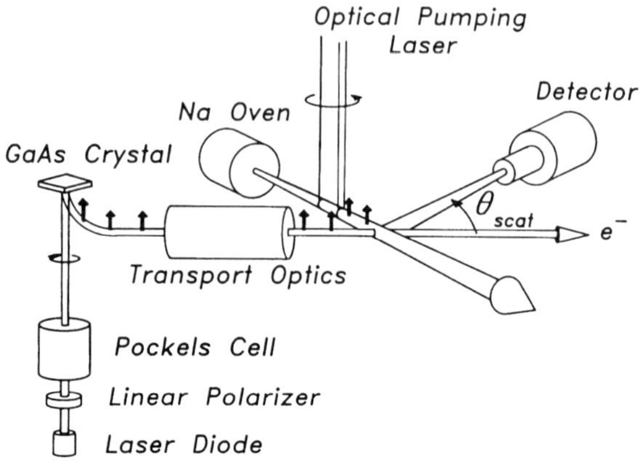


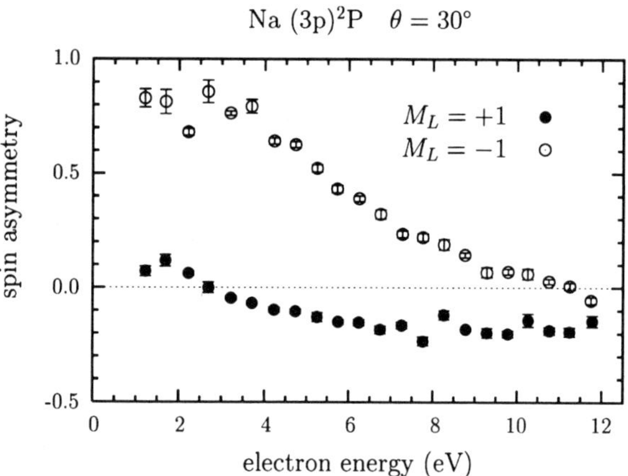


**Fig. 4.16.** Schematic diagram of the NIST experiment involving polarized electrons and polarized atoms [4.38,39]. No spin analysis is performed after the collision. Also shown are results for the spin asymmetries $A_+$ and $A_-$ for electron-impact de-excitation of laser-excited sodium atoms prepared in the $(3p)^2\mathrm{P}_{3/2}$ states with $M_L = \pm 1$ for a fixed scattering angle of 30° as a function of the incident electron energy (adapted from [4.40]).

The time-reverse version of a generalized Stokes-parameter measurement was pioneered by the NIST group [4.38,39]. The principal idea is to scatter spin-polarized electrons from spin-polarized laser-excited atoms. The setup is shown in the top part of Figure 4.16. The additional complication in this case is the laser-pumping technique to prepare the polarized target beam, but neither polarization nor coincidence analysis is performed after the collision. Consequently, this type of experiment does not suffer from the low count rate problem associated with standard electron–photon coincidence techniques.

This experimental setup has been used extensively for electron scattering from sodium atoms. Basically, spin asymmetries of the form

$$A(\theta) \equiv \frac{1}{P_e\,P_A}\,\frac{\sigma_{\uparrow\downarrow}(\theta) - \sigma_{\uparrow\uparrow}(\theta)}{\sigma_{\uparrow\downarrow}(\theta) + \sigma_{\uparrow\uparrow}(\theta)} \tag{4.4}$$

were determined with electron $(P_e)$ and atom $(P_A)$ polarizations *perpendicular* to the scattering plane. Example results for such spin asymmetries in electron-impact de-excitation of laser-excited sodium atoms in the $(3p)^2 P_{3/2}$ state are shown in the lower part of Figure 4.16. The preparation of a circular atomic state with fixed magnetic sublevel components $M_L = +1$ or $M_L = -1$ effectively corresponds to scattering from a spin-polarized target.

Since these pioneering experiments, the e–Na collision problem has been re-analyzed in great detail, resulting in a new formalism to describe the results of such experiments in terms of spin-dependent charge cloud parameters. This formalism will be discussed in Section 7.2.2.

## 4.6 Atom–Atom Collisions with Laser-Prepared Targets

An experiment for heavy-particle collisions of the type shown in Figure 1.1(c), i.c., involving optical preparation of the target *before* the collision and coincident detection of the scattered particle and the photon emitted *after* the collision, was performed by the Bielefeld group [4.41] for the process

$$H^+ + Na\,(3p) \rightarrow H\,(2p) + Na^+. \tag{4.5}$$

In their setup shown in Figure 4.17, the emitted $Ly_\alpha$ photons are detected at a fixed angle, while the scattered particles are detected at eight equally spaced azimuthal angles: $0°, 45°, ..., 315°$. This approach allows for the determination of the differential cross section and an angular correlation analysis of the photon pattern. For an incident proton energy of $1\,\mathrm{keV}$, experimental and theoretical differential cross section results for electron transfer into $H(2p)$ from the initial $Na\,(3p)$ state shown at the top of the figure are presented in the lower part [4.42]. Note that scattering angles for the process (4.5) are extremely small, typically less than $0.05°$. The angular correlation results will be discussed in Section 8.4.1.

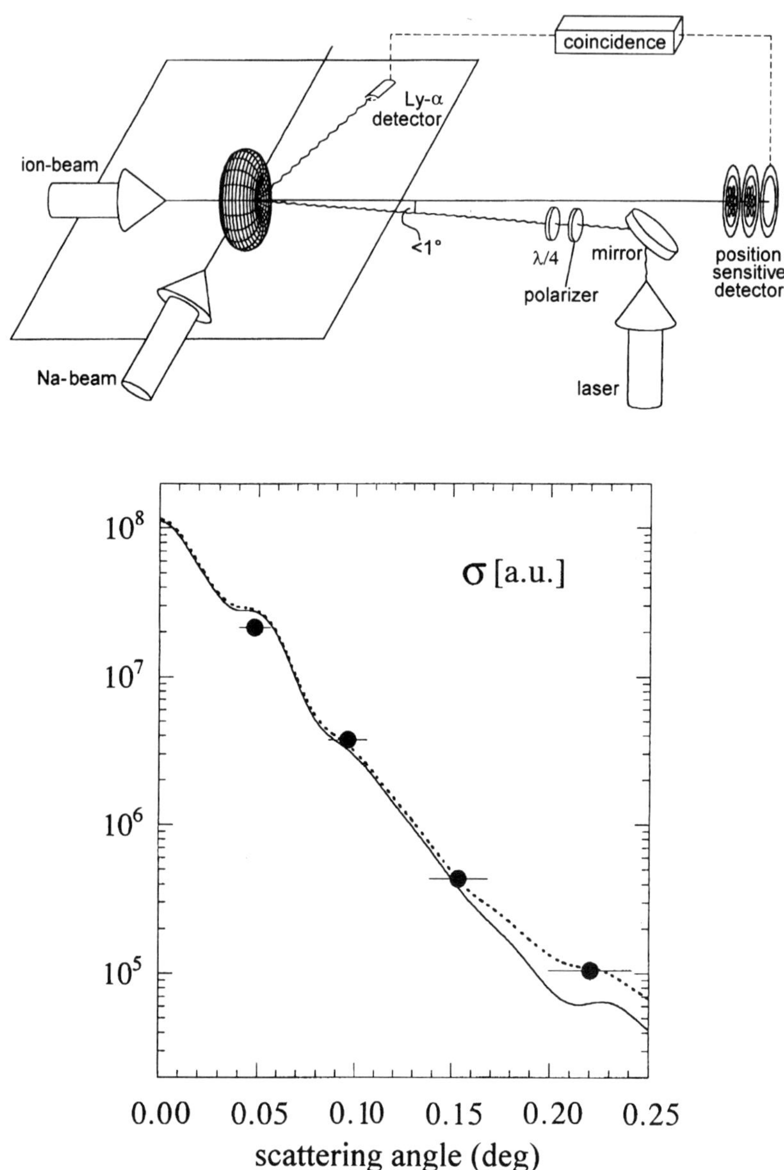



**Fig. 4.17.** The upper part is a schematic diagram of the Bielefeld experiment, involving optical preparation of a circular Na (3p) target before the collision and detection of a subsequent $Ly_\alpha$ photon from the electron transfer reaction (4.5). The lower part shows the corresponding differential cross section at an incident proton energy of 1 keV. The experimental points (•) are compared with predictions from a 36-state CC-AO calculation with (- - - - -) and without cascade (———) contributions included (adapted from [4.42].)

# References


4.1 K. Bartschat and K. Blum, Z. Phys. A **304** (1982) 85

4.2 W.R. Ott, W.E. Kauppila, and W.L. Fite, Phys. Rev. A **1** (1970) 1089.

4.3 J. van Eck, F.J. de Heer, and J. Kistemaker, Physica **28** (1962) 1184.

4.4 G.K. James, J.A. Slevin, D.E. Shemansky, J.W. McConkey, I. Bray, D. Dziczek, I. Kanik, and J.M. Allejo, Phys. Rev. A **55** (1997) 1069.

4.5 G.K. James, J.A. Slevin, D. Dziczek, J.W. McConkey, and I. Bray, Phys. Rev. A **57** (1998) 1787.

4.6 A. Werner and K.H. Schartner, J. Phys. B **29** (1996) 125.

4.7 W.B. Westerveld, K. Becker, P. Zetner, J.J. Corr, and J.W. McConkey, Appl. Opt. **24** (1985) 2256.

4.8 K. Bartschat, G.F. Hanne, and A. Wolcke, Z. Phys. A **304** (1982) 89.

4.9 D.H. Yu, P.A. Hayes, J.F. Williams, and J.E. Furst, J. Phys. B **30** (1997) 1799, and references therein.

4.10 B.G. Birdsey, H.M. Al-Khateeb, M.E. Johnston, T.C. Bowen, T.J. Gay, V. Zeman, and K. Bartschat, Phys. Rev. A **60** (1999) 1046, and references therein.

4.11 L. Vušković, M. Zuo, G.F. Chen, B. Stumpf, and B. Bederson, Phys. Rev. A **40** (1989) 133.

4.12 G. Holtkamp, K. Jost, F.J. Peitzmann, and J. Kessler, J. Phys. B **20** (1987) 4543.

4.13 D.W. Walker, Adv. Phys. **20** (1971) 257.

4.14 R. Haberland, *PhD Thesis* (1986), University of Clausthal-Zellerfeld, Germany; published as R. Haberland, L. Fritsche, and J. Noffke, Phys. Rev. A **33** (1986) 2305.

4.15 F.P. Ziemba, G.J. Lockwood, G.H. Morgan, and E. Everhart, Phys. Rev. **118** (1960) 1552.

4.16 F.P. Ziemba and E. Everhart, Phys. Rev. Lett. **118** (1959) 299.

4.17 M. Eminyan, K.B. MacAdam, J. Slevin, and H. Kleinpoppen, J. Phys. B **7** (1974) 1519.

*4.18 M.C. Standage and H. Kleinpoppen, Phys. Rev. Lett. **36** (1976) 577.

4.19 M.R. Flannery and K.J. McCann, J. Phys B **5** (1975) 1716.

4.20 I.V. Hertel and W. Stoll, J. Phys. B **7** (1974) 583.

*4.21 D.H. Jaecks, D.H. Crandall, and R.H. McKnight, Phys. Rev. Lett. **25** (1970) 491.

*4.22 G. Rahmat, G. Vassilev, J.Baudon, and M. Barat, Phys. Rev. Lett. **26** (1971) 1411.

*4.23 G. Vassilev, G. Rahmat, J. Slevin, and J. Baudon, Phys. Rev. Lett. **34** (1975) 444.

*4.24 D.H. Jaecks, F.J. Eriksen, W. de Rijk, and J. Macek, Phys. Rev. Lett. **35** (1975) 723.

4.25 L. Zehnle, E. Clemens, P.J. Martin, W. Schäuble, and V. Kempter, J. Phys. B **11** (1978) 2865.

4.26 K. Bartschat, Phys. Rep. **180** (1989) 1.

4.27 O. Berger and J. Kessler, J. Phys. B **19** (1986) 3539.

4.28 R.P. McEachran and A.D. Stauffer, J. Phys. B **19** (1986) 3523.

4.29 K. Hasenburg, K. Bartschat, R.P. McEachran, and A.D. Stauffer, J. Phys. B **20** (1987) 5165.

4.30 B. Leuer, G. Baum, L. Grau, R. Niemeyer, R., W. Raith, and M. Tondera, Z. Phys. D **33** (1995) 39.

4.31 A. Wolcke, J. Goeke, G.F. Hanne, J. Kessler, W. Vollmer, K. Bartschat, and K. Blum, Phys. Rev. Lett. **52** (1984) 1108.

4.32 M. Sohn and G.F. Hanne, J. Phys. B **25** (1992) 4627.

4.33 N. Andersen and K. Bartschat, J. Phys. B **27** (1994) 3189; corrigendum: J. Phys. B **29** (1996) 1149.

4.34 K. Blum, *Density Matrix Theory and Applications* (2nd edition), Plenum, New York 1996.

4.35 J. Goeke, G.F. Hanne, and J. Kessler, J. Phys. B **22** (1989) 1075.

4.36 N.S. Scott, P.G. Burke, and K. Bartschat, J. Phys. B **16** (1983) L361.

4.37 N. Andersen, K. Bartschat, J.T. Broad, and M. Uhrig, Phys. Rev. Lett. **76** (1996) 208.

4.38 J.J. McClelland, M.H. Kelley, and R.J. Celotta, Phys. Rev. Lett. **55** (1985) 688.

4.39 J.J. McClelland, M.H. Kelley, and R.J. Celotta, Phys. Rev. A **40** (1989) 2321.

4.40 J.J. McClelland, M.H. Kelley, and R.J. Celotta, Phys. Rev. Lett. **56** (1986) 1362.

4.41 Z. Roller-Lutz, K. Finck, Y. Wang, and H.O. Lutz, Phys. Lett. A **169** (1992) 173.

4.42 Z. Roller-Lutz, Y. Wang, H.O. Lutz, S.E. Nielsen, and A. Dubois, Phys. Rev. A **61** (2000) 022710.

# 5. Density Matrices: Connection Between Experiment and Theory

A general mathematical framework is provided. The definition of scattering amplitudes is reviewed and (reduced) density matrices are introduced for various experimental setups, such as the "generalized $STU$ parameters." The density matrix is parameterized in terms of irreducible tensor operators. The corresponding state multipoles, with special emphasis on their transformation properties and time evolution, are discussed. Finally, the general equations for the Stokes parameters are presented in this framework with S $\rightarrow$ P excitation as an example.

## 5.1 Motivation

As seen in the previous chapter, a large variety of processes can be studied experimentally to gain information about the collision dynamics in an excitation process. Hence the question arises whether it is possible to establish some connections in the description of the various experiments. In identifying such links, more systematics can be achieved both in the general formulation and in the numerical calculation of the individual observables.

As outlined in the monograph by Blum [5.1] and recent reviews [5.2–4], the density-matrix formalism is an ideal tool to achieve this goal. Although the density-matrix elements are calculated from theoretical scattering amplitudes, they can also be directly related to experimental observables introduced in the previous chapters, such as the "Stokes parameters" that describe the polarization of light emitted in optical transitions, or the "$STU$ parameters" that determine the corresponding electron spin polarization.

Although most of the theory is derived for the case where the projectile is an electron, most of the formalism, in particular the reduced density matrix approach, can directly be applied to heavy-particle impact as well.

## 5.2 Scattering Amplitudes

The scattering amplitudes

$$f(M_1 m_1, M_0 m_0; \theta) \equiv \langle J_1 M_1; \boldsymbol{k}_1 m_1 \mid \mathbf{T} \mid J_0 M_0; \boldsymbol{k}_0 m_0 \rangle \tag{5.1}$$

describe the scattering of electrons (or, more generally, spin-$\frac{1}{2}$ particles) with initial linear momentum $\boldsymbol{k}_0$ and spin component $m_0$ (with regard to a given quantization axis) from a target with total angular momentum $J_0$ and component $M_0$. The final state is characterized by the projectile momentum $\boldsymbol{k}_1$, the spin component $m_1$, and the target quantum numbers $J_1$ and $M_1$, respectively. The scattering angle $\theta$ is the angle between $\boldsymbol{k}_0$ and $\boldsymbol{k}_1$, and $\mathbf{T}$ is the transition operator, usually presented as the $\mathbf{T}$ matrix. A possible normalization factor in (5.1) is omitted for simplicity in the general formulation presented here.

### 5.2.1 Scattering amplitudes in different coordinate frames

In practical applications, it is necessary to define the scattering amplitudes with respect to a quantization axis for the angular-momentum components. A standard choice for numerical calculations (see Chapter 6) is the so-called "collision frame" where the incident beam axis is the quantization ($z^c$) axis, while the $y^c$ axis is perpendicular to the scattering plane. On the other hand, the algebra often becomes simpler and many observables can be interpreted more easily in the "natural frame" where the quantization axis coincides with the normal vector to the scattering plane and the $x^n$ axis is defined by the incident beam direction (see Chapter 2). The transformation of the scattering amplitudes from one system to another can be achieved in a straightforward way by transforming the initial and final states through standard rotation matrices and using the fact that the action of the $\mathbf{T}$ operator must be independent of the particular coordinate system.

This transformation is performed as follows. Consider states $|j, m\rangle_1$ defined in a coordinate system $(X_1, Y_1, Z_1)$. Suppose a second coordinate system $(X_2, Y_2, Z_2)$ is obtained from $(X_1, Y_1, Z_1)$ through a rotation by a set of three Euler angles $(\alpha, \beta, \gamma)$ as defined by Edmonds [5.5]. The corresponding states $|j, m'\rangle_2$ in system $(X_2, Y_2, Z_2)$ are then obtained from the $|j, m\rangle_1$ according to

$$|j, m'\rangle_2 = \sum_m D(\alpha, \beta, \gamma)^j_{mm'} |j, m\rangle_1, \tag{5.2}$$

where

$$D(\alpha, \beta, \gamma)^j_{mm'} = e^{im\alpha} \, d(\beta)^j_{mm'} \, e^{im'\gamma} \tag{5.3}$$

and $d(\beta)^j_{mm'}$ are rotation-matrix elements. Note that the angular momentum $j$ is invariant against such rotations.

As an explicit example, consider amplitudes for an $S^e \rightarrow P^o$ transition without consideration of the electron spin. Omitting the arguments $k_0$, $k_1$ and $\theta$ for simplicity, the corresponding amplitudes are

$$f_M = f(L_1 = 1, M_{L_1} = M; L_0 = 0, M_{L_0} = 0) \equiv \langle 1, M|\mathbf{T}|0,0\rangle . \qquad (5.4)$$

The natural coordinate frame evolves from the collision frame through rotation by the Euler angles $\alpha = -\pi/2$, $\beta = -\pi/2$ and $\gamma = 0$. Furthermore, only the P states have to be transformed since the spherically symmetric S states are invariant against such a rotation. Consequently, the amplitude $\langle 1, 1|\mathbf{T}|0,0\rangle_n$ in the natural frame (subscript "$n$") is related to amplitudes in the collision frame (subscript "$c$") as

$$\langle 1, 1|\mathbf{T}|0,0\rangle_n = -\frac{\mathrm{i}}{2} \langle 1, 1|\mathbf{T}|0,0\rangle_c - \frac{1}{\sqrt{2}} \langle 1, 0|\mathbf{T}|0,0\rangle_c + \frac{\mathrm{i}}{2} \langle 1, -1|\mathbf{T}|0,0\rangle_c.$$

$$(5.5)$$

It will be shown below (see also 2.28a and 2.26b) that (5.5) can be further simplified by using parity conservation and the planar symmetry of the collision process.

Before going on, however, we briefly consider a related case, namely the transformation of scattering amplitudes defined in the same coordinate system, but where the initial and final states are expanded in different basis sets. A classic example is the use of either the complex atomic or real molecular basis for P states in the natural frame introduced in Chapter 2 (see also Exercise 5.2).

Suppose the amplitude $f_{\mathrm{ba}} = \langle \mathrm{b}|\mathbf{T}|\mathrm{a}\rangle$ describes the transition from the initial state "a" to the final state "b." If these two states are expanded in a basis set $\{|n\rangle\}$ according to

$$|\mathrm{a}\rangle = \sum_n a_n |n\rangle , \qquad (5.6a)$$

$$|\mathrm{b}\rangle = \sum_m b_m |m\rangle , \qquad (5.6b)$$

we see that

$$f_{\mathrm{ba}} = \sum_{n,m} a_n b_m^* f_{mn} , \qquad (5.7)$$

where $f_{mn} = \langle m|\mathbf{T}|n\rangle$ describes the transition between the basis states "$n$" and "$m$," respectively.

### 5.2.2 Symmetry properties

An important point for the discussion of scattering amplitudes is the fact that certain symmetry properties of the projectile-target interaction lead to conservation laws through the $\mathbf{T}$ operator. These, in turn, will cause interdependences between various scattering amplitudes and even require some amplitudes to vanish.

Consider, for instance, the conservation of the total parity. This is an extremely good approximation in atomic collision physics. Consequently, computer programs in this field will generally only include projectile-target interactions that conserve the total parity, with the most important example being the Coulomb interaction.

For our case of interest, electron–atom scattering in a plane, the process must be invariant against reflection in this plane. This reflection operation can be constructed as the parity operation, followed by a 180° rotation around the normal axis $\hat{\mathbf{n}}$ of the scattering plane. If we consider a general matrix element of the form $\langle j_1 m_1 | \mathbf{T} | j_0 m_0 \rangle$, reflection invariance implies that

$$\langle j_1 m_1 | \mathbf{T} | j_0 m_0 \rangle = \langle j_1 m_1 | \mathbf{R}^\dagger \mathbf{T} \mathbf{R} | j_0 m_0 \rangle , \qquad (5.8)$$

where

$$\mathbf{R} = \mathbf{D}_{\hat{\mathbf{n}}}(180°) \circ \mathbf{P} \qquad (5.9)$$

is the operator for the reflection, constructed as a product of the parity operator $\mathbf{P}$ and the rotation operator $\mathbf{D}_{\hat{\mathbf{n}}}(180°)$, and the dagger denotes the adjoint operator.

The operation of the parity operator yields the same (eigen)state multiplied by the parity $\Pi$ of the state, and rotations are handled again through the Euler angles. Fortunately, rotations by 180° around the $z^n$ axis in the natural frame or the $y^c$ axis in the collision frame are straightforward. Applying the general transformation (5.2) one finds

$$\mathbf{R} | j_0 m_0 \rangle = \mathbf{D}(0, 0, \pi) \circ \mathbf{P} | j_0 m_0 \rangle = \Pi_0 (-1)^{m_0} | j_0 m_0 \rangle \qquad (5.10a)$$

for the natural frame and

$$\mathbf{R} | j_0 m_0 \rangle = \mathbf{D}(0, -\pi, 0) \circ \mathbf{P} | j_0 m_0 \rangle = \Pi_0 (-1)^{j_0 - m_0} | j_0 - m_0 \rangle \qquad (5.10b)$$

for the collision frame. Consequently, (5.6) becomes

$$\langle j_1 m_1 | \mathbf{T} | j_0 m_0 \rangle = \Pi_0 \Pi_1 (-1)^{m_0 - m_1} \langle j_1 m_1 | \mathbf{T} | j_0 m_0 \rangle \qquad (5.11a)$$

for the natural frame and

$$\langle j_1 m_1 | \mathbf{T} | j_0 m_0 \rangle = \Pi_0 \Pi_1 (-1)^{j_0 - m_0 + j_1 - m_1} \langle j_1 - m_1 | \mathbf{T} | j_0 - m_0 \rangle \qquad (5.11b)$$

for the collision frame, where $\Pi_0$ and $\Pi_1$ are the parities of the initial and final states, respectively.

The above symmetry relationships can be generalized to our amplitudes of interest. The results are

$$f(M_1 m_1, M_0 m_0; \theta) = \Pi_0 \Pi_1 (-1)^{M_0 - M_1 + m_0 - m_1} \, f(M_1 m_1, M_0 m_0; \theta)$$
$$(5.12a)$$

for the natural frame and

$$f(M_1 m_1, M_0 m_0; \theta) = \Pi_0 \Pi_1 (-1)^{J_1 - M_1 + J_0 - M_0 + \frac{1}{2} - m_0 + \frac{1}{2} - m_1}$$
$$\times f(-M_1 - m_1, -M_0 - m_0; \theta) \qquad (5.12b)$$

for the collision frame.

Instead of providing phase relationships ($\pm$) between amplitudes with a given set of magnetic quantum numbers and those with the sign-reversed quantum numbers in the collision frame as in (5.12b), Equation (5.12a) shows that many amplitudes simply vanish in the natural frame (namely those where the exponent is an even or odd integer, depending on the product of the parities). This fact is one of the many advantages that can be used when formulating the general theory in this frame. Numerical calculations, on the other hand, are much simpler in the collision frame. If desired, however, the resulting amplitudes can easily be transformed into the natural frame.

Finally, the corresponding form of (5.10b) for $S^e \to P^o$ transitions can be used to eliminate the amplitude $\langle 1, -1 | \mathbf{T} | 0, 0 \rangle_c$ from (5.5). This leads to the well-known results

$$\langle 1, 1 | \mathbf{T} | 0, 0 \rangle_n = -i \, \langle 1, 1 | \mathbf{T} | 0, 0 \rangle_c - \frac{1}{\sqrt{2}} \, \langle 1, 0 | \mathbf{T} | 0, 0 \rangle_c \,, \qquad (5.13a)$$

$$\langle 1, -1 | \mathbf{T} | 0, 0 \rangle_n = -i \, \langle 1, 1 | \mathbf{T} | 0, 0 \rangle_c + \frac{1}{\sqrt{2}} \, \langle 1, 0 | \mathbf{T} | 0, 0 \rangle_c \,, \qquad (5.13b)$$

$$\langle 1, 0 | \mathbf{T} | 0, 0 \rangle_n \equiv 0 \,, \qquad (5.13c)$$

for this case (see also Chapter 2, Equations 2.26–28).

### 5.2.3 Scattering amplitudes in the nonrelativistic limit

In electron scattering from light targets, it is often assumed that the total spin $S$ and the total angular momentum $L$ of the combined target + projectile system are conserved during the collision. This approximation is very good, for example, for light alkali-like targets. It is also the basis for the "fine-structure effect" which will be further discussed below

In this nonrelativistic approximation, the scattering amplitudes depend in a purely algebraic way on the spin quantum numbers, and transitions between fine-structure levels are described by standard recoupling techniques. Specifically, the scattering amplitudes (5.1) can then be expressed as

$$f(M_1 m_1, M_0 m_0; \theta)$$
$$= \sum_{\substack{L_1, M_{L_1}, \\ L_0, M_{L_0}, \\ S, M_S}} (L_0, M_{L_0}; S_0, M_{S_0} | J_0, M_0)\ (S_0, M_{S_0}; \tfrac{1}{2}, m_0 | S, M_S)$$
$$\times\ (L_1, M_{L_1}; S_1, M_{S_1} | J_1, M_1)\ (S_1, M_{S_1}; \tfrac{1}{2}, m_1 | S, M_S)$$
$$\times\ f^S(M_{L_1}, M_{L_0}; \theta), \tag{5.14}$$

where $S_0$, $S_1$, $M_{S_0}$, $M_{S_1}$, $L_0$, $L_1$, $M_{L_0}$, and $M_{L_1}$ are the spins as well as the angular momenta and the corresponding $z$ components of the initial and final target states, and

$$f^S(M_{L_1}, M_{L_0}; \theta) \equiv \langle L_1 M_{L_1}; \boldsymbol{k}_1 \,|\, \mathbf{T}^S \,|\, L_0 M_{L_0}; \boldsymbol{k}_0 m_0 \rangle \tag{5.15}$$

are nonrelativistic scattering amplitudes that describe transitions between orbital angular-momentum states.

Equation (5.14) expresses the conservation of the total spin $S$ and its component

$$M_S = M_{S_1} + m_1 = M_{S_0} + m_0 \tag{5.16}$$

through the Clebsch–Gordan coefficients $(j_1, m_1; j_2, m_2 | j_3, m_3)$. Note that this approximation will reduce the number of independent scattering amplitudes significantly, since the scattering dynamics are assumed to be identical for each member of a fine-structure multiplet.

## 5.3 Density Matrices

A thorough introduction to the theory of density matrices and their applications with emphasis on atomic physics can be found in the book by Blum [5.1], and extended applications to electron collisions have been given in [5.2–4]. As shown already in Chapters 2 and 3 for selected examples, the main advantage of the density-matrix formalism is its ability to deal with pure and mixed states in the same consistent manner. The preparation of the initial state as well as the details regarding the observation of the final state may be treated in a systematic way. In particular, averages over quantum numbers of unpolarized beams in the initial state and incoherent sums over nonobserved quantum numbers in the final state can be accounted for by the "reduced density matrix." Several examples will be presented in the case studies discussed in Part II of this book. Furthermore, expansion of the density matrix in terms of "irreducible tensor operators" and the corresponding "state multipoles" (see Section 5.5) allows for the use of advanced angular-momentum techniques.

Following Blum [5.1], the complete density operator after the collision process is given by

$$\rho_{\text{out}} = \mathbf{T}\,\rho_{\text{in}}\,\mathbf{T}^{\dagger}\,, \tag{5.17}$$

where $\rho_{\text{in}}$ is the density operator before the collision. The corresponding matrix elements are given by

$$(\rho_{\text{out}})^{M_1' M_1}_{m_1' m_1;\theta} = \sum_{m_0' m_0 M_0' M_0} \rho_{m_0' m_0}\,\rho_{M_0' M_0}$$
$$\times\; f(M_1' m_1', M_0' m_0'; \theta)\, f^*(M_1 m_1, M_0 m_0; \theta)\,, \tag{5.18}$$

where the parameters $\rho_{m_0' m_0}\rho_{M_0' M_0}$ describe the preparation of the initial state. (We assume that the projectile and target beams are prepared independently, thereby allowing for this factorization.)

As pointed out above, "reduced" density matrices account for the fact that, in practical experiments, not all quantum mechanically allowed quantum numbers are determined simultaneously. For example, if only the scattered projectiles are observed, the corresponding elements of the reduced density matrix are obtained by summing over the atomic quantum numbers as follows:

$$(\rho_{\text{out}})_{m_1' m_1;\theta} = \sum_{M_1'=M_1} (\rho_{\text{out}})^{M_1' M_1}_{m_1' m_1;\theta}\,. \tag{5.19}$$

The differential cross section is then given by

$$\frac{d\sigma(\theta)}{d\Omega} = C \sum_{m_1'=m_1} (\rho_{\text{out}})_{m_1' m_1;\theta}\,, \tag{5.20}$$

where $C$ is a constant that depends on the normalization of the continuum waves in a numerical calculation. These reduced density-matrix elements contain information about the projectile spin. The information can be extracted through a measurement of the generalized $STU$ parameters [5.4], which will be discussed as an example in the next section.

On the other hand, if only the atoms are observed (for example, by analyzing the light emitted in optical transitions), the elements

$$(\rho_{\text{out}})^{M_1' M_1} = \int d^3 k_1 \sum_{m_1'=m_1} (\rho_{\text{out}})^{M_1' M_1}_{m_1' m_1;\theta} \tag{5.21}$$

determine the "integrated Stokes parameters," i.e., the polarization of the emitted light independent of the electron scattering angle. These parameters contain information about the angular-momentum distribution in the excited target ensemble.

Finally, for electron–photon coincidence experiments without projectile spin analysis in the final state, the elements

$$\left(\boldsymbol{\rho}_{\text{out}}\right)_{\theta}^{M_1' M_1} = \sum_{m_1' = m_1} \left(\boldsymbol{\rho}_{\text{out}}\right)_{m_1' m_1 ; \theta}^{M_1' M_1} \tag{5.22}$$

*simultaneously* contain information about the projectiles and the target. This information can be extracted by measuring the angle-differential Stokes parameters.

The density-matrix formalism outlined above is very useful for obtaining a qualitative description of the geometrical and sometimes also of the dynamical symmetries of the collision process [5.4].

## 5.4 An Explicit Example: Generalized *STU* Parameters

### 5.4.1 Definition in terms of scattering amplitudes

The generalized *STU* parameters describe the scattering of a spin-polarized electron beam from unpolarized targets. They can be expressed conveniently in terms of the quantities

$$\langle m_1' m_0'; m_1 m_0 \rangle \equiv \frac{1}{2 J_0 + 1} \sum_{M_1 M_0} f(M_1 m_1'; M_0 m_0') f^*(M_1 m_1; M_0 m_0). \tag{5.23}$$

The derivation of the general formulas for these parameters proceeds as follows. First, the unpolarized target beam in the initial state is represented by matrix elements $\boldsymbol{\rho}_{M_0' M_0} = \delta_{M_0' M_0}$ in (5.18), while a spin-polarized projectile beam is represented by elements of the polarization matrix

$$\boldsymbol{\rho}_{\text{proj}} = \frac{1}{2} \begin{pmatrix} 1 + P_z & P_x - i P_y \\ P_x - i P_y & 1 - P_z \end{pmatrix}, \tag{5.24}$$

Then the reduced density-matrix elements defined in (5.19) are calculated, followed by the differential cross section given in (5.20). Finally, the spin-polarization components $P'_{x,y,z}$ after the scattering are obtained through

$$P'_{x,y,z} = \frac{tr\, \boldsymbol{\sigma}_{x,y,z} \boldsymbol{\rho}_{\text{red}}}{tr\, \boldsymbol{\rho}_{\text{red}}}, \tag{5.25}$$

where $\boldsymbol{\rho}_{\text{red}}$ is the reduced density matrix for the scattered projectiles with the matrix elements given in (5.19), $\boldsymbol{\sigma}_{x,y,z}$ are the three Pauli matrices, and $tr$ denotes the trace operation.

For the natural frame, one finds the general result (see also Figure 4.12):

$$\mathbf{P}' = \frac{(S_P + T_z P_z)\hat{z} + (T_y P_y + U_{yx} P_x)\hat{y} + (T_x P_x - U_{xy} P_y)\hat{x}}{1 + S_A P_z}. \tag{5.26}$$

The equivalent equation for the collision frame [5.4] is obtained by making the index substitutions $x \leftrightarrow z$, $y \leftrightarrow x$ and $z \leftrightarrow y$ in (5.26).

The final formulas can be simplified substantially by using the hermiticity of the density matrix and assuming parity conservation in the projectile–target interaction. The hermiticity condition yields [5.4]

$$\langle m_1' m_0'; m_1 m_0 \rangle = \langle m_1 m_0; m_1' m_0' \rangle^* \tag{5.27}$$

and parity conservation allows for the application of (5.12) to the parameters defined in (5.23). The results are

$$\langle m_1' m_0'; m_1 m_0 \rangle_n = (-1)^{m_1' - m_0' + m_1 - m_0} \, \langle m_1' m_0'; m_1 m_0 \rangle_n \,, \tag{5.28a}$$

$$\langle m_1' m_0'; m_1 m_0 \rangle_c = (-1)^{m_1' - m_0' + m_1 - m_0} \, \langle -m_1' - m_0'; -m_1 - m_0 \rangle_c, \tag{5.28b}$$

where the subscripts again denote the parameters in the natural and in the collision frame, respectively. Once more, we see that reflection invariance with respect to the scattering plane yields $\pm$ phase relationships between parameters in the collision frame, whereas half of the possible combinations of the magnetic quantum numbers will not give any contribution in the natural frame.

Omitting a possible normalization factor, the differential cross section for unpolarized projectile beams is given by

$$\sigma_u \equiv \frac{1}{2} \sum_{m_1 m_0} \langle m_1 m_0; m_1 m_0 \rangle = \frac{1}{2(2J_0 + 1)} \sum_{M_1 M_0 m_1 m_0} |f(M_1 m_1, M_0 m_0)|^2$$

$$\tag{5.29}$$

in both the natural frame and the collision frame. For the remaining seven generalized *STU* parameters one finds:

$$S_P = \frac{1}{2\sigma_u} \left[ \langle \tfrac{1}{2}\tfrac{1}{2}; \tfrac{1}{2}\tfrac{1}{2} \rangle_n + \langle \tfrac{1}{2} - \tfrac{1}{2}; \tfrac{1}{2} - \tfrac{1}{2} \rangle_n \right.$$

$$\left. - \langle -\tfrac{1}{2}\tfrac{1}{2}; -\tfrac{1}{2}\tfrac{1}{2} \rangle_n - \langle -\tfrac{1}{2} - \tfrac{1}{2}; -\tfrac{1}{2} - \tfrac{1}{2} \rangle_n \right], \tag{5.30a}$$

$$S_A = \frac{1}{2\sigma_u} \left[ \langle \tfrac{1}{2}\tfrac{1}{2}; \tfrac{1}{2}\tfrac{1}{2} \rangle_n - \langle \tfrac{1}{2} - \tfrac{1}{2}; \tfrac{1}{2} - \tfrac{1}{2} \rangle_n \right.$$

$$\left. + \langle -\tfrac{1}{2}\tfrac{1}{2}; -\tfrac{1}{2}\tfrac{1}{2} \rangle_n - \langle -\tfrac{1}{2} - \tfrac{1}{2}; -\tfrac{1}{2} - \tfrac{1}{2} \rangle_n \right], \tag{5.30b}$$

$$T_x = \frac{1}{\sigma_u} \, Re \, \{ \langle \tfrac{1}{2}\tfrac{1}{2}; -\tfrac{1}{2} - \tfrac{1}{2} \rangle_n + \langle -\tfrac{1}{2}\tfrac{1}{2}; \tfrac{1}{2} - \tfrac{1}{2} \rangle_n \}, \tag{5.30c}$$

$$T_y = \frac{1}{\sigma_u} \, Re \, \{ \langle \tfrac{1}{2}\tfrac{1}{2}; -\tfrac{1}{2} - \tfrac{1}{2} \rangle_n - \langle -\tfrac{1}{2}\tfrac{1}{2}; \tfrac{1}{2} - \tfrac{1}{2} \rangle_n \}, \tag{5.30d}$$

$$T_z = \frac{1}{2\sigma_u} \left[ \langle \tfrac{1}{2}\tfrac{1}{2}; \tfrac{1}{2}\tfrac{1}{2} \rangle_n - \langle \tfrac{1}{2} - \tfrac{1}{2}; \tfrac{1}{2} - \tfrac{1}{2} \rangle_n \right.$$

$$\left. - \langle -\tfrac{1}{2}\tfrac{1}{2}; -\tfrac{1}{2}\tfrac{1}{2} \rangle_n + \langle -\tfrac{1}{2} - \tfrac{1}{2}; -\tfrac{1}{2} - \tfrac{1}{2} \rangle_n \right], \tag{5.30e}$$

$$U_{yx} = -\frac{1}{\sigma_u}\, Im\, \{\langle \tfrac{1}{2}\tfrac{1}{2}; -\tfrac{1}{2} - \tfrac{1}{2}\rangle_n - \langle -\tfrac{1}{2}\tfrac{1}{2}; \tfrac{1}{2} - \tfrac{1}{2}\rangle_n\}, \tag{5.30f}$$

$$U_{xy} = -\frac{1}{\sigma_u}\, Im\, \{\langle \tfrac{1}{2}\tfrac{1}{2}; -\tfrac{1}{2} - \tfrac{1}{2}\rangle_n + \langle -\tfrac{1}{2}\tfrac{1}{2}; \tfrac{1}{2} - \tfrac{1}{2}\rangle_n\}, \tag{5.30g}$$

in the natural frame and

$$S_P = -\frac{2}{\sigma_u}\, Im\, \{\langle \tfrac{1}{2}\tfrac{1}{2}; -\tfrac{1}{2}\tfrac{1}{2}\rangle_c\}, \tag{5.31a}$$

$$S_A = -\frac{2}{\sigma_u}\, Im\, \{\langle \tfrac{1}{2} - \tfrac{1}{2}; \tfrac{1}{2}\tfrac{1}{2}\rangle_c\}, \tag{5.31b}$$

$$T_x = \frac{1}{\sigma_u}\, \big[\langle -\tfrac{1}{2} - \tfrac{1}{2}; \tfrac{1}{2}\tfrac{1}{2}\rangle_c + \langle -\tfrac{1}{2}\tfrac{1}{2}; \tfrac{1}{2} - \tfrac{1}{2}\rangle_c\big], \tag{5.31c}$$

$$T_y = \frac{1}{\sigma_u}\, \big[\langle -\tfrac{1}{2} - \tfrac{1}{2}; \tfrac{1}{2}\tfrac{1}{2}\rangle_c - \langle -\tfrac{1}{2}\tfrac{1}{2}; \tfrac{1}{2} - \tfrac{1}{2}\rangle_c\big], \tag{5.31d}$$

$$T_z = \frac{1}{\sigma_u}\, \big[\langle \tfrac{1}{2}\tfrac{1}{2}; \tfrac{1}{2}\tfrac{1}{2}\rangle_c - \langle \tfrac{1}{2} - \tfrac{1}{2}; \tfrac{1}{2} - \tfrac{1}{2}\rangle_c\big], \tag{5.31e}$$

$$U_{xz} = \frac{1}{\sigma_u}\, \big[\langle -\tfrac{1}{2}\tfrac{1}{2}; \tfrac{1}{2}\tfrac{1}{2}\rangle_c - \langle -\tfrac{1}{2} - \tfrac{1}{2}; \tfrac{1}{2} - \tfrac{1}{2}\rangle_c\big] = \frac{2}{\sigma_u}\, Re\, \{\langle \tfrac{1}{2}\tfrac{1}{2}; -\tfrac{1}{2}\tfrac{1}{2}\rangle_c\}, \tag{5.31f}$$

$$U_{zx} = -\frac{1}{\sigma_u}\, \big[\langle \tfrac{1}{2} - \tfrac{1}{2}; \tfrac{1}{2}\tfrac{1}{2}\rangle_c + \langle \tfrac{1}{2}\tfrac{1}{2}; \tfrac{1}{2} - \tfrac{1}{2}\rangle_c\big] = -\frac{2}{\sigma_u}\, Re\, \{\langle \tfrac{1}{2} - \tfrac{1}{2}; \tfrac{1}{2}\tfrac{1}{2}\rangle_c\}, \tag{5.31g}$$

in the collision frame.

### 5.4.2 Exact symmetry relationships

It is instructive to discuss two of the most important special cases for the generalized $STU$ parameters:

    $(i)$  the scattering from targets with zero angular momentum in both the initial and final states (i.e., $J_0 = J_1 = 0$) and

    $(ii)$ elastic scattering from targets with arbitrary angular momentum $J = J_0 = J_1$.

In the first case, it can be shown [5.4] that, in the natural frame,

$$S_P = \Pi_1\, \Pi_0\, S_A\,, \tag{5.32a}$$

$$T_z = \Pi_1\, \Pi_0\,, \tag{5.32b}$$

$$T_y = \Pi_1\, \Pi_0\, T_x\,, \tag{5.32c}$$

$$U_{ux} = \Pi_1\, \Pi_0\, U_{xy}\,, \tag{5.32d}$$

where $\Pi_0$ and $\Pi_1$ denote the parities of the atomic states involved in the transition. Hence, the number of independent $STU$ parameters is reduced to three. Note that $T_z$ can only become $+1$ or $-1$, depending on the product of the initial-state and final-state parities. If this product is $+1$ (as, for example, in elastic scattering), (5.32) yield the well-known $STU$ parameters of Kessler [5.6] by setting $S_P = S_A \equiv S$ (the so-called "Sherman function"), $T_x = T_y \equiv T$, and $U_{yx} = U_{xy} \equiv U$.

The second special case deals with *elastic* scattering from targets with arbitrary angular momentum $J_0 = J_1$. Here, time reversal invariance of the projectile–target interaction leads to

$$S_P = S_A \tag{5.33a}$$

and

$$U_{yx} - U_{xy} = (T_x - T_y)\tan\theta . \tag{5.33b}$$

Equation (5.33a) and the equivalent of (5.33b) in the collision frame were derived in [5.4]. Hence, the number of independent parameters is reduced to five relative generalized $STU$ parameters and one absolute differential cross section $\sigma_\mathrm{u}$.

### 5.4.3 An approximate symmetry: The fine-structure effect

The "fine-structure effect," in its pure form, is derived by assuming the validity of (5.14). It was discussed in detail by Hanne [5.7] for the special cases of $^2\mathrm{S^e} \rightarrow \,^2\mathrm{P^o}$ transitions in alkali-like systems and $^1\mathrm{S^e} \rightarrow \,^3\mathrm{P^o}$ transitions in quasi-two-electron targets, and a generalization of Hanne's treatment was given by Goerss *et al.* [5.8]. Using (5.14), one can obtain an interesting result for the sum over fine-structure levels of the parameters defined in (5.23). In detail, we evaluate

$$\sum_{J_1,J_0} \langle m_1' m_0'; m_1 m_0 \rangle =$$

$$\sum_{J_1,J_0,S,S',\text{all }M's} C(L_1, M_{L_1}'; S_1, M_{S_1}' | J_1, M_1) C(L_0, M_{L_0}'; S_0, M_{S_0}' | J_0, M_0)$$

$$\times\, C(S_1, M_{S_1}'; \tfrac{1}{2}, m_1' | S', M_S') C(S_0, M_{S_0}'; \tfrac{1}{2}, m_0' | S', M_S')$$

$$\times\, C(L_1, M_{L_1}; S_1, M_{S_1} | J_1, M_1) C(L_0, M_{L_0}; S_0, M_{S_0} | J_0, M_0)$$

$$\times\, C(S_1, M_{S_1}; \tfrac{1}{2}, m_1 | S, M_S) C(S_0, M_{S_0}; \tfrac{1}{2}, m_0 | S, M_S)$$

$$\times\, f^{S'}(M_{L_1}', M_{L_0}')\, f^S(M_{L_1}, M_{L_0})^* . \tag{5.34}$$

In this case, the sums over $J_1$ and $M_1$ as well as over $J_0$ and $M_0$ can be performed by using well-known relations for the Clebsch–Gordan coefficients [5.8]. The result $\delta(M_{L_1}, M_{L_1}')\,\delta(M_{L_0}, M_{L_0}')\,\delta(M_{S_1}, M_{S_1}')\,\delta(M_{S_0}, M_{S_0}')$ can then be used in the remaining Clebsch–Gordan coefficients in (5.34) to yield the selection rule

$$M_S' - M_S = m_1' - m_1 = m_0' - m_0 . \tag{5.35}$$

Hence, the sums (5.34) will be zero if $m_1' - m_1 \neq m_0' - m_0$. From (5.28) in the collision frame, it follows that the mean values

$$\langle S_P \rangle \equiv \sum_{J_1,J_0} (2J_0 + 1)\, \sigma(J_1, J_0)\, S_P(J_1, J_0)\, /\sigma_{tot}\,, \qquad (5.36a)$$

$$\langle S_A \rangle \equiv \sum_{J_1,J_0} (2J_0 + 1)\, \sigma(J_1, J_0)\, S_P(J_1, J_0)\, /\sigma_{tot}\,, \qquad (5.36b)$$

$$\langle U_{xz} \rangle \equiv \sum_{J_1,J_0} (2J_0 + 1)\, \sigma(J_1, J_0)\, U_{xz}(J_1, J_0)\, /\sigma_{tot}\,, \qquad (5.36c)$$

$$\langle U_{zx} \rangle \equiv \sum_{J_1,J_0} (2J_0 + 1)\, \sigma(J_1, J_0)\, U_{zx}(J_1, J_0)\, /\sigma_{tot}\,, \qquad (5.36d)$$

with

$$\sigma_{tot} \equiv \sum_{J_1,J_0} (2J_0 + 1)\, \sigma(J_1, J_0)\,, \qquad (5.37)$$

will vanish in the approximation of the pure fine-structure effect. With the index substitutions outlined below (5.26), these relationships will also hold for the equivalent parameters in the natural frame.

Important examples of these relationships within the approximation of the fine-structure effect are the following transitions (Exercise 5.3):

(i) $^2S_{1/2} \rightarrow\ ^2P^o_{1/2,3/2}$, for which

$$S_P(1/2) = -2\, S_P(3/2)\,, \qquad (5.38a)$$

$$S_A(1/2) = -2\, S_A(3/2)\,, \qquad (5.38b)$$

$$U_{xy}(1/2) = U_{yx}(1/2) = -2\, U_{xy}(3/2) = -2\, U_{yx}(3/2)\,, \qquad (5.38c)$$

$$T_x(1/2) = T_y(1/2) = T_z(1/2) = T_x(3/2)$$
$$= T_y(3/2) = T_z(3/2)\,; \qquad (5.38d)$$

(ii) $^1S_0 \rightarrow\ ^3P^o_{0,1,2}$, for which

$$S_P(0) = 2\, S_P(1) = -2\, S_P(2)\,, \qquad (5.39a)$$

$$S_A(0) = 2\, S_A(1) = -2\, S_A(2)\,, \qquad (5.39b)$$

$$U_{xy}(0) = U_{yx}(0) = 2\, U_{xy}(1) = 2\, U_{yx}(1)$$
$$= -2\, U_{xy}(2) = -2\, U_{yx}(2)\,, \qquad (5.39c)$$

$$\sum_{J_1} T_x(J_1) = \sum_{J_1} T_y(J_1) = \sum_{J_1} T_z(J_1) = -1\,. \qquad (5.39d)$$

(For simplicity, we have only listed $J_1$ in the arguments.)

As can be seen from (5.38,39), the nonrelativistic approximation may also lead to relationships between individual generalized $STU$ parameters for a single fine-structure level. The validity of these relationships can be checked against experimental data and either provide important information about the influence of relativistic effects in the collision or serve as a consistency check in cases where such effects should be very small.

## 5.5 Irreducible Tensor Operators and State Multipoles

The general density-matrix theory can be formulated in a very elegant fashion by decomposing the density operator in terms of irreducible components whose matrix elements become the so-called "state multipoles." In such a formulation, full advantage may be taken of the most sophisticated techniques developed in angular-momentum algebra. A thorough introduction to this method can be found, for example, in the book by Blum [5.1].

### 5.5.1 Basic definitions

The density operator for an ensemble of particles in quantum states labeled as $|J, M\rangle$ where $J$ and $M$ are the total angular momentum and its magnetic component, respectively, can be written as

$$\rho = \sum_{M'M} \rho^J_{M'M} |JM'\rangle\langle JM|, \tag{5.40}$$

where

$$\rho^J_{M'M} = \langle JM'|\rho|JM\rangle \tag{5.41}$$

are the matrix elements. Since it is sufficient for our cases of interest, we have assumed that $J$ is well defined, but coherences ($M' \neq M$) between different magnetic sublevels are possible.

Alternatively, one may write

$$\rho = \sum_{KQ} \langle T(J)^\dagger_{KQ}\rangle\, \mathbf{T}(J)_{KQ}, \tag{5.42}$$

where the irreducible tensor operators are defined as

$$\mathbf{T}(J)_{KQ} = \sum_{M'M} (-1)^{J-M}\,(J, M'; J, -M|K, Q)\,|JM'\rangle\langle JM| \tag{5.43}$$

and the "state multipoles" or "statistical tensors" are given by

$$\langle T(J)^\dagger_{KQ}\rangle = \sum_{M'M} (-1)^{J-M}\,(J, M'; J, -M|K, Q)\,\langle JM'|\rho|JM\rangle. \tag{5.44}$$

Note that the selection rules for the Clebsch–Gordan coefficients imply

$$0 \leq K \leq 2J\,, \tag{5.45a}$$
$$M' - M = Q\,. \tag{5.45b}$$

Equation (5.44) can be inverted through the orthogonality condition of the Clebsch–Gordan coefficients to give

$$\langle J'M'|\rho|JM\rangle = \sum_{KQ}(-1)^{J-M}\,(J,M';J,-M|K,Q)\,\langle T(J)^{\dagger}_{KQ}\rangle. \qquad (5.46)$$

The transformation properties of the tensor operators and the state multipoles defined in two different coordinate systems are given by expressions similar to (5.2) as

$$\mathbf{T}(J)_{KQ} = \sum_{q}\mathbf{T}(J)_{Kq}\,D(\alpha,\beta,\gamma)^{K}_{qQ}\,; \qquad (5.47a)$$

i.e., the rank $K$ of the tensor operator is invariant, and

$$\langle T(J)^{\dagger}_{KQ}\rangle = \sum_{q}\langle T(J)^{\dagger}_{Kq}\rangle\,D(\alpha,\beta,\gamma)^{K*}_{qQ}\,. \qquad (5.47b)$$

The irreducible tensor operators fulfill the orthogonality condition

$$tr\left\{\mathbf{T}(J)_{KQ}\,\mathbf{T}(J)^{\dagger}_{K'Q'}\right\} = \delta_{K'K}\,\delta_{Q'Q}. \qquad (5.48)$$

With

$$\mathbf{T}(J)_{00} = \frac{1}{\sqrt{2J+1}}\,\mathbf{1} \qquad (5.49)$$

being proportional to the unit operator $\mathbf{1}$, it follows that all tensor operators have vanishing trace, except for the monopole $\mathbf{T}(J)_{00}$.

The hermiticity condition for the density matrix implies

$$\langle T(J)^{\dagger}_{KQ}\rangle^{*} = (-1)^{Q}\langle T(J)^{\dagger}_{K-Q}\rangle. \qquad (5.50)$$

Hence, the state multipoles $\langle T(J)^{\dagger}_{K0}\rangle$ are real numbers. Furthermore, the transformation property (5.47b) of the state multipoles imposes restrictions on nonvanishing state multipoles to describe systems with given symmetry properties. In detail, one finds the following:

a)  For spherically symmetric systems,

$$\langle T(J)^{\dagger}_{KQ}\rangle = \langle T(J)^{\dagger}_{Kq}\rangle_{\mathrm{rot}} \qquad (5.51)$$

for *all* sets of Euler angles. This implies that only the monopole term $\langle T(J)^{\dagger}_{00}\rangle$ may be different from zero.

b)  For axially symmetric systems,

$$\langle T(J)^{\dagger}_{KQ}\rangle = \langle T(J)^{\dagger}_{Kq}\rangle_{\mathrm{rot}} \qquad (5.52)$$

for *all* Euler angles $\phi$ that describe a rotation around the $z$ axis. Since this angle enters through a factor $\exp\{-iQ\phi\}$ into the general transformation formula (5.47b), it follows that only state multipoles with $Q = 0$ may be different from zero in such a situation.

c)  For systems with properties invariant under reflection in the $xy$ plane,

$$\langle T(J)^\dagger_{KQ}\rangle = (-1)^Q \langle T(J)^\dagger_{KQ}\rangle. \tag{5.53}$$

In this case, state multipoles with odd component $Q$ must vanish.

d)  For systems with properties invariant under reflection in the $xz$ plane,

$$\langle T(J)^\dagger_{KQ}\rangle = (-1)^K \langle T(J)^\dagger_{KQ}\rangle^*. \tag{5.54}$$

In this case, state multipoles with even rank $K$ are real numbers, while those with odd rank are purely imaginary.

Equations (5.53) and (5.54) are important for angle-differential impact excitation studies of unpolarized targets by unpolarized projectiles or for angle-integrated experiments with spin-polarized beams. Similarly, (5.52) expresses the cylindrical symmetry of such a process in angle-integrated experiments with unpolarized beams.

### 5.5.2 Coupled systems

Tensor operators and state multipoles   for coupled systems are built up as direct products of the operators for the individual systems. For example, the combined density operator for two subsystems in basis states $|L, M_L\rangle$ and $|S, M_S\rangle$ is constructed as [5.1]

$$\rho = \sum_{KQkq} \langle T(L)^\dagger_{KQ} \times T(S)^\dagger_{kq}\rangle \, [\mathbf{T}(L)_{KQ} \times \mathbf{T}(S)_{kq}]. \tag{5.55}$$

If the two systems are uncorrelated, the state multipoles factorize as

$$\langle T(L)^\dagger_{KQ} \times T(S)^\dagger_{kq}\rangle = \langle T(L)^\dagger_{KQ}\rangle \, \langle T(S)^\dagger_{kq}\rangle. \tag{5.56}$$

Furthermore, coupled operators can be defined as

$$\mathbf{T}(J', J)_{K'Q'} = \sum_{KQkq} \hat{K}\hat{k}\hat{J}\hat{J}' \, (KQ, kq|K'Q')$$

$$\times \left\{ \begin{array}{ccc} K & k & K' \\ L & S & J' \\ L & S & J \end{array} \right\} \mathbf{T}(L)_{KQ} \times \mathbf{T}(S)_{kq}, \tag{5.57}$$

where $\hat{x} \equiv \sqrt{2x+1}$ and $\left\{ \begin{array}{ccc} j_1 & j_2 & j_3 \\ j_4 & j_5 & j_6 \\ j_7 & j_8 & j_9 \end{array} \right\}$ is a 9j-symbol.

Note that the coupling of fixed $L$ and $S$ values leads, in general, to several possible $J$ values, in accordance with the rules for angular-momentum coupling.

### 5.5.3 Time evolution of state multipoles: Quantum beats

The time evolution of the density operator is determined by the Liouville equation [5.1]

$$\rho(t) = \mathbf{U}(t)\,\rho(0)\,\mathbf{U}^{\dagger}(t)\,, \tag{5.58}$$

where $\mathbf{U}(t)$ is the time evolution operator that relates wavefunctions at times $t_0 = 0$ and $t$ according to

$$|\Psi(\mathbf{r},t)\rangle = \mathbf{U}(t)\,|\Psi(\mathbf{r},0)\rangle\,. \tag{5.59}$$

From the general expansion

$$\rho(t) = \sum_{kq} \langle T(j;t)^{\dagger}_{kq}\rangle\, \mathbf{T}(j)_{kq} \tag{5.60}$$

and (5.44) for time $t_0 = 0$, it follows that

$$\langle T(j;t)^{\dagger}_{kq}\rangle = \sum_{JKQ} \langle T(J)^{\dagger}_{KQ}\rangle\, G(J,j;t)^{Qq}_{Kk}\,, \tag{5.61}$$

where the "perturbation coefficients" are defined as

$$G(J,j;t)^{Qq}_{Kk} = \mathrm{tr}\,[\mathbf{U}(t)\mathbf{T}(J)_{KQ}\mathbf{U}(t)^{\dagger}\mathbf{T}(j)^{\dagger}_{kq}]. \tag{5.62}$$

Hence, these coefficients relate the state multipoles at time $t$ to those at $t_0 = 0$.

An important application of the perturbation coefficients is the coherent excitation of several quantum states that subsequently decay by optical transitions. Such an excitation may be performed, for example, in beam–foil experiments, charge-transfer and capture processes, or electron–atom collisions where the energy width of the electron beam is too large to resolve the fine structure (or hyperfine structure) of the target states.

Suppose, for instance, that explicitly relativistic effects, such as the spin-orbit interaction between the projectile and the target, can be neglected *during* the collision process between an incident electron and a target atom. In this case, the orbital angular momentum $(L)$ system of the collisionally excited target states may be oriented and/or aligned, depending on the scattering angle of the projectile. On the other hand, the spin $(S)$ system remains unaffected; in particular, it remains unpolarized if both the target and the projectile beams are unpolarized before the collision.

During the lifetime of the excited target states, however, the spin-orbit interaction *within* the target produces an exchange of orientation between the $L$ and the $S$ systems, which results in a net loss of orientation in the $L$ system. This effect can be observed directly through the intensity and the polarization of the light emitted from the excited target ensemble (see

Figure 2.5). The perturbation coefficients for the fine-structure interaction are found to be [5.1]

$$G(L;t)_K = \frac{\exp\{-\gamma t\}}{2S+1} \sum_{J'J} (2J'+1)(2J+1) \left\{ \begin{matrix} L & J' & S \\ J & L & K \end{matrix} \right\}^2 \cos(\omega_{J'J}t),$$

(5.63)

where $\left\{ \begin{matrix} j_1 & j_2 & j_3 \\ j_4 & j_5 & j_6 \end{matrix} \right\}$ is a standard 6j-symbol and $\omega_{J'J} = \omega_{J'} - \omega_J$ corresponds to the (angular) frequency difference between the various multiplet states with total electronic angular momenta $J'$ and $J$, respectively. Also, $\gamma$ is the natural width of the spectral line; for simplicity, the same lifetime has been assumed in (5.63) for all states of the multiplet.

Note that the perturbation coefficients are independent of the multipole component $Q$ in this case, and that there is no mixing between different multipole ranks $K$. The cosine terms represent correlations between the signals from different fine-structure states, and they lead to oscillations in the light intensity and in the measured Stokes parameters in a time-resolved experiment.

Similar results can be derived for the hyperfine interaction and also to account for the combined effects of fine structure and hyperfine structure. The result is [5.1]

$$G(J;t)_K = \frac{\exp\{-\gamma t\}}{2I+1} \sum_{F'F} (2F'+1)(2F+1) \left\{ \begin{matrix} J & F' & I \\ F' & J & K \end{matrix} \right\}^2 \cos(\omega_{F'F}t)$$

(5.64)

for the hyperfine-structure interaction after excitation of a state with well-defined total electronic angular momentum $J$; i.e, the nuclear spin $I$ replaces the electronic spin $S$ in (5.63), $J$ replaces $L$, and the total angular momentum $F$ replaces $J$. Furthermore, one obtains [5.1]

$$G(L;t)_K = \frac{\exp\{-\gamma t\}}{(2J+1)(2I+1)} \sum_{J'JF'F} (2J'+1)(2J+1)$$

$$\times (2F'+1)(2F+1) \left\{ \begin{matrix} L & J' & S \\ J & L & K \end{matrix} \right\}^2 \left\{ \begin{matrix} J & F' & I \\ F & J & K \end{matrix} \right\}^2$$

$$\times \cos(\omega_{J'F'JF}t)$$

(5.65)

for the combined effect of fine structure and hyperfine structure, provided the hyperfine splitting is much smaller than the fine-structure splitting and hence $J$ is still a good quantum number. A more elaborate treatment of the general case with comparable splittings was given by Fano and Macek [5.9].

### 5.5.4 Time integration over quantum beats

If the excitation and decay times are not resolved in a given experimental setup, the perturbation coefficients need to be integrated over time. As a result, the quantum beats will disappear, but a net effect may still be visible through a depolarization of the emitted radiation. For the case of atomic fine-structure interaction discussed above, one finds [5.1]

$$\bar{G}(L)_K = \int_0^\infty dt\, G(L;t)_K = \frac{1}{2S+1} \sum_{J'J} (2J'+1)(2J+1)$$

$$\times \left\{ \begin{matrix} L & J' & S \\ J & L & K \end{matrix} \right\}^2 \frac{\gamma}{\gamma^2 + \omega_{J'J}^2}, \qquad (5.66)$$

and similar results are obtained after time integration of (5.64) and (5.65). Note that the amount of depolarization depends on the relationship between the fine-structure splitting and the natural line width. For $|\omega_{J'J}| \gg \gamma$ (if $J' \neq J$), the terms with $J' = J$ will dominate and cause the maximum depolarization; for the opposite case $|\omega_{J'J}| \ll \gamma$, the sum rule of the 6j-symbols can be applied and no depolarization will be observed. The smooth transition between these two extremes is required by the "principle of spectroscopic stability" [5.10]. It is not fulfilled in the treatment given by Penney [5.11], who assumes from the start that $|\omega_{J'J}| \gg \gamma$. For a detailed discussion, see Bartschat and Csanak [5.12].

Table 5.1 gives a list of some practically important time-integrated perturbation coefficients, as well as the corresponding intensity reduction and depolarization factors for radiation from a P state with positive reflection symmetry introduced in Chapter 2. The latter can be verified by multiplying the state multipoles in the general formulas for the Stokes parameters given in the next section. For the fine-structure depolarization, we also list the contributions from the individual fine-structure levels to the sum (5.66) for $J' = J$, i.e., the dominant term if the fine-structure level splitting is much larger than the natural line width. This information is useful if the fine-structure is actually resolved in the optical channel; i.e., the polarization from a particular line of the multiplet is observed.

Finally, we point out that the perturbation coefficients for hyperfine-structure depolarization effects are obtained by constructing a weighted average of the results for individual nuclear spins, with the weighting factors being the relative abundances [5.13].

**Table 5.1.** Intensity reduction, light depolarization factors, and *relative* perturbation coefficients $G_K \equiv \bar{G}(L)_K / \bar{G}(L)_0$ for light emission from P states with positive reflection symmetry in a direction normal to the scattering plane for nuclear spins $I = 0$ and $I = 3/2$, respectively [5.2].* The results for $I = 3/2$ are applicable if $\omega_{\mathrm{fs}} \gg \omega_{\mathrm{hfs}} \gg \gamma$, where $\omega_{\mathrm{fs}}$ ($\omega_{\mathrm{hfs}}$) are the fine-structure (hyperfine-structure) level splittings and $\gamma$ is the natural line width. If $\omega_{\mathrm{hfs}} \ll \gamma$, the results for $I = 0$ hold for all nuclear spins, and if $\omega_{\mathrm{fs}} \ll \gamma$, there is no depolarization.

| | Transition | $s_0$ | $c_1$ | $c_2$ | $G_1$ | $G_2$ |
|---|---|---|---|---|---|---|
| $I = 0$ | $^1\mathrm{P} \to {}^1\mathrm{S}$ | 1 | 1 | 1 | 1 | 1 |
| | $^2\mathrm{P} \to {}^2\mathrm{S}$ | 7/9 | 1 | 3/7 | 7/9 | 1/3 |
| | $^2\mathrm{P}_{1/2} \to {}^2\mathrm{S}_{1/2}$ | 2/3 | 1 | 0 | 2/3 | 0 |
| | $^2\mathrm{P}_{3/2} \to {}^2\mathrm{S}_{1/2}$ | 5/6 | 1 | 3/5 | 5/6 | 1/2 |
| | $^3\mathrm{P} \to {}^3\mathrm{S}$ | 41/54 | 27/41 | 15/41 | 1/2 | 15/54 |
| | $^3\mathrm{P}_0 \to {}^3\mathrm{S}_1$ | 2/3 | 0 | 0 | 0 | 0 |
| | $^3\mathrm{P}_1 \to {}^3\mathrm{S}_1$ | 3/4 | 1/3 | 1/3 | 1/4 | 1/4 |
| | $^3\mathrm{P}_2 \to {}^3\mathrm{S}_1$ | 47/60 | 45/47 | 21/47 | 3/4 | 7/20 |
| $I = 3/2$ | $^2\mathrm{P} \to {}^2\mathrm{S}$ | 209/300 | 325/627 | 27/209 | 13/36 | 9/100 |

*The relative perturbation coefficients cannot be used if more than one line is observed and the absolute line strength must be accounted for through an averaging procedure. An important example is the $n = 3 \to 2$ transition in atomic hydrogen.

## 5.6 Stokes Parameters

The state-multipole description is also widely used for the parameterization of the Stokes parameters that describe the polarization of light emitted in optical decays of excited atomic ensembles. The general case of excitation by spin-polarized projectiles was first treated by Bartschat *et al.* [5.14], whose work was recently extended [5.15].

The photon intensity emitted in a direction $\hat{n} = (\theta_\gamma, \phi_\gamma)$, after impact excitation of an atomic state with total electronic angular momentum $J$ and an electric dipole transition to a state with $J_{\mathrm{f}}$, is given by [5.14]

$$s_0(\theta_\gamma, \phi_\gamma) = C \left( \frac{2}{3\sqrt{2J+1}} \langle T(J)_{00}^\dagger \rangle \right.$$

$$+ (-1)^{J - J_{\mathrm{f}} + 1} \begin{Bmatrix} 1 & 1 & 2 \\ J & J & J_{\mathrm{f}} \end{Bmatrix} \left[ \sqrt{\tfrac{1}{6}} \langle T(J)_{20}^\dagger \rangle \, (3\cos^2\theta_\gamma - 1) \right.$$

$$- \mathrm{Re}\{\langle T(J)_{21}^\dagger \rangle\} \, \sin 2\theta_\gamma \cos\phi_\gamma + \mathrm{Re}\{\langle T(J)_{22}^\dagger \rangle\} \, \sin^2\theta_\gamma \cos 2\phi_\gamma$$

$$\left. \left. + \mathrm{Im}\{\langle T(J)_{21}^\dagger \rangle\} \, \sin 2\theta_\gamma \sin\phi_\gamma - \mathrm{Im}\{\langle T(J)_{22}^\dagger \rangle\} \, \sin^2\theta_\gamma \sin 2\phi_\gamma \right] \right),$$

$$(5.67)$$

where

$$C = \frac{e^2\omega^4}{2\pi c^3} \, |\langle J_f||\mathbf{r}||J\rangle|^2 \tag{5.68}$$

is a constant containing the frequency $\omega$ of the transition as well as the reduced radial dipole matrix element originating from the Wigner–Eckart theorem. (For details of the derivation, see [5.1] and [5.14].)

Similarly,

$$s_3(\theta_\gamma, \phi_\gamma) = C(-1)^{J-J_f+1} \begin{Bmatrix} 1 & 1 & 1 \\ J & J & J_f \end{Bmatrix} \Bigg[ \sqrt{2}\langle T(J)^\dagger_{10}\rangle \, \cos\theta_\gamma$$

$$- \mathrm{Re}\{\langle T(J)^\dagger_{11}\rangle\} \, 2\sin\theta_\gamma \cos\phi_\gamma + \mathrm{Im}\{\langle T(J)^\dagger_{11}\rangle\} \, 2\sin\theta_\gamma \sin\phi_\gamma \Bigg], \tag{5.69}$$

so that the circular polarization can be calculated as

$$P_3(\theta_\gamma, \phi_\gamma) = s_3(\theta_\gamma, \phi_\gamma)/s_0(\theta_\gamma, \phi_\gamma). \tag{5.70}$$

Finally, the two linear polarizations can be obtained from

$$s_1(\theta_\gamma, \phi_\gamma) = C\,(-1)^{J-J_f} \begin{Bmatrix} 1 & 1 & 2 \\ J & J & J_f \end{Bmatrix} \Bigg[ \sqrt{\tfrac{3}{2}}\langle T(J)^\dagger_{20}\rangle \, \sin^2\theta_\gamma$$

$$+ \mathrm{Re}\{\langle T(J)^\dagger_{21}\rangle\} \, \sin 2\theta_\gamma \cos\phi_\gamma + \mathrm{Re}\{\langle T(J)^\dagger_{22}\rangle\} \, (1+\cos^2\theta_\gamma)\cos 2\phi_\gamma$$

$$- \mathrm{Im}\{\langle T(J)^\dagger_{21}\rangle\} \, \sin 2\theta_\gamma \sin\phi_\gamma - \mathrm{Im}\{\langle T(J)^\dagger_{22}\rangle\} \, (1+\cos^2\theta_\gamma)\sin 2\phi_\gamma \Bigg]$$

$$\tag{5.71}$$

and

$$s_2(\theta_\gamma, \phi_\gamma) = C \begin{Bmatrix} 1 & 1 & 2 \\ J & J & J_f \end{Bmatrix} (-1)^{J-J_f}$$

$$\times \Bigg[ \mathrm{Re}\{\langle T(J)^\dagger_{21}\rangle\} \sin 2\theta_\gamma \sin\phi_\gamma + \mathrm{Re}\{\langle T(J)^\dagger_{22}\rangle\} \, 2\cos\theta_\gamma \sin 2\phi_\gamma$$

$$+ \mathrm{Im}\{\langle T(J)^\dagger_{21}\rangle\} \sin 2\theta_\gamma \cos\phi_\gamma + \mathrm{Im}\{\langle T(J)^\dagger_{22}\rangle\} \, 2\cos\theta_\gamma \cos 2\phi_\gamma \Bigg].$$

$$\tag{5.72}$$

Note that each state multipole gives rise to a characteristic angular dependence in the formulas for the *absolute* Stokes parameters, i.e., the light polarizations multiplied by the intensity. Hence, (5.70) is not trivial, since the numerator and denominator have to be calculated separately. Furthermore, perturbation coefficients may need to be applied to deal, for example, with depolarization effects due to internal or external fields.

As pointed out before, some of the state multipoles may vanish, depending on the experimental arrangement. Some examples, such as the "angle-integrated" and "generalized" Stokes parameters, were introduced in Chapter 4.

## 5.7 Atomic and Photon Density Matrices for P-State Excitation

As another example, we will now apply the formalism described in this chapter to the problem discussed in detail in Chapter 2, namely $^1S \to {}^1P^o$ excitation with subsequent optical decay back to a $^1S$ state. We use the natural coordinate frame and leave the corresponding derivation in the collision frame as an exercise to the reader.

We begin with the density matrix. Since the only nonvanishing scattering amplitudes for this case are $f_{\pm 1}$, the density matrix $\rho = |\Psi\rangle\langle\Psi|$ describing the P state can be written as

$$\rho = \begin{pmatrix} f_{+1} f_{+1}^* & 0 & f_{+1} f_{-1}^* \\ 0 & 0 & 0 \\ f_{-1} f_{+1}^* & 0 & f_{-1} f_{-1}^* \end{pmatrix} . \tag{5.73}$$

Next, we need to calculate the state multipoles for $L = 1$ (instead of $J = 1$). Using (5.44), we find

$$\langle T(1)_{00}^+ \rangle = \frac{1}{\sqrt{3}} \left( |f_{+1}|^2 + |f_{-1}|^2 \right), \tag{5.74a}$$

$$\langle T(1)_{10}^+ \rangle = \frac{1}{\sqrt{3}} \left( |f_{+1}|^2 - |f_{-1}|^2 \right), \tag{5.74b}$$

$$\langle T(1)_{20}^+ \rangle = \frac{1}{\sqrt{3}} \left( |f_{+1}|^2 + |f_{-1}|^2 \right), \tag{5.74c}$$

$$\langle T(1)_{22}^+ \rangle = \frac{1}{\sqrt{3}} f_{+1} f_{-1}^* , \tag{5.74d}$$

$$\langle T(1)_{11}^+ \rangle = \langle T(1)_{21}^+ \rangle = 0 , \tag{5.74e}$$

where (5.74e) also follows directly from (5.53) owing to the planar symmetry of the problem. Note that $\langle T(1)_{20}^+ \rangle = \langle T(1)_{00}^+ \rangle$ in this special case; i.e., these two parameters are not independent here. (They are independent in the case where $f_0 \neq 0$ and states with negative reflection symmetry can be excited; see also Chapter 2.)

For the calculation of the Stokes parameters, as measured along the $z^n$ direction, we choose the polar angles $\theta_\gamma^n = \phi_\gamma^n = 0$ of the photon detector. Although $\phi_\gamma^n$ is, in principle, not defined for $\theta_\gamma^n = 0$, the above choice guarantees agreement between the results obtained in the natural and the collision frame in which $\theta_\gamma^c = \phi_\gamma^c = \pi/2$, as well as with the results from Classical Optics [5.16]. Inserting (5.74) into (5.67–72), we obtain

$$s_0 = \frac{C}{3}\left(|f_{+1}|^2 + |f_{-1}|^2\right),\tag{5.75a}$$

$$s_1 = -\frac{2C}{3}|f_{+1}|\,|f_{-1}|\cos\delta,\tag{5.75b}$$

$$s_2 = \frac{2C}{3}|f_{+1}|\,|f_{-1}|\sin\delta,\tag{5.75c}$$

$$s_3 = \frac{C}{3}\left(|f_{-1}|^2 - |f_{+1}|^2\right).\tag{5.75d}$$

These results agree with (2.33) apart from the normalization factor $C/3$.

Transforming (5.75) to the light polarizations $P_i = s_i/s_0$, $i = 1, 2, 3$, we see that the density matrix (5.73) of the excited P state can also be written as

$$\rho = \sigma\,\frac{1}{2}\begin{pmatrix} 1 - P_3 & 0 & -P_1 + iP_2 \\ 0 & 0 & 0 \\ -P_1 - iP_2 & 0 & 1 + P_3 \end{pmatrix}\tag{5.76a}$$

$$= \sigma\,\frac{1}{2}\begin{pmatrix} 1 + L_\perp & 0 & -P_\ell\,e^{-2i\gamma} \\ 0 & 0 & 0 \\ -P_\ell\,e^{2i\gamma} & 0 & 1 - L_\perp \end{pmatrix},\tag{5.76b}$$

where $\sigma$, $L_\perp$, $P_\ell$, and $\gamma$ have been defined in Chapter 2.

Comparing to the photon density matrix in the helicity frame,

$$\rho_{\mathrm{ph}}^{+z} = I\,\frac{1}{2}\begin{pmatrix} 1 - P_3 & -P_1 + iP_2 \\ -P_1 - iP_2 & 1 + P_3 \end{pmatrix},\tag{5.77}$$

which characterizes the photons emitted along the positive $z$ direction with intensity $I$ [5.1], reveals that the two matrices have the same structure in this particular case, provided the zeros in the second row and second column of (5.76a) are omitted. In other words, the structure of the *atomic* density matrix characterizing a P state with positive reflection symmetry is essentially identical to the *photon* density matrix for the light emitted along the positive $z$ direction. This results in the direct relationship between the Stokes parameters and the shape and dynamics of the atomic P-state charge cloud (see Chapter 2), since the absolute normalization (i.e., the trace of the two matrices) is irrelevant in this case. For excitation of states with angular momentum $L > 1$, however, this simple connection is no longer valid.


## Exercises

1. Write down symbolically the nonvanishing scattering amplitudes for an excited D state after impact excitation from an S state with conservation of reflection symmetry in the natural and the collision frames. Transform one set of amplitudes into the other.

2. Transform the scattering amplitudes for excitation of P and D states with conservation of reflection symmetry in the natural frame from the atomic to the molecular basis introduced in Chapter 2.

3. Prove Equations (5.38) and (5.39) for the generalized $STU$ parameters in the nonrelativistic limit.

4. Verify the items in Table 5.1. Extend the table to include nuclear spins between 1/2 and 7/2. Calculate the perturbation coefficients $G_1$ and $G_2$ for the $(6s6p)^3P_1^o \rightarrow (6s^2)^1S_0$ transition in mercury using the natural isotope mixture of this element. Compare your results with those given in [5.13].

5. By calculating the fine-structure perturbation coefficients and the proper 6j-symbols, show that a P-state density matrix, after excitation with conservation of reflection symmetry, can directly be characterized in terms of "reduced" light polarizations introduced in Chapter 2 for the following cases:

   a) $^2S \rightarrow {}^2P^o$ with subsequent optical decay to a $^2S$ state (an example would be electron-impact excitation of H $(2p)^2P^o$);

   b) $^1S \rightarrow {}^3P^o$ with subsequent optical decay to a $^3S$ state (an example would be electron-impact excitation of He $(1s3p)^3P^o$);

   c) $^1S \rightarrow {}^1P^o$ with subsequent optical decay to a $^1D$ state (an example could be electron-impact excitation of He $(1s4p)^1P^o$).

6. Derive Equation (11) of Eminyan *et al.* [5.17] for the photon angular distribution in the collision plane

   a) in terms of $\lambda$ and $\chi$,

   b) in terms of $P_\ell$ and $\gamma$.

7. Continue Exercise 1 by writing down the density matrix of an excited $^1D$ state after impact excitation from an S state with conservation of reflection symmetry in the natural frame and derive expressions for the Stokes parameter $P_1 - P_4$ for a subsequent optical decay to a P state. Express as far as possible the D-state density-matrix elements in terms of the Stokes parameters (for details, see [5.18]).

8. Consider excitation of a P state with negative reflection symmetry (i.e., $M = 0$ in the natural frame) to a D state. Such a state can be prepared, for example, by pumping with linearly polarized laser light. Assume that the interaction responsible for the excitation conserves the reflection symmetry of the target states.

   a) Which sublevels of the D states can be excited in the natural frame?

   b) Show that $P_4 \neq 0$ and explain your result.

9. Consider electron-impact excitation of a $J = 0^e$ to $J = 1^o$ transition without conservation of reflection symmetry, as in $(6s^2)^1S_0 \rightarrow (6s6p)^3P_1^o$ in Hg.

   a) Write down the nonvanishing amplitudes in the natural and the collision frames.

   b) Work out expressions for the generalized $STU$ parameters in both frames.

c) Work out expressions for the state multipoles in both frames. Compare your results with those given in [5.15] and [5.3].

d) Work out expressions for the generalized Stokes parameters in both frames. Compare your results for the natural frame with with those given in [5.3].

e) Discuss possibilities of performing a "complete experiment" by determining all independent scattering amplitudes. Compare your results with [5.3]. Comment on the structure of the resulting equations in the two frames.

# References


5.1 K. Blum, *Density Matrix Theory and Applications* (2nd edition) Plenum, New York 1996.

5.2 N. Andersen, J.W. Gallagher, and I.V. Hertel, Phys. Rep. **180** (1988) 1.

5.3 N. Andersen, K. Bartschat, J.T. Broad, and I.V. Hertel, Phys. Rep. **279** (1997) 251.

5.4 K. Bartschat, Phys. Rep. **180** (1989) 1.

5.5 A.R. Edmonds, *Angular Momentum in Quantum Mechanics*, Princeton University Press, Princeton 1957.

5.6 J. Kessler, *Polarized Electrons* (2nd edition), Springer, Berlin 1985.

5.7 G.F. Hanne, Phys. Rep. **95** (1983) 95.

5.8 H.-J. Goerss, R.-P. Nordbeck, and K. Bartschat, J. Phys. B **24** (1991), 2833.

*5.9 U. Fano and J.H. Macek, Rev. Mod. Phys. **45** (1973) 553.

*5.10 I.C. Percival and M.J. Seaton, Phil. Trans. Roy. Soc. London A **251** (1958). 113.

*5.11 W.G. Penney, Proc. Nat. Acad. Sci. **18** (1932) 231.

5.12 K. Bartschat and G. Csanak, Comments At. Mol. Phys. **32** (1996) 233.

5.13 A. Wolcke, K. Bartschat, K. Blum, H. Borgmann, G.F. Hanne, and J. Kessler, J. Phys. B **16** (1983) 639.

5.14 K. Bartschat, K. Blum, G.F. Hanne, and J. Kessler, J. Phys. B **14** (1981) 376.

5.15 N. Andersen and K. Bartschat, J. Phys. B **27** (1994) 3189; corrigendum: J. Phys. B **29** (1996) 1149.

5.16 M. Born and E. Wolf, *Principles of Optics* (4th edition), Pergamon, New York 1970.

*5.17 M. Eminyan, K.B. MacAdam, J. Slevin, and H. Kleinpoppen, J. Phys. B **7** (1974) 1519.

5.18 N. Andersen and K. Bartschat, J. Phys. B **30** (1997) 5071.

# 6. Computational Methods

The basic methods applied in numerical simulations of the collision processes discussed in the previous chapters are summarized. An introductory section compares the similarities and differences in the techniques that are used for electron or heavy-particle impact. Some details for these two cases are presented in subsequent sections.

## 6.1 Electron vs. Heavy-Particle Impact

A collision between a projectile A and a target B, prepared in initial states $A(\alpha)$ and $B(\beta)$, respectively, will typically lead to final states $A(\alpha')$ and $B(\beta')$ of the collision partners (see also Figure 1.1). This process, with a change in kinetic energy of $\Delta E$, can symbolically be described by

$$A(\alpha) + B(\beta) \longrightarrow A(\alpha') + B(\beta') - \Delta E. \tag{6.1}$$

In addition to changing the quantum states of projectile and target, such as in electron-impact excitation, rearrangement collisions that change the actual chemical behavior of the projectile or the target (or both) also deserve to be considered. Such processes include electron-impact ionization where $B(\beta')$ corresponds to one (or more) ejected electrons plus the residual ion, electron transfer that changes the ionization stage of the target and the projectile, as well as collisions that result in dissociation or (partial) recombination.

Although collisions involving arbitrary projectiles and targets are, in principle, tractable by starting with the many-particle Schrödinger equation (or the corresponding Dirac Hamiltonian in a full-relativistic approach) and introducing the proper boundary conditions for the initial and final states of the system, in practice approximate schemes have to be introduced for all but the simplest model problems. As will be shown below, the methods used for electron or heavy-particle impact differ in many important aspects, although some similarities can certainly be found, especially at high velocities.

The most important difference between the projectile being either an electron or a heavy particle is the projectile mass. For electron impact, this mass is very small compared to that of any atomic or ionic target. Consequently, the center-of-mass coordinate frame is almost identical to the laboratory frame.

In addition, the projectile electrons can only change their linear momentum (including its magnitude and therefore the energy) and their spin projection with respect to a given quantization axis. Most importantly, however, a classical-trajectory picture is typically not a good approximation to describe the scattered electron, and hence a fully quantum mechanical approach is usually necessary.

On the other hand, collisions between heavy particles, such as ion–atom, ion–molecule, or ion–surface collisions, are often well described in semi-classical approximations. Here the motion of the nuclei and all core (spectator) electrons is described via a trajectory, and only the active electrons (often just one) are described quantum mechanically. Such a trajectory picture generally leads to a time dependence in the resulting equations for the active electron(s), thereby introducing additional complexity. In contrast, the fully quantum mechanical treatment of electron collisions can, in principle, be performed in a time-independent "steady-state model," although time-dependent approaches may give additional insight into the collision dynamics. An interesting alternative for treatment of heavy-particle collisions involving, e.g., multiply charged ions, where perturbation methods fail and coupled-channel calculations are limited by basis set requirements, is the "Classical-Trajectory Monte-Carlo" (CTMC) method.

Finally, electron–atom collisions are well described by single-center expansions of the wavefunctions around the nucleus. Already in electron–molecule collisions, a choice between single-center and multiple-center expansions needs to be made or, if the problem is treated by different expansions depending on the region of space, transformations to join the results smoothly need to be introduced. Similarly, heavy-particle collisions can be described in the "atomic orbital" (AO) or the "molecular orbital" (MO) representation, and special care must be taken in the case of rearrangement collisions if, for example, an electron is transferred from one center to the other.

The ultimate goal of a numerical calculation involves the extraction of observable parameters, such as the angle-integrated or angle-differential cross sections, alignment, orientation, or spin-polarization parameters introduced in Chapter 4. In most cases, this is achieved via the calculation of scattering amplitudes that are directly related to these observables, as shown in Chapter 5. On the other hand, some quantities, such as angle-integrated cross sections, can be obtained more directly from angle-independent $\mathbf{T}$-matrix elements or via projection techniques in time-dependent approaches.

In Section 6.2, we summarize the basic ideas behind some standard methods used for the numerical simulation of elastic and inelastic electron–atom collisions in a nonrelativistic framework. Relativistic effects may be introduced either by starting with the many-particle Dirac equation or, on a perturbative level, by modifying the corresponding Schrödinger equation by including relativistic correction terms. The methods can also be extended to ionic targets by modifying the boundary conditions for the projectile wavefunction, to molecular targets by accounting for the nuclear motion, and to ionization by describing at least two outgoing electrons and the residual ion. For a more detailed discussion of these extensions, we recommend the many textbooks and reviews on this topic. In addition, many numerical aspects are discussed in [6.1].

Section 6.3 provides a summary of the theoretical methods invoked for the description of heavy-particle collisions. The main emphasis is put on describing the semiclassical approximation and the techniques that are implemented in order to evaluate collision parameters in various velocity regimes. The two basis sets chosen for solving the close-coupling equations, molecular orbitals (MO) and atomic orbitals (AO), are briefly introduced, and their advantages and limitations are pointed out. A recently developed alternative method of solving the time-dependent Schrödinger equation directly on a three-dimensional lattice is summarized. Finally, the idea behind the CTMC method, invoking the solution of Hamilton's classical equations for the system, is briefly described.

If not mentioned otherwise, atomic units are used throughout this chapter; i.e., charges are measured in units of the elementary charge $|q_e| \approx 1.602 \cdot 10^{-19}$ C, lengths in units of the Bohr radius $a_0 \approx 0.529 \cdot 10^{-10}$ m, masses in units of the electron mass $m_e \approx 9.108 \cdot 10^{-31}$ kg, and energies in Hartrees, where 1 Hartree = 2 Rydberg $\approx 27.21$ eV $\approx 4.359 \cdot 10^{-18}$ J.

## 6.2 Computational Methods for Electron Scattering

As mentioned above, most of the computational efforts to describe electron scattering have concentrated on quantal methods, although in some cases classical or semiclassical methods have been used as well. Consequently, most of this section deals with quantum mechanical attempts to simulate these collisions, based on the nonrelativistic Schrödinger equation. Note that only for atomic hydrogen and other one-electron ionic targets the problem consists of the collision part alone, since the structure part of this problem is known exactly in nonrelativistic quantum mechanics. The generalization to more complex systems, however, is straightforward, except that atomic or molecular structure codes need to be used first to describe the target states.

The numerical methods used in both electron and heavy-particle collisions generally fall into one of two categories, namely perturbation-series expansions based on variations of the Born series or the nonperturbative close-coupling approach originating from the expansion of the trial wavefunction into a set of basis functions. Typically, Born-series expansions were used successfully for high collision energies where the projectile–target interaction is a relatively small perturbation of the free-particle motion with large kinetic energy. On the other hand, close-coupling expansions were applied to simulate low-energy collisions, where the incident energy is such that only elastic scattering or at most excitation of a few low-lying target states is possible. However, this method may be extended to deal with higher collision energies as well, provided the effect of the target continuum states is accounted for properly. For details on the "convergent close-coupling" (CCC) approach and some applications, we refer to the recent reviews by Bray and Stelbovics [6.2] and Fursa and Bray [6.3].

We now give an overview of the basic equations used in electron–atom collision calculations. This overview is intended for readers with a basic knowledge of quantum mechanical scattering theory and should not be regarded as a replacement for standard textbooks on atomic collision theory. Since it seems impossible to give even a nearly complete list of general references, we only mention the recent books by McCarthy and Weigold [6.4] for a more detailed discussion of the general formulation and by Bartschat [6.1] for numerical aspects.

### 6.2.1 Potential scattering

Consider electron scattering from an atom or ion. In the simplest case, the effect of the target atom on the projectile motion is represented by a potential $V(\boldsymbol{r})$ in the Schrödinger equation for the projectile with a total energy $E$,

$$H\Psi_E^+(\boldsymbol{k},\boldsymbol{r}) = [T + V(\boldsymbol{r})]\Psi_E^+(\boldsymbol{k},\boldsymbol{r}) = E\,\Psi_E^+(\boldsymbol{k},\boldsymbol{r})\,, \tag{6.2}$$

where $T$ is the kinetic energy operator in the Hamiltonian $H$ and $\boldsymbol{k}$ is the linear momentum of the incident projectile with $k^2 = E/2$.

We simplify the situation further by specializing to the case of a short-range spherically symmetric potential $V(\boldsymbol{r}) = V(r)$ that vanishes sufficiently fast with increasing distance $r$. For such potentials, (6.2) needs to be solved subject to the boundary condition

$$\Psi_E^+(\boldsymbol{k},\boldsymbol{r}) \longrightarrow e^{i\boldsymbol{k}\cdot\boldsymbol{r}} + f(\theta)\,\frac{e^{ikr}}{r} \quad \text{for } r \longrightarrow \infty\,. \tag{6.3}$$

This boundary condition expresses the physical situation of an incident projectile beam described by a plane wave that is scattered by a potential centered at $r = 0$. The scattered wave, as seen by a detector far away from the scattering region, is an outgoing spherical wave (superscript $+$) with amplitude $f(\theta)$. This "scattering amplitude" depends on the scattering angle $\theta$; recall that this is the angle between the incident beam direction $\boldsymbol{k}$ and the direction $\boldsymbol{k}'$ where the detector is located and the scattered projectiles are detected. There is no dependence on the azimuthal angle $\phi$ in the scattering amplitude, due to the cylindrical symmetry of the problem around the incident beam ($z$) axis that is used as the quantization axis in standard numerical calculations. Modifications of (6.3) to treat long-range potentials, especially a Coulomb term proportional to $1/r$, represent no fundamental problem and will not be discussed here.

The scattering amplitude $f(\theta)$ is related to the differential cross section $\sigma(\theta)$ through

$$\frac{\mathrm{d}\sigma(\theta)}{\mathrm{d}\Omega} = |f(\theta)|^2 , \tag{6.4}$$

and the total collision cross section is given by

$$\sigma_{\text{tot}} = \int_\Omega \mathrm{d}\Omega |f(\theta)|^2 = 2\pi \int_0^\pi \mathrm{d}\theta \sin\theta |f(\theta)|^2 = \frac{4\pi}{k} \operatorname{Im}\{f(0°)\} , \tag{6.5}$$

where the second equality holds for spherically symmetric potentials and the third represents the "optical theorem" [6.4]. The latter relates the total cross section to the imaginary part (Im) of the forward scattering amplitude.

The standard way to calculate the scattering amplitude is based on the "partial wave expansion" of

$$\Psi_E^+(\boldsymbol{k},\boldsymbol{r}) = \sqrt{\frac{2}{\pi}} \frac{1}{kr} \sum_{\ell m} \mathrm{i}^\ell F_\ell(k,r) Y_{\ell m}(\hat{\boldsymbol{r}}) Y_{\ell m}^*(\hat{\boldsymbol{k}}) , \tag{6.6}$$

where $Y_{\ell m}(\hat{\boldsymbol{r}})$ is a spherical harmonic for the angles associated with the direction of the position vector $\boldsymbol{r}$, and the function $F_\ell(k,r)$ describes the radial motion of the projectile.

Inserting the expansion (6.6) into the Schrödinger equation (6.2) yields the radial Schrödinger equation

$$\left( \frac{\mathrm{d}^2}{\mathrm{d}r^2} - \frac{\ell(\ell+1)}{r^2} - 2V(r) + k^2 \right) F_\ell(k,r) = 0 . \tag{6.7}$$

In the asymptotic region, the potential is supposed to be negligible. Consequently, $F_\ell(k,r)$ in this region must be a linear combination of the regular and the irregular solutions of (6.7). Transforming to standing wave boundary conditions allows for $F_\ell(k,r)$ to be regarded as a real instead of a complex function. Its asymptotic behaviour is then given by

$$\lim_{r \to \infty} F_\ell(\rho) = A_\ell j_\ell(\rho) - B_\ell \eta_\ell(\rho) , \tag{6.8}$$

where $j_\ell(\rho)$ and $\eta_\ell(\rho)$ are regular and irregular Riccati–Bessel functions with argument $\rho = kr$.

Using the incident beam axis as the quantization axis together with the asymptotic form of the functions $j_\ell(\rho)$ and $\eta_\ell(\rho)$, we can express the scattering amplitude as

$$f(\theta) = \frac{1}{2\mathrm{i}k} \sum_{\ell=0}^{\infty} (2\ell+1) T_\ell P_\ell(\cos\theta) , \tag{6.9}$$

where

$$T_\ell = e^{2i\delta_\ell} - 1 = S_\ell - 1 \tag{6.10}$$

is the transition-matrix element, $S_\ell = e^{2i\delta_\ell}$ is the scattering-matrix element, and $\delta_\ell$ is the phase shift due to the potential $V(r)$. This phase shift is related to the asymptotic form (6.8) through

$$\tan \delta_\ell = K_\ell = B_\ell/A_\ell \,. \tag{6.11}$$

Because of the centrifugal barrier in the radial equation (6.7), the potential phase shift must converge to zero for high angular momenta and short-range potentials, thereby guaranteeing the convergence of the sum (6.9).

Elementary complex algebra reveals the connection

$$S_\ell = 1 + T_\ell = (1 + iK_\ell)/(1 - iK_\ell) \,. \tag{6.12}$$

This may be extended to give a more general relationship between the $\mathbf{S}$, $\mathbf{T}$ and $\mathbf{K}$ matrices in many-channel problems (see below).

As mentioned before, the above treatment can be generalized in a straightforward way to the case of Coulomb scattering where the potential $V(r)$ contains both a short-range term and a long-range Coulomb part of the form $Z_{\text{asym}}/r$. In this case, a Coulomb phase shift $\sigma_\ell$ needs to be introduced, and the Riccati-Bessel functions in (6.8) are replaced by regular and irregular Coulomb functions multiplied by $\rho$. In practice, the phase shift for potential scattering problems can be obtained quite easily; for details, see Chapter 1 of [6.1].

### 6.2.2 Perturbation approaches

Whereas the determination of phase shifts for potential scattering is, in principle, straightforward, the situation becomes more complicated when the target is represented by an $N$-electron wavefunction and excitation or even ionization is possible in addition to elastic scattering of the projectile. In such cases, perturbative approaches are often used to provide approximate solutions of the collision problem.

To illustrate the basic ideas within the framework of potential scattering, we follow McCarthy and Weigold [6.4] and rewrite (6.2) in the form

$$(E^{(+)} - T)\,\Psi_E^+(\mathbf{k},\mathbf{r}) = V(r)\,\Psi_E^+(\mathbf{k},\mathbf{r}) \,. \tag{6.13}$$

Equation (6.13) involves the free-particle operator $(E^{(+)} - T)$ and an inhomogeneous term on the right-hand-side. It can formally be solved by using standard Green's function approaches. The solution with the appropriate boundary condition (6.3) is given by

$$\Psi_E^+(\mathbf{k},\mathbf{r}) = e^{i\mathbf{k}\cdot\mathbf{r}} + \int d^3r' G_0(E^{(+)};\mathbf{r},\mathbf{r}')V(r')\,\Psi_E^+(\mathbf{k},\mathbf{r}') \,, \tag{6.14}$$

with the free-particle Green's function

$$G_0(E^{(+)}; r, r') = -\frac{1}{2\pi} \frac{e^{ik|r-r'|}}{|r-r'|} . \tag{6.15}$$

Equation (6.14) is the Lippmann–Schwinger equation for the exact scattering wavefunction $\Psi_E^+(k, r)$. Note that this is an integral equation, since the unknown wavefunction not only appears on the left-hand side, but also in the integral on the right-hand side. Before discussing approximate solutions, however, it is useful to investigate the asymptotic form of (6.15) for large $r$ and its consequences for (6.14). The final result is the expression

$$f(\theta) = -\frac{1}{2\pi} \int d^3r' e^{-ik \cdot r'} V(r')\Psi_E^+(k, r) \tag{6.16}$$

for the scattering amplitude [6.4]. Defining

$$|k\rangle \equiv \frac{1}{(2\pi)^{3/2}} e^{ik \cdot r} \tag{6.17a}$$

and

$$|k^+\rangle \equiv \frac{1}{(2\pi)^{3/2}} \Psi_E^+(k, r), \tag{6.17b}$$

Equation (6.16) can be abbreviated as

$$f(\theta) = -4\pi^2 \langle k'|V|k^+\rangle . \tag{6.18}$$

This equation may be used to *define* the **T** operator for potential scattering according to

$$\langle k'|V|k^+\rangle \equiv \langle k'|\mathbf{T}|k\rangle , \tag{6.19}$$

which is the momentum-space Lippmann–Schwinger equation for the transition operator.

While an accurate numerical solution of the Lippmann–Schwinger equation through basis-function expansions is, in principle, possible (see Chapter 8 of [6.1]), an important aspect of the above formalism concerns the possibility of using perturbative methods. The simplest approach consists of replacing the exact scattering wavefunction $|k^+\rangle$ by the (very) approximate plane wave $|k\rangle$ in (6.18). This yields the "first Born approximation" (FBA)

$$\langle k'|\mathbf{T}|k\rangle \approx \langle k'|V|k\rangle . \tag{6.20}$$

Equation (6.20) is expected to be a good approximation if the potential $V(r)$ is a small perturbation compared to the total energy — in other words, if the kinetic energy is much larger than the potential energy everywhere in space. (For more detailed discussions, we refer to standard textbooks on collision theory.) To obtain a better approximation, one might want to split up the potential as $V = V_1 + V_2$ and treat $V_1$ more accurately than $V_2$ by including it on the left-hand-side of (6.13); in fact, this *must* be done for any long-range Coulombic term in the potential [6.4]. This procedure corresponds to the "first-order distorted-wave approximation" (DWBA), since the plane waves are replaced by distorted waves that are calculated with the potential $V_1$. Another possibility is the next iteration in the solution of the Lippmann–Schwinger equation with the full $V$ included, a procedure called the "second Born approximation." Again, this may be improved by splitting up the potential, thereby defining the "second-order distorted-wave approximation." Partial extensions to account for even higher orders have been considered in the literature, but the computational effort becomes very extensive beyond first-order approaches.

Finally, we note that the partial-wave **T**-matrix element in this formalism is usually defined as [6.4]

$$T_\ell = -\frac{1}{2i\pi k}(S_\ell - 1)\,, \tag{6.21}$$

and thus differs from the definition in (6.10) by a complex, energy-dependent factor. The reason for the different choices is the convenience in the individual approaches, here in particular the simple form of the first-Born matrix element (6.20). This demonstrates, however, that special care must be taken when using formulas from different authors for **T**-matrix elements, scattering amplitudes, and cross sections.

Finally, we consider the use of the Born-series expansion in the concrete case of elastic and inelastic electron–atom scattering. Because of the need for fully antisymmetrized wavefunctions of the projectile + target system, one typically obtains a "direct" and an "exchange" part to the scattering amplitude. At this point, we only present one example, namely the first-order distorted-wave approximation for the transition operator. This is given by (see Chapter 4 of [6.1]):

$$\mathbf{T}_1 = (N+1)\left\langle \chi_f^-(0)\psi_f(1,\ldots,N)\,|V - U_f|\,\mathcal{A}\psi_i(1,\ldots,N)\chi_i^+(0)\right\rangle$$
$$+ \left\langle \chi_f^-(0)\psi_f(1,\ldots,N)\,|U_f|\,\psi_i(1,\ldots,N)\beta_i(0)\right\rangle. \tag{6.22}$$

Here $V$ is the interaction between the projectile and the atom, $\psi_{i,f}(1,\ldots,N)$ are the initial and final $N$-electron target states, $U_f$ is a distorting potential for the projectile, which is used to calculate the distorted wave $\chi_f^-$ by solving the equation

$$(T + U_f - E_f)\,\chi_f^- = 0\,, \tag{6.23}$$

subject to incoming wave boundary conditions (indicated by the $-$ superscript) for the final-state energy $E_f$ of the projectile, $\beta_i$ is an initial-state plane wave, and $\mathcal{A}$ is the antisymmetrizing operator for the $N+1$ electrons.

Typically, $U_f$ is chosen to be a spherically symmetric final-state approximation for $V$ since it is also the potential used to calculate the final-state wavefunction for the projectile. In this case the second term of (6.22) vanishes for inelastic scattering and orthogonal atomic wavefunctions since $U_f$

only depends on the single coordinate of the projectile. For elastic scattering, on the other hand, the second term of (6.22) is the dominant term; in fact, it is generally the only contributing term since $U_f$ is often chosen such that the matrix elements of $V - U_f$ vanish. Finally, the direct and exchange parts of the scattering amplitude are obtained by combining the **T**-matrix elements with the appropriate spherical harmonics and angular-momentum coupling coefficients to describe the directions of the incoming and outgoing projectile. For more details, we refer to Chapters 4 and 5 of [6.1] for excitation and ionization, respectively.

The first-order distorted-wave method, in both semirelativistic and full-relativistic form, as well as the related "first-order many-body theory" (FOMBT) have been used frequently for the calculation of polarization, alignment and orientation parameters. For a selection of results, see the recent reviews [6.5,6].

### 6.2.3  The close-coupling expansion

The close-coupling approximation has been a standard method of treating low-energy scattering, both elastic and inelastic, for many years. This non-perturbative method is based on an expansion of the total wavefunction for a collision system in terms of a sum of products that are constructed from target states $\Phi_i$ that diagonalize the $N$-electron target Hamiltonian

$$H_T^N = \sum_{i=1}^{N} \left[ -\frac{1}{2} \nabla_i^2 - \frac{Z}{r_i} + \frac{1}{2} \sum_{j \neq i}^{N} \frac{1}{|\boldsymbol{r}_i - \boldsymbol{r}_j|} \right] \tag{6.24}$$

according to

$$\langle \Phi_{i'} \mid H_T^N \mid \Phi_i \rangle = E_i\, \delta_{i'i}\,, \tag{6.25}$$

and unknown functions $F_{E,i}$ describing the motion of the projectile for a total (target + projectile) collision energy $E$. If relativistic effects are neglected, the wavefunction for each total orbital angular momentum $L$, total spin $S$, and parity $\pi$ is expanded as

$$\Psi^{LS\pi}(\boldsymbol{r}_1,\ldots,\boldsymbol{r}_{N+1}) = \mathcal{A} \sum_{i} \!\!\!\!\!\!\!\int \Phi_i^{LS\pi}(\boldsymbol{r}_1,\ldots,\boldsymbol{r}_N,\hat{\boldsymbol{r}}) \frac{1}{r} F_{E,i}(r)\,. \tag{6.26}$$

Here $\sum\!\!\!\!\int$ denotes a sum over all discrete and an integral over all continuum states of the target, and $\mathcal{A}$ is the antisymmetrization operator that accounts for the indistinguishability of the projectile and the target electrons. Furthermore, the angular and spin coordinates of the projectile electron (collectively denoted by $\hat{\boldsymbol{r}}$) have been coupled with the target states to produce the "channel functions" $\Phi_i^{LS\pi}(\boldsymbol{r}_1,\ldots,\boldsymbol{r}_N,\hat{\boldsymbol{r}})$.

After some algebraic manipulations, the unknown radial wavefunctions $F_{E,i}$ are determined from the solution of a system of coupled integro-differential equations given by

$$\left[\frac{\mathrm{d}^2}{\mathrm{d}r^2} - \frac{\ell_i(\ell_i+1)}{r^2} + k^2\right] F_{E,i} = 2 \sum_j V_{ij}(r) F_{E,j} + 2 \sum_j W_{ij} F_{E,j} \qquad (6.27)$$

with the direct coupling potentials

$$V_{ij}(r) = -\frac{Z}{r}\delta_{ij} + \sum_{k=1}^{N} \langle \Phi_i \mid \frac{1}{|\boldsymbol{r}_k - \boldsymbol{r}|} \mid \Phi_j\rangle \qquad (6.28)$$

and the exchange terms

$$W_{ij}F_{E,j}(r) = \sum_{k=1}^{N} \langle \Phi_i \mid \frac{1}{|\boldsymbol{r}_k - \boldsymbol{r}|} \mid (\mathcal{A}-1)\Phi_j F_{E,j}\rangle . \qquad (6.29)$$

For each "$i$", several sets of independent solutions (labeled by a second subscript "$j$") must be found, subject to the appropriate boundary conditions. For scattering from and excitation of neutral targets, these are given by [6.7]

$$F_{E,ij\,|r=0} = 0 , \qquad (6.30a)$$

$$\lim_{r\to\infty} F_{E,ij} = \delta_{ij}\sin\left(k_i r - \tfrac{1}{2}\ell_i\pi\right) + K_{ij}\cos\left(k_i r - \tfrac{1}{2}\ell_i\pi\right) ; \quad i=1,\ldots,n_{\mathrm{open}},$$
$$\qquad (6.30b)$$

$$\lim_{r\to\infty} F_{E,ij} = C_{ij}\exp(-|k_i|r); \quad i > n_{\mathrm{open}} . \qquad (6.30c)$$

In (6.30), $k_i = \sqrt{2(E - E_i)}$ is the linear momentum in channel $i$ for the total energy $E$ and the channel energy $E_i$, and $n_{\mathrm{open}}$ is the number of "open" channels for which $k_i$ is a real number. For the "closed" channels, $k_i$ is purely imaginary.

The coupled integro-differential equations (6.27) may be generalized to a relativistic framework, and the collision problem essentially consists of finding the solution to this system for each total energy. This can be achieved by various iterative, noniterative, or algebraic methods. An introductory overview is given, for example, by Burke and Seaton [6.7]. Chapters 6–8 of [6.1] are also devoted to this problem and provide simplified programs for scattering from one-electron targets. While the "convergent close-coupling" (CCC) method can presently be applied to both quasi-one-electron and quasi-two-electron targets, the standard R-matrix program of the Belfast group [6.8] may also be used to obtain the reactance-matrix element $K_{ij}$ of (6.30b) for complex atomic and ionic targets. The "R-matrix with pseudo-states" (RMPS) approach [6.9] has recently been implemented as well and extends the low-energy R-matrix method to higher energies. The principal advantage of both CCC and RMPS over standard discrete-state-only treatments lies in the fact

that the coupling effect of high-lying discrete states and the target continuum
to the states involved in the transitions of interest is accounted for by including a large number of square-integrable pseudo-states in the close-coupling
+ correlation expansion (6.25). An additional benefit of such an enlarged expansion is the possibility of partially re-interpreting the results for excitation
of the pseudo-states as ionization, thereby allowing for the treatment of such
processes as well.

The construction of the scattering amplitudes, and the observable parameters of interest, is the final step of the calculation. It involves the scattering ($\mathbf{S}$), transition ($\mathbf{T}$), and reactance ($\mathbf{K}$) matrices, which are related
through

$$\mathbf{S} = 1 + \mathbf{T} = [1 + \mathrm{i}\,\mathbf{K}]\,[1 - \mathrm{i}\,\mathbf{K}]^{-1}\,. \tag{6.31}$$

Equation (6.31) is the multi-channel generalization of (6.12). Once again,
these matrix elements have to be combined with the proper spherical harmonics and angular-momentum coupling coefficients.

This problem was discussed in Chapter 10 of [6.1], As an example, consider
transitions from an initial state denoted as $|\Gamma_0, L_0, M_{L_0}, S_0, M_{S_0}; \boldsymbol{k}_0 m_0\rangle$ to
a final state $|\Gamma_1, L_1, M_{L_1}, S_1, M_{S_1}; \boldsymbol{k}_1 m_1\rangle$ where the projectile electrons (or
other spin-$\frac{1}{2}$ particles) have initial (final) angular momentum $\boldsymbol{k}_0$ ($\boldsymbol{k}_1$) and spin
component $m_0$ ($m_1$) with respect to a given quantization axis. Furthermore,
the target states involved in the transition have total orbital and spin angular
momenta $L_0$ ($L_1$) and $S_0$ ($S_1$) with components $M_{L_0}$ ($M_{L_1}$) and $M_{S_0}$ ($M_{S_1}$),
respectively. Finally, $\Gamma_0$ and $\Gamma_1$ denote collectively all other quantum numbers
that are used to construct the target states.

Suppose that the total spin $S$, the total angular momentum $L$, and the
total parity $\Pi = (-1)^{\sum_i \ell_i}$ (the sum runs over the angular momenta $\ell_i$ of all
individual electrons) of the combined target + projectile system are conserved
during the collision. This $LS$-approximation is very well fulfilled for electron
scattering from light neutral targets such a hydrogen, helium, or the first few
alkali atoms in the periodic table. As outlined in detail by Smith [6.10], the
scattering amplitudes for this case are constructed by making the following
general assumptions:

1. When the projectile is far away from the target, the initial state may be
   written as

$$\Psi_{\mathrm{in}} = \mathrm{e}^{\mathrm{i}k_0 z}\chi(m_0)\,|L_0, M_{L_0}, S_0, M_{S_0}\rangle\,; \tag{6.32}$$

   i.e., a plane wave $\mathrm{e}^{\mathrm{i}k_0 z}$ associated with a spin function $\chi(m_0)$ is multiplied
   by the initial target state $|\Gamma_0, L_0, M_{L_0}, S_0, M_{S_0}\rangle$.

2. The plane wave is expanded in terms of partial waves, and the spherical
   harmonics and the target states are coupled together to give properly
   antisymmetrized channel functions with total orbital and spin angular
   momenta $L$ and $S$ for the combined system, respectively. Note that this
   expansion is simplified significantly if the incident plane wave is parallel

to the quantization $(z)$ axis. This is the reason why numerical calculations are usually performed in the "collision frame."

3. Again far away from the target, the scattered part in the final state wave-function $\Psi_{\text{fin}}$ is an outgoing spherical wave in the $\hat{\boldsymbol{k}}_1 \equiv (\theta, \phi)$ direction. In the coupled representation, this means

$$\Psi_{\text{scatt}} = \Psi_{\text{fin}} - \Psi_{\text{in}}$$

$$= \frac{e^{ik_1 r}}{r} \sum_{M_{L_1} M_{S_1} m_1} \chi(m_1) \, |L_1, M_{L_1}, S_1, M_{S_1}\rangle$$

$$\cdot f(\Gamma_1, L_1, M_{L_1}, S_1, M_{S_1}, \boldsymbol{k}_1, m_1; \Gamma_0, L_0, M_{L_0}, S_0, M_{0_1}, \boldsymbol{k}_0, m_0).$$

$$(6.33)$$

With this normalization and our notation, Equation (2.113) of Smith [6.10] corresponds the following expression for the scattering amplitude:

$$f(\Gamma_1, L_1, M_{L_1}, S_1, M_{S_1}, m_1; \Gamma_0, L_0, M_{L_0}, S_0, M_{S_0}, m_0; \theta, \phi)$$

$$= -\sqrt{\frac{\pi}{k_0 k_1}} \sum_{L M_L S M_S \ell_0, l_1, m_{\ell_1}} i^{\ell_0 - \ell_1 + 1} \sqrt{2\ell_0 + 1} \, Y_{\ell_1 m_{\ell_1}}(\theta, \phi)$$

$$\cdot (L_0, M_{L_0}; \ell_0, 0 | L, M_L) \, (S_0, M_{S_0}; \tfrac{1}{2}, m_0 | S, M_S)$$

$$\cdot (L_1, M_{L_1}; \ell_1, m_{\ell_1} | L, M_L) \, (S_1, M_{S_1}; \tfrac{1}{2}, m_1 | S, M_S)$$

$$\cdot \mathbf{T}^{LS\Pi}_{\Gamma_0, L_0, \ell_0; \Gamma_1, L_1, \ell_1},$$

$$(6.34)$$

where $(j_1, m_1; j_2, m_2 | j_3, m_3)$ is a standard Clebsch–Gordan coefficient and $Y_{\ell m}(\theta, \phi)$ is a spherical harmonic as defined in Edmonds [6.11]. The quantity $\mathbf{T}^{LS\Pi}_{\Gamma_0, L_0 \ell_0; \Gamma_1, L_1, \ell_1}$ in (6.3) is the energy-dependent $\mathbf{T}$-matrix element, and the superscripts indicate that the total angular momentum $L$, the total spin $S$, and the total parity $\Pi$ of the combined target + projectile system are conserved during the collision.

An important point to be made with respect to (6.34) concerns the spherical harmonics that are used to construct the coupled states. While Smith's derivation [6.10] assumes that these are of the same kind as the $Y_{\ell_1 m_{\ell_1}}(\theta, \phi)$ in (6.34), some computer codes construct these states by using the functions $\mathcal{Y}_{\ell m} \equiv i^{\ell} Y_{\ell m}$ defined by Fano and Racah [6.12]. In this case, the factor $i^{\ell_0 - \ell_1}$ must be omitted in (6.34).

Similar equations can be derived for other angular-momentum coupling schemes. The most important example is electron scattering from heavy targets where relativistic effects must be taken into account and only the total electronic (orbital + spin) angular momentum $J$ of the system and the parity are conserved (see, for example, Bartschat and Scott [6.13]). Also, the necessary extensions for scattering from charged targets can be found in Smith [6.10] who applied the results of Lane and Thomas [6.14] to this case.

Note that the dependence on the spin quantum numbers is essentially limited to the Clebsch–Gordan coefficients in (6.34). Also, for the most important case of spherically symmetric interactions, the only azimuthal dependence on the angle $\phi$ enters through the factor $e^{im\phi}$ in the spherical harmonic. One can therefore simplify this equation further by choosing the scattering plane, spanned by $k_0$ and $k_1$, as the $xz$ plane. While this choice corresponds to $\phi \equiv 0$ in (6.34), it still contains all the physically relevant information.

Consequently, we rewrite (6.34) as

$$f(\Gamma_1, L_1, M_{L_1}, S_1, M_{S_1}, m_1; \Gamma_0, L_0, M_{L_0}, S_0, M_{S_0}, m_0; \theta, \phi)$$
$$= \sum_{SM_S} (S_0, M_{S_0}; \tfrac{1}{2}, m_0 | S, M_S) \ (S_1, M_{S_1}; \tfrac{1}{2}, m_1 | S, M_S)$$
$$\cdot f^S(M_{L_1}, M_{L_0}; \theta), \tag{6.35}$$

where

$$f^S(M_{L_1}, M_{L_0}; \theta)$$
$$= -\tfrac{1}{2} \sqrt{\frac{1}{k_0 k_1}} \sum_{L, \ell_0, l_1} i^{\ell_0 - \ell_1 + 1} \sqrt{(2\ell_0 + 1)(2\ell_1 + 1)}$$
$$\cdot (L_0, M_{L_0}; \ell_0, 0 | L, M_L) \ (L_1, M_{L_1}; \ell_1, m_{\ell_1} | L, M_L)$$
$$\cdot (-1)^{m_{\ell_1}} \sqrt{\frac{(\ell - m_{\ell_1})!}{(\ell + m_{\ell_1})!}} \ P_{\ell_1 m_{\ell_1}}(\theta) \ \mathbf{T}^{LS\Pi}_{\Gamma_0, L_0, \ell_0; \Gamma_1, L_1, \ell_1}. \tag{6.36}$$

In (6.36), we have used the relationship between the spherical harmonics and the associate Legendre polynomials $P_{\ell_1 m_{\ell_1}}(\theta)$ in the form [6.15]

$$Y_{\ell m}(\theta, \phi) = (-1)^m \sqrt{\frac{2\ell + 1}{4\pi} \frac{(\ell - m)!}{(\ell + m)!}} \ P_{\ell m}(\theta) \, e^{im\phi} \tag{6.37}$$

and set $\phi = 0$. Also, we have dropped the sums over the magnetic quantum numbers $M_L$ and $m_{\ell_1}$, since the selection rules for the Clebsch–Gordan coefficients imply that

$$M_L = M_{L_0}, \tag{6.38a}$$
$$m_{\ell_1} = M_L - M_{L_1} = M_{L_0} - M_{L_1}. \tag{6.38b}$$

The simplifications seen in (6.36,38) demonstrate the advantage of the "collision frame" for explicit numerical calculations. Transformations to other coordinate systems, such as the "natural frame" (see Chapter 2), can be achieved by transforming the initial and final states through standard rotation matrices and using the fact that the action of the $\mathbf{T}$ operator must be independent of the particular coordinate frame. For more details, see Section 5.2 and also Chapter 11 of [6.1].

The above example indicates the analytical and numerical work that is involved to calculate, for example, the amplitudes $f_1^c$ and $f_0^c$ introduced in

Chapter 2. For electron-impact excitation of an S $\rightarrow$ P transition such as $(1s^2)^1$S $\rightarrow (1s2p)^1$P° in helium, these amplitudes correspond to $f^{1/2}(1,0;\theta)$ and $f^{1/2}(0,0;\theta)$ in the notation used in (6.36). Finally, we note that the differential cross section for a transition from an initial state "0" to a final state "1" in this formulation is given by

$$\frac{\mathrm{d}\sigma(\theta)}{\mathrm{d}\Omega} = \frac{k_1}{k_0}|f_{0\to 1}(\theta)|^2 \,, \tag{6.39}$$

where the factor $k_1/k_0$ accounts for the proper electron fluxes in the initial and final states, respectively.

### 6.2.4 Time-dependent approaches

With the recent development of massively parallel supercomputers, the use of more direct numerical approaches has become very popular. One example is the time-dependent close-coupling (TDCC) method described by Pindzola et al. [6.16]. For a target with one electron outside a closed shell, they derive the time-dependent close-coupling equations for the radial parts of the wavefunction describing the projectile and the active target electron. For each partial-wave symmetry, these equations are propagated in time over a two-dimensional lattice.

The initial wavefunction is usually written as the properly symmetrized product of a bound orbital, represented on the lattice, and a Gaussian wavepacket for the projectile. The width of the package is chosen in such a way that any overlap between the projectile and the target wavefunction before the collision can be neglected. The two-electron wavefunction is then propagated for a sufficiently long time to ensure that the final state after the collision has been reached. Note, however, that the propagation time has to be small enough to ensure that boundary effects, due to collisions with the wall of the lattice, do not disturb the result. Cross sections for elastic scattering, excitation and ionization are obtained from the projections of the final-state wavefunction on the individual asymptotic channel functions.

The principal advantage of the method lies in the fact that the radial part of the wavefunction, which contains all the dynamics of the collision, is not described by a finite basis set, as in the time-independent close-coupling approaches discussed above. On the other hand, the computational resources needed for such calculations are relatively extensive. Hence, the method has so far only been applied to (quasi-)one-electron targets. In addition, just the lowest few partial waves are typically treated in this way, while contributions from higher partial waves are handled by much simpler methods, such as the first-order distorted-wave Born approximation.

# 6.3 Computational Methods for Heavy-Particle Collisions

For the theoretical description, the principal new feature when going from the electron–atom collisions discussed in Section 6.2 above to heavy-particle collisions at similar velocities is the meaningful concept of a trajectory. The reason for this is the de Broglie wavelength $\lambda_{\mathrm{dB}} = h/mv$ which, due to the great increase in mass, is much smaller than typical atomic dimensions. In this semiclassical approximation, the movement of the projectile nucleus along the trajectory is treated classically, while the electrons are treated quantum mechanically [6.17].

When compared to the orbital velocity $v_e$ of the active, outer-shell electron, the range of collision velocities $v_c$ may naturally be divided into three regions. In the "velocity-matching region," $v_c \simeq v_e$, transition probabilities and corresponding cross sections may become large, and cross sections typically exhibit their maxima here. In this region, therefore, perturbation approaches will fail, and one has to resort to the solution of strongly coupled equations. Toward lower velocities, $v_c \lesssim v_e$, the collision system may be treated as a short-living quasi-molecule, and the corresponding choice of a molecular orbital (MO) basis for the calculations may be appropriate. Toward higher velocities, $v_c \gtrsim v_e$, a description in terms of an atomic orbital (AO) basis is more adequate for the theoretical analysis. When increasing the collision velocity even further, a Born approximation estimate will eventually become valid. The general theoretical framework is well developed and described in textbooks and reviews such as refs. [6.18,19]. In some cases, a Classical-Trajectory Monte-Carlo (CTMC) method is an attractive alternative. We shall now describe these approaches, their advantages, and their limitations in more detail.

### 6.3.1 Semiclassical approaches

In the semiclassical approach, and for sufficiently high velocities, the projectile trajectory is described by a straight line with constant velocity. (If necessary, a bent Coulomb-type trajectory may be invoked.) Thus the vector $\boldsymbol{R}$, locating the projectile relative to the target atom, which is at rest at the origin, may be parameterized as $\boldsymbol{R} = \boldsymbol{R}(b, t) = \boldsymbol{b} + \boldsymbol{v}t$, with $b$ being the impact parameter and $t$ the time. One thus neglects the coupling between the energy and momentum given to the excited electron(s) and the projectile motion. Using a suitable model Hamiltonian to describe the time-dependent interaction between the target and projectile cores and the interaction between the active electron and the two centers, one can solve the time-dependent Schrödinger equation (TDSE), subject to appropriate boundary conditions. If the projectile is scattered primarily by the static potential of the target, it is often possible to relate the scattering angle $\theta$ to an impact parameter $b$ and thus extract angle-differential cross sections from the $b$-dependent probabilities. At low energies, however, or for sufficiently small scattering angles, this might not work. Still, by the eikonal transformation, one may extract differential cross sections from impact parameter dependent probabilities evaluated in straight-line trajectory calculations [6.20,21].

### 6.3.1.1 Close-coupling solution of the TDSE

One approach to the solution of the TDSE is to select some basis set and expand the electronic wavefunction $\psi(r, b, t)$ in terms of this set. In the literature, this procedure is called the coupled-channels, coupled-states, or close-coupling (CC) method. With the choice of basis sets mentioned above, one then defines CC-MO or CC-AO calculations. Many studies in recent years have addressed the problem of exploring the limit toward higher energies of CC-MO calculations, which form an adequate approach at velocities somewhat below the matching velocity. Similarly, CC-AO calculations, adequate in the matching velocity region and above, break down toward lower velocities, owing to an inadequate description of the formation of a quasi-molecule at short internuclear distances. Hybrids between the two approaches (MO at short distance, AO at long distance) have thus been developed. (A comprehensive review is given in [6.22].) In Chapters 8–10, we shall encounter many examples of results from CC-MO and CC-AO calculations.

If the problem at hand only requires a single-center expansion, as is sometimes the case for target excitation, one first diagonalizes the target Hamiltonian. This yields a basis set of eigenfunctions $\phi_k^T(r)$ with energies $E_k$, and the expansion becomes

$$\psi_i(r, b, t) = \sum_k a_{ki}(b, t) e^{-\mathrm{i}(E_k - E_i)t/\hbar} \, \phi_k^T(r) \, . \tag{6.40}$$

Inserting this expansion into the TDSE yields a set of first-order equations for the time-dependent expansion coefficients. This set of equations is solved subject to appropriate boundary conditions. With the target atom initially in state $i$, the boundary condition is $a_{ki}(b, t = -\infty) = \delta_{ij}$. After solving the TDSE, the expansion coefficients $a_{fi}(b, t = +\infty)$ give the transition amplitudes from the initial state $i$ to a final state $f$.

Many processes, however, are genuine two-center phenomena, with electron transfer being an obvious example. In a two-center expansion, two sets of eigenfunctions, $\phi_k^T(r)$ centered on the target and $\phi_j^P(r)$ centered on the projectile, come into play. Hence, the expansion is written as

$$\psi_i(r, b, t) = \sum_k a_{ki}^T(b, t) e^{-\mathrm{i}(E_k - E_i)t/\hbar} \, \phi_k^T(r)$$

$$+ \sum_j a_{ji}^P(b, t) e^{-\mathrm{i}(E_j - E_i)t/\hbar} \, e^{\mathrm{i}p \cdot r} \, \phi_j^P(r) \, . \tag{6.41}$$

Here the energies $E_j$ also contain the kinetic energy of the projectile electron in the target frame because of its motion with the projectile. The electron translation factor (ETF) $e^{i\boldsymbol{p}\cdot\boldsymbol{r}}$ ensures Galilean invariance. (See [6.22] for a discussion of MO ETFs.) As above, the expansion coefficients $a_{fi}^{T,P}(b, t = +\infty)$ give the transition amplitudes from the target initial state $i$ to the final state $f$. Both the mathematical and computational challenges in the solution of the TDSE are significantly greater than for a single-center expansion. Computations involving one additional center, the center of charge, have also been explored [6.23]. As in the case of the CCC and RMPS methods discussed for electron collisions, the inclusion of ionization processes requires the introduction of additional pseudostates in order to simulate the continuum [6.22].

Results from solutions of the TDSE are particularly well suited for illustrating the collision dynamics, since they allow for a visualization of the time evolution of the electron charge cloud and the build-up of electric currents during a collision event. As a result of the calculation, one may thus produce a set of pictures, or even a "movie", to illustrate how the final state of the system, which is the one accessible to experimental investigation, develops in time for a specific impact parameter and collision velocity [6.24,25]. This aspect will be discussed further in Section 8.3.

**6.3.1.2** Lattice solutions of the TDSE

Recently, attempts have been made to circumvent the difficulties and limitations of the methods described in the previous section by directly solving the TDSE numerically on a large lattice (LTDSE). After propagating the initial state for a sufficiently long time, scattering amplitudes for the various exit channels are obtained by projecting the wavefunction on the relevant final states. The required computational effort is very substantial, but the rapid advances in computer power have made such attempts feasible. The fundamental collision system $H^+ + H$ has recently been the subject of such an investigation [6.26].

Figure 6.1 shows TDSE results for angle-integrated cross sections for H(2s) and H(2p) excitation, produced in the process

$$H^+ + H(1s) \rightarrow H^+ + H(n\ell) , \tag{6.42}$$

over the wide energy range of $1 - 1000\,\mathrm{keV}$ [6.26]. We notice pronounced maxima in the matching velocity range, $v_c \simeq 1\,v_0$, or an impact energy of $E = 25\,\mathrm{keV}$, for the H(1s) electron.

For the higher energies, only a few impact parameters needed to be considered because of the smooth variation of the excitation probabilities with impact parameter. For the lower energies, the probability curves often display several local maxima, and thus as many as 20 or 25 impact parameters were needed to obtain convergence. At the high energies, the projectile was started as close as $15\,a_0$ from the target, and the time propagation only needed to be carried out for a short time. For low collision energies, the projectile must start much farther away from the target and the final state takes longer to converge due to strong oscillations of the electron cloud induced by the slowly receding proton.

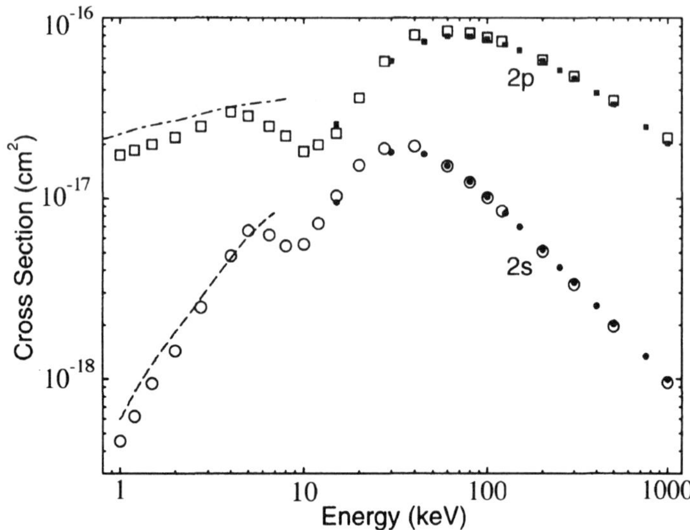

**Fig. 6.1** Comparison of LTDSE results [6.26] for 2s (circles) and 2p (squares) excitation in $H^+ + H$ collisions with other approaches applicable at low energy, CC-MO $(--)$ [6.27] and HC $(-\cdot-)$ [6.28], and intermediate to high energy, CC-AO ($\bullet$) [6.29].

The numerical grid extended over a range of $52\,a_0$ in each of the three Cartesian directions, with 135 points uniformly spaced along each direction. A complex potential, localized at the edge of the box, was used to reduce reflections and flux to the continuum or carried by the projectile. In total, the number of floating points operations for the calculation was on the order of $10^{15}$.

In Figure 6.1, the LTDSE results for H(2s) excitation are compared with the best available data from low-energy CC-MO calculations and high-energy CC-AO calculations. The CC-MO results [6.27] are expected to be accurate at low energies. A principal drawback of this approach is its reliance on electron translation factors above a few keV. The figure shows good agreement between LTDSE and CC-MO results at low energy, but also shows the failure of the CC-MO calculation to reproduce the downturn of the cross section above about 4 keV. Similarly, for H(2p) excitation, the results are compared with "hidden-crossing" (HC) results [6.28], believed to be valid in the limit of low velocities. The HC theory, too, fails to predict the downturn of the cross section above 4 keV.

The figure also shows good agreement at high velocities between LTDSE and single-center CC-AO results [6.29] for both 2s and 2p excitation, but the single-center CC-AO theory does not reproduce the oscillation in the cross section between 5 and 15 keV. A more comprehensive comparison and discussion of the LTDSE results with two- and three-center CC-AO calculations may be found in [6.26].

### 6.3.2 Classical-Trajectory Monte-Carlo approach

Another alternative to the methods outlined in Section 6.3.1 is the Classical-Trajectory Monte-Carlo (CTMC) method [6.30], essentially a computer experiment. This technique has been widely applied to atomic collision problems of strongly-coupled systems, with collisions involving multiply-charged ions as a generic example [6.31]. Recent successes of CTMC also include collisions involving Rydberg atoms and interpretation of Recoil Ion Momentum Spectroscopy (RIMS) data. Theoretical treatments based on perturbation methods fail for these processes, and basis set limitations of coupled-channel techniques prevent adequate representation of the multitude of active channels for excitation, electron transfer, and ionization. With this method, too, the rapid advances in computer power enable increasingly detailed studies of the collision processes. Originally developed for one-electron systems, the method has successfully been generalized to $n$-electron systems ($n$ CTMC [6.32]).

The starting point is the set of Hamilton's classical equations for the system. These are solved numerically, with random numbers being used to fix the impact parameter and the initial plane and eccentricity of the Kepler orbit for the active electron. A total cross section for a particular reaction $R$, such as electron transfer or ionization, is evaluated as

$$\sigma_R = \pi b_{\max}^2 \frac{N_R}{N} , \tag{6.43}$$

where $N$ is the total number of trajectories run within a given maximum impact parameter $b_{\max}$, and $N_R$ is the number of positive tests for the reaction. Generalization to cross sections, differential in angle or energy, is straightforward. As in a real experiment, the result (6.42) has a statistical uncertainty given by

$$\Delta \sigma = \sigma_R \left[ \frac{N - N_R}{N N_R} \right]^{\frac{1}{2}} , \tag{6.44}$$

which is proportional to $N_R^{-\frac{1}{2}}$ in the limit of large $N$. Small cross sections are thus subject to relatively large uncertainties unless considerable computer time is invested. Total cross sections have been evaluated with considerable success at velocities near and above the matching velocity. Differential scattering results also come out remarkably close to the quantal predictions, except in cases where quantum interference plays a dominating role. For lower collision velocities, the method also fails to account for quasi-molecular effects.

## 6.4 Visualization of Charge Clouds

As shown in Chapter 2, the angular distribution and the polarization of the light emitted from excited states are closely related to the charge cloud representing the active electron(s). A relatively straightforward way of visualizing these charge clouds, from input data given as measured or calculated scattering amplitudes or state multipoles, is possible with the help of widely available computer software packages such as *Mathematica*. A notebook to produce individual frames and eventually movies (via an angle-to-time conversion) was recently published by Loveall *et al.* [6.33].

In order to illustrate the basic procedure, let us consider a fully coherent atomic state with orbital angular momentum $L$ that was excited by projectiles scattered at an angle $\theta$. This angular part of the excited state can be written as

$$|\Psi(L)\rangle = \sum_{M} f_M(\theta)|L, M\rangle, \tag{6.45}$$

where $f_M(\theta)$ is the scattering amplitude for excitation of a magnetic sublevel with quantum number $M$. (For simplicity, we assume that the initial state has no angular momentum, and we omit the dependence on other dynamical parameters such as the collision energy.) The corresponding charge cloud density is then given by the coordinate representation of $\Psi^*\Psi$, i.e.,

$$\rho(\theta; \vartheta, \varphi) = \Psi^*\Psi = \sum_{M,M'} f_M^*(\theta)f_{M'}(\theta)Y_{LM}^*(\vartheta, \varphi)\,Y_{LM'}(\vartheta, \varphi), \tag{6.46}$$

where $Y_{LM}(\vartheta, \varphi)$ is a spherical harmonic for the polar coordinates $(\vartheta, \varphi)$ with respect to the target. Note that each scattering angle $\theta$ (and each collision energy $E$) will, in general, produce a different charge cloud.

If the excited state is not pure, for example because of averaging over spins or summing over nonobserved quantum numbers, the above formalism must be generalized. As discussed in Chapter 5, a very powerful method is the reduced density matrix description, formulated in terms of the so-called state multipoles $\langle T_{KQ}^+(L, \theta)\rangle$. Following Raeker *et al.* [6.34], the charge cloud can be expressed in terms of these multipoles as

$$\rho(\theta; \vartheta, \varphi) = \sum_{K,Q} \langle T_{KQ}^+(L, \theta)\rangle Y_{KQ}(\vartheta, \varphi), \tag{6.47}$$

where $K = 0, 2, 4, ..., 2L$, and $Q = -K, -K + 1, ..., K$.

The relationship between the state multipoles and the scattering amplitudes depends on the details of the experiment to be described. For the special case of the pure state discussed above, it is given by

$$\langle T_{KQ}^+(L, \theta)\rangle = \sum_{M,M'} (-1)^{L-M} (L, M'; L, -M|KQ) f_M^*(\theta)f_{M'}(\theta), \tag{6.48}$$

where $(L, M'; L, M | KQ)$ is a Clebsch–Gordan coefficient.

In addition to being more general, there are other advantages of the state multipole description. These include the possibility of terminating the sum (6.47) after a certain number of terms, in order to simulate what would actually be "seen" in an experiment in which only some of several photons in a cascade process are observed. A very important example of such a case is impact excitation of an S $\rightarrow$ D transition followed by a two-photon (D $\rightarrow$ P plus P $\rightarrow$ S) decay, in which only one photon is observed in coincidence with the scattered projectile. A detailed discussion of such processes will be given in Chapter 7.

Furthermore, we recall from Chapter 2 that the alignment angle $\gamma$ can be determined from

$$\gamma = \tfrac{1}{2}\, \arg\{P_1 + iP_2\}\,, \tag{6.49}$$

where $P_1$ and $P_2$ are calculated as outlined in Section 5.6. This is important, since the interpretation of $\gamma$ for excited P states as the angle between the symmetry axis of the charge cloud and the incident beam direction (see Figure 2.4(a)) is no longer obvious for excited states with angular momentum $L > 1$. Nevertheless, it is often useful to compare the result for $\gamma$ with the direction of the angular-momentum transfer $\Delta k \equiv k_0 - k_1$, where $k_0$ and $k_1$ are the initial and final linear momenta of the projectile, respectively. Note that the first Born approximation (FBA) predicts $\gamma$ to be in the direction of the momentum transfer [6.35].

Loveall *et al.* [6.33] presented several examples that were generated with their application. Figure 6.2 shows one of those cases, namely the charge clouds representing the $(1s2p)^1$P state of helium after impact excitation by 50 eV incident electrons that were scattered at various angles between $0°$ and $150°$, respectively. Note that for small scattering angles the charge cloud is indeed well represented by an $M = 0$ state aligned along the momentum-transfer direction, as predicted by the FBA. Further examples will be presented in Part II, and color movies are available on the accompanying CD-ROM.

A similar procedure can be used to represent the charge cloud as well as the current in heavy-particle collisions [6.37]. In this context, contour plots in the reaction plane are often the preferred choice, since the time evolution along the classical trajectories of the nuclei can be visualized. Examples will be presented in Part II, and a complete movie is available on the accompanying CD-ROM.

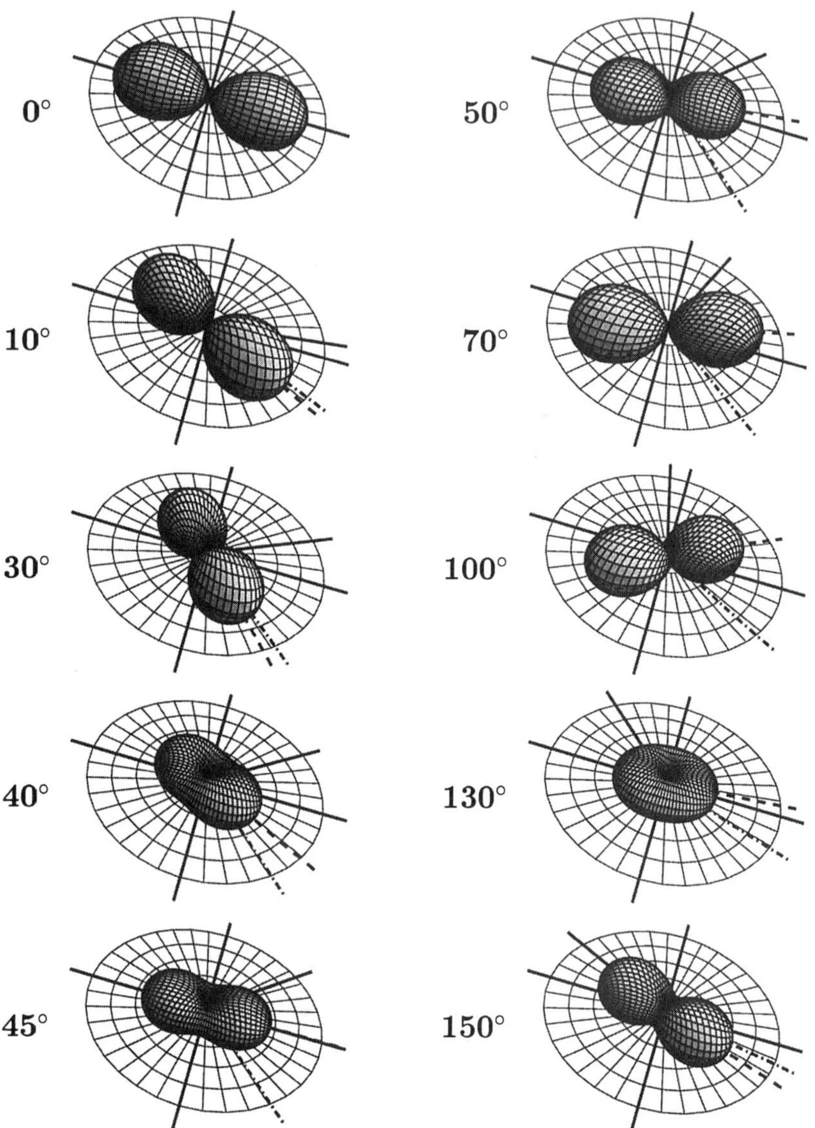

**Fig. 6.2** Charge cloud for electron-impact excitation of the $(1s2p)^1P^o$ state of helium at an incident electron energy of $50\,\mathrm{eV}$ at various scattering angles between $0°$ and $150°$. The scattering amplitudes were obtained from an RMPS calculation [6.36]. For the black-and-white printing, the thick lines were chosen as follows: solid, axes in the plane and scattering angle; dashed, alignment angle; dot-dashed, momentum transfer direction. In the color pictures created by the actual application [6.33], the solid thick lines represent the axes in the plane (red), the scattering angle (blue), the alignment angle (orange), and the momentum transfer direction (green).

# References


6.1 K. Bartschat, *Computational Atomic Physics — Electron and Positron Scattering from Atoms and Ions*, Springer, Heidelberg 1996.

6.2 I. Bray and A.T. Stelbovics, Adv. At. Mol. Opt. Phys. **35** (1995) 209.

6.3 D.V. Fursa and I. Bray, J. Phys. B **30** (1997) 757.

6.4 I.E. McCarthy and E. Weigold, *Electron–Atom Collisions*, University Press, Cambridge 1995.

6.5 N. Andersen, K. Bartschat, J.T. Broad, and I.V. Hertel, Phys. Rep. **279** (1997) 251.

6.6 N. Andersen and K. Bartschat, Adv. At. Mol. Opt. Phys. **36** (1996) 1.

6.7 P.G. Burke and M.J. Seaton, Meth. Comput. Phys. **10** (1971) 1.

6.8 K.A. Berrington, W.B. Eissner, and P.H. Norrington, Comp. Phys. Commun. **92** (1995) 290.

6.9 K. Bartschat, Comp. Phys. Commun. **114** (1998) 168.

6.10 K. Smith, *The Calculation of Atomic Collision Processes*, John Wiley, New York 1971.

6.11 A.R. Edmonds, *Angular Momentum in Quantum Mechanics*, University Press, Princeton 1957.

6.12 U. Fano and G. Racah, *Irreducible Tensorial Sets*, Academic Press, New York 1959.

6.13 K. Bartschat and N.S. Scott, Comp. Phys. Commun. **30** (1983) 369.

6.14 A.M. Lane and R.G. Thomas, Rev. Mod. Phys. **30** (1958) 257.

6.15 H. Friedrich, *Theoretical Atomic Physics* (2nd edition), Springer, Berlin 1994.

6.16 M.S. Pindzola, F. Robicheaux, N.R. Badnell, and T.W. Gorczyca, Phys. Rev. A **56** (1997) 1994.

6.17 J.M. Hansteen, Phys. Scr. **42** (1990) 299.

6.18 B.H. Bransden, *Atomic Collision Theory*, Benjamin-Cummings, Reading 1983.

6.19 J.S. Briggs and J.H. Macek, Adv. At. Mol. Opt. Phys. **28** (1991) 1.

6.20 R. Gayet and A. Salin, Nucl. Instrum. Methods B **56** (1991) 82.

6.21 A. Dubois, S.E. Nielsen, and J.P. Hansen, J. Phys. B **26** (1993) 705.

6.22 W. Fritsch and C.D. Lin, Phys. Rep. **202** (1991) 1.

6.23 T.G. Winter and C.D. Lin, Phys. Rev. A **29** (1984) 567.

6.24 J.P. Hansen, L. Kocbach, A. Dubois, and S.E. Nielsen, Phys. Rev. Lett. **64** (1990) 2491.

6.25 M. Machholm and C. Courbin, J. Phys. B **27** (1994) 4703.

6.26 D.R. Schultz, M.R. Strayer, and J.C. Wells, Phys. Rev. Lett. **82** (1999) 3976.

6.27 M. Kimura and W.R. Thorson, Phys. Rev. A **24** (1981) 1780.

6.28 R.K. Janev and P.S. Krstić, Phys. Rev. A **46** (1992) 5554.

6.29 A.L. Ford, J.F. Reading, and K.A. Hall, J. Phys. B **26** (1993) 4537.

6.30 I.C. Percival and D. Richards, Adv. At. Mol. Phys. **11** (1975) 1.

6.31 R.E. Olson and A. Salop, Phys. Rev. A **16** (1977) 531.

6.32 R.E. Olson, J. Ullrich, and H. Schmidt-Böcking, Phys. Rev. A **39** (1989) 5572.

6.33 D. Loveall, M.B. Hamley, B.J. Miller, and K. Bartschat, Comp. Phys. Commun. **124** (2000) 90.

6.34 A. Raeker, K. Blum, and K. Bartschat, *J. Phys. B* **26** (1993) 1491.

6.35 N. Andersen and K. Bartschat, J. Phys. B**30** (1997) 5071.

6.36 K. Bartschat, E.T. Hudson, M.P. Scott, P.G. Burke, and V.M. Burke, J. Phys. B **29** (1996) 2875.

6.37 M. Machholm and C. Courbin, J. Phys. B **27** (1994) 4703; corrigendum: J. Phys. B **27** (1994) 5813.

# PART II

# CASE STUDIES

# 7. Electron-Impact Excitation

The case study part of this book begins in Section 7.1 with the presentation of a few examples where either unpolarized or spin-polarized electron beams are used to excite unpolarized atoms. Only the emitted light and its polarization, but not the scattered electrons, are observed. The intensity of the light corresponds to the so-called "optical excitation function." After proper normalization and accounting for a possible anisotropy in the radiation, excitation functions are often used to determine absolute cross sections for the process of interest. Furthermore, the light polarization components correspond to the "angle-integrated Stokes parameters." They are determined by the angle-integrated state multipoles that represent alignment and orientation of the excited target state, averaged over all projectile scattering angles.

Angle-differential observation of Stokes and $STU$ parameters is discussed in Section 7.2. The discussion focuses on some benchmark cases, involving light targets such as helium and sodium and heavy targets such as mercury and cesium. Special attention is given to the noble gas targets Ne–Xe. These continue to represent a major challenge to theory, partly due to difficulties associated with the structure calculation.

## 7.1 Angle-Integrated Stokes Parameters and Cross Sections

Let us first consider the symmetry properties of the excitation process. Following Bartschat and Blum [7.1], we distinguish three basic cases of electron polarization, namely unpolarized, longitudinally polarized, and transversally polarized beams, and three directions of light observation, namely parallel and perpendicular to the incident beam direction, with the latter case being further divided into observation along or perpendicular to the electron beam polarization.

The natural choice for the coordinate system in the case of an unpolarized incident electron beam is the $z$ axis along the incident beam direction, and hence we will make this choice in all three cases. A transversal electron polarization, if present, is chosen to define the $y$ axis. Note that for unpolarized and also for longitudinally polarized incident beams, the problem exhibits cylindrical symmetry around the $z$ axis, whereas we have planar symmetry (with respect to reflection on the $xz$ plane, since the spin polarization is an *axial vector*) in case of a transversal electron polarization along the $y$ direction.

Figure 7.1 illustrates the situation for the linear light polarization $P_2$ if the light is observed along the $y$ direction. Whether or not a given light polarization may be different from zero can generally be determined by inserting fictional mirrors into the arrangement in order to see which parameters change sign under this operation and which ones remain invariant. If the problem itself is mirror-symmetric, then only those parameters that do not change sign can, in principle, be nonzero. This argument is based on parity conservation in the physical process, which will be assumed throughout this book and is a very good approximation for all our cases of interest.

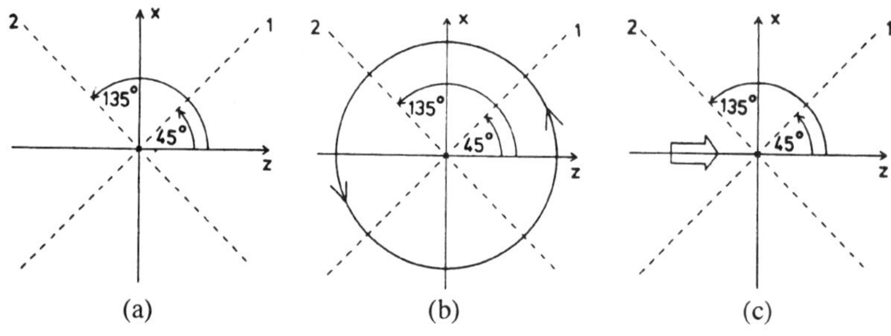

Fig. 7.1. Symmetry analysis for the linear light polarization $P_2$ for the basic cases of electron polarization discussed in the text. The light is observed along the $y$ direction. (a) Unpolarized electrons; (b) transversally polarized electrons with the polarization indicated by the circle; (c) longitudinally polarized electrons with the polarization indicated by the arrow.

Consider, for example, unpolarized (Figure 7.1a) and transversally polarized electrons (Figure 7.1b) with photon observation along the $y$ direction. Putting a mirror into the $yz$ plane transforms the axes "1" and "2" into each other. For the case of unpolarized electrons, nothing else changes and hence these two directions are physically equivalent. Consequently, the linear polarization $P_2$ measured by a photon director along the $y$ (and $x$) directions must vanish. On the other hand, this is not the case for the transversally polarized beam. Instead, the electron polarization seen in the mirror would also reverse, and thus $P_2$ can, based on *geometrical* symmetry alone, be nonzero. Furthermore, it must be proportional to the magnitude $P_y$ of the electron beam polarization. Other *dynamical* symmetries, however, may still result in a zero result for $P_2$, as will be further discussed below. Finally, inserting a

**Table 7.1** Relationship between the polarization of the incident electron beam ($P_e$) and the Stokes parameters ($P_1, P_2, P_3$) of the emitted light measured in various directions $\hat{n}$. The stars indicate the possibility of nonzero values. See text for definition of the coordinate system.

| $P_e$ | $\hat{n}$ | Stokes components | | |
|---|---|---|---|---|
| | | $P_1$ | $P_2$ | $P_3$ |
| Unpolarized | | $*$ | 0 | 0 |
| Longitudinal | $x$ | $*$ | 0 | 0 |
| Transversal | | $*$ | 0 | 0 |
| Unpolarized | | $*$ | 0 | 0 |
| Longitudinal | $y$ | $*$ | 0 | 0 |
| Transversal | | $*$ | $*$ | $*$ |
| Unpolarized | | 0 | 0 | 0 |
| Longitudinal | $z$ | 0 | 0 | $*$ |
| Transversal | | 0 | 0 | 0 |

mirror into the $xz$ plane shows that $P_2$ cannot depend on the longitudinal component $P_z$ of the electron polarization.

Using such arguments, one can derive the list of possibilities summarized in Table 7.1 [7.2,3]. Note that only three additional nonzero cases occur if the incident electron beam is polarized, namely a circular polarization $P_3$ for longitudinal and transversal electron polarization and light observation along the polarization direction, and a linear polarization $P_2$ when the incident electron beam is transversally polarized. These additional light polarizations are directly proportional to the corresponding electron polarization components, while the linear polarization $P_1$ is the same for unpolarized or polarized incident beams in all cases listed in Table 7.1.

As the first explicit example, we consider impact excitation of helium atoms from their ground state by unpolarized electrons, followed by observation of the radiation emitted in the $2^1\mathrm{P} \to 1^1\mathrm{S}$ in a direction perpendicular to the incoming beam (the $y$ direction in the definition above). Owing to the cylindrical symmetry of the problem, as well as its planar symmetry with respect to the $yz$ plane, only the following angle-integrated state multipoles can be nonzero:

$$\langle T(1)^\dagger_{00} \rangle = \frac{1}{\sqrt{3}} \left[ Q_0 + 2Q_1 \right], \tag{7.1a}$$

$$\langle T(1)^\dagger_{20} \rangle = \frac{2}{\sqrt{3}} \left[ Q_0 - Q_1 \right]. \tag{7.1b}$$

Here we have used (5.44) and the fact that the diagonal elements of the density matrix correspond to the cross sections $Q_0$ and $Q_1$ for excitation of the $M_L = 0, \pm 1$ magnetic sublevels. Note that $M_L = M$ for singlet–singlet transitions, and that $Q_1 \equiv Q_{-1}$ in the above choice of the coordinate system.

Inserting this result into (5.67) and (5.71) yields the simple formula

$$P_1 = \frac{Q_0 - Q_1}{Q_0 + Q_1} \tag{7.2}$$

for this case. Despite its simplicity, several important conclusions can already be drawn from this result:

$(i)$    The denominator of (7.2) is similar, but not identical to the total cross section $Q = Q_0 + 2Q_1$.

$(ii)$    The "optical excitation function," which is proportional to the intensity of the emitted light and hence to the above denominator, can be used to give the *relative* total cross section (i.e., its shape as a function of the incident electron energy), once the ratio $Q_1/Q_0$ has been determined from a $P_1$ measurement. An independent *normalization* is required to obtain *absolute* cross sections.

$(iii)$    Near the excitation threshold, excitation of the $M_L = 0$ sublevel is much more likely than excitation of $M_L \neq 0$, whereas the opposite happens for high energies. Consequently, the expected near-threshold result for the above case is $P_1 \approx 1$. A reduction of the measured $P_1$ values and even a change in sign is likely for higher energies. Also, as seen from Figure 4.2, fine-structure depolarization effects reduce the minimum and maximum values of $P_1$.

Figure 7.2 shows results for the light polarization $P_1$ and the angle-integrated magnetic sublevel cross sections for electron-impact excitation of the $2^1$P state in helium. The experimental data of Norén *et al.* [7.4] are compared with predictions from a relatively simple 11-state R-matrix (close-coupling) calculation. Based on the excellent agreement between theory and experiment in the shape of the curves in the near-threshold regime below the $n = 3$ thresholds, the above theoretical model was also used to normalize the experimental data. Note the verification of the threshold result discussed above, the effect of resonances, which is partly smoothed out by the convolution of the theoretical prediction with the experimental energy resolution (160 meV FWHM), and the depolarization effect (not included in the theory) caused by indirect population of the $2^1$P state through cascading, i.e., radiative transitions from the $n = 3$ and higher lying states that can also be excited at sufficiently high incident energies.

Such measurements have been performed for numerous targets, as discussed in detail in the review by Heddle and Gallagher [7.5]. We will not attempt to update the reference list here, but summarize a few important points:

$(i)$    These experiments represent a standard method for obtaining accurate results for the energy dependence of total cross sections as well as for the alignment parameter.

$(ii)$    Cascade effects can either be estimated or, in principle, be accounted for in an *ab initio* way by obtaining the equilibrium solutions of the corresponding rate equations for population of magnetic sublevels

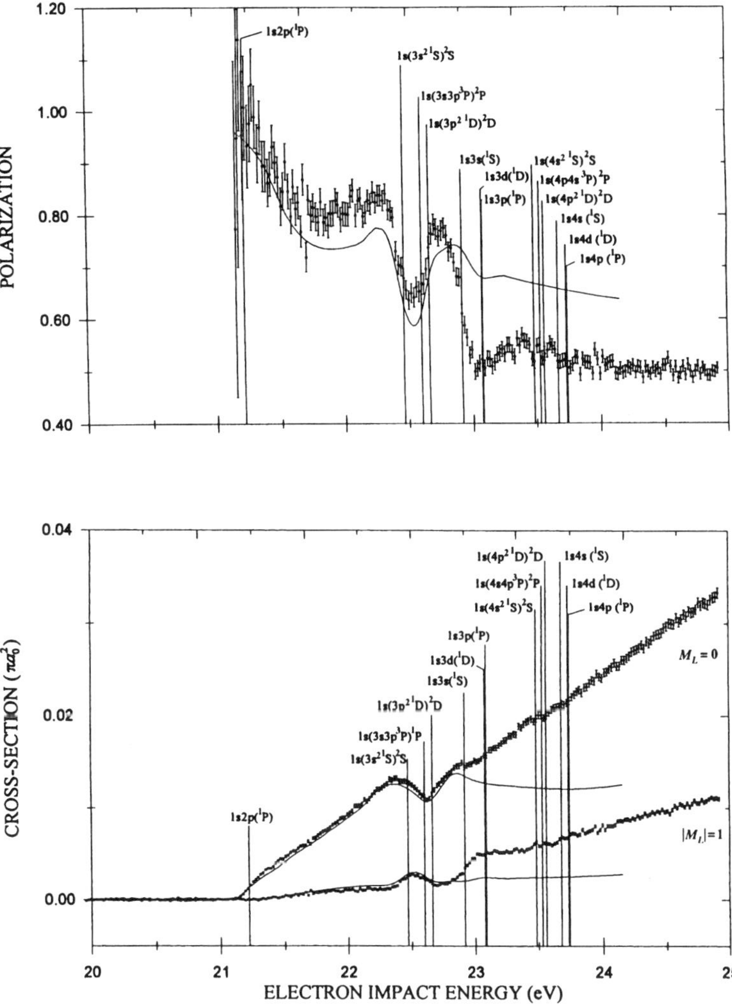



**Fig. 7.2.** Linear polarization $P_1$ and angle-integrated magnetic sublevel cross sections for impact excitation of the $2^1$P state of helium by unpolarized electrons. The vertical lines indicate target thresholds and resonance positions. The experimental data are compared with predictions from an 11-state R-matrix calculation (from [7.4]).

by electron and photon impact [7.6]. The latter treatment is non-trivial, since it requires input of all relevant cross sections and optical transition strengths.

(*iii*) Important practical applications of such measurements, often supplemented by theoretical results, are found in plasma polarization spectroscopy [7.7] (see also Section 11.8). For example, data from experiments in electron beam ion traps (EBITs) provide valuable insights for plasma diagnostics, particularly with respect to the direction of electron currents [7.8].

Studies of this kind may be extended from S $\rightarrow$ P transitions to higher angular-momentum states as well as spin-exchange transitions in the electron-impact excitation process. As shown in the fundamental paper by Percival and Seaton [7.9], the final results can be expressed in relatively simple form in terms of the magnetic sublevel cross sections, with the details depending on the target angular momenta involved, as well as on the resolution (or lack thereof) of the fine-structure and hyperfine-structure. A modern version of the Percival–Seaton treatment, based on the density-matrix formalism outlined in Chapter 5, was given by Bartschat and Csanak [7.10] and Bartschat [7.11].

If the excited state is well described in $LS$-coupling, the threshold polarization may be expressed in terms of angular-momentum coefficients, since only the $M_L = 0$ magnetic sublevel can be excited and thus will cancel in the formula for the light polarization $P_1$. For example, for excitation of a state with orbital angular momentum $L$, which optically decays into a state with angular momentum $L_\mathrm{f}$, the threshold polarization value for light observation perpendicular to the incident beam direction is given by [7.12]

$$
P_T = \frac{3\,G_2 \begin{Bmatrix} 1 & 1 & 2 \\ L & L & L_\mathrm{f} \end{Bmatrix} \begin{pmatrix} L & L & 2 \\ 0 & 0 & 0 \end{pmatrix}}{\sqrt{\dfrac{8}{15}}\dfrac{(-1)^{L_\mathrm{f}}}{2L+1} + G_2 \begin{Bmatrix} 1 & 1 & 2 \\ L & L & L_\mathrm{f} \end{Bmatrix} \begin{pmatrix} L & L & 2 \\ 0 & 0 & 0 \end{pmatrix}}, \tag{7.3}
$$

where $G_2$ corresponds to the perturbation coefficient defined in Section 5.5.4. Such considerations provide important consistency checks for both experimental and theoretical data, although the results may be modified dramatically in the presence of resonances. In fact, these modifications can sometimes be used to classify the resonances [7.4,13–16] or the dielectronic satellite lines in complex ionic recombination spectra [7.8].

As seen from Table 7.1, nonvanishing values of $P_2$ and $P_3$ require the use of a spin-polarized electron beam and a component of this spin polarization along the direction of light observation. Under such circumstances, the circular polarization $P_3$ measures the angle-integrated orientation of the excited target state while the linear polarization $P_2$ determines a second independent component of the alignment tensor.

Before discussing explicit examples, let us investigate the *dynamical* mechanisms that can cause such effects, in addition to the purely *geometric* symmetry properties outlined in Table 7.1. Beginning with the orientation, one might intuitively guess that the target can become oriented if its spin system becomes polarized due to exchange processes. This, however, is not the complete story since electric dipole radiation effectively investigates the *orbital* angular-momentum system. Hence, the polarization of the target spin system needs to be made visible in the emitted radiation. There are two principal mechanisms by which this can occur, namely:

(a) One can observe radiation between well-defined fine-structure levels, such as $^3\mathrm{P}_1 \to {}^3\mathrm{S}_1$, $^3\mathrm{S}_1 \to {}^3\mathrm{P}_0$, or $^3\mathrm{D}_3 \to {}^3\mathrm{P}_2$, to name a few.
(b) Alternatively, the spin-orbit interaction, acting *within* the target between excitation and decay, may transfer spin orientation into orbital angular-momentum orientation. This, in turn, results in circular light polarization *without* resolving the fine structure.

Based on mechanism (a), Farago and Wykes [7.17,18] suggested the measurement of an electron polarization via the optical method of observing the circular light polarization. The feasibility of the approach was demonstrated by Eminyan and Lampel [7.19], but owing to various experimental problems, partly related to the generally heavy targets needed to separate the fine structure, the method did not become competitive to Mott scattering as the standard electron polarization analyzer.

This situation changed dramatically when Gay [7.20] proposed a method based on mechanism (b) that could use a helium target. If an electron beam with polarization $P_e$ excites the helium atom, he showed that the circular polarization $P_3$ of the light emitted in the $3^3\mathrm{P} \to 2^3\mathrm{S}$ transition of helium is given by

$$P_3 = -\tfrac{1}{2}\left(1 - \tfrac{1}{6}P_1\right) P_e \equiv A P_e \,. \tag{7.4}$$

Although $P_1$ has to be determined to obtain very high precision, its value is generally small and its effect is further reduced by the coefficient $\tfrac{1}{6}$ in (7.4). In other words, the "analyzing power" $A$ is nearly independent of the electron energy.

Figure 7.3 shows the result of an experiment by the Münster group [7.21] who used Gay's method to calibrate a Mott detector to approximately one percent accuracy. Although this was a very desirable result for high-precision experiments, they did not advocate the method for routine measurements, owing to the relatively low efficiency of the helium polarimeter. This point was addressed by further work in Gay's group [7.22] who went back to fine-structure resolved measurements and investigated the $(2p^5 3p)^3\mathrm{D}_3 \to (2p^5 3s)^3\mathrm{P}_2$ transition in neon, as well as the corresponding transitions in Ar, Kr, and Xe. They showed that the figure of merit, namely the product of the analyzing power squared and the excitation cross section, could be increased by almost two orders of magnitude, thereby making the optical polarimeter a viable alternative to the standard Mott apparatus.

We now move on to the linear light polarization $P_2$, which, according to the geometrical symmetry properties listed in Table 7.1, may also be expected to be nonzero for the case of transversal electron polarization and light observation along the polarization direction. According to the discussion in Chapter 2, a nonvanishing value of $P_2$ corresponds to an average charge cloud

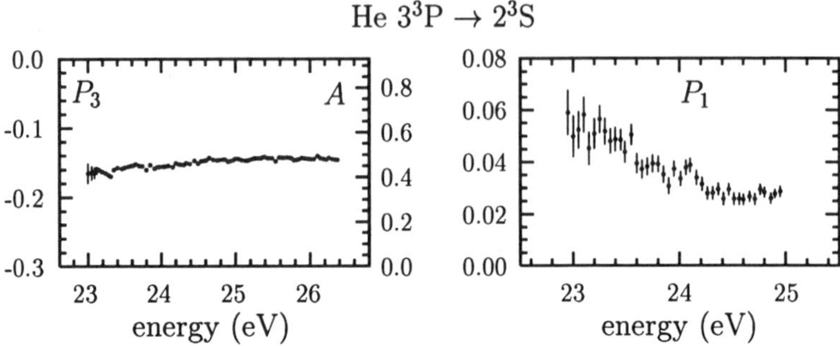



**Fig. 7.3.** Circular polarization $P_3$ and linear polarization $P_1$ of the $3^3$P $\rightarrow$ $2^3$S (388.9 nm) line in helium as function of the incident electron energy. The scale on the right side of the circular polarization figure shows the analyzing power of approximately 0.50, nearly independent of the energy [7.21].

that is aligned but also *rotated* by an angle $\gamma$ with respect to the incident beam axis. However, it was shown by Bartschat and Blum [7.1] that neither of the two mechanisms discussed above can lead to such a rotation; exchange alone is not sufficient, and the spin-orbit interaction within the target cannot produce such an effect either, as long as the excited target state is well described by $LS$-coupling.

Two other mechanisms have therefore been suggested that might yield nonzero values of $P_2$, namely:

(c) The spin-orbit interaction within the target, combined with configuration mixing, makes it impossible to describe the excitation mechanism in $LS$-coupling. Instead, an intermediate coupling scheme, i.e., combinations of $LS$-states with different values of $L$ and $S$ but the same value of the total electronic angular momentum $J$ must be used.

(d) Coupling of the continuum electron spin to the target angular momenta is so strong that it can effectively rotate the charge cloud. This mechanism represents the optical analogue to Mott scattering, but this time affecting the *target* without even detecting a specific electron scattering angle.

Figure 7.4(a) shows the first measurement of such a nonzero $P_2$ in an angle-integrated experiment with spin-polarized electrons, carried out by Bartschat *et al.* [7.23] for electron-impact excitation of the $(6s6p)^3$P$_1$ state in mercury. The same parameter, with much improved counting statistics [7.13], is shown in Figure 7.4(b) in the near-threshold regime, where the results are also compared with predictions from a 5-state Breit–Pauli R-matrix calculation carried out in the Belfast group [7.24]. There is excellent agreement between theory and experiment for $P_2$, and the results can be understood in terms of the intermediate-coupling nature (singlet + triplet mixture) of the excited state. Furthermore, the existence of resonances was believed to en-

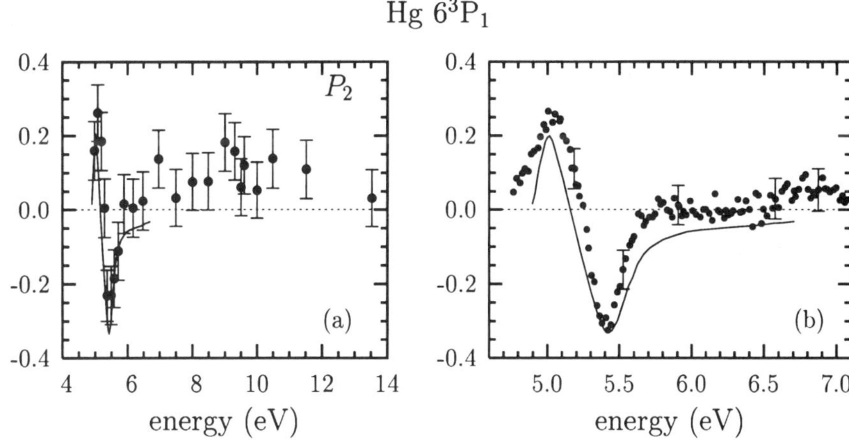



**Fig. 7.4.** Linear polarization $P_2$ (normalized to 100% incident electron polarization) for impact excitation of the $6^3P_1 \rightarrow 6^1S_0$ (254 nm) radiation in mercury, plotted as a function of the incident electron energy. The original results of Bartschat *et al.* [7.23] are shown on the left, and the near-threshold data of Wolcke *et al.* [7.13] are compared with theoretical predictions from a Breit–Pauli R-matrix approach [7.24] on the right.

hance the effect which has also been seen for noble gases by the Lincoln [7.25] and the Perth [7.26,27] groups.

Despite several attempts on different targets [7.25], however, the optical detection of Mott scattering has not been successful to date. Figure 7.5 shows the most recent results of Gay's group [7.28] for the linear polarization $P_2$ in the $(4p^55p)^3D_3 \rightarrow (4p^55s)^3P_2$ transition in krypton, after impact excitation by spin-polarized electrons. Although krypton is a fairly heavy target ($Z = 36$), the states of interest are well $LS$-coupled, and hence mechanism (c) is not expected to be relevant. Note that two different ways of determining $P_2$ did not yield an experimental result that could be interpreted as nonzero in a statistically meaningful way. On the other hand, theoretical predictions from two semirelativistic 15-state and 31-state Breit–Pauli R-matrix calculations suggest positive values of $P_2$ in the order of a few percent over an energy range that is free of cascades and several times wider than the energy width of the electron beam. Although the differences between the two theoretical curves clearly indicate a potential lack of convergence with respect to the number of states included in the calculation, the qualitative disagreement between theory and experiment regarding the $P_2$ parameter is somewhat surprising, since the theoretical model was otherwise very successful in describing near-threshold electron-impact excitation of the noble gases [7.29–31]. The reason for the discrepancy is currently unknown, and hence this problem represents an important open question in the field of polarized electron physics.

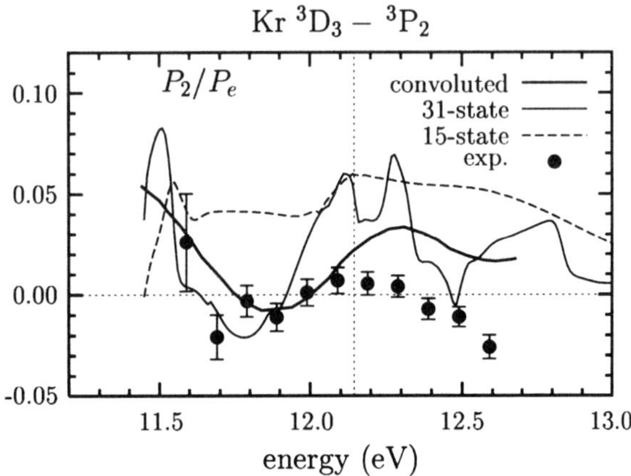

Fig. 7.5. Stokes parameter $P_2$ (normalized to 100% incident electron polarization) for impact excitation of the $(4p^5 5p)^3 D_3 \rightarrow (4p^5 5s)^3 P_2$ transition in krypton, plotted as a function of the incident electron energy. The experimental results of Gay's group [7.28] are compared with theoretical predictions based on 31-state and 15-state Breit–Pauli R-matrix models [7.31]. The thick line represents the 31-state results after convolution with the energy width of the electron beam.

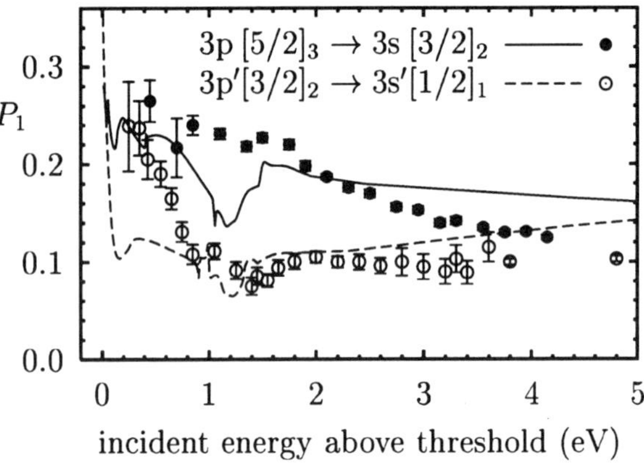

Fig. 7.6. Linear polarization $P_1$ after electron-impact excitation of the $3p[5/2]_3$ ($^3 D_3$) and $3p'[3/2]_2$ ($^3 D_2$) states in neon [7.31]. The experimental data were taken at $\theta = 135°$.

We finish this section with one more example of the $P_1$ parameter, this time for excitation of the $(2p^5 3p)$ manifold of states in neon, with subsequent decay to various states of the $(2p^5 3s)$ manifold. Figure 7.6 shows $P_1$ values for excitation of the $3p[5/2]_3$ and the $3p'[3/2]_2$ states in neon. The former corresponds to the $(2p^5 3p)^3 D_3$ state in $LS$-notation, while the latter is predominantly described as $(2p^5 3p)^3 P_2$ but contains significant admixtures from $^1\mathrm{D}_2$ and $^3\mathrm{D}_2$ symmetries of the same configuration and thus is not a pure Russell-Saunders state. The data were taken in Gay's group with the photons being observed at a polar angle of $\theta = 135°$ [7.31] relative to the incident beam direction.

In this example, the 31-state Breit–Pauli R-matrix model [7.31] is very successful in describing the observed energy dependence and magnitude of the data in the near-threshold regime. Since cascading from higher levels begins to be appreciable about 2 eV above threshold, this will likely account for the increasing discrepancy between theory and experiment at higher energies. Second, the $^3\mathrm{D}_3$ data approach the required kinematic threshold value of $P_1 = 0.28$ [7.32] for the particular experimental setup. This value is determined for a well $LS$-coupled $^3\mathrm{D}_3$ state by the fact that ($i$) it must be excited by exchange and ($ii$) only $M_L = 0$ magnetic sublevels can be populated at threshold (see above). Since the $^3\mathrm{P}_2$ level is not well $LS$-coupled, it does not have a kinematically defined threshold value. The calculation also indicates that negative-ion resonances [7.4,14,15,33], which decay into the $^3\mathrm{D}_3$ state, should affect $P_1$. These are not seen as clearly in the experimental data because of the energy width of the electron beam in the experiment ($\approx 0.3\,\mathrm{eV}$).

Qualitatively, the most interesting aspect of the $^3\mathrm{D}_3$ data is the fact that they do not drop as rapidly from their threshold value as do the $^3\mathrm{P}_2$ results. The calculations indicate that the $^3\mathrm{P}_2$ values of $P_1$ drop to less than 50% of their threshold value within 0.1 eV of threshold. (The differences between theory and experiment below 1 eV for this state are due, at least in part, to the high-energy "tail" of the electron beam.) In contrast, the $^3\mathrm{D}_3$ values do not decrease to half their initial value until 6 eV above threshold. An interesting difference between the $^3\mathrm{D}_3$ and $^3\mathrm{P}_2$ states concerns the angular momentum $j_c$ of their respective cores. While they both have their excited-electron orbital angular momentum, their excited-electron spins, and $j_c$ "lined up" to give the maximal $J$ for the state, the $^3\mathrm{D}_3$ state has an alignable ($j_c = 3/2$) core while the $^3\mathrm{P}_2$ state does not ($j_c = 1/2$). The tendency of the atomic $^3\mathrm{D}_3$ alignment to remain high well above threshold has thus in the past been interpreted in terms of an alignment "flywheel" model [7.34,35]: the "storage" of alignment in the $^3\mathrm{D}_3$ core could reduce both the depolarizing influence of negative-ion decay and the ordinary fall of $P_1$ from its threshold value as sublevels with $M_L > 0$ begin to be excited.

This type of physical information, made apparent by the comparison of states with different angular-momentum coupling schemes in the same manifold, points out the utility of such comparative studies. Consequently, we now look more broadly at the dynamical parameter that uniquely determines the value of $P_1$: the relative alignment

$$\langle t(J)_{20}^+\rangle \equiv \frac{\langle T(J)_{20}^+\rangle}{\langle T(J)_{00}^+\rangle}, \tag{7.5}$$

where the integrated state multipoles are defined according to (5.44). Owing to the cylindrical and planar symmetry of the problem, we recall that $\langle t(J)_{20}^+\rangle$

is the dynamical parameter that fully determines the anisotropy in the emitted dipole radiation. For emission from a state with angular momentum $J$ to one with $J_{\mathrm{f}}$ and photon observation at an angle $\theta$ with respect to the incident beam axis, $P_1$ can be expressed as (see Section 5.6)

$$
P_1 = \frac{3\, G_2 \begin{Bmatrix} 1\,1\,2 \\ J\,J\,J_{\mathrm{f}} \end{Bmatrix} \langle t_{20}^+ \rangle \sin^2 \theta}{(-1)^{J+J_{\mathrm{f}}} \sqrt{\dfrac{8}{3\,(2J+1)}} - G_2 \begin{Bmatrix} 1\,1\,2 \\ J\,J\,J_{\mathrm{f}} \end{Bmatrix} \langle t_{20}^+ \rangle (3\cos^2 \theta - 1)} , \tag{7.6}
$$

where $G_2$ is again a perturbation coefficient to account, if necessary, for depolarization of the radiation due to atomic hyperfine structure. (Note that the fine-structure is already resolved in this case.)

Figure 7.7 shows results for $\langle t_{20}^+ \rangle$ as a function of incident electron energy for the eight neon states with configuration $2p^5 3p$ and $J \neq 0$. (For $J = 0$ states, $\langle t_{20}^+ \rangle$ is identically zero, and the light intensity can be used directly to obtain the total excitation cross section [7.31].) For energies of only a few tenths of an eV above threshold, we note a striking systematic effect: the results cluster by $J$ value of the excited state. Although the agreement between theory and experiment (and between different sets of experimental data) is not perfect, the general trend is clearly confirmed.

One can qualitatively explain this clustering with arguments based on angular-momentum coupling. Defining once again the angle-integrated magnetic sublevel cross sections $Q_M = \langle JM|\rho|JM \rangle$ as the diagonal elements of the density matrix, the relative alignment parameters for the various $J$ values can be written as [7.31]

$$
J = 1 : \langle t_{20}^+ \rangle = \sqrt{2}\, \frac{Q_1 - Q_0}{2Q_1 + Q_0} ; \tag{7.7a}
$$

$$
J = 2 : \langle t_{20}^+ \rangle = \sqrt{\frac{10}{7}}\, \frac{2Q_2 - Q_1 - Q_0}{2Q_2 + 2Q_1 + Q_0} ; \tag{7.7b}
$$

$$
J = 3 : \langle t_{20}^+ \rangle = \frac{1}{\sqrt{3}}\, \frac{5Q_3 - 3Q_1 - 2Q_0}{2Q_3 + 2Q_2 + 2Q_1 + Q_0} . \tag{7.7c}
$$

Furthermore, conservation of the total angular momentum of the collision system implies the selection rule

$$
M + m_{\ell_{\mathrm{f}}} + m_{\mathrm{f}} = m_{\mathrm{i}} , \tag{7.8}
$$

where $m_{\ell_{\mathrm{f}}}$ is the orbital angular-momentum component of the scattered electron while $m_{\mathrm{f}}$ ($m_{\mathrm{i}}$) is its final (initial) spin projection with respect to the quantization axis. Note that (7.8) holds for an initial atomic state with $J = 0$ and our choice of quantization axis along the incident beam direction.

It follows from (7.8) that excitation processes without spin-flip ($m_{\mathrm{f}} = m_{\mathrm{i}}$) require $M = -m_{\ell_{\mathrm{f}}}$, while excitation processes with spin-flip ($m_{\mathrm{f}} = -m_{\mathrm{i}}$) require $M = -m_{\ell_{\mathrm{f}}} \pm 1$. Consequently, there are no contributions from projectile

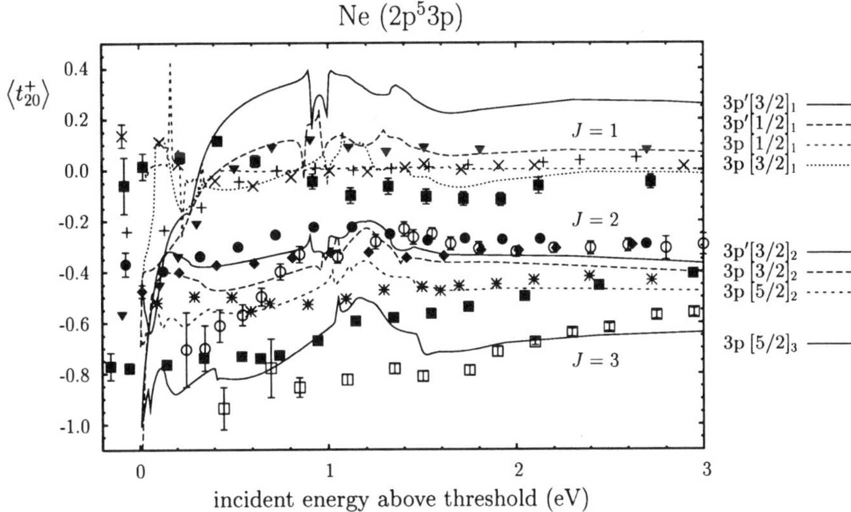


**Fig. 7.7.** Relative alignment parameter $\langle t_{20}^+ \rangle$ after electron-impact excitation of the Ne 2p$^5$3p manifold. For incident energies more than 0.2 eV above threshold, the top four curves belong to states with $J=1$, the next three to $J=2$, and the bottom one to the $J=3$ state. In detail, the curves and symbols are as follows: $J=1$: solid line, $\times$, 3p[3/2]$_1$; long dashes, squares, 3p[1/2]$_1$; short dashes, triangles, 3p'[3/2]$_1$; dots, $+$, 3p'[1/2]$_1$. $J=2$: solid line, $*$, 3p[5/2]$_2$; long dashes, diamonds, 3p[3/2]$_2$; short dashes, $\bullet$ and o, 3p'[3/2]$_2$. $J=3$: solid line, open and solid squares, 3p[5/2]$_3$. The experimental data are taken from Yu *et al.* [7.26,27] and from Zeman *et al.* [7.31] ($\square$ for 3p[5/2]$_3$ and o for 3p'[3/2]$_2$).

partial waves with $\ell_f = 0, 1$ to $Q_3$ at all, while $Q_2$ contains only an exchange contribution from $\ell_f = 1$. On the other hand, optically forbidden transitions, like the ones discussed in this example, are strongly affected by partial waves with small angular momenta. Hence, one can expect $Q_0$ and $Q_1$ to be significantly larger than $Q_2$ and $Q_3$, respectively, especially near threshold. According to (7.7), this result will indeed lead to a large negative alignment for states with $J \geq 2$, and a slower increase of its value with increasing energy for the $J = 3$ state than for the $J = 2$ state. Furthermore, the alignment parameter for the $J = 1$ states contains the *difference* between $Q_1$ and $Q_0$ in the numerator, and thus one would expect a smaller alignment value in this case. This grouping of alignment by $J$ had not been observed before because of the lack of comprehensive data sets for a given atom, and because much of the previous alignment data was taken with unresolved fine structure.

In summary, these comprehensive studies revealed an interesting grouping of alignment according to the total electronic angular momentum $J$ of the excited state. Since the findings can be qualitatively explained using angular-momentum coupling rules, they are expected to be a general feature for similar excitation processes in Ar, Kr, and Xe. While some evidence for the validity of the "flywheel model," i.e., the storage of alignment in the core of the excited atom was found, any such effect is not as important as the overall dynamical clustering with $J$ [7.31].

## 7.2 Angle-Differential Stokes and *STU* Parameters

We now move on to the next level of detail, namely the angle-resolved study of electron-impact excitation. The first case study deals with the helium target, more specifically P-state and D-state excitation of the first few states. In this case, the projectile spin is essentially irrelevant, since singlet states can only be excited without changing this spin, whereas triplet states can only be excited by exchange.

### 7.2.1 Electron-impact excitation of helium

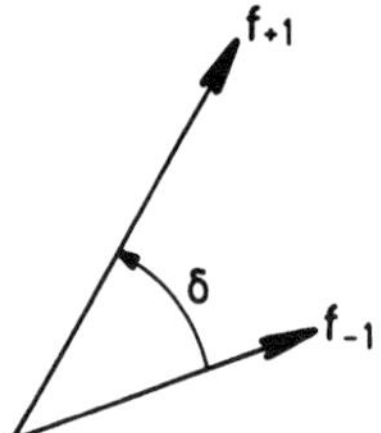


**Fig. 7.8.** The amplitudes $f_{+1}$ and $f_{-1}$ for impact S $\rightarrow$ P excitation without accounting for electron spin.

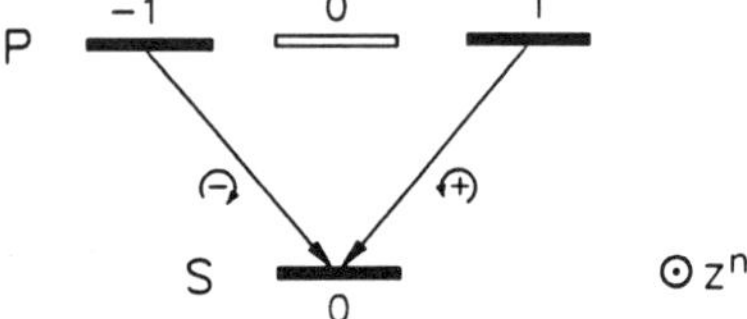


**Fig. 7.9.** Atomic P-state excitation and P $\rightarrow$ S photon emission along the quantization axis in the natural coordinate frame. Only the magnetic sublevels with $M = \pm 1$ will be populated.

We start with the case of $^1$S $\rightarrow$ $^1$P$^\circ$ transitions for which the framework outlined in Chapter 2 can be applied directly. Recall that this case is described by the two scattering amplitudes $f_{-1}$ and $f_{+1}$ shown in Figure 7.8. Transitions to the $M_L = 0$ state are forbidden (see Figure 7.9) due to conservation of reflection symmetry in the scattering plane. Hence, this process is determined by three parameters: the absolute differential cross section

$$\sigma = |f_{+1}|^2 + |f_{-1}|^2 , \tag{7.9}$$

as well as a parameter describing the relative size of the two amplitudes and an angle that fixes their relative phase $\delta$.

Following the discussion outlined in Chapter 2, the first of the latter two dimensionless parameters may be chosen as the angular-momentum transfer $\langle \boldsymbol{L} \rangle = (0, 0, L_\perp)$ with

$$L_\perp = \frac{|f_{+1}|^2 - |f_{-1}|^2}{|f_{+1}|^2 + |f_{-1}|^2} = \frac{|f_{+1}|^2 - |f_{-1}|^2}{\sigma} \equiv w^+ - w^- , \tag{7.10}$$

where we have defined the weighting factors $w^\pm = |f_{\pm1}|^2/\sigma$ with $w^+ + w^- = 1$. As the second parameter, we choose the alignment angle $\gamma$ of the major symmetry axis of the charge cloud in the scattering plane. This angle, defined mod $\pi$ within the interval $-\frac{\pi}{2} \le \gamma \le \frac{\pi}{2}$, is related to the phase angle $\delta$ (cf. Figure 7.8) through (2.39),

$$\delta = \arg\left(f_{+1} f_{-1}^*\right) = -2\gamma \pm \pi. \tag{7.11}$$

Determination of the parameter set $(\sigma; L_\perp; \gamma)$ thus constitutes a perfect scattering experiment, as can be seen from the relationship between the density matrices describing the excited P state and the emitted photons in the optical decay of this state (see Section 5.7).

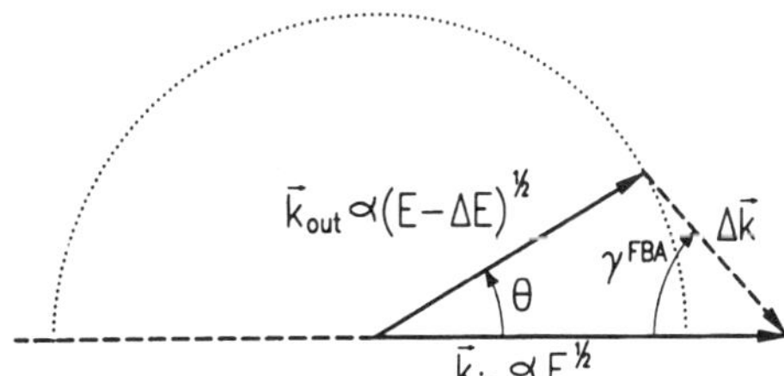


**Fig. 7.10.** Diagram corresponding to (7.12). Note the sense of rotation for $\gamma^{FBA}$, which is defined mod $\pi$.

The first Born approximation (FBA) is a useful starting point for the discussion of experimental and theoretical data. For electron scattering, the selection rule $\Delta M = 0$ holds along the direction of momentum transfer [7.36]. Furthermore, $L_\perp^{FBA} = 0$. Denoting the incident energy by $E$ and the energy loss by $\Delta E$, the relation between the scattering angle $\theta$ and the alignment angle $\gamma^{FBA}$ is directly read from the momentum vector diagram in Figure 7.10, namely

$$\tan \gamma^{FBA} = \sin\theta / (\cos\theta - x) , \tag{7.12}$$

where $x = [E/(E - \Delta E)]^{1/2}$. For $\Delta E > 0$, $\gamma^{\mathrm{FBA}}$ is always negative (note the sense of rotation in the definition of $\gamma$ in Figure 7.10) and assumes its minimum value, $\gamma_{\min}^{\mathrm{FBA}}$, at the scattering angle $\theta_{\min}$, where $\Delta k$ is perpendicular to $k_{\mathrm{out}}$; i.e.,

$$\gamma_{\min}^{\mathrm{FBA}} = -\arcsin(1/x) = \theta_{\min} - \frac{\pi}{2}\,. \tag{7.13}$$

The three Stokes parameters for S $\to$ P excitation and subsequent P $\to$ S decay are

$$(P_1^{\mathrm{FBA}}, P_2^{\mathrm{FBA}}, P_3^{\mathrm{FBA}}) = (\cos 2\,\gamma^{\mathrm{FBA}}, \sin 2\,\gamma^{\mathrm{FBA}}, 0)\,. \tag{7.14}$$

The results become even simpler in the so-called "atomic frame" which is obtained from the natural frame by rotation around the $z$ axis such that the new $\hat{x}$ axis points in the $\gamma$ direction. Quantities in this frame will be labeled by a "hat" below. By definition, $\hat{\gamma} = 0$, and therefore

$$(\hat{P}_1^{\mathrm{FBA}}, \hat{P}_2^{\mathrm{FBA}}, \hat{P}_3^{\mathrm{FBA}}) = (1, 0, 0)\,. \tag{7.15}$$

For the scattering amplitude (again except for an overall phase), we obtain

$$(\hat{f}_{+1}^{\mathrm{FBA}}, \hat{f}_{-1}^{\mathrm{FBA}}) = \sqrt{\frac{\sigma}{2}}\,(1, -e^{2\mathrm{i}\gamma})\,. \tag{7.16}$$

FBA results for higher angular momenta will be discussed below.

For later reference, we recall in Figures 7.11 and 7.12 how the parameter set $(L_\perp, \gamma)$ can be used to visualize the outcome of the electron-impact induced He ($1^1$S $\to$ $2^1$P) excitation process [7.37]. The upper part of Figure 7.11 shows experimental results at $80\,\mathrm{eV}$ impact energy for the orientation parameter $L_\perp$, while the lower part presents data for the alignment angle $\gamma$, here compared to the value predicted by the first Born approximation. The orientation parameter exhibits a characteristic dependence on the scattering angle $\theta$: starting at zero, positive values grow with increasing $\theta$ until an almost circular state is observed, with the alignment angle following the FBA prediction until near its minimum. From then on, the orientation decreases rapidly, changes sign, and approaches its other extremum value. This corresponds again to a circular state, but with opposite sense of rotation (i.e., clockwise) of the excited electron around the atomic core. In the above angular range, the $\gamma$ angle is almost perpendicular to the FBA prediction, with another rapid change near the minimum value of $L_\perp$. At even larger scattering angles, the orientation and the alignment angle converge back to zero until, for backward scattering, the excited state assumes once again its initial shape, a P orbital aligned along the beam axis. Figure 7.12(a) shows the corresponding evolution of the charge cloud density in the collision plane, with the upper strip corresponding to $L_\perp > 0$ and the lower one to $L_\perp < 0$. The three-dimensional clouds are shown in Figure 7.12(b).

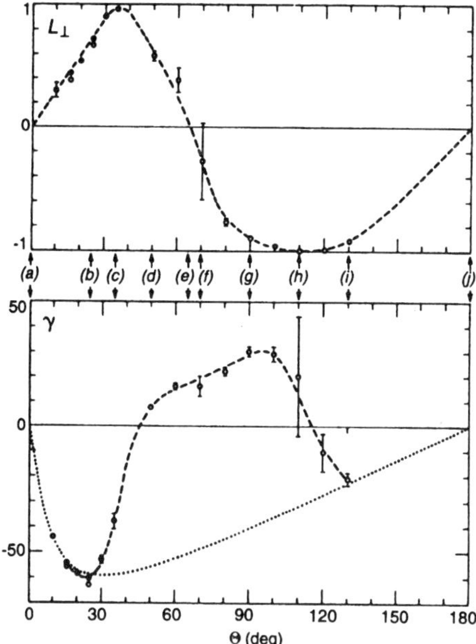

**Fig. 7.11.** Angular-momentum transfer $L_\perp$ and alignment angle $\gamma$ for electron-impact excitation of the $2^1\mathrm{P}^\circ$ state in helium for an incident energy of 80 eV [7.37]. The dashed line is a smooth fit through the experimental data of [7.36] (•) and [7.38] (o), while the dotted line for $\gamma$ represents the FBA prediction. The labels (a)–(j) are referred to in Figure 7.12.

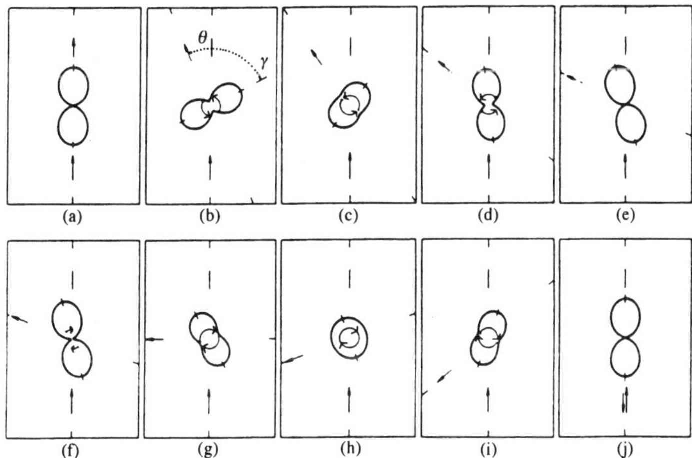



**Fig. 7.12.** (a) Shape and dynamics of the He $(1s2p)^1\mathrm{P}^\circ$ state corresponding to Figure 7.11 [7.37]. The scattering angles are $0^\circ$(a), $25^\circ$(b), $35^\circ$(c), $50^\circ$(d), $65^\circ$(e), $70^\circ$(f), $90^\circ$(g), $110^\circ$(h), $130^\circ$(i), and $180^\circ$(j). The upper row (a–e) corresponds to counterclockwise rotation of the electron around the atomic core, whereas the lower row (f–j) corresponds to clockwise rotation.

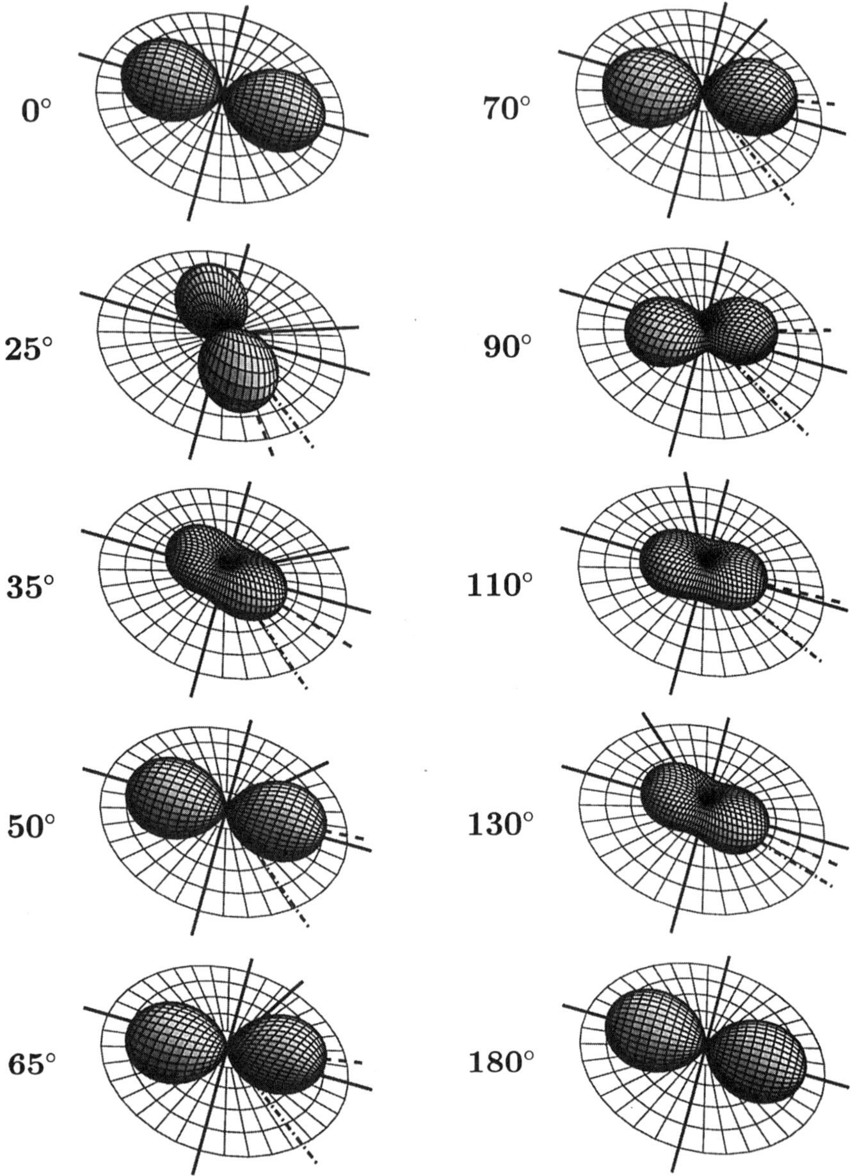

**Fig. 7.12.** (b) Three-dimensional charge cloud pictures corresponding to the scattering angles selected in Figures 7.11 and 7.12(a). The dashed line represents the alignment angle and the dash-dotted line the momentum-transfer direction.

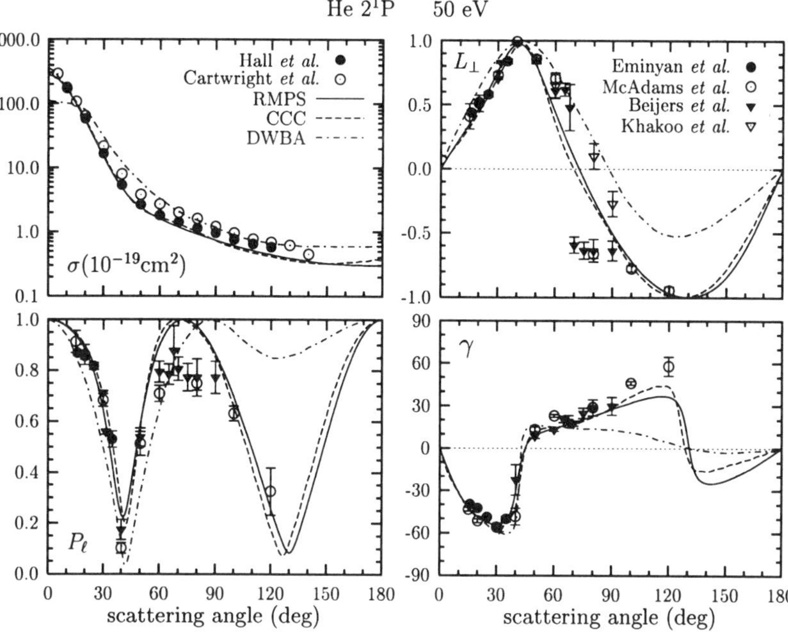


**Fig. 7.13.** Differential cross section (a) and electron-impact coherence parameters $L_\perp$ (b), $\gamma$ (c), and $P_\ell$ (d) for electron-impact excitation of the $2^1$P$^\circ$ state in helium from the ground state $1^1$S at an incident electron energy of 50 eV. The theoretical curves correspond to ——, RMPS [7.39]; – – –, CCC [7.40]; – · – ·, DWBA [7.41]. The experimental data are from [7.42–46].

Figures 7.13 and 7.14 demonstrate how well the S → P excitation process is now understood, both experimentally and theoretically, and how the detailed electron-impact coherence parameters (EICP) can serve as benchmark tests of theoretical models. As seen in Figure 7.13, there is very good agreement between various experimental datasets and theoretical predictions from the "R-matrix with pseudostates" (RMPS) [7.39] and the "convergent-close-coupling" (CCC) [7.40] models for excitation of the $2^1$P in helium at an incident electron energy of 50 eV. A much simpler first-order distorted-wave approach [7.41], formulated many years ago by Madison and Shelton [7.47], does very well in reproducing the differential cross section (DCS) and also the EICPs for small scattering angles. Not surprisingly, however, it fails for the large scattering angles if the very detailed sublevel and phase resolved information is compared. Similar problems, at comparable impact energies, are found with predictions from the "first-order many-body theory" (FOMBT) [7.48].

At lower collision energies, such as 30 eV, simpler close-coupling-type methods including only discrete target states can also be expected to predict the outcome of the collision process with reasonable accuracy. This is

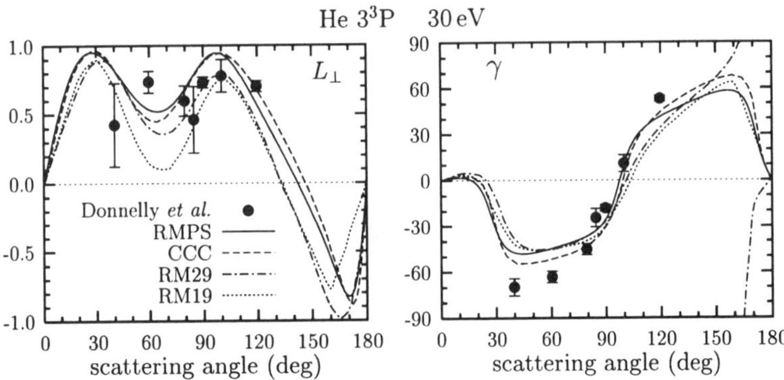

**Fig. 7.14.** Electron-impact coherence parameters $L_\perp$ and $\gamma$ for electron-impact excitation of the $3^3P^\circ$ state in helium from the ground state $1^1S$ at an incident electron energy of 30 eV. The experimental data of Donnelly *et al.* [7.49] are compared with results from various theoretical models: solid line, CCC [7.50]; long-dashed line, RM29 [7.51]; short-dashed line, RM19 [7.52].

shown in Figure 7.14 where experimental EICP results [7.49] for excitation of the $3^3P^\circ$ state in helium are compared with results from CCC [7.50] as well as 29-state [7.51] and 19-state [7.52] R-matrix calculations. The overall agreement between experiment and theory is satisfactory, slightly favoring (as expected) the CCC model. Note that the observed radiation ($3^3P^\circ \rightarrow 2^3S^e$) is depolarized in this case, but the EICPs directly after excitation have been recovered (see also Chapter 2).

Having summarized the good experimental and theoretical understanding of S $\rightarrow$ P excitation processes in helium, we now move on to the case of electron induced S $\rightarrow$ D transitions. This case has been discussed in the literature over the years [7.53–55]. Our summary here is based on a recent review [7.56] to which we refer for further details.

The three independent amplitudes are shown in Figure 7.15. A complete analysis of the optical decay pattern requires a triple-coincidence experiment involving the scattered particle and the two photons from the subsequent D $\rightarrow$ P and P $\rightarrow$ S transitions. The decay pattern for observation of both photons along the positive $z$ direction is detailed in Figure 7.16, where the helicities of the individual photons and the branching ratios for the sublevel transitions are indicated as well. A characteristic distinction of this observation direction is the fact that photons from transitions through the $M = 0$ sublevel of the P state, $|2\,0\rangle \rightarrow |1\,0\rangle$ and $|1\,0\rangle \rightarrow |0\,0\rangle$ (dashed lines), are not emitted and detected in the $z$ direction, owing to the properties of $\Delta M = 0$ electric dipole radiation.

Because of the extreme difficulty of triple-coincidence experiments, only double-coincidences between the scattered particle and one of the emitted photons [7.41,54,57–59], or the two photons without observation of the scat-

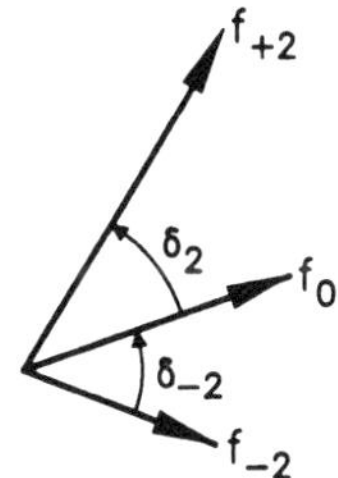

**Fig. 7.15.** The amplitudes $f_{+2}$, $f_0$ and $f_{-2}$ for impact S $\rightarrow$ D excitation without accounting for electron spin.

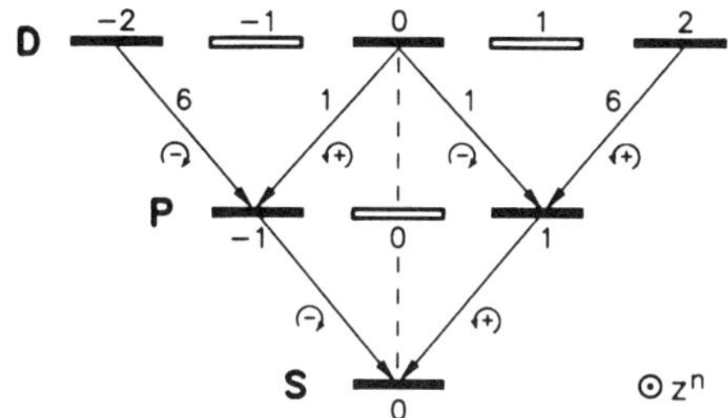

**Fig. 7.16.** Branching ratios for decay of an atomic D state through photon emission along the quantization axis. S $\rightarrow$ D impact excitation will only populate the magnetic sublevels with $M = 0, \pm 2$. Transitions through the $M = 0$ sublevel of the P state (dashed lines) are not visible for photon observation along the $z$ direction.

tered particle [7.60,61] have been observed in the majority of experimental studies until today. However, a pioneering triple-coincidence experiment was recently performed [7.62], thereby demonstrating the possibility (and experimental difficulty) of such studies.

Although the observation geometry was somewhat different in the latter experiment [7.62] we first summarize the analysis for the case described above owing to its conceptional simplicity. The usual coincidence experiment without observation of the cascade photon is then discussed in terms of this framework.

In Figure 7.16, two decay paths can be distinguished by the sublevel ($M_L = +1$ or $M_L = -1$) of the intermediate P level. (Recall that transitions to or from the state with $M_L = 0$ are not observable along the $\pm z$ directions.) These paths will be labeled by the helicity ($+$ or $-$) of the subsequent P $\rightarrow$ S cascade photon. The two channels may be studied individually if observed in coincidence with the corresponding cascade photon, detected along the positive $z$ direction through a circular polarization analyzer. If only the D $\rightarrow$ P photons are observed without regard of the cascade, the total signal is the weighted, incoherent sum of the two channels.

We now evaluate the signals observed for the two channels individually. The three scattering amplitudes are parameterized as

$$f_2 = \alpha_2 e^{i\phi_{+2}}, \tag{7.17a}$$

$$f_0 = \alpha_0 e^{i\phi_0}, \tag{7.17b}$$

$$f_{-2} = \alpha_{-2} e^{i\phi_{-2}}, \tag{7.17c}$$

and the differential cross section is given by

$$\sigma = \alpha_2^2 + \alpha_0^2 + \alpha_{-2}^2 . \tag{7.18}$$

Note (cf. Figure 7.16) that the $M_L = 2$ sublevel of the excited D state only contributes to the "+" channel, the $M_L = -2$ sublevel only contributes to the "−" channel, and the contribution from the $M_L = 0$ sublevel is equally shared between the two. We therefore define

$$\tilde{\sigma}^{\pm} = \alpha_{\pm 2}^2 + \frac{1}{2}\alpha_0^2 \tag{7.19}$$

and the relative size parameters

$$\tilde{w}^{\pm} = \tilde{\sigma}^{\pm}/\sigma \tag{7.20}$$

with

$$\tilde{w}^+ + \tilde{w}^- = 1 \tag{7.21}$$

and

$$\tilde{h}^{\pm} = \alpha_0^2/(3\tilde{\sigma}^{\pm}) \tag{7.22a}$$

$$\tilde{L}_\perp^{\pm} = \pm 2 \, \frac{\alpha_{\pm 2}^2 - \alpha_0^2/6}{\alpha_{\pm 2}^2 + \alpha_0^2/2} . \tag{7.22b}$$

A tilde is used to label all quantities that are related to the two decay channels rather than to the charge cloud of the excited D state (see below). We note the relations

$$\tilde{h}^{\pm} = \frac{1}{4}(2 \mp \tilde{L}_\perp^{\pm}), \tag{7.23a}$$

$$\tilde{w}^{\pm} = \frac{2 \pm \tilde{L}_\perp^{\mp}}{4 - \tilde{L}_\perp^+ + \tilde{L}_\perp^-} = \frac{\tilde{h}^{\mp}}{\tilde{h}^+ + \tilde{h}^-} . \tag{7.23b}$$

Starting from the relative phase angles (cf. Figure 7.15)

$$\delta_2 \equiv \phi_2 - \phi_0 , \tag{7.24a}$$

$$\delta_{-2} \equiv \phi_0 - \phi_{-2} , \tag{7.24b}$$

we define the alignment angles $\tilde{\gamma}^{\pm}$

$$\delta_{\pm 2} = -2\tilde{\gamma}^{\pm} \pm \pi . \tag{7.25}$$

The (coincidence) Stokes parameters for the two channels are

$$P_1^{\pm} + iP_2^{\pm} = P_\ell^{\pm}\, e^{2i\tilde{\gamma}^{\pm}} = -\frac{\sqrt{\tfrac{2}{3}}\alpha_0\alpha_{\pm 2}}{\alpha_{\pm 2}^2 + \alpha_0^2/6}\, e^{-i\delta_{\pm 2}}\;; \qquad (7.26a)$$

$$P_3^{\pm} = \mp\frac{\alpha_{\pm 2}^2 - \alpha_0^2/6}{\alpha_{\pm 2}^2 + \alpha_0^2/6}\,. \qquad (7.26b)$$

It is important to note that the vectors $\boldsymbol{P}^{\pm}$ are *unit vectors*, i.e.,

$$P_\ell^{\pm\,2} + P_3^{\pm\,2} = 1\,. \qquad (7.27)$$

Furthermore, we obtain

$$\tilde{L}_\perp^{\pm} = -2\,P_3^{\pm}(1 - \tilde{h}^{\pm})\,. \qquad (7.28)$$

Although the relative size parameters $\tilde{L}_\perp^{\pm}$ and $\tilde{h}^{\pm}$ are related to circular polarizations in a mathematical form analogous to P-state excitation (see equa-
tion (3.3.12) of Andersen *et al.* [7.63]), these quantities, individually, *neither
represent atomic angular momenta nor heights of charge clouds.* However,
their weighted sums

$$h = \tilde{w}^{+}\tilde{h}^{+} + \tilde{w}^{-}\tilde{h}^{-} = \frac{2}{3}\alpha_0^2/\sigma = \frac{2}{3}\rho_{00} \qquad (7.29)$$

and

$$L_\perp = \tilde{w}^{+}\tilde{L}_\perp^{+} + \tilde{w}^{-}\tilde{L}_\perp^{-} = 2\,(\alpha_2^2 - \alpha_{-2}^2)/\sigma = 2\,(\tilde{w}^{+} - \tilde{w}^{-}) = -2P_3\,(1 - h)\,, \qquad (7.30)$$

as measured in the scattered-particle–one-photon coincidence setup discussed
below, correspond to the angular momentum and the height of the D-state
charge cloud.

A complete experiment is then defined by the set

$$(\sigma; \tilde{L}_\perp^{+}, \tilde{L}_\perp^{-}; \tilde{\gamma}^{+}, \tilde{\gamma}^{-})\,, \qquad (7.31)$$

with the corresponding wavefunction given by

$$|\psi\rangle = f_2\,|2+2\rangle + f_0\,|2\,0\rangle + f_{-2}\,|2-2\rangle\,. \qquad (7.32)$$

Apart from an overall phase, the scattering amplitudes can be expressed as

$$(f_{+2}, f_0, f_{-2}) =$$
$$\frac{\sqrt{\sigma}}{2}\,(-\sqrt{2 + L_\perp - 3h}\, e^{-2i\tilde{\gamma}^{+}}, \sqrt{6h}, -\sqrt{2 - L_\perp - 3h}\, e^{2i\tilde{\gamma}^{-}})\,, \qquad (7.33)$$

and the ensemble of atoms that were excited by particles scattered at an
angle $\theta$ are in the pure quantum state
Interestingly, the analysis can equally well be made in the opposite order,
i.e., Stokes-vector analysis of the P $\to$ S cascade decay in coincidence with

helicity-selected D → P photons, with identical outcome. The optimal choice of order may thus be based on experimental convenience, such as photon wavelengths [7.64,56]. Also note that photon observation in only one direction is, in principle, sufficient to obtain a complete set of parameters.

The previous definitions can be used to obtain the formulas for the scattered-particle–one-photon cascade case by summing the contributions from the two channels (+ and −) with proper weights according to the photon intensities. This yields the following expression for the one-photon coincidence Stokes vector $\boldsymbol{P}$:

$$(1-h)\,\boldsymbol{P} = \tilde{w}^+\,(1-\tilde{h}^+)\,\boldsymbol{P}^+ + \tilde{w}^-\,(1-\tilde{h}^-)\,\boldsymbol{P}^- \tag{7.34}$$

or, splitting into linear and circular polarizations,

$$(1-h)\,P_\ell e^{2i\gamma} = \tilde{w}^+\,(1-\tilde{h}^+)\,P_\ell^+\,e^{2i\tilde{\gamma}^+} + \tilde{w}^-\,(1-\tilde{h}^-)\,P_\ell^-\,e^{2i\tilde{\gamma}^-}\,; \tag{7.35a}$$

$$(1-h)\,P_3 = \tilde{w}^+\,(1-\tilde{h}^+)\,P_3^+ + \tilde{w}^-\,(1-\tilde{h}^-)\,P_3^-\,. \tag{7.35b}$$

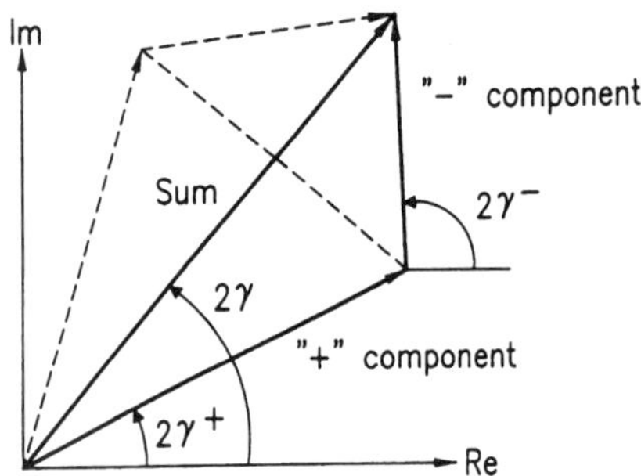


**Fig. 7.17.** Vector diagram corresponding to (7.35a). Instead of the values indicated for $2\gamma^\pm$, the ones obtained using the long dashed lines in the kite would also fulfill the equation.

Since $\boldsymbol{P}^+$ and $\boldsymbol{P}^-$ in general are not parallel, $\boldsymbol{P}$ is not a unit vector. The degree of polarization $P = |\boldsymbol{P}|$ is thus not unity, although the excitation process is fully coherent. Equation (7.35a) corresponds to a triangle in the complex plane shown in Figure 7.17. For this triangle, the lengths of all sides can be evaluated from scattered-particle–one-photon coincidence data, and the same holds for the direction $2\gamma$ of the sum vector. This information, however, is not sufficient to uniquely determine the directions of the "+" and "−" components, since, as the figure suggests, the mirror triangle has the same properties.

Before we analyze the corresponding ambiguity further, we note that additional information about the excited D state, namely the height parameter $h$, can still be obtained using the scattered-particle–one-photon coincidence setup. This is achieved by photon observation in a second direction, traditionally chosen along the $y$ axis, i.e., in the collision plane perpendicular to the incident beam direction. The Stokes vector in this direction is $(P_4, 0, 0)$. Here, light emitted with linear polarization perpendicular to the scattering plane arises from $\Delta M = 0$ optical decays of the $|2\,0\rangle$ state through the $|1\,0\rangle$ state which has negative reflection symmetry with respect to the scattering plane. The height parameter $h$ can then be expressed as the following function of $P_1$ and $P_4$:

$$h = \frac{(1 + P_1)(1 - P_4)}{4 - (1 - P_1)(1 - P_4)} \,. \tag{7.36}$$

The traditional (incomplete) parameterization for this case is the set [7.63]

$$(\sigma; L_\perp, \rho_{00} = \tfrac{3}{2}\, h; \gamma, P_\ell) \tag{7.37}$$

with the latter four evaluated from the four Stokes parameters $(P_1, P_2, P_3, P_4)$. One obtains the following expressions for the Stokes parameters in terms of the scattering amplitude parameters:

$$\sigma(1 - h)P_1 = -\sqrt{\frac{2}{3}}\,\alpha_0(\alpha_2 \cos \delta_2 + \alpha_{-2} \cos \delta_{-2}) \,, \tag{7.38a}$$

$$\sigma(1 - h)P_2 = \sqrt{\frac{2}{3}}\,\alpha_0(\alpha_2 \sin \delta_2 + \alpha_{-2} \sin \delta_{-2}) \,, \tag{7.38b}$$

$$\sigma(1 - h)P_3 = \alpha_{-2}^2 - \alpha_2^2 \,, \tag{7.38c}$$

where

$$\sigma(1 - h) = \alpha_2^2 + \alpha_{-2}^2 + \alpha_0^2/3 \,. \tag{7.39}$$

In the atomic frame (see Chapter 2) these equations simplify further to

$$\hat{P}_1 = P_\ell = \sqrt{\frac{2}{3}}\,\alpha_0\sqrt{\alpha_2^2 + \alpha_{-2}^2 + 2\alpha_2\alpha_{-2} \cos(\delta_2 - \delta_{-2})}/\sigma(1 - h); \tag{7.40a}$$

$$\hat{P}_2 = 0\,; \tag{7.40b}$$

$$\hat{P}_3 = P_3 = (\alpha_{-2}^2 - \alpha_2^2)/\sigma(1 - h)\,. \tag{7.40c}$$

Finally,

$$P_4 = \frac{\alpha_2^2 + \alpha_{-2}^2 - \alpha_0^2 - \sqrt{\frac{2}{3}}\,\alpha_0(\alpha_2 \cos \delta_2 + \alpha_{-2} \cos \delta_{-2})}{\alpha_2^2 + \alpha_{-2}^2 + \frac{5}{3}\alpha_0^2 - \sqrt{\frac{2}{3}}\,\alpha_0(\alpha_2 \cos \delta_2 + \alpha_{-2} \cos \delta_{-2})} \,. \tag{7.41}$$

The three probabilities for excitation of the individual magnetic sublevels are then expressed in terms of the parameter sets introduced above as

$$\frac{\alpha_2^2}{\sigma} = \frac{1}{4}(2 + L_\perp - 3h) = \tilde{w}^+ \frac{1}{4}(2 + \tilde{L}_\perp^+ - 3\tilde{h}^+) + \tilde{w}^- \frac{1}{4}(2 + \tilde{L}_\perp^- - 3\tilde{h}^-) ;$$

$$(7.42a)$$

$$\frac{\alpha_{-2}^2}{\sigma} = \frac{1}{4}(2 - L_\perp - 3h) = \tilde{w}^+ \frac{1}{4}(2 - \tilde{L}_\perp^+ - 3\tilde{h}^+) + \tilde{w}^- \frac{1}{4}(2 - \tilde{L}_\perp^- - 3\tilde{h}^-) ;$$

$$(7.42b)$$

$$\frac{\alpha_0^2}{\sigma} = \frac{3}{2}h = \tilde{w}^+ \frac{3}{2}\tilde{h}^+ + \tilde{w}^- \frac{3}{2}\tilde{h}^- . \qquad (7.42c)$$

We are now in a position to further address the ambiguity mentioned in Figure 7.17 above. If the atoms are in the pure quantum state given by (7.32), the angular part of the charge cloud density of the excited electron can be obtained as

$$\begin{aligned}
\Psi\Psi^*(\Theta,\Phi) = \frac{5}{16}\pi\Big\{ & \frac{3}{2}\left(|f_2|^2 + |f_{-2}|^2\right)\sin^4\Theta + |f_0|^2\left(3\cos^2\Theta - 1\right)^2 \\
& + \sqrt{6}\left(3\cos^2\Theta - 1\right)\sin^2\Theta\left[Re(f_2 f_0^* + f_0 f_{-2}^*)\cos 2\Phi\right. \\
& \qquad\qquad\qquad\qquad\left. - Im(f_2 f_0^* + f_0 f_{-2}^*)\sin 2\Phi\right] \\
& + 3\,Re(f_2 f_{-2}^*)\sin^4\Theta\cos\left(4\Phi + \delta_2 + \delta_{-2}\right)\Big\} \\
= \frac{5}{16}\pi\Big\{ & \frac{3}{2}\left(\alpha_2^2 + \alpha_{-2}^2\right)\sin^4\Theta + \alpha_0^2\left(3\cos^2\Theta - 1\right)^2 \\
& + \sqrt{6}\left(3\cos^2\Theta - 1\right)\sin^2\Theta\,\alpha_0\left[(\alpha_2\cos\delta_2 + \alpha_{-2}\cos\delta_{-2})\cos 2\Phi\right. \\
& \qquad\qquad\qquad\qquad\left. - (\alpha_2\sin\delta_2 + \alpha_{-2}\sin\delta_{-2})\sin 2\Phi\right] \\
& + 3\,\alpha_2\alpha_{-2}\sin^4\Theta\cos\left(4\Phi + \delta_2 + \delta_{-2}\right)\Big\} , \qquad (7.43)
\end{aligned}$$

As in the case of excited P states, the alignment angle $\gamma$ is obtained by choosing $\hat{P}_2 \equiv 0$ or, equivalently, the coefficient of $\sin 2\Phi$ in (7.43) to vanish in the atomic frame. The corresponding condition is

$$\tan 2\gamma = -\frac{Im(f_2 f_0^* + f_0 f_{-2}^*)}{Re(f_2 f_0^* + f_0 f_{-2}^*)} = -\frac{\alpha_2\sin\delta_2 + \alpha_{-2}\sin\delta_{-2}}{\alpha_2\cos\delta_2 + \alpha_{-2}\cos\delta_{-2}} . \qquad (7.44)$$

The inversion yields solutions for $2\gamma \bmod \pi$, i.e., $\gamma$ and $\gamma+90°$ will fulfill (7.44). These two directions correspond to maximum and minimum light intensity transmitted along the direction of the $z$ axis through a linear polarizer. The polarizer direction corresponding to the *maximum* intensity then defines the $\gamma$ angle.

Inserting this result into (7.43) yields

$$\begin{aligned}
\Psi\Psi^*(\Theta,\Phi) = \frac{15\sigma}{32\pi}\Big\{ & 1 - \frac{3}{2}h + h(3\cos^2\Theta - 1)^2 \\
& + (1 - h)\big[2(1 - 3\cos^2\Theta)\sin^2\Theta\,P_\ell\cos 2(\Phi - \gamma) \\
& + \sin^4\Theta\sqrt{(1 - \tfrac{1}{2}h)^2 - P_3^2}\,\cos 4(\Phi - \eta)\big]\Big\} \qquad (7.45)
\end{aligned}$$

with

$$\eta = -\frac{1}{4}(\delta_2 + \delta_{-2})\,. \tag{7.46}$$

It is this angle $\eta$, however, that cannot be defined uniquely in terms of the four Stokes parameters $(P_1, P_2, P_3, P_4)$. By transforming the problem into the "atomic frame" and defining (up to mod $\pi/4$)

$$\hat{\eta} = -\frac{1}{4}(\hat{\delta}_2 + \hat{\delta}_{-2}) = \frac{1}{2}(\hat{\gamma}^+ + \hat{\gamma}^-) = \frac{1}{2}(\tilde{\gamma}^+ + \tilde{\gamma}^-) - \gamma = \eta - \gamma\,, \tag{7.47}$$

it can be shown that two D states that only differ by the sign of $\hat{\eta}$ will yield *an identical set of Stokes parameters*. The two solutions merge to one if, and only if, all three amplitudes are real in the atomic frame.

This implies that the term depending on $4(\phi - \eta)$ in (7.43) for the electron charge cloud is determined except for a mirror operation in the plane defined by the $\hat{z}$ axis and the $\hat{x}$ axis. The electron charge clouds for the two possible states are thus mirror images of each other in the $(\hat{x}, \hat{z})$ plane. The proper sign is revealed by a coincidence analysis with the cascade photons.

To illuminate further the fundamental differences between P-state and D-state excitation and the increasing complexity of the latter, it is instructive to compare the matrices characterizing the photons emitted along the positive $z$ direction in a scattered-particle–single-photon coincidence experiment with the density matrix of the excited D state [7.56]. Although there are similarities between these two matrices, their structures for D-state excitation are *not identical*, in contrast to the situation for P-state excitation with positive reflection symmetry. Hence, it is understandable that not all parameters of the atomic density matrix can be extracted from the set of Stokes parameters measured in a scattered particle single-photon coincidence experiment alone.

As an explicit example to illustrate the concepts outlined above, we select the He $1^1$S $\rightarrow$ $3^1$D excitation process at 40 eV for which experimental electron–single-photon coincidence results with good quality are available in a wide angular range [7.57,59]. The theories selected for comparison are the FBA, the "convergent close-coupling" (CCC) calculation of Fursa and Bray [7.40,50], and the "R-matrix with pseudostates" (RMPS) results of Bartschat [7.65].

In analogy to the S $\rightarrow$ P excitation case, the FBA prediction for S $\rightarrow$ D excitation can be read from Figure 7.10, keeping the $\Delta M = 0$ selection rule along $\Delta \boldsymbol{k}$ in mind. The excited D state thus corresponds to an $M_L = 0$ sublevel with quantization axis in this direction. The two "mirror" states are degenerate in this case. Furthermore

$$L_\perp = 0; \tag{7.48a}$$

$$h = \frac{1}{6}; \tag{7.48b}$$

$$\tilde{w}^+ = \tilde{w}^- = \frac{1}{2}; \tag{7.48c}$$

$$\tilde{h}^+ = \tilde{h}^- = \frac{1}{6}; \tag{7.48d}$$

$$\tilde{L}_\perp^+ = -\tilde{L}_\perp^- = \frac{4}{3}; \tag{7.48e}$$

$$\tilde{\gamma}^+ = \tilde{\gamma}^- = \gamma; \tag{7.48f}$$

$$\hat{\eta} = 0. \tag{7.48g}$$

and thus

$$(P_1^{FBA}, P_2^{FBA}, P_3^{FBA}) = \frac{3}{5}(\cos 2\gamma^{FBA}, \sin 2\gamma^{FBA}, 0) \tag{7.49}$$

and

$$P_\ell = P = \frac{3}{5}. \tag{7.50}$$

The corresponding scattering amplitudes are

$$(f_2^{FBA}, f_0^{FBA}, f_{-2}^{FBA}) = \frac{\sqrt{\sigma}}{2}\left(-\sqrt{\frac{3}{2}}e^{-2i\gamma}, 1, -\sqrt{\frac{3}{2}}e^{2i\gamma}\right). \tag{7.51}$$

Figure 7.18 shows results for the four Stokes parameters $(P_1, P_2, P_3, P_4)$. These, in turn, are used to produce the standard incomplete parameter set (7.37) displayed in Figure 7.19, together with the total degree of light polarization $P$. There are several interesting features to be noticed. To begin with, the overall agreement between the CCC predictions and experiment is very good, except maybe for $P_2$ in the angular range between 60° and 100°. This discrepancy is propagated into the results for the degree of linear polarization $P_\ell$, as seen in Figure 7.19. Furthermore, the alignment angle $\gamma$ follows the prediction of the first Born approximation (i.e., $\gamma$ along the direction of the linear momentum transfer) approximately to the scattering angle where the angular-momentum transfer (which we recall to be identically zero in the Born approximation) reaches a maximum. A rapid increase in $\gamma$ follows, leading to the major axis of the linear light polarization being almost perpendicular to the direction of momentum transfer until a second rapid change occurs near the scattering angle where the angular-momentum transfer changes sign. This qualitative result is analogous to that encountered for P-state excitation (see Figure 7.11), and it shows the importance of the momentum transfer direction also for excitation of higher angular-momentum states.

Another interesting feature can be seen in the results for $\rho_{00} = \frac{3}{2}h$ and $P$. The height of the charge cloud is very small in the angular range between

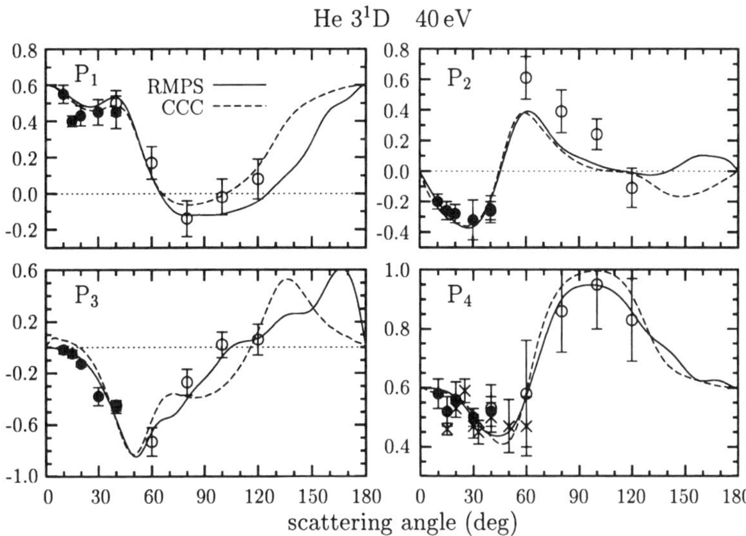


**Fig. 7.18.** Stokes parameters $(P_1, P_2, P_3, P_4)$ for electron-impact excitation of the He $1^1$S $\rightarrow 3^1$D transition at an incident electron energy of 40 eV. The experimental data of Beijers *et al.* [7.41] ($\times$), McLaughlin *et al.* [7.57] (o) and Mikosza *et al.* [7.59] ($\bullet$) are compared with predictions from CCC [7.50] and RMPS [7.65] calculations.

$80°$ and $120°$, indicating that excitation of the $M = 0$ sublevel is almost negligible in this range. In analogy to the excitation of P states in heavy rare gases, one might perhaps expect fully coherent light, since the excitation process is completely coherent. Instead, the observed radiation in a scattered-electron–single-photon coincidence is almost unpolarized ($P_1, P_2, P_3 \approx 0.0$) for scattering angles in the range between $100°$ and $120°$.

These findings become more transparent in the parameters introduced in (7.19–24). We first discuss the relative sizes. Figure 7.20 displays the results for the size parameters $(L_\perp, \tilde{L}_\perp^\pm)$, $(h, \tilde{h}^\pm)$, and $\tilde{w}^\pm$. The first interesting feature occurs near a scattering angle of $50°$ where both $\tilde{L}_\perp^-$ and $\tilde{h}^-$ exhibit rapid variations, with $\tilde{L}_\perp^-$ even becoming positive. On the other hand, due to the small value of $\alpha_{-2}^2$, the weight factor $\tilde{w}^-$ reaches its minimum in this region, and thus these structures are much less pronounced in the $(L_\perp, h)$ set. The second important feature arises for $\theta \approx 100°$ where $\alpha_0^2 \approx 0$, thereby resulting in a vanishing height parameter. This leads to the extreme values of $\tilde{L}_\perp^\pm \approx \pm 2$, whereas $L_\perp \approx 0$ because of $\tilde{w}^+ \approx \tilde{w}^-$. Also note that the individual $\tilde{L}_\perp^\pm$ do *not* vanish for scattering in the forward and backward directions.

Next we discuss the phase parameters $(\gamma, \tilde{\gamma}^\pm, \hat{\eta})$. Since no experimental results for the complete set $(\tilde{\gamma}^+, \gamma^-)$ are available, we invert the experimental data for $(P_1, P_2, P_3, P_4)$ by using the geometry introduced in Figure 7.17 and resolve the sign ambiguity for the pair $(\tilde{\gamma}^+, \tilde{\gamma}^-)$ — and thereby for $\hat{\eta}$

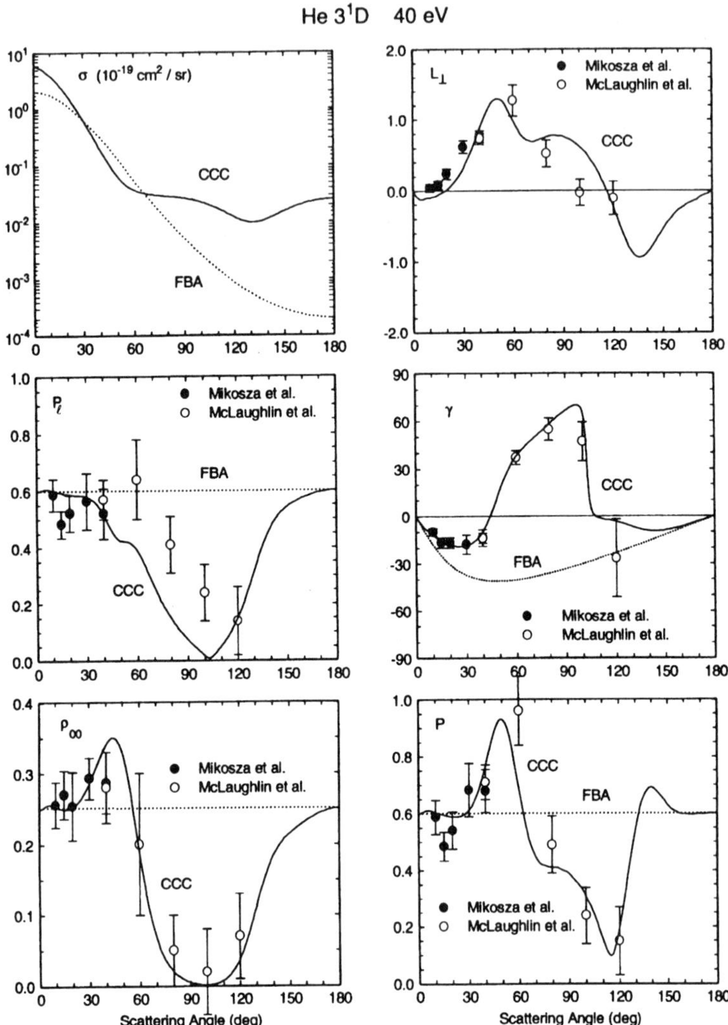



**Fig. 7.19.** Differential cross section $\sigma$ and coherence parameters ($L_\perp, P_\ell, \gamma, \rho_{00}, P$) for electron-impact excitation of the He $1^1$S $\rightarrow$ $3^1$D transition at an incident electron energy of 40 eV. The experimental data of Mikosza *et al.* [7.59] and McLaughlin *et al.* [7.57] are compared with CCC calculations (solid line) of Fursa and Bray [7.50] and with the predictions from the first Born approximation (dotted line).

— by comparison with theoretical predictions. The results are shown in Figure 7.21(a). Problems in the $(P_\ell, \gamma) \leftrightarrow (\tilde{\gamma}^+, \tilde{\gamma}^-)$ inversion for scattering angles of 60°, 80°, and 100° can, in part, be traced back to the relatively large magnitude of $P_\ell$ shown in Figure 7.19. (Mathematically, this results in one side of the triangle in Figure 7.17 being longer than the sum of the other two sides,

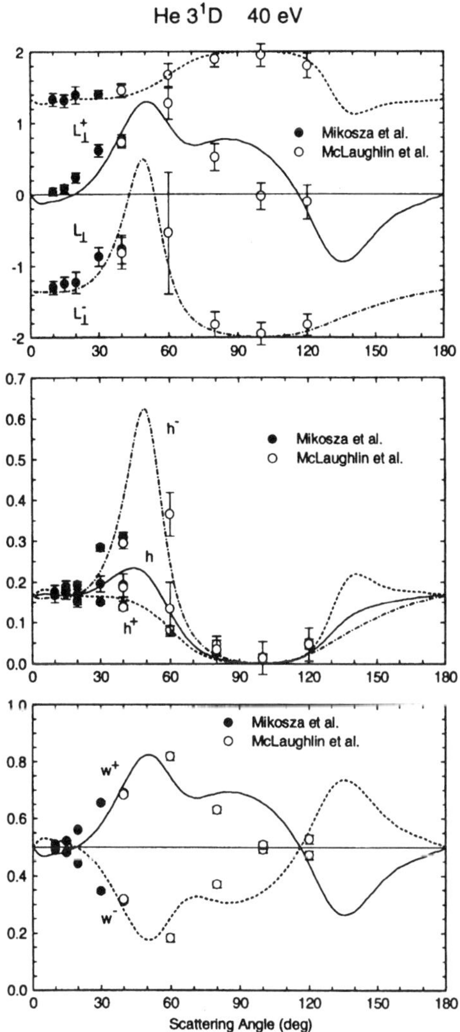



**Fig. 7.20.** The size parameters $(L_\perp, \tilde{L}_\perp^{\pm})$, $(h, \tilde{h}^{\pm})$, and $\tilde{w}^{\pm}$ for electron-impact excitation of the He $1^1$S $\rightarrow$ $3^1$D transition at an incident electron energy of 40 eV. The experimental points have been calculated from the data of Mikosza *et al.* [7.59] and McLaughlin *et al.* [7.57]. They are compared with results generated from CCC scattering amplitudes of Fursa and Bray [7.50].

in which case we set $\tilde{\gamma}^+ = \tilde{\gamma}^- = \gamma$.) A potential reason for the above problem in the experimental data may be the nearly circular nature of the emitted radiation in the individual channels that adds up incoherently in the cascade but is used to determine the $\gamma$ angle. For the same reason, the alignment angles $\tilde{\gamma}^{\pm}$ cannot be distinguished in this angular range. Based on the good

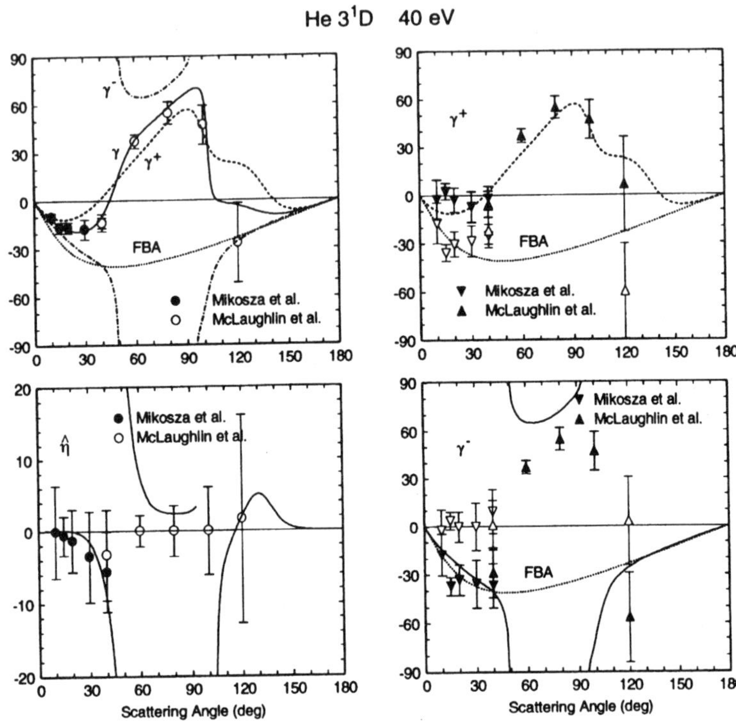


**Fig. 7.21.** (a) The alignment parameters $\gamma$, $\tilde{\gamma}^{\pm}$, and $\hat{\eta}$ for electron-impact excitation of the He $1^1$S $\rightarrow$ $3^1$D transition at an incident electron energy of 40 eV. The experimental points have been calculated from the data of Mikosza *et al.* [7.59] and McLaughlin *et al.* [7.57]. They are compared with results generated from CCC scattering amplitudes (solid line) of Fursa and Bray [7.50] and with the predictions from the first Born approximation (dotted line). The solid (open) symbols represent the "true" ("ghost") values for the set $(\tilde{\gamma}^{+}, \tilde{\gamma}^{-}, \hat{\eta})$, as suggested by using the theoretical results as a guide.

agreement between theory and experiment for the parameter set $(\tilde{L}_{\perp}^{+}, \tilde{L}_{\perp}^{-}, \gamma)$ and the fact that the theoretical data are internally consistent (even if they do not describe Nature perfectly), one might suspect that the experimental data for $P_\ell$ are slightly overestimated. If this were true indeed, the agreement between experiment and theory could further improve.

Similar experiments were recently performed in Crowe's group for excitation of the $3^3$D state of helium [7.66–68]. Again, comparison with CCC prediction allowed for the elimination of the "ghost solution" with a high degree of confidence. Figure 7.21(b) shows the form of the charge cloud predicted by the CCC calculation for various scattering angles of 40°, 60°, 80°, 100°, and 120°. The full picture is shown on the left, whereas the form derived from a one-photon angular distribution is shown on the right. Note the

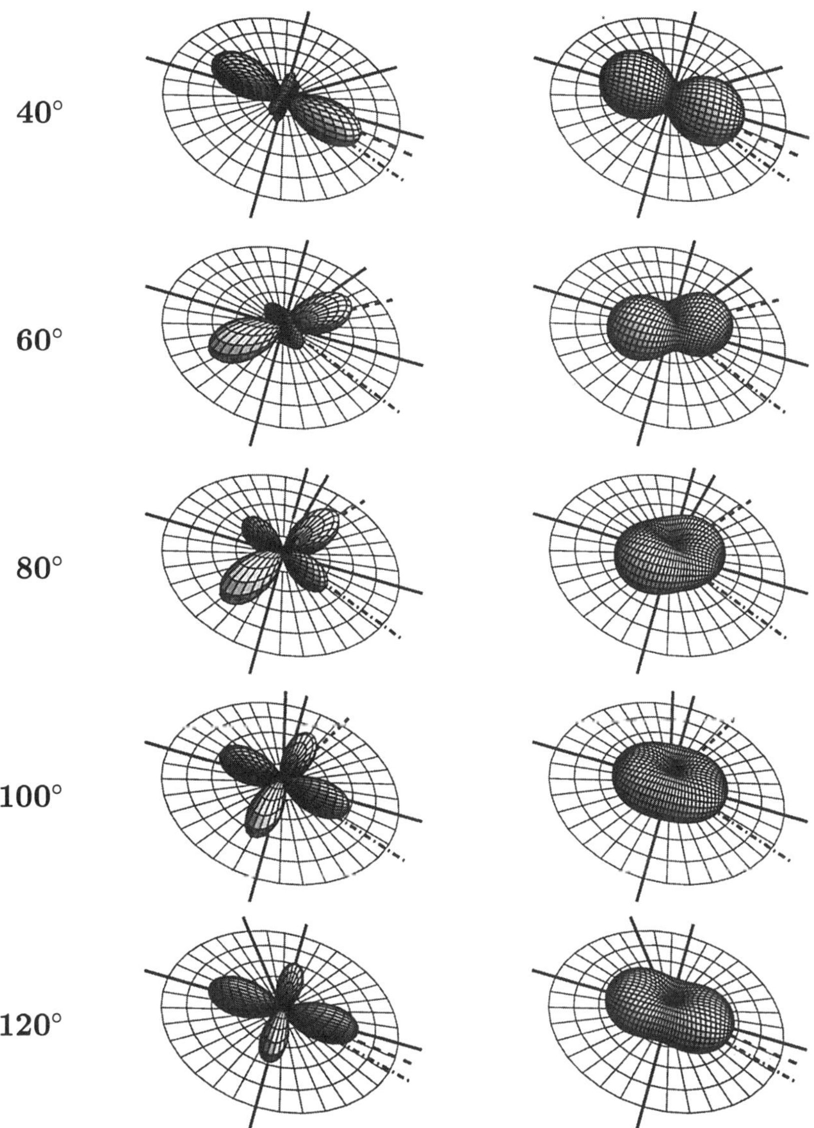

**Fig. 7.21.** (b) The form of the charge cloud predicted by the CCC calculation for electron-impact excitation of the He $1^1S \rightarrow 3^1D$ transition at an incident electron energy of 40 eV. The full picture for selected scattering angles is shown on the left, whereas the result derived from a one-photon angular distribution is shown on the right. The dashed line represents the alignment angle and the dash-dotted line the momentum-transfer direction.

almost pure four-leaf-clover shape around 100°, emitting nearly unpolarized light perpendicular to the scattering plane.

Since the scattered-particle–one-photon coincidence approach only leaves the sign ambiguity of $\hat{\eta}$ to be resolved in order to establish a complete experiment, one might ask whether perhaps an experimental shortcut exists to this information, compared to the full scattered-particle–two-photon coincidence approach. Whereas a general answer to this important question is difficult to formulate, a few specific cases can be pointed out.

If only neutral particles are involved and the active electron experiences internal forces, such as the spin-orbit coupling, a level-crossing experiment may yield the missing information. As demonstrated by Neitzke and Andersen [7.69] for Li ($2^2$S $\rightarrow$ $3^2$D) excitation by collisions with He atoms, appropriately chosen external fields influence the time evolution of the excited state, between the time of impact and the optical decay, differently for the two alternative states $|\Psi_{\rm D}^{(1)}\rangle$ and $|\Psi_{\rm D}^{(2)}\rangle$. Thus, a study of the signal as function of the external fields may distinguish the two alternatives and thus provide the missing sign. See also the discussion of Figure 8.5 in the following chapter.

More recently, Wang and Williams [7.70], in their theoretical analysis of He ($1^1$S $\rightarrow$ $3^1$D) electron-impact excitation, discussed which additional information may be obtained in a geometry where, using the scattered-electron–two-photon coincidence approach, only the cascade intensity in a selected direction away from the normal to the scattering plane is monitored. From a technical point of view, this is a simpler procedure to perform than the coincidence experiment with circular polarization analysis of the cascade photon. The results can then be transformed into the language and the parameter set outlined here, an this has indeed been done in a recent experiment of the Perth group [7.62]. The results for electron-impact excitation of the $3^1$D state, this time at an incident energy of 60 eV, are displayed in Figure 7.22. At a scattering angle of 40°, a triple-coincidence measurement was able to distinguish *experimentally* between the true and the "ghost" solution, both of which are consistent with the data obtained from a standard scattered-particle–one-photon double-coincidence setup. The good agreement between the corresponding CCC predictions and the two sets of data provide further support for the simpler method of using a reliable theory to distinguish between the two possibilities.

Based on the systematics outlined for P-state and D-state excitation, it is relatively straightforward to derive the corresponding formulas for excitation of states with higher angular momentum for both the scattered-particle–multi-photon as well as the more restricted scattered-particle–one-photon coincidence setups [7.56]. Even for electron-impact excitation of F states in helium, however, the important effect of singlet–triplet mixing of the atomic states on the photon signal must be taken into account. These complications

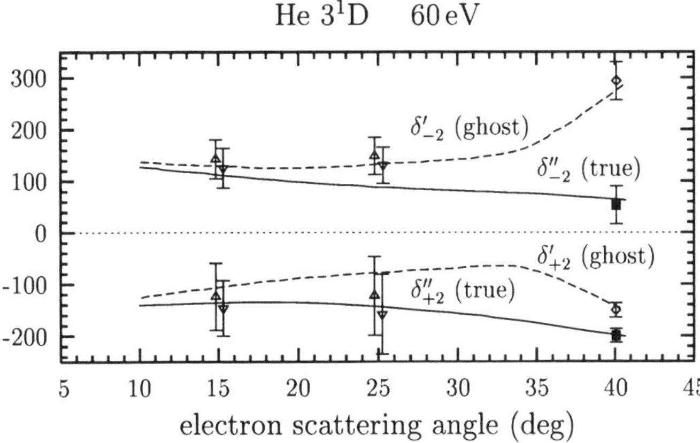



**Fig. 7.22.** The true and ghost solutions for the relative phase $\delta_{\pm 2}$ for electron-impact excitation of the He $1^1$S $\rightarrow 3^1$D transition at an incident electron energy of 60 eV obtained by Mikosza *et al.* [7.62] in scattered-electron–single-photon (double) coincidence measurements. By using a scattered-electron–two-photon (triple) coincidence setup, it was possible to distinguish the two solutions experimentally for a scattering angle of 40°. Also shown are the corresponding predictions obtained from the CCC calculation of Fursa and Bray [7.50].

go beyond the scope of the present book, and we refer the reader to the literature for details [7.71,72].

## 7.2.2 Electron-impact excitation of hydrogen, lithium, and sodium

We now move on to the case of light (quasi-)one-electron targets. In contrast to excitation processes involving at least one singlet state, such as the $(1s^2)^1$S ground state of helium, the projectile and target spins of $s = \frac{1}{2}$ double the number of independent scattering amplitudes, because we now have the possibility of triplet (t) and singlet (s) scattering, i.e., two independent channels for the combined spin $S$ of the projectile + target system. Figure 7.23 illustrates the situation for the most important case, $(n\,s)^2$S $\rightarrow (n'\,p)^2$P°.

We parameterize the amplitudes $f^S(M_{L_f}, M_{L_i} = 0; \theta) \equiv f^S_{M_{L_f}}$ as

$$f^t_{+1} = \alpha_+ e^{i\phi_+}\,; \tag{7.52a}$$

$$f^t_{-1} = \alpha_- e^{i\phi_-}\,; \tag{7.52b}$$

$$f^s_{+1} = \beta_+ e^{i\psi_+}\,; \tag{7.52c}$$

$$f^s_{-1} = \beta_- e^{i\psi_-}\,. \tag{7.52d}$$

Neglecting an overall phase, we thus need to determine seven independent parameters for a complete experiment. In addition to the differential cross

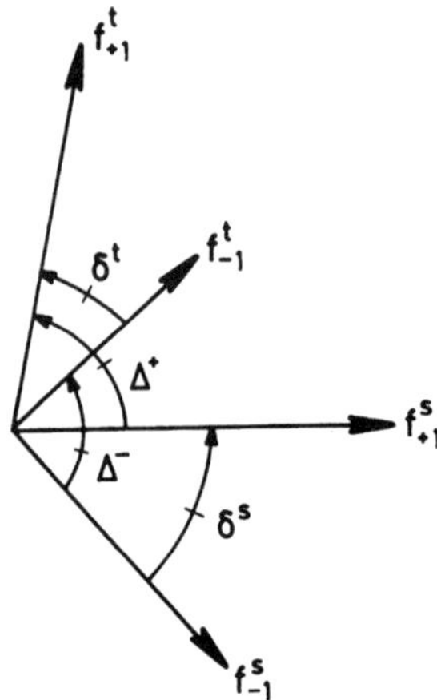



**Fig. 7.23.** Schematic diagram of triplet (t) and singlet (s) scattering amplitudes in the natural frame for $^2S \to {}^2P$ transitions by electron impact. Note that $\Delta^+ + \delta^s = \Delta^- + \delta^t$.

section $\sigma_u$, six dimensionless parameters must be defined, three to characterize the relative lengths of the four vectors, and three to define their relative phase angles.

Next, we parameterize the density matrices for the individual spin channels in analogy to (5.76) as

$$\rho^t = \sigma^t \, \frac{1}{2} \begin{pmatrix} 1 + L^t_\perp & 0 & -P^t_\ell \, e^{-2i\gamma^t} \\ 0 & 0 & 0 \\ -P^t_\ell \, e^{2i\gamma^t} & 0 & 1 - L^t_\perp \end{pmatrix} \tag{7.53a}$$

and

$$\rho^s = \sigma^s \, \frac{1}{2} \begin{pmatrix} 1 + L^s_\perp & 0 & -P^s_\ell \, e^{-2i\gamma^s} \\ 0 & 0 & 0 \\ -P^s_\ell \, e^{2i\gamma^s} & 0 & 1 - L^s_\perp \end{pmatrix}, \tag{7.53b}$$

where

$$\sigma^t = \alpha^2_+ + \alpha^2_-; \tag{7.54a}$$

$$\sigma^s = \beta^2_+ + \beta^2_-; \tag{7.54b}$$

$$L^t_\perp = \frac{1}{\sigma^t} \, (\alpha^2_+ - \alpha^2_-); \tag{7.54c}$$

$$L^s_\perp = \frac{1}{\sigma^s} \, (\beta^2_+ - \beta^2_-); \tag{7.54d}$$

$$P^t_\ell \, e^{2i\gamma^t} = P^t_1 + i\,P^t_2 = -\frac{2\alpha_+\alpha_-}{\sigma^t} \, e^{-i\delta^t}; \tag{7.54e}$$

$$P^s_\ell \, e^{2i\gamma^s} = P^s_1 + i\,P^s_2 = -\frac{2\beta_+\beta_-}{\sigma^s} \, e^{-i\delta^s}. \tag{7.54f}$$

with $\delta^t = \phi_+ - \phi_-$ and $\delta^s = \psi_+ - \psi_-$. In the case where unpolarized beams are used, the total density matrix becomes the weighted sum of the two matrices $\rho^s$ and $\rho^t$, i.e.,

$$\rho_u = \sigma_u \frac{1}{2} \begin{pmatrix} 1 + L_\perp & 0 & -P_\ell\, e^{-2i\gamma} \\ 0 & 0 & 0 \\ -P_\ell\, e^{2i\gamma} & 0 & 1 - L_\perp \end{pmatrix}$$

$$= 3w^t\, \sigma_u \frac{1}{2} \begin{pmatrix} 1 + L_\perp^t & 0 & -P_\ell^t\, e^{-2i\gamma^t} \\ 0 & 0 & 0 \\ -P_\ell^t\, e^{2i\gamma^t} & 0 & 1 - L_\perp^t \end{pmatrix}$$

$$+ w^s\, \sigma_u \frac{1}{2} \begin{pmatrix} 1 + L_\perp^s & 0 & -P_\ell^s\, e^{-2i\gamma^s} \\ 0 & 0 & 0 \\ -P_\ell^s\, e^{2i\gamma^s} & 0 & 1 - L_\perp^s \end{pmatrix}, \tag{7.55}$$

where

$$w^t = \frac{\sigma^t}{\sigma^s + 3\sigma^t} = \frac{\sigma^t}{4\sigma_u}, \tag{7.56a}$$

$$w^s = \frac{\sigma^s}{\sigma^s + 3\sigma^t} = \frac{\sigma^s}{4\sigma_u} = 1 - 3w^t, \tag{7.56b}$$

$$\sigma_u = [3w^t + w^s]\,\sigma_u = \tfrac{3}{4}\sigma^t + \tfrac{1}{4}\sigma^s. \tag{7.56c}$$

The parameters $w^t$ and $w^s$ are related to the parameter $r - \sigma^t/\sigma^s$ used by Hertel *et al.* [7.73] through

$$r = \frac{w^t}{w^s} = \frac{w^t}{1 - 3w^t} = \frac{1 - w^s}{3w^s}, \tag{7.57}$$

but we prefer the use of $w^{t,s}$ to $r$ for reasons of mathematical symmetry and simplicity.

At this point we have thus introduced a total of six parameters, namely $\sigma_u, w^t, L_\perp^t, L_\perp^s, \gamma^t$, and $\gamma^s$, leaving one parameter, a relative phase, still to be chosen. Inspection of Figure 7.23 suggests, for example, the angle $\Delta^+$. Note that the fourth angle, $\Delta^-$, is then fixed through the relation

$$\Delta^+ - \Delta^- = \delta^t - \delta^s = 2\left(\gamma^s - \gamma^t\right). \tag{7.58}$$

The first equality sign follows from inspection of Figure 7.23, the second one from (7.11), applied individually to the singlet and triplet components. We thus use the following complete set:

$$(\sigma_u; w^t, L_\perp^t, L_\perp^s; \gamma^t, \gamma^s, \Delta^+). \tag{7.59}$$

These parameters

- allow for a complete description of the scattering process;
- are a generalization of the parameters used for unpolarized beams;

- can be interpreted in simple physical pictures; and
- are accessible in "partial" (i.e., noncomplete) experiments.

The reduced Stokes vector $\boldsymbol{P}$ of an unpolarized beam experiment is given by the weighted sum of the singlet and triplet (unit) Stokes vectors $\boldsymbol{P}^{s,t}$ as

$$\boldsymbol{P} = 3\,w^t\,\boldsymbol{P}^t + w^s\,\boldsymbol{P}^s, \tag{7.60}$$

from which the set of parameters $(L_\perp, \gamma, P_\ell)$ for an unpolarized beam experiment may be evaluated. In particular,

$$L_\perp = 3\,w^t\,L_\perp^t + w^s\,L_\perp^s. \tag{7.61}$$

Since, in general, $L_\perp^t \neq L_\perp^s$ and $\gamma^t \neq \gamma^s$, this causes the (reduced) degree of polarization $P$ to be smaller than unity, i.e., $P \leq 1$ for an unpolarized electron beam (see also Section 2.2.3).

Before we discuss experiments involving spin-polarized electron and target beams, we show two sets of results obtained with unpolarized particles. The first one concerns the famous case of electron-impact excitation of atomic hydrogen, H(1s) $\rightarrow$ H(2p), at an incident electron energy of 54.4 eV. For many years, theorists struggled with the fact that the most sophisticated methods agreed very well with each other but were unable to reproduce the parameters extracted from angular correlation measurements, especially at large scattering angles [7.74,75]. When the correlation experiments were eventually repeated [7.76–78] and supplemented by sets of light polarization measurements [7.78,79], however, the large-angle problem was resolved.[5]

Figure 7.24 shows several sets of results for the parameters $(\sigma, P_\ell, \gamma, L_\perp)$.[6] The overall agreement between the experimental data and theoretical predictions is now satisfactory, except for the small-angle polarization results of O'Neill *et al.* [7.79] that deviate by several standard deviations from the theoretical values. Based on the well-known dominance of direct over exchange scattering at small angles and relatively high energies, one might suspect that significant deviations of $P_\ell$ from unity are due to remaining systematic problems in the experiment, rather than a physically meaningful effect.

Many of the difficulties associated with experiments on the 1s$\rightarrow$2p transition in atomic hydrogen can be overcome by investigating the 2s $\rightarrow$ 2p resonance transition in lithium. One of the biggest advantages lies in the wavelength of this transition, which is long enough to allow for employing the time-reverse approach, i.e., studying the superelastic scattering of electrons from laser-prepared targets. Figure 7.25 shows essentially perfect agreement between the measurements and calculations of the Flinders group [7.84],

---

[5] The original data of Williams [7.75] should actually be corrected for scattering angles $\theta \geq 90°$. The revised data agree well with the theoretical predictions, and also with new data [7.78] plotted in Figure 7.24.

[6] Data for this problem are often published in terms of the parameter set $\lambda \equiv (1 + \bar{P}_1)/2$, $R \equiv -\bar{P}_2/2\sqrt{2}$, $I \equiv \bar{P}_3/2\sqrt{2}$ which, unfortunately, does not reveal directly the connection to the form of charge cloud.

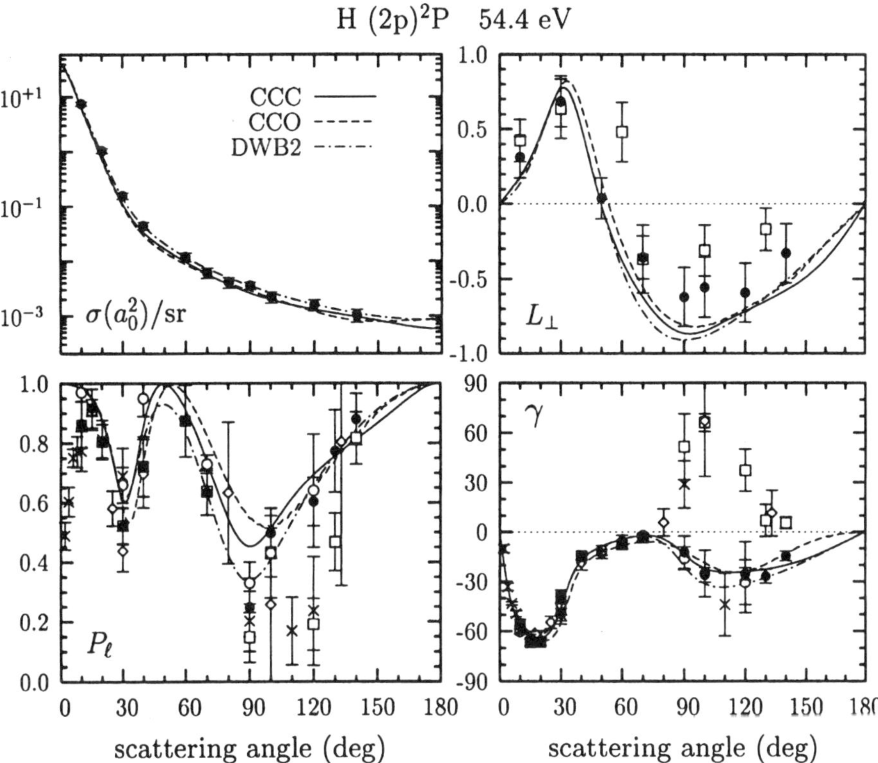



**Fig. 7.24.** Measured and calculated charge cloud parameters for the 1s → 2p electron-impact induced transition in atomic hydrogen at an incident electron energy of 54.4 eV. Predictions from three representative theories, namely a second-order distorted-wave (DWBA2) model [7.81], a close-coupling plus optical-potential method [7.82], and a convergent close-coupling (CCC) approach [7.83], are compared with the following sets of experimental data: • [7.75] for DCS, [7.78,80] for $(L_\perp, \gamma, P_\ell)$; ◇ [7.74] for $(\gamma, P_\ell)$; □ [7.75] for $(\gamma, P_\ell)$; o [7.76,77] for $(\gamma, P_\ell)$; × [7.79] for $(\gamma, P_\ell)$.

thereby giving confidence in both the experimental and theoretical methods as a common starting point for investigating more complicated systems in the future.

Experiments with polarized electron $(P_e)$ and target $(P_A)$ beams have been performed in the NIST and the Münster groups for electron-impact excitation and de-excitation of sodium. We start the analysis by expressing the generalized *STU* parameters discussed in Section 4.4 in terms of the complete parameter set (7.59). In addition, we introduce the parameter

$$V \equiv \frac{P_e'}{P_e P_A} \tag{7.62}$$

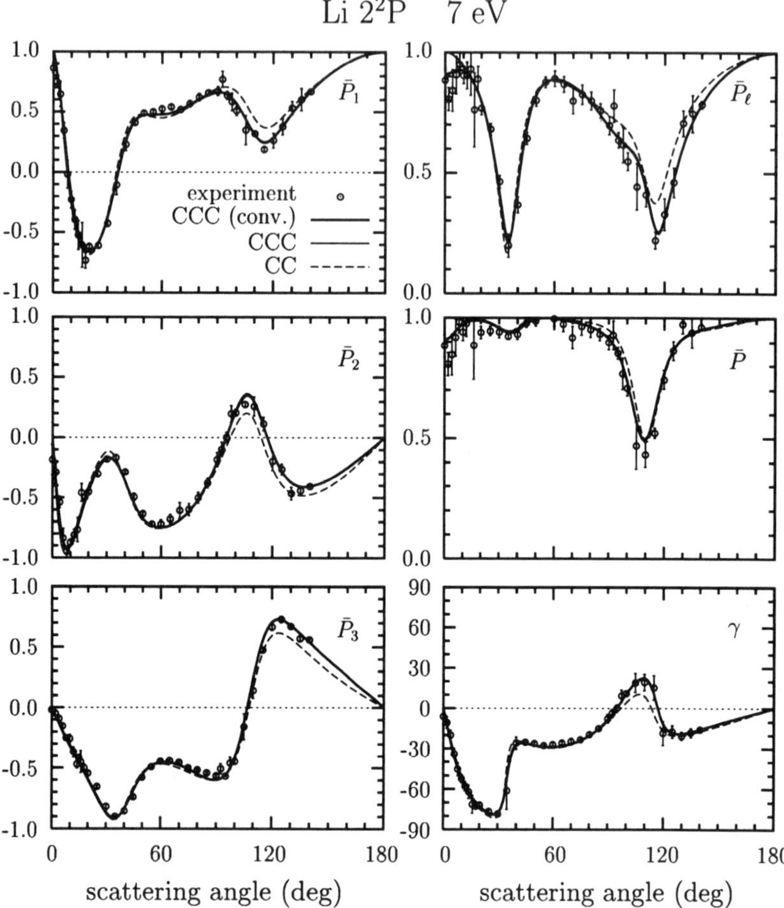



**Fig. 7.25.** Measured and calculated reduced Stokes and charge cloud parameters for the 2s $\rightarrow$ 2p electron-impact induced transition in lithium at an incident electron energy of 7 eV. The measured data are compared with predictions from a CCC and a CC model; the latter neglects the effect of the target continuum states [7.84].

that describes the final electron spin polarization component $P'_e$ *perpendicular* to the scattering plane for *in-plane* initial polarizations, namely $P_e$ in the forward direction and $P_A$ perpendicular to the beam [7.85]. As will be shown below, measurements of the latter kind would be necessary to remove otherwise remaining ambiguities in a truly complete experiment.

Provided that electron exchange is the only spin-dependent effect of importance for the excitation, and that the fine-structure energy splitting is negligible compared to the initial and final energies of the projectile, the *STU* parameters get vastly simplified, as shown in (5.38). The seven polarization, asymmetry, contraction, and rotation parameters for each fine-structure level reduce to the following set of only four independent parameters (the superscripts denote the $J$ value of the excited target state):

$$S_P \equiv S_P^{1/2} = -2\,S_P^{3/2}, \tag{7.63a}$$

$$S_A \equiv S_A^{1/2} = -2\,S_A^{3/2}, \tag{7.63b}$$

$$T \equiv T_x^{1/2} = T_y^{1/2} = T_z^{1/2} = T_x^{3/2} = T_y^{3/2} = T_z^{3/2}, \tag{7.63c}$$

$$U \equiv U_{xy}^{1/2} = U_{yx}^{1/2} = -2\,U_{xy}^{3/2} = -2\,U_{yx}^{3/2}. \tag{7.63d}$$

The results for the parameters of interest are

$$S_P = -w^t\,L_\perp^t + w^s\,L_\perp^s, \tag{7.64a}$$

$$S_A = \frac{1}{2\sigma_u}\left[\alpha_+\beta_+\cos\Delta^+ - \alpha_-\beta_-\cos\Delta^-\right] - 2w^t\,L_\perp^t, \tag{7.64b}$$

$$T = \frac{1}{2\sigma_u}\left[\alpha_+\beta_+\cos\Delta^+ + \alpha_-\beta_-\cos\Delta^-\right] + 2w^t, \tag{7.64c}$$

$$U = \frac{1}{2\sigma_u}\left[\alpha_+\beta_+\sin\Delta^+ - \alpha_-\beta_-\sin\Delta^-\right], \tag{7.64d}$$

$$V = \frac{1}{2\sigma_u}\left[\alpha_+\beta_+\sin\Delta^+ + \alpha_-\beta_-\sin\Delta^-\right]. \tag{7.64e}$$

The amplitude sizes $\alpha_\pm$ and $\beta_\pm$ may be eliminated from (7.64) by using

$$\frac{\alpha_\pm\beta_\pm}{2\sigma_u} = \sqrt{w^t w^s\,(1 \pm L_\perp^t)(1 \pm L_\perp^s)}. \tag{7.65}$$

We now investigate to what extent the perfect scattering experiment has been achieved to date. We begin with the key experiment performed by the NIST group [7.86] and shown in Figure 4.16, but with the beam overlap modified so that scattering may take place also from the excited state. Spin-polarized electrons with polarization vector perpendicular to the scattering plane were scattered superelastically from spin-polarized sodium atoms in the $3^2$P state. This state was produced by pumping with circularly polarized laser light. By reversing the directions of the two polarizations individually, the experiment allows for the determination of $L_\perp^t$, $L_\perp^s$, and $w^t$ (for details, see Hertel *et al.* [7.73]).

This experiment does not determine the alignment angles $\gamma^t$ and $\gamma^s$. However, the off-diagonal elements of (7.55) show that

$$P_\ell\,\mathrm{e}^{2\mathrm{i}\gamma} = 3w^t\,P_\ell^t\,\mathrm{e}^{2\mathrm{i}\gamma^t} + w^s\,P_\ell^s\,\mathrm{e}^{2\mathrm{i}\gamma^s}. \tag{7.66}$$

As illustrated in Figure 7.26, this complex equation corresponds to addition of the two vectors $\boldsymbol{P}_\ell^t$ and $\boldsymbol{P}_\ell^s$, multiplied by weighting factors $3w^t$ and $w^s$, respectively, to form the resulting vector $\boldsymbol{P}_\ell$. Hence, elementary geometry can be applied to obtain *two* pairs of solutions, $(\gamma^t,\gamma^s)_{\mathrm{true}}$ (the true solution) and $(\gamma^t,\gamma^s)_{\mathrm{ghost}}$ (the other possibility) as

$$\gamma^t = \gamma \mp \chi/2, \tag{7.67a}$$

$$\gamma^s = \gamma \pm \psi/2, \tag{7.67b}$$

where the angles $\chi$ and $\psi$ are defined in the figure. Provided experimental data are available for the parameter set $(P_\ell, \gamma, w^t, L^t_\perp, L^s_\perp)$ at a given collision energy and scattering angle, two sets of possible angles $(\gamma^t, \gamma^s)$ can be determined.

As pointed out by Hertel *et al.* [7.73], this ambiguity is mathematically identical to the one for S $\rightarrow$ D excitation processes discussed above. The physical origin is the same in the two cases, namely the incoherent addition in the experiment of two channels that are, in principle, distinguishable. It could have been resolved with the following modified version of the NIST experiment. If the laser light propagates *in* the collision plane perpendicular to the electron beam direction, with the electron spin polarization still parallel or antiparallel to the atomic spin polarization, one can determine an asymmetry parameter $B$ analogous to the $A$ parameter of (4.4). The result is [7.85]

$$B = -\frac{w^t\left(1 - P_1^t\right) - w^s\left(1 - P_1^s\right)}{\left(1 - P_1\right)} . \tag{7.68}$$

Thus, a consistency check between the results of (7.67) and (7.68) will eliminate the ghost solution. Although such an experiment has not been performed yet, the ambiguity for the alignment angles can, with some confidence, be removed with help from state-of-the-art theory. While data for all the parameters necessary for determination of the two sets of angles $(\gamma^t, \gamma^s)$ have indeed been obtained for electron–sodium (de-)excitation, the energy and scattering angle combinations investigated by the Adelaide [7.87] and the NIST [7.88,89] groups do not overlap. By replacing the missing experimental data for $P_\ell$ and $\gamma$ with theoretical "close-coupling plus optical potential" (CCO) results of Bray [7.90] at a total energy of 4.1 eV, however, it was possible to "invert" the NIST data at this energy. The approach seemed justified in light of the excellent agreement between the CCO predictions and experiment for these observables at the energies for which the experiment was performed.

The results of the inversion are shown in Figure 7.27 for the angles $\gamma^t$ and $\gamma^s$ [7.85]. Since the true and the ghost solutions evaluated from theory are in very good agreement with the experimental values, the "true" experimental data were selected as those that follow the true theoretical solution. Note that the two sets of solutions can cross each other, and that it is essentially impossible to stay on the "true" experimental curve by assuming, for example, a smooth angle and energy dependence of the phase angles.

Next, we note from inspection of (7.64) that information on the still missing phase difference between a singlet and triplet amplitude can only be obtained from *STU* parameters. Owing to the lack of data from different

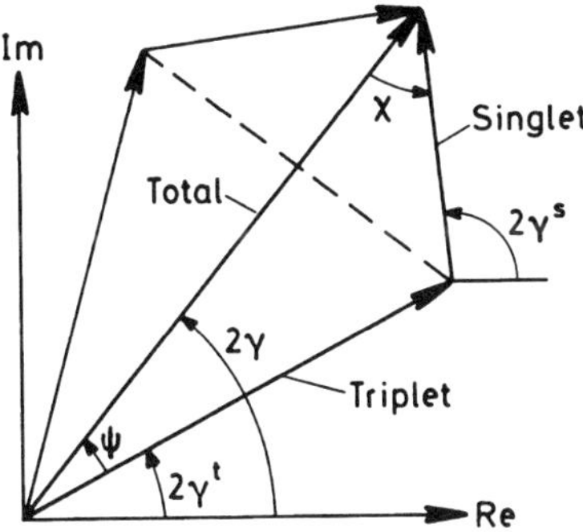


**Fig. 7.26.** Vector diagram corresponding to (7.66). Note the analogy to Figure 7.17.

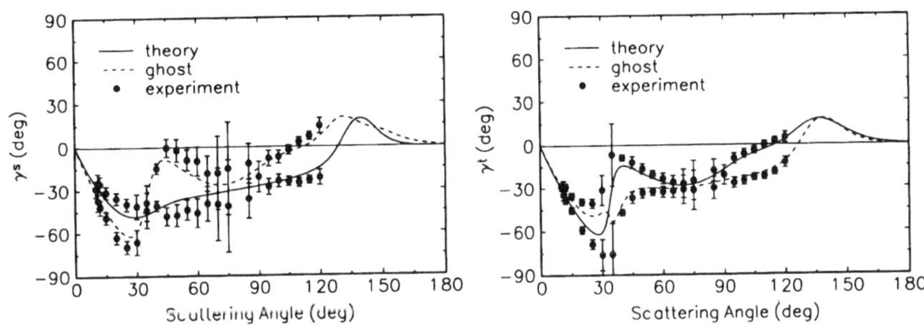


**Fig. 7.27.** Alignment angles $\gamma^t$ and $\gamma^s$ calculated from the NIST data [7.86] for $(w^t, L_\perp^t, L_\perp^s)$ and theoretical results for $(P_\ell, \gamma)$ from scattering amplitudes of Bray [7.90] for electron-impact excitation of the $3\,^2$P state of sodium at an incident electron energy of 4.1 eV; •, two sets of inverted experimental data as well as true (———) and ghost (– – –) theoretical solutions [7.85].

experiments at the same energy, an inversion method for the $\Delta$ angles [7.85] required theoretical results for the parameter $T$ at 4.1 eV.[7]

The idea of the second inversion procedure is illustrated in Figure 7.28. Equation (7.64c) for the $T$ parameter corresponds to a nonlinear equation for $\Delta^+$ and $\Delta^-$. In addition, the difference between these two angles is related to the difference between the alignment angles $\gamma^t$ and $\gamma^s$ through (7.58). Consequently, solutions for $\Delta^+$ and $\Delta^-$ can be found by searching for crossings between the lines determined by

$$A \, \cos \Delta^+ + B \, \cos \Delta^- = C \,, \tag{7.69}$$

---

[7] This parameter was actually measured by Hegemann *et al.* [7.91,92], but at total collision energies of 4.0 eV and 12.1 eV compared to the NIST experiments at 4.1 eV and 10.0 eV. Of course, if experimental results for the $U$ or $V$ parameter were available, evaluation of the missing phase angle is straightforward from (7.64)

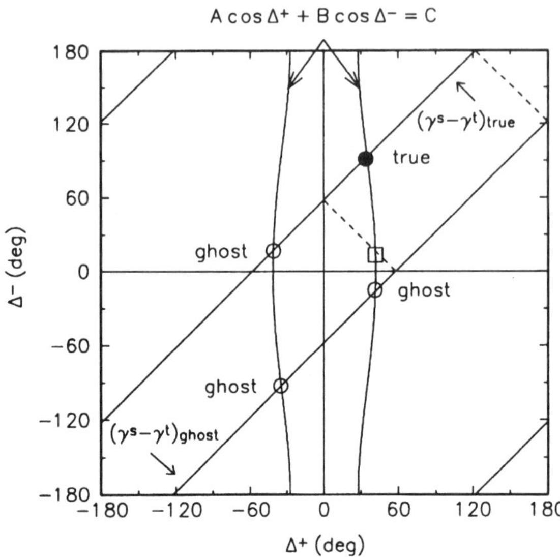


**Fig. 7.28.** Determination of the singlet-triplet phase angles $\Delta^+$ and $\Delta^-$ from experimentally observable parameters [7.85].

where the constants $A$, $B$, and $C$ are evaluated from (7.64c), and the lines defined by (7.58) and labeled $(\gamma^s-\gamma^t)_{\text{true,ghost}}$ in Figure 7.28. Because of the ambiguity in the sign of the arguments in the cosines and the ambiguity in the pair $(\gamma^t, \gamma^s)$, one will usually find *four* solutions, only one of which is correct. This is illustrated in Figure 7.28 for inversion of the theoretical data at a scattering angle of 40° and a total collision energy of 4.1 eV [7.85].

Note that the problem can be reduced to searches in the first quadrant, since the "ghosts" in the second and fourth quadrants may be found via intersections with the dashed lines in Figure 7.28 that are mirror images of the difference lines in those quadrants seen in the first quadrant. Since the slopes of the dashed mirror lines are reversed compared to the original difference lines, the actual crossing points in the second and fourth quadrants can easily be reconstructed, while the only remaining crossing, in the third quadrant, is related to the one in the first quadrant through a simultaneous sign change in $\Delta^+$ and $\Delta^-$.

The results for $\Delta^+$ and $\Delta^-$ as a function of the scattering angle for a collision energy of 4.1 eV are shown in Figure 7.29. For simplicity, only one ghost solution (where $\Delta^+ > 0$) is shown. Again, the theoretical results help to identify, in most cases unambiguously, the true solution among the possibilities obtained from an inversion of the experimental data alone.

As an example for the importance of such detailed benchmark measurements, followed by a thorough analysis of the data, Figure 7.30 shows the full set of results obtained at a collision energy of 10 eV [7.85]. Looking at

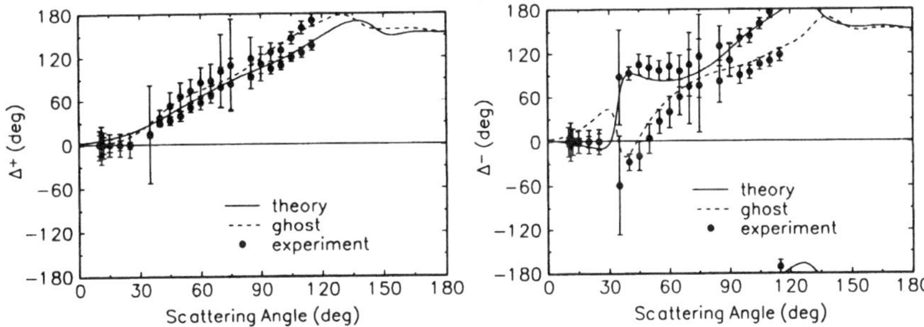

**Fig. 7.29.** Singlet-triplet phase angles $\Delta^+$ and $\Delta^-$ calculated from data for $(w^t, L_\perp^t, L_\perp^s)$ [7.86], the corresponding alignment angles $(\gamma^t, \gamma^s)$ presented in Fig. 7.27, and theoretical $T$ parameter results [7.90] for electron-impact excitation of the $3\,^2\mathrm{P}$ state of sodium at an incident electron energy of 4.1 eV; •, two sets of inverted experimental data as well as the true (——) and one ghost (– – –) solution [7.85].

the parameter $L_\perp^s$ reveals that only the most sophisticated CCC theory of Bray [7.93] can reproduce the experimental results for this observable; in contrast, the parameter $L_\perp^t$ is much less sensitive to the quality of the theoretical model.

Although a complete set of parameters presently can only be extracted from available experimental data after two guesses guided by theory, the discussion shows that a complete experiment is within reach for this case, in particular if the experimental programs of the participating groups aim at common choices of angles and energies. Indeed, the somewhat complicated inversion procedure could be avoided and the perfect scattering experiment be achieved directly, for example, as follows: a measurement of the in-plane asymmetry parameter $B$ would resolve the ambiguity in the two $(\gamma^t, \gamma^s)$ pairs. Furthermore, since the equations for the pairs $(S_A, U)$ and $(T, V)$ can be recast in a form similar to (7.60), determination of one of the pairs gives a geometrical ambiguity in the $(\Delta^+, \Delta^-)$ pair, as discussed for $(\gamma^t, \gamma^s)$. This ambiguity can be removed by measurement of one more of the remaining $STUV$ parameters. If $(\gamma, P_\ell)$ are known from unpolarized beam experiments and $(L_\perp^t, L_\perp^s, w^t)$ from a NIST-type experiment, *any three* of the five parameters $B$, $S_A$, $T$, $U$, and $V$ will suffice to achieve a perfect scattering experiment.

The scattering-amplitude information contained in the atomic density matrix (i.e., the Stokes parameters) and the reduced density matrix of the scattered electrons (i.e., the $STU$ parameters) is illustrated in Figure 7.31. From a Stokes-parameter analysis, one obtains information about the relative phase between the two $f_{+1}^t$ and $f_{-1}^t$ amplitudes and the relative phase between the two $f_{+1}^s$ and $f_{-1}^s$ amplitudes, as well as the relative sizes of all four amplitudes. However, the Stokes parameters do not depend on the relative phases between any triplet and singlet amplitude. The $STU$ parameters, on

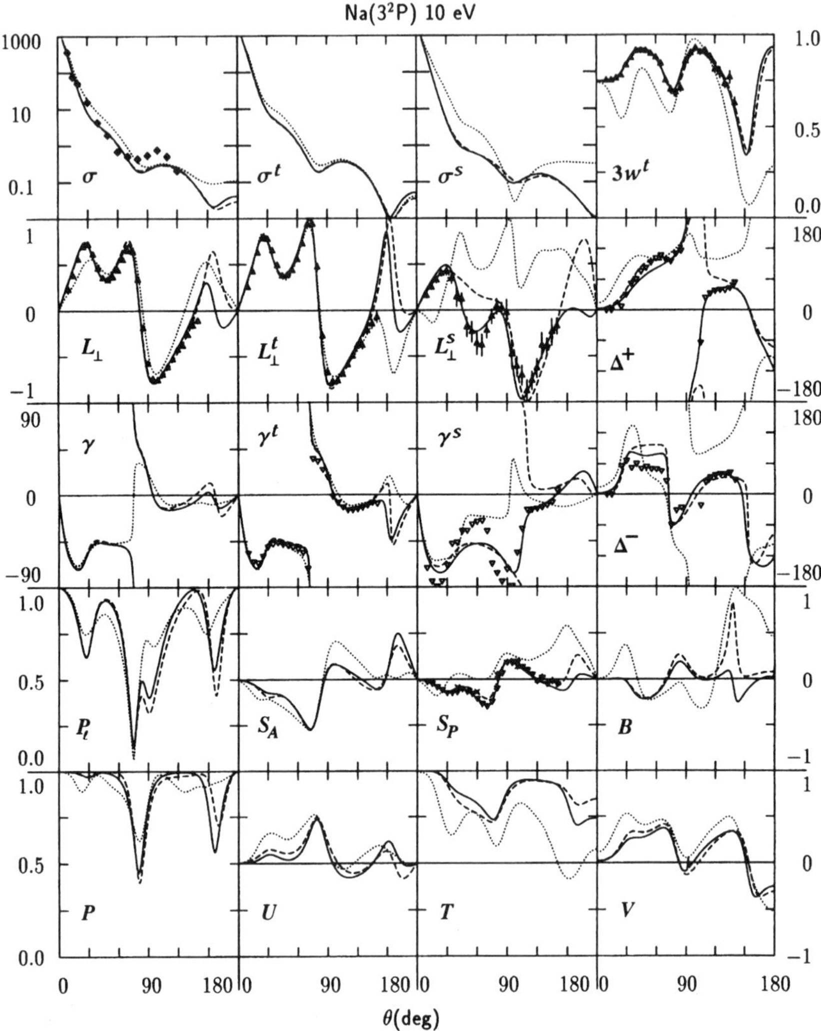


**Fig. 7.30.** Survey of alignment and orientation parameters for excitation of the Na $3\,^2$P state by spin-polarized electrons at an incident electron energy of 10 eV. The differential cross sections are given in units of $a_0^2$/sr. The experimental data of the NIST and Münster groups have been transformed to the parameter set (7.44); they are compared with CCC (——— [7.93]), CCO (– – – [7.90]), and DWBA2 ($\cdots$) [7.94] results. The differential cross sections are from Srivastava and Vuskovic [7.95].

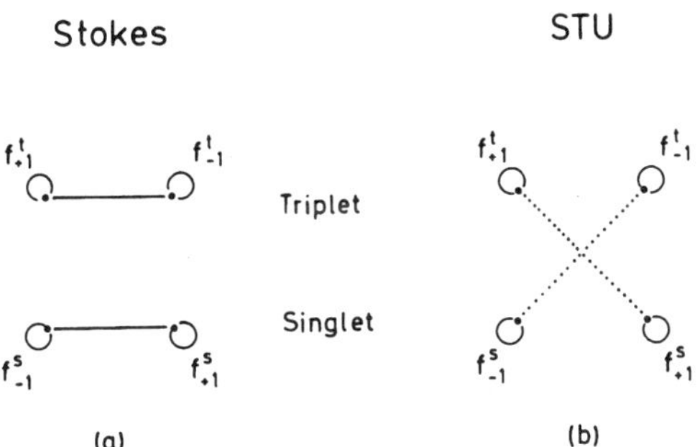



**Fig. 7.31.** Relative amplitude sizes and phases that can be evaluated from an analysis of the Stokes (a) and *STU* (b) parameters.

the other hand, can be used to determine the relative phase $\Delta^+$ between the two $f^t_{+1}$ and $f^s_{+1}$ amplitudes and the relative phase $\Delta^-$ between the two $f^t_{-1}$ and $f^s_{-1}$ amplitudes, provided that the relative sizes of all four amplitudes are known from a Stokes parameter measurement.

The discussion above also demonstrates how the inversion procedures may serve as consistency checks among separate experimental data sets. Consistent experimental data should always allow for inversion within experimental uncertainties. Consider, for example, how (7.64a) points to an interesting link between Stokes parameters and *STU* parameters. For Na $3^2$P excitation, Nickich *et al.* (1990) [7.96] performed a measurement in a similar geometry to the NIST setup. The essential difference was the choice of a target density high enough to ensure unpolarized target atoms owing to the depolarization effect of radiation trapping. For each fine-structure level, the asymmetry function

$$S_A(\theta) \equiv \frac{1}{P_\perp} \frac{\sigma_{\text{left}}(\theta) - \sigma_{\text{right}}(\theta)}{\sigma_{\text{left}}(\theta) + \sigma_{\text{right}}(\theta)} \tag{7.70}$$

was measured for the superelastic transition $3^2$P $\to 3^2$S.

Because of time-reversal invariance of the interaction, however, the $S_A$ measurement for the *de-excitation* process is equivalent to determination of the polarization function $S_P$ for *excitation*. One can thus use (7.64a) to *predict* the results of this experiment from the data set $(L^t_\perp, L^t_\perp, w^t)$ of the NIST experiment [7.73,97]. The data of Nickich *et al.* [7.96] and the results predicted from the NIST experiment are shown together with theoretical predictions in Figure 7.32. The consistency between the two independent experimental data sets and their agreement with the theoretical prediction underlines the power of the present formalism.

We finish this section with a few examples of the relatively rare cases where transitions other than S $\to$ P have been investigated. Since the number of independent scattering amplitudes is significantly larger in these cases, a complete experiment is currently unrealistic. Also, the interpretation of

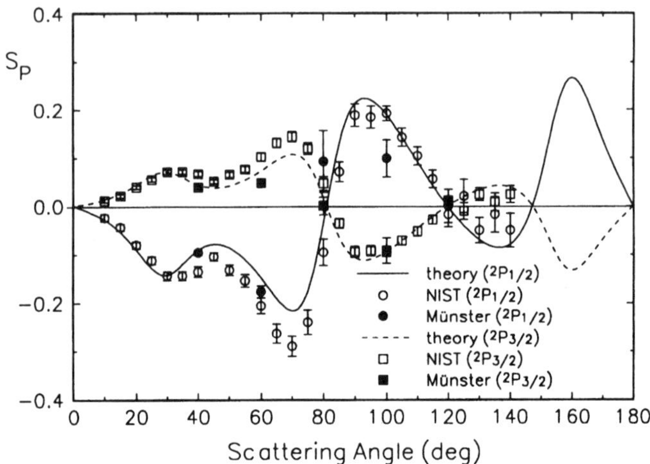

**Fig. 7.32.** Spin-polarization function $S_P$ for electron-impact excitation of the $(3p)^2P_{1/2,3/2}$ states of sodium at an incident electron energy of $10\,\mathrm{eV}$; data of Nickich *et al.* [7.96] for $S_P(^2P_{1/2})$ (●) and $S_P(^2P_{3/2})$ (■); prediction of $S_P(^2P_{1/2})$ (○) and $S_P(^2P_{3/2})$ (□) from the data of Scholten *et al.* [7.89] for $(w^t, L_\perp^t, L_\perp^s)$. The theoretical curves for $S_P(^2P_{1/2})$ (———) and $S_P(^2P_{3/2})$ (– – –) were calculated from CCO scattering amplitudes [7.90].

the results is not straightforward. On the other hand, such investigations may provide valuable benchmark data for testing theoretical models for more complicated situations.

Figure 7.33 shows results from the Brisbane group [7.98] for the $3p \to 3d$ transition in sodium. The experiments were performed by laser-preparation of the $(3p)^2P^{\circ}$ state (using particular hyperfine transitions), followed by observation of the energy loss peak corresponding to the $3p \to 3d$ transition.[8] The signal depends on the polarization of the pumping light and is expressed in terms of so-called "pseudo-Stokes parameters." These, in turn, are then transformed into the standard set $(L_\perp, \gamma, P_\ell, P^+)$ of correlation parameters using standard angular-momentum algebra.

As can be seen from Figure 7.33, the agreement between the experimental data and the CCC predictions is quite satisfactory over the small angular range in which the measurements were performed. The small values of $P_\ell$ and $P^+$ in this case are likely due to the many paths that can be followed in the optical decay but are not separated in this experiment (see also the previous discussion for the D states in He). Finally, we note that $L_\perp$ starts off with positive values for small scattering angles, indicating the dominance of $\Delta M_L = +1$ transitions. Positive small-angle results for the

---

[8] This group also investigated the $3p \to 4s$ transition, which will be further discussed in Chapter 9.

$$\text{Na } (3p)^2P \to (3d)^2D$$

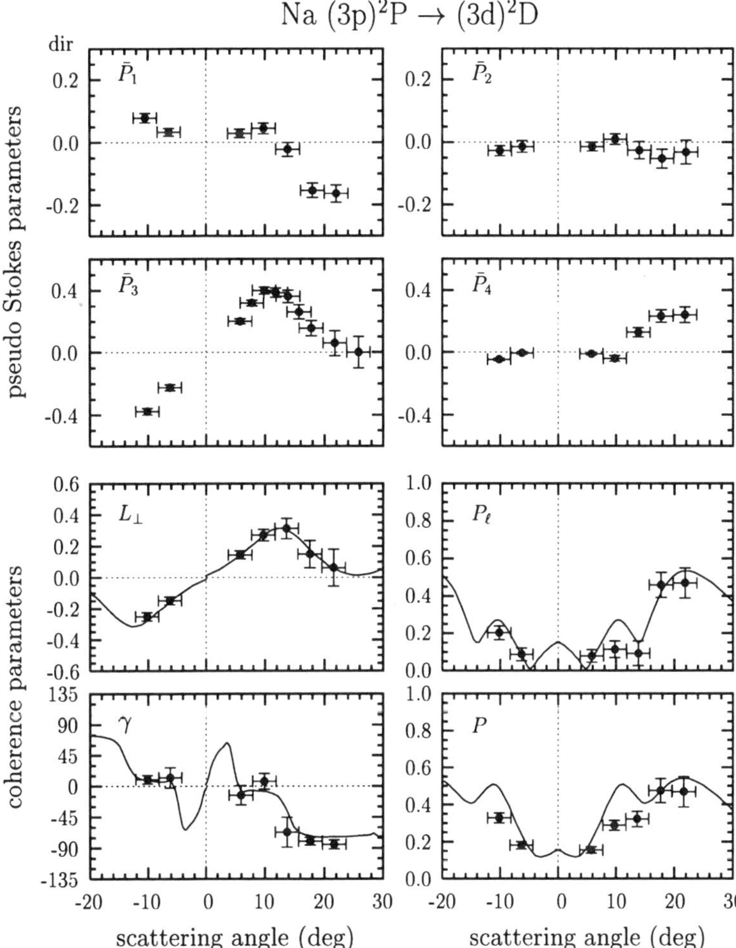

Fig. 7.33. Pseudo-Stokes parameters and the corresponding set of correlation parameters for the $(3p)^2P^\circ \to (3d)^2D$ transition of sodium at an incident electron energy of 30 eV. The experimental data of Shurgalin *et al.* [7.98] are compared with predictions based on the CCC calculations of Bray [7.93].

angular-momentum transfer are also seen in all the examples for S $\to$ P transitions shown so far (cf. Figures 7.7,11,13,14,24,25,30), thereby indicating the possibility of a propensity rule. Such rules will be further discussed in Chapter 9.

An experimental investigation of *elastic* electron scattering from the laser excited $(3p)^2P^\circ$ state of sodium was performed by Shi *et al.* [7.99]. The principal idea is to study the orbital angular-momentum asymmetry, i.e., the possible dependence of the scattering intensity at a certain angle on the orbital angular-momentum orientation of the initial state. For a fixed scattering

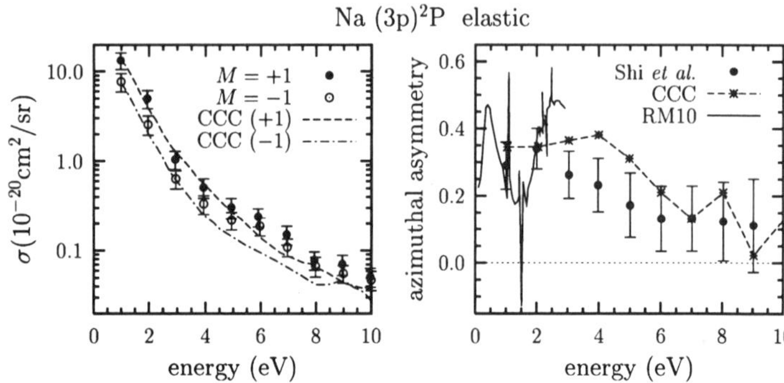



**Fig. 7.34.** Differential cross sections and orbital asymmetry for elastic electron scattering from the $(3p)^2P^\circ$ state of sodium. For a fixed scattering angle of 135°, the experimental results of Shi *et al.* [7.99] are compared with predictions based on various close-coupling models [7.100,101].

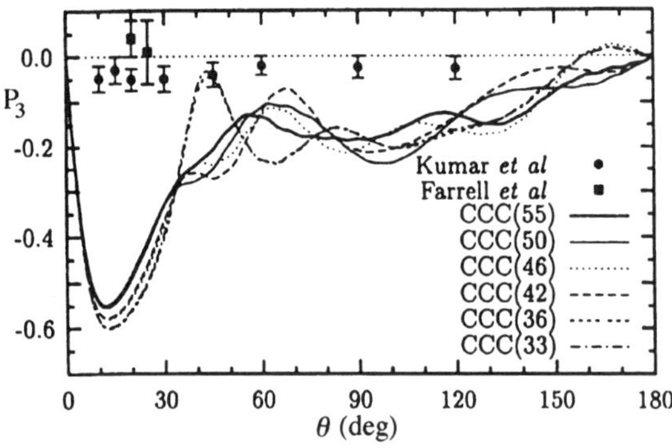



**Fig. 7.35.** Degree of circular polarization measured for 54.4 eV electron-impact excitation of the H $(n = 3)$ states in the experiments of Farrell *et al.* [7.102] and Kumar *et al.* [7.103] in comparison with CCC-type calculations of Bray and Stelbovics [7.104]. The brackets indicate the number of states included in the close-coupling expansion.

angle of 135°, Figure 7.34 shows experimental results for the cross sections $\sigma_{3p}^\pm$ and the corresponding asymmetry $A \equiv (\sigma_{3p}^+ - \sigma_{3p}^-)/(\sigma_{3p}^+ + \sigma_{3p}^-)$, where the superscript indicates the angular-momentum orientation of the target state, as a function of the incident electron energy. Again, the agreement between the measured data and theoretical predictions from various close-coupling approaches is quite good. On the other hand, the difficulties of the experiment lead to uncertainties of such magnitude that stringent tests of theoretical models are very limited.

Our last example deals again with electron-impact excitation of atomic hydrogen, specifically with the circular polarization of light emitted in the $3d \rightarrow 2p$ transition after excitation from the 1s ground state. Figure 7.35

shows two sets of experimental data [7.102,103] at 54.4 eV incident energy in comparison with predictions from various close-coupling calculations [7.104].[9] Given the severe discrepancy between theory and experiment, particularly at small angles where the calculation seems to have converged with the number of states included in the expansion, and the success of the CCC theory in many other cases, it is not clear at the present time whether the discrepancy is due to an experimental or a theoretical problem.

### 7.2.3 Electron-impact excitation of heavy noble gases

As pointed out in Section 2.2.4, impact excitation of these systems by unpolarized electron beams is generally characterized by a nonvanishing height of the charge cloud in S $\rightarrow$ P or, more precisely, $J = 0^+ \rightarrow 1^-$ transitions, where the superscript indicates the parity of the state. This height is caused by the possibility of spin-flips due to relativistic effects in the target as well as in the projectile–target interaction. These, in turn, make possible the excitation of the $M_J = 0$ sublevel. As a consequence, the atomic reflection symmetry is violated, and the Stokes parameter $P_4$ measured *in the scattering plane* (along the $y^n$ direction) may differ from unity.

Most of the electron–photon coincidence experiments involving unpolarized beams were performed on the $(n\,\mathrm{p}^6)^1\mathrm{S}_0 \rightarrow (n\,\mathrm{p}^5[n{+}1]\mathrm{s})^{3,1}\mathrm{P}_1^\mathrm{o}$ transitions in Ne ($n = 2$), Ar ($n = 3$), Kr ($n = 4$), and Xe ($n = 5$). They were reviewed by Andersen and et al. [7.63] and Becker et al. [7.105]. Figure 7.36 shows an example of the set of Stokes parameters $(P_1, P_2, P_3, P_4)$ for electron-impact excitation of the $(5\mathrm{p}^5[3/2]6\mathrm{s})$ ("$^3\mathrm{P}_1$") state in xenon at an incident energy of 30 eV.[10] After accounting for the angular acceptance of the photon detector and the depolarization of the radiation due to hyperfine-structure effects, there is excellent agreement between experiment [7.106] and the first-order distorted-wave (DWBA) results of Bartschat and Madison [7.107] for the set of Stokes parameters. Hence, one might think that this relatively simple, semirelativistic theoretical approach is sufficient to describe the collision process.

However, this assessment clearly needs to be revised after examining Figure 7.37, which displays results for the spin-asymmetry function $S_A$ for impact excitation of the same state as well as the $(5\mathrm{p}^5[1/2]6\mathrm{s})$ ("$^1\mathrm{P}_1$") state. Investigating an independent observable at the same collision energy apparently reveals major deficiencies in the theoretical model, and the discrepancies between theory and experiment become even more visible owing to the wide

---

[9] The signal also contains an unpolarized part from the 3s $\rightarrow$ 2p transition which was accounted for in the theory.

[10] The inverted commas indicate that the $LS$-notation for this state is only an approximation, since it is heavily mixed and must be described by an intermediate coupling scheme. The number in brackets denotes the angular momentum of the Xe$^+$ core.

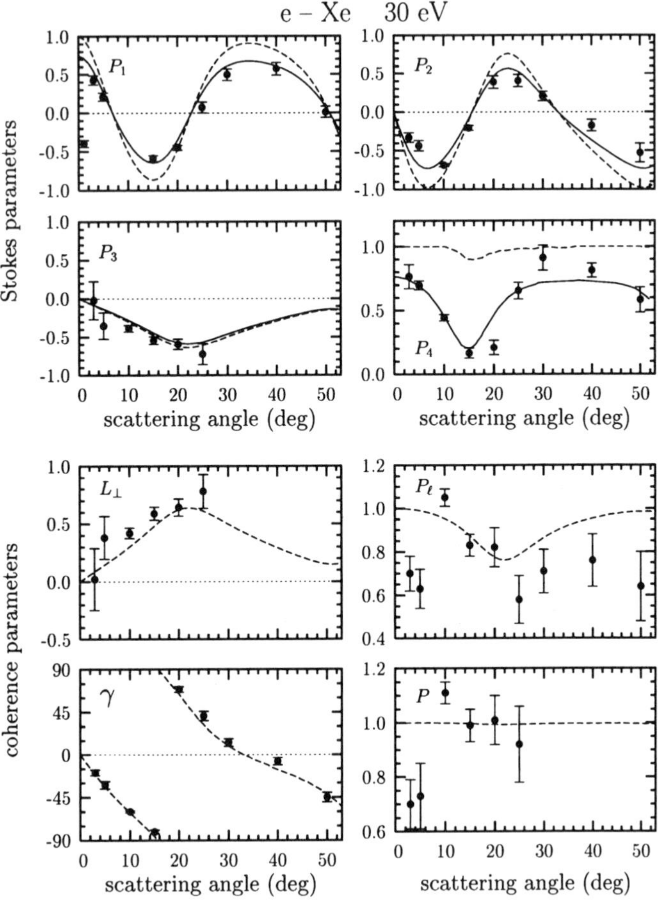


**Fig. 7.36.** Stokes parameters $(P_1, P_2, P_3, P_4)$ and the corresponding charge cloud parameters $(L_\perp, \gamma, P_\ell, P^+)$ for electron-impact excitation of the $(5p^5[3/2]6s)$ ($\text{``}^3P_1\text{''}$) state in xenon at an incident energy of 30 eV. The experimental data of Corr *et al.* [7.106] are compared with semirelativistic DWBA results of Bartschat and Madison [7.107]. The solid line includes the depolarization effects due to the hyperfine structure and the acceptance angles of the photon detector.

range of scattering angles for which the experiment was performed. We also note that predictions from a full-relativistic distorted-wave calculation by the Toronto group [7.108] agrees neither with the semirelativistic DWBA results nor with experiment over a large range of energies for these and also for other excited states. For more details, see the paper by Dümmler *et al.* [7.109].

These results raise questions regarding the sensitivity of the various observables regarding the details in a theoretical model, and hence the value of experimental benchmark data to test such models. Figure 7.38 compares the parameters $P_1$ and $S_A$ for impact excitation of the $(3p^5[3/2]4s)$ ($\text{``}^3P_1\text{''}$)

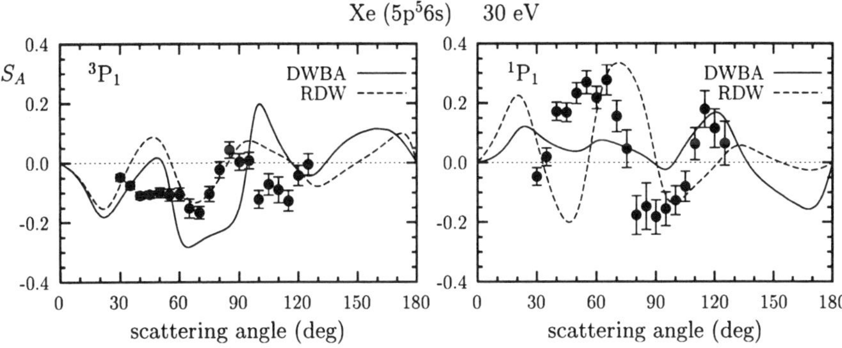



**Fig. 7.37.** Spin asymmetry function $S_A$ for electron-impact excitation of the $(5p^5[3/2]6s)$ ("$^3P_1$") and $(5p^5[1/2]6s)$ ("$^1P_1$") states in xenon at an incident energy of 30 eV. The experimental data of Dümmler *et al.* [7.109] are compared with semi-relativistic (DWBA) and full-relativistic (RDW) first-order distorted-wave results of Bartschat and Madison [7.107] and of Zuo *et al.* [7.108].

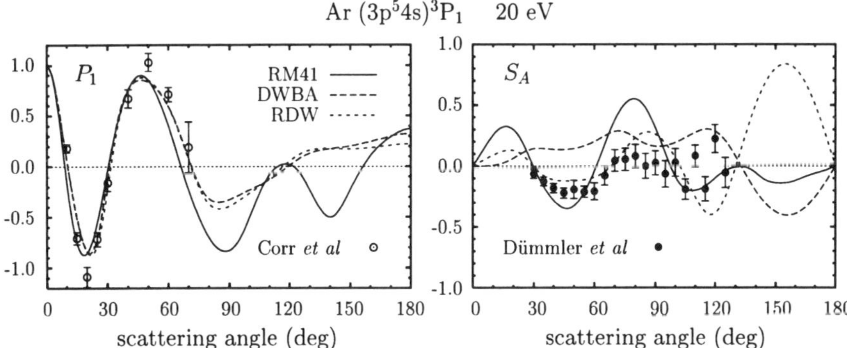



**Fig. 7.38.** Stokes parameter $P_1$ and spin asymmetry function $S_A$ for electron-impact excitation of the $(3p^5[3/2]4s)$ ("$^3P_1$") in argon at an incident energy of 20 eV. The experimental data of Dümmler *et al.* [7.109] are compared with semi-relativistic (DWBA) and full-relativistic (RDW) first-order distorted-wave results of Bartschat and Madison [7.107] and of Zuo *et al.* [7.108], and with predictions from a 41-state Breit–Pauli R-matrix model [7.110].

in argon. Both distorted-wave models [7.107,108], as well as a recent 41-state Breit–Pauli R-matrix (close-coupling) approach [7.110], are able to predict the experimental data for $P_1$ very well, but neither model does particularly well for the spin asymmetry function $S_A$. Although part of the problems lies in the limited angular range of the $P_1$ data ($10° - 70°$), the predictions for $S_A$ are clearly more sensitive to the theoretical model than those for $P_1$.

The need for a more sophisticated treatment of electron collisions with the heavy noble gases is further demonstrated in Figure 7.39, which again compares distorted-wave and close-coupling predictions for $S_A$, this time for impact excitation of the manifold of four states with configuration $4p^5 5s$ in krypton. Clearly, the 31-state Breit–Pauli model [1.109] agrees much better with the experimental data than either of the distorted-wave models. Nevertheless, the agreement is still not perfect and quickly deteriorates in the "intermediate energy regime" between approximately 15 and 50 eV incident energy.

Electron–photon coincidence experiments involving noble-gas targets and spin-polarized electron beams have been very rare. Figure 7.40 shows experimental results [7.111] for the asymmetry

$$A^{zz}(0^\circ) \equiv \frac{1}{P_z} \frac{I(0^\circ, +P_z) - I(0^\circ, -P_z)}{I(0^\circ, +P_z) + I(0^\circ, -P_z)}, \tag{7.71}$$

where $I(0^\circ, \pm P_z)$ is the light intensity observed along the $z$ direction through a linear polarization filter with transmission axis along the incident beam direction, with electron spin polarization in the $\pm z$ direction as well. The experiment was performed in the Münster group for electron-impact excitation of the $(5p^5[3/2]6s)$ state in xenon at an incident energy of 25 eV [7.111]. Given the size of the asymmetry and the small scattering angles to which the investigation was limited, this experiment is not likely to provide a very detailed test of theoretical models.

Finally, experiments on members of the $(2p^5 3p)$ manifold of states in neon were performed by the Utrecht group [7.112,113]. Their data, in combination with predictions from a FOMBT model, allowed for a detailed discussion of the various excitation mechanisms, especially for states that need to be described by an intermediate coupling scheme.

### 7.2.4 Electron-impact excitation of mercury

In contrast to the noble gas targets discussed in the previous section, the $6^1S_0 \rightarrow 6^3P_1^o$ excitation process has been investigated in detail using spin-polarized electron beams. Consequently, we will now discuss this special case of a transition from an initial state with total electronic angular momentum $J_i = 0$ and even parity to a final state with $J_f = 1$ and odd parity in some detail.

Based on the arguments presented in Chapter 5, this process is described by six independent scattering amplitudes, thereby requiring the determination of one *absolute* differential cross section, five *relative* magnitudes, and five *relative* phases. The large number of independent parameters reflects the additional degrees of freedom that the problem presents. Figure 7.41, for instance, shows that for electron-spin polarization in the scattering plane the charge-cloud symmetry axis is no longer restricted to this plane; instead, it may tilt away from the plane and even twist (Raeker *et al.* [7.114]). This will necessarily lead to considerable complications in the algebra.

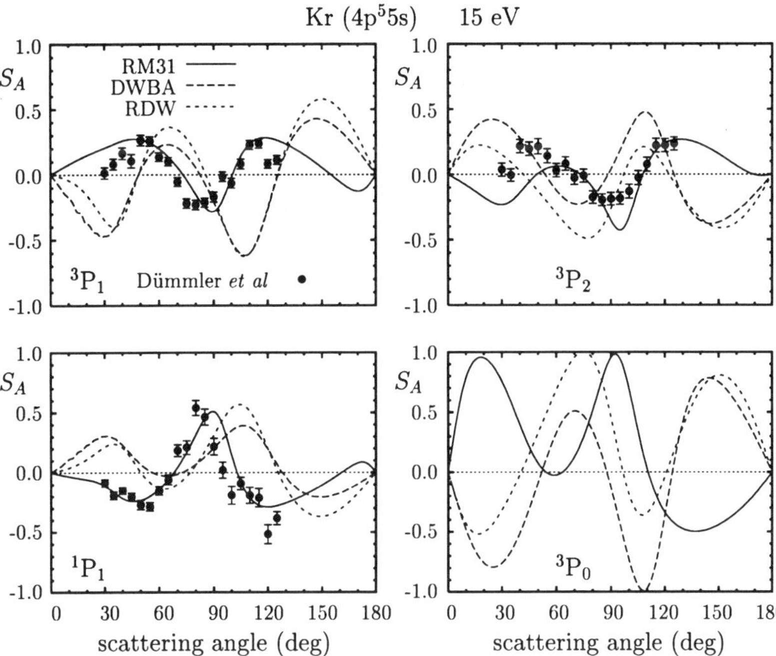



**Fig. 7.39.** Spin asymmetry function $S_A$ for electron-impact excitation of the $(4p^5 5s)$ states in krypton at an incident energy of 15 eV. The experimental data of Dümmler *et al.* [7.109] are compared with semirelativistic (DWBA) [7.107] and full-relativistic (RDW) [7.108] first-order distorted-wave results, and with predictions from a 31-state Breit–Pauli R-matrix model [7.110].

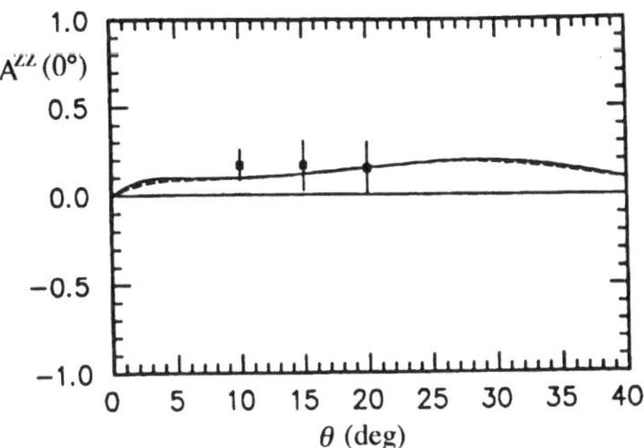



**Fig. 7.40.** Asymmetry $A^{zz}(0°)$ for electron-impact excitation of the $(5p^5[3/2]6s)$ ("³P₁") state in xenon at an incident energy of 25 eV. The experimental data of Uhrig *et al.* [7.111] are compared with semirelativistic (DWBA) and full-relativistic (RDW) first-order distorted-wave results of Bartschat and Madison [7.107] and of Zuo *et al.* [7.108].

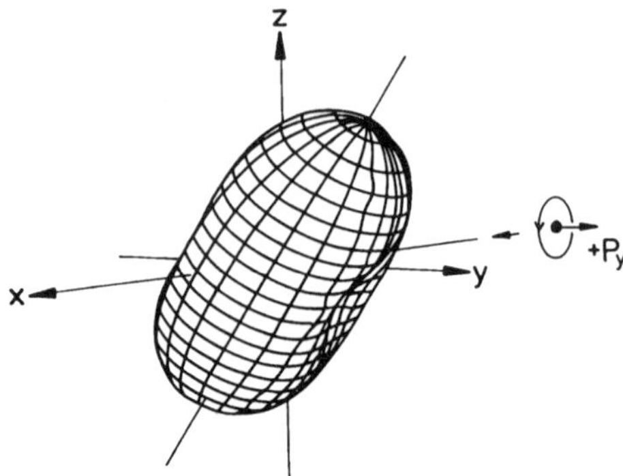



**Fig. 7.41.** Example of a tilted and twisted charge cloud resulting from an in-plane spin-polarized incident electron beam (from Raeker *et al.* [7.114]). The figure is for Hg $6^3P_1$ excitation with impact energy $8\,\mathrm{eV}$, initial spin polarization $P_y = 1$, and scattering angle $\theta = 30°$.

Omitting $J_f = 1$, $J_i = M_i = 0$, and $\theta$, we parameterize the six scattering amplitudes $f(M_f, m_f, m_i)$ in Figure 7.42 as

$$f(\ \ 1,\ \tfrac{1}{2},\ \tfrac{1}{2}) \equiv f^{\uparrow}_{+1} = \alpha_+ e^{i\phi_+}\,, \tag{7.72a}$$

$$f(\ \ 1,-\tfrac{1}{2},-\tfrac{1}{2}) \equiv f^{\downarrow}_{+1} = \beta_+ e^{i\psi_+}\,, \tag{7.72b}$$

$$f(-1,\ \tfrac{1}{2},\ \tfrac{1}{2}) \equiv f^{\uparrow}_{-1} = \alpha_- e^{i\phi_-}\,, \tag{7.72c}$$

$$f(-1,-\tfrac{1}{2},-\tfrac{1}{2}) \equiv f^{\downarrow}_{-1} = \beta_- e^{i\psi_-}\,, \tag{7.72d}$$

$$f(\ \ 0,-\tfrac{1}{2},\ \tfrac{1}{2}) \equiv f^{\uparrow}_{0} = \alpha_0 e^{i\phi_0}\,, \tag{7.72e}$$

$$f(\ \ 0,\ \tfrac{1}{2},-\tfrac{1}{2}) \equiv f^{\downarrow}_{0} = \beta_0 e^{i\psi_0}\,. \tag{7.72f}$$

The quantities (7.72a–d) represent *no-flip* amplitudes that leave the projectile spin unchanged while (7.72e,f) describe the cases where the electron spin is *flipped*. The up/down arrows correspond to the initial spin projection in the natural frame.

As for e–Na excitation, we want to express the photon and electron polarization properties in terms of the amplitude parameters, as well as the density matrix for the excited state. The analysis is done most systematically in the reduced density matrix formalism, with an expansion of the density-matrix elements in terms of state multipoles. Since details of the derivations can be found in ref. [7.115], we will only summarize the results here.

We begin with the photon pattern and parameterize the density matrix for the classical oscillator density by further generalizing the method used for sodium above. For excitation with an unpolarized electron beam, the density matrix for heavy atoms is decomposed into a pair of matrices, one having positive reflection symmetry with respect to the scattering plane and the other having negative reflection symmetry, respectively. Recall that excitation

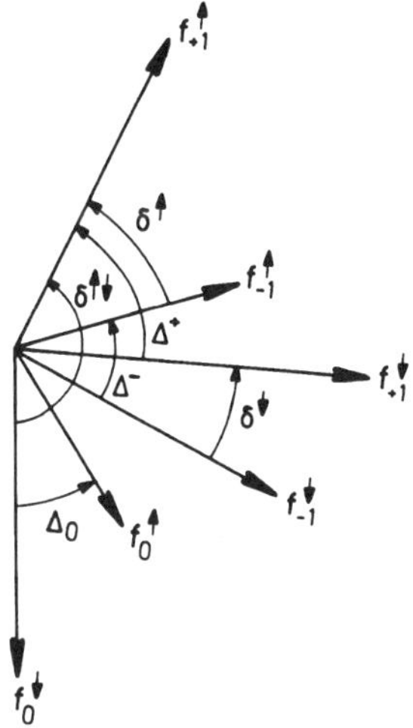



**Fig. 7.42.** Schematic diagram of scattering amplitudes in the natural frame for $J = 0 \rightarrow J = 1$ transitions by electron impact [7.115]. Note that $\Delta^+ + \delta^\downarrow = \Delta^- + \delta^\uparrow$.

of states with negative atomic reflection symmetry and the corresponding spin-flip processes are measured by the height parameter $h$. We consider the two cases:

(*i*) *Electron beam polarization perpendicular to the scattering plane*, i.e., along the $z$ direction of the natural coordinate system. The extension of the earlier decomposition is a pair of density matrices, one for spin-up electron-impact excitation and one for spin-down excitation, where "up" and "down" correspond to the initial spin component orientation with respect to the scattering plane. Hence,

$$\rho_u = \sigma_u \left[ (1-h)\frac{1}{2} \begin{pmatrix} 1 + L^+_\perp & 0 & -P^+_\ell\, e^{-2i\gamma} \\ 0 & 0 & 0 \\ -P^+_\ell\, e^{2i\gamma} & 0 & 1 - L^+_\perp \end{pmatrix} + h \begin{pmatrix} 0 & 0 & 0 \\ 0 & 1 & 0 \\ 0 & 0 & 0 \end{pmatrix} \right]$$

$$= w^\uparrow \rho^\uparrow + w^\downarrow \rho^\downarrow$$

$$= w^\uparrow \sigma_u \left[ (1-h^\uparrow)\frac{1}{2} \begin{pmatrix} 1 + L^{+\uparrow}_\perp & 0 & -P^{+\uparrow}_\ell\, e^{-2i\gamma^\uparrow} \\ 0 & 0 & 0 \\ -P^{+\uparrow}_\ell\, e^{2i\gamma^\uparrow} & 0 & 1 - L^{+\uparrow}_\perp \end{pmatrix} + h^\uparrow \begin{pmatrix} 0 & 0 & 0 \\ 0 & 1 & 0 \\ 0 & 0 & 0 \end{pmatrix} \right]$$

$$+ w^\downarrow \sigma_u \left[ (1-h^\downarrow)\frac{1}{2} \begin{pmatrix} 1 + L^{+\downarrow}_\perp & 0 & -P^{+\downarrow}_\ell\, e^{-2i\gamma^\downarrow} \\ 0 & 0 & 0 \\ -P^{+\downarrow}_\ell\, e^{2i\gamma^\downarrow} & 0 & 1 - L^{+\downarrow}_\perp \end{pmatrix} + h^\downarrow \begin{pmatrix} 0 & 0 & 0 \\ 0 & 1 & 0 \\ 0 & 0 & 0 \end{pmatrix} \right].$$

$$(7.73)$$

Here we have defined

$$L_{\perp}^{+\uparrow} = \frac{\alpha_+^2 - \alpha_-^2}{\alpha_+^2 + \alpha_-^2} = -P_3^{\uparrow} \,, \tag{7.74a}$$

$$L_{\perp}^{+\downarrow} = \frac{\beta_+^2 - \beta_-^2}{\beta_+^2 + \beta_-^2} = -P_3^{\downarrow} \,, \tag{7.74b}$$

$$P_{\ell}^{+\uparrow} \, \mathrm{e}^{2i\gamma^{\uparrow}} = P_1^{\uparrow} + iP_2^{\uparrow} = -\frac{2\alpha_+\alpha_- \, \mathrm{e}^{i(\phi_- - \phi_+)}}{\alpha_+^2 + \alpha_-^2} \,, \tag{7.74c}$$

$$P_{\ell}^{+\downarrow} \, \mathrm{e}^{2i\gamma^{\downarrow}} = P_1^{\downarrow} + iP_2^{\downarrow} = -\frac{2\beta_+\beta_- \, \mathrm{e}^{i(\psi_- - \psi_+)}}{\beta_+^2 + \beta_-^2} \,, \tag{7.74d}$$

$$\sigma^{\uparrow} = \alpha_+^2 + \alpha_-^2 + \alpha_0^2 \,, \tag{7.74e}$$

$$\sigma^{\downarrow} = \beta_+^2 + \beta_-^2 + \beta_0^2 \,, \tag{7.74f}$$

$$\sigma_u = \tfrac{1}{2} \left( \alpha_+^2 + \alpha_-^2 + \alpha_0^2 + \beta_+^2 + \beta_-^2 + \beta_0^2 \right) = \tfrac{1}{2} \left( \sigma^{\uparrow} + \sigma^{\downarrow} \right) \,, \tag{7.74g}$$

$$h^{\uparrow} = \alpha_0^2 / \sigma^{\uparrow} \,, \tag{7.74h}$$

$$h^{\downarrow} = \beta_0^2 / \sigma^{\downarrow} \,, \tag{7.74i}$$

$$w^{\uparrow} = \sigma^{\uparrow} / (2\,\sigma_u) \,, \tag{7.74j}$$

$$w^{\downarrow} = \sigma^{\downarrow} / (2\,\sigma_u) = 1 - w^{\uparrow} \,. \tag{7.74k}$$

From these definitions, it follows that

$$(1 - h)\, L_{\perp}^{+} = w^{\uparrow} (1 - h^{\uparrow})\, L_{\perp}^{+\uparrow} + w^{\downarrow} (1 - h^{\downarrow})\, L_{\perp}^{+\downarrow} \,, \tag{7.75a}$$

$$(1 - h)\, P_{\ell}^{+} \mathrm{e}^{2i\gamma} = w^{\uparrow} (1 - h^{\uparrow})\, P_{\ell}^{+\uparrow} \, \mathrm{e}^{2i\gamma^{\uparrow}} + w^{\downarrow} (1 - h^{\downarrow})\, P_{\ell}^{+\downarrow} \, \mathrm{e}^{2i\gamma^{\downarrow}} \,, \tag{7.75b}$$

$$h = w^{\uparrow} h^{\uparrow} + w^{\downarrow} h^{\downarrow} = (\alpha_0^2 + \beta_0^2)/(2\sigma_u) \,, \tag{7.75c}$$

$$P_{\ell}^{+\uparrow} = \sqrt{1 - \left( L_{\perp}^{+\uparrow} \right)^2} = \sqrt{1 - P_3^{\uparrow\,2}} \,, \tag{7.75d}$$

$$P_{\ell}^{+\downarrow} = \sqrt{1 - \left( L_{\perp}^{+\downarrow} \right)^2} = \sqrt{1 - P_3^{\downarrow\,2}} \,, \tag{7.75e}$$

$$2\gamma^{\uparrow} = \phi_- - \phi_+ \pm \pi = -\delta^{\uparrow} \pm \pi \,, \tag{7.75f}$$

$$2\gamma^{\downarrow} = \psi_- - \psi_+ \pm \pi = -\delta^{\downarrow} \pm \pi \,. \tag{7.75g}$$

Consequently, the decomposition of the density matrix (7.73) is described in terms of the absolute cross section $\sigma_u$ and seven dimensionless independent parameters:

$$(\sigma_u; \; w^{\uparrow}, \; L_{\perp}^{+\uparrow}, \; L_{\perp}^{+\downarrow}, \; h^{\uparrow}, \; h^{\downarrow}; \gamma^{\uparrow}, \; \gamma^{\downarrow}) \,. \tag{7.76}$$

This set leaves three relative phases still unknown. Before searching for additional information about these phases using electron polarization in the scattering plane (to be discussed in paragraph (*ii*) below), we first map the results of the radiation pattern analysis with electron polarization $P_z$ by showing the relationship between the parameter set (7.76) and the generalized Stokes parameters. For an electron beam polarized along the $z$ axis, the generalized Stokes matrix for observation in the $+z$ direction is given in terms of the density-matrix parameters by [7.115]:

$$\frac{2}{3}\left(IQ_{ij}^{zz}\right) =$$

$$\begin{pmatrix}
(1-h)\,P_1;\, w^\uparrow\,(1-h^\uparrow)\,P_1^\uparrow - w^\downarrow\,(1-h^\downarrow)\,P_1^\downarrow;\, \frac{2}{3}\left[\,w^\uparrow\,(1-3h^\uparrow) - w^\downarrow\,(1-3h^\downarrow)\right] \\
(1-h)\,P_2;\, w^\uparrow\,(1-h^\uparrow)\,P_2^\uparrow - w^\downarrow\,(1-h^\downarrow)\,P_2^\downarrow;\, \frac{2}{3}\left[\,w^\uparrow\,(1-3h^\uparrow) - w^\downarrow\,(1-3h^\downarrow)\right] \\
(1-h)\,P_3;\, w^\uparrow\,(1-h^\uparrow)\,P_3^\uparrow - w^\downarrow\,(1-h^\downarrow)\,P_3^\downarrow;\, \frac{2}{3}\left[\,w^\uparrow\,(1-3h^\uparrow) - w^\downarrow\,(1-3h^\downarrow)\right]
\end{pmatrix}$$

$$(7.77a)$$

with the (normalized) light intensity

$$I = I_u^z = \tfrac{3}{2}(1 - h) \tag{7.77b}$$

and

$$(1 - h)\,P_i = w^\uparrow\,(1 - h^\uparrow)\,P_i^\uparrow + w^\downarrow\,(1 - h^\downarrow)\,P_i^\downarrow; \quad i = 1, 2, 3\,. \tag{7.77c}$$

If the height $h$ of the charge cloud for unpolarized electron impact is known from a standard $P_4$ measurement, one may use the sum and the difference of the elements in the first two columns (i.e., six parameters) of (7.77a) to obtain $w^\uparrow\,(1 - h^\uparrow)\,\boldsymbol{P}^\uparrow/(1 - h)$ and $w^\downarrow\,(1 - h^\downarrow)\,\boldsymbol{P}^\downarrow/(1 - h)$ where $\boldsymbol{P}^\uparrow = (P_1^\uparrow, P_2^\uparrow, P_3^\uparrow)$ and $\boldsymbol{P}^\downarrow = (P_1^\downarrow, P_2^\downarrow, P_3^\downarrow)$, respectively. Since the degrees of polarization $P^{+\uparrow,\downarrow} \equiv |\boldsymbol{P}^{\uparrow,\downarrow}| = 1$ for the two initial spin projections, one can first extract $c^\uparrow \equiv w^\uparrow\,(1 - h^\uparrow)/(1 - h)$ and $c^\downarrow \equiv w^\downarrow\,(1 - h^\downarrow)/(1 - h)$ from the sum of the squares of the individual components, and subsequently $L_\perp^{+\uparrow}$, $L_\perp^{+\downarrow}$, $\gamma^\uparrow$, and $\gamma^\downarrow$ from the Stokes vectors $\boldsymbol{P}^\uparrow$ and $\boldsymbol{P}^\downarrow$. (Since $c^\uparrow + c^\downarrow = 1$ theoretically, any measured deviation from this relationship should be remedied by renormalizing all elements of the generalized Stokes matrix by a common factor.) The last column determines $[w^\uparrow\,(1-3h^\uparrow) - w^\downarrow\,(1-3h^\downarrow)]/(1-h)$ which, when combined with $c^\uparrow$ and $c^\downarrow$, allows for determination of $w^\uparrow$, $h^\uparrow$ and $h^\downarrow$. Thus, knowing $h$, the seven dimensionless parameters of (7.76) can be determined from $Q_{ij}^{zz}$. Note that the independent determination of $h$ can *not* be replaced by (7.77c); this equation is not independent from the others and can only serve as a consistency check.

Switching now to the matrices measured in the two remaining directions $y$ and $x$, equations for the nonvanishing elements in the first row of the matrices $(Q_{ij}^{yz})$ and $(Q_{ij}^{xz})$ are, written here as a *column* for convenience [7.115]:

$$\frac{4}{3}\begin{pmatrix} IQ_{11}^{yz} \\ IQ_{12}^{yz} \\ IQ_{13}^{yz} \end{pmatrix} =$$

$$\begin{pmatrix} -[w^{\uparrow}\,(1-3h^{\uparrow})+w^{\downarrow}\,(1-3h^{\downarrow})] - [w^{\uparrow}\,(1-h^{\uparrow})\,P_1^{\uparrow}+w^{\downarrow}\,(1-h^{\downarrow})P_1^{\downarrow}] \\ -[w^{\uparrow}\,(1-3h^{\uparrow})-w^{\downarrow}\,(1-3h^{\downarrow})] - [w^{\uparrow}\,(1-h^{\uparrow})\,P_1^{\uparrow}-w^{\downarrow}\,(1-h^{\downarrow})P_1^{\downarrow}] \\ [w^{\uparrow}\,(1+h^{\uparrow})-w^{\downarrow}\,(1+h^{\downarrow})] + [w^{\uparrow}\,(1-h^{\uparrow})\,P_1^{\uparrow}-w^{\downarrow}\,(1-h^{\downarrow})P_1^{\downarrow}] \end{pmatrix}$$

$$(7.78a)$$

with the normalized light intensity

$$I = I_u^y = \frac{3}{4}[(1+h)+(1-h)\,P_1)] \,, \tag{7.78b}$$

and

$$\frac{3}{4}\begin{pmatrix} IQ_{11}^{xz} \\ IQ_{12}^{xz} \\ IQ_{13}^{xz} \end{pmatrix} =$$

$$\begin{pmatrix} -[w^{\uparrow}\,(1-3h^{\uparrow})+w^{\downarrow}\,(1-3h^{\downarrow})] + [w^{\uparrow}\,(1-h^{\uparrow})\,P_1^{\uparrow}+w^{\downarrow}\,(1-h^{\downarrow})P_1^{\downarrow}] \\ -[w^{\uparrow}\,(1-3h^{\uparrow})-w^{\downarrow}\,(1-3h^{\downarrow})] + [w^{\uparrow}\,(1-h^{\uparrow})\,P_1^{\uparrow}-w^{\downarrow}\,(1-h^{\downarrow})P_1^{\downarrow}] \\ [w^{\uparrow}\,(1+h^{\uparrow})-w^{\downarrow}\,(1+h^{\downarrow})] - [w^{\uparrow}\,(1-h^{\uparrow})\,P_1^{\uparrow}-w^{\downarrow}\,(1-h^{\downarrow})P_1^{\downarrow}] \end{pmatrix} \,,$$

$$(7.79a)$$

with the normalized light intensity

$$I = I_u^x = \frac{3}{4}[(1+h)-(1-h)\,P_1)] \,. \tag{7.79b}$$

Except for the $P_4$ measurement, we see from inspection that no additional information can be obtained from generalized Stokes parameters observed in other directions (such as $x$ or $y$) with electron beam polarization vector perpendicular to the scattering plane. On the other hand, such additional measurements with photon detectors in the scattering plane can provide valuable consistency checks.

As found above, three more relative independent phases are needed to determine the scattering amplitudes uniquely. In analogy to the sodium case, we define

$$\Delta^+ \equiv \phi_+ - \psi_+, \tag{7.80a}$$

$$\Delta^- \equiv \phi_- - \psi_-, \tag{7.80b}$$

$$\Delta^0 \equiv \phi_0 - \psi_0. \tag{7.80c}$$

Inspection of Figure 7.42, however, shows that only two of these are independent, since

$$\Delta^+ - \Delta^- = \delta^{\uparrow} - \delta^{\downarrow} = 2\,(\gamma^{\downarrow} - \gamma^{\uparrow}) \tag{7.81}$$

in analogy to (7.58). Therefore, it remains to fix the phase of the spin-flip amplitudes $f_0^{\uparrow,\downarrow}$ relative to the nonflip amplitudes. A convenient choice for the remaining phase angle is

$$\delta^{\uparrow\downarrow} \equiv \phi_+ - \psi_0 \, . \tag{7.82}$$

A complete set of independent parameters is then given by

$$(\sigma_u; w^\uparrow, \, L_\perp^{+\uparrow}, \, L_\perp^{+\downarrow}, \, h^\uparrow, \, h^\downarrow; \, \gamma^\uparrow, \, \gamma^\downarrow, \, \Delta^+, \, \Delta^0, \, \delta^{\uparrow\downarrow}) \, , \tag{7.83}$$

i.e., one absolute cross section, five relative sizes and five relative phases. Information about the remaining three phase angles may be sought for in experiments with *in-plane* spin polarization, a possibility that we will now explore.

($ii$) *Electron beam polarization in the scattering plane:* No additional information is obtained with such an electron polarization if the photons are observed in the $z$ direction [7.115]. However, in the directions $x$ and $y$ one obtains eight nontrivial components, namely $IQ_{22}^{\hat{m}\hat{n}}$ and $IQ_{32}^{\hat{m}\hat{n}}$ with $\hat{m}\hat{n} = xx$, $xy$, $yx$ and $yy$. To clarify the algebraic structure of the general expressions for these components, we introduce the following abbreviations:

$$A_1 = \alpha_+ \beta_0, \tag{7.84a}$$

$$A_2 = \beta_+ \alpha_0, \tag{7.84b}$$

$$A_3 = \alpha_- \beta_0, \tag{7.84c}$$

$$A_4 = \beta_- \alpha_0, \tag{7.84d}$$

$$\omega_1 = \phi_+ - \psi_0 = \delta^{\uparrow\downarrow}, \tag{7.84e}$$

$$\omega_2 = \psi_+ - \phi_0 = \delta^{\uparrow\downarrow} - \Delta^0 - \Delta^+, \tag{7.84f}$$

$$\omega_3 = \phi_- - \psi_0 = \delta^{\uparrow\downarrow} - \delta^\uparrow, \tag{7.84g}$$

$$\omega_4 = \psi_- - \phi_0 = \delta^{\uparrow\downarrow} - \Delta^0 - \Delta^+ - \delta^\downarrow, \tag{7.84h}$$

where the phases are defined in Figure 7.42. With this notation, we find

$$\begin{pmatrix} IQ_{22}^{yy} \\ IQ_{32}^{yy} \end{pmatrix} = \frac{3}{2} \begin{pmatrix} -A_1 \sin\omega_1 + A_2 \sin\omega_2 + A_3 \sin\omega_3 - A_4 \sin\omega_4 \\ -A_1 \cos\omega_1 + A_2 \cos\omega_2 + A_3 \cos\omega_3 - A_4 \cos\omega_4 \end{pmatrix}, \tag{7.85a}$$

$$\begin{pmatrix} IQ_{22}^{yx} \\ IQ_{32}^{yx} \end{pmatrix} = \frac{3}{2} \begin{pmatrix} A_1 \cos\omega_1 - A_2 \cos\omega_2 + A_3 \cos\omega_3 - A_4 \cos\omega_4 \\ -A_1 \sin\omega_1 + A_2 \sin\omega_2 - A_3 \sin\omega_3 + A_4 \sin\omega_4 \end{pmatrix}, \tag{7.85b}$$

$$\begin{pmatrix} IQ_{22}^{xy} \\ IQ_{32}^{xy} \end{pmatrix} = \frac{3}{2} \begin{pmatrix} -A_1 \cos\omega_1 - A_2 \cos\omega_2 + A_3 \cos\omega_3 + A_4 \cos\omega_4 \\ A_1 \sin\omega_1 + A_2 \sin\omega_2 - A_3 \sin\omega_3 - A_4 \sin\omega_4 \end{pmatrix}, \tag{7.85c}$$

$$\begin{pmatrix} IQ_{22}^{xx} \\ IQ_{32}^{xx} \end{pmatrix} = \frac{3}{2} \begin{pmatrix} A_1 \sin\omega_1 + A_2 \sin\omega_2 + A_3 \sin\omega_3 + A_4 \sin\omega_4 \\ A_1 \cos\omega_1 + A_2 \cos\omega_2 + A_3 \cos\omega_3 + A_4 \cos\omega_4 \end{pmatrix}. \tag{7.85d}$$

Inspection shows that a measurement of the generalized Stokes parameters in the $y$ and $x$ directions with in-plane electron beam polarizations $P_y$ and $P_x$

will provide the four relative magnitudes and the four relative phase angles defined in (7.84). If analysis in the $z$ direction with polarization $P_z$ has been completed, all relative magnitudes are already known, and the four $A$ parameters may serve as consistency checks. For example, the parameters $L_\perp^{+\uparrow}$ and $L_\perp^{+\downarrow}$ can be derived as

$$L_\perp^{+\uparrow} = \frac{A_1^2 - A_2^2}{A_1^2 + A_2^2}, \tag{7.86a}$$

$$L_\perp^{+\downarrow} = \frac{A_3^2 - A_4^2}{A_3^2 + A_4^2}. \tag{7.86b}$$

Additional information may be obtained from the $\omega$ angles. They determine the relative phase angles within the amplitude triplets $(f_{+1}^\uparrow, f_{-1}^\uparrow, f_0^\downarrow)$ and $(f_{+1}^\downarrow, f_{-1}^\downarrow, f_0^\uparrow)$, corresponding to *final* spin up and down, respectively. However, the phase of the two triplets with respect to each other can *not* be obtained from the $\omega$ angles, since

$$\omega_1 - \omega_3 = \delta^\uparrow = -2\gamma^\uparrow + \pi, \tag{7.87a}$$
$$\omega_2 - \omega_4 = \delta^\downarrow = -2\gamma^\downarrow + \pi. \tag{7.87b}$$

Provided that $\gamma^\uparrow$ and $\gamma^\downarrow$ are already known, these relations allow us to extract *two* additional angles only, e.g., $\delta^{\uparrow\downarrow}$ and $(\Delta^0 + \Delta^-)$. This is possible by photon observation in the $y$ direction only, as can be seen from (7.85a,b) after eliminating $\omega_3$ and $\omega_4$ with (7.87). For a perfect experiment, a determination of any of the three angles $\Delta^+$, $\Delta^-$ or $\Delta^0$ will suffice. This can only be achieved by using the $STU$ parameters discussed below since, in the photon analysis, the incoherent summation over the unobserved final electron spins destroys the phase information between the amplitude triplets that belong to final electron spins up and down, respectively.

Before we turn to the experimental situation, we summarize the discussion above. Figure 7.43 shows the three generalized Stokes-parameter matrices that can be measured in each of the three orthogonal coordinate directions. In the $z$ direction, apart from the Stokes parameters $P_1, P_2, P_3$ (the first column) corresponding to an unpolarized beam, new information is obtained only with electron spin polarization in the $z$ direction. From this matrix, provided $P_4$ is also measured, the dimensionless parameter set (7.76) can be evaluated. For this spin direction, the three numbers in the third column should be identical and may thus serve as consistency checks for the three polarizer settings. Given this information, observation in the $y$ direction with polarization $P_z$ adds nothing new, except for the parameter $P_4$ that can be obtained without spin polarization. The two other nonzero components in the first row can be derived from the data obtained already, but their measurement may serve as another consistency check. For the spin directions $P_x$ and $P_y$, however, new information is derived from the two last elements of the second column, sufficient to evaluate two additional angles.

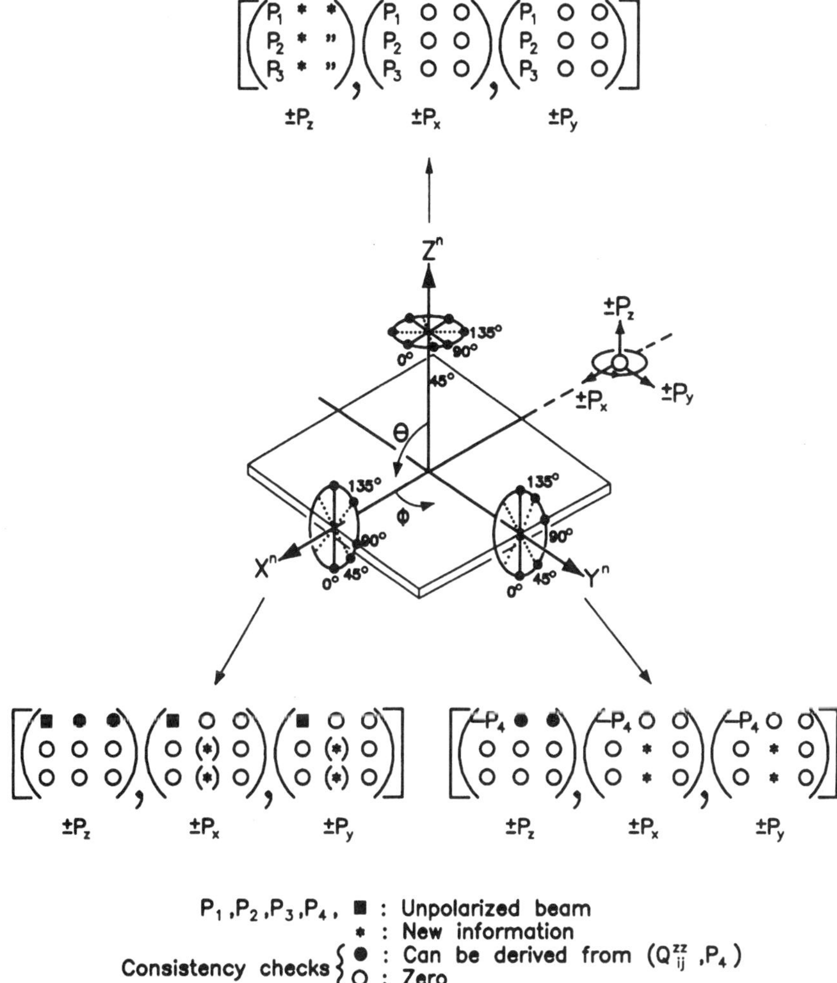



$P_1 , P_2 , P_3 , P_4 ,$  ■ : Unpolarized beam

           ✱ : New information

Consistency checks $\Big\{$ ● : Can be derived from $(Q_{ij}^{zz} , P_4 )$

           ○ : Zero

**Fig. 7.43.** Summary of the information that can be obtained from generalized Stokes parameter measurements.

At this point, we have thus extracted all information that is available in the photon polarization pattern. Observation in a third direction in the collision plane, such as $x$, is equivalent to observation in the $y$ direction. Thus it can either replace this, or serve as consistency check, since all elements may be predicted from the results obtained in the $z$ and $y$ directions.

Regarding the experimental situation, results from an electron – polarized-photon coincidence experiment with a spin-polarized incident electron beam were reported by Goeke *et al.* [7.116] and by Sohn and Hanne [7.117] for the present case of interest, namely electron-impact excitation of the $(6s6p)^3P_1^o$

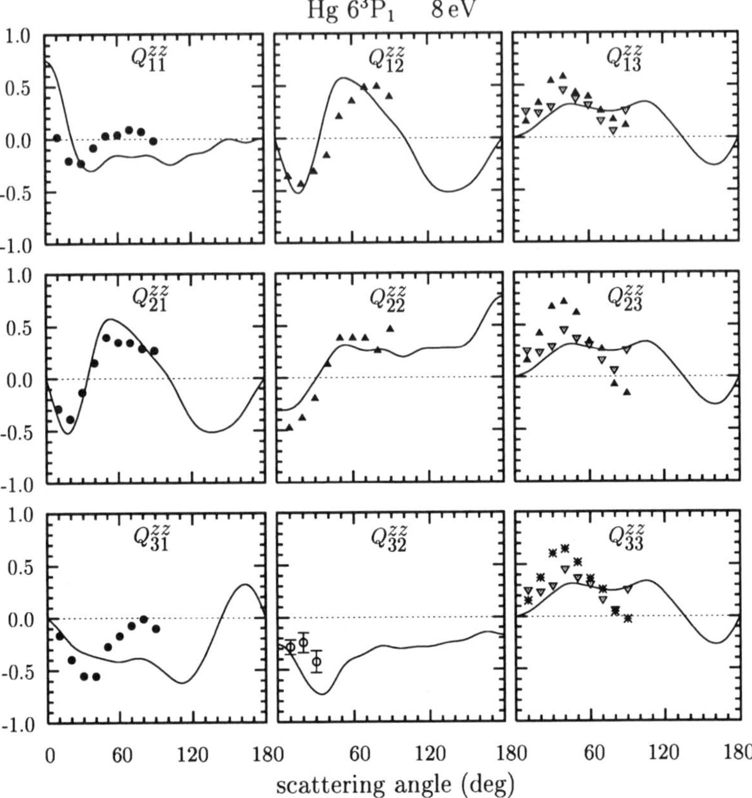



**Fig. 7.44.** Generalized Stokes parameters for electron-impact excitation of the Hg(6s6p)$^3$P$_1$ state at an incident electron energy of 8 eV with light observation along the $z$ direction. $\Diamond$ , published results [7.116,117]; $\square$, unpublished data of Sohn and Hanne [7.117]; $\triangle$, derived from measured polarization and asymmetry data of Goeke *et al.* [7.116] and Sohn and Hanne [7.117]; $\times$, average of measured ($\triangle$) $Q_{13}^{zz}$ and $Q_{23}^{zz}$ data; $\triangledown$, prediction based on properties of the generalized Stokes parameters (see text). The experimental results are compared with predictions from a five-state Breit–Pauli R-matrix calculation based on Scott *et al.* [7.118].

state in mercury. Their data can be used to calculate the generalized Stokes parameters for the case of initial electron spin polarization perpendicular to the scattering plane. Experimental results for an incident electron energy of 8 eV are shown in Figure 7.44 together with theoretical results based on a five-state Breit–Pauli R-matrix (close-coupling) calculation [7.118]. Unfortunately, the full set of generalized Stokes parameters for the $z$ direction could only be obtained for three scattering angles (10°, 20°, and 30°), owing to some missing data for the circular polarization parameters. Although $Q_{13}^{zz} = Q_{23}^{zz} = Q_{33}^{zz}$ according to (7.77a), the experimental data for $Q_{13}^{zz}$ and $Q_{23}^{zz}$ do not agree within the specified error bars, thus indicating some internal

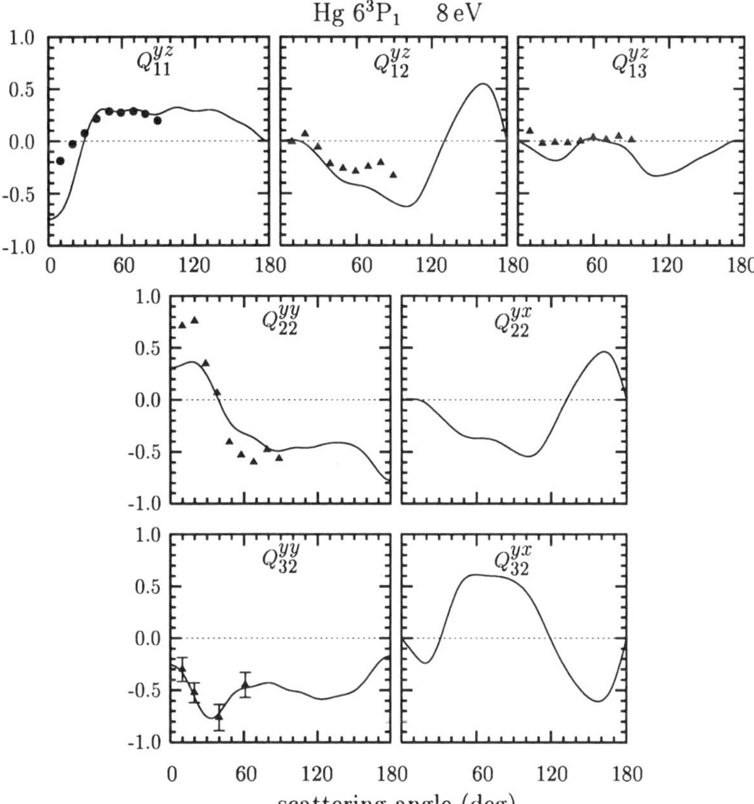



Fig. 7.45 Same as Figure 7.44 for light observation along the $y$ direction.

inconsistency between them. Therefore, these results were averaged and plotted as $\times$ in the field for $Q_{33}^{zz}$. Further consistency checks may be performed by using data from the $y$ direction (shown in Figure 7.45) to evaluate, for example, the third column of $(Q_{ij}^{zz})$ in yet another way. Inspection of (7.77,78) shows that

$$2\,Q_{12}^{zz} + 3\,Q_{13}^{zz} + 4\,\frac{1+P_1}{1+P_4}\,Q_{12}^{yz} = 0\,.\tag{7.88}$$

Solving for $Q_{13}^{zz}$ and plotting the results as "$\triangledown$" in the third column of Figure 7.44 reveals that the trend is correct, but the data are not entirely consistent.

Finally, using the available experimental data, one can *predict* experimental results for generalized Stokes parameters that have *not been measured to date*, such as those that would be observed in the $x$ direction, i.e., along the incident beam axis. Inspection of (7.77–79) yields

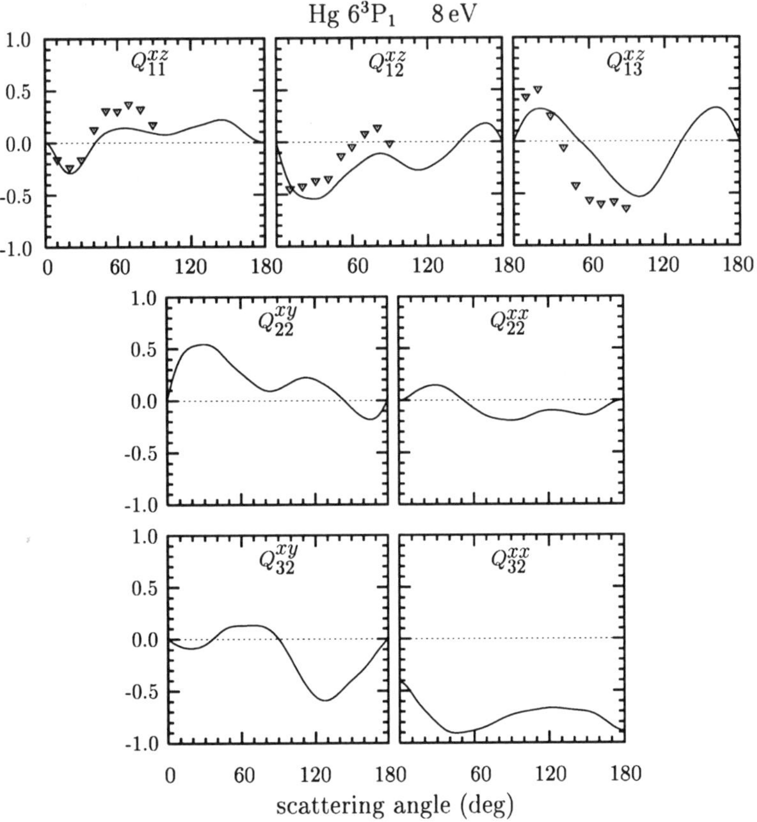



**Fig.** 7.46 Same as Figure 7.44 for light observation along the $x$ direction.

$$Q_{11}^{xz} = \frac{1+P_1}{1-P_1P_4}\, Q_{11}^{yz} + \frac{1+P_4}{1-P_1P_4}\, Q_{11}^{zz} = \frac{P_1-P_4}{1-P_1P_4}, \tag{7.89a}$$

$$Q_{12}^{xz} = \frac{1+P_1}{1-P_1P_4}\, Q_{12}^{yz} + \frac{1+P_4}{1-P_1P_4}\, Q_{12}^{zz}, \tag{7.89b}$$

$$Q_{13}^{xz} = \frac{1+P_1}{1-P_1P_4}\, Q_{13}^{yz} - \frac{1+P_4}{1-P_1P_4}\, Q_{12}^{zz}, \tag{7.89c}$$

where we have used that $Q_{11}^{zz} = P_1$ and $Q_{11}^{yz} = -P_4$. The predictions are plotted in Figure 7.46.

The agreement between the actual and derived experimental results in Figures 7.44–46 and the theoretical predictions is satisfactory, bearing in mind the complexity of the collision problem, the level of detail in the comparison, and the remaining inconsistencies within the experimental data set. The agreement between experiment and theory is good for the in-plane polarization parameters $Q_{22}^{yy}$ and $Q_{32}^{yy}$ shown in Figure 7.45, except for a theoretical underestimation of $Q_{22}^{yy}$ at small scattering angles.

In the concrete case of the mercury target, extraction of the set (7.76) of spin-dependent density matrix parameters (except for the cross section $\sigma_u$) from the generalized Stokes parameters is further complicated by depolarization effects from hyperfine-structure interaction in mercury isotopes with nonvanishing nuclear spin [7.119]. The two sets of results plotted in Figure 7.47(a) were calculated from published data ($\times$) [7.116,117], together with unpublished results of Sohn and Hanne ($+$) [7.117] for scattering angles of $10°$, $20°$, and $30°$, respectively. The important point to be made is the fact that only the latter set corresponds to a full determination of the generalized Stokes parameters shown in Figures 7.44 and 7.45, i.e., all parameters that can be determined from the photon radiation pattern with electron polarization perpendicular to the scattering plane. On the other hand, the first set lacks data for $Q_{32}^{zz}$ and $Q_{33}^{zz}$.

While data for $Q_{33}^{zz}$ are desirable for a consistency check between the three elements in the third column of the $(Q_{ij}^{zz})$ matrix, data for $Q_{32}^{zz}$ are *mandatory* for performing the evaluation procedure outlined following (7.77). In particular, the normalization $P^{+\uparrow} = P^{+\downarrow} = 1$, ensuring full coherence in the individual spin channels, was found to be *crucial* for the numerical stability of the inversion [7.119]. Only a complete set of generalized Stokes parameters can be tested for this normalization and, consequently, be renormalized to fulfill this fundamental condition exactly. If any element in the first two columns of the $(Q_{ij}^{zz})$ matrix is missing, the normalization condition can be used *without verification* to calculate this element (except for the sign). Obviously, this is a much less satisfactory procedure and leads to significantly larger systematic errors, thereby illuminating the importance of complete data sets. In the above case, the additional data for $Q_{32}^{zz}$ not only provide more information about the spin-dependent orientation, but also stabilize the inversion procedure for *all other* parameters through built-in internal consistency checks.

Interesting features can be seen in Figure 7.47(a) after transformation of the generalized Stokes parameters into the spin-dependent coherence parameters defined in (7.74) [7.119]. The differential cross sections for the two spin directions are fairly similar at all angles. The coherence parameters, however, are vastly different for the two incident spin projections. Note, for example, that $L_\perp^{+\uparrow} > 0$ while $L_\perp^{+\downarrow} < 0$ at small scattering angles, a result that will be further discussed in Chapter 9. While the positive value of $L_\perp^{+\uparrow}$ is in agreement with well-established propensity rules, the negative value of $L_\perp^{+\downarrow}$ might seem surprising. Both $L_\perp^{+\uparrow}$ and $L_\perp^{+\downarrow}$ are nonzero for forward scattering, and $L_\perp^{+\uparrow}(0°) = -L_\perp^{+\downarrow}(0°)$ by symmetry requirements. The alignment angles $\gamma^\uparrow$ and $\gamma^\downarrow$ show no similarities, with the directions of the major axes often being perpendicular to each other $(\gamma_\perp^{+\uparrow}(0°) = -\gamma_\perp^{+\downarrow}(0°) \neq 0)$. There is a large difference between the height parameters $h^\uparrow$ and $h^\downarrow$ (which measure the relative importance of spin-flips) in this angular range, with $h^\downarrow$ assuming a maximum value of 75% near a scattering of $40°$. This means that spin-flips are very likely

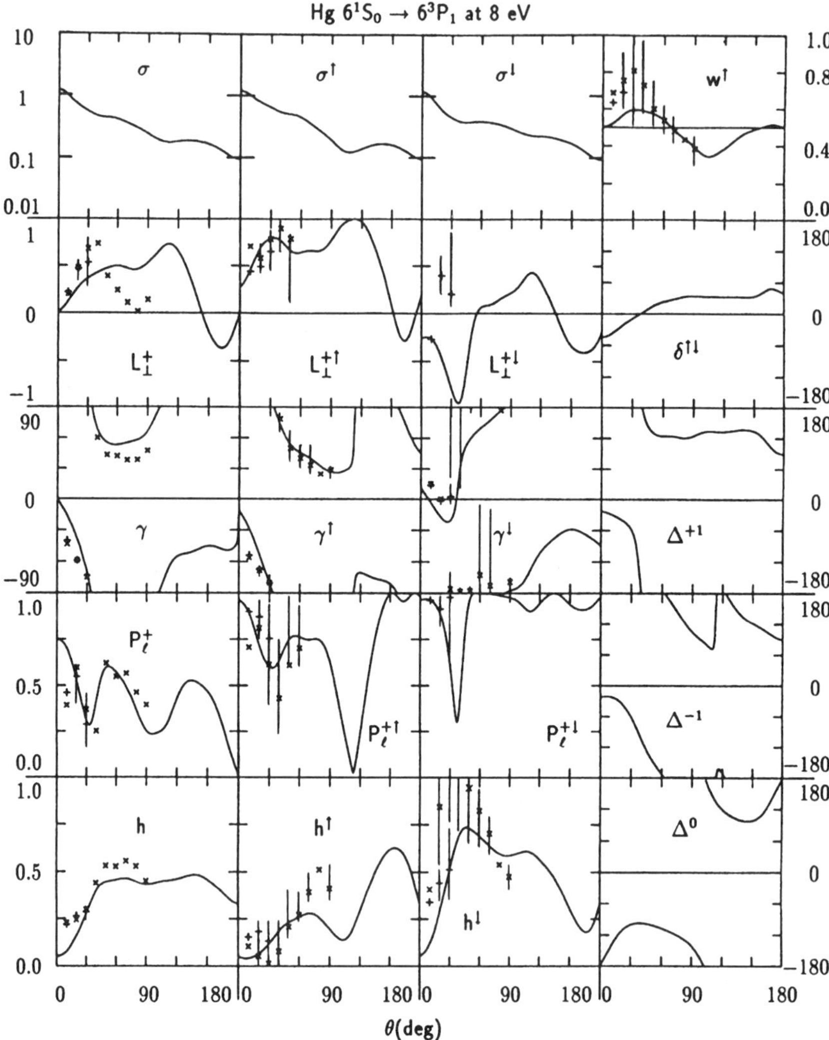



**Fig. 7.47.** (a) Spin-dependent coherence parameters for electron-impact excitation of Hg(6s6p)$^3$P$_1$ at an incident electron energy of 8 eV [7.85,119]. The experimental data were calculated from the generalized Stokes parameters presented in Figures 7.44,45 (see text). ×, prediction based on data of Goeke *et al.* [7.116] and Sohn and Hanne [7.117]; +, prediction based on unpublished data of Sohn and Hanne. The experimental data are compared with the results from a five-state Breit–Pauli R-matrix calculation based on Scott *et al.* [7.118].

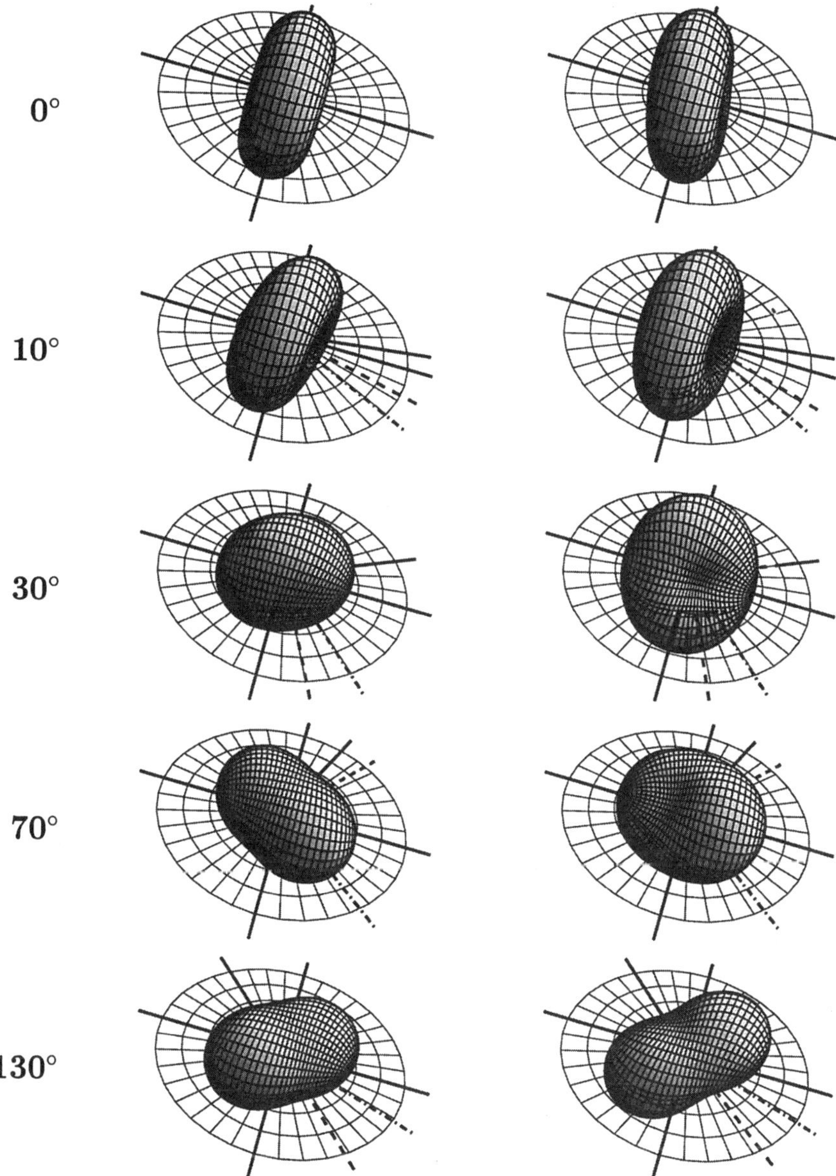


**Fig. 7.47.** (b) Charge cloud representing the excited $6p$ electron after impact excitation of $\mathrm{Hg(6s6p)}^3\mathrm{P}_1$ at an incident electron energy of $8\,\mathrm{eV}$. The left column is for unpolarized incident electrons, the right column for an in-plane transversal spin polarization $P_y = 1$. The dashed line represents the alignment angle and the dash-dotted line the momentum-transfer direction. The predictions are based on a five-state Breit–Pauli R-matrix calculation [7.118].

for incoming spin-down electrons, but those spin-down electrons, whose spin is not flipped, tend to transfer a *negative* angular momentum to the atom.

Figure 7.47(b) shows the charge clouds representing the excited $6p$ electron. The graphs on the left correspond to unpolarized incident electrons, whereas those on the right correspond to an in-plane transversal spin polarization $P_y = 1$. Note how the charge cloud tilts and twists out of the reaction plane in the latter case.

We complete the analysis with a discussion of the generalized $STU$ parameters for the spin polarization of the scattered electron beam. In terms of our parameter set, they can be written as [7.115]

$$S_A = w^\uparrow - w^\downarrow ,  \tag{7.90a}$$

$$S_P = w^\uparrow(1 - 2h^\uparrow) - w^\downarrow(1 - 2h^\downarrow) ,  \tag{7.90b}$$

$$T_z = w^\uparrow(1 - 2h^\uparrow) + w^\downarrow(1 - 2h^\downarrow) = 1 - 2h ,  \tag{7.90c}$$

$$T_y = \frac{1}{\sigma_u} \left\{ \alpha_+\beta_+ \cos \Delta^+ + \alpha_-\beta_- \cos \Delta^- - \alpha_0\beta_0 \cos \Delta^0 \right\} ,  \tag{7.90d}$$

$$T_x = \frac{1}{\sigma_u} \left\{ \alpha_+\beta_+ \cos \Delta^+ + \alpha_-\beta_- \cos \Delta^- + \alpha_0\beta_0 \cos \Delta^0 \right\} ,  \tag{7.90e}$$

$$U_{yx} = -\frac{1}{\sigma_u} \left\{ \alpha_+\beta_+ \sin \Delta^+ + \alpha_-\beta_- \sin \Delta^- - \alpha_0\beta_0 \sin \Delta^0 \right\} ,  \tag{7.90f}$$

$$U_{xy} = -\frac{1}{\sigma_u} \left\{ \alpha_+\beta_+ \sin \Delta^+ + \alpha_-\beta_- \sin \Delta^- + \alpha_0\beta_0 \sin \Delta^0 \right\} .  \tag{7.90g}$$

For brevity, we kept the products of the amplitude magnitudes in the $T$ and $U$ parameters. They can be expressed in terms of density matrix parameters by using

$$\frac{\alpha_\pm\beta_\pm}{\sigma_u} = \sqrt{w^\uparrow w^\downarrow (1 - h^\uparrow)(1 - h^\downarrow)(1 \pm L_\perp^{+\uparrow})(1 \pm L_\perp^{+\downarrow})} ,  \tag{7.91a}$$

$$\frac{\alpha_0\beta_0}{\sigma_u} = \sqrt{w^\uparrow w^\downarrow h^\uparrow h^\downarrow} .  \tag{7.91b}$$

Note that the three parameters $S_A$, $S_P$, and $T_z$ may be *predicted* from coherence parameters extracted in electron–photon coincidence experiments and vice versa; the parameter $T_z = 1 - 2h$ can already be obtained from the coincidence setup with an unpolarized electron beam [7.120], while prediction of $S_{A,P}$ requires an electron beam polarized perpendicular to the scattering plane. On the other hand, measurements of the set $(S_P, S_A, T_z)$ allows for the evaluation of the three relative sizes $h^\uparrow$, $h^\downarrow$, and $w^\uparrow = 1 - w^\downarrow$. The two missing relative sizes $L_\perp^{+\uparrow}$ and $L_\perp^{+\downarrow}$ can only be obtained by circular polarization measurements.

The parameters $T_y$, $T_x$, $U_{yx}$, and $U_{xy}$ contain new information, namely the three phase angles $(\Delta^+, \Delta^0, \Delta^-)$. We see that linear combinations of the form $T_y \pm T_x$ and $U_{yx} \pm U_{xy}$ allow for determination of the two complex

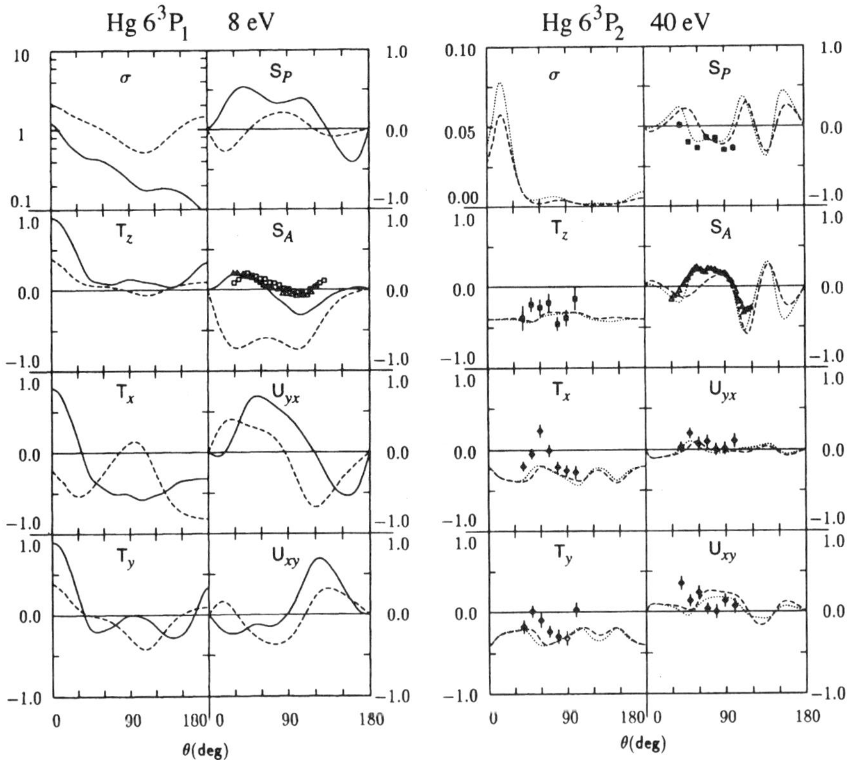



**Fig. 7.48.** Differential cross section and generalized *STU* parameters for electron-impact excitation of Hg $6\,^3P_1$ at an incident energy of 8 eV (a) and $6\,^3P_2$ at an incident energy of 40 eV (b). In (a), the experimental data for $S_A$ are measurements by Borgmann *et al.* [7.121] ($\triangle$) and predictions from electron–photon coincidence data by Goeke *et al.* [7.116] ($\square$). In (b), the experimental data are from Borgmann *et al.* [7.121] ($\triangle$) and Dümmler *et al.* [7.123] ($\square$) for $S_A$, Müller and Kessler [7.124] ($\square$) for $S_P$ and $T_z$, and Klose [7.125] ($\Diamond$) for $T_x$, $T_y$, $U_{xy}$, and $U_{yx}$. The theoretical curves are from a five-state Breit–Pauli R-matrix calculation based on Scott *et al.* [7.118] (——), a first-order semirelativistic distorted-wave calculation by Bartschat and Madison [7.122] (– – –), and a first-order full-relativistic distorted-wave calculation by Srivastava *et al.* [7.126] ($\cdots\cdots$).

numbers $\alpha_+\beta_+ e^{i\Delta^+} + \alpha_-\beta_- e^{i\Delta^-}$ and $\beta_0\alpha_0 e^{i\Delta^0}$, thereby providing the crucial phase difference $\Delta^0$ not obtainable with the electron–polarized-photon coincidence technique without simultaneous electron spin analysis in the exit channel. Hence, the *combination of techniques* for the generalized Stokes and the generalized *STU* parameters constitutes the *ultimate goal of a perfect scattering experiment* for the important $J = 0^+ \rightarrow J = 1^-$ excitation problem in its most general form.

The left side of Figure 7.48(a) shows results for the generalized $STU$ parameters at 8 eV. Unfortunately, experimental data only exist for the $S_A$ parameter. This reflects the fact that their determination is a major experimental effort. That a full determination is, however, within the capabilities of present-day technology is illustrated on the right side of Figure 7.48, which presents a complete set of generalized $STU$ parameters, measured in Münster at 40 eV for Hg $6^1S_0 \to 6^3P_2$ excitation.

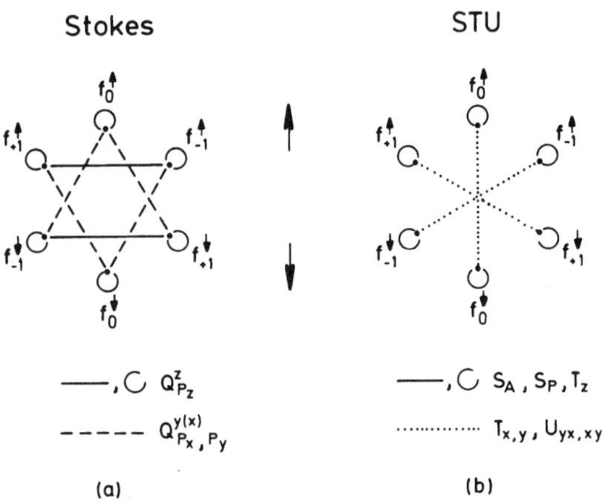


**Fig. 7.49.** Determination of relative amplitude sizes and phases from a generalized Stokes parameter analysis (a) in the $z$ direction with electron polarization $P_z$ and from in-plane measurements with polarizations $P_x$ and $P_y$. Also shown are the relative amplitude sizes and phases that enter into (7.90) for the generalized $STU$ parameters (b). Note that $S_A$, $S_P$, and $T_z$ only depend on the relative sizes of the amplitudes and can, therefore, be predicted from a generalized Stokes parameter analysis.

These considerations and the fundamental difference between the information extracted from generalized Stokes and $STU$ parameters are summarized in Figure 7.49. Figure 7.49(a) illustrates that generalized Stokes parameter analysis in the $z$ direction with electron spin polarization $P_z$ perpendicular to the scattering plane determines all relative amplitude magnitudes, and the phase relationship between the $f_{+1}$ and $f_{-1}$ amplitudes — provided $P_4$ is also known. Additional analysis in the $y$ (or $x$) direction with in-plane spin polarization $P_y$ or $P_x$ yields the four relative phases within the two triples $(f_{+1}^\downarrow, f_{-1}^\downarrow, f_0^\uparrow)$ and $(f_{+1}^\uparrow, f_{-1}^\uparrow, f_0^\downarrow)$, respectively. None of the relative phases $(\Delta^+, \Delta^0, \Delta^-)$ between up/down amplitudes for the same magnetic sublevel enter. For the generalized $STU$ parameters, Figure 7.49(b) illustrates that $S_A$, $S_P$, and $T_z$ depend only on the relative sizes of the amplitudes, while $T_x$, $T_y$, $U_{yx}$, and $U_{xy}$ also depend on the relative phases $(\Delta^+, \Delta^0, \Delta^-)$ *within* the three amplitude pairs $(f_{+1}^\uparrow, f_{+1}^\downarrow)$, $(f_{-1}^\uparrow, f_{-1}^\downarrow)$, and $(f_0^\uparrow, f_0^\downarrow)$. No information on the relative phase angles *between* these pairs can be extracted from a generalized Stokes parameter analysis.

### 7.2.5 Elastic electron scattering from cesium

Electron collisions with cesium atoms have been of interest for a long time, since spin-dependent effects due to *both* exchange and the spin-orbit interaction may play a role. For example, based on a thorough analysis of the *elastic scattering problem* by Burke and Mitchell [7.127], Farago predicted the possibility of a left-right asymmetry $A_1$ in the differential cross section for scattering of unpolarized electrons from spin-polarized target atoms [7.128].[11] This asymmetry should only occur if both electron exchange and the spin-orbit interaction between the projectile electron and the target nucleus are important in the process — either mechanism alone would not be sufficient.

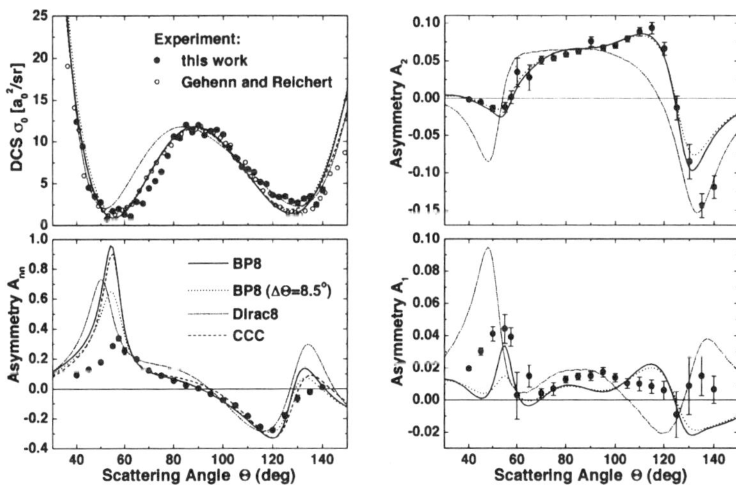


**Fig. 7.50.** Differential cross section $\sigma_0$(a), normalized to the BP8 theory at a scattering angle of 90°, and spin asymmetries $A_{nn}$(b), $A_2$(c), and $A_1$(d) for elastic electron scattering from cesium atoms at an energy of 3 eV. The experimental results [7.129] are compared with various theoretical predictions [7.130–132]. In order to compare the shapes of the differential cross sections with the measurements, the CCC and Dirac8 results were multiplied by 0.82 and 1.12, respectively.

Using spin-polarized electron and atom beams, the Bielefeld group recently produced benchmark data for elastic e–Cs collisions at an energy of 3 eV for various spin asymmetries [7.129]. Figure 7.50 shows their results

---

[11]Owing to the lack of angle-differential experimental data for excitation processes, we concentrate on elastic scattering here

for $A_1$ together with those for the differential cross section $\sigma_0$ (for unpolarized beams) as well as the asymmetry function $A_2$ ($\equiv S_A$) and a generalized exchange asymmetry $A_{nn}$.[12] Comparison of the experimental data with theoretical predictions from a nonrelativistic CCC model [7.130] (only $\sigma_0$ and $A_{nn}$ are nonzero in this model), an eight-state Breit–Pauli R-matrix model (BP8 [7.131]), and an eight-state Dirac–Breit calculation (Dirac8 [7.132]) shows qualitative and sometimes quantitative agreement between experiment and theory. On the other hand, there remain significant discrepancies among the results from different the theoretical models and, particularly for the small asymmetry $A_1$, the measured data. Further experimental benchmark data are highly desirable and are expected to present significant challenges but also opportunities for further theoretical development toward a relativistic formulation of the convergent close-coupling approach.

Inelastic transitions in heavy alkalis, such as $(4s)^2S_{1/2} \rightarrow (4p)^2P^\circ_{1/2,3/2}$ in K, $(5s)^2S_{1/2} \rightarrow (5p)^2P^\circ_{1/2,3/2}$ in Rb, and $(6s)^2S_{1/2} \rightarrow (6p)^2P^\circ_{1/2,3/2}$ in Cs, would also be ideal candidates for future benchmark studies. In fact, results for the set $(L_\perp, \gamma, P_\ell, P)$ have now been published for K [7.133], and experiments are in progress for Rb in Brisbane [7.134] and are planned for Cs in Bielefeld [7.135] and Adelaide [7.136]. So far the experiments indicate that a nonrelativistic CCC theory without accounting for core excitation is sufficient to describe the principal features of these processes in K and Rb. Particularly for the Cs target, the general description outlined in this book should serve as a starting point for parameterizing the measured quantities, for example, in a further generalization of the Stokes parameters by accounting for both target and projectile polarizations in the initial state.

## 7.3 Conclusions

Although the case studies presented in this chapter only present a small amount of the work that has been carried out in this field, they can be used to assess the current level of understanding:

a) Dominant transitions in relatively simple systems, such as excitation of H(2p), Li(2p), Na(3p), and He($n = 2, 3$) from their respective ground states, are fairly well understood.

b) However, challenges remain for higher angular-momentum states, the heavy alkalis (K,Rb,Cs), nobles gases (Ne,Ar,Kr,Xe), and other quasi-two-electron systems like Hg.

---

[12]If the spin-orbit interaction is important, the exchange asymmetry not only depends on the relative orientation of the projectile and target polarizations, but also on their orientation with respect to the scattering plane. For $A_{nn}$, both initial polarizations are chosen perpendicular to the scattering plane.

c)  Angle-integrated measurements can provide a valuable alternative to double- or even triple-coincidence setups for testing theoretical models, particularly if advantage is taken of the increased signal rate to produce benchmark data of high accuracy.

d)  The visualization of the charge cloud shapes and currents using state-of-the-art technology is a powerful tool to enhance our understanding of the collision processes.

# References



7.1 K. Bartschat and K. Blum, Z. Phys. A **304** (1982) 85.

7.2 K. Bartschat, *Diploma Thesis*(1981), University of Münster, Germany.

7.3 J. Kessler, *Polarized Electrons* (2nd edition), Springer, Heidelberg 1985.

7.4 C. Norén, J.W. McConkey, P. Hammond, and K. Bartschat, Phys. Rev. A **53** (1996) 1559.

7.5 D.W.O. Heddle and J.W. Gallagher, Rev. Mod. Phys. **61** (1989) 221.

7.6 P. Hakel, R.C. Mancini, and G. Csanak, Bull. Am. Phys. Soc. **43** (1998) 1299.

7.7 S.A. Kazantsev and J.-C. Hénoux, *Plasma Polarization Spectroscopy of Ionized Gases*, Kluwer, Boston 1995.

7.8 A.S. Shlyaptseva, R.C. Mancini, P. Neill, P. Beiersdorfer, J.R. Crespo Lopéz-Urrutia, and K. Widmann, Phys. Rev. A **57** (1998) 888.

*7.9 I.C. Percival and M.J. Seaton, Phil. Trans. Roy. Soc. A **251** (1958) 113.

7.10 K. Bartschat and G. Csanak, Comm. At. Mol. Phys. **32** (1996) 233.

7.11 K. Bartschat in: *Selected Topics on Electron Physics*, D.M. Campbell and H. Kleinpoppen (eds.), Plenum, New York 1996.

7.12 K. Blum, *Density Matrix Theory and Applications* (2nd edition), Plenum, New York 1996.

7.13 A. Wolcke, K. Bartschat, K. Blum, H. Borgmann, G.F. Hanne, and J. Kessler, J. Phys. B **16** (1983) 639.

7.14 C. Norén and J.W. McConkey, Phys. Rev. A **53** (1996) 1559.

7.15 C. Norén, W.L. Karras, J.W. McConkey, and P. Hammond, Phys. Rev. A **54** (1996) 510.

7.16 V. Zeman, K. Bartschat, C. Norén, and J.W. McConkey, Phys. Rev. A **58** (1998) 1275.

7.17 P.S. Farago and J.S. Wykes, J. Phys. B **2** (1969) 747.

7.18 J.S. Wykes, J. Phys. B **4** (1971) L91.

7.19 M. Eminyan and G. Lampel, Phys. Rev. Lett. **45** (1980) 1171.

7.20 T.J. Gay, J. Phys. B **16** (1983) L553.

7.21 M. Uhrig, A. Beck, J. Goeke, F. Eschen, M. Sohn, G.F. Hanne, K. Jost, and J. Kessler, Rev. Sci. Instrum. **60** (1988) 872.

7.22 T.J. Gay, J.E. Furst, K.W. Trantham, and W.M.K.P. Wijayaratna, Phys. Rev. A **53** (1996) 1623.

7.23 K. Bartschat, G.F. Hanne, and J. Kessler, Z. Phys. A **304** (1982) 89.

7.24 K. Bartschat, N.S. Scott, K. Blum, and P.G. Burke, J. Phys. B **17** (1984) 269.

7.25 J.E. Furst, T.J. Gay, W.M.K.P. Wijayaratna, K. Bartschat, H. Geesmann, M.A. Khakoo, and D.H. Madison, J. Phys. B **25** (1992) 1089.

7.26 D.H. Yu, P.A. Hayes, J.F. Williams, and J.E. Furst, J. Phys. B **30** (1997) 1799.

7.27 D.H. Yu, P.A. Hayes, J.E. Furst, and J.F. Williams, Phys. Rev. Lett. **78** (1997) 2724.

7.28 B.G. Birdsey, H.M. Al-Khateeb, M.E. Johnson, T.C. Bowen, T.J. Gay, V. Zeman, and K. Bartschat, Phys. Rev. A **60** (1999) 1046.

7.29 V. Zeman and K. Bartschat, J. Phys. B **30** (1997) 4609.

7.30 M.J. Brunger, S.J. Buckman, P.J.O. Teubner, V. Zeman, and K. Bartschat, J. Phys. B **31** (1998) L387.

7.31 V. Zeman, K. Bartschat, T.J. Gay, and K.W. Trantham, Phys. Rev. Lett. **79** (1997) 1825.

7.32 K.W. Trantham, T.J. Gay, and R.J. Vandiver, Rev. Sci. Instrum. **67** (1996) 4103.

7.33 S.J. Buckman and C.W. Clark, Rev. Mod. Phys. **66** (1994) 539.

7.34 P.A. Hayes, D.H. Yu, J.E. Furst, M. Donath, and J.F. Williams, J. Phys. B **29** (1996) 3989.

7.35 K.W. Trantham, M.E. Johnston, and T.J. Gay, Bull. Am. Phys. Soc. **41** (1996) 1135.

7.36 M. Eminyan, K.B. MacAdam, J. Slevin, M.C. Standage, and H. Kleinpoppen, J. Phys. B **8** (1975) 2058.

7.37 N. Andersen, I.V. Hertel, and H. Kleinpoppen, J. Phys. B **17** (1984) L901.

7.38 M.T. Hollywood, A. Crowe, and J.F. Williams, J. Phys. B **12** (1979) 819.

7.39 K. Bartschat, E.T. Hudson, M.P. Scott, P.G. Burke, and V.M. Burke, J. Phys. B **29** (1996) 2875.

7.40 D.V. Fursa and I. Bray, Phys. Rev. A **52** (1995) 1279.

7.41 J.P. Beijers, D.H. Madison, J. van Eck, and H.D.M. Heideman, J. Phys. B **20** (1987) 167.

7.42 R.I. Hall, G. Joyez, Y. Mazeau, J. Reinhard, and C. Schermann, J. Physique **34** (1973) 827.

7.43 D.C. Cartwright, G. Csanak, S. Trajmar, and D.F. Register, Phys. Rev. A **45** (1992) 1602.

7.44 R. McAdams R, M.T. Hollywood, A. Crowe A, and J.F. Williams, J. Phys. B **13** (1980) 3691.

*7.45 M. Eminyan, K.B. MacAdam, J. Slevin, and H. Kleinpoppen, J. Phys. B **7** (1974) 1519.

7.46 M.A. Khakoo, K. Becker, J.L. Forand, and J.W. McConkey, J. Phys. B **19** (1986) L209.

7.47 D.H. Madison and W.N. Shelton, Phys. Rev. A **7** (1973) 499.

7.48 D.C. Cartwright and G. Csanak, Phys. Rev. A **38** (1988) 2740.

7.49 B.P. Donnelly, P.A. Neill, and A. Crowe, J. Phys. B **21** (1988) L321.

7.50 D.V. Fursa and I. Bray, J. Phys. B **30** (1997) 757.

7.51 W.C. Fon, K.P. Lim, K.A. Berrington, and T.G. Lee, J. Phys. B **28** (1995) 1569.

7.52 W.C. Fon, K.A. Berrington, and A.E. Kingston, J. Phys. B **24** (1991) 2161.

7.53 G. Nienhuis, in *Coherence and Correlation in Atomic Collisions*, H. Kleinpoppen and J.F. Williams (eds.), Plenum, New York 1980.

7.54 N. Andersen, T. Andersen, J.S. Dahler, S.E. Nielsen, G. Nienhuis, and K. Refsgaard, J. Phys. B **16** (1983) 817.

7.55 N. Andersen and K. Bartschat, Adv. At. Mol. Opt. Phys. **36** (1996) 1.

7.56 N. Andersen and K. Bartschat, J. Phys. B **30** (1997) 5071.

7.57 D.T. McLaughlin, B.P. Donnelly, and A. Crowe, Z. Phys. D **29** (1994) 259.

7.58 D.T. McLaughlin, B.P. Donnelly, and A. Crowe, Phys. Rev. A **49** (1994) 2545.

7.59 A.G. Mikosza, R. Hippler, J.B. Wang, and J.F. Williams, Z. Phys. D **30** (1994) 129.

7.60 A.G. Mikosza, R. Hippler, J.B. Wang, and J.F. Williams, Phys. Rev. Lett. **71** (1993) 235.

7.61 A.G. Mikosza, R. Hippler, J.B. Wang, and J.F. Williams, Phys. Rev. A **53** (1996) 3287.
7.62 A.G. Mikosza, J.F. Williams, and J.B. Wang, Phys. Rev. Lett. **79** (1997) 3375.
7.63 N. Andersen, J.W. Gallagher, and I.V. Hertel, Phys. Rep. **165** (1988) 1.
7.64 H.B. van Linden van den Heuvell, G. Nienhuis, J. van Eck, and H.G.M. Heideman, J. Phys. B **14** (1981) 2667.
7.65 K. Bartschat, J. Phys. B **32** (1999) L355.
7.66 A. Crowe, B.P. Donnelly, D.T. McLaughlin, I. Bray, and D.V. Fursa, J. Phys. B **27** (1994) L795.
7.67 D.V. Fursa, I. Bray, B.P. Donnelly, D.T. McLaughlin, and A. Crowe, J. Phys. B **30** (1997) 3459.
7.68 D.V. Fursa, I. Bray, B.P. Donnelly, D.T. McLaughlin, and A. Crowe, Phys. Rev. A **56** (1997) 4606.
7.69 H.P. Neitzke and T. Andersen, J. Phys. B **17** (1984) 1559.
7.70 J.B. Wang and J.F. Williams, Comp. Phys. Commun. **75** (1993) 275.
7.71 J.B. Wang, J.F. Williams, A.T. Stelbovics, J.E. Furst, and D.H. Madison, Phys. Rev. A **52** (1995) 2885.
7.72 D. Cvejanović and A. Crowe, Phys. Rev. Lett. **80** (1998) 3033.
7.73 I.V. Hertel, M.H. Kelley, and J.J. McClelland, Z. Phys. D **6** (1987) 163.
7.74 E. Weigold, L. Frost, and K.J. Nygaard, Phys. Rev. A **21** (1980) 1950.
7.75 J.F. Williams, J. Phys. B **14** (1981) 1997.
7.76 H.A. Yalim, D. Cvejanović, and A. Crowe, Phys. Rev. Lett. **79** (1997) 2951.
7.77 H.A. Yalim, D. Cvejanović, and A. Crowe, J. Phys. B **32** (1999) 3437.
7.78 J.F. Williams and A.G. Mikosza (1999), private communication.
7.79 R.W. O'Neill, P.J.M. van der Burgt, D. Dziczek, P. Bowe, S. Chwirot, and J.A. Slevin, Phys. Rev. Lett. **80** (1998) 1630.
7.80 J.F. Williams, Austr. J. Phys. **39** (1986) 621.
7.81 D.H. Madison, I. Bray, and I.E. McCarthy, J. Phys. B **24** (1991) 3861.
7.82 I. Bray, D.A. Konovalov, and I.E. McCarthy, Phys. Rev. A **44** (1991) 5586.
7.83 I. Bray and A.T. Stelbovics, Phys. Rev. A **46** (1992) 6995.
7.84 V. Karaganov, I. Bray, and P.J.O. Teubner, J. Phys. B **31** (1998) L187.
7.85 N. Andersen, K. Bartschat, J.T. Broad, and I.V. Hertel, Phys. Rep. **279** (1997) 251.
7.86 J.J. McClelland, M.H. Kelley, and R.J. Celotta, Phys. Rev. A **40** (1989) 2321.
7.87 P.J.O. Teubner and R.E. Scholten, J. Phys. B **25** (1992) L301.
7.88 J.J. McClelland, M.H. Kelley, and R.J. Celotta, Phys. Rev. Lett. **55** (1985) 688.
7.89 R.E. Scholten, S.R. Lorentz, J.J. McClelland, M.H. Kelley, and R.J. Celotta, J. Phys. B **24** (1991) L653.
7.90 I. Bray, Phys. Rev. Lett. **69** (1992) 1908.
7.91 T. Hegemann, M. Oberste-Vorth, R. Vogts, and G.F. Hanne, Phys. Rev. Lett. **66** (1991) 2968.
7.92 T. Hegemann, S. Schroll, and G.F. Hanne, J. Phys. B **26** (1993) 4607.
7.93 I. Bray, Phys. Rev. A **49** (1994) 1066.
7.94 D.H. Madison, K. Bartschat, and R.P. McEachran, J. Phys. B **25** (1992) 5199.
7.95 S.K. Srivastava and L. Vuskovic, J. Phys. B **13** (1980) 2633.
7.96 V. Nickich, T. Hegemann, M. Bartsch, and G.F. Hanne, Z. Phys. D **16** (1990) 261.
7.97 V.V. Balashov and A.N. Grum-Grzhimailo, Z. Phys. D **23** (1991) 127.
7.98 M. Shurgalin, A.J. Murray, W.R. MacGillivray, and M.C. Standage, J. Phys. B **31** (1998) 4205.
7.99 Z. Shi, C.H. Ying, and L. Vusković, Phys. Rev. A **54** (1996) 480.
7.100 I. Bray, D.V. Fursa, and I.E. McCarthy, Phys. Rev. A **49** (1994) 2667.

7.101 W.K. Trail, M.A. Morrison, H.L. Zhou, M.A. Morrison, K. Bartschat, K.B. MacAdam, T.L. Goforth, and D.W. Norcross, Phys. Rev. A **49** (1994) 3620.

7.102 D. Farrell, S. Chwirot, R. Srivastava, and J.A. Slevin, J. Phys. B **23** (1990) 315.

7.103 M. Kumar, A.T. Stelbovics, and J.F. Williams, J. Phys. B **26** (1993) 2165.

7.104 I. Bray and A.T. Stelbovics, J. Phys. B **30** (1997) L493.

7.105 K. Becker, A. Crowe, and J.W. McConkey, J. Phys. B **25** (1992) 3885.

7.106 J.J. Corr, P. Plessis, and J.W. McConkey, Phys. Rev. A **42** (1990) 5240.

7.107 K. Bartschat and D.H. Madison, J. Phys. B **20** (1987) 5839.

7.108 T. Zuo, R.P. McEachran, and A.D. Stauffer, J. Phys. B **25** (1992) 3393.

7.109 M. Dümmler, G.F. Hanne, and J. Kessler, J. Phys. B **28** (1995) 2985.

7.110 K. Bartschat, in *Atomic Physics 16*, G.W.F. Drake (ed.), American Institute of Physics (1999), 254

7.111 M. Uhrig, G.F. Hanne, and J. Kessler, J. Phys. B **27** (1994) 4007.

7.112 A.W. Baerveldt, W.B. Westerveld, J. van Eck, and H.G.M. Heidemann, Can. J. Phys. **74** (1996) 897.

7.113 A.W. Baerveldt, W.B. Westerveld, J. van Eck, H.G.M. Heidemann, G.D. Meneses, G. Csanak, R.E.H. Clark, and J. Abdallah Jr, J. Phys. B **31** (1998) 573.

7.114 A. Raeker, K. Blum, and K. Bartschat, J. Phys. B **26** (1993) 1491.

7.115 N. Andersen and K. Bartschat, J. Phys. B **27** (1994) 3189; corrigendum, J. Phys. B **29** (1996) 1149.

7.116 J. Goeke, G.F. Hanne, and J. Kessler, J. Phys. B **22** (1989) 1075.

7.117 M. Sohn and G.F. Hanne, J. Phys. B **25** (1992) 4627.

7.118 N.S. Scott, P.G. Burke, and K. Bartschat, J. Phys. B **16** (1983) L361.

7.119 N. Andersen, K. Bartschat, J.T. Broad, G.F. Hanne, and M. Uhrig, Phys. Rev. Lett. **76** (1996) 208.

7.120 N. Andersen, K. Bartschat, and G.F. Hanne, J. Phys. B **28** (1995) L29.

7.121 H. Borgmann, J. Goeke, G.F. Hanne, J. Kessler, and A. Wolcke, J. Phys. B **20** (1987) 1619.

7.122 K. Bartschat and D.H. Madison, J. Phys. B **21** (1988) 2621.

7.123 M. Dümmler, M. Bartsch, H. Geesmann, G.F. Hanne, and J. Kessler, J. Phys. B **23** (1990) 3407.

7.124 H. Müller and J. Kessler, J. Phys. B **27** (1994) 5933; corrigendum, J. Phys. B **28** (1995) 911.

7.125 M. Klose, *PhD Thesis* (1995), University of Münster, Germany.

7.126 R. Srivastava, T. Zuo, R.P. McEachran and A.D. Stauffer, J. Phys. B **25** (1992) 2409.

7.127 P.G. Burke and J.F.B. Mitchell, J. Phys. B **7** (1974) 214.

7.128 P.S. Farago, J. Phys. B **7** (1974) L28.

7.129 G. Baum, W. Raith, B. Roth, M. Tondera, K. Bartschat, I. Bray, S. Ait-Tahar, I.P. Grant, and P.H. Norrington, Phys. Rev. Lett. **83** (1999) 1128.

7.130 K. Bartschat and I. Bray, Phys. Rev. A **54** (1996) 1723.

7.131 K. Bartschat, J. Phys. B **26** (1993) 3695.

7.132 S. Ait-Tahar, I.P. Grant, and P.H. Norrington, Phys. Rev. Lett. **79** (1997) 2955.

7.133 K.A. Stockman, V. Karaganov, I. Bray, and P.J.O. Teubner, J. Phys. B **32** (1999) 3003.

7.134 M.C. Standage (1999), private communication.

7.135 G. Baum (1999), private communication.

7.136 P.J.O. Teubner (1999), private communication.

# 8. Ion- and Atom-Impact Excitation

The first case studies on ion- and atom-impact excitation are examples of scattering angle-differential S $\rightarrow$ P, D excitation and electron transfer studies in which radiation from the final state is polarization analyzed. This is followed in Section 8.2 by a discussion of angle-integrated initial-state and final-state alignment studies using targets in optically prepared P states, and by a description of angle-differential studies using optically prepared targets in Section 8.3. The closing section 8.4 addresses P $\rightarrow$ P electron transfer processes in which radiation from the final state is also polarization analyzed. A comprehensive review of results up to 1996 can be found in [8.1].

## 8.1. Angle-Differential S $\rightarrow$ P, D Excitation and Transfer

From a fundamental viewpoint, one would have liked to start this section by showing systematic, scattering angle-differential results for collision systems such as $H^+ - H(1s)$ or $He^{2+} - H(1s)$ in which the active electron is excited or transferred to the 2p state, with a full Stokes-parameter analysis of the subsequently emitted photon pattern. However, no such data exist, owing to a long series of very severe experimental difficulties related to such an endeavor. Furthermore, no such results can be expected within the foreseeable future, so for these systems one has to rely on purely theoretical data [8.2]. However, extensive experimental data are available for a series of other cases, including simple, so-called quasi-one-electron systems that are characterized by the presence of an isolated, loosely bound electron outside two closed shells.

### 8.1.1. S $\rightarrow$ P, D excitation in $Mg^+-$ and Li$-$rare-gas systems

Excitation processes in the $Mg^+-$rare-gas systems have been some of the most extensively investigated cases, experimentally as well as theoretically, at the level of first-, second-, and third-generation studies. Together with systems such as $Be^+-$rare-gas, and the iso-electronic neutral systems Li, Na, K$-$rare-gas, they belong to the group of quasi-one-electron systems,

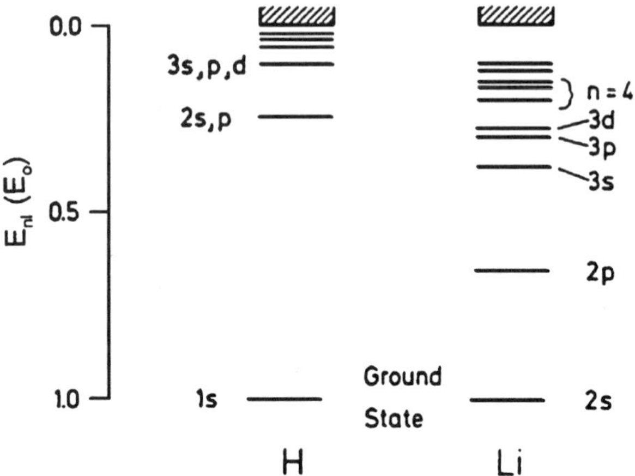


**Fig. 8.1.** Comparison of hydrogenic energy levels with those of a typical alkali-like atom, Li. Energies are measured in units of the ground state energy.

consisting of a single valence electron loosely bound to an ionic core that is colliding with a rare gas. Such cases turn out to be simpler in many respects than genuine one-electron systems involving H or $He^+$. One key reason for this may be gleaned from inspection of Figure 8.1, which shows binding energies of some excited levels in H compared to a typical alkali atom, Li, in units of the ground state energy $E_0$. The graph illustrates that for hydrogenic and hydrogen-like ions, the 1s ground state is strongly bound compared to the excited states, which are situated in a relatively narrow band below the continuum and, furthermore, are degenerate with respect to the orbital angular-momentum quantum number. This structure may be contrasted to alkali-like atoms, where the first excited level, $n$p, is well separated from both the $n$s ground state and the higher excited levels, and in particular the continuum. One would thus conclude that for excitation in the quasi-one-electron systems the most prominent inelastic process is the $n$s − $n$p resonance transition and that, theoretically, just a few states will form a sufficient basis for a fair description of the process. Furthermore, experimentally, the corresponding prominent spectral lines are situated in or near the visible range where determination of absolute cross sections, polarizations, etc., is possible with high efficiency and precision.

Studies at the level of first- and second-generation experiments have established the picture illustrated in Figure 8.2 [8.3]. A summary of the early work may be found in [8.4], while the systematics of alignment effects for all quasi-one-electron systems studied at the level of second-generation experiments were presented in [8.5]. The graph illustrates the relatively large extension of the loosely bound valence electron centered on the small ionic

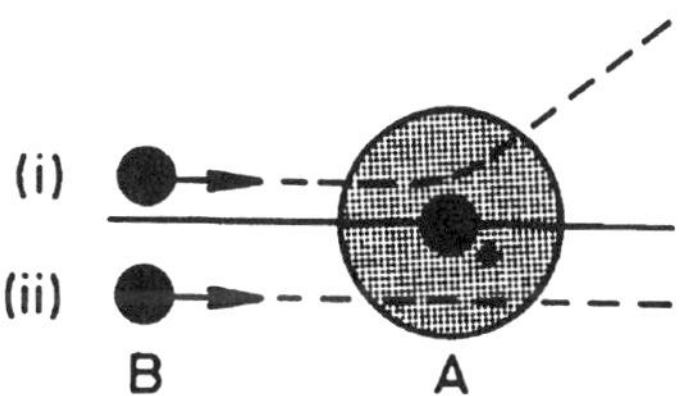

**Fig. 8.2.** Schematic diagram showing a collision between an alkali-like atom with a large valence electron loosely bound to a small ionic core A and a rare-gas atom B. Two types of collisions are indicated for ($i$) small impact parameters, where the two cores interact strongly, leading to a large scattering angle, and ($ii$) large impact parameters, where the important interaction is between B and the alkali valence electron only, leading to a very small scattering angle.

core A (at rest for convenience), colliding with a rare-gas atom B, also of relatively small extension. The active excitation mechanisms may be classified according to whether the impact parameter is smaller than [case ($i$)] or larger than [case ($ii$)] the sum of the radii of the cores A and B. The corresponding excitation mechanisms ($i$) and ($ii$) are of very different complexity:

($i$)   For *violent*, small impact-parameter collisions, the two cores interpenetrate significantly, and the corresponding scattering angles are relatively large. The excitation takes place at well-localized quasi-molecular curve crossings, following the general correlation rules outlined by Barat and Lichten [8.6]. Thus the rare-gas electrons play an active role during the formation and break-up of the quasi-molecule, and the outcome of a collision will depend primarily on the impact parameter.

($ii$)   For *soft*, large impact-parameter collisions, corresponding to small deflection angles, the two cores will overlap insignificantly. Here, the rare-gas electrons remain relatively unperturbed during the collision and excitation is mainly induced by the direct interaction between the valence electron and the rare-gas atom B. In this case, the excitation of the valence electron is induced by a nonlocalized interaction. We expect the excitation to depend primarily on the duration and strength of this interaction and only weakly on the impact parameter.

Several experimental and theoretical studies have addressed the exploration of this picture at the level of third-generation experiments. Here we shall restrict the presentation to a few selected, typical results for low and high energies, beginning with — somewhat arbitrarily — the cases $Mg^+ - He$ and $Mg^+ - Ar$, and introduce the new concepts that emerge from these studies.

Results from planar scattering experiments are shown in Figure 8.3 for (a) $Mg^+ - He$ at $b \simeq 1.0\,a_0$, and (b) $Mg^+ - Ar$ at $b \simeq 1.8\,a_0$, as a function of energy [8.7]. Striking about the alignment and orientation parameters is

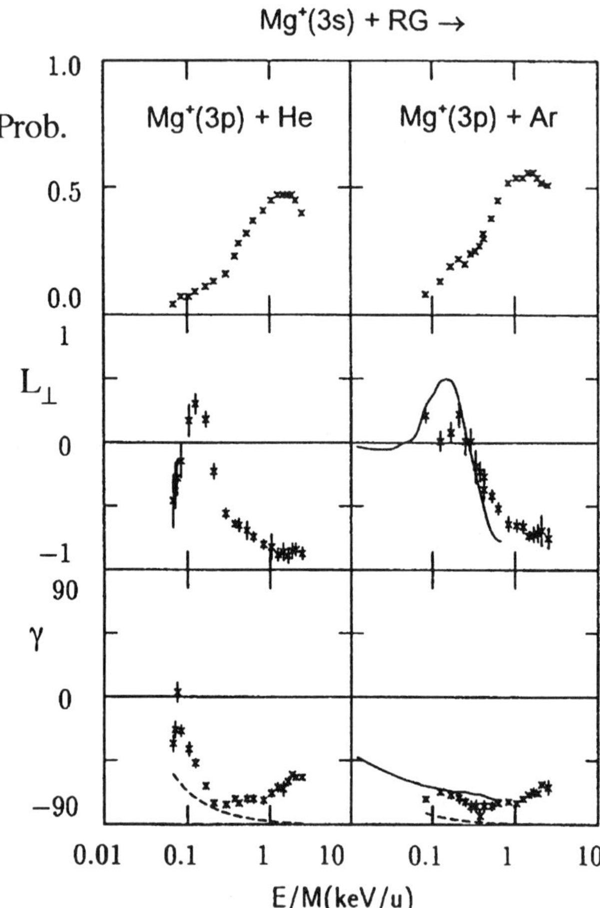



**Fig. 8.3.** $Mg^+(3s \rightarrow 3p)$ excitation probability, and orientation and alignment parameters $(L_\perp, \gamma)$ versus collision energy for (a) $Mg^+$–He collisions with $b \simeq 1.0\,a_0$ and (b) $Mg^+$–Ar collisions with $b \simeq 1.8\,a_0$. The experimental results are from [8.7]. The dashed line for $\gamma$ shows the direction perpendicular to the internuclear axis, $\theta_{cm} + 90°$.

*the strong propensity to populate the* $p_{-1}$ *state near the excitation maximum,* where in both cases $L_\perp \simeq -1$. This is a common feature observed for the direct excitation process in the impact-energy region of maximum excitation. At energies below the maximum region, this propensity is no longer seen. For $Mg^+$ – He the orientation even changes sign, with a $p_{+1} : p_{-1}$ population ratio of 2:1 near $0.15\,keV/u$ (atomic mass unit), and switching once more to predominant $p_{-1}$ population at the lowest energies measured. For both the He and Ar targets, the alignment angle stays in the neighborhood of $-70°$ in most of the energy region studied, not far from the perpendicular direction, which is the FBA high-energy limit.

A derivation of the propensity rule for the orientation in the region of maximum excitation was given by Andersen and Nielsen [8.8], actually before the experimental verification. We shall now analyze the collision dynamics in more detail, following their line of thought. Expressing the state $|\psi\rangle$ in terms of the three states involved,

$$|\psi\rangle = a_{\mathrm{s}}|\mathrm{s}\rangle + a_{-1}|\mathrm{p}_{-1}\rangle + a_{+1}|\mathrm{p}_{+1}\rangle , \tag{8.1}$$

the time-dependent Schrödinger equation may be cast in the form

$$i\,v\,\hbar\,\frac{d}{dx}\begin{pmatrix} a_{\mathrm{s}} \\ a_{-1} \\ a_{+1} \end{pmatrix} = \mathbf{A}\begin{pmatrix} a_{\mathrm{s}} \\ a_{-1} \\ a_{+1} \end{pmatrix} , \tag{8.2}$$

where the matrix $\mathbf{A}$ has the very simple structure

$$\mathbf{A} = F_{\mathrm{sp}}\begin{pmatrix} 0 & c.c. & c.c. \\ e^{\mathrm{i}(\frac{\Delta E\,x}{\hbar v} - \phi)} & 0 & 0 \\ -e^{\mathrm{i}(\frac{\Delta E\,x}{\hbar v} + \phi)} & 0 & 0 \end{pmatrix} + F_{\mathrm{pp}}\begin{pmatrix} 0 & 0 & 0 \\ 0 & 0 & e^{-\mathrm{i}2\phi} \\ 0 & e^{\mathrm{i}2\phi} & 0 \end{pmatrix} , \tag{8.3}$$

with

$$F_{\mathrm{sp}} = F_{\mathrm{sp}}(R) = \langle \mathrm{s}|V|\mathrm{p}_{-1}\rangle = -\langle \mathrm{s}|V|\mathrm{p}_{+1}\rangle \tag{8.4}$$

and

$$F_{\mathrm{pp}} = F_{\mathrm{pp}}(R) = \langle \mathrm{p}_{+1}|V|\mathrm{p}_{-1}\rangle. \tag{8.5}$$

Here $V$ is the interaction between the active electron and particle B and $\Delta E = \Delta E(R(x))$ is the S$-$P energy difference. Thus, the first part of $\mathbf{A}$ governs the flow of probability between the ground state and the excited state, while the second part determines the flow between the two P states. The phases entering the exponentials are easily identified: The $\Delta E\,x/\hbar v$ term is the phase difference that develops in time between two states with an energy difference $\Delta E$, while the $\pm\phi$ term is due to the transformation properties of the spherical harmonics $Y_{1\pm1}(\theta,\phi)$ under rotation by an angle $\phi$ around the $z$ axis. From this, one may easily obtain the following first-order perturbation expressions for the two excited-state probability amplitudes after the collision, using $(a_{\mathrm{s}}, a_{-1}, a_{+1}) = (1, 0, 0)$ as initial conditions:

$$\begin{cases} a_{-1} = \frac{1}{v}\int_{-\infty}^{\infty} F_{\mathrm{sp}}\,e^{\mathrm{i}(\frac{\Delta E\,x}{\hbar v} - \phi)}\,dx \\[2mm] a_{+1} = -\frac{1}{v}\int_{-\infty}^{\infty} F_{\mathrm{sp}}\,e^{\mathrm{i}(\frac{\Delta E\,x}{\hbar v} + \phi)}\,dx. \end{cases} \tag{8.6}$$

We now explore under which conditions these amplitudes may become large. The function $F_{\mathrm{sp}}$ generally is a smooth, bell-shaped function with maximum at $x = 0$. Thus, the size of $a_{-1}$ is mainly determined by the behavior of the phase of the exponential function. This phase consists of two parts. The latter one, the rotation angle $\phi$ of the internuclear axis, changes by about $\pi$ during a full collision. The variation of the first one depends on the size of the velocity $v$. If $v$ is small, this phase term may accumulate many multiples of $\pi$ during the collision, leading to important cancellation in the integral and a correspondingly small $a_{-1}$ value. On the other hand, if $v$ is very large, this phase term stays small. However, if $v$ is adjusted so that this term also increases by $\pi$ during the collision, the two terms may effectively cancel, leading to a stationary phase and thereby yielding the maximum value of $a_{-1}$. Furthermore, at this velocity, the phase term of the $a_{+1}$ integral will accordingly vary by about $2\pi$ during the collision, causing this amplitude to become very small.

In summary, in a velocity range near $v_{\mathrm{max}}$ determined by the criterion

$$\frac{\Delta\epsilon\, a}{\hbar\, v_{\mathrm{max}}} \simeq \pi\,, \tag{8.7}$$

the S $\rightarrow$ P excitation probability has its maximum. Here, $a$ is the effective interaction length of the collision, and $\Delta\epsilon$ is the effective energy defect, in most cases close to its asymptotic value. Thus, the propensity for orientation is seen when the phase build-up due to the S$-$P energy difference matches the rotation angle of the internuclear axis. The overwhelming part of the probability goes to the $p_{-1}$ state, causing an orientation parameter $L_\perp \simeq -1$. The velocity range for which the above argument applies is $\frac{1}{2} < v/v_{\mathrm{max}} < 2$, and the most favorable impact parameter may be estimated by $b \simeq a/\pi$. From the above formulas, the subsequent oscillations in $L_\perp$ for increasing collision time are also readily understood. Finally, for low energies, truly in the quasi-molecular energy region, the phases vary so rapidly that no prediction of $L_\perp$ is possible from this argument, while at the high-energy limit the Born approximation value $L_\perp = 0$ is approached.

Note the importance of $\Delta\epsilon > 0$. If $\Delta\epsilon < 0$, i.e., in a *de*-excitation process, the role of $p_{+1}$ and $p_{-1}$ will be reversed, and $L_\perp \simeq +1$ near the maximum. Further interesting consequences are discussed in [8.9].

The rule is readily generalized to transitions other than S $\rightarrow$ P. In the general form, the stationary-phase argument above may be stated as [8.10]

$$\frac{\Delta\epsilon\, a}{\hbar\, v_{\mathrm{max}}} + \Delta m \cdot \pi = 0\,, \tag{8.8}$$

where $\Delta m = m_{\mathrm{final}} - m_{\mathrm{initial}}$ is the difference between the magnetic quantum numbers of final and initial states. Figure 8.4 shows excitation probabilities and orientation and alignment parameters $L_\perp$ and $\gamma$ for 2s $\rightarrow$ 2p [8.11] and 2s $\rightarrow$ 3d [8.12] excitation in the Li-He system. The Li(2p) results provide another nice example of the propensity rule for orientation presented above, with $L_\perp$ being very close to $-1.0$ in the region of the maximum excitation

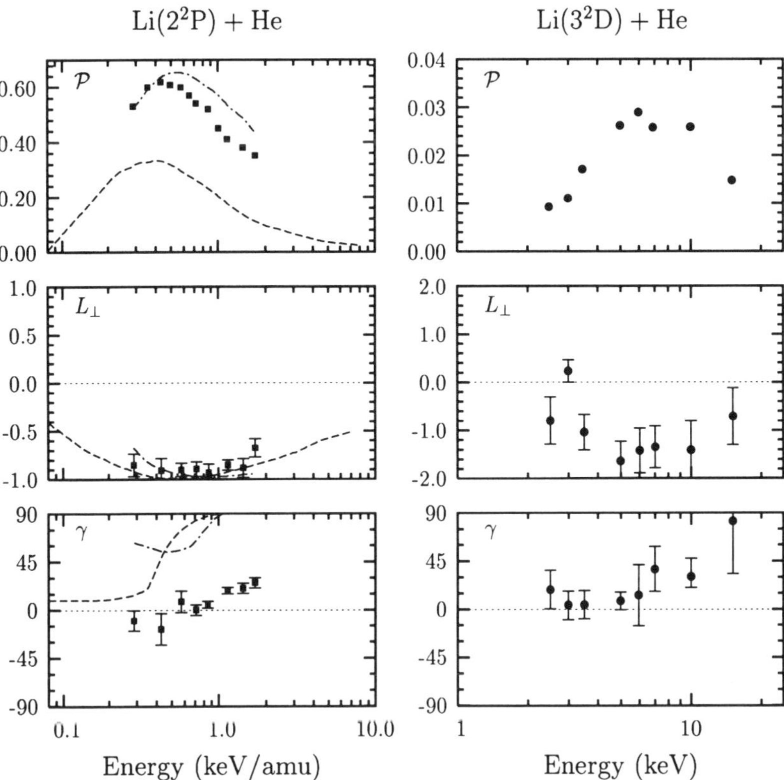



**Fig. 8.4.** Li (2s–$n\ell$) excitation probability, and orientation and alignment parameters versus collision energy for (a) Li (2p) excitation with $b \simeq 1.25a_0$. Experimental data [8.11] are compared with 9-state AO predictions [8.10] and MO results including ETFs [8.13]; (b) Li (3d) excitation with $b \simeq 0.95\ a_0$ [8.11].

probability. The experimental results are compared with theoretical close-coupling predictions using an AO basis [8.10] and an MO basis [8.13] including electron translation factors. The theoretical results for the alignment angle exhibit the correct trend, but the actual values predicted differ dramatically from the experimental ones, thereby illustrating the difficulties in a precise estimate of the alignment angle for a nearly circular state.

The data for Li (3d) excitation also display the predicted orientation propensity near the excitation maximum. (Note the dramatic decrease in probability when going from the 2p state to the 3d state.) In terms of the set of nonzero scattering amplitudes $(f_{-2}, f_0, f_2)$, we read from the figure

$$L_\perp = 2(|f_2|^2 - |f_{-2}|^2) \simeq -1.4 \,. \tag{8.9}$$

Furthermore, since [8.12]

$$\rho_{00} = |f_0|^2 \simeq 0.3 \,, \tag{8.10a}$$

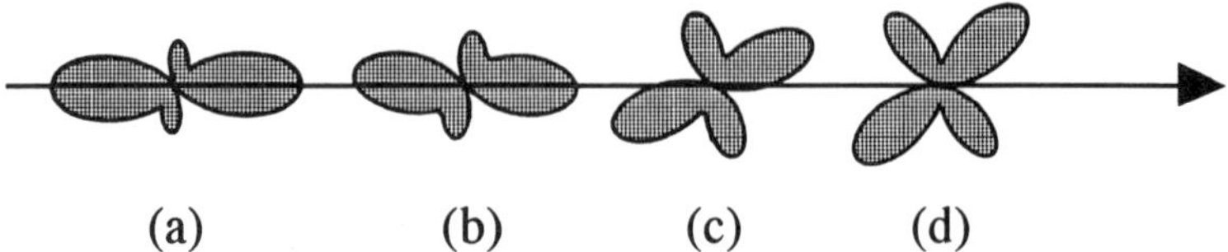

(a)          (b)          (c)          (d)

**Fig. 8.5.** Cut in the collision plane of the shape of the Li (3d) charge cloud produced in Li (2s)–He collisions as function of impact energy $E$ =3.5 (a), 5 (b), 10 (c), and 15 keV(d) for an impact parameter $b \simeq 0.95\, a_0$ [8.14].

this means that

$$|f_{-2}|^2 \simeq 0.7 \tag{8.10b}$$

and

$$|f_2|^2 \simeq 0.0 , \tag{8.10c}$$

so that $|d_{-2}\rangle$ is indeed the preferentially populated state near the maximum.

As discussed in Section 7.2, the complete set of four Stokes parameters characterizing the D $\rightarrow$ P decay is not sufficient to completely characterize the D state. In general, two solutions are consistent with the observed set of Stokes parameters. However, the two Li (3d) states will develop differently in an external magnetic field, and one may thus distinguish between the two cases by an angle-differential level-crossing experiment. In this way, the missing sign of the angle $\hat{\eta}$ may be determined, as described and achieved for this case in [8.14]. At the time of writing, this remains the only complete experiment for S $\rightarrow$ D excitation by heavy-particle impact. Based on the results, one may reconstruct the shape of the Li (3d) charge cloud as function of impact energy, as illustrated in Figure 8.5.

There have been no further experiments on higher-rank alignment or orientation in this system. A single theoretical study exists of excitation probabilities and alignment and orientation parameters for the processes Li (2p$_0$,3p$_0$)+ He $\rightarrow$ Li (3d$_{\pm 1}$)+ He, i.e., for transitions among states with negative reflection symmetry in the scattering plane [8.15].

In conclusion, the active excitation mechanisms in the quasi-one-electron systems and the general behavior of the alignment and orientation parameters are now well understood, in particular the orientation propensities observed near the excitation maximum at higher energies. A similar understanding of the behavior in the molecular regime is not yet available for these, here relatively complex, collision systems. Based on the qualitative discussion related to Figure 8.3 this will require a major theoretical effort, which may or may not reveal recognizable general trends in the low velocity/small impact parameter region.

## 8.1.2.  S → P transfer excitation in $B^{3+}$–He, Ne collisions

Following the discussion in the previous section, it appears natural to proceed to a case where the excitation takes place at a well-defined potential energy curve-crossing, i.e., $\Delta E(R) = 0$, and with as little interference from additional mechanisms as possible. An early discussion of such a geometry was given by Russek *et al.* [8.16]. An appropriate choice is the $B^{3+}$–He system for which the molecular curves of importance have the very simple structure shown in Figure 8.6.

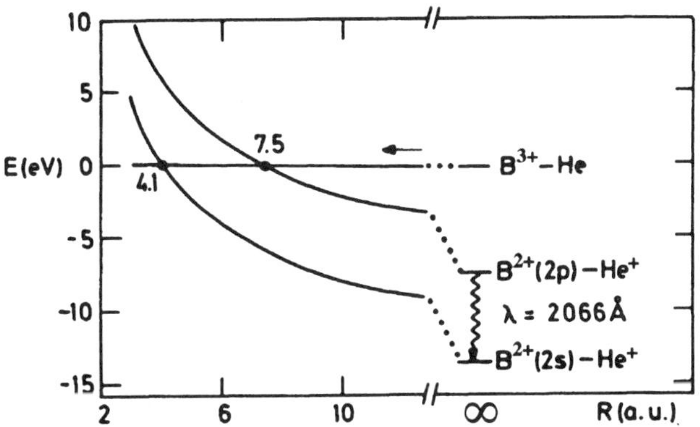


**Fig. 8.6.** Molecular potential energy curves for the $B^{3+}$–He collision. Only two curves cross the incoming channel, with the Coulomb curves leading to charge exchange into $B^{2+}(2s)$ and $B^{2+}(2p)$, respectively.

The essentially flat incoming $B^{3+}$–He channel crosses the two Coulomb curves leading to electron transfer into the $B^{2+}(2s)$–$He^+$ and $B^{2+}(2p)$–$He^+$ channels at internuclear distances of approximately 4.1 and $7.5\,a_0$, respectively. No other channels interfere. The $B^{2+}(2p)$ level decays to the ground state by emission of a 2066 Å photon, i.e., a wavelength for which linear and circular polarization analysis is readily performed with standard techniques.

The first report on alignment and orientation for this process was a theoretical analysis [8.17] predicting strong orientation effects, with a sense of orientation following the rotation direction of the internuclear axis for the main, forward peak of the charge exchange cross section. The paper also illustrated the evolution of the electron charge cloud along the trajectory, and in particular the build-up of the $B^{2+}(2p)$ component, as shown in Figure 8.7. We see population of a 2p orbital, initially aligned in the direction of the He target, but later staying almost space-fixed until, on the way out, additional population builds up to create a 2p charge cloud with a large (negative) orientation and an alignment perpendicular to the trajectory.

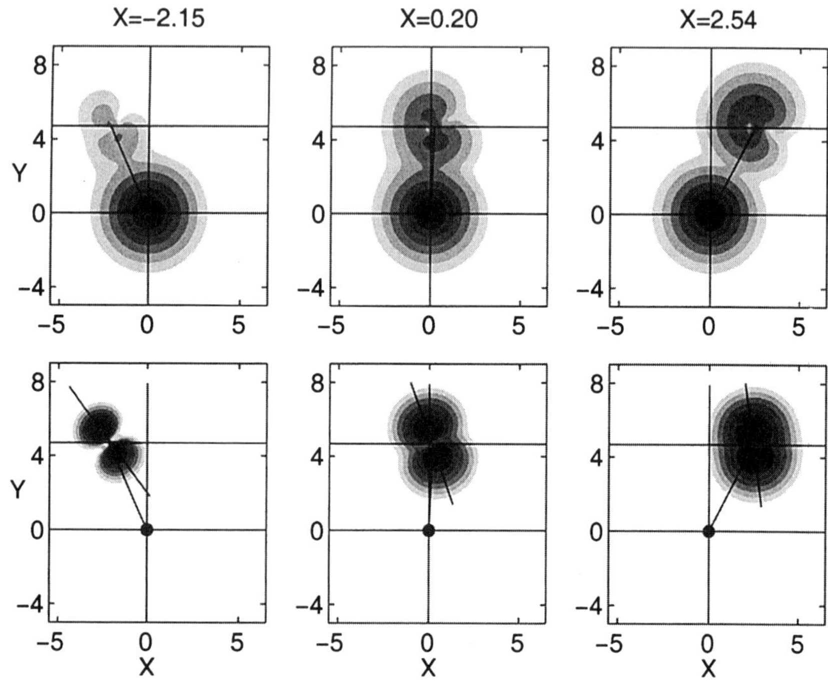



**Fig. 8.7.** The upper panel shows $B^{3+}$–He electron charge cloud densities in the $xy$ collision plane at three different positions, $x = -2.15$, $0.20$, and $2.54\,a_0$, along the trajectory for an impact parameter $b = 4.7a_0$ and impact energy $E = 25\,\mathrm{keV}$. The lower panel shows the corresponding $B^{2+}(2p)$ component. The position of the He target is marked by a black dot. Also shown are the instantaneous position of the internuclear axis and the alignment angle, respectively [8.17].

A subsequent experimental investigation mapped the double-differential cross section in scattering angle and energy loss, as illustrated in Figure 8.8 [8.18,19]. The data confirm the expectation from Figure 8.6 of electron transfer into the $B^{2+}(2s)$ and $B^{2+}(2p)$ levels, with capture into the ground state being the dominant channel. A photon polarization study of the $B^{2+}(2p)$ channel yielded an orientation with a sign confirming the theoretical prediction, as shown in Figure 8.9(a) together with theoretical predictions [8.20]. We also note a sign reversal when going to larger scattering angles, corresponding to the tail of the differential cross section. For the $B^{3+}-$ Ne collision the $B^{2+}(2p)$ channel dominates $B^{2+}(2s)$, so it is possible to obtain better photon counting statistics. Figure 8.9(b) shows the results in this case, revealing the same sign of the orientation for forward scattering as for the He target, and a clearly pronounced, regular oscillatory structure in the tails.

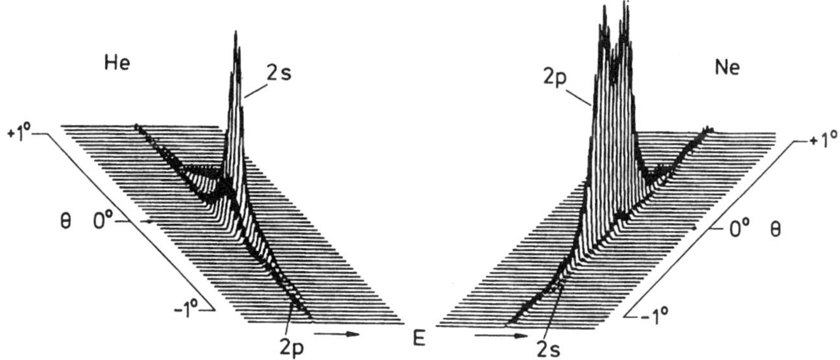



**Fig. 8.8.** Double-differential cross sections in energy loss $\Delta E$ and scattering angle $\theta$ for the charge exchange channels, clearly separating electron transfer into the $B^{2+}(2s)$ and $B^{2+}(2p)$ levels, respectively, for He and Ne targets. The impact energy is 1.8 keV [8.18,19].

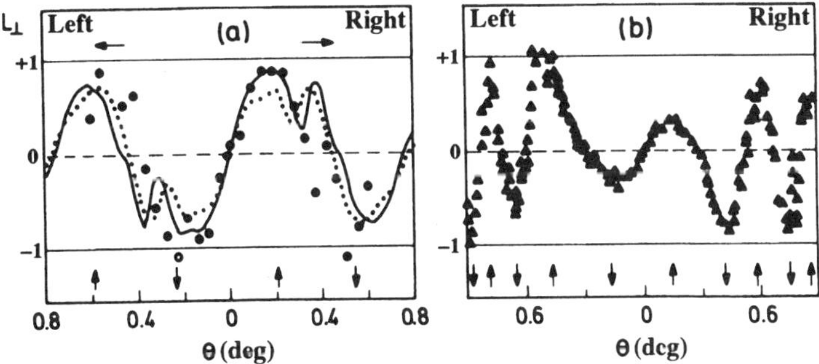



**Fig. 8.9.** Orientation parameter $L_\perp$ as function of scattering angle for the $B^{2+}(2p)$ state produced in 1.5 keV (a) $B^{3+}$−He [8.18], and (b) $B^{3+}$−Ne [8.19] collisions, respectively.

The principal features of these observations may be understood from the following analysis. Figure 8.10(a) shows how the incident $\Sigma$ potential curve crosses the two degenerate $\Sigma$ and $\Pi$ curves of the outgoing channel. Excitation takes place by $\Sigma - \Sigma$ radial coupling on the way in (1) or on the way out (2). After excitation, efficient rotational $\Sigma - \Pi$ coupling will cause the excited orbital to remain space-fixed. Thus, the collision geometry will be as outlined in Figure 8.10(b), in which two $\sigma$-orbitals are drawn, (1) corresponding to excitation on the way in, and (2) to excitation at the symmetric point on the way out. The rotation angle of the internuclear axis from (1) to (2) is $\phi$. The total amplitude for excitation is obtained by coherently adding the two corresponding dipoles, which differ in two ways: (2) is rotated by an angle

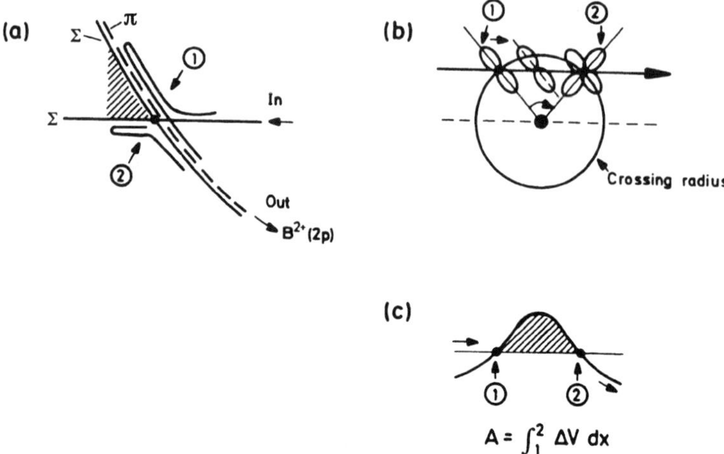



**Fig. 8.10.** Collision geometry and energy curves for interpretation of $L_\perp$ oscillations in a quasi-molecular picture. (a) The incoming $\Sigma$ potential curve crosses the outgoing $\Sigma$ and $\Pi$ curves. (b) The corresponding orbital geometry along the trajectory. Compare with Figure 8.7, lower panel. (c) Molecular potential energy curves along the trajectory.

$\phi$ with respect to (1); (1) has accumulated an additional phase $\Phi = A/\hbar v$, where $A$ is the area shown in (c) corresponding to the energy difference between the two alternative potential-curve paths that can be followed along the trajectory. Assuming that the transition probability at the crossings is small, a superposition of the two contributions yields the following expression for the excitation amplitudes:

$$\begin{cases} a_{-1} = \cos \tfrac{1}{2}\left(\tfrac{A}{\hbar v} - \phi\right) = \cos \tfrac{1}{2}(\Phi - \phi)\,, \\[2ex] a_{+1} = \cos \tfrac{1}{2}\left(\tfrac{A}{\hbar v} + \phi\right) = \cos \tfrac{1}{2}(\Phi + \phi)\,. \end{cases} \tag{8.11}$$

These particular formulas were derived by Ostrovsky [8.21], but the line of thought outlined is more general and can be traced much further back in the literature [8.22–24]. The corresponding orientation is derived as

$$L_\perp = -\frac{\sin \tfrac{A}{\hbar v} \sin \phi}{1 + \cos \tfrac{A}{\hbar v} \cos \phi}\,. \tag{8.12}$$

It may also be concluded that in this simple model the alignment angle $\gamma$ is equal to $0°$ or $90°$, depending on whether the major axis of the polarization ellipse is parallel or perpendicular to the incident beam direction. An equivalent statement is $P_2 = 0$ for the second Stokes-vector component, or $\chi = 90°$ in the "$(\lambda, \chi)$-language," as first noted in [8.22]. From the equation above it is clear that, with this geometry of the potential curves, $L_\perp$ always starts out *negative* at small scattering angles, i.e., it follows the rotation direction of the internuclear axis. Furthermore, sign reversals in $L_\perp$ will develop with increasing scattering angle, due to an increase of the area $A$ as progressively shorter internuclear distances are reached. It can also be seen that, if the geometry is changed so that the potential curve difference changes sign, the corresponding orientation will change sign as well because of the sign change of $A$.

It is appropriate to end this section by comparing the present picture with the one outlined for quasi-one-electron systems in Section 8.1.1. Although the geometric details of the two cases are quite different, the underlying physics responsible for the orientation effects is, by closer inspection, very similar. In both cases, the orientation propensity, which for the main scattering component follows the rotation direction of the internuclear axis, arises from an interplay between a phase $\Phi$, proportional to the energy difference $\Delta E$ between the incoming and outgoing potential curves, and the rotation angle $\phi$ of the internuclear axis. In both cases, an oscillatory pattern in $L_\perp$ may develop due to an increase of the first phase term, $\Phi$, in case ($i$) by decreasing the collision velocity and keeping the impact parameter fixed, in case ($ii$) by decreasing the impact parameter and keeping the collision velocity fixed. In both cases, the orientation propensity reverses if the sign of $\Delta E$ is reversed as well. Indeed, the pair of equations (8.11) may be derived from the pair (8.6) by using a sum of two $\delta$-functions for the interaction term $\Gamma_{\text{sp}}(R)$, one centered at the point (1) on the way in and one centered at the point (2) on the way out. This alternative derivation of the Ostrovsky formulas (8.11) highlights the fact that the underlying dynamical effects responsible for the observed orientation phenomena in the two cases have the same physical origin.

### 8.1.3. S → P transfer in small-angle H$^+$, Li$^+$–Na (3s) collisions

In the next class of collision processes to be addressed, an ion collides with an alkali-atom target and the valence electron is transferred to the projectile, thereby creating a neutral atom in an excited state. Over most of the energy range, the favored channels have a small energy defect and the electron transfer may take place at large impact parameters, $10 - 20\, a_0$, comparable to or larger than the orbital radius of the alkali valence electron. The corresponding scattering angles are extremely small, typically a fraction of a degree, and the produced neutrals are focused in a narrow, forward cone, often comparable to the collimation of the incident beam. The precise determination of such small scattering angles poses considerable challenges.

The solution is to map in detail the complete spot of neutral particles using a position-sensitive detector located directly in the beam path, as shown in Figure 8.11 [8.25]. In order to achieve sufficient angular resolution, the particle detector is typically placed several meters downstream from the collision region. Photons emitted from the collision region in a fixed, conveniently chosen direction are detected in coincidence with the scattered neutral atoms.

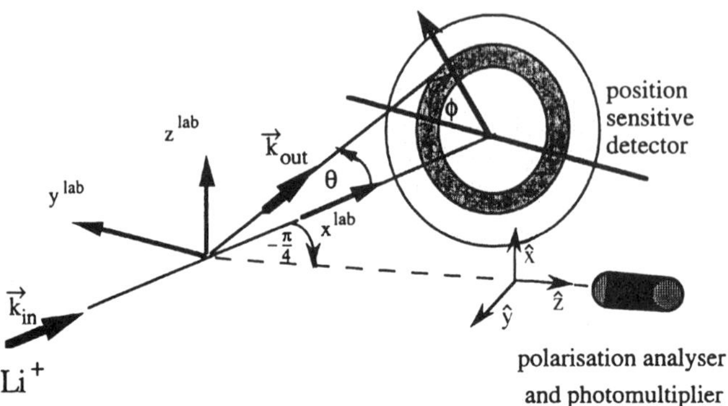


**Fig. 8.11.** The experimental geometry of a setup used for angle-differential excitation studies at very small scattering angles. The intensity pattern $I(\theta,\phi)$ of forwardly scattered projectiles is mapped by a position-sensitive detector located a few meters downstream from the collision region. The particles are measured in coincidence with (polarization-analyzed) photons emitted in a fixed direction. Subsequently, the recorded intensity pattern is Fourier analyzed according to (8.13) [8.25].


Depending on wavelength, the photons may be filtered with a linear or circular polarizer before detection. The location on the position sensitive detector corresponding to zero scattering angle is determined through symmetry considerations, and the recorded intensity pattern $I(\theta,\phi)$ is then fitted to a Fourier series with the generic structure

$$I(\theta,\phi)\sin\theta = \sum_m C_m(\theta)\cos m\phi + \sum_m S_m(\theta)\sin m\phi. \tag{8.13}$$

The factor $\sin\theta$ is included in order to facilitate subsequent integration of the counts over the solid angle $\sin\theta\,d\theta\,d\phi$ and retrieval of the expansion coefficients $C_m$ and $S_m$. The number of terms entering the expression depends on the process studied and the kind of polarizer (if any) that is invoked. For example, circular-polarization analysis is normally necessary in order to obtain nonzero $S_m(\theta)$ coefficients.

As an example, Figure 8.12 shows recorded spectra using a circular polarizer for photon analysis [8.25]. For brevity we will often suppress the variable $\theta$ in the Fourier coefficients, but when appropriate add a label specifying the setting of the photon polarization analyzer, e.g., $C_0(90°)$, $S_1(\mathrm{RHC})$, .... . The Fourier coefficients may be expressed in terms of the scattering amplitudes referring to a specific coordinate frame and basis set chosen, or the standard Stokes parameters for the problem. The structural details of the expression (8.12) depend on the observation direction, the polarizer used, and possible depolarization due to fine and hyperfine interactions in the final state before the optical decay.

Here we shall restrict ourselves to the discussion of example results for P states produced by electron transfer from an S state target. Furthermore,

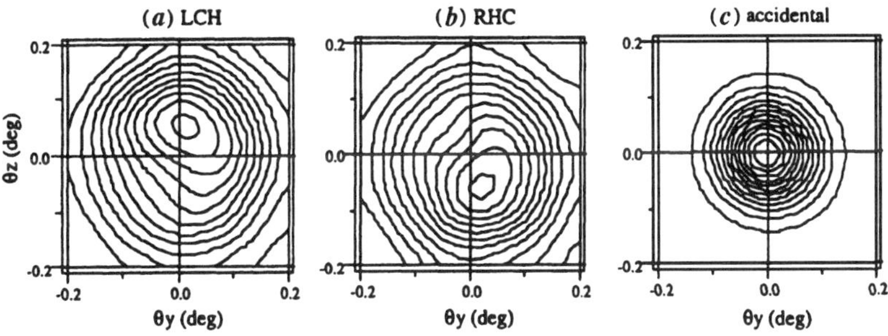



**Fig. 8.12.** Contour plots for Li(2p) formed in the $\mathrm{Li^+ - Na\,(3s) \to Li\,(2p) - Na^+}$ process for polarizer settings (a) LHC and (b) RHC in the setup of Figure 8.11. The beam spots are Fourier resolved according to (8.13). An up-down asymmetry due to a $\sin\phi$ component is clearly visible. The pattern for accidental coincidences is shown in (c). The $^6\mathrm{Li^+}$ ion impact energy is $1\,\mathrm{keV}$ [8.25].

we shall omit effects due to internal forces and instead refer to the literature for details on how these may be incorporated [8.25]. It should, however, be pointed out that from an experimental point of view these depolarization effects are very important for the signal-to-noise ratios and they should be carefully considered in the planning stage of an actual experiment. Due to conservation of reflection symmetry, only orbitals in the 2D-space spanned by, e.g., the two states $|\sigma\rangle$ and $|\pi^+\rangle$ may be populated, while there is no component along $|\pi^-\rangle$. The population of the two allowed components may be monitored directly through the intensity measured from a direction perpendicular to the collision plane, using a linear polarizer with transmission angles of $0°$ and $90°$, respectively, with respect to the direction of the incident beam. It is convenient to introduce names for the orthogonal orbital pairs corresponding to the alternative polarizer settings $\pm 45°$ and RHC/LHC, respectively. We shall use $\kappa^\pm$ and $\lambda^\pm$ for the p orbitals tilted by $\pm 45°$ and the circular states with angular-momentum projection $\pm\hbar$, respectively, i.e.,

$$|\kappa^+\rangle = \tfrac{1}{\sqrt{2}}(|\sigma\rangle + |\pi^+\rangle), \tag{8.14a}$$

$$|\kappa^-\rangle = \tfrac{1}{\sqrt{2}}(|\sigma\rangle - |\pi^+\rangle), \tag{8.14b}$$

and

$$|\lambda^+\rangle = -\tfrac{1}{\sqrt{2}}(|\sigma\rangle + i|\pi^+\rangle), \tag{8.14c}$$

$$|\lambda^-\rangle = \tfrac{1}{\sqrt{2}}(|\sigma\rangle - i|\pi^+\rangle). \tag{8.14d}$$

These basis sets are particularly useful for analysis of left-right scattering asymmetries for alignment and orientation, respectively. The excited P state may thus be represented in three equivalent ways, namely

$$|p\rangle = f_\sigma|\sigma\rangle + f_{\pi+}|\pi^+\rangle = f_{\kappa+}|\kappa^+\rangle + f_{\kappa-}|\kappa^-\rangle = f_{\lambda+}|\lambda^+\rangle + f_{\lambda-}|\lambda^-\rangle. \tag{8.15}$$

It is convenient to introduce the corresponding total and partial differential cross sections. With obvious notation,

$$\sigma = |f_\sigma|^2 + |f_{\pi+}|^2 = |f_{\kappa+}|^2 + |f_{\kappa-}|^2 = |f_{\lambda+}|^2 + |f_{\lambda-}|^2 \tag{8.16a}$$

$$\equiv \sigma_\sigma + \sigma_{\pi+} = \sigma_{\kappa+} + \sigma_{\kappa-} = \sigma_{\lambda+} + \sigma_{\lambda-} \,. \tag{8.16b}$$

The relationships between the standard Stokes-vector components $(P_1, P_2, P_3)$ and the partial cross sections are

$$\sigma_{\sigma/\pi+} = \tfrac{1}{2}\sigma(1 \pm P_1), \tag{8.17a}$$

$$\sigma_{\kappa\pm} = \tfrac{1}{2}\sigma(1 \pm P_2), \tag{8.17b}$$

$$\sigma_{\lambda\pm} = \tfrac{1}{2}\sigma(1 \mp P_3), \tag{8.17c}$$

where the signs in (8.17c) correspond to the convention of classical optics [8.26] that RHC/LHC polarized photons have negative/positive helicity, respectively (see also Chapter 2). The Stokes-vector components relate directly to real and imaginary parts of the density matrix components. It may be instructive to see this explicitly for the three alternative basis sets introduced above. Using the basis $(\sigma, \pi^+, \pi^-)$, one obtains

$$\rho = \begin{pmatrix} f_\sigma f_\sigma^* & f_\sigma f_{\pi+}^* & 0 \\ f_{\pi+} f_\sigma^* & f_{\pi+} f_{\pi+}^* & 0 \\ 0 & 0 & 0 \end{pmatrix} \tag{8.18a}$$

$$= \sigma \tfrac{1}{2} \begin{pmatrix} 1 + P_1 & P_2 + iP_3 & 0 \\ P_2 - iP_3 & 1 - P_1 & 0 \\ 0 & 0 & 0 \end{pmatrix}. \tag{8.18b}$$

Using the basis $(\kappa^+, \kappa^-, \pi^-)$, one obtains

$$\rho = \begin{pmatrix} f_\kappa f_{\kappa+}^* & f_\kappa f_{\kappa-}^* & 0 \\ f_\kappa f_{\kappa+}^* & f_\kappa f_{\kappa-}^* & 0 \\ 0 & 0 & 0 \end{pmatrix} \tag{8.19a}$$

$$= \sigma \tfrac{1}{2} \begin{pmatrix} 1 + P_2 & P_1 + iP_3 & 0 \\ P_1 - iP_3 & 1 - P_2 & 0 \\ 0 & 0 & 0 \end{pmatrix}. \tag{8.19b}$$

Finally, using the basis $(\lambda^+, \pi^-, \lambda^-)$, one obtains

$$\rho = \begin{pmatrix} f_\lambda f_{\lambda+}^* & 0 & f_\lambda f_{\lambda-}^* \\ 0 & 0 & 0 \\ f_\lambda f_{\lambda+}^* & 0 & f_\lambda f_{\lambda-}^* \end{pmatrix} \tag{8.20a}$$

$$= \sigma \tfrac{1}{2} \begin{pmatrix} 1 - P_3 & 0 & -P_1 + iP_2 \\ 0 & 0 & 0 \\ -P_1 - iP_2 & 0 & 1 + P_3 \end{pmatrix}. \tag{8.20b}$$

It is convenient to express the Fourier decomposition of the scattering signal in terms of the Stokes parameters, thereby avoiding reference to a specific choice of basis set. With the geometry of Figure 8.11, one finds [8.25]

$$I = \sigma[(5 - P_1) + 4P_2 \cos \phi - (1 - P_1) \cos 2\phi], \tag{8.21a}$$

$$I(0°) = \sigma[2(1 - P_1) - 2(1 - P_1) \cos 2\phi], \tag{8.21b}$$

$$I(90°) = \sigma[(3 + P_1) + 4P_2 \cos \phi + (1 - P_1) \cos 2\phi], \tag{8.21c}$$

$$I(\text{RHC/LHC}) = \sigma[\tfrac{1}{2}(5 - P_1) + 2P_2 \cos \phi - \tfrac{1}{2}(1 - P_1) \cos 2\phi$$
$$\pm 2\sqrt{2} P_3 \sin \phi]. \tag{8.21d}$$

Recall that all parameters $\sigma$, $P_1$, $P_2$, and $P_3$ depend on the scattering angle $\theta$. Note also that

$$I = I(0°) + I(90°) = I(\text{RHC}) + I(\text{LHC}) \tag{8.22}$$

and that the intensity pattern obtained using a circular polarizer, (8.21d), is the only one whose Fourier components contain complete information on the set $(\sigma, P_1, P_2, P_3)$ or $(\sigma, L_\perp, \gamma)$. From the photon angular correlation pattern (8.21a) obtained without polarization analysis, one may extract $\sigma$, $P_1$, and $P_2$, or $(\sigma, \gamma, P_\ell)$, only. Thus, this approach does not give access to the sign of $L_\perp$ nor to the consistency check that the degree of light polarization $P$ is unity (see also Section 4.3.1).

We now proceed to specific results obtained using the techniques outlined above. The Bielefeld group [8.27] has studied photons from the process

$$\text{H}^+ - \text{Na}\,(3\text{s}) \rightarrow \text{H}\,(2\text{p}) - \text{Na}^+. \tag{8.23}$$

Since the polarization of the VUV $\text{Ly}_\alpha$ photons is difficult to measure with high efficiency, only the photon angular correlation pattern was mapped and the parameters $(\sigma, P_1, P_2)$ or $(\sigma, \gamma, P_\ell)$ were extracted. (Since the experimental geometry was slightly different from the one shown in Figure 8.11, the corresponding algebraic expression for the correlation pattern differs from (8.21a)). Experimental results at proton impact energies of 1 and 2 keV [8.27] are shown in Figure 8.13 together with theoretical predictions [8.28,29], convoluted with the finite experimental angular resolution. We note a rather strong variation of the results with proton energy. The overall agreement between theory and experiment is satisfactory.

The Orsay group [8.25] studied photons from the process

$$\text{Li}^+ - \text{Na}\,(3\text{s}) \rightarrow \text{Li}\,(2\text{p}) - \text{Na}^+. \tag{8.24}$$

The polarization of the red LiI $2^2\text{P} \rightarrow 2^2\text{S}$ photons at 6707 Å is readily measured with conventional techniques. Over a wide energy range, this channel is second only to Li(2s) in importance. Figure 8.12 shows patterns of scattered Li(2p) neutrals recorded with a position sensitive detector and the photon

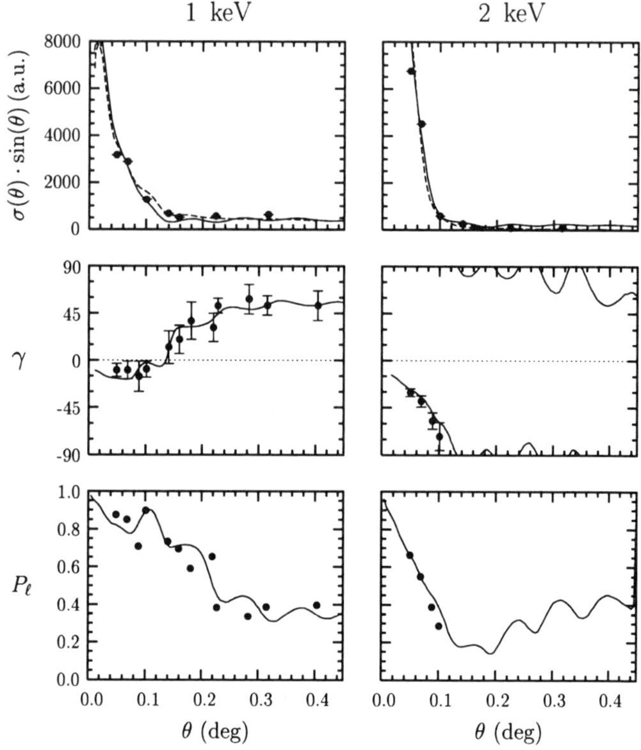



**Fig. 8.13.** Differential cross section $\sigma$ and alignment parameters $\gamma$ and $P_\ell$ for H(2p) formed in the process $\mathrm{H^+ - Na\,(3s) \rightarrow H\,(2p) - Na^+}$ for proton impact energies of 1 and 2 keV (adapted from [8.27]).

polarizer settings LHC/RHC. The isotope $^6\mathrm{Li}$ is used as projectile to minimize hfs depolarization effects in the Li(2p) final state. The impact energy is 1 keV, or $v = 0.081\,v_0$.

Figure 8.14 shows the sets of Fourier coefficients extracted using polarizer settings of $0°$, $90°$, and LHC, respectively. Although complete information on the process (8.24) is, in principle, contained in the four Fourier coefficients obtained for LHC polarization alone, see (8.21d), additional spectra were recorded with linear polarizer settings of $0°$ and $90°$. In this way consistency checks between various sets of coefficients may be performed. Also, because of a strong attenuation of the higher-order coefficients due to the finite angular width of $0.08°$ of the incident beam, some coefficients may in practice be more adequate for extraction of the coherence parameters than others. After proper convolution, the sign and size of all coefficients are excellently reproduced by the MO theory [8.30].

Based on these coefficients, one may extract the Stokes parameters $(P_1, P_2, P_3)$, $\sigma$, and the complete set of partial cross sections (8.17) shown

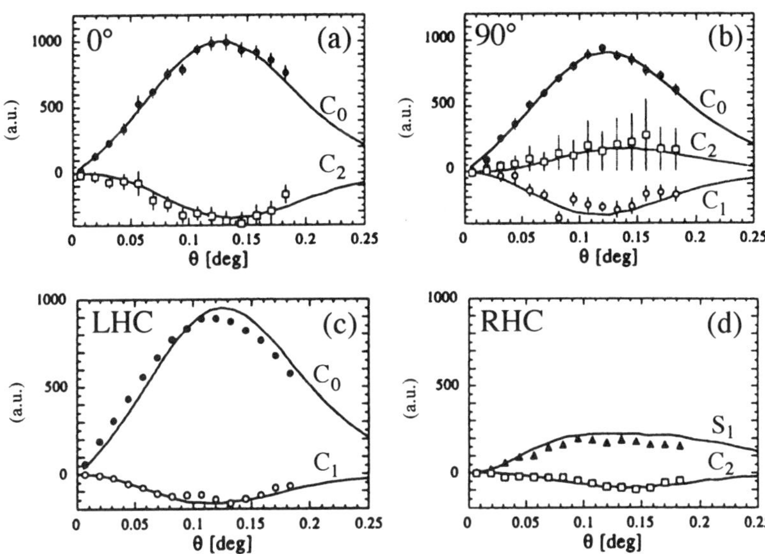


**Fig. 8.14.** Complete sets of Fourier coefficients for polarizer settings (a) $0°$ ($C_0, C_2$), (b) $90°$ ($C_0, C_1, C_2$), and (c,d) LHC ($C_0, C_1, C_2, S_1$) extracted from an analysis of the 2D coincidence patterns using (8.21b–d). The FWHM of the incident beam is $0.08°$. Experimental data points [8.25] are compared with MO predictions (full lines) [8.30].

in Figure 8.15. Recall that these quantities depend on $\theta$, the angle of the outgoing Li(2p) atom scattered to the left side in the $xy$ scattering plane. Scattering to the right side instead produces the mirror-symmetric state in the $xy$ plane. The first two partial cross sections corresponding to scattering into the $\sigma$ and $\pi^+$ states will thus be symmetric, while the cross sections for scattering into the two states $\kappa^+$ and $\lambda^+$ will exhibit a left-right scattering asymmetry. We note the relationships $\sigma_{\kappa+}(\text{right}) = \sigma_{\kappa-}(\text{left})$ and $\sigma_{\lambda+}(\text{right}) = \sigma_{\lambda-}(\text{left})$. The lower two panels thus represent the asymmetric patterns of Li(2p) particles in the scattering plane corresponding to the $\kappa^+$ and $\lambda^+$ final states, respectively. The corresponding Stokes parameters $P_1$, $P_2$, and $P_3$ are shown in the right column of the figure. Again, we note an excellent agreement with the predictions of the MO theory [8.30] for all parameters.

The Stokes parameters have been converted into the parameters $(L_\perp, \gamma)$ characterizing the nascent Li(2p) charge cloud, which are shown together with $\sigma$ in Figure 8.16. We first notice that $L_\perp$ grows to an almost constant value of $+0.3\hbar$ beyond $\theta = 0.05°$. This implies a correspondingly large value of the parameter $P_\ell$, indicating that the charge cloud exhibits a shape close to that of a dumbbell-shaped p orbital. The orbital is aligned at an angle $\gamma$ typically near $-65°$ in the scattering plane, almost independent of the scat

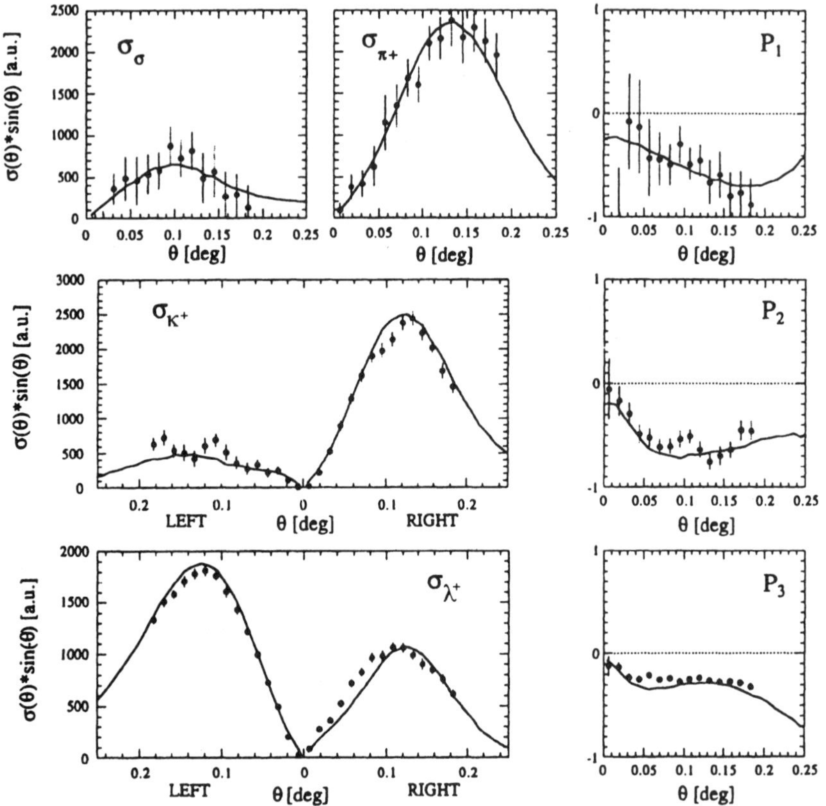

**Fig. 8.15.** The six partial cross sections of (8.17) are shown on the left, while the right column presents the corresponding three Stokes parameters. Experimental data points [8.25] are compared with MO predictions (full lines) [8.30].

tering angle $\theta$. The weak variation with $\theta$ for this system may be contrasted with the strong variations of the analogous parameters observed for H (2p) (cf. Figure 8.13).

A representative Li (2p) charge cloud is shown on the right side of Figure 8.16. We shall now explore how a collision event may result in this outcome, and how the picture depends on collision velocity. The very detailed, quantitative agreement between experiment and the MO theory encountered for all parameters displayed in Figures 8.15,16 encourages a closer, systematic look at the theoretical predictions that can provide access to regions that have not been accessible experimentally. Figure 8.17 presents electron transfer probabilities $b \cdot P(b)$ to the three orthogonal pairs of final states $(\sigma, \pi^+)$, $(\kappa^+, \kappa^-)$, and $(\lambda^+, \lambda^-)$ corresponding to the cross sections $\sigma_{\sigma, \pi+}$, $\sigma_{\kappa\pm}$, and $\sigma_{\lambda\pm}$ as a function of impact parameter $b = 0 - 20\, a_0$ and velocity

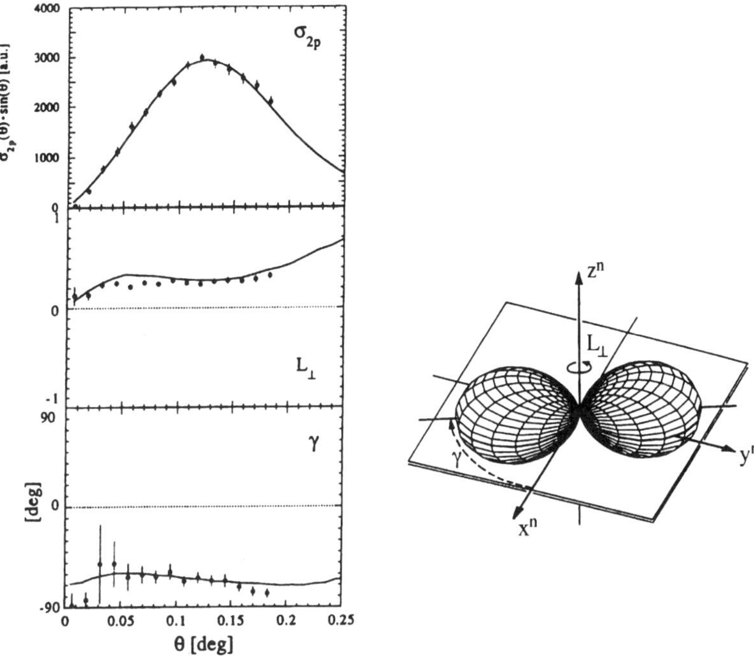

**Fig. 8.16.** The parameters $(\sigma, L_\perp, \gamma)$ characterizing the nascent Li(2p) charge cloud are shown on the left. Experimental data points [8.25] are compared with MO predictions (full lines) [8.30]. On the right is a representative Li(2p) electron charge cloud with $L_\perp = 0.3\hbar$ and $\gamma = -65°$.

$v = 0.05 - 0.30\, v_0$ [8.30]. The impact parameters considered correspond to a left passage of the projectile in the scattering plane. The experimental results presented above correspond to the low-velocity part of the graphs in Figure 8.17. We notice a distinct change in the patterns when going from low to high velocities. First, at low velocities, electron transfer takes place at relatively small impact parameters, $b \simeq 5\, a_0$. The $P(b)$ maximum is narrow, a fingerprint of a localized, short range transition [8.30]. There is a strong preference for the states $\pi^+$, $\kappa^-$, and $\lambda^+$ shown in the right column. When going toward higher velocities, transfer takes place also at large impact parameters, $10-15\, a_0$, here with a preferential population of $\sigma$, $\kappa^+$ (also increasing), and in particular $\lambda^-$ (see left column). The preference for $\lambda^-$ at higher velocities is in accordance with the propensity rule for orientation encountered in Section 8.1.1, which can readily be generalized to transfer processes [8.31]. The high-velocity alignment propensity for $\sigma$ population will be addressed in Section 8.2.1 below.

Solving the time-dependent Schrödinger equation offers the unique possibility of following the evolution of the collision complex along a specific trajectory. To illustrate the development of quantum states similar to the

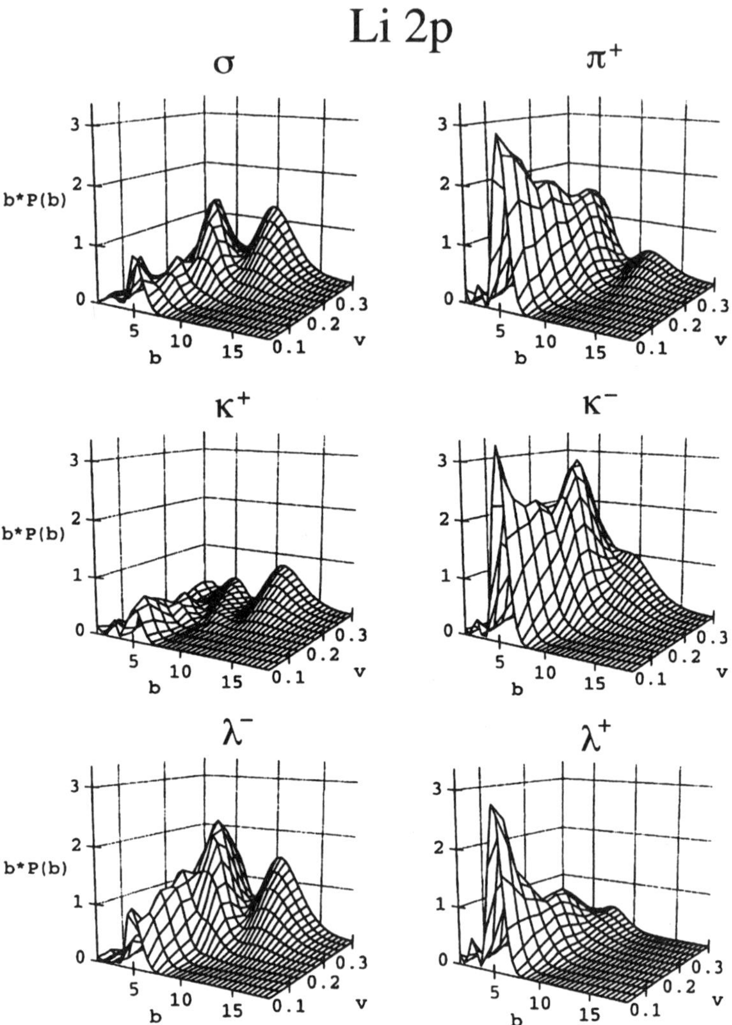



**Fig. 8.17.** Electron transfer probabilities $b \cdot P(b)$ to the three alternative, final pairs of orthogonal states $(\sigma, \pi^+)$, $(\kappa^+, \kappa^-)$, and $(\lambda^-, \lambda^+)$ as function of impact parameter $b = 0 - 20\, a_0$ and velocity $v = 0.05 - 0.30\, v_0$ [8.30].

one shown on the right in Figure 8.16, an animation of the evolution of the density and current of the electronic wavefunction along the straight-line trajectory was produced by plotting the probability density in the scattering plane, $|\Psi(x = vt, y = b, z = 0)|^2$, at a series of positions along the trajectory [8.30]. For clarity, only the partial density $|\Psi_{2p}|^2$ given by the projection of the wavefunction on the two $\mathrm{Li}\,(2p\Sigma)$ and $\mathrm{Li}\,(2p\Pi)$ molecular components, is shown. Parallel to this, the electron current was calculated to illustrate the

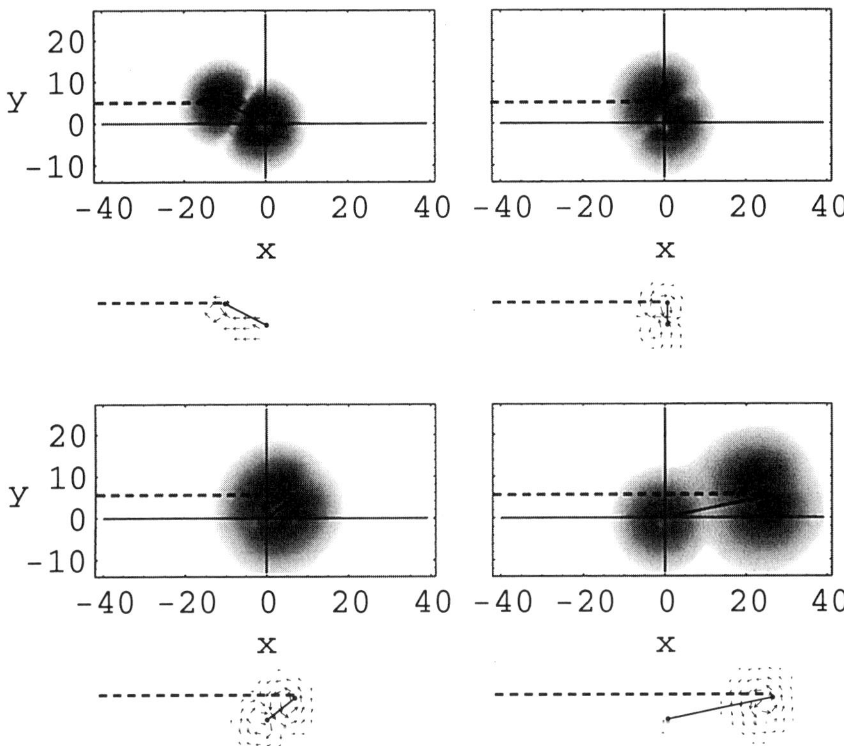

**Fig. 8.18.** Electron density for the Li (2p) wavefunction component in the scattering plane. Below are shown the corresponding currents. The impact parameter is $b = 5\,a_0$ and the collision velocity is $v = 0.1\,v_0$. The four panels correspond to positions $x = vt = -10,\ 0,\ 6.5,$ and $25\,a_0$, respectively, along the trajectory (adapted from [8.30]).

build-up of electron-cloud orientation along the trajectory. A typical impact parameter of $b = 5\,a_0$ and collision velocity of $v = 0.1\,v_0$ (i.e., slightly higher than the $0.08\,v_0$ of Figure 8.16) was selected. Figure 8.18 shows the result at four different times corresponding to the positions $x = vt = -10,\ 0,\ 6.5,$ and $25\,a_0$ relative to the point of closest approach to the Na (3s) center. We note that, at the closest approach, Li $(2p\Pi)$ is clearly the most populated molecular state. After this, the electron cloud rotates with the internuclear axis, the p orbital staying almost perpendicular to the internuclear axis until a certain distance, the locking radius [8.32], where the motion of the electron cloud decouples from the molecular axis. From here on, the electron cloud stays fixed in the laboratory frame. The fourth picture shows a final state aligned at an angle $\gamma \simeq -75°$. Depending on which basis set the asymptotic 2p wavefunction is projected on, one sees that the $\pi^+$ state is preferred to the $\sigma$ state, or the $\kappa^-$ state is preferred to the $\kappa^+$ state. Furthermore, the current illustrates that the final circulation is counterclockwise, i.e., positive orientation, so that $\lambda^+$ is preferred to $\lambda^-$.

All this is consistent with the picture presented in Figure 8.16. The characteristic features described above become particularly distinct to the eye in animations with a finer time grid than that used in Figure 8.18. Such a movie is provided in connection with this book. Finally, the potential-curve geometry responsible for the Li(2p) population at low energy is similar to the one encountered for production of $B^{2+}(2p)$ in Section 8.1.2, except that $\Delta E(R)$, and thereby the orientation, here has the opposite sign [8.30].

## 8.2 Angle-Integrated Alignment Studies Using Optically Prepared Targets

It is evident from the previous section that detailed scattering-angle resolved experimental tests of theoretical predictions for the electron-transfer processes become increasingly difficult with increasing velocity, since the electron transfer takes place at larger distances and the corresponding scattering angles become too small to be accessible. This statement applies even to the sophisticated experimental techniques shown in Figure 8.11, mainly because the requirements on initial-beam collimation become prohibitively severe. Tests of theoretical estimates over a wide velocity range may, however, be performed at the level of scattering-angle *integrated* Stokes parameters.

Figure 8.19 shows an example for the process (8.24) just considered [8.33], namely the scattering-angle integrated linear polarization (the only component surviving the angular integration) for the Li(2p $\rightarrow$ 2s)transition as function of projectile energy $E/M$ and collision velocity $v$. The polarization was measured with the $^6$Li isotope and corrected for final-state depolarization. The experimental results have small error bars and compare well with predictions based on a 26-state AO theory above a velocity of $0.2\,v_0$ and with a 28-state MO theory below $0.4\,v_0$. Average electron cloud shapes, exhibiting rotational symmetry with respect to the incident beam axis, are shown on the right of Figure 8.19. They correspond to the experimental results at velocities $v = 0.08, 0.14, 0.20$, and $0.36\,v_0$, respectively. The evolution of the charge cloud with velocity clearly confirms the theoretical predictions of Figure 8.17 for a gradual domination of the Li(2p) $\sigma$ orbital, which has momentum components along the incident beam direction, with increasing velocity.

Integral alignment effects can also be studied effectively in the reverse scheme of (8.23,24). A large number of experiments have been performed in which the target has been optically prepared, typically in the Na(3p) state, prior to the binary encounter. The details of the Na(3p) state can be controlled by the laser light polarization. The variation in the exit cannel with orbital shape or orientation of the initial state may then be studied in detail. This kind of investigation can be carried out at several levels of refinement,

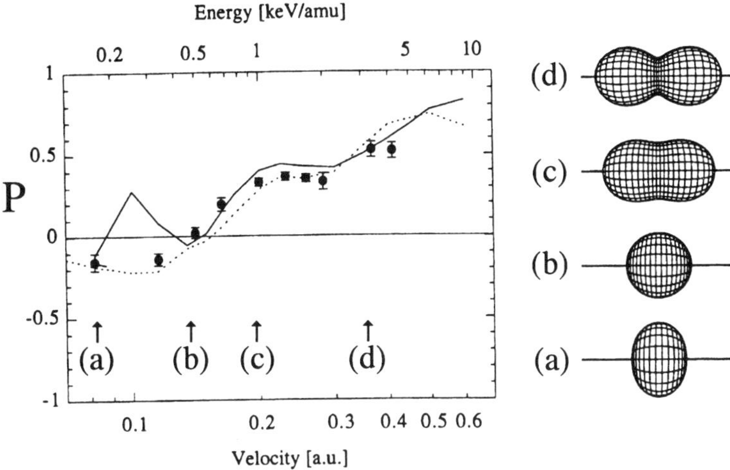



**Fig. 8.19.** The figure shows the scattering-angle integrated linear polarization $P$ for the Li(2p→2s) transition as function of projectile energy $E/M$ and collision velocity. Experimental results (•) are compared to predictions based on a 26-state AO theory (full line) and a 28-state MO theory (dotted line) [8.33]. Average electron cloud shapes corresponding to the experimental results at $v = 0.08$ (a), 0.14 (b), 0.20 (c), and 0.36 $v_0$ (d), respectively, are shown on the right.

such as scattering-angle integrated or differential studies, and with various degrees of sophistication for the final-state analysis. We shall now present examples of such studies at various levels of detail, still with emphasis on the physical pictures of the collision dynamics that have emerged from the results.

### 8.2.1 Alignment effects in H$^+$, Li$^+$–Na(3p) collisions

We first address integral alignment effects in the electron transfer process

$$\text{H}^+ + \text{Na}(3\text{p}) \rightarrow \text{H}(n) + \text{Na}^+. \tag{8.25}$$

Figure 8.20 shows hydrogen energy-loss spectra for the process (8.25) at an impact energy of 2 keV, recorded for an optically prepared Na(3p) target. The latter was created by pumping with laser light linearly polarized parallel and perpendicular, respectively, to the incident beam direction [8.34]. For comparison, spectra for a ground-state Na(3s) target are also shown.

In the analysis of such data, several aspects must be carefully considered. First, the fraction $\alpha$ of excited states in such a target is not 100%, but rather 10–15% [8.35]. This fraction can, however, be increased to the 40–50% range by means of an electro-optical modulator (EOM) [8.36]. Second, when pumping with linearly polarized light, the presence of Na fine and hyperfine structure generally leads to an incoherent mixture of $\sigma$ and $\pi$ orbitals in

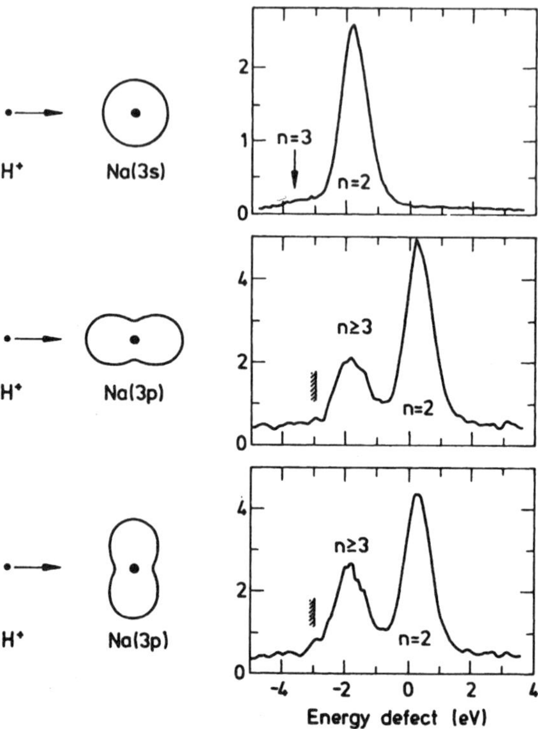



**Fig. 8.20.** Orbital alignment anisotropies for $H^+ + Na\,(3s,3p) \rightarrow H(n) + Na^+$ for 2 keV proton impact energy [8.34]. At this energy, the parallel orbital alignment for $n = 2$ is more effective for electron transfer than the perpendicular one. The opposite statement holds for $n \geq 3$.

the target. The weights depend on the hyperfine transition used, the extent to which steady-state pumping conditions have been reached, etc. In practice, the degree of target anisotropy is measured by rotating the plane of polarization of the exciting photon beam and measuring the corresponding anisotropy of the emitted radiation in a direction perpendicular to the photon beam. The intensity variation $I(\phi)$ is then fitted to an expression of the form $I(\phi)/I_0 = 1 + f \cos \phi$, with the contrast parameter $f = \infty$ for a pure p orbital. For the optimal choice of hyperfine pumping transition, however, $f = 0.75$ at best, and in practice often in the range 0.4–0.6, depending on experimental details. For measured values of $\alpha$ and $f$, the target therefore consists of a fraction $1 - \alpha$ of atoms in the $Na\,(3s)$ ground state and a fraction $\alpha$ in the $Na\,(3p)$ state. The excited fraction may be described as an incoherent superposition of three orthogonal p orbitals with relative weights $(2f + 1) : 1 : 1$, with the first, dominating one being aligned parallel to the linear polarization direction of the pump beam.

The situation is somewhat simpler for pumping with circularly polarized light. For a proper choice of transition, the optically prepared part of the target may approach an ensemble in a pure, circular state, with helicity determined by the helicity of the pump beam. However, in this process, the valence electron in both the $Na(3s)$ and $Na(3p)$ states becomes spin-polarized along the direction of the laser beam. This has important consequences for the light emission from the exit channels, as we shall discuss further below.

In the spectra for $Na(3p)$ targets in Figure 8.20, the (large) background from $Na(3s)$ states has been subtracted for clarity, while the effect of fine structure and hyperfine structure has not, as indicated by the shape of the orbitals shown. However, from the data one may derive the signals corresponding to the pure atomic orbitals, $3p\sigma$ and $3p\pi$, aligned parallel and perpendicular to the beam direction, respectively. With obvious notation we get for the two polarizer settings

$$I(0°) = \frac{2f+1}{2f+3}\,\sigma(3p\sigma) + \frac{2}{2f+3}\,\sigma(3p\pi)\,, \tag{8.26a}$$

$$I(90°) = \frac{2f+2}{2f+3}\,\sigma(3p\pi) + \frac{1}{2f+3}\,\sigma(3p\sigma)\,, \tag{8.26b}$$

with $\sigma(3p\pi) = \frac{1}{2}\left[\sigma(3p\pi^+) + \sigma(3p\pi^-)\right]$. (For brevity we use the notation $\sigma$ for scattering-angle integrated cross sections.) One may now define a dimensionless alignment anisotropy parameter $A$ as

$$A = \frac{\sigma(3p\sigma) - \sigma(3p\pi)}{\sigma(3p\sigma) + \sigma(3p\pi)}\,. \tag{8.27}$$

This quantity is equal to the reduced polarization of the $Na(3p{\rightarrow}3s)$ decay in the (somewhat unrealistic) reverse experiment, in which the $Na(3p)$ atoms are produced by collisions of an $Na^+$ beam with the target consisting of the proper isotropic, incoherent superposition of $H(n\ell)$ states.

If sufficient energy loss resolution can be achieved, $n$-resolved alignment anisotropy parameters $A(n)$ may be extracted. If not, only the parameter $A = A(\Sigma n)$ corresponding to the total transfer cross sections is obtained. We shall now study more closely the velocity dependence of the alignment anisotropy parameter.

### 8.2.1.1 Initial-state alignment dependence

Results for the alignment anisotropy parameter $A$ for $H^+$ impact [8.37] are shown in Figure 8.21, which also includes data for $He^+$ [8.38] and $Li^+$ [8.39] ions. Collision velocities are comparable to the orbital velocity of the active electron, the so-called "velocity-matching region." We note a fairly dramatic velocity dependence, with increasingly positive $A$ values beyond the matching velocity. This strong velocity dependence might be expected, with electron transfer being favored for the orbital geometry with maximum momentum distribution along the direction of the incident beam, as illustrated

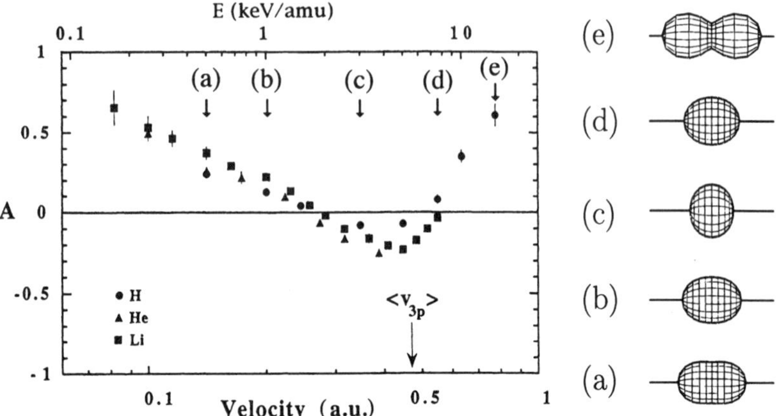



**Fig. 8.21.** The orbital alignment anisotropy parameter $A$ as function of collision velocity for the ions H$^+$(•) [8.37], He$^+$($\triangle$) [8.38], and L$^+$($\square$) [8.39] colliding with an optically prepared Na(3p) target. For reference, the mean orbital velocity $< v_{3p} >= 0.47$ a.u. of the Na(3p) electron is indicated. Also shown are average electron cloud shapes for the reverse process corresponding to the experimental results for H$^+$ at $v=0.14$ (a), 0.20 (b), 0.35 (c), 0.55 (d), and 0.77 $v_0$ (e), respectively.

by the electron clouds in the right part of Figure 8.21. However, the large similarity of the $A$ curves for the three projectiles, including the minimum near $v_c = 0.4$ a.u., where the perpendicular orbital is the favored geometry for transfer, is surprising, although generally in good agreement with numerical predictions based on close-coupling [8.37,40–43] or CTMC codes [8.44]. We will return to this feature in the discussion below and here just note that analogous electron-transfer studies using He$^{2+}$ [8.45,46], Ne$^+$[8.47], Na$^+$[8.48], Ar$^+$[8.47], and K$^+$[8.49] all yield anisotropy parameters *not* coinciding with the group in Figure 8.21. The propensity for perpendicular alignment of the Na(3p) orbital for electron transfer at a collision velocity near 0.4 a.u. is thus not universal.

### 8.2.1.2 Final-state alignment dependence

A step further in probing the transfer dynamics is achieved by including also polarization analysis of the final state. This has been done for the dominant electron transfer channel using a $^6$Li$^+$ beam,

$$\text{Li}^+ + \text{Na}(3p) \rightarrow \text{Li}(2p) + \text{Na}^+, \tag{8.28}$$

for which further information may be obtained from a determination of the integrated Stokes parameters of the final Li(2p) state as function of the initial optical preparation of the Na(3p) target atom. For a P–P transition, such as (8.28), a complete set of Stokes parameters consists of eight independent components [8.50].

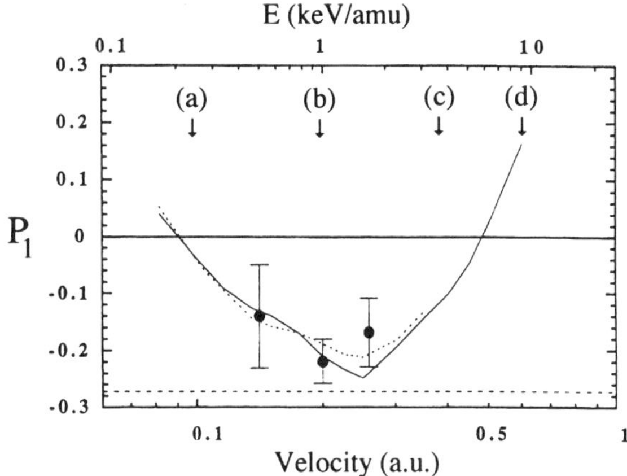

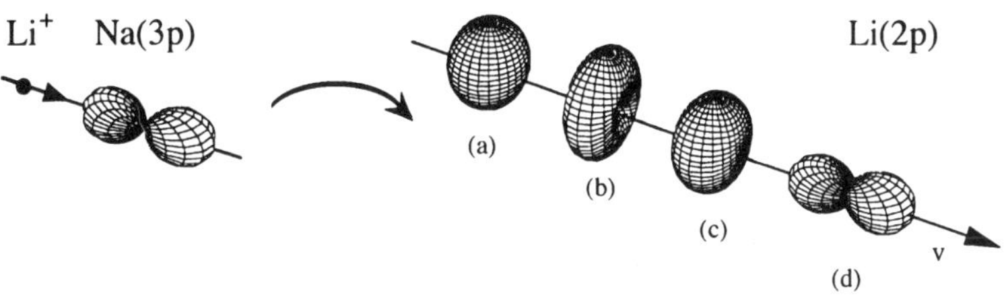



**Fig. 8.22.** The integrated Stokes parameter $P_1$ for light emitted from the Li (2p) state, measured perpendicular to the ion-beam direction, for optical preparation of the Na (3p) target with linearly polarized light along the beam direction. Experimental results are shown together with theoretical predictions [8.42,51]. The lower limit of $-3/11$, due to fine-structure and hyperfine-structure depolarization in the initial and final states, is indicated by a dashed line. Theoretical average electron cloud shapes corresponding to $v = 0.1$ (a), 0.2 (b), 0.4 (c), and $0.6\,v_0$ (d), respectively, are shown below.

As a characteristic example, Figure 8.22 shows results for the integrated Stokes parameter $P_1$ for light emitted from the Li (2p) state, measured perpendicular to the ion-beam direction, for optical preparation of the Na (3p) target with linearly polarized light along the ion-beam direction, thus preparing primarily an orbital along the incident beam. Experimental results are shown together with theoretical predictions [8.51]. The lowest possible value of $-3/11$, indicated by a dashed line, is due to fine-structure and hyperfine-structure depolarization in the initial and final states when the isotopes $^6$Li and $^{23}$Na are used in the standard pumping scheme [8.51]. (Without these depolarizations, the $P_1$ range would be the usual interval $[-1.0, 1.0]$.) We note that $P_1$ is close to its minimum value near a collision velocity $v_c = 0.2$ a.u., corresponding to an electron-cloud density for the excited Li (2p) electron with a very small component along the direction of the incident beam. Remarkably, *this observation is essentially independent of the initial preparation of the target state* [8.51]. In other words, at this collision velocity, there is a universal, strong propensity for the final P state to be aligned perpendicular to the beam direction. Predictions of close-coupling calculations [8.51] are in excellent agreement with this finding but give no clue concerning an interpretation of this striking observation.

### 8.2.1.3 The Sidky–Simonsen velocity-matching model

The velocity-matching model of Sidky and Simonsen [8.52] offers a convenient framework for understanding the observed integral-alignment propensities for the initial and final P states outlined in the previous two paragraphs. Building on earlier work of Brinkmann and Kramers [8.53] and Schippers *et al.* [8.54], they present the following picture, which holds to a good approximation for large impact-parameter collisions at electron velocities near or beyond the matching velocity.

The first step is to transform the scattering problem from configuration space $(x, y, z)$ to a mixed configuration–momentum space, $V_{\mathrm{mixed}}$ or $(p_x, y, z)$, by Fourier transforming the coordinate $x$ in the direction of motion of the projectile to the momentum component $p_x$. In this frame, the transition amplitude $f_{\mathrm{P \leftarrow T}}$ for electron transfer from the target T to the projectile P may be written as (in atomic units)

$$f_{\mathrm{P \leftarrow T}} = C \cdot \int_{V_{\mathrm{mixed}}} y \cdot \mathrm{O_{PT}} \cdot \delta \left( p_x - \frac{-\Delta E + \frac{1}{2} v^2}{v} \right) dp_x \, dy \, dz \,, \qquad (8.29)$$

where $C$ is a normalization factor. The $\delta$-function expresses energy conservation [8.52]. The overlap of the mixed-space wavefunctions (labeled by a tilde) for the projectile final state $\tilde{\psi}_{\mathrm{P}}$ and target initial state $\tilde{\psi}_{\mathrm{T}}$, the former one being shifted in the $p_x$ direction by the collision velocity $v$ and in the $y$ direction by the impact parameter $b$, is denoted by

$$\mathrm{O_{PT}} = \tilde{\psi}_{\mathrm{P}}^*(p_x - v, y - b, z) \cdot \tilde{\psi}_{\mathrm{T}}(p_x, y, z) \,. \qquad (8.30)$$

Figure 8.23 illustrates schematically how the integral (8.29) is evaluated. Two situations are of particular interest:

- (*i*) The $\delta$-function is centered on the *target* T, i.e., $p_x = 0$. This is the case for $\Delta E > 0$, i.e., if the process is *endoergic*, at a critical velocity $v_c = \sqrt{2\Delta E}$.
- (*ii*) The $\delta$-function is centered on the *projectile* P, i.e., $p_x = v$. This is the case for $\Delta E < 0$, i.e., if the process is *exoergic*, at a critical velocity $v_c = \sqrt{-2\Delta E}$.

Note that a sign change in $\Delta E$ interchanges the role of target and projectile, as should be the case for the reaction if the direction of time is reversed.

Application of these results to the two situations presented above may now explain the observed propensities:

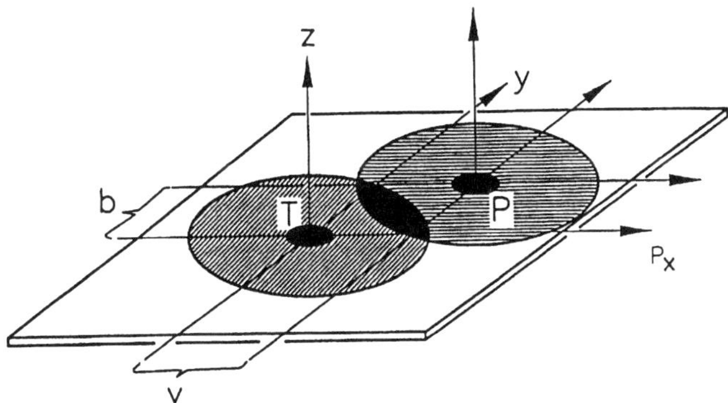


**Fig. 8.23.** In the velocity-matching model, the electron transfer scattering amplitude is obtained from an integral in the mixed configuration-momentum space $(p_x, y, z)$ over the overlap of the mixed space wavefunctions of projectile P and target T, taken at the value of $p_x$ fixed by the $\delta$-function in (8.29).

($i$) *Initial-state alignment dependence:* In the collisions of $H^+, He^+, Li^+$ with Na(3p), $\Delta E > 0$ for the $n = 3$ levels with a critical velocity $v_c \approx 0.4$ a.u. At this velocity, the mixed space wavefunction corresponding to a *target* p orbital aligned along the beam direction will exhibit negative reflection symmetry with respect to the plane defined by the $\delta$-function. Hence, the corresponding scattering amplitude is zero in the velocity-matching model, causing the observed minimum in the $A$ curve with the negative value corresponding to a propensity for the perpendicular orbital geometry. Experiment [8.34] and theoretical calculations [8.37,40–44] confirm that this minimum is indeed primarily due to the $n = 3$ levels.

($ii$) *Final-state alignment dependence:* For the process (8.28), $\Delta E < 0$ with a critical velocity $v_c \approx 0.2$ a.u. At this velocity, the mixed-space wavefunction corresponding to a *projectile* p orbital aligned along the beam direction will exhibit negative reflection symmetry with respect to the plane defined by the $\delta$-function. Hence, the corresponding scattering amplitude is zero in the velocity-matching model, giving rise to the observed propensity for alignment of the transferred electron charge cloud perpendicular to the incident beam direction.

Armed with this insight, we shall now briefly explore to which extent other data in the literature on integral alignment effects may be interpreted within this framework.

## 8.2.2  Alignment effects in $He^{2+}-Na\,(3p)$ collisions

Extensive experimental and theoretical results on the alignment anisotropy parameter for the process

$$He^{2+} + Na\,(3p) \rightarrow He^{+}(n) + Na^{+} \tag{8.31}$$

have been presented by the Vienna [8.45] and Groningen [8.46] groups. The
experimental results by Aumayr et al. [8.45] for $n = 3, 4$ were obtained from
analysis of Translational Energy Spectra (TES) which may resolve the $n$, but
in this case not the $l$, quantum number. Schlatmann et al. [8.46] used Photon
Emission Spectroscopy (PES) and recorded the HeII $n=4 \rightarrow n=3$ line emission at 468.6 nm as function of the optical target preparation. Strictly speaking, their anisotropy parameter should thus be denoted $A(4 \rightarrow 3)$ since it is not
exactly identical to $A(4)$. However, analysis of the relative cross sections and
branching ratios involved shows [8.46] that the $He^{+}(4f)$ level by far dominates
both parameters, so for the present purpose a distinction is not necessary.
The experimental results together with theoretical 38-state CC-AO predictions of Gieler et al. [8.55], CTMC results [8.46], and the Momentum-Space
Overlap (MSO) model of Schippers et al. [8.54], are shown in Figure 8.24. In
the MSO model, electron transfer cross sections are estimated by evaluating
the overlap of the projectile (P) and target (T) wavefunctions in momentum
space [8.54],

$$\sigma = \pi R_{c}^{2}\,\Big|\int d\mathbf{k}\,\Psi_{T}^{*}(\mathbf{k}+\mathbf{v})\,\Psi_{P}(\mathbf{k})\Big|^{2}, \tag{8.32}$$

where $R_{c}$ is the maximum internuclear distance at which an electron can
classically cross the Coulomb barrier between the two atomic cores. This
model has an appealing simplicity, but is essentially a zero impact-parameter
approximation, and it does not respect energy conservation.

The velocity-matching model presented in Section 8.2.1.3 may also be
invoked, but in a less straightforward way in this case of different charges
on the two centers (here $+2$ and $+1$, respectively) than for the symmetric
case discussed above, since now the critical velocity depends on the impact
parameter and the size of the interaction range [8.52]. Using a typical impact
parameter for electron transfer of $15\,a_0$ and a Bohr–Lindhard estimate of $27\,a_0$
for the interaction range gives $v_c = 0.27\,v_0$ for the endoergic channel $n = 4$, as
indicated by the arrow on Figure 8.24. (Using instead an impact parameter of
$20\,a_0$ gives $v_c = 0.23\,v_0$.). In this model, $H(n=3)$ does not exhibit a minimum
for the anisotropy parameter since this channel is exoergic.

Comparison of the theoretical predictions with the experimental results
in Figure 8.24 shows that the dramatic variation of the anisotropy parameter $A(4)$ with velocity is well reproduced by the CC-AO calculation. The
CTMC curve also follows the experimental data and correctly predicts the
minimum of the $A(4)$ curve, while the agreement is less satisfactory at higher

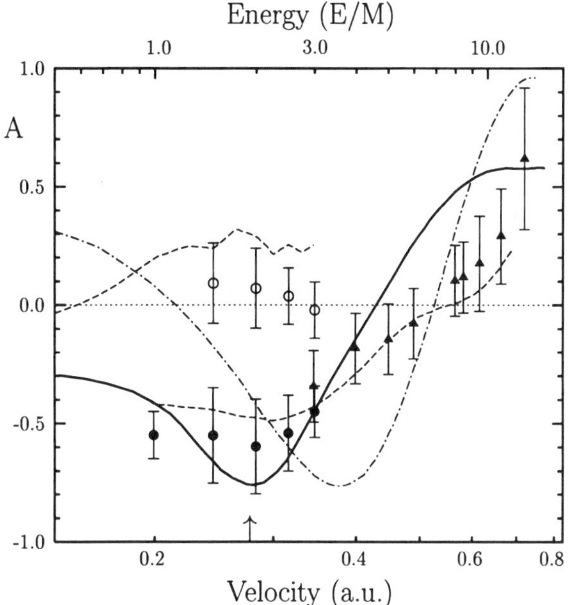

**Fig. 8.24.** Orbital alignment anisotropy parameters $A(n)$ as function of collision velocity for $He^{2+}$ ions colliding with an optically prepared $Na(3p)$ target. Experimental data are from the groups in Vienna for $n = 3$ (o) and $n = 4$ (•) [8.45], and Groningen for $n = 4$ (triangles) [8.46]. Theoretical predictions are CC ($- - -$, $n = 3, 4$) [8.43], CTMC (———, $n = 4$) [8.46], and MSO ($- \cdot - \cdot$, $n = 4$) [8.54]. The arrow at a collision velocity of 0.27 a.u. indicates the predicted position of the $A(4)$ minimum in the velocity-matching model [8.55].

velocities. The MSO model predicts a curve with roughly the correct shape, but the minimum is displaced toward higher velocities than found in the experiment. Finally, the Sidky-Simonsen velocity-matching model provides a reasonable estimate of the position of the minimum of the $A(4)$ curve also in this case, and the prediction of no pronounced minimum for $A(3)$ is in agreement with the (limited) experimental data available.

In closing we note that alignment anisotropies have also been studied for electron transfer to multiply charged ions such as $O^{6+}$ [8.56]. For the $O^{6+} + Na(3p)$ system, the experimental results for the important endoergic channel $n = 10$, studied via the anisotropy parameter $A(10 \rightarrow 8)$ for the $O^{5+}$ 450 nm emission, exhibits a variation with velocity opposite to the one just discussed for $He^{2+}$ impact, in the same velocity range. The experimental data, however, are too limited to enable an unambiguous location of an $A$-minimum. For highly charged ions such as $O^{6+}$, CC-AO calculations are presently not feasible because of practical limitations on the size of the basis set. The CTMC approach, however, is well suited for electron-transfer studies of these systems, and the theoretical curve follows nicely the experimental data. The MSO predictions, on the contrary, are very different from the experimental findings, as are those from the Oppenheimer-Brinkman-Kramers (OBK) theory [8.56]. The velocity-matching model, using an impact parameter for electron transfer of $15\,a_0$ and a Bohr–Lindhard estimate of $42\,a_0$ for the interaction range, gives $v_c = 0.53\,v_0$ for the $n = 10$ channel, in close agreement with the CTMC result [8.56].

## 8.3 Angle-Differential Studies Using Optically Prepared Targets

We now proceed to scattering angle-differential studies using optically prepared targets. As expected from the discussions above, all experimental studies have been performed at relatively low velocities.

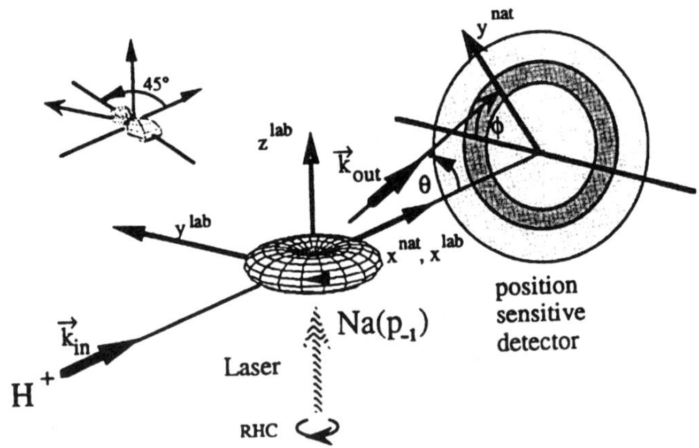


**Fig. 8.25.** Schematic diagram of the experimental scattering geometry used for angle-differential studies of an optically prepared target at very small scattering angles. The intensity pattern $I(\theta, \phi)$ of forwardly scattered projectiles is mapped by a position-sensitive detector located a few meters downstream from the collision region. The recorded intensity pattern is subsequently Fourier analyzed according to (8.33) [8.57].

### 8.3.1 Level populations in $H^+$–$Na\,(3p) \to H\,(n = 2, 3)$–$Na^+$ scattering

Figure 8.25 shows an experimental scattering geometry used for angle-differential studies of an optically prepared target at very small scattering angles [8.57] (compare with Figure 8.11). A sodium-atom beam, a laser beam, and a proton beam intersect each other at right angles. The effusive sodium beam is collimated to a small diameter and a small angular divergence. The polarization of the laser pump beam for the Na target can be manipulated by linear or circular polarizers. The incident proton beam is collimated to a small angular divergence, and chopped in order to separate the various reaction channels by time-of-flight (TOF) spectroscopy. For each channel, the intensity pattern $I(\theta, \phi)$ of forwardly scattered neutrals is mapped by a position-sensitive detector located a few meters downstream from the collision region. After subtracting the signal component from the Na(3s) ground state, the recorded intensity pattern is Fourier analyzed according to

$$I(\theta, \phi) \sin \theta = \sum_{m}^{m_{\max}} C_m(\theta) \cos m\phi \qquad (8.33)$$

(compare with (8.13)), where $m_{\max} = 2L$ for an angular-momentum quantum number $L$ of the optically prepared state. The Fourier coefficients $C_m(\theta)$ depend on the actual optical preparation, as will be detailed below.

For the process $H^+ - Na(3p)$ at $1\,\mathrm{keV}$ impact energy, the near-resonant $H(n = 2)$ channels dominate the transfer [8.34]. At higher impact energies, however, $H(n > 2)$ channels also become important. Here we compare results for $H(n = 2)$ with those for an exit channel with a nonnegligible energy defect, the $H(n = 3)$ channel, at an impact energy of $2\,\mathrm{keV}$, i.e.,

$$H^+ + Na(3p) \rightarrow H(n = 2) + Na^+ , \qquad \Delta E = -0.36\,\mathrm{eV} , \qquad (8.34)$$

$$H^+ + Na(3p) \rightarrow H(n = 3) + Na^+ , \qquad \Delta E = +1.52\,\mathrm{eV} . \qquad (8.35)$$

Intensity patterns $I(\theta, \phi)$ are accumulated for the three standard pairs $(a, b)$ of linear or circular light polarizations with respect to the incident beam direction, $(a, b) = (0°, 90°)$, $(+45°, -45°)$, and $(\mathrm{RHC}, \mathrm{LHC})$.

The theory of measurement for such an experiment was developed in [8.57,58]. Formulas equivalent to (8.21) may be derived in terms of scattering amplitudes for various choices of basis sets, or in terms of Stokes parameters. The formalism is more complicated, however, than that presented in (8.21) for an S-state target, since now also amplitudes corresponding to states with negative reflection symmetry come into play. Here we restrict the presentation to formulas connecting the scattering amplitudes $f_{\mathrm{fi}}(\theta)$ to the intensity distributions for the various target preparations, and how these in turn are related to the standard set of differential cross sections $\sigma(\theta)$ and dimensionless parameters $(P_1, P_2, P_3, P_4)$. A determination of this set is equivalent to a complete determination of the atomic density matrix. Finally, these may be converted to the perhaps more illuminating parameters $(L_\perp^+, \gamma, P_\ell^+, h)$, conventionally used for discussion of such experiments. If we select the set $(|\sigma\rangle, |\pi^+\rangle, |\pi^-\rangle)$ of real p orbitals as the basis for the three-dimensional manifold of initial Na(3p) states, the density matrix corresponding to (8.18) becomes

$$\rho = \begin{pmatrix} f_\sigma f_\sigma^* & f_\sigma f_{\pi+}^* & 0 \\ f_{\pi+} f_\sigma^* & f_{\pi+} f_{\pi+}^* & 0 \\ 0 & 0 & f_\pi- f_{\pi-}^* \end{pmatrix} \tag{8.36a}$$

$$= \sigma \tfrac{1}{2}(1-h) \begin{pmatrix} 1+P_1 & P_2+iP_3 & 0 \\ P_2-iP_3 & 1-P_1 & 0 \\ 0 & 0 & 0 \end{pmatrix} + h \begin{pmatrix} 0 & 0 & 0 \\ 0 & 0 & 0 \\ 0 & 0 & 0 \end{pmatrix}, \tag{8.36b}$$

where we have used the abbreviations

$$f_\sigma f_\sigma^* = \sum_f |f_{f\sigma}|^2 \equiv \sigma_\sigma, \tag{8.37a}$$

$$f_{\pi+} f_{\pi+}^* = \sum_f |f_{f\pi+}|^2 \equiv \sigma_{\pi+}, \tag{8.37b}$$

$$f_{\pi-} f_{\pi-}^* = \sum_f |f_{f\pi-}|^2 \equiv \sigma_{\pi-}, \tag{8.37c}$$

$$f_\sigma f_{\pi+}^* = \sum_f f_{f\sigma} f_{f\pi+}^*, \tag{8.37d}$$

$$\mathrm{Re}\,(f_\sigma f_{\pi+}^*) = \sum_f \mathrm{Re}\,(f_{f\sigma} f_{f\pi+}^*) \equiv \tfrac{1}{2}(\sigma_{\kappa+} - \sigma_{\kappa-}), \tag{8.37e}$$

$$\mathrm{Im}\,(f_\sigma f_{\pi+}^*) = \sum_f \mathrm{Im}\,(f_{f\sigma} f_{f\pi+}^*) \equiv \tfrac{1}{2}(\sigma_{\lambda+} - \sigma_{\lambda-}). \tag{8.37f}$$

Recall that all these parameters depend on the scattering angle $\theta$, omitted here to simplify notation. In (8.37), "f" is one of the four final states of the $H(n=2)$ manifold, or one of the nine final states of the $H(n=3)$ manifold. The two equations (8.37e,f) originate from the expressions for the cross sections for in-plane scattering from states $\kappa^\pm$ and $\lambda^\pm$, respectively, namely

$$\sigma_{\kappa\pm}(\phi = 0) = \frac{1}{2}\left[\sigma_\sigma + \sigma_{\pi+}\right] \pm \mathrm{Re}\,(f_\sigma f_{\pi+}^*), \tag{8.38a}$$

$$\sigma^{\lambda\pm}(\phi = 0) = \frac{1}{2}\left[\sigma_\sigma + \sigma_{\pi+}\right] \pm \mathrm{Im}\,(f_\sigma f_{\pi+}^*), \tag{8.38b}$$

and the relation

$$\sigma_\sigma + \sigma_{\pi+} = \sigma_{\kappa+} + \sigma_{\kappa-} = \sigma_{\lambda+} + \sigma_{\lambda-}. \tag{8.39}$$

The cross-section terms may be parameterized as one average cross section, $\sigma$, and four dimensionless asymmetry parameters, $(P_1, P_2, P_3, P_4)$, all functions of $\theta$, and related to relative cross sections for the states $\sigma$, $\pi^\pm$, $\kappa^\pm$, and $\lambda^\pm$ in the following way:

$$\sigma = \frac{1}{3}(\sigma_\sigma + \sigma_{\pi+} + \sigma_{\pi-}) \tag{8.40}$$

and

$$P_1 = \frac{\sigma_\sigma - \sigma_{\pi+}}{\sigma_\sigma + \sigma_{\pi+}} \, , \tag{8.41a}$$

$$P_2 = \frac{\sigma_{\kappa+} - \sigma_{\kappa-}}{\sigma_{\kappa+} + \sigma_{\kappa-}} = \frac{2\,\mathrm{Re}\,(f_\sigma\, f_{\pi+}^*)}{\sigma_\sigma + \sigma_{\pi+}} \, , \tag{8.41b}$$

$$P_3 = \frac{\sigma_{\lambda-} - \sigma_{\lambda+}}{\sigma_{\lambda-} + \sigma_{\lambda+}} = -\frac{2\,\mathrm{Im}\,(f_\sigma\, f_{\pi+}^*)}{\sigma_\sigma + \sigma_{\pi+}} \, , \tag{8.41c}$$

$$P_4 = \frac{\sigma_\sigma - \sigma_{\pi-}}{\sigma_\sigma + \sigma_{\pi-}} \, . \tag{8.41d}$$

For settings of the linear polarizer angle of $0°$, $90°$, $\pm45°$ and LHC/RHC polarization, we obtain the corresponding double differential cross sections as Fourier series in $\cos\phi$ and $\cos 2\phi$ [8.56]:

$$I(0°) = \frac{1}{3+2f}\left[(1+2f)\sigma_\sigma + \sigma_{\pi+} + \sigma_{\pi-}\right] \, , \tag{8.42a}$$

$$I(90°) = \frac{1}{3+2f}\left[\sigma_\sigma + (1+f)(\sigma_{\pi+} + \sigma_{\pi-})\right.$$

$$\left. + (\sigma_{\pi+} - \sigma_{\pi-})\cos 2\phi\right] \, , \tag{8.42b}$$

$$I(\pm45°) = \frac{1}{3+2f}\left[(1+f)\sigma_\sigma + (1+\frac{f}{2})(\sigma_{\pi+} + \sigma_{\pi-})\right.$$

$$\pm f(\sigma_{\kappa+} - \sigma_{\kappa-})\cos\phi$$

$$\left. + \frac{f}{2}(\sigma_{\pi+} - \sigma_{\pi-})\cos 2\phi\right] \, , \tag{8.42c}$$

$$I(\mathrm{LHC/RHC}) = \frac{1}{4}\left[2\sigma_\sigma + (\sigma_{\pi+} + \sigma_{\pi-})\right.$$

$$\pm 2(\sigma_{\lambda+} - \sigma_{\lambda-})\cos\phi$$

$$\left. + (\sigma_{\pi+} - \sigma_{\pi-})\cos 2\phi\right] \, . \tag{8.42d}$$

The standard parameters $(L_\perp^+, \gamma, P_\ell^+, h)$ are evaluated from the four asymmetry parameters by the relations

$$L_\perp^+ = -P_3 \, , \tag{8.43a}$$

$$P_\ell^+ e^{2i\gamma} = P_1 + i\,P_2 \, , \tag{8.43b}$$

$$h = \frac{(1+P_1)(1-P_4)}{4 - (1-P_1)(1-P_4)} \, . \tag{8.43c}$$

Although the choice of labels $P_1, \dots, P_4$ above for the dimensionless parameters (8.41) nicely brings out the analogy to the case considered in Section 8.1.3, this notation may also be seducing in the sense that the parameters are not Stokes parameters measured in the usual way, but instead the (reduced) Stokes parameters that would be measured in the reverse experiment, i.e., by analysis of the $\mathrm{Na}(3\mathrm{p}\rightarrow 3\mathrm{s})$ emission resulting from an $\mathrm{Na}^+$ ion beam impinging on an isotropic, incoherent target of $\mathrm{H}(n=2,3)$ atoms.

We note the following relationships, analogous to (8.17),

$$\sigma_{\sigma/\pi+} = 3\sigma\,(1-h)\,\tfrac{1}{2}\,(1 \pm P_1)\,, \tag{8.44a}$$

$$\sigma_{\kappa\pm} = 3\sigma\,(1-h)\,\tfrac{1}{2}\,(1 \pm P_2)\,, \tag{8.44b}$$

$$\sigma_{\lambda\pm} = 3\sigma\,(1-h)\,\tfrac{1}{2}\,(1 \mp P_3)\,, \tag{8.44c}$$

$$\sigma_{\sigma/\pi-} = 3\sigma\,[(1+h) + P_1(1-h)]\,\tfrac{1}{4}\,(1 \pm P_4)$$

$$= 3\sigma\,\frac{2(1+P_1)}{4-(1-P_1)(1-P_4)}\,\tfrac{1}{2}\,(1 \pm P_4)\,. \tag{8.44d}$$

One may insert these equations into the expressions (8.42) for the scattering intensities and thus obtain expressions in terms of the basis-set independent cross section $\sigma$, the Stokes parameters $P_i$, and $h$ in analogy to (8.21). However, owing to the increased complexity of this case, the expressions become correspondingly more complicated. Consequently, the actual extraction of the parameters of interest from Fourier decomposition of the spatial patterns of scattered neutrals is a difficult and tedious procedure, and its reliability requires much attention to detail [8.56,57].

Figure 8.26 compares experimental results for the reduced differential cross sections with theoretical predictions. The theoretical approach was based on a one-electron model and a two-center, electron translation factor modified, 36-state atomic basis expansion of the electronic state using the coupled-channel impact parameter method. The 36-state basis consisted of 23 projectile states, $\mathrm{H}(n=1,2,3; 4\mathrm{s}, 4\mathrm{p}, 4\mathrm{d})$, and 13 $\mathrm{Na}\,(n=3; 4\mathrm{s}, 4\mathrm{p})$ target states. From the semiclassical probability amplitudes, assuming a straight-line projectile trajectory and constant relative velocity, the quantal scattering amplitudes, $f_{\mathrm{fi}}(\theta)$, are obtained with the Eikonal method, and the differential cross sections for capture into $\mathrm{H}(n=2)$ and $\mathrm{H}(n=3)$ follow as the appropriate sums (8.37a–f) over final states of $|f_{\mathrm{fi}}(\theta)|^2$. The theoretical curves were convoluted with the experimental divergence determined by the beam apertures. ($\Delta\theta=0.04°$ for $\mathrm{Na}\,(3\mathrm{s})$ and the circular $\mathrm{Na}\,(3\mathrm{p})$ state, and $0.06°$ for the $\mathrm{Na}\,(3\mathrm{p})$ data obtained with linearly polarized light.) There is a satisfactory overall agreement between experimental and theoretical results concerning the relative sizes and shapes of the cross sections. A striking feature is the opposite behavior of most of the cross sections for the various target preparations for the $n=2$ and $n=3$ channels, respectively, with the circular target state preparation being the exception. Some details of the experimental results are not reproduced by theory; among these we note the oscillations observed in the wings of the differential cross sections, particularly pronounced for $\kappa^+$ state, for which no counterpart is found in the theory curves.

The asymmetries are quantified by the four asymmetry parameters shown in Figure 8.27. The parameters $P_4$, $P_1$, and $P_2$ all show close to opposite behavior for $n=2$ and $n=3$, respectively, in the angular range up to scattering

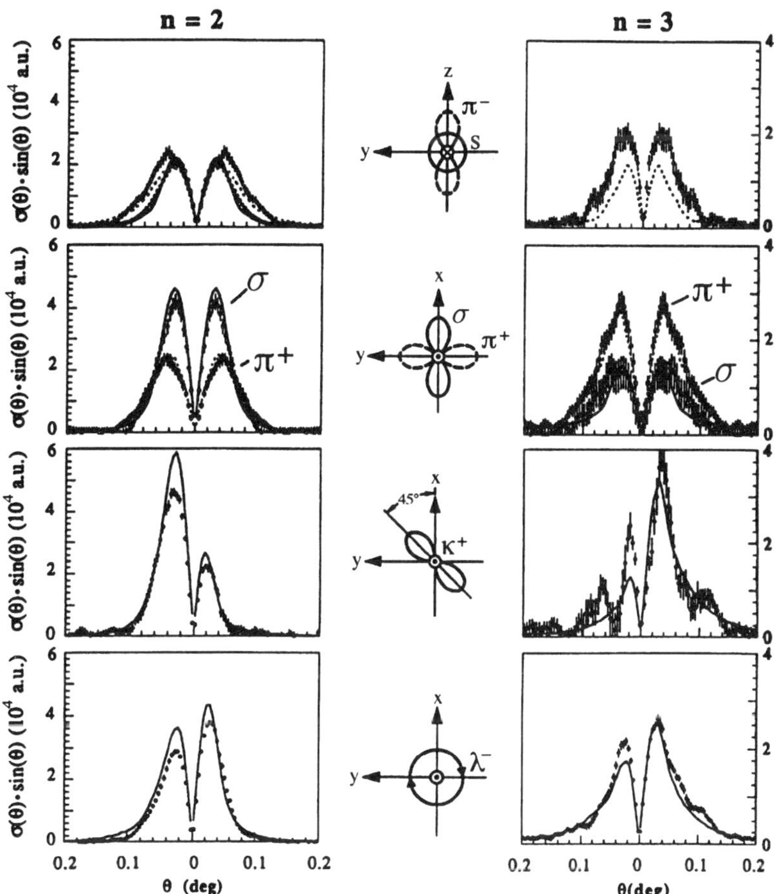



**Fig. 8.26.** Comparison of measurements and convoluted theoretical results for the reduced differential cross sections for electron transfer into $H(n = 2, 3)$. The beam divergence was $\Delta\theta = 0.04°$ for $Na(3s)$ and the circular $Na(3p)$ state, and $0.06°$ for the remaining $Na(3p)$ data [8.57].

angles of $\theta = 0.05°$, inside which the dominant part of the cross sections is located. These propensities reflect corresponding effects in the electron-transfer probability curves as function of impact parameter [8.56]. Only the parameter $P_3$, which measures the relative capture from the oppositely oriented circular states, shows similar angular behavior for the $n = 2$ and $n = 3$ channels. However, inspection of the impact parameter-dependent probabilities [8.56] reveals that this similarity, too, corresponds to *opposite* left-right asymmetries when translated into impact parameter dependence.

Figure 8.28 shows the differential cross sections and coherence parameters for the $n = 2$ and $n = 3$ channels, respectively. We note again the generally

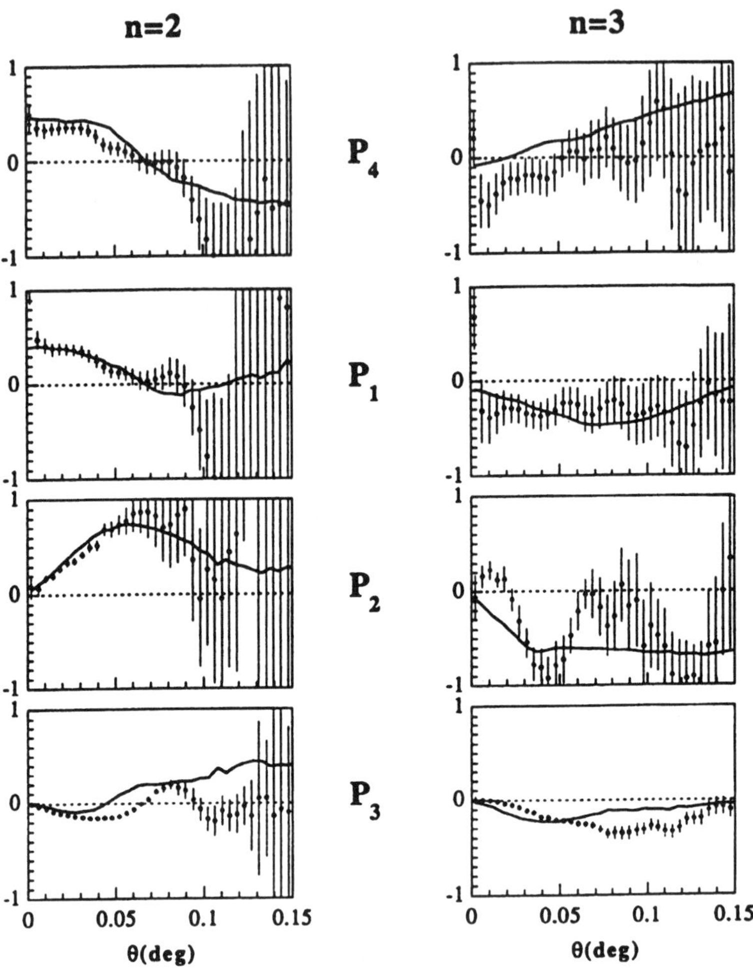



**Fig. 8.27.** The complete set of dimensionless asymmetry parameters $(P_1, P_2, P_3, P_4)$ for electron transfer into $H(n=2,3)$ [8.57].

good agreement between theory and experiment, taking the complexity and detailed level of investigation into account. The parameter $h$, which measures the relative electron-transfer cross section for preparation of the $\pi^-$ target orbital, shows opposite variation with scattering angle for $n=2$ and $n=3$. For $n=2$, it has its minimum value near forward scattering and increases significantly with scattering angle, with the $\pi^-$ cross section being the dominant one beyond $\theta = 0.10°$. For $n=3$, on the other hand, $h$ has its maximum value near forward scattering, where the $\pi^-$ cross section dominates, and then decreases gradually with scattering angle.

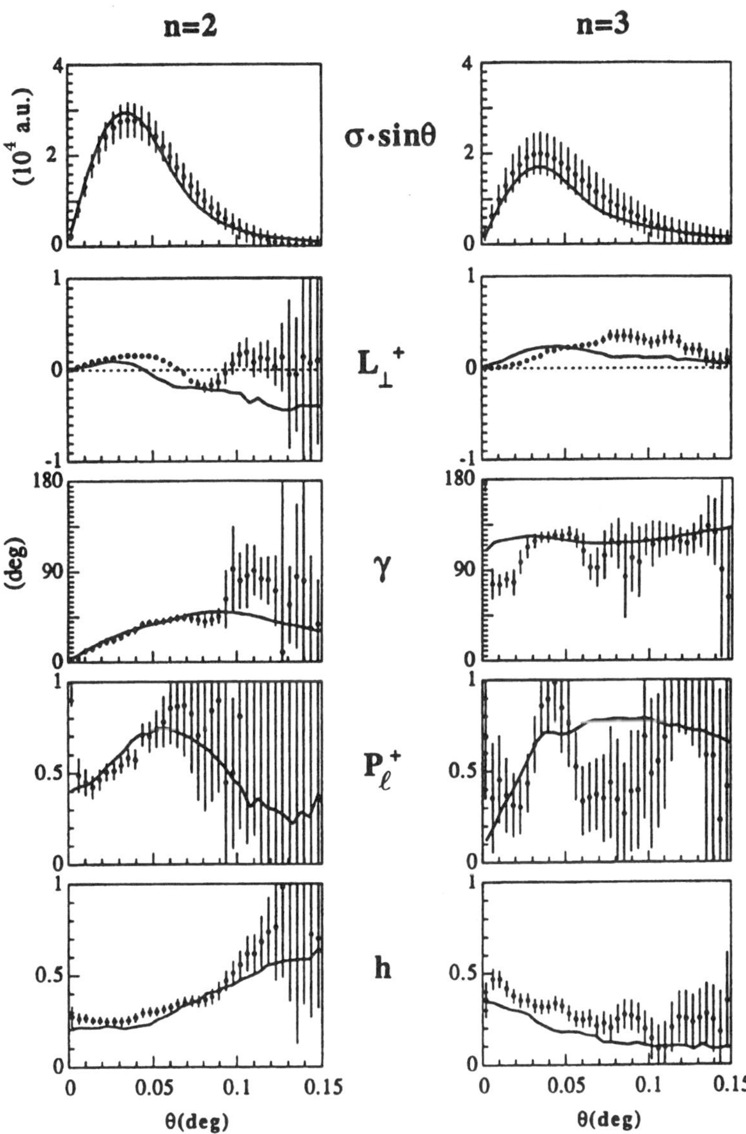



**Fig. 8.28.** Cross sections $\sigma$ and coherence parameters $(L_\perp^+, \gamma, h)$ for electron transfer into $\mathrm{H}(n=2,3)$ [8.57].

Figure 8.29 summarizes in a qualitative way some of the principal observations for the two channels (8.35a,b). It shows the Na(3p) target preparations which are most efficient for electron transfer into (upper row) the four-dimensional $\mathrm{H}(n=2)$ manifold and (lower row) the nine-dimensional

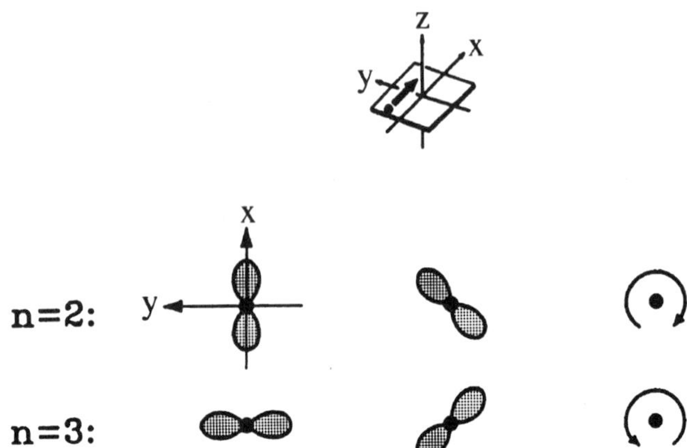

Fig. 8.29. Summary of initial Na(3p) target preparations that favor electron transfer into the H($n\!=\!2$) and H($n\!=\!3$) channels, respectively [8.57].

H($n = 3$) manifold, respectively, in most of the angular range contributing to electron transfer. Here the standard geometry is assumed; i.e., the proton passes the Na(3p) orbital on the left side in the $xy$ plane along the direction of the $x$ axis, as shown at the top. For the first two target geometries, these propensities prevail in the scattering angle dependence of the cross sections, as seen in Figures 8.26 and 8.27. They are manifested by the almost 90° difference of the $\gamma$ angles for $n\!=\!2$ and $n\!=\!3$, for which a satisfactory agreement between theory and experiment is obtained. For the circular state preparation, however, the opposite $b$ dependence for $n = 2$ and $n = 3$, respectively, is neither reflected in the differential cross sections nor in the $P_3$ asymmetry parameters. In addition to the propensities shown in Figure 8.29, preparation of a p orbital along the $z$ axis, $\pi^-$, is increasingly favorable for electron transfer into the H($n\!=\!2$) manifold at larger scattering angles, while the opposite is true for the H($n\!=\!3$) manifold.

Further unraveling of the underlying dynamics of the transfer process will require probing the individual states of the two manifolds. Experimentally, this information is in principle accessible through a full Stokes parameter analysis of the photons emitted when the excited states in the exit channel decay. This is, however, a difficult task, as demonstrated by the Bielefeld group in a pioneering experimental study [8.27] of the linear light polarization of the Ly$_\alpha$ photons from H(2p) by correlation techniques, to be described below. Theoretically, however, the information is accessible through the individual scattering amplitudes, $a_{fi}(b)$ and $f_{fi}(\theta)$. Such an analysis indeed confirms the velocity-matching picture of Kohring $et\ al.$ [8.59] for the quasi-resonant process (8.35a), as discussed in more detail for resonant electron transfer in the Na$^+$ $-$ Na(3p) system by Campbell $et\ al.$ [8.60]. For the present system, it manifests itself as the electron-transfer channel Na($3p_{-1}$)$\rightarrow$H($2p_{-1}$)

being dominant for left-side passage. When discussing the endoergic channel (8.35b), we first note that, for quantization axis perpendicular to the scattering plane, a left-side passage of an $\mathrm{Na}(3\mathrm{p}_m)$ state is equivalent to a right-side passage of an $\mathrm{Na}(3\mathrm{p}_{-m})$ state, as is evident from looking at the collision event in a mirror. For collision velocities at which a specific electron-transfer channel becomes important, the criterion (8.8) for orientation may be invoked. When this equation is satisfied for transfer into $\mathrm{H}(n=3)$, $\Delta\epsilon > 0$ implies $\Delta m < 0$. For the $\mathrm{Na}(3\mathrm{p}_{-1})$ state, only the $\mathrm{H}(3\mathrm{d}_{-2})$ exit channel fulfills this criterion, while the $\mathrm{H}(3\mathrm{s}_0, 3\mathrm{p}_{-1}, 3\mathrm{d}_0, 3\mathrm{d}_{-2})$ manifold is accessible for $\mathrm{Na}(3\mathrm{p}_1)$. Inspection of the theoretical data shows indeed that at this velocity $\mathrm{Na}(3\mathrm{p}_1) \rightarrow \mathrm{H}(3\mathrm{p}_{-1})$ transfer dominates. Further efforts to elucidate the electron-transfer dynamics may, however, concentrate on the conceptually and experimentally simpler system $\mathrm{Li}^+ - \mathrm{Na}(3\mathrm{p})$ for which electron transfer into the three-dimensional $\mathrm{Li}(2\mathrm{p})$ manifold dominates completely over a wide energy range.

### 8.3.2 Level populations in $\mathrm{Li}^+ - \mathrm{Na}(3\mathrm{p}) \rightarrow \mathrm{Li}(2\mathrm{p}) - \mathrm{Na}^+$ scattering

At $1\,\mathrm{keV}$ impact energy, the energy loss spectrum [8.61] shows that the near-resonant (slightly exoergic) process

$$\mathrm{Li}^+ + \mathrm{Na}(3\mathrm{p}) \rightarrow \mathrm{Li}(2\mathrm{p}) + \mathrm{Na}^+, \quad \Delta E = +0.51\,\mathrm{eV} \tag{8.45}$$

is by far the dominating channel. Experiments may thus be performed with a continuous $^6\mathrm{Li}^+$ beam, which improves counting statistics significantly, as will be seen from the data below. Thus, the scattering patterns of $\mathrm{Li}(2\mathrm{p})$ atoms have been mapped for an optically prepared $\mathrm{Na}(3\mathrm{p})$ target, using a full set of polarizer settings $(0°, 90°)$, $(+45°, -45°)$, and $(\mathrm{RHC}, \mathrm{LHC})$. Information on the cross sections for the corresponding $\sigma$, $\pi^+$, $\kappa^\pm$, and $\lambda^\pm$ target states may subsequently be derived from (8.42a–d) [8.61].

The left column of Figure 8.30 shows experimental and theoretical results for the reduced differential cross section $\sigma(\theta)\sin(\theta)$ corresponding to the set of five pure state target orbital geometries $(\pi^-, \sigma, \pi^+, \kappa^+, \lambda^-)$. The theoretical results presented are from a 28-state molecular basis calculation [8.62] (MO) and a 26-state atomic basis calculation [8.61] (AO). The theoretical curves were convoluted with the experimental divergence determined by the beam apertures, $\Delta\theta = 0.04°$. The experimental results were brought to an absolute scale by requiring that the integrals of the measured and convoluted theoretical cross section $\sigma_{3\mathrm{s}}(\theta)$ be identical in the angular range $\theta \leq 0.15°$, properly taking into account the effective fraction of excited states $\alpha$.

The results display two strong anisotropies of the scattering process (8.45). First, electron transfer from an initial state $\sigma$ dominates strongly compared to capture from an initial $\pi^+$ or $\pi^-$ state. This is consistent with the results for the angle-integrated cross sections shown in Figure 8.21, but in addition the present data display the angular dependence and the different behavior

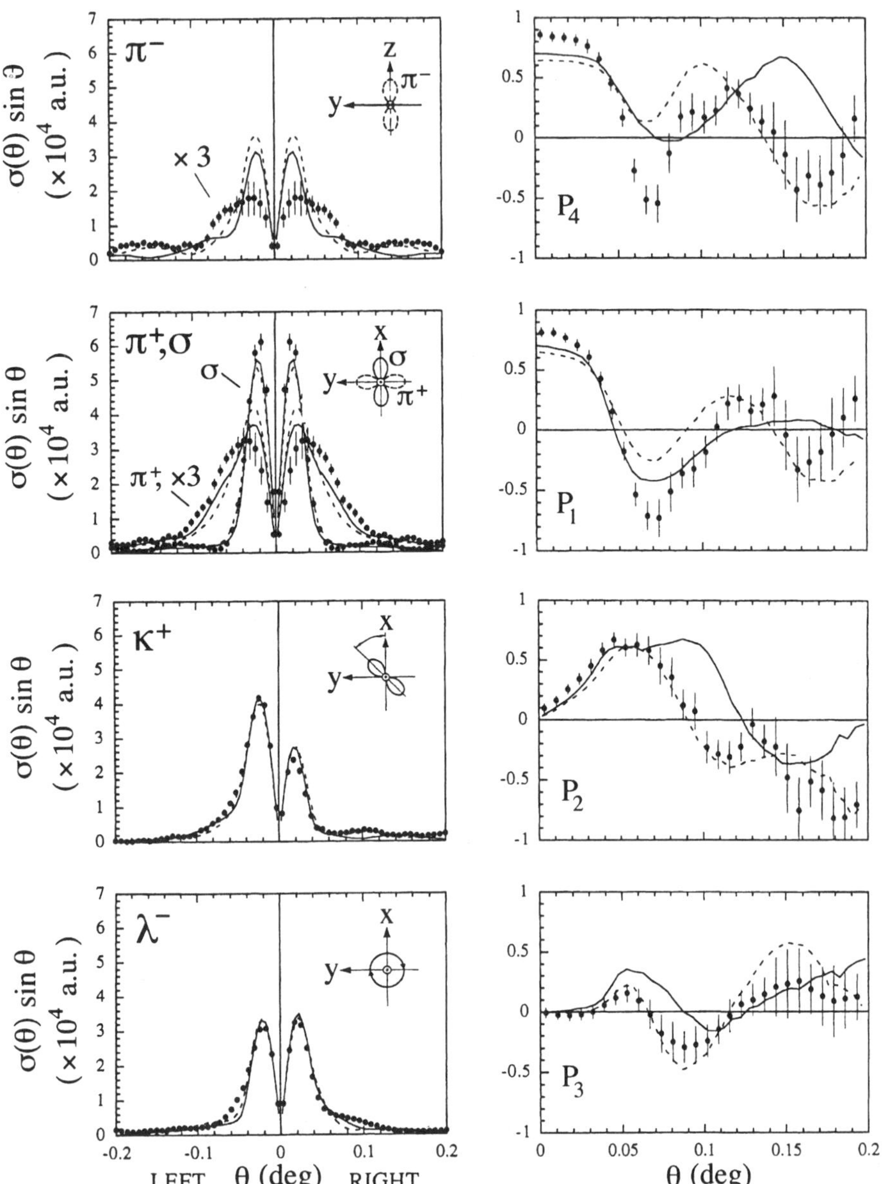



**Fig. 8.30.** Comparison of measurements and convoluted theoretical results for the reduced differential cross sections (left) and the complete set of dimensionless asymmetry parameters $(P_4, P_1, P_2, P_3)$ for electron transfer into Li (2p). Experiment (•) [8.61], MO theory (− − −) [8.62], AO theory (———) [8.61].

of the $\pi^+$ and $\pi^-$ states. Second, a strong left-right scattering anisotropy is observed for the scattering from an initial $\kappa^+$ state. A weaker left-right anisotropy is observed for the $\lambda^-$ state.

Except for the shape of the $\pi$ state cross sections, we notice a satisfactory agreement between experimental and theoretical results concerning the relative sizes and shapes of the cross sections. Agreement is good with both the AO and MO predictions at small scattering angles, where, however, the smoothing of finer details of the theoretical curves due to the finite angular beam divergence is very important because of azimuthal $\phi$-mixing. At larger angles, some details of the experimental results are not reproduced by theory, especially for the asymmetry parameters; among these discrepancies we note the oscillations observed in the wings of the differential cross sections, particularly pronounced for the $\kappa^+$ orbital, which are not reproduced by the AO theory.

The asymmetries are quantified by the four asymmetry parameters shown in the right column of Figure 8.30. We note a similarity in shape of the measured $P_i$ parameters with the corresponding set for H (2p). Although theory reproduces the general trends of the scattering symmetries, the experimentally determined anisotropies are often even more pronounced than theoretically predicted.

The left column of Figure 8.31 shows the corresponding reduced differential cross section $\sigma_{3p}$ and the complete set of coherence parameters versus scattering angle. Recall the interpretation of these parameters as characterizing the Na (3p) charge cloud created in the reverse process to (8.45), using an incoherent, isotropic Li (2p) target. We note good agreement between the two theories and experiment concerning the cross section. Only the trends are reproduced for most of the other parameters, which may also be considered satisfactory in light of the complexity of the analysis and the detailed level of investigation. Again, at larger scattering angles corresponding to the tail of the cross sections, the MO results tend to agree better with experiment than the AO predictions. The rich structures seen in the four coherence parameters contrast the much smoother behavior of the differential cross section and illustrate the challenges facing the theoretician when attempting to simulate the reaction (8.45) at the level defined by the present experimental approach. The right column of Figure 8.31 visualizes the parameters in terms of charge clouds at selected scattering angles. We note a very distinct dependence on the scattering angle for both the cloud shape and the alignment angle, even for an angular step size of only $0.03°$ — close to the angular resolution of the experiment.

When comparing with the results presented for $H^+$ impact in Section 8.3.1, we note that the present data are very similar to the findings for the analogous, exoergic process for capture into H($n = 2$) (see Figures 8.26–28) and thereby different from the endoergic H($n = 3$) channel. At the present level of investigation, which sums over the magnetic substates of the Li (2p)

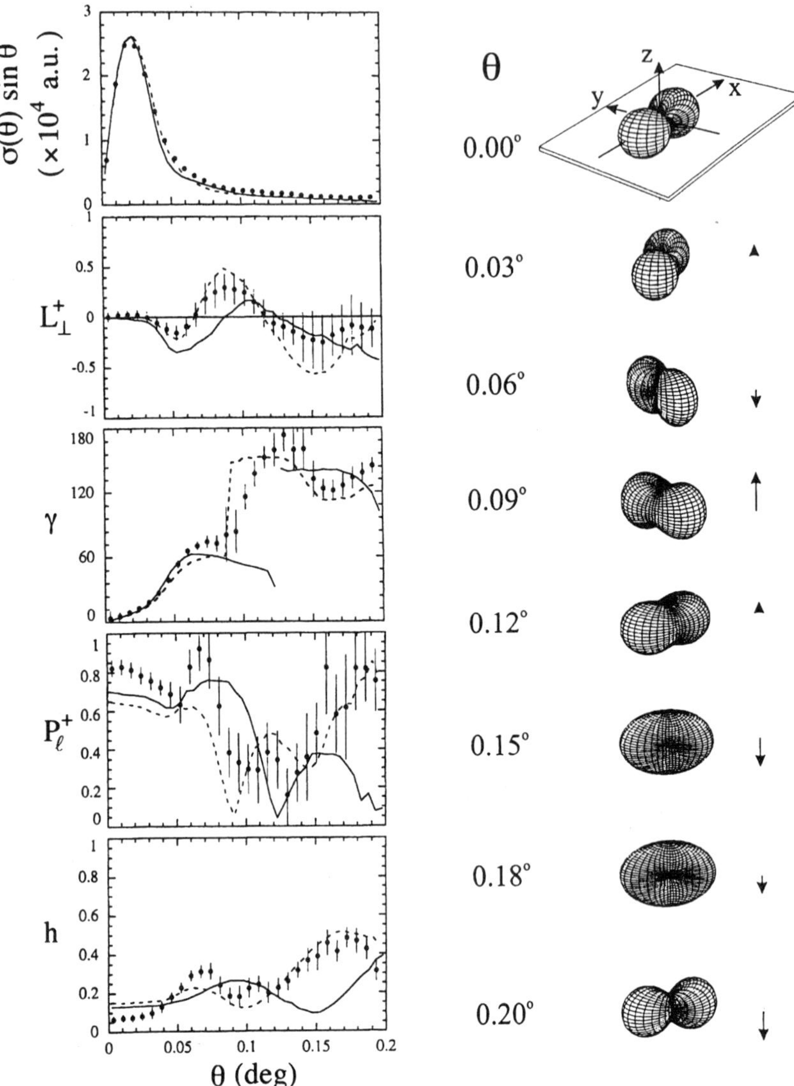

**Fig. 8.31.** The coherence parameters $(\sigma, L_\perp^+, \gamma, P_\ell^+, h)$ for electron transfer into Li (2p). The corresponding electron charge clouds for angular steps of $0.03°$ are shown on the right. Experiment ($\bullet$) [8.61], MO theory ($- - -$) [8.62], AO theory (———) [8.61].

final level, both theoretical calculations, MO and AO, are reliable concerning prediction of the electron transfer for the various initial Na(3p) states. The theories may be put to further test by analyzing the shape of and current circulation in the final-state electron orbital, as discussed in Section 8.4.2 below.

## 8.4 Angle-Differential Studies Using Optically Prepared Targets and Optical Final-State Analysis

We now advance one level further in the study of excitation and transfer processes, state-to-state transitions, by addressing processes of the generic type

$$A^+ + B(n\text{p}) \rightarrow A(n'\text{p}) + B^+ , \tag{8.46}$$

in which the initial target state $B(n\text{p})$ is prepared optically in a specific P state, and the final state $A(n'\text{p})$ is studied by photon polarization or correlation analysis of the subsequent optical decay. In the present discussion, we neglect effects of fine structure and hyperfine structure, which may cause depolarization in the initial and final states, as well as possible effects of electron spin polarization caused by pumping with circularly polarized light. We thus assume that reflection symmetry with respect to the collision plane is conserved. A P$-$P transition process is also interesting from the point of view that it is, in some sense, its own time-reverse, contrary to the S$-$P (or S$-$D) transitions discussed earlier.

Accounting for conservation of reflection symmetry with respect to the scattering plane, Figure 8.32 shows the nonzero transition amplitudes $f_1 - f_5$ connecting the initial and final P states. We have here, somewhat arbitrarily, selected the basis of real-valued orbitals. The actually preferred choice of basis set may depend on the problem considered, but in practical applications the algebra in this basis often leads to shorter analytical expressions than using, for example, the helicity basis.

With five independent transition amplitudes, a quantum-mechanically complete scattering experiment should thus determine a total of nine independent parameters, such as one absolute cross section, four relative amplitude sizes, and four relative angles.[13] It is of interest to address the general question of which measurement strategies could preferably be applied for this purpose, such as the choice of suitable observation directions and proper collision parameters to determine. In this context, it is useful to have an experimental

---

[13]In case the two P states are identical, as in elastic scattering from a P state or in resonant electron transfer, the number of parameters is reduced to seven; here the two amplitudes $f_2$ and $f_3$ are no longer independent, but one is the complex conjugate of the other.

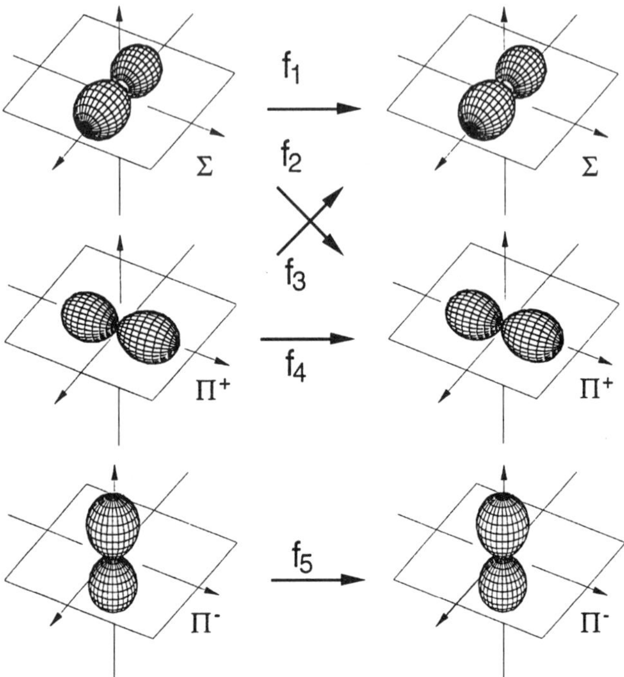



**Fig. 8.32.** The five nonzero P−P transition amplitudes, here shown in the basis of real-valued orbitals, in an experiment that conserves reflection symmetry in the scattering plane of the atomic wavefunction.

approach in mind. One setup of this kind from Bielefeld [8.63], using optical target preparation and subsequent photon correlation analysis, has already been presented in Figure 4.17. Another one, from Orsay, which invokes Stokes parameter analysis of the emitted photons, is shown in Figure 8.33 [8.64]. This setup unifies the two approaches presented in Figures 8.11 and 8.25. A laser prepares the target atoms in specific states, as controlled by a polarizer $P_i$. The intensity pattern of forwardly scattered particles is measured in coincidence with photons emitted in a specific direction and polarization analyzed with the polarizer $P_f$. As before, the recorded intensity pattern $I(\theta, \phi)$ is then fitted to a Fourier series with the generic structure

$$I(\theta, \phi) \sin \theta = \sum_{m=0}^{N} C_m(\theta) \cos m\phi + \sum_{m=1}^{N} S_m(\theta) \sin m\phi, \qquad (8.47)$$

where $N = 2(L_i + L_f) = 4$ in the present case of a P $\rightarrow$ P transition with $L_i = L_f = 1$ [8.50]. The Fourier coefficients $C_m(\theta)$ and $S_m(\theta)$ also depend on the initial and final states selected, and thereby on the settings of the polarizers $P_i$ and $P_f$. They may be evaluated in terms of the real and imaginary

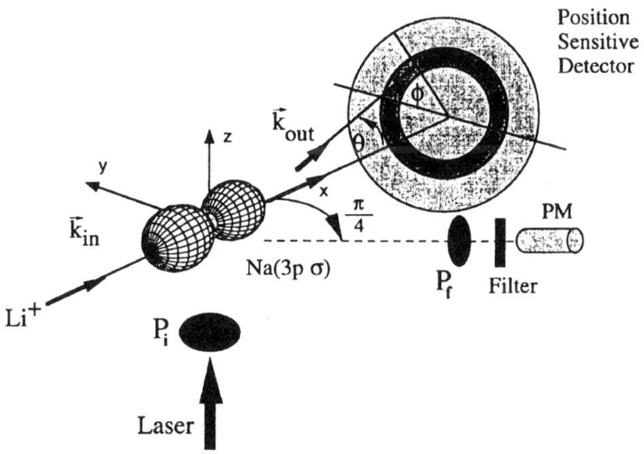


**Fig. 8.33.** An experimental setup invoked for the study of P–P transitions. The optical preparation of the target states is controlled by the polarizer $P_i$. The intensity pattern $I(\theta, \phi)$ of forwardly scattered projectiles is measured in coincidence with photons emitted in a fixed direction, polarization analyzed with the polarizer $P_f$. The recorded intensity pattern is Fourier analyzed according to (8.47) for selected pairs of polarizer settings $(P_i, P_f)$ [8.64].

parts of the scattering amplitudes $f_1, \ldots, f_5$. The complete analysis of this problem is a straightforward but somewhat elaborate affair [8.50] and will not be attempted here. Only the reasoning necessary for understanding the following examples will be presented.

We define partial cross sections $r_{pp}$ and coherence terms $r_{pq}$ and $i_{pq}$ by

$$r_{pq} = Re(f_p f_q^*), \tag{8.48a}$$
$$i_{pq} = Im(f_p f_q^*). \tag{8.48b}$$

With the help of (8.47), one may evaluate the Stokes parameters in any direction for an arbitrary preparation of the initial target state. For the present purpose, the results listed in Table 8.1 suffice.

We first consider the case of initial preparation of a $\sigma$ state, i.e., a p orbital aligned along the laboratory $x$ axis in Figure 8.33. With this preparation, only the amplitudes $f_1$ and $f_2$ come into play (cf. Figure 8.32). For observation in the laboratory $z$ direction, the Fourier components $C_0(s_{0z}^\sigma)$ and $C_0(s_{1z}^\sigma)$ determine the (relative) lengths, or partial cross sections $r_{11}$ and $r_{22}$, of the amplitude pair $(f_1, f_2)$, while the relative angle may be evaluated from $C_1(s_{2z}^\sigma)$ and $C_1(s_{3z}^\sigma)$, respectively.

Proceeding to initial preparation of a $\pi^+$ state, i.e., a p orbital aligned perpendicular to the incident beam direction along the laboratory $y$ axis, further inspection of Table 8.1 shows that the (relative) lengths of the amplitude triple $(f_3, f_4, f_5)$ may be obtained, for example, from a determination of the four independent parameters $C_0(s_{0z}^{\pi^+})$, $C_0(s_{0y}^{\pi^+})$, $C_2(s_{0z}^{\pi^+})$, and $C_2(s_{1z}^{\pi^+})$,

**Table 8.1** Overview of Fourier components $C_m$, $S_m$ up to order $m = 2$ of Stokes parameters $s^i_{j\hat{n}}$ ($j = 0-3$) in the direction $\hat{n}$ for initial state preparation "i," extracted from [8.50]. For $\sigma$ preparation, higher-order terms are zero. For $\pi^+$ preparation, the higher-order terms yield no additional information.

| $s^i_{m\hat{n}}$ | $C_0$ | $C_1$ | $S_1$ | $C_2$ | $S_2$ |
|---|---|---|---|---|---|
| $s^\sigma_{0z}$ | $4(2r_{11}+r_{22})$ | $0$ | $0$ | $4r_{22}$ | $0$ |
| $s^\sigma_{1z}$ | $4(2r_{11}-r_{22})$ | $0$ | $0$ | $-4r_{22}$ | $0$ |
| $s^\sigma_{2z}$ | $0$ | $16r_{12}$ | $0$ | $0$ | $0$ |
| $s^\sigma_{3z}$ | $0$ | $16i_{12}$ | $0$ | $0$ | $0$ |
| $s^{\pi^+}_{0z}$ | $4r_{33}+3r_{44}+3r_{55}+2r_{45}$ | $0$ | $0$ | $4(r_{33}+r_{44}-r_{55})$ | $0$ |
| $s^{\pi^+}_{1z}$ | $4r_{33}-3r_{44}-3r_{55}-2r_{45}$ | $0$ | $0$ | $4(r_{33}-r_{44}+r_{55})$ | $0$ |
| $s^{\pi^+}_{2z}$ | $0$ | $4(3r_{34}+r_{35})$ | $0$ | $0$ | $0$ |
| $s^{\pi^+}_{3z}$ | $0$ | $4(3i_{34}+i_{35})$ | $0$ | $0$ | $0$ |
| $s^{\pi^+}_{0y}$ | $4r_{33}+r_{44}+r_{55}-2r_{45}$ | $0$ | $0$ | $4r_{33}$ | $0$ |
| $s^{\pi^+}_{1y}$ | $-4r_{33}+r_{44}+r_{55}-2r_{45}$ | $0$ | $0$ | $-4r_{33}$ | $0$ |
| $s^{\pi^+}_{2y}$ | $0$ | $0$ | $-4(r_{34}-r_{35})$ | $0$ | $0$ |
| $s^{\pi^+}_{3y}$ | $0$ | $0$ | $4(i_{34}-i_{35})$ | $0$ | $0$ |

which in addition will yield the term $r_{45}$. (We note that this step requires knowledge of the relative detector efficiencies along the $y$ and $z$ directions. These, however, can be obtained from observations of the $\sigma$ preparation for which the signals in the two directions have equal strengths due to rotational symmetry.) The four Fourier components $C_1(s^{\pi^+}_{2z})$, $C_1(s^{\pi^+}_{3z})$, $S_1(s^{\pi^+}_{2y})$, and $S_1(s^{\pi^+}_{3y})$ can be converted into $r_{34}$, $r_{35}$, $i_{34}$, and $i_{35}$, from which the relative angles may be evaluated. Determination of the other Fourier coefficients in Table 8.1 is not strictly necessary, but they may serve as important consistency checks, or alternative ways of determining the parameters, in these highly sophisticated experiments.

The relative length scale connecting the amplitude sets $(f_1, f_2)$ and $(f_3, f_4, f_5)$ may be established from the relative $\sigma$ and $\pi^+$ signal strengths, as measured in the $z$ direction. This leaves only one angle determining the relative position of the two sets to be established. Inspection of the expressions for the Fourier components for preparation of the $\kappa^+$ state [8.50] shows that

$$C_0(s^{\kappa^+}_{2z}) = 4(r_{14} + r_{23} + r_{15}), \quad C_0(s^{\kappa^+}_{3z}) = 4(i_{14} - i_{23} + i_{15}), \tag{8.49a}$$

$$C_2(s^{\kappa^+}_{2z}) = 4(r_{14} + r_{23} - r_{15}), \quad C_2(s^{\kappa^+}_{3z}) = 4(i_{14} - i_{23} - i_{15}). \tag{8.49b}$$

From these expressions, $r_{15}$ and $i_{15}$, and thereby the angle between $f_1$ and $f_5$, may be extracted. Alternatively, the same phase information can be obtained from preparation of a $\lambda^+$ state [8.50].

The analysis described above is thus one possible recipe for determining the nine independent parameters required for *the complete scattering experiment* in the case of a P$-$P transition. Depending on practical limitations, other routes may be preferable and other choices of parameterization may be more convenient. This will be further discussed in two examples below, which to date are the only experimental studies in this direction for heavy-particle collisions.

### 8.4.1  $H^+-Na\,(3p) \rightarrow H\,(2p)-Na^+$ scattering experiments

With the setup of Figure 4.17, a photon correlation measurement was carried out for the process [8.63]

$$H^+ - Na\,(3p) \;\rightarrow\; H\,(2p) - Na^+ \tag{8.50}$$

at a proton impact energy of 1 keV. The initial state prepared with circularly polarized light propagating collinearly with the ion beam had $M_L = +1$ with respect to the beam axis. It is thus natural to discuss this experiment in terms of the helicity basis in the collision frame, labeled by "$c$." With an initial state $\lambda_c^+$ and photon observation ($\hat{n}$) at the magic angle $\theta_\gamma = 54.7°$ with respect to the beam axis, one obtains the following five Fourier coefficients for the photon intensity $s_{0\hat{n}}^{\lambda_c^+}$ in terms of the collision frame amplitudes $f_{\mathrm{fi}}$ [8.63]:

$$C_0(s_{0\hat{n}}^{\lambda_c^+}) = 3(|f_{-11}^c|^2 + |f_{01}^c|^2 + |f_{11}^c|^2)\,, \tag{8.51a}$$

$$C_1(s_{0\hat{n}}^{\lambda_c^+}) = 2\,\mathrm{Re}\,(f_{01}^c f_{11}^{c*})\,, \tag{8.51b}$$

$$S_1(s_{0\hat{n}}^{\lambda_c^+}) = 2\,\mathrm{Im}\,(f_{-11}^c f_{11}^{c*})\,, \tag{8.51c}$$

$$C_2(s_{0\hat{n}}^{\lambda_c^+}) = 2\,\mathrm{Re}\,(f_{-11}^c f_{11}^{c*})\,, \tag{8.51d}$$

$$S_2(s_{0\hat{n}}^{\lambda_c^+}) = 2\,\mathrm{Im}\,(f_{-11}^c f_{11}^{c*})\,. \tag{8.51e}$$

All these may be extracted from the experiment. In the derivation of (8.51), it is important to take into account that, with the optical pumping scheme of [8.63], the transferred electron is spin-polarized. Due to spin-orbit coupling, this spin-polarization affects the subsequent optical decay pattern. Using theoretical results to bring the data to an absolute scale and to resolve an ambiguity in the extraction of $|f_{11}^c|^2$, one may obtain complete information on the amplitude triple $(f_{-11}^c, f_{-11}^c, f_{11}^c)$.

With this information in hand, it is possible to evaluate the shape of the final state electron charge H$\,(2p)$ charge cloud shown in Figure 8.34 [8.63]. For forward scattering, the orbital is transferred without change in shape. For the larger scattering angles, $\theta = 0.05°$ and $0.10°$, both theory and experiment indicate dramatic changes in the shape of the electron cloud. For this choice of initial state, the angular momentum of the final state may point in *any direction in space*, since all three states in the H$\,(2p)$ subspace can be excited

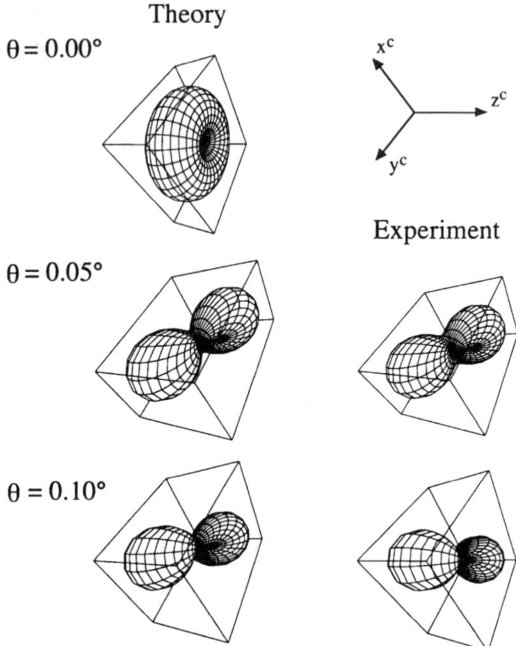



**Fig. 8.34.** Theoretical and experimental results as function of scattering angle for the final H(2p) state, produced by electron transfer from a circular Na(3p) state orbiting in the $x^c y^c$ plane to a 1 keV proton incident along the $z^c$ axis [8.63].

coherently. Interestingly, since the final state is a pure state, a coordinate frame exists in which the shape and dynamics of the electron charge cloud may be described by the usual two parameters $(L_\perp, \gamma)$. This frame, however, is both tilted and twisted with respect to the collision frame, as clearly seen in Figure 8.34. For further discussion of this aspect, see [8.50].

## 8.4.2 Li$^+$–Na(3p) → Li(2p)–Na$^+$ scattering experiments

With the setup of Figure 8.33, final-state polarization analysis was carried out for the process

$$^6\text{Li}^+ - \text{Na}(3\text{p}) \rightarrow {}^6\text{Li}(2\text{p}) - \text{Na}^+ \tag{8.52}$$

at an impact energy of 1 keV [8.63]. An observation direction $\hat{\boldsymbol{u}} = \frac{1}{\sqrt{2}}(1, -1, 0)$ was chosen for reasons of instrumental design. We first verify that a complete scattering experiment may be performed with this setup. According to the general theory of measurement [8.50], one obtains expressions for the Stokes parameters in the direction $\hat{\boldsymbol{u}}$ given in Table 8.2 [8.64].

Inspection of Table 8.2 shows that preparation of a $\sigma$ state and determination of the $C_0$ coefficients for $s^\sigma_{0u}$ and $s^\sigma_{1u}$ allows for derivation of the

**Table 8.2** Overview of Fourier components $C_m$, $S_m$ up to order $m = 2$ of Stokes parameters $s^i_{j\hat{u}}$ $(j = 0 - 3)$ in the direction $\hat{u}$ for initial state preparation "i" [8.64]. For $\sigma$ preparation, higher order terms are zero. For $\pi^+$ preparation, the higher-order terms yield no additional information.

| $s^i_{m u}$ | $C_0$ | $C_1$ | $S_1$ | $C_2$ | $S_2$ |
|---|---|---|---|---|---|
| $s^{\sigma}_{0u}$ | $2(\,2r_{11} + 3r_{22})$ | $8r_{12}$ | $0$ | $-2r_{22}$ | $0$ |
| $s^{\sigma}_{1u}$ | $2(-2r_{11} + r_{22})$ | $-8r_{12}$ | $0$ | $-6r_{22}$ | $0$ |
| $s^{\sigma}_{2u}$ | $0$ | $0$ | $8\sqrt{2}r_{12}$ | $0$ | $2\sqrt{2}r_{22}$ |
| $s^{\sigma}_{3u}$ | $0$ | $0$ | $-8\sqrt{2}i_{12}$ | $0$ | $0$ |
| $s^{\pi^+}_{0u}$ | $\frac{1}{2}(5(r_{44}+r_{55})+4r_{33}-2r_{45})$ | $2(3r_{34}+r_{35})$ | $0$ | $2(r_{33}+r_{44}-r_{55})$ | $0$ |
| $s^{\pi^+}_{1u}$ | $\frac{1}{2}(-(r_{44}+r_{55})-4r_{33}-6r_{45})$ | $-2(3r_{34}+r_{35})$ | $0$ | $2(r_{33}+r_{44}-r_{55})$ | $0$ |
| $s^{\pi^+}_{2u}$ | $0$ | $0$ | $2\sqrt{2}(r_{34}-r_{35})$ | $0$ | $2\sqrt{2}(r_{44}-r_{55})$ |
| $s^{\pi^+}_{3u}$ | $0$ | $0$ | $-2\sqrt{2}(i_{34}-i_{35})$ | $0$ | $2\sqrt{2}i_{45}$ |

amplitude moduli of $f_1$ and $f_2$, while their relative phase may be extracted from $S_1(s^{\sigma}_{3u})$, combined with any one of the remaining nonzero first-order Fourier coefficients. No additional information can be derived from the other nonzero coefficients, but their measurement may serve as a valuable consistency check.

Preparation of a $\pi^+$ state and determination of all moduli and relative phases corresponding to the triple $(f_3,f_4,f_5)$ is a more complicated task. From Table 8.2 we see that the four parameters $r_{33}$, $r_{44}$, $r_{55}$, and $r_{45}$ may be derived, for example, from $C_0(s^{\pi^+}_{0u})$, $C_0(s^{\pi^+}_{1u})$, $C_2(s^{\pi^+}_{1u})$, and $S_2(s^{\pi^+}_{2u})$. Concerning the relative phases, $C_1(s^{\pi^+}_{0u})$ and $S_1(s^{\pi^+}_{2u})$ yield $r_{34}$ and $r_{35}$. Furthermore, $S_1(s^{\pi^+}_{3})$ determines $i_{35} - i_{34}$. This, together with $S_2(s^{\pi^+}_{3u})$, i.e., $r_{45}$, the already known quantities, and the identity $r_{33}i_{45} = r_{34}i_{35} - r_{35}i_{34}$, is in principle sufficient for determination of $i_{34}$ and $i_{35}$, and therefore of all remaining relative phase angles in the triple $(f_3,f_4,f_5)$. Thus, no third-order or fourth-order Fourier coefficients are needed for a complete determination of the triple, but once again they may serve as consistency checks.

The only missing piece of information on the complete set of five scattering amplitudes $f_1,\ldots,f_5$ is one angle between the pair $(f_1,f_2)$ and the triple $(f_3,f_4,f_5)$. Inspection of the Fourier coefficients corresponding to preparation of a $\kappa^+$ or $\lambda^+$ state (not tabulated) reveals several ways of extracting this information, for example, by using the pair $C_0(s^{\kappa^+}_{0u})$ and $C_0(s^{\lambda^+}_{0u})$. In this way, nine independent amplitude parameters, and thereby a complete experiment, may be performed with this geometry.

In an actual experiment, however, many practical obstacles effectively block this approach. First, due to fine-structure and hyperfine-structure couplings, the anisotropies of both the initial $\mathrm{Na}(3p)$ state and the final $\mathrm{Li}(2p)$ state are significantly reduced. Second, owing to the very small scattering angles of typically less than $0.05°$ (cf. Figure 8.31), it is necessary to use highly collimated beams, with a correspondingly drastic reduction of count rates and increase in data accumulation time. Even then, convolution with the finite angular beam divergence noticeably affects the Fourier coefficients, an effect that increases with increasing order. In the present case, $C_1$ and $S_1$ are affected near the center, and $C_2$ and $S_2$ are severely distorted by convolution. In addition, the smallness of the $C_2$ and $S_2$ coefficients makes them very sensitive to any ellipticity of the initial beam. Consequently, alternative strategies for data collection and evaluation need to be carefully considered, economizing on accumulation time and making maximal use of already existing data, such as results from Section 8.3.2.

We now present the outcome of such an approach [8.64]. The target was prepared with linearly polarized light propagating perpendicular to the ion beam, with polarization axis parallel or perpendicular to the beam. The emitted photons were analyzed with linear polarizer settings parallel and perpendicular to the beam, and with a circular polarization analyzer. Taking all the complications listed above into consideration in the data analysis, making use of the four zero-order Fourier coefficients $C_n(P_i, P_f) = C_0(0°, 0°)$, $C_0(90°, 0°)$, $C_0(0°, 90°)$, and $C_0(90°, 90°)$ for extraction of the partial cross sections, and the four first-order coefficients $C_1(0°, 90°)$, $C_1(90°, 90°)$, $S_1(0°, \mathrm{LHC})$, and $S_1(90°, \mathrm{LHC})$ for extraction of phase information, together with the total $\mathrm{Li}(2p)$ transfer data of Section 8.3.2, enables extraction of the parameters presented in Figure 8.35. (See [8.64] for details on the procedure.)

Figure 8.35(a) presents the five partial cross sections $r_{11}$–$r_{55}$ for electron transfer from $\mathrm{Na}(3p)$ to $\mathrm{Li}(2p)$. At this impact energy, the $\mathrm{Na}(3p\sigma)$ preparation is the most efficient one for electron transfer, with the $\mathrm{Li}(2p\pi^+)$ channel as the most important final state, but $\mathrm{Li}(2p\sigma)$ is also important. The $\mathrm{Na}(3p\pi^+)$ state preparation is less effective, with about equal probabilities for the two final states. Preparation of the state $\mathrm{Na}(3p\pi^-)$, having negative reflection symmetry with respect to the scattering plane, is the least effective, with the corresponding exit channel $\mathrm{Li}(2p\pi^-)$ being the only accessible one. (The remaining four panels in Figure 8.35(a) are empty owing to the selection rule of conserving the reflection symmetry of the atomic wavefunction with respect to the scattering plane during the collision event.)

Figure 8.35(b) presents the measured six independent real and imaginary parts of coherence terms for electron transfer from $\mathrm{Na}(3p)$ to $\mathrm{Li}(2p)$, which may be converted into three phase angles. The values of these terms are seen to be generally small, compared to the partial cross sections of Figure 8.35(a). Thus, eight out of nine independent parameters have been extracted from the experiment, leaving the phase angle of the amplitude $f_5$ as the only undetermined parameter.

Comparing the experimental results [8.64] with theoretical MO [8.62] and AO [8.64] predictions, we note a satisfactory overall agreement for both the-

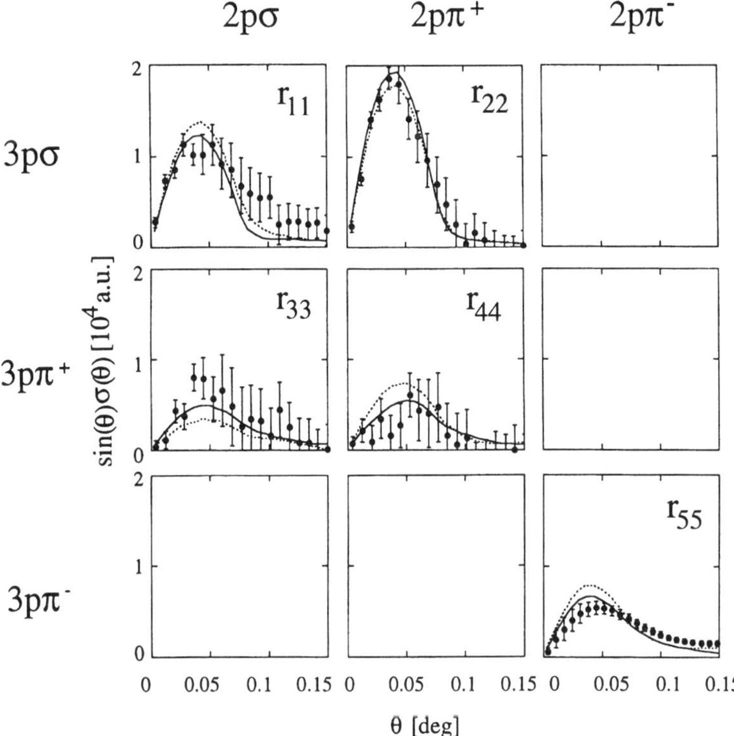


**Fig. 8.35(a).** The five partial cross sections for electron transfer from Na(3p) to Li(2p). Experiment (•) [8.64], MO theory (– – –) [8.62], AO theory (———) [8.64].

ories, with the quantitative agreement often being less satisfactory near $0°$ scattering angle for the small terms that are derived from the first-order Fourier components of the polarization signals. Again, however, considering the level of detail of the investigation and the many effects that may influence the data acquisition, one may consider the degree of success of the two theoretical approaches as an accomplishment, giving confidence in their predictive power also in areas not yet accessible to experimental scrutiny.

As an example, Dowek *et al.* [8.64] investigated theoretically the validity of orientation propensities for electron-transfer processes near the matching velocity, formulated first by Campbell *et al.* [8.60] in a discussion of resonant Na(3p) → Na(3p) electron transfer. Figure 8.36 shows the impact-parameter dependence for left-side trajectory electron capture from circular Na(3pλ±) states, resolved into the two circular Na(3pλ±) basis state components of the final states (cf. the graph at the top of the figure). The six panels show the dependence on impact velocity for several values ranging from $v = 0.08$ a.u. (or a $^6$Li$^+$ impact energy of 1 keV, as in the experiment

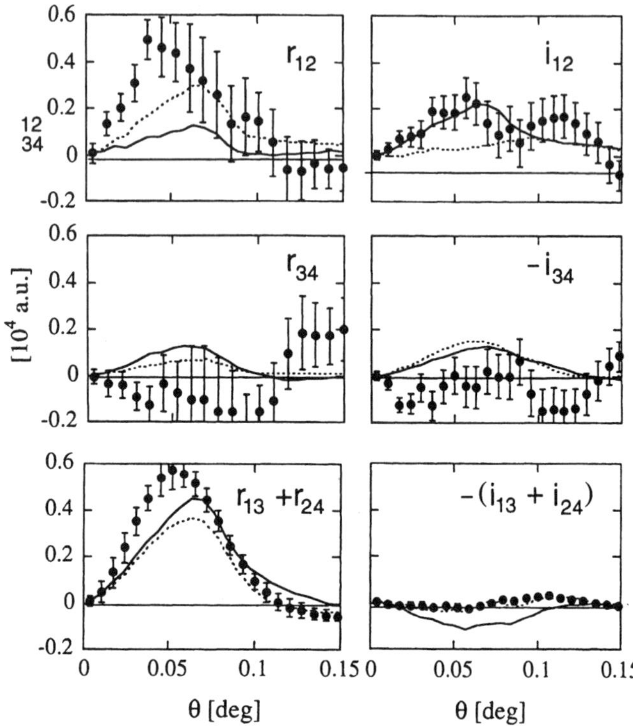

**Fig. 8.35(b).** The six coherence terms for electron transfer from Na (3p) to Li (2p). Experiment (•) [8.64], MO theory (− − −) [8.62], AO theory (———) [8.64].

just discussed) up to and beyond the matching velocity of $v = 0.46$ a.u., i.e., the mean orbital velocity of the Na (3p) electron. Around the maximum of the electron-transfer curves, and for all velocities, we may conclude that transfer is favored when the velocity vector of the incoming ion is parallel (as opposed to antiparallel) to the velocity vector of the target electron. Furthermore, at lower velocities up to about half of the matching velocity, the dominant channel corresponds to *orientation reversal*. In the subsequent graphs where the matching velocity is approached and exceeded, the channel with *orientation conservation* becomes strongly dominating, so that, in this impact velocity range, the circulation of the active electron is the same before and after the transfer.

More details of this picture are shown in Figure 8.37. The shape and angular momentum of the Li (2p) state produced by electron transfer from a circular Na (3p$\lambda^-$) state for a left-side impact parameter are calculated for the same series of velocities as used in Figure 8.36. The impact parameters were chosen at the maximum of the transfer probability curves. It is clearly seen that, close to the matching velocity of 0.46 a.u., the orbital shape of the

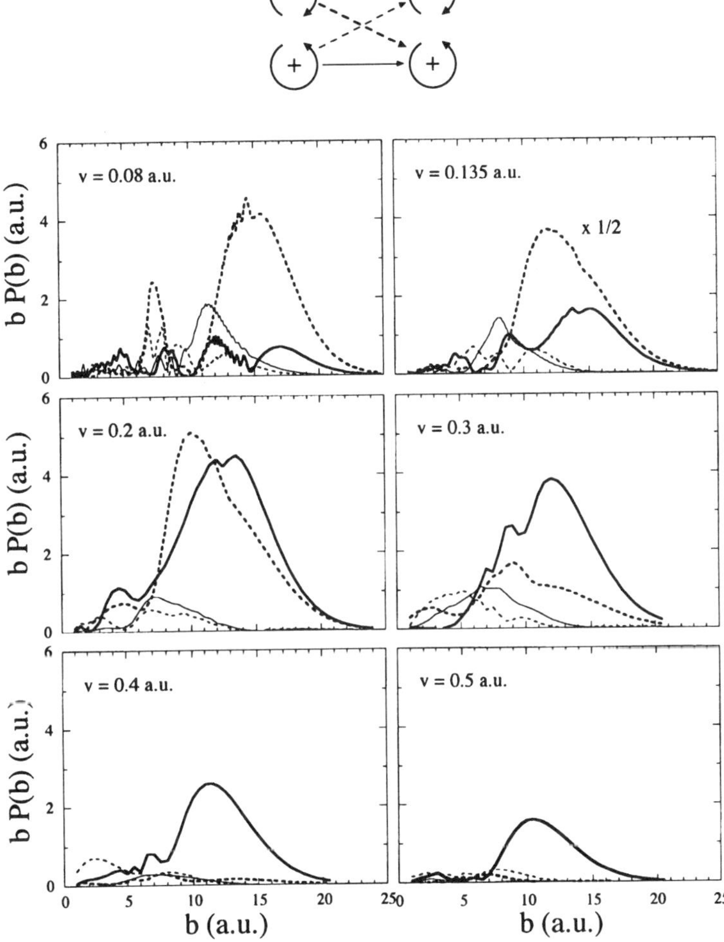



**Fig. 8.36.** The impact parameter dependence of the four transition probabilities for electron transfer from $\mathrm{Na}\,(3p)$ to $\mathrm{Li}\,(2p)$ connecting the circular states in the collision plane, evaluated for left-side impact with impact velocities in the range $0.08 - 0.5$ a.u. [8.64]. The graph at the top defines the notation.

electron cloud and the sense of circulation of the active electron around the atomic core are nearly conserved after the transfer. Decreasing the velocity has a dramatic influence on the shape and the orientation of the orbital of the transferred electron. The intuitive picture of orientation conserving transitions being favored is thus restricted in validity to a narrow velocity range around the matching velocity. Particularly near a velocity of $0.2$ a.u., we notice that the transferred electron cloud has hardly any component along the direction of the incident ion, as predicted by the (impact parameter indepen-

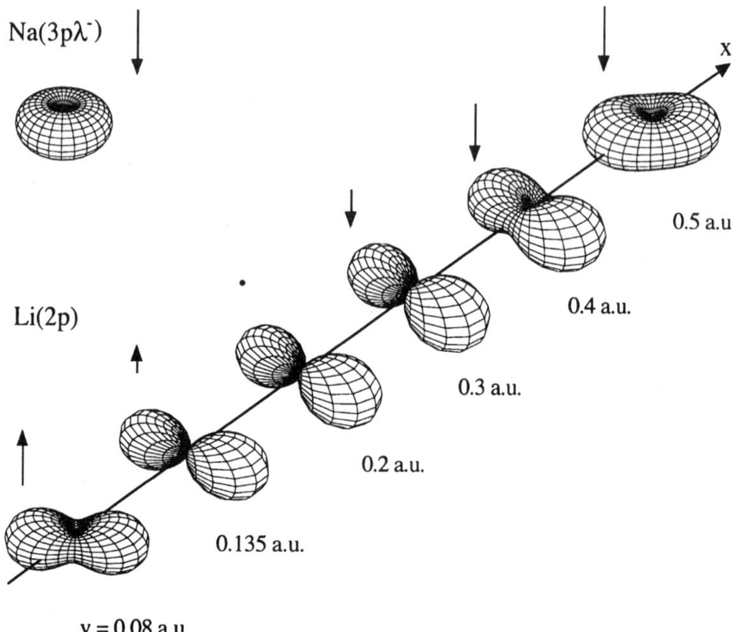


**Fig. 8.37.** The shape and angular momentum of the Li (2p) state produced by electron transfer from a circular Na (3pλ⁻) state, upper left, for impact on the left side, as calculated for selected velocities [8.64]. The impact parameters chosen correspond to maximum transfer probability (see Figure 8.36). The incident beam direction is along $x$.

dent) Sidky–Simonsen alignment propensity rule presented in Section 8.2.1 on the integral alignment for this system.

### 8.4.3 Future directions

The results presented in the two preceding sections demonstrate how new, illuminating details on the collision dynamics of the various reaction channels may be extracted from combined experimental and theoretical struggles toward the realization and understanding of complete scattering experiments. It is also clear from the discussion that further progress in the understanding of electron-transfer collision dynamics will require a rather dramatic improvement in angular resolution to enable further differential scattering studies, preferably at even higher collision velocities, near or beyond the matching velocity. At the same time, a drastic increase in data accumulation rates is strongly called for, in order to retrieve the desired information within a reasonable measurement time. It is also evident that the present experimental approaches, as illustrated in Figures 4.17 and 8.32, have been stretched to their very limits. One attractive path forward would be to introduce the so-called COLTRIMS (COLd Target Recoil Ion Momentum Spectroscopy) technique [8.65,66] for the investigation of these processes. In the present context, this will require implementation of the technology of laser-cooled targets for the experimental studies. Such developments are currently in progress in several laboratories for the study of electron transfer phenomena in collisions involving singly or multiply charged ions. See, for example, [8.67].

# References


8.1 N. Andersen, J.T. Broad, E.E.B. Campbell, J.W. Gallagher, and I.V. Hertel, Phys. Rep. **278** (1997) 107.

8.2 See Section 3.1 of [8.1] for an overview.

8.3 J. Fayeton, N. Andersen, and M. Barat, J. Phys. B **9** (1976) L149.

8.4 N. Andersen and S.E. Nielsen, Adv. At. Mol. Phys. **18** (1982) 265.

8.5 N. Andersen, T. Andersen, and J.Ø. Olsen, J. Phys. B **12** (1979) 3723.

8.6 M. Barat and W. Lichten, Phys. Rev. A **6** (1972) 211.

8.7 G.S. Panev, N. Andersen, T. Andersen, and P. Dalby, Z. Physik D **5** (1987) 331.

8.8 N. Andersen and S.E. Nielsen, Europhys. Lett. **1** (1986) 15.

8.9 N. Andersen, J.W. Gallagher, and I.V. Hertel, Phys. Rep. **165** (1988) 1.

8.10 S.E. Nielsen and N. Andersen, Z. Physik D **5** (1987) 321.

8.11 T. Andersen and J.E. Pedersen, J. Phys. B **22** (1989) 617.

8.12 N. Andersen, T. Andersen, J.S. Dahler, S.E. Nielsen, G. Nienhuis, and K. Refsgaard, J. Phys. B **16** (1983) 817.

8.13 B. Archer, N.F. Lane, and M. Kimura, Phys. Rev. A **42** (1990) 6379.

8.14 H.-P. Neitzke and T. Andersen, J. Phys. B **17** (1984) 1559.

8.15 S.E. Nielsen and N. Andersen, Z. Physik D **11** (1989) 123.

8.16 A. Russek, D.B. Kimball, and M.J. Cavagnero, Phys. Rev. A **23** (1981) 139.

8.17 J.P. Hansen, L. Kocbach, A. Dubois, and S.E. Nielsen, Phys. Rev. Lett. **64** (1990) 2491.

8.18 P. Roncin, C. Adjouri, M.N. Gaboriaud, L. Guillemot, M. Barat, and N. Andersen, Phys. Rev. Lett. **65** (1990) 3261.

8.19 C. Adjouri, P. Roncin, M.N. Gaboriaud, M. Barat, and N. Andersen, J. Phys. B **27** (1994) 3093.

8.20 C. Adjouri, *PhD Thesis* (1993), Université Paris-Sud, France.

8.21 V. Ostrovsky, J. Phys. B **24** (1991) L507.

*8.22 D.H. Jaecks, F.J. Eriksen, W. de Rijk, and J. Macek, Phys. Rev. Lett. **35** (1975) 723.

8.23 F.J. Eriksen, D.H. Jaecks, W. de Rijk, and J. Macek, Phys. Rev. A **14** (1976) 119.

8.24 J.P. Gauyacq, J. Phys. B **11** (1978) 85.

8.25 J.W. Thomsen, I. Reiser, N. Andersen, J.C. Houver, J. Salgado, E. Sidky, A. Svensson, and D. Dowek, J. Phys. B **29** (1996) 5459.

8.26 M. Born and E. Wolf, *Principles of Optics* (4th edition), Pergamon Press, New York 1970.

8.27 Z. Roller-Lutz, Y. Wang, K. Finck, and H.O. Lutz, J. Phys. B **26** (1993) 2697.

8.28 R. Shingal and B.H. Bransden, J. Phys. B **20** (1987) 4815.

8.29 A. Dubois, private communication in [8.27].

8.30 C. Courbin, M. Machholm, I. Reiser, D. Dowek, and J.C. Houver, J. Phys. B **31** (1998) 2305.

8.31 S.E. Nielsen, J.P. Hansen, and A. Dubois, J. Phys. B **23** (1990) 2595.

8.32 J. Grosser, J. Phys. B **14** (1981) 1449.

8.33 S. Grego, J. Salgado, J.W. Thomsen, M. Machholm, S.E. Nielsen, and N. Andersen, J. Phys. B **31** (1998) 3419.

8.34 D. Dowek, J.C. Houver, J. Pommier, C. Richter, T. Royer, N. Andersen, and B. Pálsdottir, Phys. Rev. Lett. **64** (1990) 1713.

8.35 A. Fischer and I.V. Hertel, Z. Physik A **304** (1982) 103.

8.36 J.F. Kelly and A. Gallagher, Rev. Sci.Instrum. **58** (1987) 563.

8.37 U. Müller, H.A.J. Meijer, N.C.R. Holme, M. Kmit, J.H.V. Lauritsen, J.O.P. Pedersen, C. Richter, J.W. Thomsen, N. Andersen, and S.E. Nielsen, Z. Physik D **33** (1995) 187.

8.38 J.W. Thomsen, N. Andersen,D. Dowek, J.C. Houver, M.O. Larsson, J.H.V. Lauritsen, U. Müller, J.O.P. Pedersen, J. Salgado, and A. Svensson, J. Phys. B **28** (1995) L93.

8.39 J.H.V. Lauritsen, J.W. Thomsen, N. Andersen, D. Dowek, J.C. Houver, J.O.P. Pedersen, J.Salgado, and A. Svensson, J. Phys. B **29** (1996) 1093.

8.40 S.E. Nielsen, J.P. Hansen, and A. Dubois, J. Phys. B **28** (1995) 5295; corrigendum, J. Phys. B **29** (1996) 1419.

8.41 T.H. Rod and S.E. Nielsen, J. Phys. B **28** (1995) L607.

8.42 M. Machholm and C. Courbin, J. Phys. B **29** (1996) 1079.

8.43 A. Jain and T.G. Winter, Phys. Rev. A **51** (1995) 2963.

8.44 C.J. Lundy and R.E. Olson, J. Phys. B **29** (1996) 1723.

8.45 F. Aumayr, M. Gieler, J. Schweinzer, H. Winter, and J.P. Hansen, Phys. Rev. Lett. **68** (1992) 3277.

8.46 A.R. Schlatmann, R. Hoekstra, R. Morgenstern, R.E. Olson, and J. Pascale, Phys. Rev. Lett. **71** (1993) 513.

8.47 J.W. Thomsen, N. Andersen, D. Dowek, J.C. Houver J.H.V. Lauritsen, U. Müller, J.O.P. Pedersen, J. Salgado, and A. Svensson, Z. Physik D **37** (1996) 133.

8.48 J.W. Thomsen, N. Andersen, E.E.B. Campbell, I.V. Hertel, and S.E. Nielsen, J. Phys. B **31** (1998) 3429.

8.49 J.W. Thomsen, Can. J. Phys. **74** (1996) 950.

8.50 E. Sidky, S. Grego, D. Dowek, A. Svensson, and N. Andersen (2000), in preparation.

8.51 S. Grego, J. Salgado, P. Borel, S.E. Nielsen, and N. Andersen (2000), in preparation.

8.52 E.Y. Sidky and H.J.T. Simonsen, Phys. Rev. A **54** (1996) 1417, and (2000), in preparation.

8.53 H.C. Brinkmann and H.A. Kramers, Proc. Acad. Sci. Amsterdam **33** (1930) 973.

8.54 S. Schippers, A.R. Schlatmann, and R. Morgenstern, Phys. Lett. A **181** (1993) 80.

8.55 M. Gieler, F. Aumayr, J. Schweinzer, W. Koppensteiner, W. Husinsky, H.P. Winter, K. Lozhkin, and J.P. Hansen, J. Phys. B **26** (1993) 2137.

8.56 S. Schippers, A.R. Schlatmann, W.P. Wiersema, R. Hoekstra, R. Morgenstern, R.E. Olson, and J. Pascale, Phys. Rev. Lett. **72** (1994) 1628.

8.57 J. Salgado, J.W. Thomsen, N. Andersen, D. Dowek, A. Dubois, J.C. Houver, S.E. Nielsen, I. Reiser, and A. Svensson, J. Phys. B **30** (1997) 3059.

8.58 C. Richter, N. Andersen, J.C. Brenot, D. Dowek, J.C. Houver, J. Salgado, and J.W. Thomsen, J. Phys. B **26** (1993) 723.

8.59 G.A. Kohring, A.E. Westmore, and R.E. Olson, Phys. Rev. A **28** (1983) 2526.

8.60 E.E.B. Campbell, I.V. Hertel, and S.E. Nielsen, J. Phys. B **24** (1991) 3825.

8.61 J.W. Thomsen, J. Salgado, N. Andersen, D. Dowek, A. Dubois, J.C. Houver, S.E. Nielsen, and A. Svensson, J. Phys. B **32** (1999) 5189.

8.62 M. Machholm, E. Lewartowski, and C. Courbin, J. Phys. B **27** (1994) 4681; corrigendum, J. Phys. B **28** (1995) 1419.

8.63 Z. Roller-Lutz, Y. Wang, H.O. Lutz, S.E. Nielsen, and A. Dubois, Phys. Rev. A **61** (2000) 022710.

8.64 D. Dowek, I. Reiser, S. Grego, N. Andersen, J.C. Houver, S.E. Nielsen, C. Richter, J. Salgado, A. Svensson, and J.W. Thomsen (2000), in preparation.

8.65 J. Ullrich, R. Moshammer, R. Dörner, O. Jagutzki, V. Mergel, H. Schmidt-Böcking, and L. Spielberger, J. Phys. B **30** (1997) 2917.

8.66 S. Wolf and H. Helm, Phys. Rev. A **56** (1997) R4385.

8.67 M. van der Poel, M.-A. Gearba, C.V. Nielsen, J.W. Thomsen, N. Andersen, V. Mergel, R. Dörner, and H. Schmidt-Böcking, *Book of Abstracts XXI. ICPEAC* (1999) 527.

# 9. Propensity Rules

Orientation propensities in electron/positron–atom and proton/antiproton–atom collisions are discussed. The current understanding of the origin and the range of validity of these rules is investigated. It is shown how this analysis may yield deeper insight into the collision dynamics responsible for electron excitation, de-excitation, and transfer in atomic collision processes.

## 9.1 Orientation for S → P Impact Excitation by Electrons and Positrons

In the previous two chapters some examples of propensities for orientation and alignment in electron–atom and ion/atom–atom collisions were encountered. In this chapter, additional results for orientation propensities in electron/positron–atom and proton/antiproton–atom collisions are presented. The concept of a propensity rule has, in the words of Fano, "been introduced with reference to a transition — or to a class of transitions — which is much more likely than alternative but accessible ones. Propensity thus amounts to an attenuated version of a selection rule. Selection rules result from exact symmetries or other properties of a system, *propensities from less clearly identified circumstances*" [9.1].

Orientation and alignment parameters essentially characterize, respectively, the circulation of the active electron around the atomic core and the shape of the excited electron cloud and its direction in space, thereby allowing new kinds of questions to be addressed. An early example of this is illustrated in Figure 9.1, taken from the paper *"Orientation by Collision"* of Kohmoto and Fano [9.2]. It serves to illustrate how, in a classical trajectory picture, attractive and repulsive forces lead to collisions with opposite values of $[\boldsymbol{p}_i \times \boldsymbol{p}_f] \cdot \boldsymbol{J}$.

We first treat the simple case of S → P excitation in electron–atom collisions. As discussed in Chapter 7 (see Figure 7.10), within the first Born approximation (FBA), the excitation is described as a transition transferring only linear momentum along the direction of momentum transfer $\Delta \boldsymbol{k}$. Thus

$$L_\perp^{\mathrm{FBA}} = 0, \tag{9.1}$$

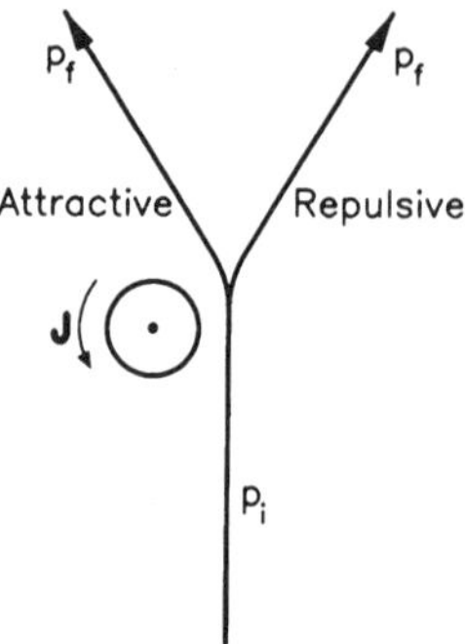



**Fig. 9.1.** Diagram showing how attractive or repulsive forces lead to collisions with opposite values of $[\boldsymbol{p}_i \times \boldsymbol{p}_f] \cdot \boldsymbol{J}$ [9.2].

and the alignment angle $\gamma^{\mathrm{FBA}}$ can be calculated by simple trigonometry from (7.12). The classic case of He $2\,^1$P excitation ($\Delta E = 21.2\,\mathrm{eV}$) at an impact energy of 80 eV was shown in Figure 7.11. Except for forward and backward scattering, the orientation parameter $L_\perp$ deviates dramatically from the FBA prediction, approaching its maximum value of $+1$ near a scattering angle of $35°$ and its minimum value of $-1$ near a scattering angle of $110°$. At small scattering angles, the alignment angle $\gamma$ follows the FBA predictions, but dramatic deviations occur beyond an angle of $30°$.

The pattern shown for $L_\perp$ in Figure 7.11, namely that the value is positive for small scattering angles and increases with scattering angle to a maximum beyond which the variation becomes more individual, has turned out to be a very general observation for electron-impact excitation, almost without exceptions. Therefore, considerable efforts have been put into finding a solid theoretical foundation for this empirical propensity rule, the Kohmoto–Fano paper cited in connection with Figure 9.1 being an early example. A significant step further was the analysis by Madison and Winters [9.3]. After pointing out a phase problem in the Kohmoto–Fano paper, they discussed the orientation parameter in terms of the charge $q$ of the projectile, with $q=\pm 1$ for positron and electron impact, respectively, by expanding the scattering amplitude in a Born series as

$$f_{\mathrm{P\leftarrow S}} = \langle \Phi_P | V | \Phi_S \rangle + \langle \Phi_P | V G^+ V | \Phi_S \rangle + \ldots = q\, t^{(1)} + q^2\, t^{(2)} + \ldots , \quad (9.2)$$

where $V$ is the interaction potential and $G^+$ is the free-particle Green's function with appropriate boundary conditions. This analysis yields the following expression for the transferred angular momentum:

$$L_\perp \propto q^3 \,\mathrm{Im}\{t_0^{(1)}\}\mathrm{Re}\{t_1^{(2)}\} + q^4 \,\mathrm{Im}\{t_0^{(2)}t_1^{(2)*}\} + \ldots \quad (9.3)$$

where the subscript indicates the magnetic quantum number. The first term in (9.2), involving a product of first- and second-order amplitudes, dominates

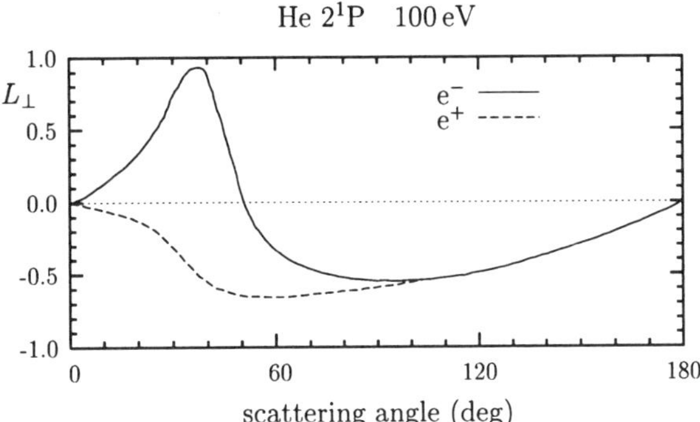


**Fig. 9.2.** The orientation parameter $L_\perp$ for He $2\,^1$P excitation at an impact energy of 100 eV for electron and positron projectiles, respectively [9.3].

at small scattering angles, whereas the second one, with a product of two second-order amplitudes, takes over at large angles. Thus, for small scattering angles, the orientation parameter should change sign when switching from electron to positron impact, while both projectiles are expected to produce orientation parameters of the same sign at larger angles.

Figure 9.2 shows the Madison and Winters results for the He $2\,^1$P state after electron and positron impact, respectively, for an incident energy of 100 eV. Concerning the actual sign of $L_\perp$ at small scattering angles, however, no definitive, general answer could be derived.

A semiclassical argument for the sign of the orientation parameter, valid for small scattering angles, was outlined by Andersen and Hertel [9.4] and is summarized in Figure 9.3. The starting point is the fact that the electrostatic interaction between an electron and the target atom is always attractive when the electron penetrates the atomic electron cloud (left column), while the force is repulsive for positrons (right column).[14] As shown schematically, analyzing the motion of the active electron around the atomic core suggests $L_\perp > 0$ for electron impact and $L_\perp < 0$ for positron impact. For de-excitation, Andersen and Hertel pointed out that a change in sign of the momentum-transfer vector $\Delta\mathbf{k}$ reverses the sign of the first-order term in (9.2), while the sign of the second-order term remains unchanged. Thus, for S → P de-excitation at small scattering angles, the sign of $L_\perp$ should change when compared to the excitation process. (Note the analogy to the discussion of direct excitation in heavy-particle collisions in Section 8.1.1. Reversing the sign of the energy transfer in (8.6), i.e., going from excitation to de-excitation, also reverses the sign of the angular-momentum transfer.)

No decisive improvement has been achieved since then concerning a theoretical, quantum-mechanical foundation of the propensity rule. However,

---

[14]This simple argument neglects the attractive polarization force for electron and positron scattering from neutral targets in their ground state.

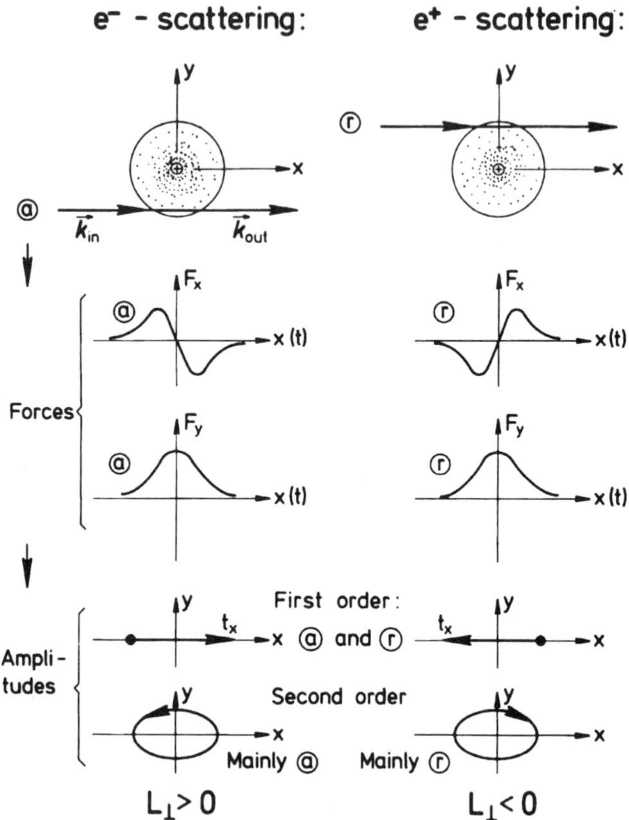



**Fig. 9.3.** A schematic diagram showing the forces acting on the atomic electron charge cloud for small scattering angles of electrons and positrons, together with the resulting excitation amplitudes of the equivalent classical oscillator in first- and second-order treatments [9.4].

the present satisfactory situation concerning the numerical predictions for electron-impact excitation of the lighter elements presented in Chapter 7 might, however, enable further understanding of this problem. Two example cases will be discussed in Section 9.3 below.

Experimental and theoretical progress in the study of collisions involving spin-polarized electron beams has allowed for the extraction of spin-dependent orientation and alignment parameters for impact excitation of the Hg $6^1S_0 \rightarrow 6^3P_1$ transition [9.5]. For heavy atoms, explicitly spin-dependent forces, such as the spin-orbit interaction or other relativistic effects, must be taken into account. As discussed in Section 7.2.4, six scattering amplitudes come into play for the above transition, and a complete scattering experiment becomes a complicated affair [9.6]. Here we focus on the spin-resolved orienta-

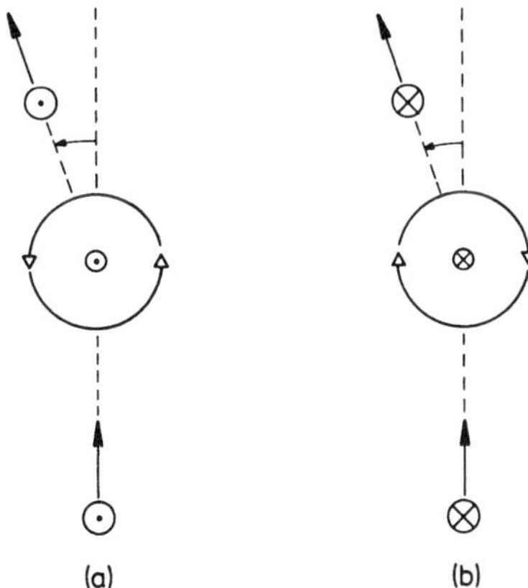

**Fig. 9.4.** Illustration of angular-momentum transfer by spin-up (a) and spindown (b) electrons, as predicted for electron-impact excitation of $\mathrm{Hg}(6s6p)^3\mathrm{P}_1$ for small scattering angles at an incident electron energy of 8 eV [9.5].

tion parameters only, for which the general results at 8 eV impact energy (cf. Figure 7.47) are illustrated in Figure 9.4, the spin-dependent analogue of the Kohmoto–Fano graph (Figure 9.1).

Figure 7.47 shows that $L_\perp^{+\uparrow} > 0$ while $L_\perp^{+\downarrow} < 0$ at small scattering angles. While the positive value of $L_\perp^{+\uparrow}$ is in agreement with the propensity rule, the negative value of $L_\perp^{+\downarrow}$ might seem surprising. Both $L_\perp^{+\uparrow}$ and $L_\perp^{+\downarrow}$ are nonzero for forward scattering, and $L_\perp^{+\uparrow}(0°) = -L_\perp^{+\downarrow}(0°)$ by symmetry requirements. There is also a large difference between the relative importance of spin-flips in this angular range, as seen from the height parameters $h^{\uparrow,\downarrow}$ in Figure 7.47. Spin-flips are relatively unimportant for "spin-up" incoming electrons (small $h^\uparrow$), but very likely for "spin-down" electrons (large $h^\downarrow$). Those spin-down electrons, however, whose spin is not flipped, tend to transfer a *negative* angular momentum to the atom, as indicated in Figure 9.4.

The presently available data for spin-resolved electron-impact excitation of heavy atoms are too sparse to reveal if the behavior shown in Figure 9.4 is a singular case or part of a more regular pattern. Further studies are therefore highly desirable.

## 9.2 Orientation for S → P Impact Excitation by Protons and Antiprotons

The collision dynamics responsible for the propensities outlined in Chapter 8 for heavy-particle collisions, including the propensity rule for orientation induced by direct excitation expressed in Eqs. (8.7,8), was further elucidated by Lin and collaborators [9.7]. Comparing hydrogen and helium excitation by protons, antiprotons, electrons, and positrons at impact velocities of $1.0 - 2.5\,v_0$, including electron transfer where appropriate, they searched for general trends and similarities in the calculated parameters in order to assess the possibility of simple models for interpreting the variation in the sign of the orientation parameter for different projectiles.

Using Figure 9.2 as starting point for the discussion, Lin *et al.* [9.7] discussed $L_\perp$ in a classical picture, where the sign is determined from consideration of repulsive and attractive forces. The effective interaction between an electron and an atom in its ground state is always attractive, which supports the positive sign for electron impact at small scattering angles. Furthermore, the negative sign calculated for positron impact is ascribed to the fact that the effective interaction between a positron and an atom is always repulsive [9.7], provided that polarization forces can be neglected. Note, however, that the sign reversal for electron impact at larger angles cannot be interpreted in this simple picture, since this would imply a repulsive force. Although arguments using such a repulsive force have been used in the literature to explain the sign reversal, it was pointed out by Madison *et al.* [9.8] that the electron–neutral-atom interaction is *always attractive*. For large scattering angles, therefore, a fully quantum mechanical treatment accounting for interference effects between various partial waves is necessary.

Bearing in mind that $L_\perp$ represents the transferred angular momentum, Lin *et al.* [9.7] also discussed scaling properties of this parameter, which shows a similar behaviour at different electron-impact energies, except for shifts in the angles where maxima and minima occur. A problem here lies in the fact that the concept of an impact parameter is not well defined for most of the impact energies of interest. However, from the relation

$$b = \frac{Z}{2E}\cot(\tfrac{1}{2}\theta) \tag{9.4}$$

for Coulomb interaction with a charge $Z$ to attribute an impact parameter value $b$ to a scattering angle $\theta$, Lin *et al.* [9.7] demonstrated an approximate scaling of $L_\perp$ results for various energies $E$ by using the parameter $vb$ as the independent variable. Figure 9.5 shows an example of this scaling.

Proton and antiproton impact excitation of He $2^1$P and He $3^1$P was calculated at 100, 200, and 400 keV, respectively, using a one-center atomic-orbital close-coupling expansion method and a large basis set, including S-, P-, and D-type bound states, together with pseudostates to represent the effect of

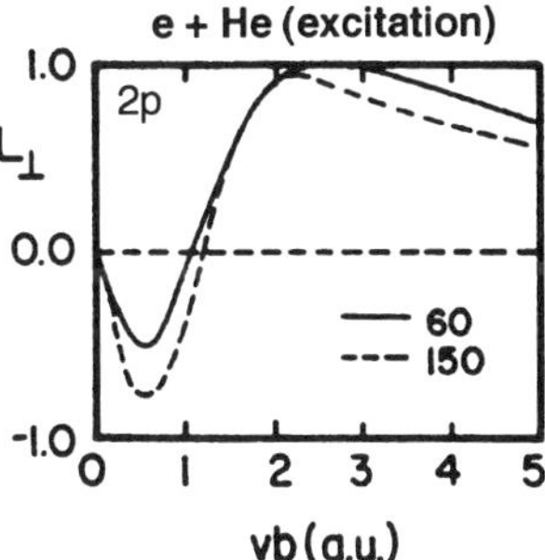


Fig. 9.5. Scaling of the parameter angular-momentum transfer with respect to $vb$ for electron-impact excitation of the He $2^1$P state at 60 and 150 eV. The $L_\perp$ values were obtained from the distorted-wave Born calculations by Madison (from [9.7]).

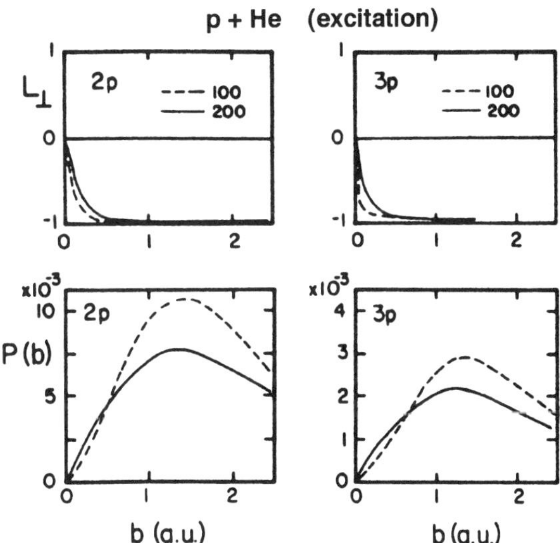


Fig. 9.6. The angular-momentum transfer and the excitation probabilities vs. impact parameter for excitation to the He $2^1$P and $3^1$P states by proton impact at 100 and 200 keV incident energy (from [9.7]).

the continuum. The helium atom was represented by a model potential. The results are shown in Figures 9.6 and 9.7. For proton impact we note that the calculated $L_\perp$ is in agreement with the propensity rule (8.7) for S → P excitation. Its value is close to −1 for proton impact on helium where the effective interaction is repulsive.

In the corresponding plots for antiproton impact, where the effective interaction is attractive, the calculated value of $L_\perp$ is very close to +1, except

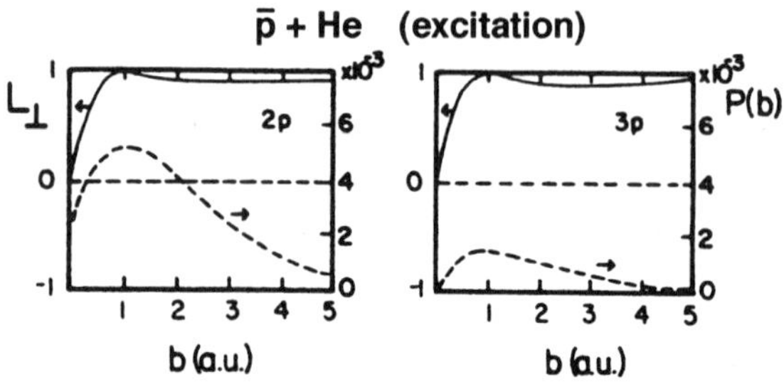



**Fig. 9.7.** Angular-momentum transfer (solid lines) and excitation probabilities (dashed lines) vs. impact parameter for excitation to the He $2^1$P and $3^1$P states by proton impact at an incident energy of 400 keV. The right-hand scale is for the probabilities [9.7].

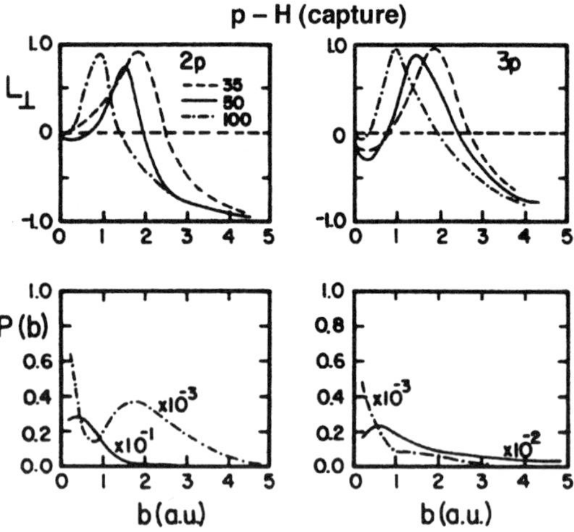



**Fig. 9.8.** Electron capture probabilities and $L_\perp$ vs. impact parameter for electron capture into H(2p) and H(3p) in H$^+$-H(1s) collisions at 35, 50, and 100 keV [9.7].

at very small impact parameters. This result is consistent with the propensity rule for an attractive potential.

The orientation parameter was also addressed in a study of electron transfer to H(2p) and Ps(2p) for proton and positron impact, respectively. Figure 9.8 shows electron capture probabilities and $L_\perp$ vs. impact parameter for electron capture into H(2p) and H(3p) in H$^+$ − H(1s) collisions at various energies. The results were obtained using a two-center atomic-orbital expansion method with 22 atomic states, including some pseudostates on each center. The angular-momentum transfer $L_\perp$ is small and mostly negative at small impact parameters, reaching a maximum value close to +1 before it drops to

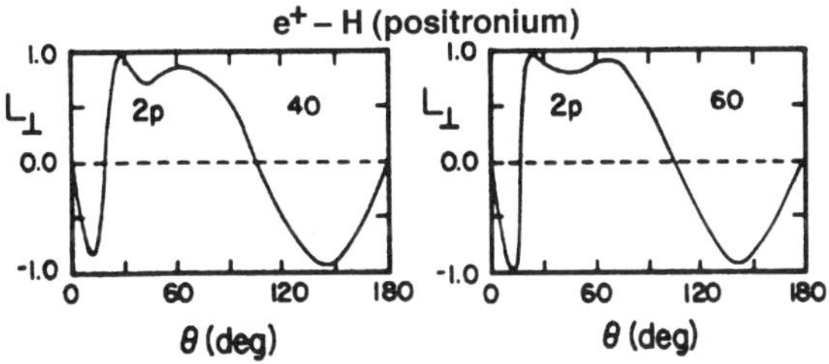


**Fig. 9.9.** The $L_\perp$ parameter for electron capture into Ps(2p) states formed in $e^+$-H(1s) collisions at 40 and 60 eV [9.7].

negative values at large impact parameters. Furthermore, the $L_\perp$ values for capture to 2p and 3p are very similar at each impact energy.

Figure 9.9 shows the orientation of positronium 2p states formed in $e^+ - \mathrm{H}(1s)$ collisions at 40 and 60 eV. The scattering amplitudes were obtained from a distorted-wave Born approximation. We note that $L_\perp$ is nearly insensitive to the incident energy, assuming large negative values at both small and large angles, but becoming positive in between. Its angular dependence is similar to that shown for capture to 2p states in $\mathrm{H}^+ - \mathrm{H}(1s)$ collisions at comparable velocities. It thus seems that $L_\perp$ is not very sensitive to the theoretical details, so long as the distortion effect is included. The authors point out, however, that more calculations or experiments are needed before one can conclude whether the general shape of $L_\perp$ seen here is an indication of a "propensity rule" for rearrangement collisions to 2p states.

## 9.3 Orientation for Excitation and De-excitation by Electrons and Positrons

As mentioned in the discussion of Figure 9.3, Andersen and Hertel [9.4] suggested that the angular dependence of the angular-momentum transfer *at small scattering angles* should be opposite for electron and positron impact, and also opposite for a positive (excitation) or negative (de-excitation) energy transfer from the projectile to the target. Owing to the prohibitively small signal rates expected in positron experiments of this kind, they suggested investigating the 3s → 3p and 4s → 3p electron-impact induced transitions in sodium, preferably using the time-reversed scheme of scattering from the laser-prepared excited $(3p)^2\mathrm{P}$ state.

This study was undertaken in a pioneering experiment by the Brisbane group [9.9]. As suggested by Andersen and Hertel, the experiment was performed by measuring the "pseudo" Stokes parameters, i.e., the dependence of the scattered electron intensity on the optical preparation of the initial 3p state for the two final states 3s and 4s, respectively. Using time reversal

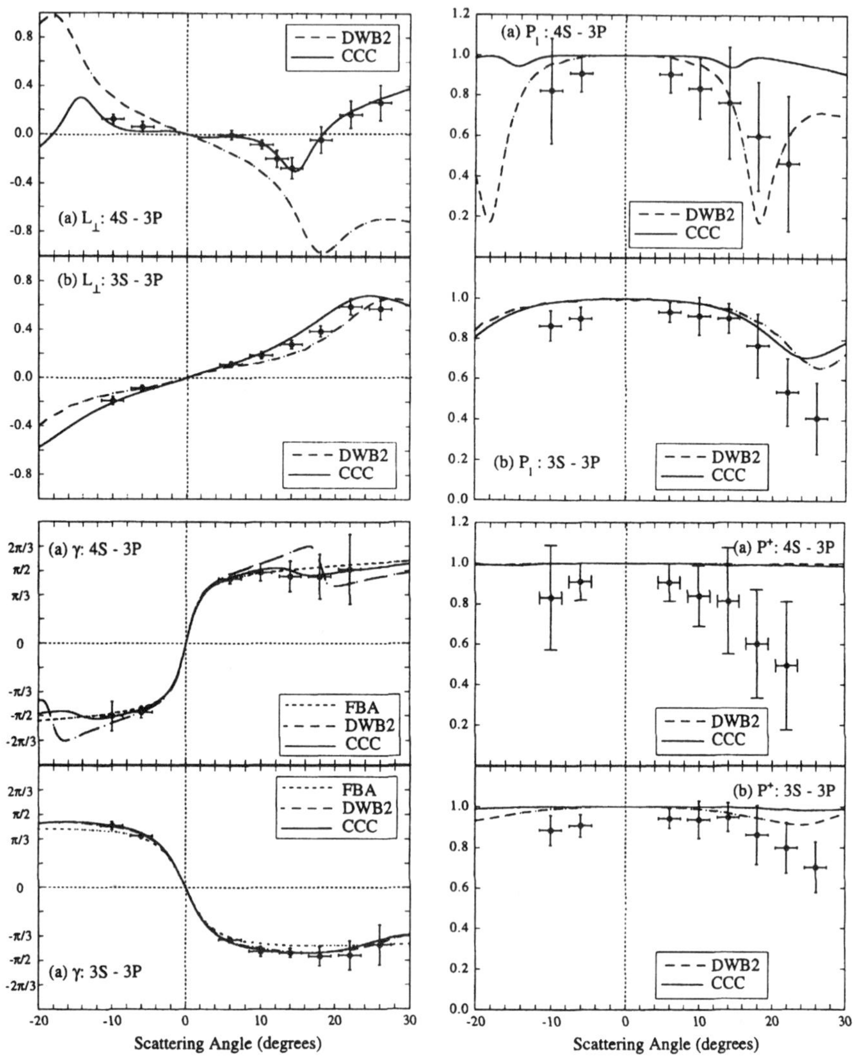

**Fig. 9.10.** Angular-momentum transfer $L_\perp$, alignment angle $\gamma$, and the degrees of linear ($P_\ell$) and total ($P$) light polarization for electron-impact excitation ($3s \rightarrow 3p$) and de-excitation ($4s \rightarrow 3p$) of sodium as a function of the scattering angle for an impact energy of $22.0\,\mathrm{eV}$. The experimental data are compared with predictions from convergent close-coupling (CCC) and second-order distorted-wave (DWBA) calculations. Also shown is the first-order Born (FBA) result for the alignment angle, corresponding to the momentum transfer direction (from [9.9]).

invariance of the interaction, the corresponding parameters for the respective
$S \to P$ transitions were obtained. Comparing the results for electron scattering under reversal of the energy transfer therefore allowed, for the first time,
to study the Andersen–Hertel predictions.

Figure 9.10 shows the result of the Brisbane experiment for incident energies of 19.9 eV and 23.1 eV on the excited $(3p)^2 P$ state. This corresponds
to the same energy of 22.0 eV for the excitation and de-excitation processes,
respectively.[15] Note that the experimental data indicate indeed a negative
angular-momentum transfer in the 4s $\to$ 3p de-excitation process for scattering angles below $18°$. For larger angles, however, a significant increase to
positive values is observed, thereby severely limiting the range of validity for
the propensity rule at this energy.

Whereas the experiment demonstrated the feasibility of such studies, several open questions remained. To begin with, the experiment was performed
for three relatively high incident energies. (In addition to the 22 eV data
shown in Figure 9.10, similar results were obtained for incident energies of
30 eV and 50 eV, respectively [9.10].) Furthermore, the data were restricted to
the angular range $4° \leq \theta \leq 26°$. And finally, the experimental results, as well
as those for the "control transition" 3s $\to$ 3p, agreed very well with "convergent close-coupling" (CCC) predictions [9.11], whereas significant deviations
were found with those based on a "second-order distorted-wave" (DWB2)
approach [9.12]. In light of the satisfactory performance of the DWB2 model
to describe the 3s $\to$ 3p transition, the latter finding was somewhat surprising, since the energy difference of 1.1 eV between the 4s and 3p levels is only
about 5% of the incident energy. Hence, one might have expected a perturbative treatment, especially when carried out to second order, to be sufficiently
accurate.

Following up on the Brisbane work, a more extensive study of the
Andersen–Hertel predictions was performed by Bartschat *et al.* [9.13], who
investigated theoretically the above transitions over the full range of scattering angles and a large number of incident energies. They performed
several R-matrix calculations, ranging from standard 3-state (3s, 3p, 4s),
5-state (3s, 3p, 4s, 3d, 4p), and 7-state (3s, 3p, 4s, 3d, 4p, 4d, 4f) models to an
"R-matrix with pseudostates" (RMPS) approach. As pointed out in Section 6.3, the latter method is essentially equivalent to the CCC method,
except that it is formulated in an R-matrix framework in coordinate space
rather than as a close-coupling method in momentum space.

Figure 9.11 shows $L_\perp$ results for electron-impact excitation (3s $\to$ 3p) and
de-excitation (4s $\to$ 3p) of sodium as a function of the scattering angle for
various collision energies. Whereas reasonable agreement between the experimental data and the theoretical predictions for the excitation process can

---

[15]The total collision energy, given by the energy of the projectile plus the target
relative to the target ground state, is different in the two cases, namely 22.0 eV
for 3s $\to$ 3p and 25.1 eV for 4s $\to$ 3p.

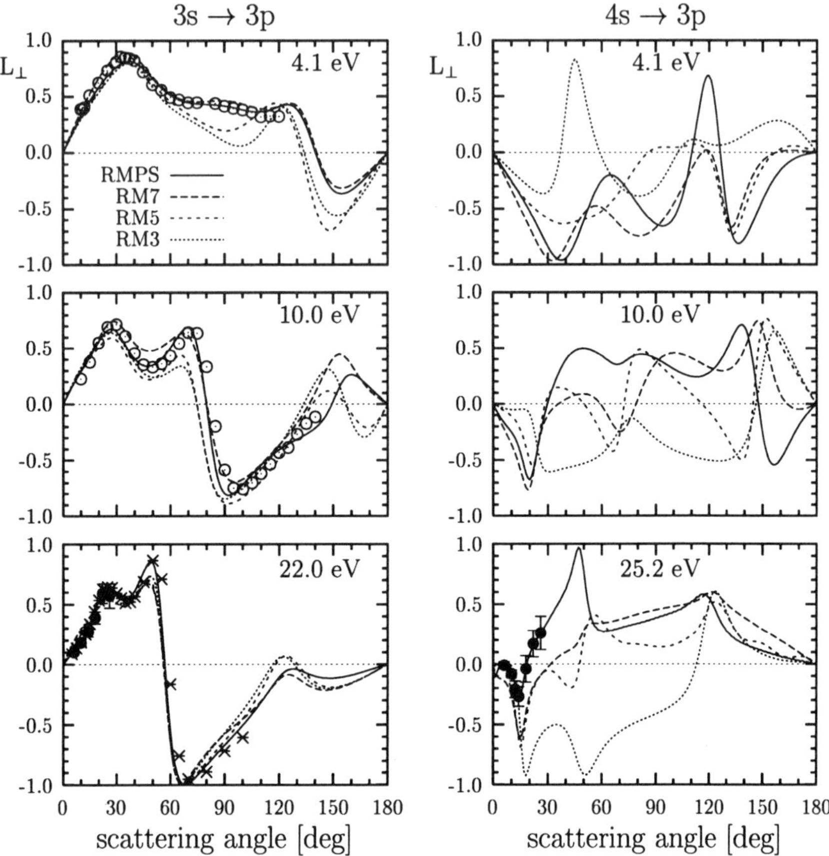



**Fig. 9.11.** Angular-momentum transfer $L_\perp$ for electron-impact excitation (3s → 3p) and de-excitation (4s → 3p) of sodium as a function of the scattering angle for various collision energies. The predictions from different R-matrix models (see text) are compared with experimental data of the NIST (o) [9.14,15], Adelaide (×) [9.16], and Brisbane (●) [9.9] groups.

already be achieved with a simple three-state close-coupling model (RM3), only the RMPS model is able to reproduce the experimental data for the 4s → 3p transition. The slow convergence of the close-coupling models for the latter case is consistent with the problems in the perturbative DWB1 and DWB2 models found by Shurgalin *et al.* [9.9,10]. A detailed partial-wave study suggests [9.13] that the very large dipole polarizability of the 4s state (a value of approximately $+3,600\,a_0^3$ was obtained in the RMPS model) effectively prevents the convergence of a perturbative treatment at the second-order level of the projectile–target interaction.

Figure 9.12 displays $L_\perp$ results for electron-impact excitation (3s → 3p) and de-excitation (4s → 3p) of sodium at fixed scattering angles of 10° and 20°

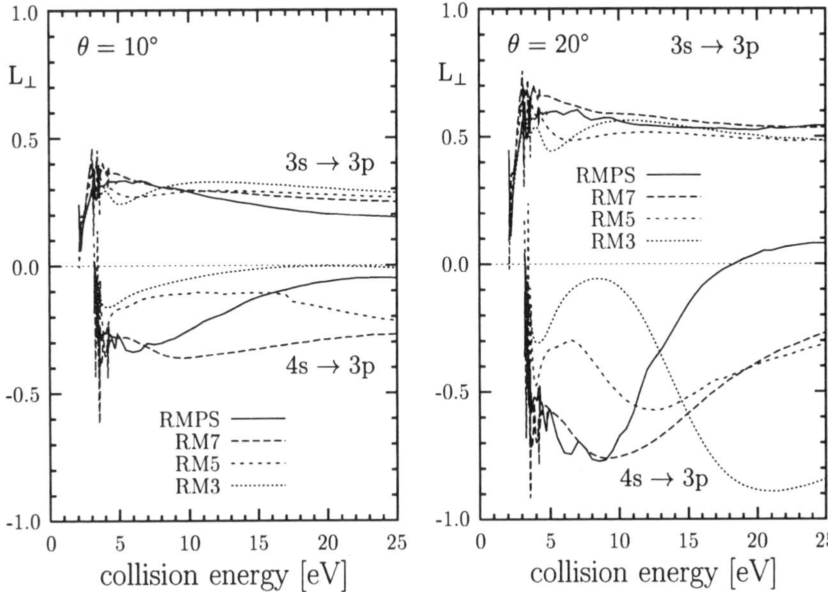

Fig. 9.12. Angular-momentum transfer $L_\perp$ for electron-impact excitation (3s $\to$ 3p) and de-excitation (4s $\to$ 3p) of sodium at fixed scattering angles of 10° and 20° as a function of the collision energy.

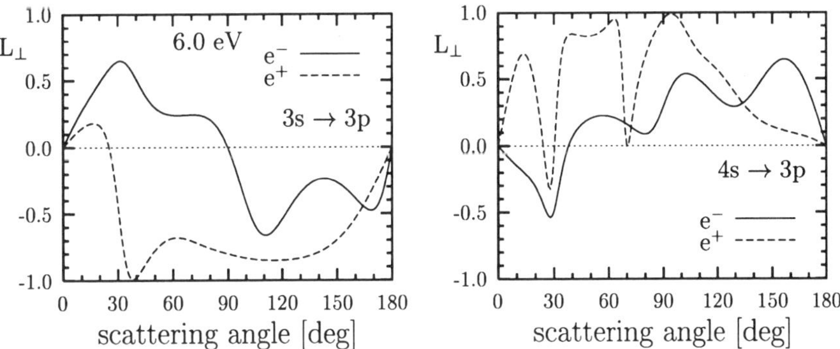

Fig. 9.13. Angular-momentum transfer $L_\perp$ for excitation (3s $\to$ 3p) or de-excitation (4s $\to$ 3p) of sodium by electron or positron impact as a function of the scattering angle. The results are from five-state R-matrix and close-coupling calculations.

as a function of the collision energy. As in Figure 9.11, we see the fast (3s $\to$ 3p) and slow (4s $\to$ 3p) convergence of the results with the number of states in the close-coupling expansion. Clearly, the RMPS results support the Andersen–Hertel propensity rule at these small scattering angles and sufficiently low collision energies where the high partial waves are relatively less important than at high energies. The Brisbane result at 20° for 25.2 eV (cf. Figure 9.10) is indeed an exception to the rule because of the high energy.

Figure 9.13 shows a comparison of five-state close-coupling results for the two transitions of interest, induced by electron or positron impact. The latter results were obtained with the computer code of McEachran [9.17]. They should be regarded as a purely qualitative indication, owing to the possible lack of convergence in the close-coupling expansion and because of the missing positronium formation channel. On the other hand, this channel is also neglected in the Andersen–Hertel argument, and the collision energy of 6.0 eV is low enough to give some confidence in the interpretation of the results. Here the small-angle propensity rule only works for positron-impact de-excitation of the 4s → 3p transition. As for the electron-induced 3s → 3p transition, the projectile charge, the energy transfer, and the attractive long-range interaction all suggest a positive angular-momentum transfer at small scattering angles, as supported by the figure. In the other two cases, arguments can be made for a sign change due to the reversal of the projectile charge or the energy transfer, provided polarization is neglected. A partial-wave analysis of the positron results for the 3s → 3p transition showed indeed that omitting the partial waves with $\ell \geq 10$, i.e., the terms affected most strongly by the long-range *attractive* polarization terms, *reverses* the sign of the small-angle results for $L_\perp$.

Figure 9.14 shows the full set of the RMPS results for the angular-momentum transfer by electron impact of sodium in three-dimensional and contour plots as a function of both the scattering angle and the total collision energy for electron impact. Up to scattering angles of about 60° and energies of nearly 20 eV, the Andersen–Hertel propensity rule is quite well fulfilled. For higher energies, however, including those at which the Brisbane experiment was performed, the validity range of the propensity rule diminishes quickly. The principal reason for this finding seems to be the dominant influence of the partial waves with high angular momentum that are strongly affected by the projectile-target distortion.

Contour plots of the theoretical results from the five-state close-coupling approximations for electron and positron impact are shown in Figure 9.15. Although a tendency of support for the propensity rule can be seen, no definite conclusions can be drawn owing to the importance of coupling to other discrete and continuum channels, as well as the currently unknown effect of the positronium formation channel on the angular-momentum transfer. Further studies of these effects are certainly desirable.

In summary, predictions from RMPS calculations show excellent agreement with the experimental data for the angular-momentum transfer in electron-impact induced 3s → 3p and 4s → 3p transitions in sodium. The results from these calculations support the validity of the Andersen–Hertel propensity rule, provided the impact energy is sufficiently low and the angular range is chosen such that (*i*) partial waves with high angular momenta

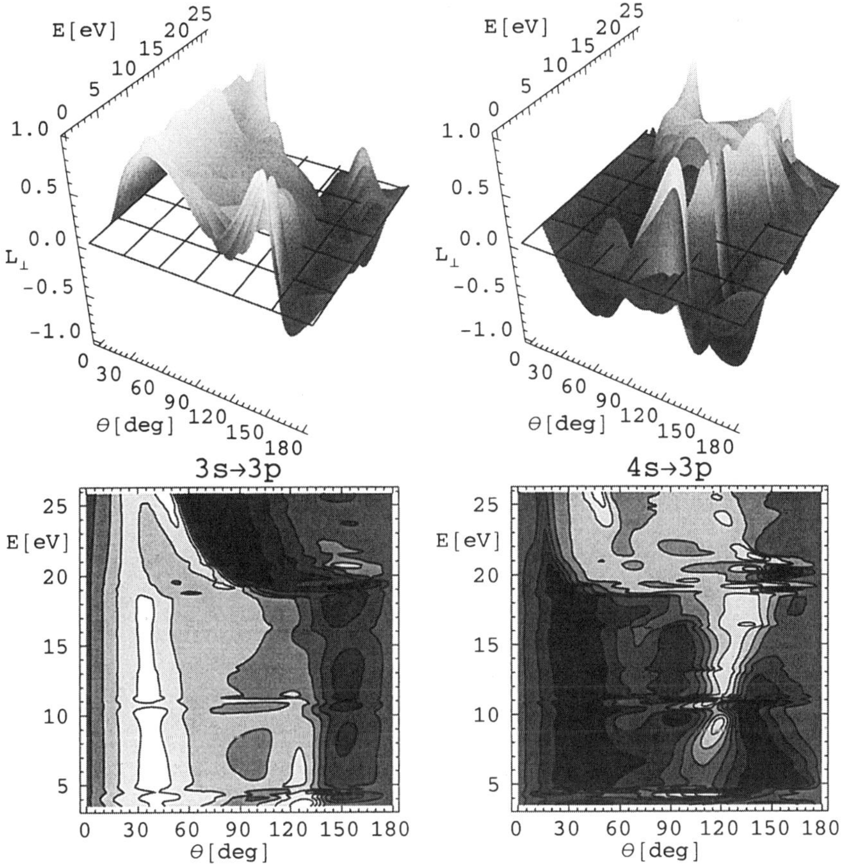

**Fig. 9.14.** 3D and contour plots of the RMPS results for $L_\perp$ for electron-impact excitation (3s $\rightarrow$ 3p) and de-excitation (4s $\rightarrow$ 3p) of sodium as a function of the scattering angle and the collision energy.

do not dominate the outcome of the collision, and (*ii*) the penetration of the target electron cloud by the projectile is insufficient to exhibit the rapidly varying dependence of the results on the scattering angle.

As a second example, Figure 9.16 shows results for the angular-momentum transfer in electron collisions with barium. Specifically, the data for the inelastic (excitation) transitions $(6s^2)^1S_0 \rightarrow (6s6p)^1P_1$ and $(6s5d)^1D_2 \rightarrow (6s6p)^1P_1$ are compared with those for the corresponding superelastic (de-excitation) transitions $(6s7s)^1S_0 \rightarrow (6s6p)^1P_1$ and $(6s6d)^1D_2 \rightarrow (6s6p)^1P_1$. The experiment was performed in Zetner's group [9.18–20] by laser-preparing the $6^1P_1$ state, and the results were re-interpreted for the time-reverse process. We see once again that the angular-momentum transfer at small angles changes sign when the sign of the energy transfer is reversed. However, whereas the results

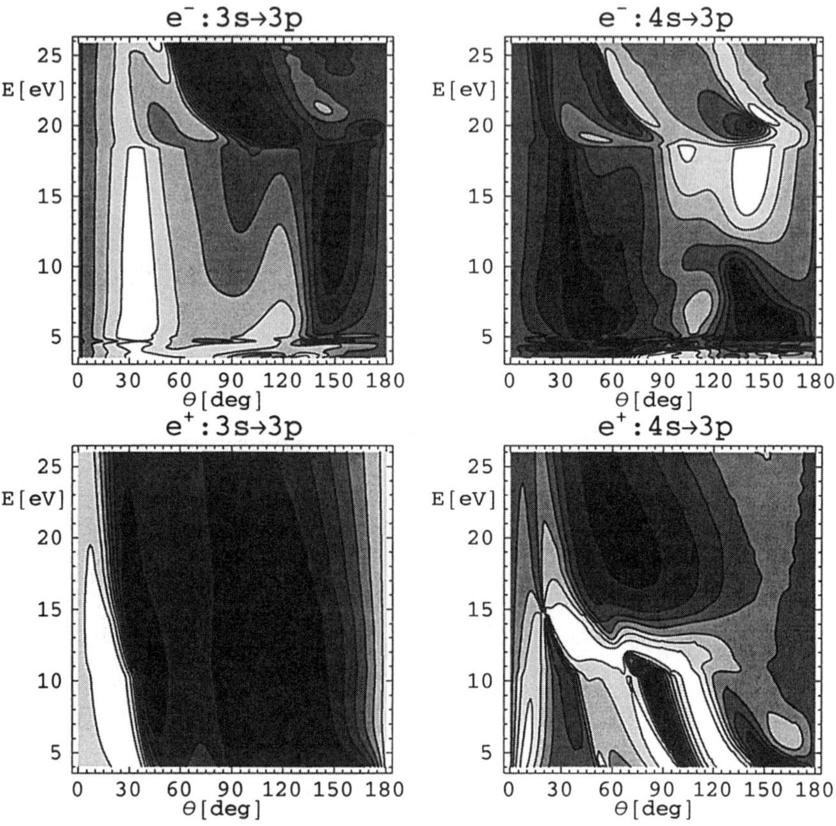

**Fig. 9.15.** Contour plots for $L_\perp$, as predicted in five-state R-matrix and close-coupling calculations for electron-impact and positron-impact excitation ($3s \rightarrow 3p$) and de-excitation ($4s \rightarrow 3p$) of sodium as a function of the scattering angle and the collision energy.

for the $S \rightarrow P$ transitions also exhibit the sign predicted by the Andersen–Hertel propensity rule, the situation is reversed for the $D \rightarrow P$ transitions. Even for this heavy target ($Z = 56$), there is reasonable, though not perfect, agreement between the experimental data and the predictions from a non-relativistic CCC calculation by Fursa and Bray [9.21]. Consequently, a more detailed study of the propensity rule, particularly with regard to the sign difference in the $S \rightarrow P$ vs. $D \rightarrow P$ case, might be possible based on the theoretical predictions.

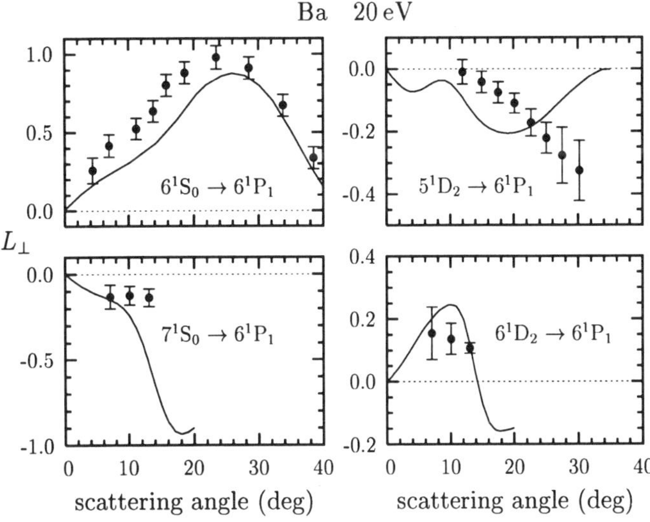


**Fig. 9.16.** Angular-momentum transfer $L_\perp$ for electron-impact excitation and de-excitation of barium. Experimental data for the $6^1S \rightarrow 6^1P$ [9.18], $5^1D \rightarrow 6^1P$ [9.19], $7^1S \rightarrow 6^1P$ [9.20], and $6^1D \rightarrow 6^1P$ [9.20] transitions in barium as a function of the scattering angle for a total collision energy of 20 eV are compared with CCC predictions [9.19,21].

## 9.4 Principal Quantum Number Dependence of Orientation and Alignment Parameters

We finish this chapter with a discussion of the principal quantum number dependence of the orientation and alignment parameters. The basic idea was put forward by Csanak and Cartwright [9.22], who investigated this problem within the first-order many-body theory (FOMBT) approach. They concluded that the principal quantum number dependence of the **T**-matrix elements can essentially be factored out for transitions in which the orbital of the active electron in the final state has a significantly larger range than the corresponding orbital in the initial state (or vice versa). Csanak and Cartwright illustrated the idea with the example of the $(1s^2)^1S \rightarrow (1snp)^1P$ $(n = 2, \ldots, 8)$ transitions in helium, and the results for incident energies of 40 eV and 81.6 eV are shown in Figure 9.17. Indeed, the predictions for the (relative) coherence parameters are very similar for all $n$ values, whereas the differential cross sections scale approximately with $1/n^3$. These results can be understood from the fact that the $np$ orbitals for small radii (i.e., approximately the range of the 1s orbital) scale with a factor $(n^*)^{-3/2}$ [9.23]. Here $n^*$ is the effective quantum number, which is close to the principal quantum number $n$ in the above case. Note, however, that this argument is only valid for the *direct first-order* matrix element. Nevertheless, some experimental

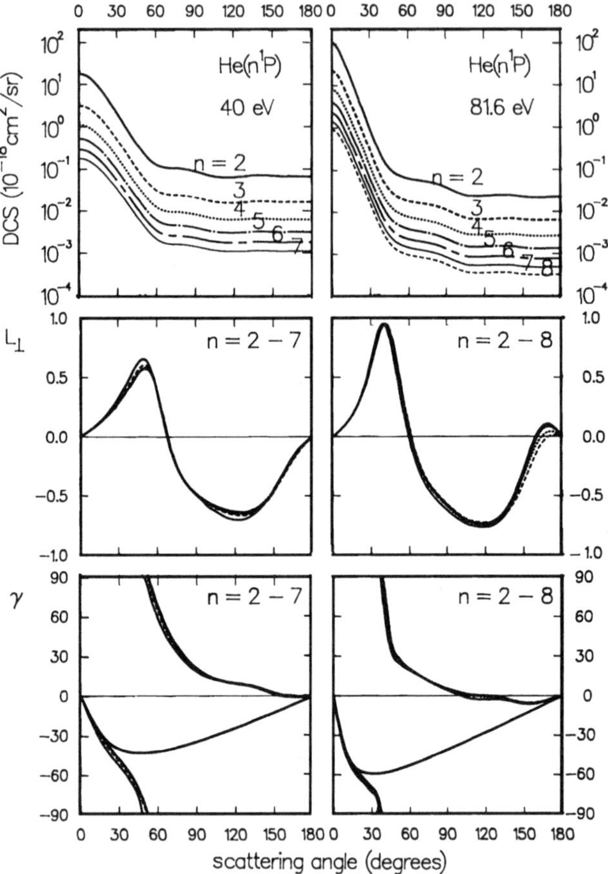

**Fig. 9.17.** Differential cross section, angular-momentum transfer, and alignment angle for electron-impact excitation of the $(1s^2) \rightarrow (1snp)^1\mathrm{P}$ transitions in helium as a function of the scattering angle for impact energies of 40 eV and 81.6 eV. The predictions are based on first-order many-body theory calculations by Csanak and Cartwright [9.22].

support for the $n$-independence in the alignment and orientation parameters for e–He excitation was found by Hammond *et al.* [9.24].

Figure 9.18 shows the corresponding results for electron-impact excitation of the 2p, 3p, and 4p states in atomic hydrogen. Results from CCC calculations [9.25] again predict angular-momentum transfers and alignment angles that are nearly independent of the principal quantum number, while the cross sections decrease with increasing $n$. Also, the similarity between the predictions in the triplet and singlet spin channels indicate a small effect of electron exchange.

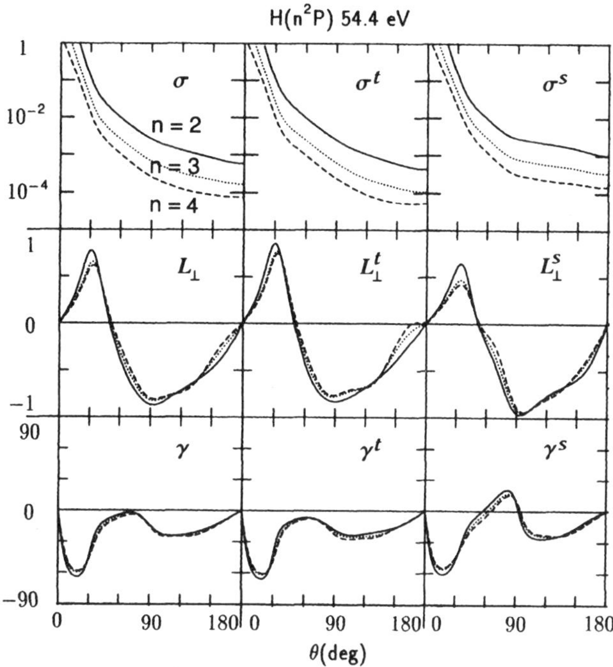



**Fig. 9.18.** Differential cross section, angular-momentum transfer, and alignment angle for electron-impact excitation of the $1s \rightarrow np$ transitions in atomic hydrogen as a function of the scattering angle for an impact energy of 54.4 eV. The predictions are based on CCC calculations by Bray and Stelbovics [9.25].

Whereas the argument apparently works very well for these two targets, the validity of the scaling laws is certainly not general. As an example, Figure 9.19 shows the differential cross section and the spin-averaged angular-momentum transfer $L_\perp$ for electron-impact excitation of the $(np)^2\mathrm{P}$ states of sodium, as calculated by Bubelev *et al.* [9.12] in the DWB2 approach. Although similarities can be recognized in the results for higher $n$ and the dependence of the results on $n$ seems to weaken with increasing energy, the scaling certainly does not work as well in this case as it did for the hydrogen and helium targets. In general, Bubelev *et al.* [9.12] concluded that the results for the lowest $n$ ($\equiv n_g$) of the excited states in Li, Na, and K differ significantly from those for states with higher principal quantum number. Approximate scaling starts at $n_g+2$ and requires sufficiently high energies as well as relatively large angles.

The principal difference between these alkali targets and hydrogen or helium is the fact that the resonance transitions from the ground state involve orbitals with the *same* principal quantum number, i.e., orbitals whose ranges are much more similar than is the case for excitation of hydrogen and helium

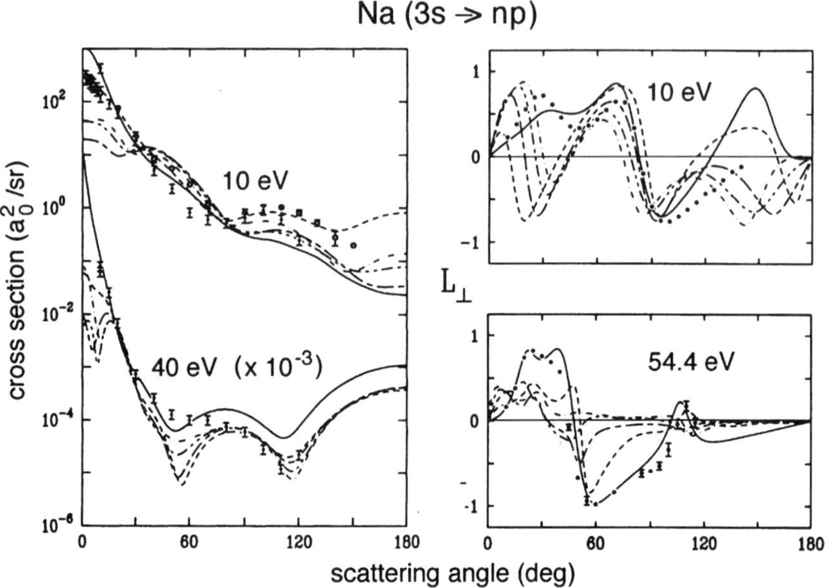



**Fig. 9.19.** Differential cross section and angular-momentum transfer for electron-impact excitation of the 3s →np transitions in sodium as a function of the scattering angle for various impact energies, as predicted in the DWBA2 calculations by Bubelev *et al.* [9.12]: ——; $n = 3$; – – – , $n = 4$; – – – – , $n = 5$; – – – –, $n = 6$; – · – ·, $n = 7$. The experimental data are for the 3s → 3p transition (DCS: [9.26,27]; $L_\perp$ [9.15].

from their respective ground states. As a result, the cross sections for these resonance transitions are often orders of magnitude larger than those for the higher $n$ states, and particularly the excitation of the states with $n_g + 1$ can be strongly suppressed. Consequently, it is not surprising that neither scaling laws nor propensity rules work very well for such transitions. Furthermore, as pointed out by Bubelev *et al.* [9.12], the validity of such rules will also be diminished if electron exchange or higher-order effects are important. In either case, the argument of Csanak and Cartwright for the *first-order direct amplitude* cannot be applied.

# References


9.1 U. Fano, Phys. Rev. A **32** (1985) 617.
9.2 M. Kohmoto and U. Fano, J. Phys. B **14** (1981) L447.
9.3 D.H. Madison and K.H. Winters, Phys. Rev. Lett. **47** (1981) 1885.
9.4 N. Andersen and I.V. Hertel, Comments At. Mol. Phys. **19** (1986) 1.
9.5 N. Andersen, K. Bartschat, J.T. Broad, G.F. Hanne, and M. Uhrig, Phys. Rev. Lett. **76** (1996) 208.
9.6 N. Andersen and K. Bartschat, Adv. At. Mol. Opt. Phys. 36 (1996) 1.
9.7 C.D. Lin, R. Singhal, A. Jain, and W. Fritsch, Phys. Rev. A **39** (1989) 4455.
9.8 D.H. Madison, G. Csanak, and D.C. Cartwright, J. Phys. B **19** (1986) 3361.
9.9 M. Shurgalin, A.J. Murray, W.R. MacGillivray, M.C. Standage, D.H. Madison, K.D. Winkler, and I. Bray, Phys. Rev. Lett. **81** (1998) 4604.
9.10 M. Shurgalin, A.J. Murray, W.R. MacGillivray, M.C. Standage, D.H. Madison, K.D. Winkler, and I. Bray, J. Phys. B **32** (1999) 2439.
9.11 I. Bray, Phys. Rev. A **49** (1994) 1066.
9.12 V.E. Bubelev, D.H. Madison, and M.A. Pinkerton, J. Phys. B **29** (1996) 1751.
9.13 K. Bartschat, N. Andersen, and D. Loveall, Phys. Rev. Lett. **83** (1999) 5254.
9.14 J.J. McClelland, M.H. Kelley, and R.J. Celotta, Phys. Rev. A **40** 2321 (1989).
9.15 R.E. Scholten, S.R. Lorentz, J.J. McClelland, M.H. Kelley, and R.J. Celotta, J. Phys. B **24** (1991) L653.
9.16 R.E. Scholten, G.F. Shen, and P.J.O Teubner, J. Phys. B **26** (1993) 987.
9.17 R.P. McEachran, in *Computational Atomic Physics — Electron and Positron Scattering from Atoms and Ions*, K. Bartschat (ed.), Springer, Heidelberg 1996.
9.18 Y. Li and P.W. Zetner, Phys. Rev. A **49** (1994) 950.
9.19 P.V. Johnson, B. Eves, P.W. Zetner, D.V. Fursa, and I. Bray, Phys. Rev. A **59** (1999) 439.
9.20 P.W. Zetner (1999), private communication
9.21 D.V. Fursa and I. Bray, Phys. Rev. A **59** (1999) 282.
9.22 G. Csanak and D.C. Cartwright, Phys. Rev. A **34** (1986) 93.
9.23 M.J. Seaton, Rep. Prog. Phys. **46** (1983) 167.
9.24 P. Hammond, M.A. Khakoo, and J.W. McConkey, Phys. Rev. A **36** (1987) 5127.
9.25 I. Bray and A.T. Stelbovics, Phys. Rev. A **46** (1992) 6995.
9.26 B. Marinkovic, V. Pejcev, D. Filipovic, I. Cadez, and L. Vuskovic, J. Phys. B **25** (1992) 5179.
9.27 S.K. Srivastava and L. Vuskovic, J. Phys. B **13** (1980) 2633.

# 10. Impact Ionization

The discussion of case studies for elastic scattering and excitation processes
is extended to ionization, including ionization and simultaneous excitation.
Examples are given for electron and heavy-particle impact, both for angle-
integrated and angle-differential studies.


## 10.1 Ionization by Electron Impact

Electron-impact ionization of atoms has been of great interest for many years.
In most cases, experimental efforts concentrated on the measurements of
ionization probabilities, with the most detailed information being obtained
from triple-differential cross section (TDCS) studies, for which the kinematics
of the incident, scattered, and ejected electrons are fully defined. Figure 10.1
shows the generic scheme of the most complex process that will be discussed
in this chapter. A possibly spin-polarized electron beam is used to ionize
a possibly spin-polarized target. In addition, the residual ion may end up
in an excited state, which subsequently can decay by photon emission.[16]
Alternatively, the ionization process may follow the excitation of a quasi-
discrete autoionizing state.

Even if the projectile and target spins are essentially irrelevant (as in
ionization of helium atoms in their ground state), the full information about
an ionization–excitation process, e.g.,

$$e(\boldsymbol{k}_i) + \mathrm{He}(1\mathrm{s}^2)^1\mathrm{S} \;\longrightarrow\; \mathrm{He}^+(2\mathrm{p})^2\mathrm{P} + e_e(\boldsymbol{k}_e) + e_s(\boldsymbol{k}_s)$$

$$\downarrow$$

$$\mathrm{He}^+(1\mathrm{s})^2\mathrm{S} + h\nu \; (30.4 \; nm)\,, \tag{10.1}$$

can only be obtained in a triple-coincidence experiment between the two
outgoing electrons and the photon which, in turn, also needs to be analyzed
with respect to its polarization. In (10.1), $\boldsymbol{k}_i$ is the momentum of the incident
electron while $\boldsymbol{k}_s$ and $\boldsymbol{k}_e$ denote the momenta of the scattered and ejected

---

[16]We limit the discussion to single-ionization and single-photon decay. For even
more complicated processes such as multiple-ionization and/or multiple-photon
emission, we refer to the current literature.

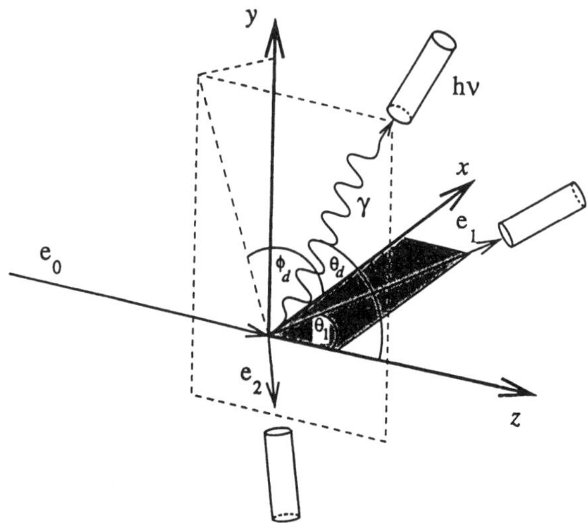



**Fig. 10.1.** Scheme of an (e, 2eγ) process for ionization–excitation.

electrons, respectively. Although the two outgoing electrons are, in principle, indistinguishable, the faster (slower) one of these is usually referred to as the scattered (ejected) electron.

Although pioneering triple-coincidence setups between one electron and two photons [10.1] or all three outgoing electrons in double-ionization studies [10.2] have been developed, the above two-electron–one-photon coincidence experiment has not been performed to date. Instead, the investigations have been simplified in the following ways:

($i$)   Only the emitted photon is observed to perform an (optical) measurement of the angle-integrated ionization–excitation cross section [10.3,4] and, if the polarization or the angular distribution of the light is determined, to obtain additional information about the magnetic sublevel cross sections for the residual ions [10.5,6]. Such experiments yield no information about the outgoing electrons and, therefore, the phases between the scattering amplitudes.

($ii$)   Only one electron is observed to obtain double-differential cross sections (DDCS) [10.7] or, in the decay of autoionizing states [10.8], information about the magnetic sublevels of such a state.

($iii$)   One electron and the photon are observed in coincidence. This problem exhibits exactly the same symmetry properties as the electron–photon coincidence setup for excitation [10.9–11] discussed in Chapters 4 and 7, although the number of independent parameters may be larger than in the excitation-only case, owing to the incoherence associated with the undetected second electron.

(*iv*)  Both electrons are observed in coincidence to obtain triple-differential cross sections (TDCS) [10.12–14]. Such studies yield the most (and in some cases the complete) information about these electrons, but they do not reveal any details regarding the magnetic sublevels of the residual ion.

As mentioned above, it is also possible to prepare spin-polarized projectile and/or target beams. In such experiments one can study angle-integrated as well as angle-differential ionization asymmetries defined according to (4.4) [10.15,16]. Furthermore, by using a spin-polarized electron beam and resolving the fine-structure of the residual ion in the energy loss of the scattered electron (for example, $Xe(5p^5)^2P_{3/2}$ vs. $Xe(5p^5)^2P_{1/2}$), an angle-differential up-down spin-asymmetry may be observed in the relative TDCS [10.17–19].

### 10.1.1 Angle-integrated studies

We begin with examples of the ionization spin asymmetry for quasi-one-electron targets, i.e.,

$$A = \frac{1}{P_e P_A} \frac{N_{\uparrow\downarrow} - N_{\uparrow\uparrow}}{N_{\uparrow\downarrow} + N_{\uparrow\uparrow}} = \frac{\sigma^s - \sigma^t}{\sigma^s + 3\sigma^t} , \tag{10.2}$$

where $N_{\uparrow\downarrow}$ ($N_{\uparrow\uparrow}$) is the count rate for ionization with antiparallel (parallel) projectile ($P_e$) and target ($P_A$) spin polarizations, while $\sigma^s$ ($\sigma^t$) is the contribution from the singlet (triplet) total spin channel, with $\sigma^s + 3\sigma^t = 4\sigma$. Figure 10.2 shows results for this asymmetry as well as the total ionization cross section $\sigma$ for electron impact on the $(1s)^2S$ ground state of atomic hydrogen, from threshold (13.6 eV) to 100 eV incident energy. The convergent close-coupling (CCC) [10.20] and R-matrix with pseudostates (RMPS) [10.21] results are compared with the measurements of Shah *et al.* [10.22] for the cross section and of Fletcher *et al.* [10.23] and Crowe *et al.* [10.24] for the spin asymmetry. The agreement between the two theoretical predictions is excellent over the entire energy range, as is the agreement with the experimental cross section and, given the scatter in the experimental data, with the spin asymmetry as well.

Measurements for the ionization spin asymmetry were also performed in Bielefeld for the alkalis Li, Na, K [10.15], and Cs [10.25], as well as for the metastable excited $(1s2s)^3S$ state of helium [10.26]. Once again, CCC calculations [10.27–29] reproduced the experimental results in a very satisfactory way.

As mentioned above, the alignment of the residual ion in ionization–excitation processes can be studied by observing the photons emitted in an optical decay, either in an angular-distribution or a polarization experiment. Figure 10.3 shows the results of Götz *et al.* [10.5] for the alignment of the

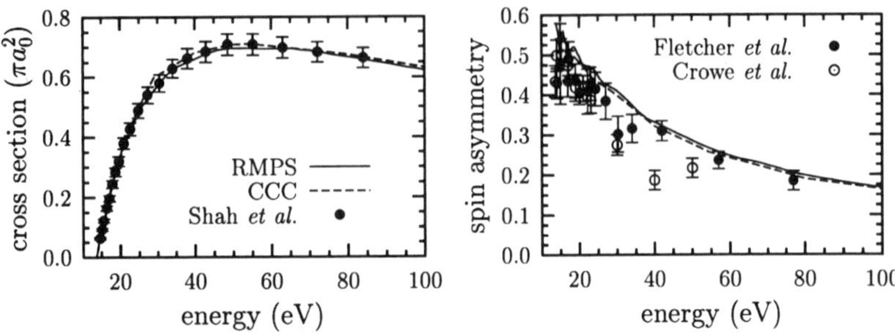

**Fig. 10.2.** Total cross section (left) and spin asymmetry (right) for electron-impact ionization of H(1s) as a function of the projectile energy. The experimental data [10.22–24] are compared with predictions from CCC [10.20] and RMPS [10.21] calculations.

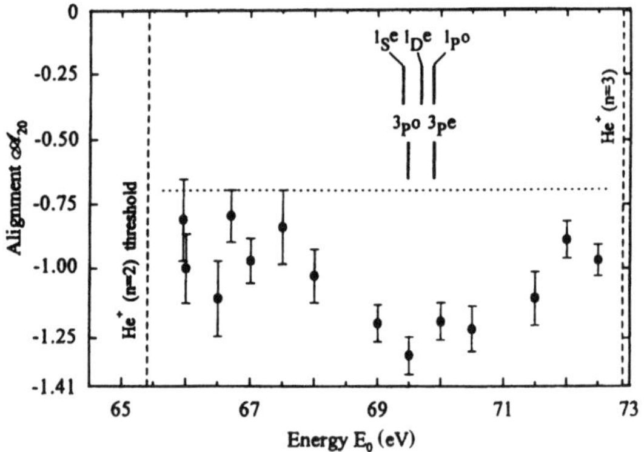

**Fig. 10.3.** Alignment of the $He^+(2p)$ ionic state as a function of the incident energy in ionization–excitation of $He\,(1s^2)^1S$ (from [10.5]). The vertical dashed lines indicate the $n=2$ and $n=3$ thresholds of $He^+$, and some $3\ell3\ell'$ doubly excited states are indicated as well.

$He^+(2p)$ state produced after ionization of the $(1s^2)^1S$ ground state. Although there is qualitative agreement between the experimental results and the theoretical prediction, based on a hybrid model using a distorted-wave description for the "fast" electron and a three-state R-matrix (close-coupling) model for the initial bound state and the interaction between the "slow" ejected electron and the residual ion, the process is apparently affected by resonances not included in the theoretical description.

A similar process was studied by Hayes $et\ al.$ [10.30], who investigated the ionization of $Kr\,(4p^6)^1S_0$. Using spin-polarized electrons for the ionizing impact, the alignment and the orientation of the residual ions left in the $Kr^+(4p^45p)^4P_{3/2,5/2}$ or $^4D_{7/2}$ excited states were studied through the polarization of the light emitted in the subsequent decay to the $\mathrm{Kr}^+(4\mathrm{p}^45\mathrm{s})^4\mathrm{P}_{5/2}$ state.

As another example of angle-integrated measurements, Figure 10.4 shows the results of the Freiburg group [10.31,32] for the electron-collision-induced alignment of the $\mathrm{Na}^*(2\mathrm{p}^53\mathrm{s}^2)^2\mathrm{P}_{3/2}$ autoionizing state. It was determined experimentally via the anisotropic emission of the autoionization electrons in the impact energy range from threshold to $100\,\mathrm{eV}$. In contrast to studies of the fluorescence polarization that are often strongly influenced by secondary excitation processes such as cascade population of the excited state, as well as depolarization effects due to the atomic fine and hyperfine structure, the nonradiative decay of such an autoionizing state via the Coulomb interaction by ejection of an autoionizing electron is so fast that a direct comparison between theoretical and experimental values of the alignment is possible at essentially arbitrary incident energies.

As seen from Figure 10.4, negative-ion resonance structures have a major effect on the alignment of the autoionizing state near threshold. They were observed in this context for the first time in the above study and, for impact energies up to $80\,\mathrm{eV}$, the experimental results were successfully reproduced by a 26-state R-matrix close-coupling calculation. This work resolved a major discrepancy between earlier experiments [10.33] and predictions from calculations based on perturbative methods [10.34,35].

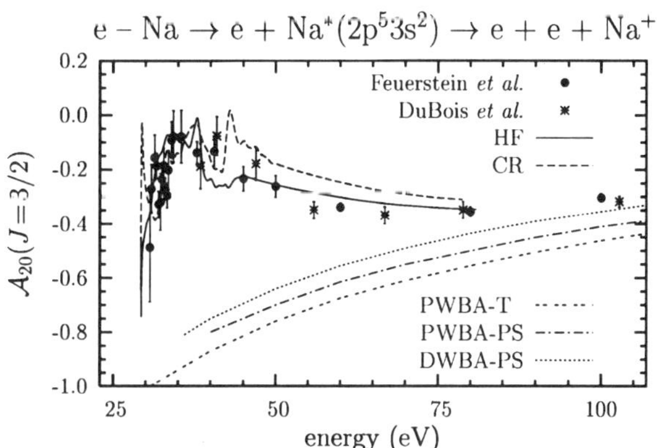



**Fig. 10.4.** Alignment of the $\mathrm{Na}\,(2\mathrm{p}^53\mathrm{s}^2)^2\mathrm{P}_{3/2}$ autoionizing state [10.31,32]. The solid and short-dashed lines represent the R-matrix calculations with two different sets of single-electron orbitals marked as "HF" and "CR," convoluted with the energy width of the incident beam. The other lines represent previous calculations in the plane-wave and distorted-wave Born approximations by Theodosiou [10.34] (PWBA-T) and Pangantiwar and Srivastava [10.35] (PWBA-PS and DWBA-PS). The circles represent the Freiburg measurements [10.31,32] and the stars those of DuBois et al. [10.33].

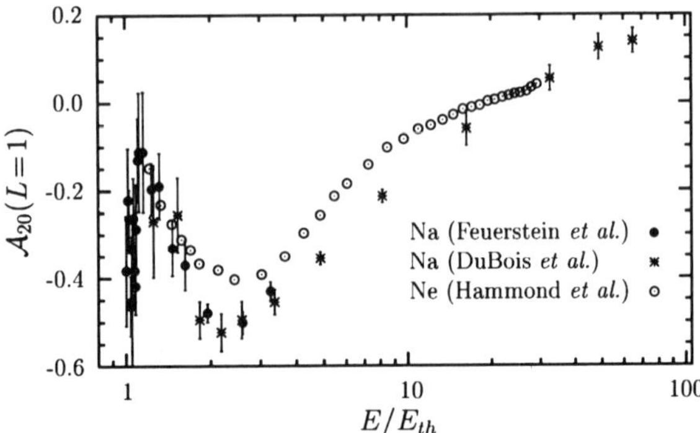


**Fig. 10.5.** Orbital angular-momentum alignment of the autoionizing $(2p^5 3s^2)^2 P_{3/2}$ state in Na and the $(2p^5 3s)^1 P_1$ state of Ne as function of incident energy over threshold energy. The alignment data for the Ne state were evaluated from the linear light polarization measurements of Hammond *et al.* [10.36], according to $\mathcal{A}_{20}(L) = 2\sqrt{2}P/(P-3)$.

Finally, when the results for the orbital angular-momentum alignment[17] of the autoionizing $(2p^5 3s^2)^2 P$ state in sodium were compared with those for the $(2p^5 3s)^1 P$ discrete state in neon for incident energies above the resonance region, a striking similarity was discovered between the alignment data for the two states which differ by a 3s electron. This is shown in Figure 10.5, indicating a potentially general form of the alignment as a function of the ratio between incident energy and threshold energy.

## 10.1.2 Angle-differential studies

We begin the angle-differential studies again with a spin-asymmetry for impact ionization of spin-polarized atoms by spin-polarized electrons. Figure 10.6 shows results of the Bielefeld group for the lithium target at an incident electron energy of 54.4 eV [10.16]. In a coplanar geometry, one electron is detected at a fixed scattering angle $\theta_A$ while the detector for the other electron is moved through angles $\theta_B$. Clearly, such an experiment is much more difficult to perform than an angle-integrated measurement, thereby resulting in substantial statistical and possibly also systematic uncertainties. The overall agreement between the experimental data and theoretical predictions based on distorted-wave (DWBA) and convergent close-coupling (CCC) models is satisfactory. The CCC predictions agree very well with experiment for the case of asymmetric sharing of the excess energy, while the experiment seems to slightly favor the DWBA model for the symmetric energy sharing case. Note that the Pauli exclusion principle requires $A_{\mathrm{TDCS}} \equiv 1$ in the fully

---

[17]The orbital angular momentum $\mathcal{A}_{20}(L=1)$ is related to $\mathcal{A}_{20}(J=3/2)$ of the $^2P_{3/2}$ state by $\mathcal{A}_{20}(L=1) = \sqrt{2}\mathcal{A}_{20}(J=3/2)$.

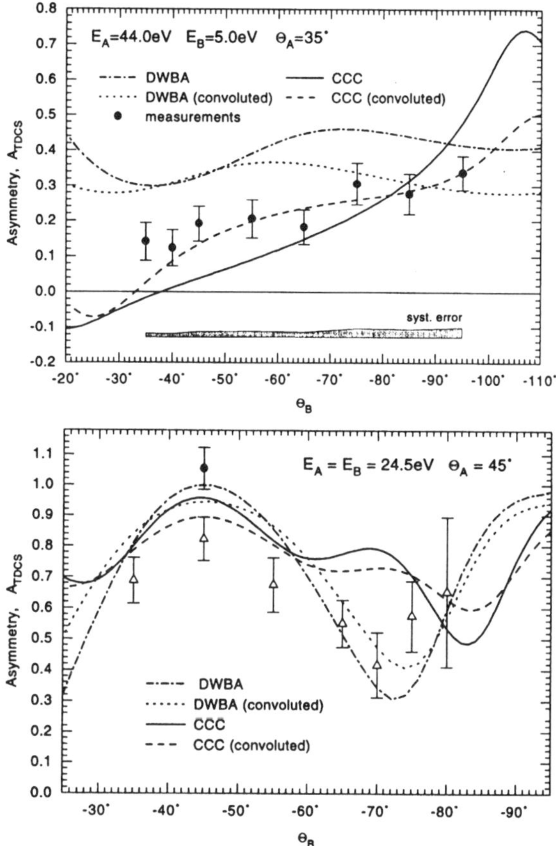

**Fig. 10.6.** Spin asymmetry of the TDCS as a function of the scattering angle $\theta_B$ for impact ionization of polarized Li atoms by spin-polarized electrons at an incident electron energy of 54.4 eV. The excess energy of 49.0 eV is shared either asymmetrically (top) or symmetrically (bottom) between the two outgoing electrons. The experimental data of the Bielefeld group are compared with theoretical predictions from DWBA and CCC calculations (from Streun *et al.* [10.16]).

symmetric case of equal energy sharing and $\theta_A = -\theta_B = 45°$; i.e., the scattering amplitude for the triplet spin channel must vanish. The deviations from this relationship seen in Figure 10.6 are due to finite energy and angle resolution in the experiment, and to minor numerical convergence problems in the CCC calculation even before convolution.

Electron-impact excitation of laser-excited sodium atoms has also been studied [10.37]. Figure 10.7 shows the main result, namely a strong orientational dichroism, i.e., a dependence of the TDCS on the orientation of the 3p state. For an incident electron energy of 150 eV, theoretical predictions from a first Born approximation (FBA) and a much more sophisticated

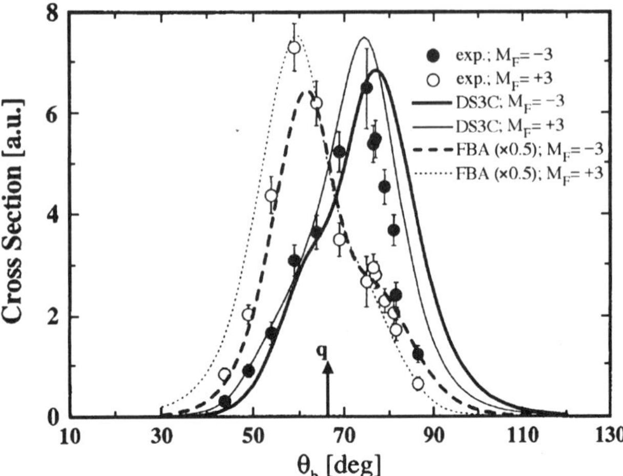



**Fig. 10.7.** Orientational dichroism observed in the TDCS for electron-impact ionization of laser-prepared sodium atoms in the 3p ($M_F = \pm 1$) states of sodium. The incident electron energy is 150 eV and the fast outgoing electron is observed at $\theta_a = 20°$, while the angle $\theta_b$ of the slow (20 eV) electron is varied. The experimental data, normalized to the DS3C results for $M_F = -1$ are compared with predictions from various theoretical models (adapted from [10.37]).

DS3C model [10.38], which accounts for a dynamical screening and incorporates the correct three-body asymptotic boundary condition, predict the effect very well. Not surprisingly, however, the FBA predictions become significantly worse at the lower energy of 60 eV [10.37].

Another possibility for a spin-dependent effect in electron-impact ionization was suggested by Hanne [10.39]: If a spin-polarized electron beam is used to ionize a heavy closed-shell atom, such as xenon in its ground state $(5p^6)^1S_0$, a spin-up/down asymmetry should be detectable if the $J$ value of the residual ion, i.e., $Xe^+(5p^5)^2P_{3/2}$ or $Xe^+(5p^5)^2P_{1/2}$, is resolved in the energy loss of the incident electron. This is the extension of the fine-structure effect [10.40] discussed already in Chaps. 5 and 7 to excitation processes.[18]

After being verified numerically in a model calculation [10.17], the latter effect was confirmed experimentally [10.18,19]. Figure 10.8 shows theoretical results [10.41] for this spin asymmetry, indicating the sensitivity of the predictions on the details of the numerical model. Consequently, experimental and theoretical benchmark studies of such processes may provide further advances in the treatment of spin-dependent electron-impact ionization.

We now return to coincidence studies involving one of the outgoing electrons and the photon (cf. Figure 10.1) in the schematic process described

---

[18] A similar effect was predicted by Macek and Shakeshaft (Phys. Rev. Lett. **27** [1971] 1487; Phys. Rev. A**6** [1972] 1876.) for capture of the 5p electron in proton impact of Xe.

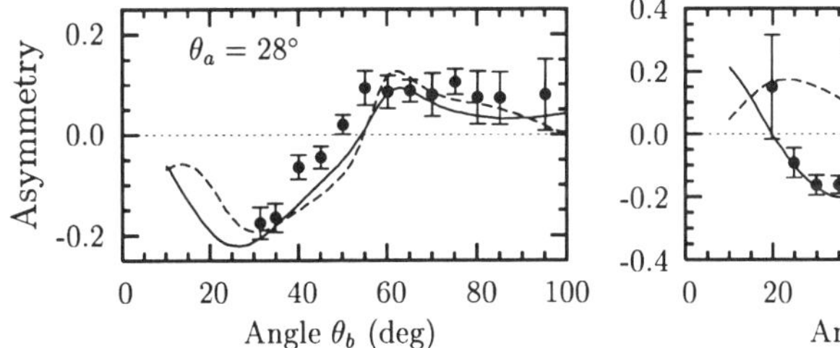

**Fig. 10.8.** Spin-up/down asymmetry in the TDCS as a function of the scattering angle $\theta_b$ for impact ionization of Xe atoms by unpolarized electrons at an incident electron energy of 147 eV. The faster electron with an energy of 100 eV is observed at a fixed angle $\theta_a$ while the detection angle of the slow electron with energy 34.9 eV corresponding to the final ionic state $\mathrm{Xe}^+(5p^5)^2\mathrm{P}_{3/2}$ is varied. The experimental data of Granitza *et al.* [10.19] are compared with predictions from DWBA models including (solid) or neglecting (dashed) exchange between the incident electron and the atomic charge cloud (adapted from Madison *et al.* [10.41]).

by (10.1). Figure 10.9 presents measured angular correlations of the photon emitted in the scattering plane for ionization–excitation of $\mathrm{He}\,(1s^2)$ to the excited ionic state $\mathrm{He}^+(2p)^2\mathrm{P}$. Recall from (2.48b) that the correlation amplitude is equivalent to half the degree of linear polarization $P_\ell$, which would be measured in scattered-electron–polarized-photon coincidence setups, and that the alignment angle $\gamma$ corresponds to the minimum in the correlation signal. These parameters are also shown in Figure 10.9.

For an incident electron energy of 200 eV with the detection of a 133.4 eV fast electron, corresponding to an (undetected) ejected electron energy of 1.2 eV, the experimental data of Hayes and Williams [10.10] (at $\theta = 8°$) and of Dogan and Crowe [10.11] are compared with predictions from various theoretical models based on the computer program **RMATRX-ION** of Bartschat [10.42]. Note that the theoretical predictions center at nearly constant values of $P_\ell \approx 0.3$ in the angular range investigated, in agreement with the Newcastle measurements [10.11], while the correlation amplitude obtained by the Perth group [10.10] is considerably smaller. The alignment angle follows the momentum-transfer direction for small scattering angles, as suggested by the plane-wave Born-approximation (PWBA). Not surprisingly, the agreement between experiment and the theoretical predictions deteriorates with increasing scattering angle, indicating the need for further improvement of the theoretical model. On the other hand, the relatively small angular range covered in the correlation measurements leads to substantial uncertainties in the fitting procedure.

Figure 10.10 shows predictions for the parameters that could be observed in a "complete" (e,2e$\gamma$) experiment for ionization–excitation of $\mathrm{He}(1s^2)$ to the ionic state $\mathrm{He}^+(2p)^2\mathrm{P}$ in a coplanar geometry for the two outgoing electrons, as calculated in a "PWBA-RMPS" model, corresponding to a plane-wave description of the fast electron combined with an RMPS-type treatment of

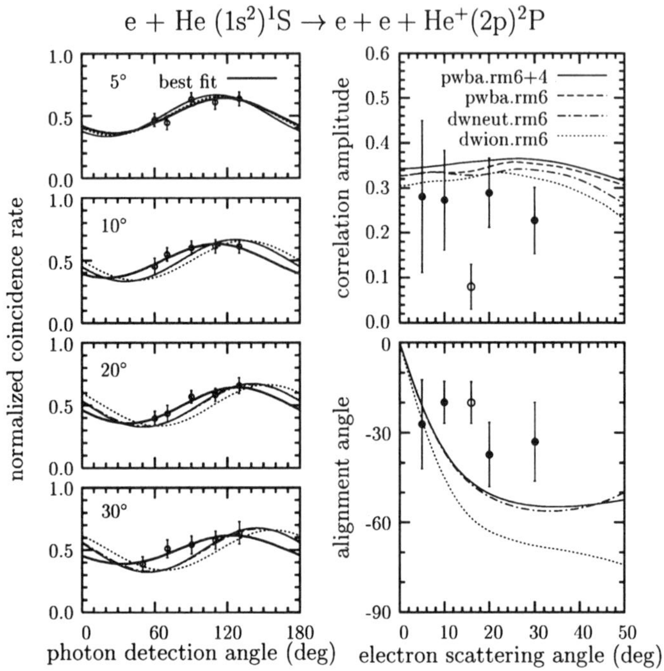



**Fig. 10.9.** Measured angular correlation signals and the corresponding parameters $P_\ell$ and $\gamma$ for ionization–excitation of He(1s$^2$) $\rightarrow$ He$^+$(2p). The thick solid lines in the left column are the least-squares fits to the experimental correlation signal, while the other lines represent predictions from various theoretical models that differ in the choice of distortion potential for the fast electron as well as the treatment of the ejected-electron–residual-ion interaction. The curves labeled "pwba" indicate no distortion, while "dwneut" and "dwion" correspond to the distortion potentials of the neutral atom and the residual ion, respectively. Furthermore, "rm6" indicates a 6-state R-matrix model that was extended with an additional four pseudostates in the "rm6+4" approach (adapted from [10.11]).

the initial bound state and the interaction between the residual ion and the ejected electron [10.43,44]. Note that the predicted degree of light polarization, $P = \sqrt{P_1^2 + P_2^2 + P_3^2}$, is less than unity, due to fine-structure depolarization effects in the linear polarizations $P_1$ and $P_2$. Furthermore, if the electron is ejected parallel or antiparallel to the direction of the momentum transfer (66° or 246°, respectively), the symmetry properties of the PWBA force the angular-momentum transfer $L_\perp (= -P_3)$ to vanish and the charge cloud to be aligned along this direction. On the other hand, the symmetry may be broken further in nonplanar observation geometries, thereby resulting in twisted and tilted charge clouds similar to those discussed in Chapter 7 for electron-impact excitation of mercury with (in-plane) spin-polarized electrons. Examples of such twisted charge clouds are shown in Figure 10.11,

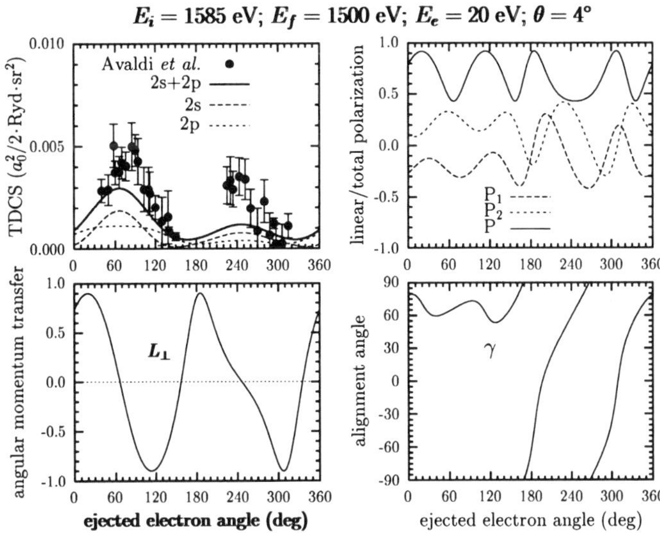



**Fig. 10.10.** TDCS, light polarizations, angular-momentum transfer, and alignment angle for ionization–excitation of $\mathrm{He}(1s^2) \to \mathrm{He}^+(2p)$ [10.43]. A fast incident electron of energy $1,585\,\mathrm{eV}$ is observed at a scattering angle of $4°$ and an energy of $1,500\,\mathrm{eV}$, corresponding to an energy of $20\,\mathrm{eV}$ for the slow electron. The experimental data for the TDCS ($n = 2$) are taken from Avaldi et al. [10.14].

where the ejected electron is observed at an angle of $45°$ above the plane defined by the incident and the scattered electron directions.

## 10.2 Ionization with Excitation by Heavy-Particle Impact

Little has been done for heavy-particle collisions in the field of ionization with simultaneous excitation, particularly if one restricts the focus to cases where photon emission data are available for both angle-integrated and angle-differential studies. One system, however, for which data exist that nicely extend the line of thought outlined in Chapter 8, is the process

$$\mathrm{Li}^- + \mathrm{He} \to \mathrm{Li}\,(2p) + \mathrm{He} + e \tag{10.3a}$$

with an energy loss given by

$$\Delta E = 2.47\,\mathrm{eV} + \epsilon\,(e). \tag{10.3b}$$

Here $\epsilon(e)$ is the kinetic energy of the detached electron, typically around $1\,\mathrm{eV}$. It is interesting to compare results for this reaction with data for the corresponding excitation process in the neutral system,

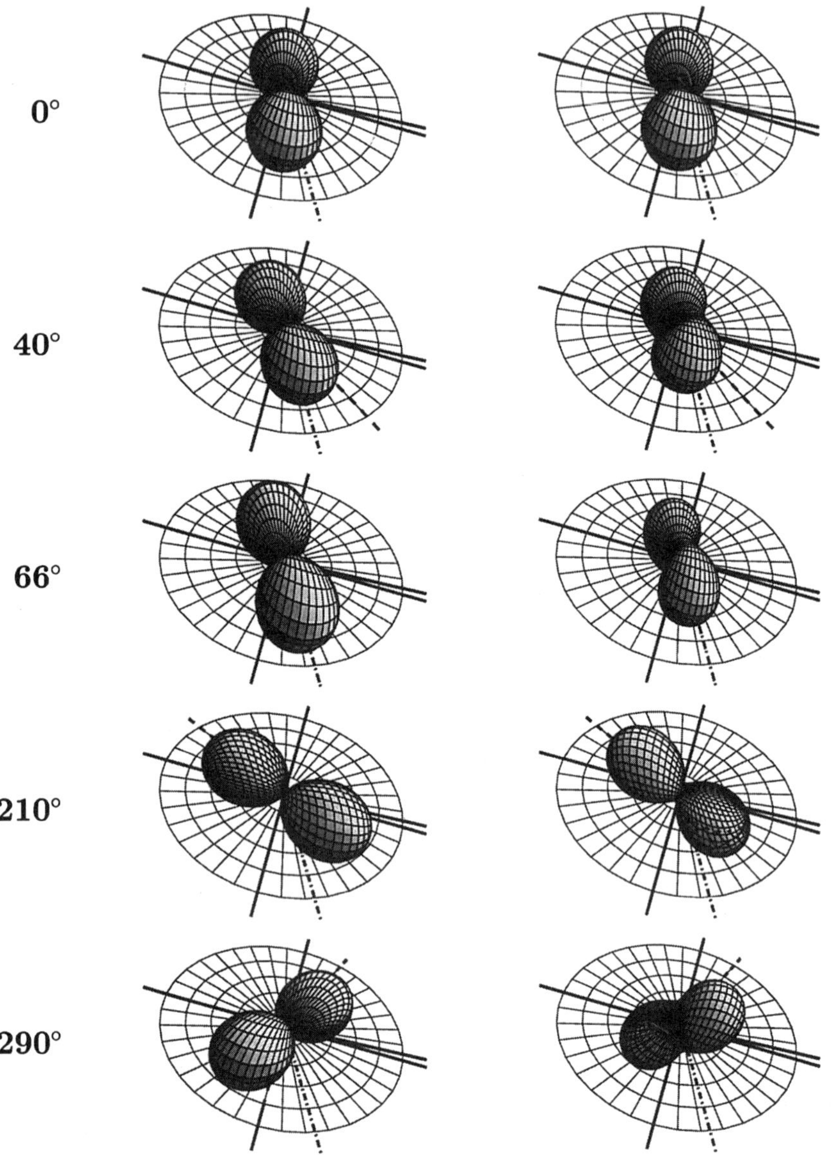

**Fig. 10.11.** Charge cloud describing the $\mathrm{He}^+(2\mathrm{p})^2\mathrm{P}$ state for the kinematical situation shown in Figure 10.10 for various angles of the ejected electron, if this electron is observed in the reaction plane (left) or at 45° out of the plane (right). The dashed line represents the ejected electron and the dash-dotted line the momentum-transfer direction.

$$\mathrm{Li} + \mathrm{He} \rightarrow \mathrm{Li}\,(2\mathrm{p}) + \mathrm{He} \qquad (10.4)$$

with $\Delta E = 1.85\,\mathrm{eV}$. This comparison will be discussed below.

### 10.2.1 Angle-integrated studies

Figure 10.12 compares angle-integrated emission cross sections for the LiI $2^2\mathrm{S} - 2^2\mathrm{P}$ transition for Li$^-$ [10.45] and Li [10.46] colliding with He as a function of impact energy $E$. According to the analysis of direct excitation processes given in Section 8.1, the bell-shaped curves exhibit a maximum in the velocity range near $v_{\max}$, determined by the criterion (8.7)

$$\frac{\Delta E\, a}{\hbar\, v_{\max}} \simeq \pi\,. \qquad (10.5)$$

Here, $\Delta E$ is the energy loss and $a$ is the effective interaction length of the collision. Including the electron kinetic energy $\epsilon\,(\mathrm{e})$, the combined energy loss for the process involving Li$^-$ is about twice the value for Li impact. Assuming comparable interaction lengths for the two systems, the Li$^-$ cross section maximum should thus be shifted up by about a factor of four in impact energy compared to Li, a result that is indeed seen in the figure. Furthermore, the broadening of the peak for Li$^-$ compared to the one for Li can be understood from the distribution of the electron kinetic energies $\epsilon\,(\mathrm{e})$.

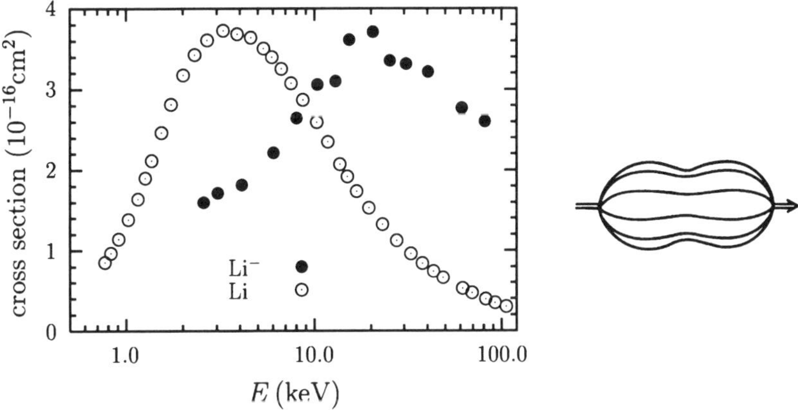


**Fig. 10.12.** Angle-integrated LiI $2^2\mathrm{S}-2^2\mathrm{P}$ emission cross sections for Li$^-$ ($\bullet$) [10.45] and Li (o) [10.46] colliding with He. The shape of the average Li(2p) electron cloud for Li$^-$ impact is shown on the right [10.45].

The angle-integrated alignment for the emitted light, using $^7$Li$^-$ as projectile, was found to be $4.0\pm0.8\%$ [10.45], with negligible energy dependence in the range studied. This observation is very similar to the finding of a small, positive value for the light polarization in the neutral system Li + He, again almost independent of energy around the maximum [8.41]. Taking fine-structure and hyperfine-structure depolarization into account, the light polarization may be converted into the cross section ratio for the magnetic sublevels $\sigma_{M_L}$ given by $\sigma_0 : \sigma_1 = 2.2 \pm 0.2$ [10.45]. The corresponding angle-averaged electron cloud is seen on the right of Figure 10.12.

Tuan and Gauyacq [10.47] conducted a theoretical study of the total electron detachment for the Li$^-$ + He system using an "independent-scattering model". In this model, the outer electron and the neutral core of the Li$^-$ ion are assumed to collide independently with the target. They found that the total detachment data could not be reproduced if detachment of the outer electron is the only process taken into account. Both the outer electron and the neutral Li core contribute to the detachment. However, no analysis of the specific channel (10.3) for simultaneous detachment and excitation, accounting for about 30% of the cross section, was performed in this theoretical study.

### 10.2.2 Angle-differential studies

Further insight into the collision dynamics for the process (10.3) can be gleaned from angle-differential data [10.48]. Figure 10.13 shows the excitation probability $\mathcal{P}$, charge cloud orientation $L_\perp$, alignment angle $\gamma$, and degree of polarization $P$ for Li$^-$ and Li for a fixed impact parameter $b$ as a function of collision energy. The angular momentum was evaluated from the formula

$$L_\perp = -D_1 P_3 / P \, , \tag{10.6}$$

where $D_1$ is a depolarization factor [10.48]. (This formula is based on the assumption that cascade processes from higher lying levels have the net effect of an isotropic, unpolarized background [10.49].)

Note that the shapes and relative sizes of the excitation probabilities exhibit a striking similarity to the corresponding total cross sections shown in Figure 10.12, thereby confirming that the key parameter governing the collision dynamics for direct excitation is the collision velocity, or time, but not the impact parameter. Furthermore, there is a strong similarity between the Li$^-$ and Li results for the parameter set $(L_\perp, \gamma)$ characterizing the shape and dynamics of the Li(2p) electron cloud. The measured values of $L_\perp \simeq -1.0$ are a strong confirmation of the propensity rule for orientation by direct excitation, discussed in Section 8.1, that is expected to be valid near the excitation probability maximum. The results shown in Figure 10.13 were found to depend only weakly on the impact parameter [10.48]. Thus, the average electron cloud shape seen on the right of Figure 10.12, with a length-to-width ratio of 2:1, is precisely what is obtained by azimuthal angular averaging of the circular states observed in the angle-differential studies.

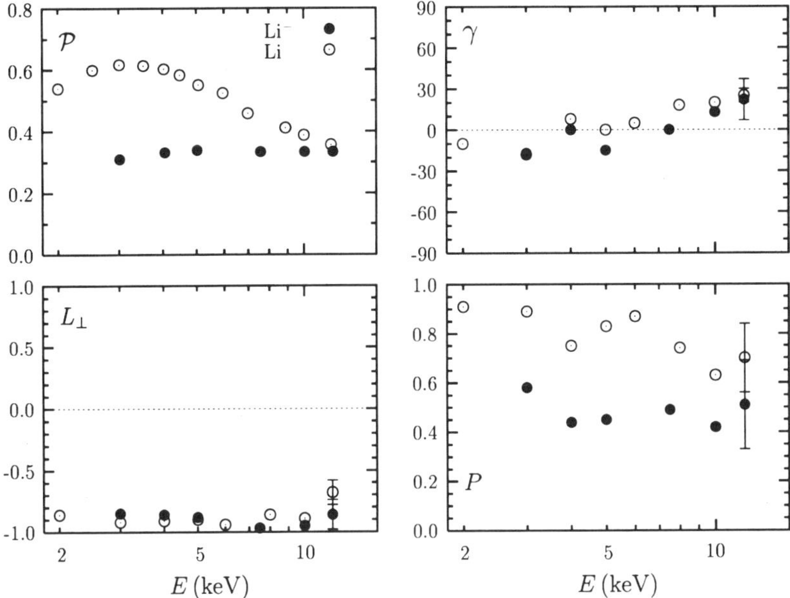

**Fig. 10.13.** Excitation probability $\mathcal{P}$, charge cloud orientation $L_\perp$, alignment angle $\gamma$, and degree of polarization $P$ for $\mathrm{Li}^-$ ($\bullet$) and Li ($\circ$) colliding with He. The impact parameters were $b \simeq 1.07$ ($\mathrm{Li}^-$) and 1.25 (Li) a.u. [10.48].

Finally, the degree of polarization is significantly different for the two systems. It is reduced from values of about 0.8 for Li to 0.5 for $\mathrm{Li}^-$. (The deviation from unity in Li is mainly caused by cascade contributions from the 3d state [8.43].) This reduction in the degree of polarization for $\mathrm{Li}^-$ may be interpreted as an effect of electron correlation between the detached and the excited electrons, since an increase of cascade effects from higher lying levels can be ruled out [10.48]. For $\mathrm{Li}^-$, the experiment sums over the energy and direction of the outgoing electron, which in principle can be observed. The measured Stokes vector for $\mathrm{Li}^-$ is thus the result of an incoherent, weighted sum of Stokes-vector components for these channels, causing a reduction of the effective length of the resulting Stokes vector, and hence of the polarization. A triple-coincidence experiment, mapping also the direction and energy of the outgoing electron, would restore the Stokes vector to a unit vector and thus bring the measured degree of polarization to a value near unity.

# References


10.1 A.G. Mikosza, J.F. Williams, and J.B. Wang, Phys. Rev. Lett. **79** (1997) 3375.

10.2 I. Taouil, A. Lahmam-Bennani, A. Duguet, and L. Avaldi, Phys. Rev. Lett. **81** (1998) 4600.

10.3 E.W.O. Bloemen, H. Winter, T.D. Märk, D. Dijkkamp, D. Barends, and F.J. de Heer, J. Phys. B **14** (1981) 717.

10.4 J.J. Forand, K. Becker, and J.W. McConkey, J. Phys. B **18** (1985) 1409.

10.5 A. Götz, W. Mehlhorn, A. Raeker, and K. Bartschat, J. Phys. B **29** (1996) 4699.

10.6 H. Merabet, M. Bailey, R. Bruch, I. Bray, D.V. Fursa, J.W. McConkey, and P. Hammond, Phys. Rev. A **60** (1999) 1187.

10.7 R. Müller-Fiedler, K. Jung, and H. Ehrhardt, J. Phys. B **19** (1986) 1211.

10.8 B. Matterstock, R. Huster, B. Paripas, A.N. Grum-Grzhimailo, and W. Mehlhorn, J. Phys. B **28** (1995) 4301.

10.9 R. Schwienhorst, A. Raeker, K. Bartschat, and K. Blum, J. Phys. B **29** (1996) 2305.

10.10 P.A. Hayes and J.F. Williams, Phys. Rev. Lett. **77** (1996) 3098.

10.11 M. Dogan, A. Crowe, K. Bartschat, and P.J. Marchalant, J. Phys. B **31** (1998) 1611.

10.12 R. Müller-Fiedler, P. Schlemmer, K. Jung, and H. Ehrhardt, Z. Phys. A **19** (1986) 1211.

10.13 C. Dupré, A. Lahmam-Bennani, A. Duguet, F. Mota-Furtado, P.F. O'Mahony, and C. Dal Cappello, J. Phys. B **25** (1992) 259.

10.14 L. Avaldi, R. Camilloni, R. Multari, G. Stefani, J. Langlois, O. Robaux, R.J. Tweed, and G.N. Vien, J. Phys. B **31** (1998) 2981.

10.15 G. Baum, M. Roede, W. Raith, and W. Schröder, J. Phys. B **18** (1985) 531.

10.16 M. Streun, G. Baum, W. Blask, J. Rasch, I. Bray, D.V. Fursa, S. Jones, D.H. Madison, H.R.J. Walters, and C.T. Whelan, J. Phys. B **31** (1998) 4401.

10.17 S. Jones, D.H. Madison, and G.F. Hanne, Phys. Rev. Lett. **72** (1994) 2254.

10.18 G.F. Hanne, Can. J. Phys. **74** (1996) 811.

10.19 B. Granitza, X. Guo, J. Hurn, J. Lower, S. Mazevet, I.E. McCarthy, Y. Shen, and E. Weigold, Aust. J. Phys. **49** (1996) 383.

10.20 I. Bray and A.T. Stelbovics, Phys. Rev. Lett. **70** (1993) 746.

10.21 K. Bartschat and I. Bray, J. Phys. B **29** (1996) L577.

10.22 M.B. Shah, D.S. Elliott, and B.H. Gilbody, J. Phys. B **20** (1987) 3501.

10.23 G.D. Fletcher, M.J. Alguard, T.J. Gay, P.F. Wainwright, M.S. Lubell, W. Raith, and V.W. Hughes, Phys. Rev. A **31** (1985) 2854.

10.24 D.M. Crowe, X.Q. Guo, M.S. Lubell, J. Slevin, and M. Eminyan, J. Phys. B **23** (1990) L325.

10.25 G. Baum, M. Fink, W. Raith, H. Steidl, and J. Taborski, Phys. Rev. A **40** (1989) 6734.

10.26 G. Baum, B. Granitza, L. Grau, B. Leuer, W. Raith, K. Rott, M. Tondera, and B. Witthuhn, J. Phys. B **26** (1993) 331.

10.27 I. Bray, J. Phys. B **28** (1995) L247.

10.28 K. Bartschat and I. Bray, Phys. Rev. A **54** (1996) 1723.

10.29 I. Bray, Can. J. Phys. **74** (1996) 875.

10.30 P.A. Hayes, D.H. Yu, and J.F. Williams, J. Phys. B **31** (1998) L193.

10.31 B. Feuerstein, A.N. Grum-Grzhimailo, K. Bartschat, and W. Mehlhorn, J. Phys. B **32** (1999), 3727.

10.32 A.N. Grum-Grzhimailo, K. Bartschat, B. Feuerstein, and W. Mehlhorn, Phys. Rev. A **60** (1999) R1751.

10.33 R.D. DuBois, L. Mortensen, and M. Rødbro, J. Phys. B **14** (1981) 1613.

10.34 C.E. Theodosiou, Phys. Rev. A **36** (1987) 3138.

10.35 A.W. Pangantiwar and R. Srivastava, J. Phys. B **20** (1987) 5881.

10.36 P. Hammond, W. Karras, A.G.McConkey, and J.W. McConkey, Phys. Rev. A **40** (1989) 1804.

10.37 A. Dorn, A. Elliott, J. Lower, E. Weigold, J. Berakdar, A. Engelns, and H. Klar, Phys. Rev. Lett. **80** (1998) 257.

10.38 J. Berakdar, Phys. Rev. A **53** (1996) 2314.

10.39 G.F. Hanne, in *Correlations and Polarization in Electronic and Atomic Collisions and (e,2e) Reactions*, Institute of Physics Conference Series 122, P.J.O. Teubner and E. Weigold (eds.), Bristol 1991.

10.40 G.F. Hanne, Phys. Rep. **95** (1983) 95.

10.41 D.H. Madison, V.D. Kravtsov, and S. Mazevet, J. Phys. B **31** (1998) L17.

10.42 K. Bartschat, Comp. Phys. Commun. **75** (1993) 219.

10.43 K. Bartschat, J. Phys. IV France **9** (1999) Pr6-17.

10.44 A.S. Kheifets, I. Bray, and K. Bartschat, J. Phys. B **32** (1999) L433.

10.45 N. Andersen, T. Andersen, L. Jepsen, and J. Macek, J. Phys. B **17** (1984) 2281.

10.46 J.Ø. Olsen, N. Andersen, and T. Andersen, J. Phys. B **10** (1977) 1723.

10.47 V.N. Tuan and J.P. Gauyacq, J. Phys. B **20** (1987) 3843.

10.48 T. Andersen and J. Engholm-Pedersen, J. Phys. B **22** (1989) 617.

10.49 G.S. Panev, N. Andersen, T. Andersen, and P. Dalby, Z. Phys. D **5** (1987) 331.

# 11. Related Topics and Applications

The importance of polarization, alignment, and orientation studies in various fields of physics is illustrated with a few selected examples. Within the limitations of this book, only a very small number of topics from the large amount of possible applications can be discussed. For more details, we refer to the original literature.

## 11.1 Photoionization

Similar to ionization by charged-particle impact, the photoionization of atoms and molecules has been studied extensively over the past years. The major difference — and simplification — compared to charged-particle impact processes lies in the fact that photoionization is governed by selection rules that severely reduce the number of independent parameters. For the case of dipole radiation (the only one considered here), the well-known selection rules for the state of the ejected electron are

$$\Delta\ell = \pm 1; \tag{11.1a}$$

$$\Delta j = \pm 1, 0. \tag{11.1b}$$

Here $\ell$ and $j$ are the orbital and total (spin plus orbital) angular-momentum quantum numbers. In fact, if a target electron with angular-momentum value $(\ell_t, j_t)$ is ejected, there is a maximum of *three* independent dipole matrix elements that determine the outcome of the process.

To begin with, consider the simplest case of photoionization of a single outer-shell electron, such as H (1s), Na (3s), or Cs (6s). In this case, the selection rules (11.1) require that the photoelectron be represented by a continuum p-wave. If all explicitly spin-dependent forces, with the spin-orbit interaction being the most important one, are neglected, the process is completely determined by a single matrix element of the electric dipole operator between the initial bound state and the final continuum state. The magnitude of this matrix element can be determined by an absolute cross-section measurement, which then represents the simplest form of a complete photoionization experiment.

The situation becomes significantly more complicated, however, if the spin-orbit interaction cannot be neglected. This is the case, for example, in photoionization of unpolarized Cs (6s), especially near the Cooper minimum of the cross section. As a result, the matrix elements between the initial bound state and the continuum p-wave go through minima at slightly different incident photon energies, thereby resulting in a dominant portion of $p_{1/2}$ or $p_{3/2}$ photoelectrons. This $j$-dependence is the basis of the "Fano effect" [11.1] and results in a high longitudinal spin-polarization of the photoelectrons if the incident light is circularly polarized and all the photoelectrons are collected. This transfer of light polarization into spin polarization was first verified indirectly by Lubell and Raith [11.2] and shortly thereafter in a direct electron polarization measurement by Kessler and Lorenz [11.3]. After first being used as a low-intensity, high-polarization source of spin-polarized electrons from photoionization of Cs atoms in the gas phase, the same basic physical mechanism is taken advantage of in today's GaAs-type sources [11.4] that are essential for many of the experiments described in previous chapters.

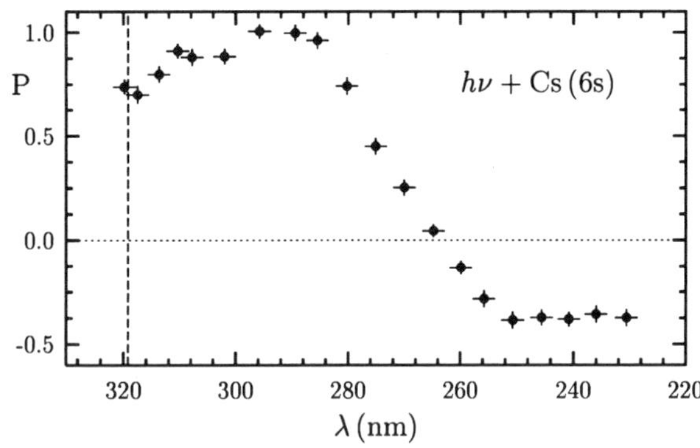


**Fig. 11.1.** Experimental data for the photoelectron spin polarization as a function of wavelength for photoionization of Cs (6s) atoms by circularly polarized light. The vertical line marks the ionization threshold (adapted from [11.5]).

Figure 11.1 shows experimental results for photoionization of Cs atoms by circularly polarized light. Indeed, the Fano effect leads to a nearly complete spin polarization around a wavelength of 290 nm [11.5]. For further illustrations of the Fano effect, we refer to the book by Kessler [11.6] and the original references given therein.

It is also possible to generate spin-polarized electrons from unpolarized atoms if the incident light beam is linearly polarized or even unpolarized. However, in this case, it is necessary to resolve the angle of the ejected electron, and parity conservation requires that the spin-polarization only have a nonvanishing component perpendicular to the reaction plane defined by the incident light beam and the ejected electron. Also, the spin-orbit interaction

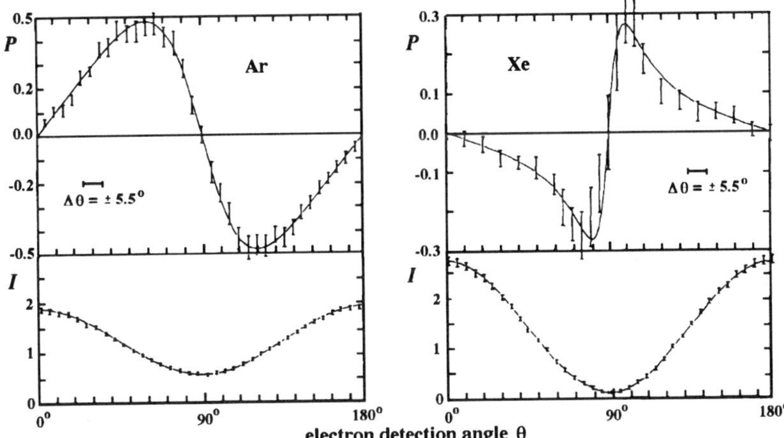



**Fig. 11.2.** Angular distribution of the photoelectron spin polarization $P$ and the intensity $I$ for photoionization of Ar (3p) and Xe (6p) by linearly polarized light of energy 21.22 eV (adapted from [11.7]).

must play a role and lead to different matrix elements for $j = \ell \pm \frac{1}{2}$ of the continuum electron.

Figure 11.2 shows experimental results of Schönhense [11.7] for photoionization of Ar and Xe, leaving the residual ion in the $\mathrm{Ar}^+(^2\mathrm{P}_{1/2})$ and $\mathrm{Xe}^+(^2\mathrm{P}_{1/2})$ ionic states. The angular distribution $I(\theta)$ and the polarization component $P(\theta)$ are shown as a function of the ejected electron angle $\theta$ defined in the reaction plane relative to the axis of the linearly polarized incident light beam. For this arrangement, the spin-polarization component is given by [11.8]

$$P_\perp(\theta) = \frac{2\xi \sin(2\theta)}{1 + \frac{\beta}{2}(3\cos^2\theta - 1)}. \tag{11.2}$$

A similar formula holds for unpolarized incident light, except that the denominator is replaced by $1 - \frac{\beta}{4}(3\cos^2\theta - 1)$, where the asymmetry parameter $\beta$ describes the anisotropy in the angular distribution of the photoelectrons.

Having already mentioned four experimental observables related to the photoionization process, namely the absolute cross section $\sigma$, the angle-integrated polarization transfer (usually described by the dynamical parameter $A$) in the Fano effect of ionizing unpolarized electrons by circularly polarized light, the angular distribution parameter $\beta$ for the photoelectrons, and the dynamical spin-polarization parameter $\xi$, suggests the possibility of "perfect photoionization experiments." Since at most three complex parameters, namely the transition-matrix elements, determine the outcome of the photoionization process in closed-shell atoms, a full determination of three magnitudes and two relative phase angles would be sufficient. In general, one might think that five independent measurements could give the absolute cross section, two relative size parameters, and two phases — except for possible ambiguities in the phases due to the fact that they usually appear as differences in arguments of sine and cosine functions. Therefore, we need at least one more independent observable that, for unpolarized atoms, is usually chosen as the polarization component *in the reaction plane* for circularly polarized incident light. Indeed, this polarization component is given by [11.8][19]

$$P_{\|}(\theta) = \frac{3\gamma \sin(2\theta)}{1 - \frac{\beta}{4}(3\cos^2\theta - 1)}. \tag{11.3}$$

For photoionization of the Xe (5p) subshell, a complete set of parameters $(\sigma; \beta, A; \xi, \gamma)$ describing the emitted photoelectron was analyzed by Heckenkamp *et al.* [11.9]. Using additional criteria based on a smooth transition from quantum defects in the discrete spectrum to phase shifts in the continuum to resolve the known ambiguities in the relative phases, they obtained a set of matrix elements and relative phases that was consistent with the set of measured dynamical parameters. Recently, however, it was discovered [11.10] that this set is apparently not unique. Some interdependence remains between the observables that results in an infinite manifold of matrix-element sets compatible with the measurements.

A complete photoionization experiment for the most general case of three independent matrix elements is therefore not possible if the atoms are unpolarized in the initial state and only the photoelectron but not the residual ion is observed. In principle, this problem can be overcome by removing either one of these restrictions, i.e., analyzing a case where fewer matrix elements determine the process, using polarized atoms, or investigating further details of an Auger decay in the ion.

An example of the first kind is the photoionization of the Xe (5p) shell with the residual ion being left in the $^2\mathrm{P}_{1/2}$ state. Since the selection rule (11.1b) prohibits a photoelectron in a $\mathrm{d}_{5/2}$ state, the number of independent parameters is reduced from five to three, and these can indeed be determined from a measurement of the photoelectron alone [11.9].

Another possibility is the introduction of additional approximations, with the most important one being a nonrelativistic model in which the phase difference between continuum waves for the same orbital angular momentum $\ell_f$ but different $j_f = \ell_f \pm \frac{1}{2}$ is neglected, and the magnitudes of the matrix elements are related by the statistical branching ratio. This extreme case of the *LS*-approximation yields a three-parameter model which, however, may have to be modified by introducing the experimental branching ratio $\rho$ instead of the statistical ratio as a fourth independent parameter.

---

[19]The parameter $\gamma$ is not related to the alignment angle in the previous chapters.

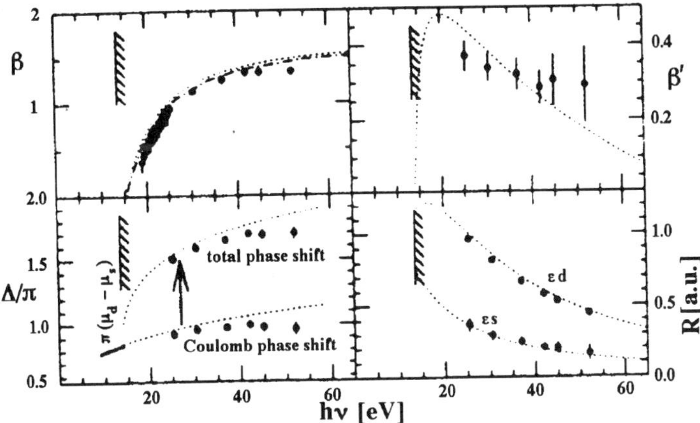

**Fig. 11.3.** Angular distribution parameters $\beta$ and $\beta'$, together with the magnitudes and relative phase $\Delta = \delta_\mathrm{d} - \delta_\mathrm{s}$ of the matrix elements for photoionization of atomic oxygen [11.11]. Theoretical results represent calculations performed with the codes RCN and RCG of Cowan [11.12] ($\cdots\cdots$) and predictions from Starace *et al.* [11.13] ($-\cdot-\cdot$); the $\beta$ results below 25 eV are from van der Meulen *et al.* [11.14].

An example of a complete experiment, within the limits of the three-parameter $LS$-model, is the photoionization of polarized oxygen atoms investigated by Plotzke *et al.* [11.11]. In this particular case, the oxygen atoms in the ground state $(2p^4)^3P_2$ were polarized through a hexapole magnet that focused the various $M_J$ components differently. Figure 11.3 shows their results for the values of matrix elements to generate photoelectrons with $\ell_\mathrm{f} = 0$ and $\ell_\mathrm{f} = 2$, respectively, as well as the relative phase between the two matrix elements. Combined with information on the *absolute* cross section, this experiment is indeed complete in the nonrelativistic approximation.

A recent example of a complete experiment within the four-parameter model is the investigation by Snell *et al.* [11.15] for photoionization of the Xe (4d) subshell. Their experimental data for the transferred spin polarization after ionizing by circularly polarized light, the dynamical spin polarization after ionizing by linearly polarized light, and the ratio of the Auger spin polarizations (see also Section 11.2), together with known data for the angular distribution parameter and the cross section, allowed for a complete determination of the three magnitudes and the one remaining independent relative phase. The results are shown in Figure 11.4, in comparison with predictions based on the relativistic random-phase approximation [11.17,18]. The agreement is very satisfactory, and the small theoretical values of the relative phases between ejected electron waves with the same orbital angular momentum supports the validity of the four-parameter approximation.

As a final example of alignment studies in photoionization experiments, Figure 11.5 shows the linear polarization observed in the optical decay

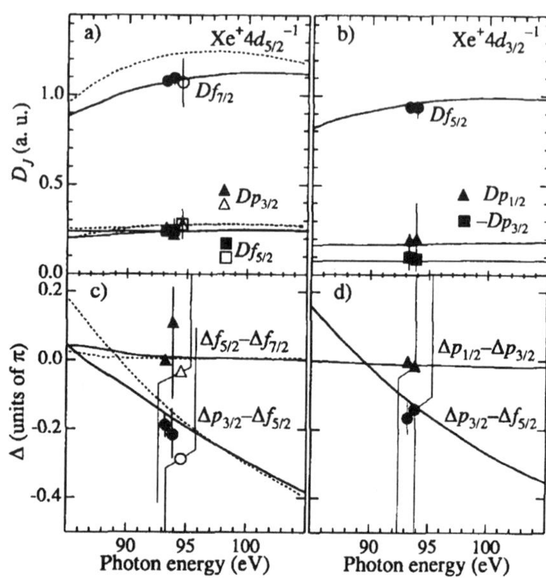

**Fig. 11.4.** Matrix elements and relative phases for photoionization of the 4d-electron in Xe. The closed symbols represent data from Snell *et al.* [11.15], evaluated under the assumption of a vanishing phase difference between matrix elements involving continuum electrons with the same $\ell$ but different $j$. Also shown are experimental results of Kämmerling and Schmidt [11.16] (open symbols) and theoretical results from a relativistic random-phase approximation model by Johnson and Cheng [11.17,18].

of the $\mathrm{Ar}^+[3\mathrm{p}^4(^3\mathrm{P})4\mathrm{p}]^2\mathrm{D}_{3/2}$ and $\mathrm{Ar}^+[3\mathrm{p}^4(^3\mathrm{P})4\mathrm{p}]^2\mathrm{P}_{3/2}$ satellite lines to the final states $\mathrm{Ar}^+[3\mathrm{p}^4(^3\mathrm{P})4\mathrm{s}]^2\mathrm{P}_{3/2,1/2}$, respectively, after production of the doubly excited $\mathrm{Ar}^+[3\mathrm{p}^4(^3\mathrm{P})4\mathrm{p}(^2\mathrm{S})n\,\mathrm{s}]$ Rydberg series by linearly polarized synchrotron radiation. According to the analysis presented by McLaughlin *et al.* [11.19], the linear polarization detected in the radiation corresponds to spin-aligned autoionizing states. The necessary spin-flips for creating these states are experimental evidence for a violation of the Wigner spin-conservation rule [11.20,21]. Also, the apparent transformation of the spin alignment in the autoionizing state to the orbital alignment observed in the polarization of the satellite indicates a strong spin-orbit interaction in the decay of an autoionizing state.

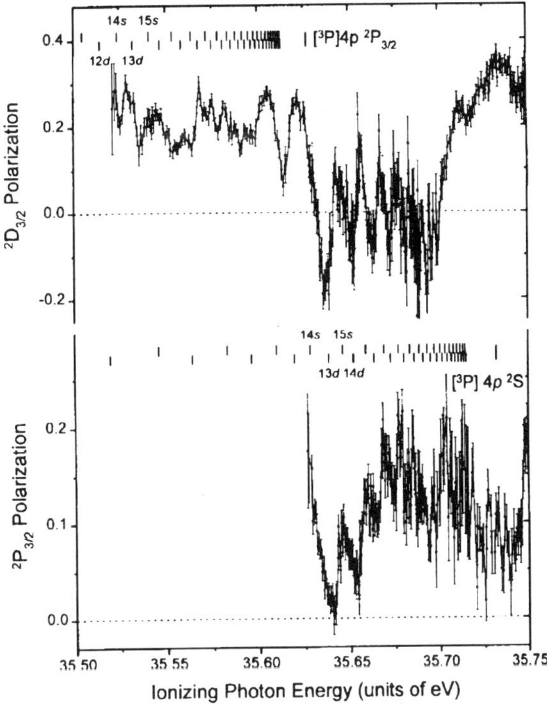



**Fig. 11.5.** Linear polarization of the $\mathrm{Ar^+[3p^4(^3P)4p]^2D_{3/2} \to [3p^4(^3P)4s]^2P_{3/2}}$ and $\mathrm{Ar^+[3p^4(^3P)4p]^2P_{3/2} \to [3p^4(^3P)4s]^2P_{1/2}}$ lines as a function of the incident photon energy [11.19].

## 11.2 Spin-Polarized Auger Electrons

In addition to studying the spin polarization of valence photoelectrons, it was predicted theoretically [11.22,23] that the Auger electrons could be spin-polarized as well. As with the discussion of the valence case, one has to distinguish between a *transfer-type* of polarization induced by creating a hole with circularly polarized light and a *dynamically induced* polarization that already occurs if the hole is only aligned.

An early attempt to measure a dynamical spin polarization of Auger electrons from Kr, in which the hole was created by unpolarized 1.5 keV incident electrons, indicated only a very small polarization effect [11.24], and attempts with a Ba target yielded a zero result within the experimental uncertainty [11.25]. On the other hand, studies of the transferred polarization have been more successful in generating spin-polarized Auger electrons. Figure 11.6 shows an example of the latter case for the Xe $M_{4,5}N_{4,5}N_{4,5}$ lines, excited by ionization of Xe with circular polarized light of energy 834.5 eV [11.26].

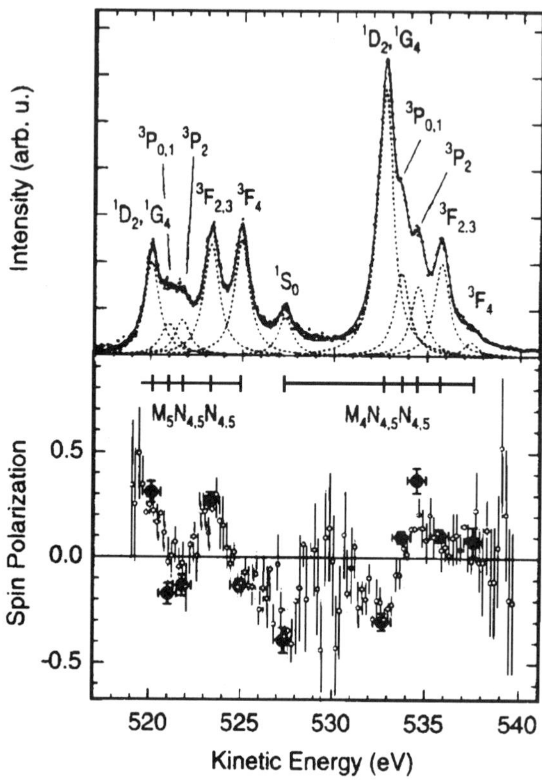

**Fig. 11.6.** Intensity and Auger spin polarization of the Xe $M_{4,5}N_{4,5}N_{4,5}$ lines for circularly polarized incident light of energy 834.5 eV [11.26]. The open symbols show the measured spin polarization, whereas the solid symbols represent results for the individual lines after fitting them to Voigt profiles.

A significant *dynamical* spin polarization of Auger electrons was measured recently in the special case of a "resonant Auger decay," where the inner-shell excited *ionic* hole state was substituted by an excited state of the *neutral* system. In this respect, the situation is similar to the decay of a doubly excited autoionizing state with outer-shell electrons. Selected results of Hergenhahn *et al.* [11.27] are shown in Figure 11.7.

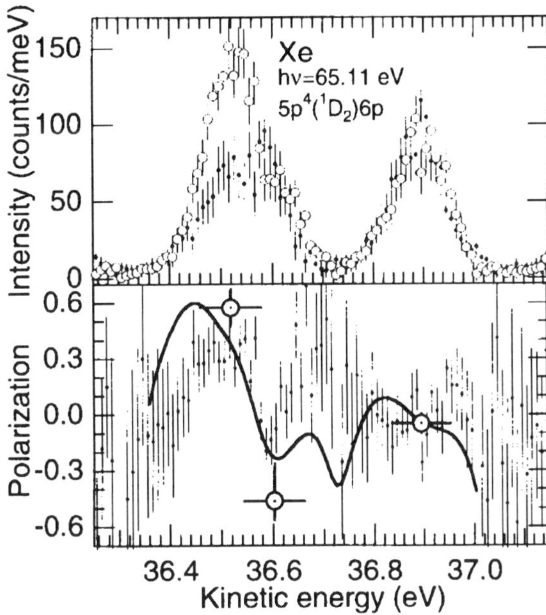


**Fig. 11.7.** Intensity and Auger spin polarization of the $\mathrm{Xe}\,4\mathrm{d}_{5/2}^{-1}6\mathrm{p}\,(J=1)$ to the multiplet of $5\mathrm{p}^4(^1\mathrm{D}_2)6\mathrm{p}$ states. The open and solid symbols in the top panel correspond to signal rates for spin-up and spin-down electrons that translate into the thin solid symbols for the spin polarization in the bottom part of the figure. The solid line and the large symbols are obtained from a least-squares analysis of individual lines (adapted from [11.27]).

## 11.3 Autoionization Anisotropies in Heavy-Particle Collisions

Similar to the case of photon emission, information may be obtained on scattering amplitudes and collision dynamics from a study of ejected-electron emission patterns when autoionizing states decay. Below we present a few illuminating examples of heavy-particle collisions where this technique has been used.

Boskamp *et al.* [11.28] studied the collision process

$$\mathrm{He}^+ + \mathrm{He}\,(1\mathrm{s}^2)^1\mathrm{S} \rightarrow \mathrm{He}^+ + \mathrm{He}\,(2\mathrm{p}^2)^1\mathrm{D} \qquad (11.4)$$

by analyzing the energy and angular distributions of the electrons emitted when the $\mathrm{He}\,(2\mathrm{p}^2)^1\mathrm{D}$ state decays back to the $\mathrm{He}^+$ ground state. Figure 11.8 shows angular distributions of such electrons from $2\,\mathrm{keV}\ \mathrm{He}^+-\mathrm{He}$ collisions at a laboratory scattering angle of $10°$ for which the $(2\mathrm{p}^2)^1\mathrm{D}$ channel accounts for 95% of the inelastic cross section. The electron-energy spectrum

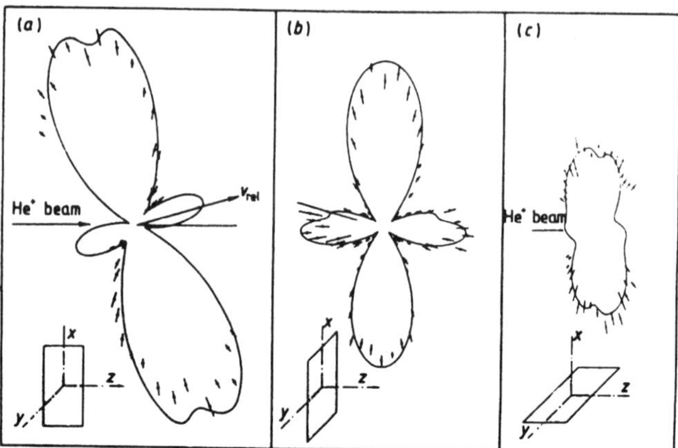


**Fig. 11.8.** Angular distributions of autoionizing $He\,(2p^2)^1D$ electrons from $2\,keV$ $He^+$–He collisions at a scattering angle of $10°$, measured in the three orthogonal planes indicated in the figure. The electron energy spectrum is integrated over target and projectile contributions. The solid curve represents fitting calculations, in which the complex population amplitudes for the magnetic sublevels of $He\,(2p^2)^1D$ are taken as fit parameters (from Boskamp *et al.* [11.28]).

plotted is obtained by integrating over target and projectile contributions. Since the collision complex may be treated as a quasi-molecule at this low energy, the probabilities for exciting the target or the projectile are the same. Consequently, the ejected electrons may come from either target or projectile, but their origin can generally be determined by measuring the Doppler shift. This distinction, however, cannot be made at observation angles perpendicular to the final relative velocity vector. Consequently, the contributions from the two emission centers add coherently, giving rise to the interference pattern observed in the figure. This feature is seen even better in Figure 11.9, which exhibits a three-dimensional picture of the emission pattern, obtained by fitting the complex scattering amplitudes for the $He\,(2p^2)^1D$ state to the observed autoionization decay pattern. Using these amplitudes, one may also calculate the shape of the excited electron cloud, shown for the same fit to the right in Figure 11.9. A theoretical description of the experimental findings has not yet been achieved from first principles, but it was found that the dominant features of the observations can be accounted for by an independent-electron two-step rotational $2p\sigma_u$–$2p\pi_u$ coupling model [11.28].

Insight into the collision dynamics of a different kind has been obtained from a study of the electron energy spectra emitted from the slightly asymmetric system

$$Li^+ + He\,(1s^2)^1S \rightarrow Li^+ + He\,(2p^2)^1D \text{ or } He\,(2s2p)^1P\,. \tag{11.5}$$

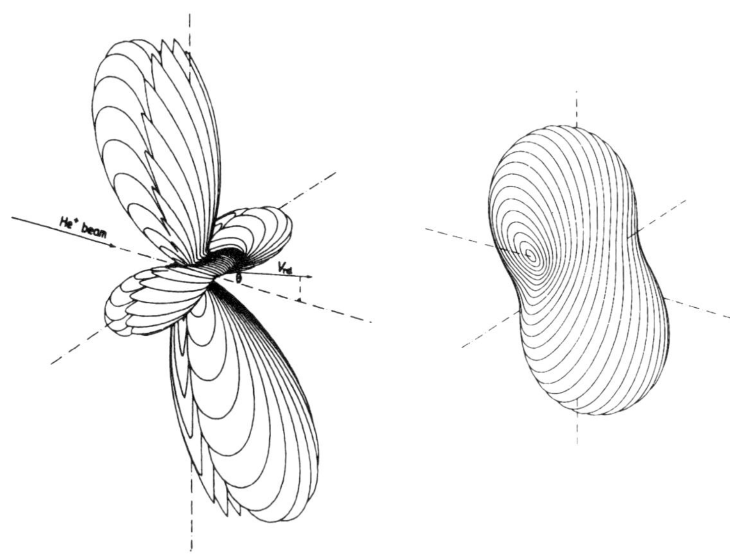


**Fig. 11.9.** A three-dimensional view of the complete distribution corresponding to the measured data of Figure 11.8 [11.28] is shown on the left. The corresponding shape of the collisionally excited atom is seen on the right [11.29].

Figure 11.10 shows energy spectra of autoionization electrons from this collision, measured in the scattering plane at opposite positions of the electron detector, at an impact energy of 2 keV and a laboratory scattering angle of 10°, corresponding to an impact parameter of $b = 0.22\,a_0$. The electron spectra in the two directions are very different. Note, for example, that at emission energies where the electron intensity has a maximum in one direction, it exhibits a minimum in the opposite direction [11.29]. The oscillations are caused by interferences of autoionizing transitions from different states, $\mathrm{He}\,(2p^2)^1\mathrm{D}$ and $\mathrm{He}\,(2s2p)^1\mathrm{P}$. These interferences depend on the relative phase of the transition amplitudes, which in turn depends on the time of transition. These times again depend on the internuclear separation of the collision partners and thereby correspond to different electron energies, since the transitions end on a repulsive Coulomb potential curve. The energy scale of Figure 11.10 therefore resembles a scale of emission times, where higher energies correspond to later emission times. Hence, the structures shown in the spectra reflect time-dependent oscillations of the atomic charge cloud. A proper data analysis eventually yields the relative scattering amplitudes which allows determination of the electron charge cloud [11.29].

The resulting picture is shown in Figure 11.11. The oscillation period $T = 17$ femtoseconds is determined by the energy separation of the two states involved, $\mathrm{He}\,(2p^2)^1\mathrm{D}$ and $\mathrm{He}\,(2s2p)^1\mathrm{P}$, respectively. Note that the oscillations seen here differ from the typical quantum beats observed in beam-foil

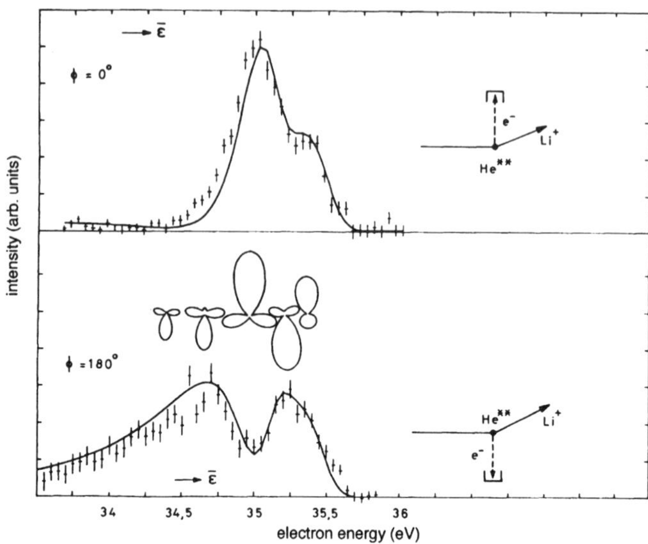

Fig. 11.10. Energy spectra of autoionization electrons originating from 2 keV Li$^+$ + He collisions, measured in the scattering plane at opposite positions of the electron detector, as shown. The electron energies refer to the emitter frame, and the solid lines represent fit calculations [11.29].

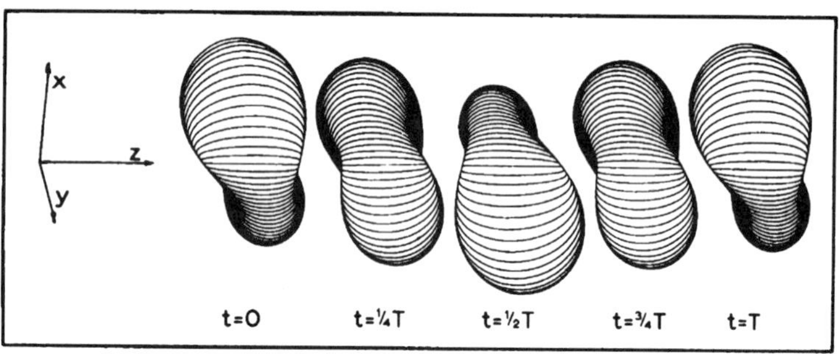


Fig. 11.11. Three-dimensional view of the shape of collisionally excited He$^{**}$ at different times after the collision. The oscillation period of $T = 17$ femtoseconds is determined by the energy separation of the two states involved, He $(2p^2)^1$D and He $(2s2p)^1$P [11.29].

spectroscopy (see Section 11.5 below) caused by interferences between fine and hyperfine structure states. The present case involves two states of opposite parity. Observation of quantum beats with a characteristic time in the femtosecond regime is quite unique in atomic collision physics.

## 11.4 Collisions with Molecules

### 11.4.1 Electron collisions with molecules

A theoretical framework for alignment and orientation studies involving molecular targets was proposed by Jakubovicz and Blum [11.30], and the treatment of Fano and Macek [11.31] was used as a starting point in the work of Greene of and Zare [11.32]. An extensive discussion of the role of orientation and alignment in molecular processes is also provided in Chapter 7 of Blum [11.33].

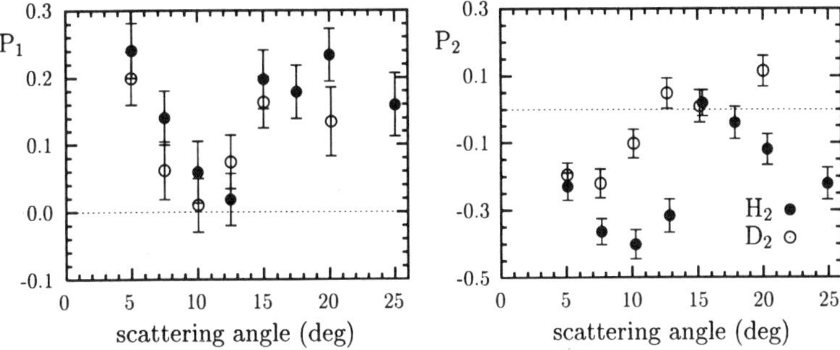

**Fig. 11.12** Linear polarizations $P_1$ and $P_2$ for electron-impact excitation of the $X^1\Sigma_g^+ \rightarrow C^1\Pi_u$ transition in $H_2$ and $D_2$ at an incident electron energy of 50 eV as a function of the scattered electron angle $\theta_e$. The open circles represent the experimental data of Becker *et al.* [11.34], while the squares show earlier results of Malcolm and McConkey [11.35].

Figure 11.12 shows an example of a scattered-electron–polarized photon coincidence study performed by the Windsor group [11.34,35]. They measured the linear polarizations $P_1$ and $P_2$ for electron-impact excitation of the $X^1\Sigma_g^+ \rightarrow C^1\Pi_u$ transition in $H_2$ and $D_2$, with subsequent decay back to $X^1\Sigma_g^+$. Although the experiment was only performed over a small range of scattering angles, interesting differences can already be seen between the results for the two molecules. These differences were attributed to the different rotational structure of the two molecules.

Figure 11.13 shows results of Hegemann *et al.* [11.36], who measured the $T$-parameter, i.e., the ratio of electron polarization after ($P'$) and before ($P$) when sent through a beam of $O_2$ and NO molecules, respectively. As seen from the figure, exchange effects, resulting in a significant depolarization or even reversal of the initial polarization, are apparently much smaller in the molecules compared to the case of the atomic Na target, even at such low energies as 5 eV. Based on the theoretical predictions of da Paixao *et al.* [11.38] and Fullerton *et al.* [11.39], the principal reason is likely the fact that the experiment averages over all directions of the molecular axis and sums over

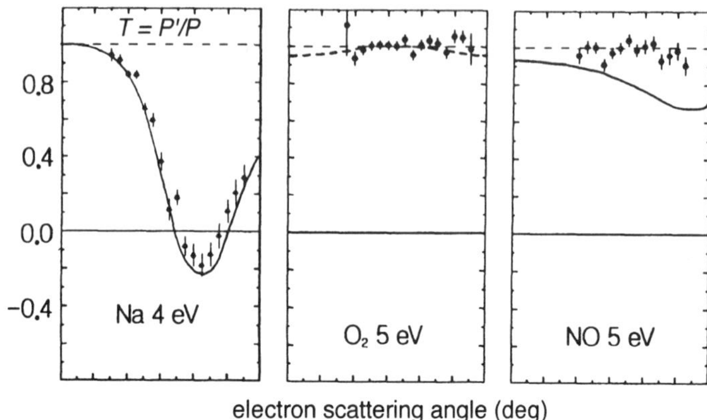


**Fig. 11.13** Polarization fraction $T = P'/P$ for elastic electron scattering from Na atoms and $O_2$ and NO molecules, respectively. The experimental data of Hegemann *et al.* [11.36] are compared with theoretical predictions by Bray and McCarthy [11.37] for Na, Fullerton *et al.* [11.39] for $O_2$, and da Paixao *et al.* [11.38] for NO (adapted from [11.40]).

various rotational states. Significantly bigger effects may be expected if such experiments are performed with oriented molecules [11.40].

The last example concerns the electron optic dichroism, i.e., an asymmetry measured in the transmission rate of longitudinally spin-polarized electrons passing through a beam of chiral molecules. Such an experiment was first suggested by Farago [11.41], and a nonzero result was indeed reported by Campbell and Farago [11.42]. However, later studies in the Lincoln [11.43] and Münster [11.44] groups indicated that the effect for unoriented camphor molecules used by Campbell and Farago is well below the current experimental detection limit — in agreement with a theoretical estimate by Blum and Thompson [11.45]. Nevertheless, if chiral molecules containing at least one heavy atom are used, the effect was expected to be larger. Indeed, Figure 11.14 shows the high-precision results of Mayer and Kessler [11.44] for L- and D-handed $Yb(hfc)_3$ molecules, i.e., an ytterbium atom surrounded by three camphor-like ligands. The reversal of the measured asymmetry for the two target varieties, resulting in a mirror pattern with respect to the zero axis, provides convincing evidence that a spurious contribution to the asymmetry due to experimental asymmetries is significantly smaller than the true physical asymmetry. Although a heavy constituent in the molecule seems essential for the effect to be detectable, we emphasize that it is *not* caused by the standard Mott mechanism, because of the *longitudinal* spin polarization.

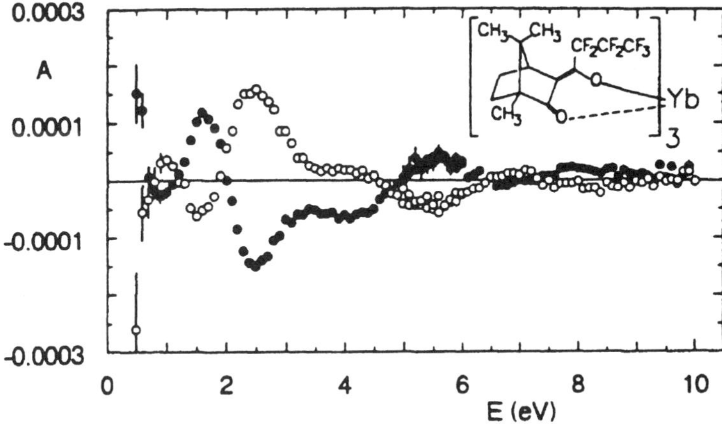

Fig. 11.14 Transmission asymmetry $A = [I(P) - I(-P)]/]I(P) + I(-P)]$ for D-Yb(hfc)₃ (filled symbols) and L-Yb(hfc)₃ (open symbols) as a function of the incident electron energy. The statistical uncertainty is only indicated by the error bars if it is larger than the symbol size. The structure of D-Yb(hfc)₃ is shown in the inset. (From [11.44].)

## 11.4.2 Heavy-particle collisions with molecules

Similarly to electron–molecule collisions, alignment and orientation studies involving heavy particles scattered from a molecular target are complicated by the fact that in most cases there is no control of the spatial direction of the internuclear axis in the molecule. Therefore, the photons or electrons emitted will originate from an isotropic distribution of target molecule directions, which furthermore are in more or less uncontrolled vibrational and rotational states. This summation over degrees of freedom, which are in principle observable, is a potential source for loss of coherence.

The first angular differential study of this kind was performed by Eriksen and Jaecks [11.46] on the $He^+ - H_2$ system. They investigated the production of $He(3\,^3P)$ states by analyzing the linear polarization of the $3889\,\text{Å}$ decay radiation emitted perpendicular to the scattering plane. Their results at $He^+$ impact energies of 1.5 and 3.0 keV and various scattering angles are shown in Figure 11.15. Typical data accumulation times were 2–4 days per point. The analysis of the variation in the transmitted intensity with polarizer angle, taking the fine-structure depolarization into account, showed a nonvanishing alignment parameter with the largest value occurring at 1.5 keV and a scattering angle of 1.5°, for which $P_\ell = 0.68 \pm 0.09$. This result is surprisingly large in light of the expectations stated above. Furthermore, the alignment angles (shown as the major axis of the elliptic curve fits in Figure 11.15) are close to the direction of momentum transfer, as calculated with the assumption that the entire $H_2^+$ molecule recoils as a single particle. The only exception occurs

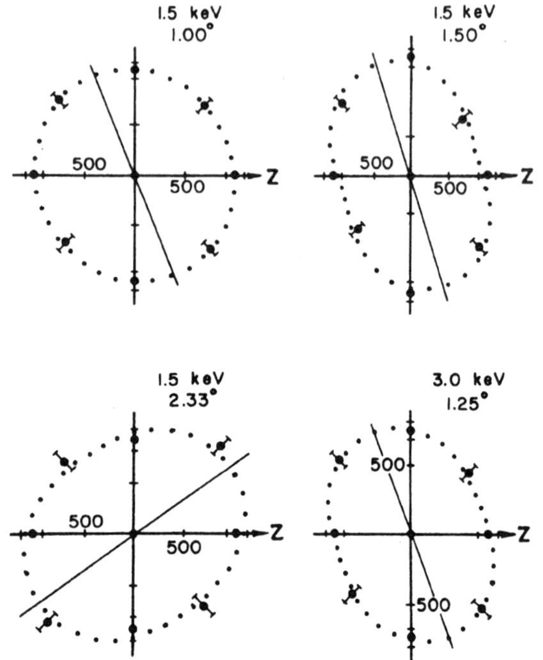



**Fig. 11.15.** Polar plots of the average number of real coincidences per $10^9$ scattered neutral particles versus polarizer angle for three scattering angles at 1.5 keV and one angle at 3.0 keV. The data are shown by solid dots with error bars and the dots without error bars are best fits to the data as described in the text. In all plots, the incident-beam direction is along the $z$ axis (to the right in the figure) and the neutral particles are detected above the $z$ axis [11.46].

at the largest angle of 2.33°, where $\gamma$ is almost perpendicular to $\Delta k$. This pioneering experiment therefore showed that for this system excitation of the He$(3\,^3\mathrm{P})$ state by electron transfer depends little on the initial direction of the H$_2$ molecule and the ro-vibrational state of the H$_2^+$ molecule.

Polarization analysis may also be used as a powerful tool to obtain a detailed understanding of bond-rearranging processes, as was recently demonstrated for the

$$\mathrm{He^+ + H_2 \rightarrow HeH^+}(v) + \mathrm{H\,(2s, 2p)} \tag{11.6}$$

"hot" proton-transfer reaction at an energy $\mathrm{E_{cm}} = 20\,\mathrm{eV}$ [11.47]. The total number of polarization studies of systems involving a molecular species is still relatively limited; see ref. [11.48] for another example and further references.

## 11.5 Collisions with Surfaces and Foils

In the late 1960s, beam-foil spectroscopy was developed, initially as a tool for the determination of lifetimes of excited atomic and ionic states. In this technique, an ion beam is accelerated and sent through a thin carbon foil, oriented perpendicular to the beam direction. The intensity of the various spectral lines emitted from the particles excited in the foil passage is monitored as a function of the distance downstream from the foil, or equivalently (knowing the energy) as a function of time after excitation. From the decay curve of this emission, the corresponding lifetime may be extracted, provided that cascade effects can be neglected or corrected for. Shortly after the introduction of the technique, Macek [11.49] pointed out that the atomic state can be created with an initial alignment. Furthermore, since the excitation may be considered impulsive and independent of spin forces, the observed photon emission should subsequently be modulated in time according to

$$I(t) = (a + \Sigma_i \, b_i \cos \omega_i t)e^{-\gamma t}, \tag{11.7}$$

with the frequencies $\omega_i$ determined by the splittings of the coherently excited fine- and hyperfine-structure levels. Shortly after this suggestion, these "zero-field quantum beats" were indeed observed experimentally by Andrä [11.50]. The early work on quantum beat spectroscopy is thoroughly discussed by Fano and Macek [11.31].

A particularly clear example of zero-field quantum beats was presented by Burns and Hancock [11.51]. Figure 11.16 shows their results for the two polarization components of the $3889\,\text{Å}$ line emitted from the $\text{He}\,(3\,^3\text{P})$ state, parallel and perpendicular to the beam direction, respectively. The middle curve, recorded with the linear polarizer set at "the magic angle" of 54.4°, corresponding to determination of the quantity $I_\parallel + 2I_\perp$, demonstrates that the exponential decay of the total light intensity emitted is not modified, but only its angular distribution.

A natural next step is to break the cylindrical symmetry of the setup by tilting the foil and looking for a possible circular polarization. This was done for the first time by Berry $et\ al.$ [11.52], who detected elliptical polarization of the $5016\,\text{Å}$ photons emitted from the $\text{He}\,(3^1\text{P})$ state. For a beam energy of $135\,\text{keV}$ and a foil tilt angle of 45°, they measured (in the standard geometry) $P_1 = 0.08 \pm 0.03$, $P_2 = 0.14 \pm 0.02$, and $P_3 = 0.11 \pm 0.01$. These results correspond to a total degree of polarization $P = 0.19 \pm 0.04$ and an alignment angle $\gamma = 30° \pm 5°$. Thus, a significant circular polarization is seen, but the excitation is far from fully coherent. In this geometry, the degree of circular polarization generally increases with increasing tilt angle of the foil.

Shortly thereafter, the idea was further generalized to the study of the circular polarization of light emitted from particles scattered off solid surfaces at grazing incidence. A comparison of the "tilted-foil" and the "grazing-incidence" geometries is shown in Figure 11.17 [11.53], in which the most

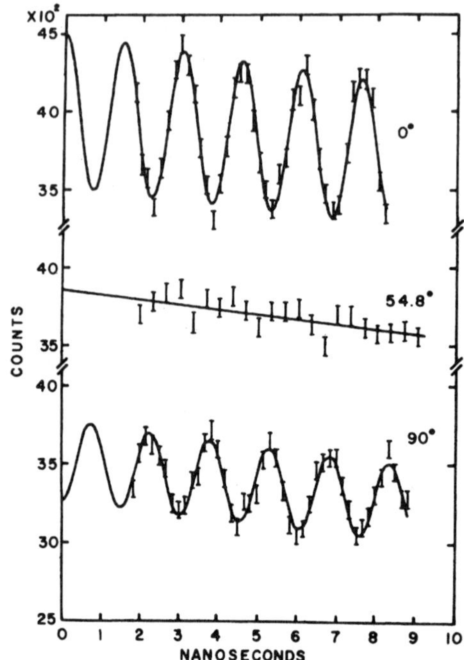



**Fig. 11.16.** Zero-field modulations in the decay of the foil-excited He ($3\,^3$P) state measured for polarizer settings parallel (top) and perpendicular (bottom) to the incident beam direction. The middle curve shows the absence of structure for a polarizer setting at "the magic angle" [11.51].

common photon helicity observed is indicated. Andrä *et al.* [11.54] measured the circular polarization of light emitted from a series of ArII levels excited by letting a 300 keV Ar$^+$ beam scatter off a solid Cu surface at incident angles of a few degrees. Large circular polarizations were observed for some of the transitions. For an angle of incidence of 1.5°, $P_3$ values in the range from $-0.14$ to $+0.76$ were measured, depending on the spectral line selected. Since these first results, the experimental developments were pioneered by Andrä, Winter, and collaborators at Münster, as demonstrated in the review by Andrä [11.53]. Parallel to the experimental developments, theoretical work was initiated with the aim of linking the observations with characteristic surface parameters [11.55]. The subsequent development of this technique into a surface probe can be traced in the literature.

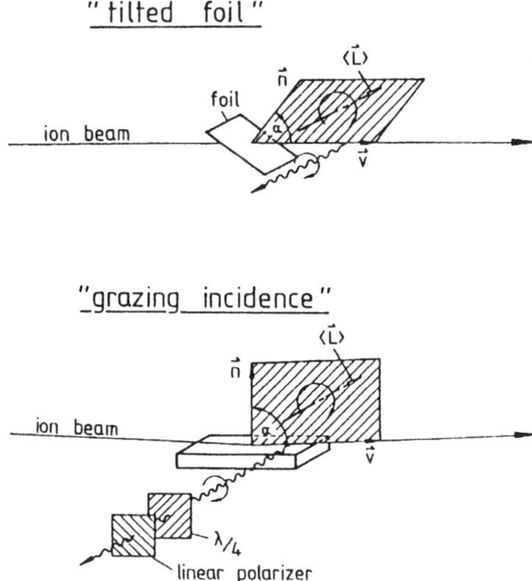


**Fig. 11.17.** Tilted-foil and grazing-incidence geometries [11.53]. See text for details.

# 11.6 Polarization in Collisional Broadening and Redistribution

The shape of atomic spectral lines and its relationship to the atom's sur-
roundings is a key question in fields such as astronomy [11.56] and plasma
physics [11.57], and the fields of atomic line broadening and collisional redistribution have played a distinguished role in the development of atomic
spectroscopy and collision physics [11.58]. Collision-broadening experiments
are concerned with the probability of absorption or emission of a photon during a collision and are capable of yielding information about the adiabatic
interaction of one atom with another.

A typical redistribution scattering experiment, on the other hand, is illustrated in Figure 11.18(a), which shows light of frequency $\omega_L$ and linear polarization $\hat{\epsilon}_L$ incident on an atom undergoing collisions with nearby perturbers.
The light emitted perpendicular to the direction of the incident light may
be analyzed in terms of intensity and polarization. If $\omega_L$ is near a resonant
frequency $\omega_0$ of the atom and the incident fields are weak, the scattered-light
spectrum typically consists of a Rayleigh peak centered at $\omega_L$ and a fluorescence peak centered near $\omega_0$, as seen in Figure 11.18(b). The corresponding
linear polarizations shown in Figure 11.18(c) exhibit a peak value of unity
for the Rayleigh peak in $J = 0 \rightarrow 1$ transitions, while the fluorescence peak in
general is not completely polarized and may even be unpolarized. Its value
is determined largely by the collision dynamics, and it is this feature that
makes it highly interesting in the context of the present book. Below we

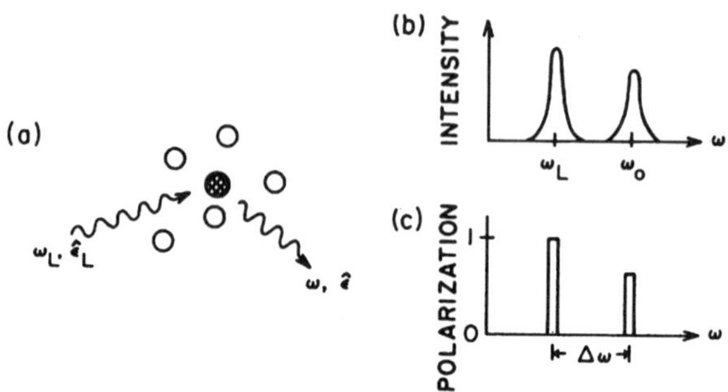

**Fig. 11.18** (a) Light of frequency $\omega_L$ and polarization $\hat{\epsilon}_L$ is scattered off an atom surrounded by perturbers. Parts (b) and (c) show the intensity of the polarization redistribution spectra of the scattered light. Note that (c) assumes scattering off a $^1S$ ground state.

shall briefly outline the qualitative physics of this phenomenon and illustrate the results with an example. For simplicity, we restrict the perturbers to be neutral atoms, different in species from the radiating atoms. Nevertheless, many of the following results are applicable to broadening by other types of perturbers.

A discussion of line broadening is pertinent to the problem of redistribution, because the physics is essentially the same in both fields. A typical absorption line-broadening experiment would send light on the system in Figure 11.18(a) and observe the spectrum of the transmitted light. Clearly, the two experiments are similar since any light redistributed in frequency (fluorescence peak) or redistributed in direction (Rayleigh peak) will show up as a loss in the transmitted light. Both line broadening and redistribution depend on the physics of the collision between the radiator and perturber. For comprehensive treatments of spectral line broadening and collisional redistribution we refer to the reviews by Allard and Kielkopf [11.59], Gallagher [11.60], Peach [11.61], and Burnett [11.62], which also contain extensive references to the original literature.

For detunings $\Delta\omega = \omega_L - \omega_0 > 1/\tau_c$, where $\tau_c$ is the typical duration of a collision ($\sim 1$ picosecond for thermal collisions between neutrals), the appearance of fluorescence light in the redistribution experiment in Figure 11.18(a) is interpreted in the way illustrated in Figure 11.19. When the atom and perturber are far away from each other, the photon energy of the incoming light beam, $\hbar\omega_L$, does not match the energy separation between the ground and excited states of the atom. However, if an internuclear distance $R_C(\omega_L)$ exists at which the energy difference between the two molecular potential curves matches the photon energy, the system may be transferred from the ground-state potential curve to the potential curve of the excited state. When the two particles separate, the atom may thus end up in an excited state, with the energy difference $\hbar\,\Delta\omega$ being compensated for by a corresponding increase or

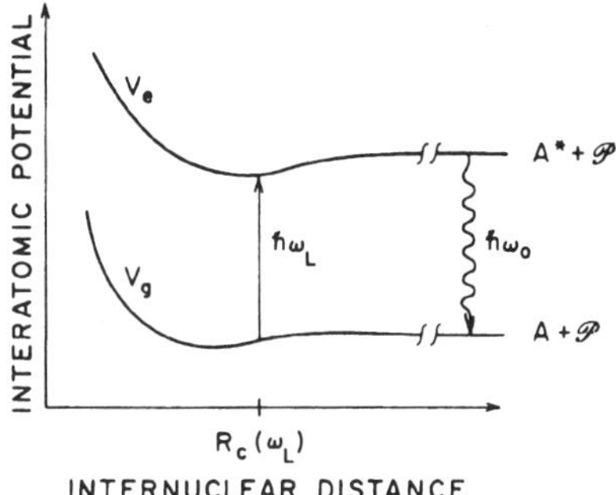


**Fig. 11.19** Hypothetical interatomic potentials between atom A and perturber $\mathcal{P}$, with A* referring to the atom in an excited state. The distance $R_C(\omega)$ is called the Condon point.

decrease in kinetic energy. Traditionally, $R_C(\omega_L)$ is called the Condon point. Thus, through the frequency of the external field, one has a handle on the internuclear distance at which the transition can take place. There is, however, no guarantee that such a point exists for a given external field, but one may also imagine cases with several such points.

We now address how a collision-induced alignment may accompany such an excitation. With emphasis on which quasi-molecular parameters control the resulting polarization, we shall here only outline the essential qualitative aspects for the simple case of a $^1S + {}^1S$ ground state system giving rise to a $\Sigma$ potential curve, while the outgoing channel may have either $\Sigma$ or $\Pi$ symmetry, corresponding to the $^1P + {}^1S$ asymptotic states. A rigorous treatment of the "reorientation model" may be found in the literature [11.63-66]. Several cases are considered in Figure 11.20 that will now be discussed in some detail.

$\Sigma \to \Sigma$ excitation: Figure 11.20(a) shows schematically the collision plane with the atom, initially in its ground state, moving along a straight-line trajectory. At the Condon point $R_C$, the quasi-molecule may be transferred from its ground state to the excited $\Sigma$ state. This state exhibits the usual dumb-bell shape of a p orbital, oriented along the internuclear axis as shown. At short distances, where the atom–perturber interaction is strong, the orientation of the orbital is locked to the internuclear axis and undergoes a rotation by an angle $\phi$ from the point of excitation (i) to a point (ii) further along the trajectory. Here the interaction becomes so weak that the motion of the orbital is decoupled from the internuclear axis from then on. This decoupling radius $R_{\text{dec}}$ is related to the distance where the energy difference between the two merging $\Sigma$ and $\Pi$ potential curves is comparable to the coupling by the rotation of the internuclear axis. For simplicity, $R_{\text{dec}}$ is assumed to be independent of the impact parameter and the point of excitation along the trajectory, and thus independent of detuning. We also assume that

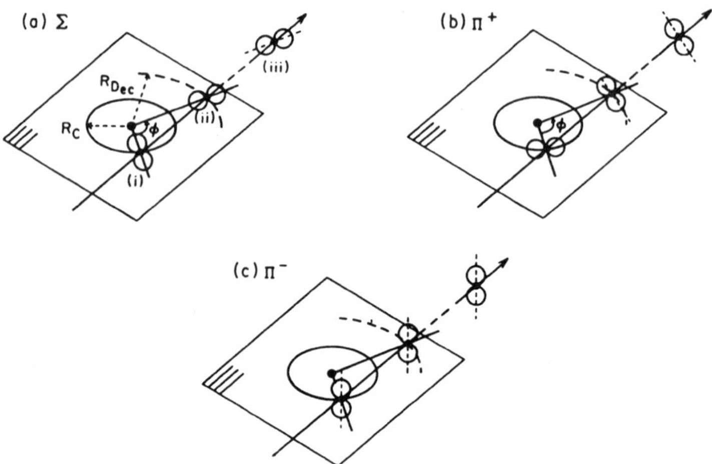

**Fig. 11.20** Qualitative illustration of the time evolution in the orientation of the excited-state orbital from (i) the point of excitation at a radius $R_C$ to (ii) the point of decoupling at a radius $R_{dec}$. The three cases of excitation are (a) the $\Sigma$ state, (b) the $\Pi^+$ state, and (c) the $\Pi^-$ state. The optical decay takes place at point (iii).

$R_C \leq R_{dec}$. From point (ii) onward, the orbital is assumed to stay fixed in space without change of shape until the radiative decay at point (iii).

Assuming straight-line trajectories and an excitation probability amplitude proportional to the component of the electric vector along the internuclear axis, one obtains the fluorescence polarization as

$$P_\Sigma = \frac{9x^2}{25 + 3x^2}, \tag{11.8}$$

where $x = R_C/R_{dec}$. Also, all impact parameters and all orientations of the collision plane with respect to the direction of the (linear) laser polarization have been averaged over. From (11.8), we see that the polarization assumes a maximum of about 32% at $R_C = R_{dec}$ and decreases to 0% in the limit where $R_C \ll R_{dec}$.

$\Sigma \rightarrow \Pi$ excitation: The discussion of polarization resulting from a $\Sigma \rightarrow \Pi$ excitation follows the same lines, but the quantitative result is different since a $\Pi$ orbital may be decomposed in two components, $\Pi^+$ and $\Pi^-$, where the index refers to the reflection symmetry with respect to the scattering plane. Figure 11.20(b) shows excitation of a $\Pi^+$ orbital. Apart from the initial rotation by 90° in the scattering plane, the reorientation of the orbital during the collision is the same as that encountered in the $\Sigma \rightarrow \Sigma$ case discussed above, and thus the depolarizing influence is identical for the same excitation radius. However, a $\Pi^-$ orbital, oriented perpendicular to the scattering plane, keeps its orientation in space during the whole period from creation to decay, and thus suffers no depolarization at all. Under the same assumptions as above, (11.8) is replaced by

$$P_\Pi = \frac{15 + 15x + 9x^2}{55 + 5x + 3x^2} \cdot \qquad (11.9)$$

In this case, the maximum polarization is about 62% (at $R_C = R_{\mathrm{dec}}$), while the minimum polarization is about 27% (for $R_C \ll R_{\mathrm{dec}}$).

To summarize the analysis, we see that the polarization will decrease with decreasing excitation radius for both $\Sigma$ and $\Pi$ excitation, as long as the trajectories are straight. For the same excitation radius, $\Sigma$ excitation will yield a smaller polarization than $\Pi$ excitation since the $\Pi$ component perpendicular to the scattering plane is not rotated. In the large detuning limit, $\Sigma$ excitation may yield 0% polarization, while pure $\Pi$ excitation will never result in a polarization below 27%. We note that the assumption of a straight-line trajectory breaks down for large detunings. Analyzing the effect of bent trajectories indicates that they always tend to *increase* the polarization [11.64,66]. Finally, the above discussion has been restricted to a single Condon point. Depending on the shape of the potential curves, several Condon points may actually occur for a given detuning $\Delta\omega$. The resulting polarization then has to be calculated by proper weighting of the individual contributions.

The first redistribution experiment was carried out at JILA by Carlsten *et al.* [11.67] who studied the Sr Ar system. Subsequent investigations there focused on alkaline-earth atoms perturbed by rare gases [11.66,68–70]. The first redistribution scattering experiments with polarization analysis were performed by Thomann *et al.* [11.68] for Sr–Ar and later extended by Alford *et al.* to Sr [11.69] and Ba [11.66] perturbed by other rare gases. They also demonstrated that additional information may be gained from scattering of circularly polarized light [11.70]. Behmenburg and coworkers performed extensive studies of selected alkali–rare-gas systems (see, for example, [11.71]).

As an example, we now discuss results for the Ba–Xe case. It is relatively simple, since there is no fine structure present and hyperfine-structure effects are easily corrected for, and it displays strong polarization effects [11.66]. Figure 11.21 shows red wing (left) and blue wing (right) results for the polarization (upper panels) and lineshape (lower panels), respectively. (Note that the units are chosen in such a way that a Lorentzian lineshape will correspond to a horizontal line [11.66]). For the Ba–Xe system at thermal energies, the upper limit of the so-called "impact region," $1/\tau_c$, may be estimated to about $3\,\mathrm{cm}^{-1}$ [11.66]. Hence all the data displayed are within the so-called "quasistatic region" for which the above analysis is valid. We notice a general trend of decreasing polarization with increasing $|\Delta\omega|$, followed by an increase at even larger detunings. Furthermore, we note a correlation between structures in the polarization and structures in the lineshape, especially clear around the red-wing lineshape "satellite."

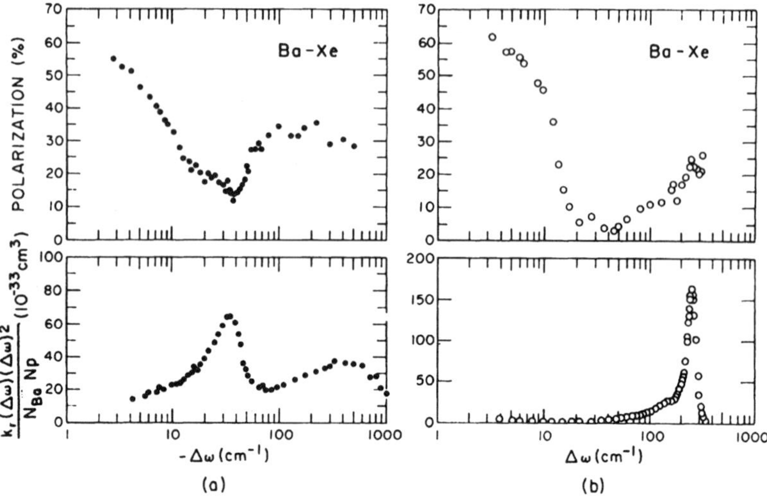

Fig. 11.21 Polarizations (upper row) for the Ba–Xe red (left) and blue (right) wings. The corresponding values of the lineshape intensity profiles, $k_r(\Delta\omega)\Delta\omega^2$, are plotted below. (From Alford *et al.* [11.66].)

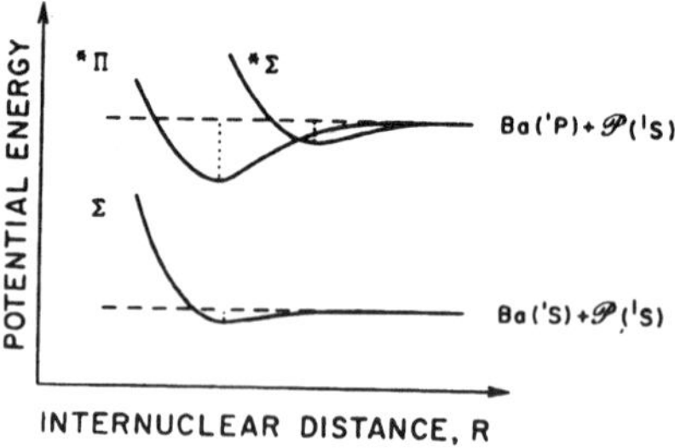


Fig. 11.22 Qualitative sketch of the interatomic potentials for the Ba–rare-gas systems.

These results can be interpreted in terms of the topology of the appropriate interatomic potential curves. Figure 11.22 shows a qualitative sketch of the interatomic potentials for the Ba–Xe system. As confirmed by recent state-of-the-art calculations by Czuchaj and coworkers [11.72], however, the general structure is believed to be representative for two-electron systems (Mg, Ca, Sr, Ba, Zn, Cd, Hg) being perturbed by the heavy rare gases (Ar, Kr, Xe) [11.73–77]. The ground state has a shallow well and becomes repulsive at small internuclear distances, while the excited $*\Sigma$ state has a long-range van-der-Waals potential, which becomes repulsive at smaller $R$ due to the electron overlap of the $^1$P state with the rare-gas electron cloud.

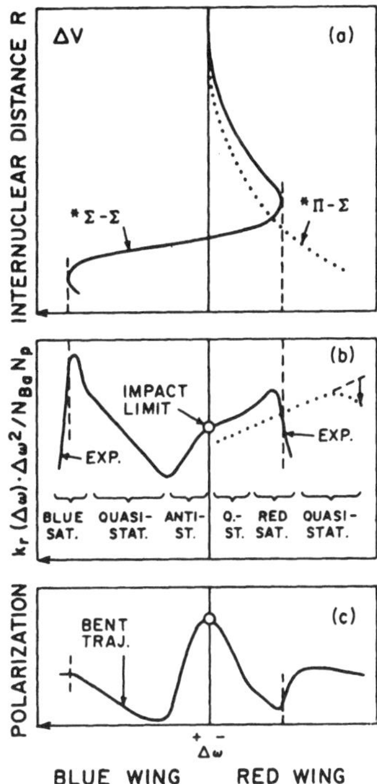


**Fig. 11.23** Qualitative sketch of the excited-state–ground-state potential difference curves versus internuclear distance (a), together with the corresponding blue-wing and red-wing profiles (b), and the polarizations (c). Exponential satellite tails are denoted by EXP.

Figure 11.23 summarizes how the principal features of the experimental polarization and lineshape data of Figure 11.21, schematically outlined in the bottom two panels of Figure 11.23, are consistent with the picture outlined in Figure 11.22. The potential curve differences for the latter are shown at the top of Figure 11.23, rotated by 90°, since the experimentally controllable parameter is the detuning, used as abscissa. The "satellite lines" in the red and blue wings are associated with maximum or minimum values of the relevant potential-curve differences. The $^*\Sigma$ well depth must be less than or approximately $35\,\mathrm{cm}^{-1}$ given the red-wing satellite position. Furthermore, the $^*\Pi$ state has a long-range van-der-Waals potential, which crosses the $^*\Sigma$ curve, and the depth of the $^*\Pi$ state well is probably between 300 and $2000\,\mathrm{cm}^{-1}$.

The red-wing polarization drops from large values at small detunings to a pronounced minimum below 15% near the top of the red-wing satellite, unambiguously labeling the origin of the satellite to be a $^*\Sigma - \Sigma$ excitation. Beyond the satellite the contribution from the $^*\Sigma - \Sigma$ potential difference curve drops exponentially, leaving only the $^*\Pi - \Sigma$ contribution to lead to a corresponding rise in the polarization. The origin of the observed feature at the largest detunings is not entirely clear, but a likely interpretation of the drop in intensity is depletion of the ground-state potential, which turns repulsive at short distances.

The blue-wing first shows very low intensities and a drop in polarization to values close to zero, where quasi-static excitation to the short-range repulsive part of the $^*\Sigma - \Sigma$ difference potential (labeled from the low value of the polarization) becomes dominant. At larger detunings, the excitation radius is quite small and trajectory effects become important, causing a gradual, but pronounced rise in the polarization. Even further out we reach the blue-wing satellite due to the local maximum in the $^*\Sigma - \Sigma$ difference potential. Beyond the peak of the blue satellite, the intensity drops exponentially.

From the measured polarizations and the equations for the polarizations given above, a corresponding decoupling radius $R_{\mathrm{dec}}$ may be estimated as about $12\,a_0$ [11.66]. This nicely demonstrates how dynamical information, not available from lineshape measurements, may be extracted from redistribution polarization studies. The use of a definite decoupling radius is in fact only a simple way of approximating the transition from molecular eigenstates to atomic eigenstates, in order to describe the final orientation of the p orbital after a collision. For an introduction to the quantitative treatment of this interesting problem, we refer to the pioneering work of Grosser [11.78].

## 11.7 Alignment and Orientation Studies at Thermal Energies

### 11.7.1 Alignment studies involving an optically prepared atom

A particularly instructive case for demonstration of alignment effects in thermal collisions is the exothermic energy-transfer process

$$\mathrm{Ca}\,(4s5p\,^1\mathrm{P}_1) + \mathrm{He} \rightarrow \mathrm{Ca}\,(4s5p\,^3\mathrm{P}_1) + \mathrm{He} + \Delta\mathrm{E} \tag{11.10}$$

($\Delta E = 177\,\mathrm{cm}^{-1}$), as first studied by Hale $et\ al.$ [11.79]. The alkaline earths are especially well suited for alignment studies since they have (almost) no nuclear spin, and, for the singlet states, one can in principle prepare the orbital in a completely aligned state. The transition between singlet and triplet systems is possible since neither the singlet nor the triplet states are purely $LS$-coupled. Each contains an admixture of the other one so that rotational coupling may mix $\Sigma$ and $\Pi$ states of both systems.

The reaction (11.10) was studied in a crossed-beam experiment, in which the $\mathrm{Ca}\,(4s5p\,^1\mathrm{P}_1)$ state was optically prepared by linearly polarized laser light of wavelength $2721\,\text{\AA}$. The production of $\mathrm{Ca}\,(4s5p\,^3\mathrm{P}_1)$ was monitored by

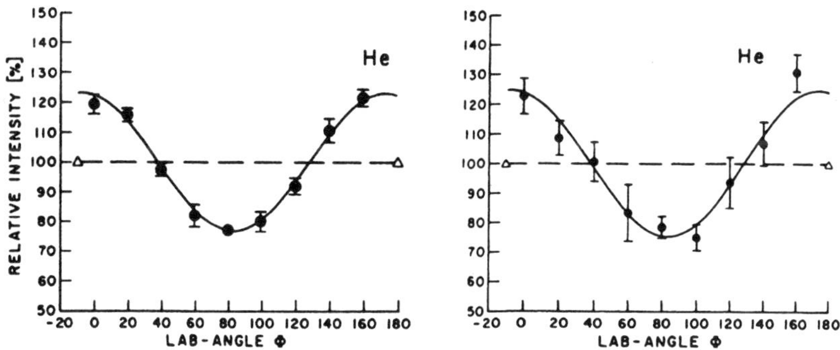


**Fig. 11.24.** The left graph shows final alignment signals after normalization and background subtraction for the forward $^1P_1 \rightarrow {}^3P_1$ transfer, as function of polarizer angle. The right graph shows the same for the reverse process, $^3P_1 \rightarrow {}^1P_1$. From Bussert *et al.* [11.80].

analyzing the optical decay transitions. Similarly, the reverse process could be investigated by optically preparing the $Ca(4s5p^3P_1)$ state with linearly polarized laser light of wavelength $2735\,\text{Å}$, again with the production of $Ca(4s5p^1P_1)$ being monitored by the subsequent optical decay [11.80]. The initial alignment of the excited P state with respect to the internuclear axis could be varied by rotating the linear polarizer.

Figure 11.24 shows results for the forward $^1P_1 \rightarrow {}^3P_1$ transfer (left). A pronounced alignment anisotropy is observed. Taking the various angles in the experiment into account, the signal corresponds to a maximum-to-minimum cross section ratio of $1.60 \pm 0.03$ [11.80]. Similarly, for the reverse direction, $^3P_1 \rightarrow {}^1P_1$, a ratio of $1.65 + 0.10$ was found.

A qualitative interpretation of these findings may be obtained from inspection of Figure 11.25, which shows schematic potential-energy curves for the $Ca - He$ system [11.79]. The excited Ca atom is initially aligned with respect to the collision frame. This initial alignment is partially preserved when the calcium and helium atoms first form molecular wavefunctions around a distance $R_L$, from which on the orbital stays locked to the rotating internuclear axis. At smaller distances, the curve-crossing region is reached at a distance $R_C$. Different initial alignments give rise to different cross sections for energy transfer. Proper analysis of the various angles determining the collision geometry shows that the $\pi$ orbital alignment with respect to the internuclear axis is more efficient (by a factor of 1.60) than the parallel $\sigma$ alignment. This result may be expected from the potential curves which show a crossing of the slightly attractive $^1\Pi$ curve with the repulsive $^3\Sigma$ curve.

The measured integral alignment for the reverse reaction, $^3P_1 \rightarrow {}^1P_1$, shown to the right in Figure 11.24, also exhibits a strong $\pi$ dominance for collisions with He. Naively, one might expect from Figure 11.25 that a $\sigma$ orbital preparation would give the maximum cross section since the $^3\Sigma$ poten-

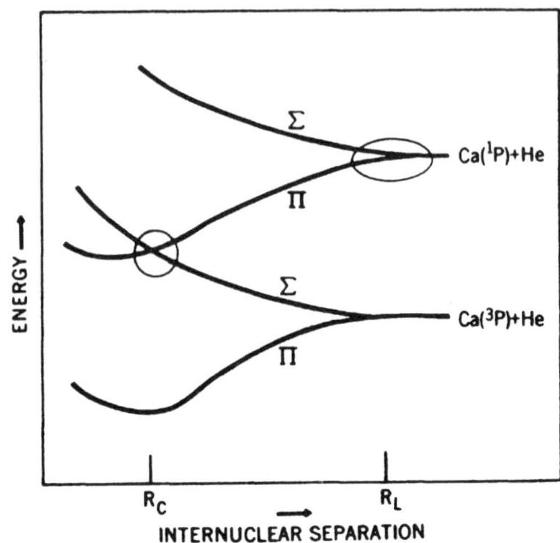


**Fig. 11.25.** Schematic potential-energy curves for the Ca − He system. The curve crossing radius R$_C$ and the region of molecular orbital formation R$_L$ are indicated.

tial curve needs to be populated in order to reach the crossing leading to the $^1\Pi$ potential curve. Closer inspection shows, however, that the optical preparation of the $^3$P state with linearly polarized light parallel to a given axis prepares this state to 100% in the $\pi$ states. For cylindrical symmetry, therefore, one obtains an average of 50% in each of the $\pi^+$ and $\pi^-$ states, whereas the perpendicular polarization leads to 50% $\sigma$ and 50% $\pi$ (i.e., 25% each in $\pi^+$ and $\pi^-$). This is a consequence of the Clebsch–Gordan coefficients describing these states in the $LS$-coupling scheme. Thus, the correct expectation in this picture is a $\pi$ dominance for the reverse reaction as well, as confirmed by experiment.

These pioneering experiments with He were followed up with studies of alignment effects for other rare-gas and alkaline-earth targets [11.80,81], and with investigations of collisions involving higher angular-momentum states (D,F) by using two or three photons for the optical preparation [11.82,83]. Parallel to the experimental progress, several theoretical studies have addressed these systems, eventually resulting in a very satisfactory, quantitative understanding of these processes [11.84–88]. The studies yield a value for $R_\mathrm{L} \approx 9\,a_0$. For more details, see ref. [11.89].

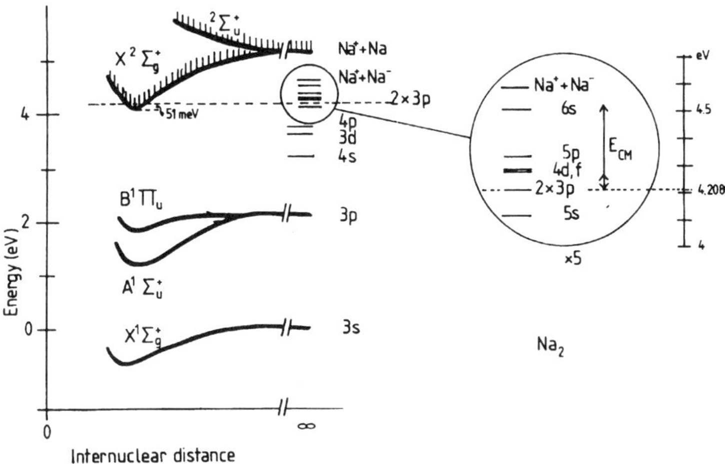


**Fig. 11.26.** A schematic diagram of relevant potential curves and important dissociation limits for the outcome of a Na (3p) − Na (3p) collision. The arrows indicate the lower and upper limits for the center-of-mass energy in the experiments of Meijer *et al.* [12.15] shown in Figure 11.27 below.

## 11.7.2 Alignment and orientation studies involving two optically prepared atoms

A process that has been subject of intensive experimental and theoretical studies by many groups and that involves two optically prepared atoms is the associative ionization of two Na (3p) atoms

$$\mathrm{Na}\,(3\mathrm{p})^2\mathrm{P}_{3/2} + \mathrm{Na}\,(3\mathrm{p})^2\mathrm{P}_{3/2} \rightarrow \mathrm{Na}_2^+ + \mathrm{e}^-. \tag{11.11}$$

The minimum of the $\mathrm{Na}_2^+(\mathrm{X}\ ^2\Sigma_g^+)$ ground-state potential curve shown in Figure 11.26 corresponds to an energy of 51 meV below the combined energy of the two excited atoms. Kircz *et al.* [11.90] were the first to demonstrate that the probability of this process depends strongly on the polarization properties of the exciting laser light. Subsequent studies also involved the energy pooling process

$$\mathrm{Na}\,(3\mathrm{p})^2\mathrm{P}_{3/2} + \mathrm{Na}\,(3\mathrm{p})^2\mathrm{P}_{3/2} \rightarrow \mathrm{Na}\,(n\ell)^2L_J + \mathrm{Na}\,(3\mathrm{s})^2\mathrm{S}_{1/2}, \tag{11.12}$$

where several states $\mathrm{Na}\,(n\ell)^2L_J$ are energetically possible, as shown by the insert in Figure 11.26. Since space limitations do not allow for a full exposition of all interesting aspects of these competing processes, we refer to the papers by Weiner *et al.* [11.91], by Masnou-Seeuws *et al.* [11.92], and the references therein. In the context of this book, it is interesting to ask to which extent the full density matrix for associative ionization can be extracted experimentally, and which experimental geometries and optical reparations of

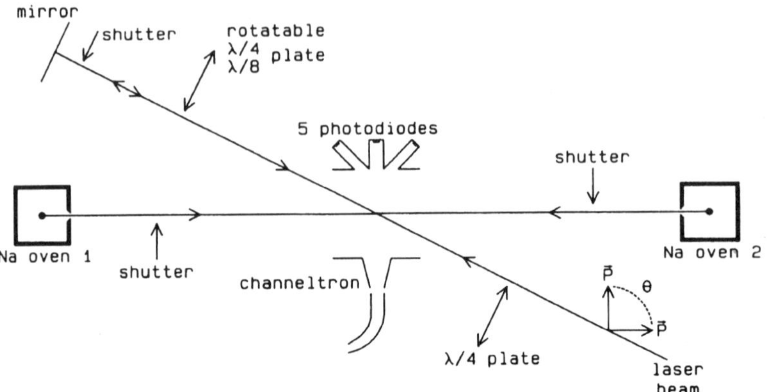



**Fig. 11.27.** Two counter-propagating thermal sodium beams from ovens 1 and 2 are crossed by a laser beam at an angle of 87°. Optical elements allow for an independent preparation of the two beams. Photodiodes monitor the optical preparation. The $Na_2^+$ ions produced are detected by a channeltron. From Meijer *et al.* [11.93].

the atoms should be exploited in order to achieve this goal. Here we briefly outline a possible approach and refer to the work of Meijer *et al.* [11.93] for a full discussion.

Figure 11.27 shows the experimental setup used by Meijer *et al.* [11.93] for this purpose. Two counter-propagating sodium beams are optically prepared in the Na (3p) state by a laser crossing the beams at an angle of 87°. The small deviation from perpendicular incidence allows for the possibility of addressing different velocity classes in the two beams, depending on the laser detuning. Furthermore, insertion of proper optical elements in the retroreflected laser beam makes it possible for the two beams to be in different states whose shape and orientation may be varied independently. Characterizing the preparation by the polarization state $P = \theta, \pm$ of the corresponding laser beam, where $\theta$ is the angle between the plane of linear polarization and the atomic beam direction and $\pm$ labels the photon helicity for circularly polarized light, the rate of production $R = R^{(P_1, P_2)}$ may be expressed as

$$R = R_0 + R_1 \cos 2\theta + R_2 \cos 4\theta + R_3 \sin 2\theta . \tag{11.13}$$

A general analysis shows that one may extract a total of eight independent parameters from the experiment for each collision velocity.

Figure 11.28 contains an overview of a set of detection schemes from which these eight parameters may be extracted, including some consistency checks. (In addition to the setup shown in Figure 11.27, a varying magnetic field in the detection region is required to resolve an otherwise remaining ambiguity [11.93]). Figures 11.29 and 11.30 show examples of how dramatic the molecular-ion production rate depends on the optical preparation and

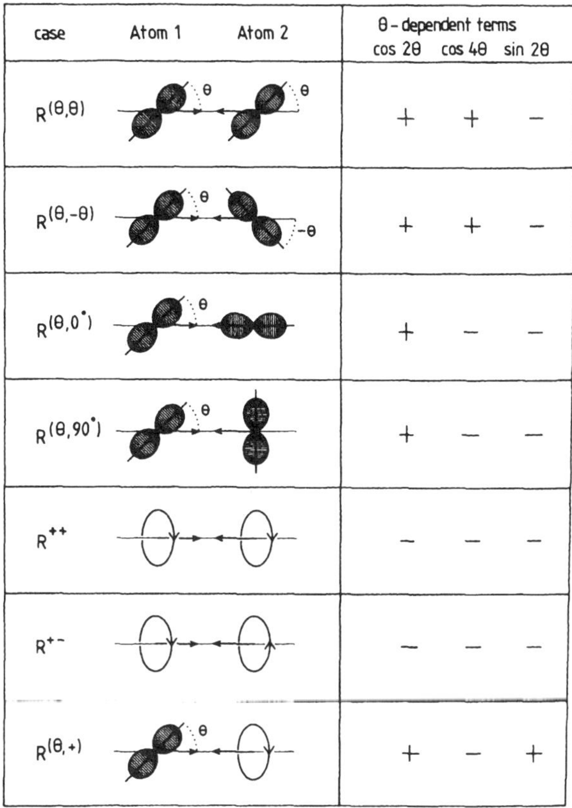

**Fig. 11.28.** Overview of the set of detection schemes that may be used for extraction of a complete set of parameters [11.93]. The column to the right refers to (11.13).

the collision velocity. Although we refrain from a detailed discussion of these curves, it is worth pointing out that the symmetries of the various preparation geometries translate into different populations of the many molecular potential curves that connect to the entrance channel. Of these potential curves, only one or two are typically efficient for molecular-ion production, connecting into the ionic continuum below the energy threshold. The identification of these curves and the evaluation of their interaction with the other potential curves, some of which lead to energy pooling, is a delicate and channeling question that is still undergoing detailed scrutiny [11.92].

We end the discussion of this system by briefly returning to the question to which extent the set of experimental parameters extracted above allows for the determination of the *complete density matrix*. The answer to this question depends on the collision velocity. For sufficiently high veloci-

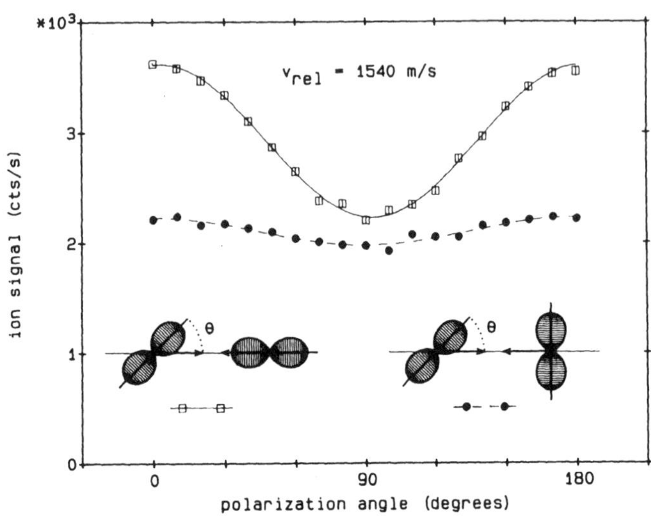



**Fig. 11.29.** Polarization dependence of the molecular ion production rate $R$ for the cases $(P_1, P_2) = (\theta, 0°)$ and $(\theta, 90°)$ at a relative collision velocity of 1540 m/s [11.93]. The curves are fits to (11.13) with $R_2 = 0$.

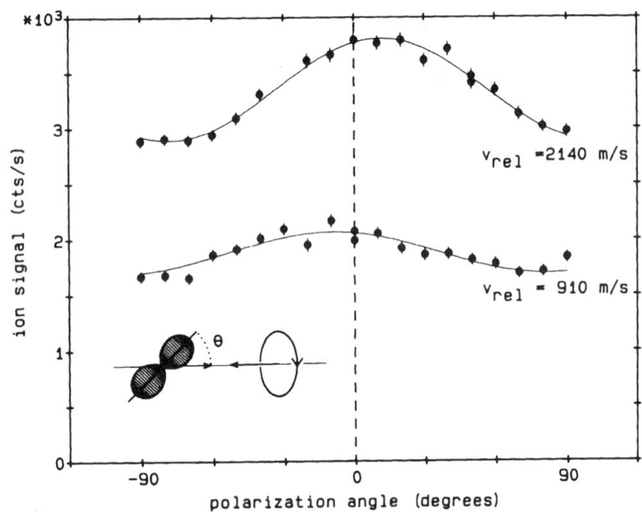



**Fig. 11.30.** Polarization dependence of the molecular ion production rate $R$ for the case $(P_1, P_2) = (\theta, +)$ at relative collision velocities of 910 m/s and 2140 m/s, respectively [11.93]. The curves are fits to (11.13) with $R_2 = R_3 = 0$.

ties, the electron spin effectively stays fixed in space, and the system may be described in terms of $M_L$−dependent scattering amplitudes and the corresponding density-matrix elements $\rho(M_L, M_{L'})$. In this velocity regime, the approach described above may yield the complete density matrix. If, however, the velocity is reduced into a region where the spin-orbit interaction may no longer be neglected during the collision, the scattering amplitudes depend on $M_J$ and the correspondingly larger density matrix is no longer fully determined.

Reducing the collision velocity even further, as in cold and ultracold collisions, will finally necessitate an analysis in terms of $M_F$−dependent amplitudes [11.94]. Such a situation opens up a new world of alignment and orientation effects in cold-collision studies. This field is still in its infancy, partly because most experimental studies of cold and ultracold collisions to date have been performed in traps in which the parameters defining the orbital alignment and orientation are not well defined. It may be expected, however, that polarization studies will play a central role in the next generation of experiments in this rapidly growing field.

## 11.8 Plasma Polarization Spectroscopy

The general concepts of plasma polarization spectroscopy (PPS) have been reviewed by Fujimoto and Kazantsev [11.95]. The principal idea concerns the search for a linear polarization of the light emitted from excited atoms or ions in the plasma, i.e., for an alignment in the (angle-integrated) magnetic sublevel population (see also Section 4.1). If such a polarization is detected, it indicates some nonisotropy in the plasma that can be traced back to fields and/or directional beams in the plasma — provided the underlying processes are sufficiently well understood.

As pointed out by Fujimoto and Kazantsev, the effects of magnetic and electric fields on the light polarization have been investigated since the pioneering works of Hanle [11.96] and Mark and Wierl [11.97]. They are very well understood at the present time. However, the same cannot be said for the creation of collision-induced alignment by electron or heavy-particle impact. Additional complications include the need to account for cascade contributions, typically resulting in a depolarization of the radiation, and the need to implement atomic collision data obtained with relatively well-defined incident energies of the beams into plasma modeling codes involving velocity distributions. Nevertheless, it is clear that benchmark data for atomic collisions are a critical ingredient for an improved understanding of the plasma properties. In addition to the most important case of inelastic S → P transitions, we note that alignment may also be created in elastic collisions from excited states with nonzero angular momentum. An example was presented

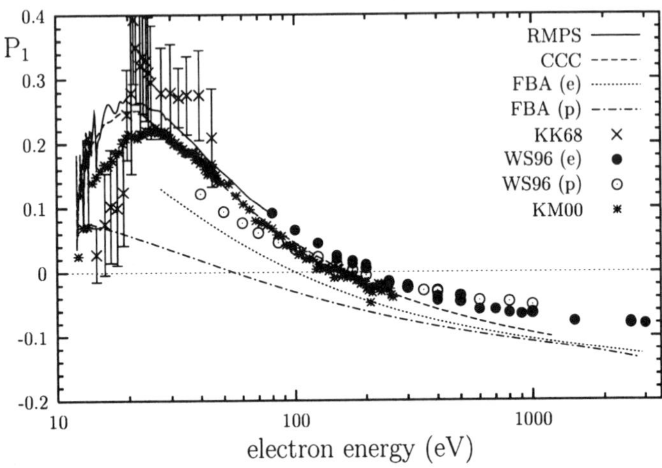

**Fig. 11.31.** Polarization of the H $(n=3) \rightarrow (n=2)$ line after impact excitation by electrons and protons of equal velocity, plotted as a function of the corresponding *electron* energy. The experimental data for electron impact are from Kleinpoppen and Kraiss [11.99] (KK68), Werner and Schartner [11.100] (WS96), and Kedzierski and McConkey [11.101] (KM99). Also shown are the results of Werner and Schartner [11.100] for proton impact (energy $\times$ 1836). The theoretical predictions are from FBA [11.100], CCC [11.101], and RMPS [11.101] calculations.

by Trajmar *et al.* [11.98] for elastic electron scattering from the laser-excited $\mathrm{Ba}\,(6s6p)^1\mathrm{P}_1$ state.

From the many investigations of the alignment and the corresponding angle-integrated linear light polarization $P_1$, we show three representative examples below. Figure 11.31 exhibits the polarization of the Balmer $\alpha$ radiation in atomic hydrogen, excited either by electron or proton impact of the same velocity. The polarization of this line after electron impact was first investigated by Kleinpoppen and Kraiss [11.99], and more recently by Werner and Schartner [11.100] and by Kedzierski and McConkey [11.101]. In addition, Werner and Schartner obtained results for proton impact. As can be seen from the figure, the uncertainty in the experimental data has now been reduced to a few percent, and the accuracy is sufficiently high to demonstrate the need for a theoretical model beyond the first Born approximation (FBA). For electron impact, predictions from the CCC and RMPS models (see Chapter 6 and the many comparisons with experimental data in previous chapters) agree very well with the experimental data. Consequently, predictions from these models are currently used with confidence in many applications.

Figure 11.32 shows a less satisfactory situation, namely electron-impact ionization–excitation of helium from the ground state to the $\mathrm{He}^+(2p)^2\mathrm{P}$ state (see also Section 10.1.1). There are major differences between the experimental data of Götz *et al.* [11.102] and Merabet *et al.* [11.103] in the

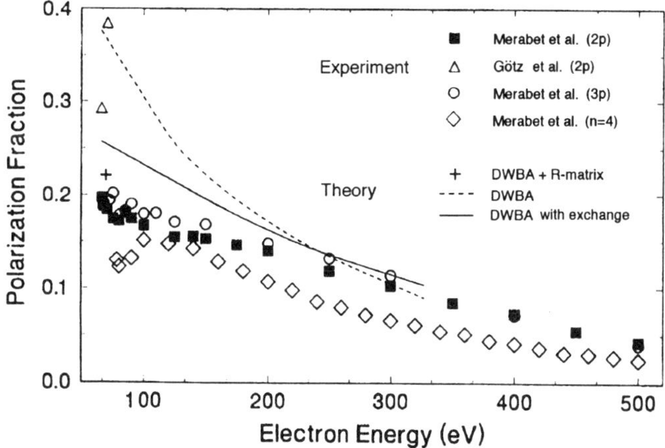


**Fig. 11.32.** Polarization of the $He^+(2p)^2P \to (1s)^2S$ and $He^+(3p)^2P \to (1s)^2S$ lines for ionization–excitation of $He(1s^2)^1S$ as a function of the incident energy. The theoretical predictions are from refs. [11.102] (DWBA+R-matrix), [11.104] (DWBA), and [11.105] (DWBA with exchange). Adapted from [11.103].

near-threshold region, and first-order distorted-wave theories [11.104,105] seem inadequate to reproduce the experimental data for incident energies below 200 eV. The DWBA+R-matrix model, based on the computer code of Bartschat [11.106] and used by Götz *et al.* [11.102], seems more reliable in the low-energy regime, but it is computationally very intensive and still needs substantial improvement, for example, with respect to autoionizing resonances and the influence of second-order effects.

Finally, Figure 11.33 shows an example of work performed at the electron beam ion trap (EBIT) at Livermore. It exhibits intensity spectra of lines in Li-like and Be-like iron excited by incident electrons with energies of 4.698, 4.717, and 4.762 keV, respectively [11.107]. The upper panel corresponds to the signal recorded with a crystal that selects a light polarization parallel to the electron beam, while the bottom panel represents the signal obtained without polarization selection. The differences in the spectra indicate strong polarization effects that are well reproduced by theoretical models. For details, see [11.107].

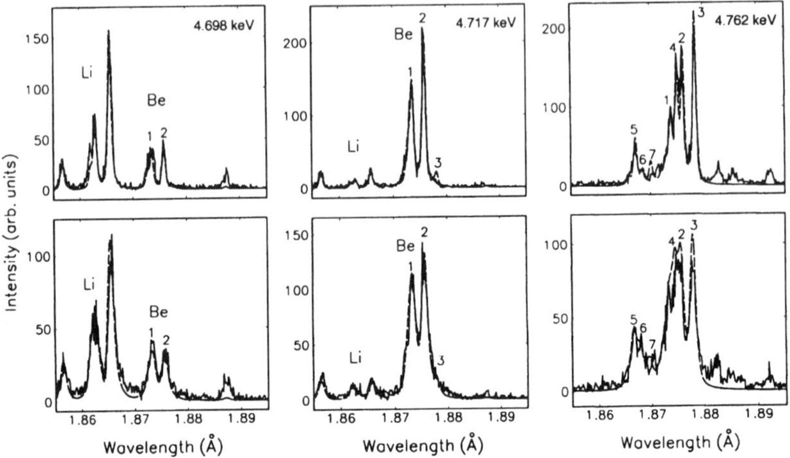

**Fig. 11.33.** Comparison of experimental (——) and theoretical (– – –) spectra of Li-like and Be-like iron for incident electron energies of 4.698 keV (a), 4.717 keV (b), and 4.762 keV (c) for a nearly pure parallel polarizer setting (top panel) and a polarization mixture (bottom panel). Adapted from [11.107].

## 11.9 Spin-Polarized Beams for Nuclear and Particle Physics

The production of beams and targets with spin-polarized nuclei is of great current interest for many applications in nuclear and particle physics, solid state physics, astrophysics, and biomedical fields. An early, simple technique to generate such beams was to send unpolarized beams through a tilted foil [11.108]. Several of the powerful sources in use today rely on a detailed understanding of collision physics, often combined with optical pumping techniques. In particular, there is a need for further development of polarized proton, deuterium and $^3$He ion beams for use with high-energy particle accelerators. The optically pumped polarized-ion source (OPPIS), as used for example at TRIUMF [11.109], works in the following way [11.110,111]: An $H^+$ beam with an energy of a few keV is partially neutralized in an optically pumped electron-spin-polarized alkali target. Capture occurs primarily into the H $(n=2)$ levels (see Section 8.2.1) and a large magnetic field is required at the alkali target to avoid depolarization of the electron spin due to the $2p \rightarrow 1s$ radiation. The resulting *electron-spin-polarized* $H^0$ atomic beam is converted into a *nuclear-spin-polarized* $H^0$ beam by the so-called Sona effect [11.112]. The $H^0$ atoms leave the alkali target with approximately equal populations in the two high-field states $|m_J, m_I\rangle = |\frac{1}{2}, \frac{1}{2}\rangle$ and $|\frac{1}{2}, -\frac{1}{2}\rangle$ corresponding to complete electron-spin polarization and zero nuclear-spin polarization. The atoms then pass into a low-field region, now predominantly in the low-field states $|Fm_F\rangle = |11\rangle$ and $|10\rangle$. When they pass rapidly through zero field into a region where the field is reversed, the atoms are left predominantly in the states $|Fm_F\rangle = |1-1\rangle$ and $|10\rangle$. As the field increases, the atoms populate the high-field states $|m_J, m_I\rangle = |\frac{1}{2}, -\frac{1}{2}\rangle$ and $|-\frac{1}{2}, -\frac{1}{2}\rangle$ corresponding to zero electron-spin polarization and full nuclear-spin polarization. The nuclear-spin-polarized $H^0$ beam is then ionized, either in a second alkali-vapor target to form a nuclear-spin-polarized $H^-$ ion beam, or in a He gas to form a nuclear-spin-polarized $H^+$ beam. Among other parameters, a detailed knowledge of the size and energy dependence of the relevant electron-transfer cross sections is a key prerequisite for the optimization of the source parameters.

Currently, there are strong efforts directed toward developing an OPPIS-type source for $^3$He, as can be seen in the proceedings of a recent conference [11.113]. A full discussion of this complex question is outside the scope of this book, but one of the many necessary ingredients is again a detailed knowledge of the pertinent cross sections for electron transfer into the individual excited states of He in $He^+$−alkali-atom collisions for kinetic energies around 1 keV. The theoretical methods developed for description of the electron-transfer processes described in Sections 8.2–4 are able to provide this information, as we now illustrate by an example.

**Fig. 11.34.** Geometries for the optically prepared Na (3p) targets in Figure 11.35.

Figure 11.34 shows the scheme for optically preparing an alkali atom in an excited P state, either by linearly or by circularly polarized laser light incident perpendicular to the direction of an incoming ion beam. (The "tilted orbital" shown to the left may be considered as an analogue to the "tilted foil" geometry mentioned previously.) For both preparations, circular polarization may evolve in the final state. Similar to $H^+$ impact, capture into the four $He(n=2)$ states is prominent for collisions involving $He^+$ and alkali-atom targets as well. For $He^+$ impact on a Na (3s) target at an impact energy around 1 keV, the electron is predominantly transferred to the $He(2^3S)$ state. For a Na (3p) target, however, at the same impact energy, transfer to the $He(2^1P)$ and $He(2^3P)$ states dominates. If the light used for preparation is linearly polarized, the electron spin is initially isotropic, but if the light is circularly polarized instead, the initial electron spin polarization is asymptotically 100 % for the standard pumping scheme. Furthermore, for states with orbital angular momentum $L > 0$, the expectation value of the angular-momentum projection, $L_\perp$, is nonzero and may depend sensitively on the op-

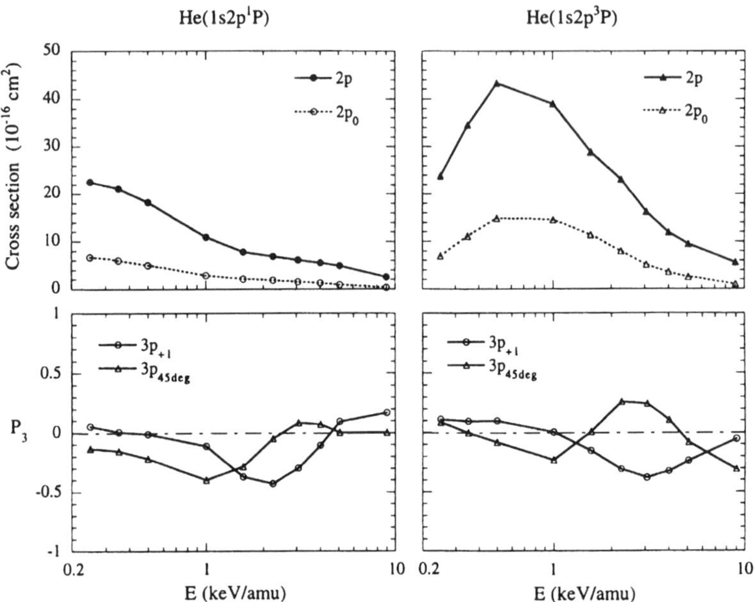

**Fig. 11.35.** Statistically weighted electron-transfer cross sections and circular polarizations for the He $(2^1\mathrm{P})$ and He $(2^3\mathrm{P})$ states in He$^+$–Na $(3\mathrm{p})$ collisions as function of impact energy. The target atom is the Na $(3\mathrm{p})$ tilted or circular orbital shown in Figure 11.34.

tical preparation and collision energy. After the collision, internal dynamical couplings, such as fine-structure and hyperfine-structure effects, redistribute the total angular momentum among the various individual components of the total angular momentum $\mathbf{F}=\mathbf{L}+\mathbf{S}+\mathbf{I}$. Hence, $F_\perp = L_\perp + S_\perp + I_\perp$, according to standard angular-momentum algebra [11.33].

Here we shall focus on the component that depends on the collision dynamics, $L_\perp$, and the related Stokes parameter, the circular polarization $P_3$ for the nascent charge cloud. It is measured in a direction perpendicular to the ion beam but parallel to the laser beam. Using the latter direction as the quantization axis, we obtain the electron-transfer cross section $\sigma$ and the Stokes parameter $P_3$ from the cross section components $\sigma_{M_L}$ as

$$\sigma = \sigma_{+1} + \sigma_0 + \sigma_{-1}; \tag{11.14}$$

$$P_3 = \frac{\sigma_{-1} - \sigma_{+1}}{\sigma_{-1} + \sigma_{+1}}. \tag{11.15}$$

Figure 11.35 shows the total cross sections and circular polarizations for the He $(2^1\mathrm{P})$ and He $(2^3\mathrm{P})$ states corresponding to the target preparations of Figure 11.34, i.e., the tilted Na $(3\mathrm{p}_{45^\circ})$ orbital and the Na $(3\mathrm{p}_{M_L=+1})$ state [11.114]. The total transfer cross section $\sigma$, as well as the $\sigma_0$ component, are the same for the two preparations, but the orientations of the nascent charge clouds, and therefore the circular polarizations, differ and vary significantly with impact energy. At higher impact energies, the $He(n = 3)$ states also acquire significant cross sections. We shall here refrain from further discussion, but draw attention to the fact that reliable and detailed cross sections and angular-momentum variables are now available for the $He^+ - Na$ system from state-of-the-art theoretical codes for this energy range [11.115]. These allow for the calculation of the time evolution and asymptotic polarization properties of the beam for given, well-defined experimental conditions. Interestingly, an optically prepared Li target might provide a more suitable choice than Na, since the $Li(2p)$ state is closer to resonance with the $He(2p)$ states than is $Na(3p)$, thereby enhancing key cross section ratios in the lower keV energy range.

# References


11.1 U. Fano, Phys. Rev. **178** (1969) 131.

11.2 M.S. Lubell and W. Raith, Phys. Rev. Lett. **23** (1969) 211.

*11.3 J. Kessler and J. Lorenz, Phys. Rev. Lett. **24** (1970) 87.

11.4 D.T. Pierce, in *Experimental Methods in the Physical Sciences — Atomic, Molecular and Optical Physics: Charged Particles*, F.B. Dunning and R.G. Hulet (eds.), Academic Press, New York 1995.

*11.5 U. Heinzmann, J. Kessler, and J. Lorenz, Z. Phys. **240** (1970) 42.

11.6 J. Kessler, *Polarized Electrons* (2nd edition), Springer, Heidelberg 1985.

11.7 G. Schönhense, Phys. Rev. Lett. **44** (1980) 640.

11.8 C.M. Lee, Phys. Rev. A **10** (1974) 1598.

11.9 Ch. Heckenkamp, F. Schäfers, G. Schönhense, and U. Heinzmann, Z. Phys. D **32** (1974) 751.

11.10 B. Schmidtke, M. Drescher, and U. Heinzmann, *Book of Abstracts XXI. ICPEAC* (1999) 36.

11.11 O. Plotzke, G. Prümper, B. Zimmermann, U. Becker, and H. Kleinpoppen, Phys. Rev. Lett. **77** (1996) 2642.

11.12 R.D. Cowan, *The Theory of Atomic Structure and Spectra*, University of California Press, Berkeley 1981.

11.13 A.F. Starace, S.T. Manson, and D.J. Kennedy, Phys. Rev. A **9** (1974) 2453.

11.14 P. van der Meulen, M.O. Krause, and C.A. de Lange, Phys. Rev. A **43** (1991) 5997.

11.15 G. Snell, B. Langer, M. Drescher, N. Müller, B. Zimmermann, U. Hergenhahn, J. Viefhaus, U. Heinzmann, and U. Becker, Phys. Rev. Lett. **82** (1999) 2480.

11.16 B. Kämmerling and V. Schmidt, Phys. Rev. Lett. **69** (1992) 1145; J. Phys. B **26** (1993) 1141.

11.17 W.R. Johnson and K.T. Cheng, Phys. Rev. Lett. **69** (1992) 1144.

11.18 W.R. Johnson and K.T. Cheng, Phys. Rev. A **46** (1992) 2952.

11.19 K.W. McLaughlin, O. Yenen, and D.H. Jaecks, Phys. Rev. Lett. **81** (1998) 289.

11.20 E. Wigner, Nachr. Akad. Wiss. Gött. Math.-Physik. K1. **IIa** (1927) 375.

11.21 J.H. Moore, Jr., Phys. Rev. A **8** (1973) 2359.

11.22 H. Klar, J. Phys. B **13** (1980) 4741.

11.23 N.M. Kabachnik and O.V. Lee, J. Phys. B **22** (1989) 2705.

11.24 U. Hahn, J. Semke, H. Merz, and J. Kessler, J. Phys. B **18** (1985) L417.

11.25 R. Kuntze, M. Salzmann, N. Böwering, U. Heinzmann, V.K. Ivanov, and N.M. Kabachnik, Phys. Rev. A **50** (1994) 489.

11.26 G. Snell, M. Drescher, N. Müller, U. Heinzmann, U. Hergenhahn, J. Viefhaus, F. Heiser, U. Becker, and N.B. Brookes, Phys. Rev. Lett. **76** (1996) 3923.

11.27 U. Hergenhahn, G. Snell, M. Drescher, B. Schmidtke, N. Müller, U. Heinzmann, M. Wiedenhöft, and U. Becker, Phys. Rev. Lett. **82** (1999) 5020.

11.28 E. Boskamp, R. Morgenstern, P. van der Straten, and A. Niehaus, J. Phys. B **17** (1984) 2823.

11.29 P. van der Straten and R. Morgenstern, Comments At. Mol. Phys. **17** (1986) 243.

11.30 K. Blum and H. Jakubovicz, J. Phys. B **11** (1978) 909.

*11.31 U. Fano and J. Macek, Rev. Mod. Phys. **45** (1973) 553.

11.32 C.H. Greene and R.N. Zare, J. Chem. Phys. **78** (1983) 6741.

11.33 K. Blum, *Density Matrix Theory and Applications* (2nd edition), Plenum, New York 1996.

11.34 K. Becker, H.W. Dassen, and J.W. McConkey, J. Phys. B **17** (1984) 2535.

11.35 I.C. Malcolm and J.W. McConkey, J. Phys. B **12** (1979) L67.

11.36 T. Hegemann, S. Schroll, and G.F. Hanne, J. Phys. B **26** (1993) 4607.

11.37 I. Bray and I.E. McCarthy, Phys. Rev. A **47** (1993) 317.

11.38 F. da Paixao, M.A.P. Lima, and V. McCoy, Phys. Rev. Lett. **68** (1992) 1698.

11.39 C.M. Fullerton, G. Wöste, D.G. Thompson, K. Blum, and C.J. Noble, J. Phys. B **27** (1994) 185.

11.40 G.F. Hanne, in *Photon and Electron Collisions with Atoms and Molecules*, P.G. Burke and C.J. Joachain (eds.), Plenum Press, New York 1997.

11.41 P.S. Farago, J. Phys. B **13** (1980) L567.

11.42 D.M. Campbell and P.S. Farago, J. Phys. B **20** (1987) 5133.

11.43 K.W. Trantham, M.E. Johnston, and T.J. Gay, J. Phys. B **28** (1995) L543.

11.44 S. Mayer and J. Kessler, Phys. Rev. Lett. **74** (1995) 4803.

11.45 K. Blum and D.G. Thompson, J. Phys. B **22** (1989) 1823.

11.46 F.J. Eriksen and D.H. Jaecks, Phys. Rev. Lett. **36** (1976) 1491.

11.47 D. Dhuicq, B. Jugi, C. Benoit, and V. Sidis, J. Chem. Phys. **109** (1998) 512.

11.48 H. Hülser, M. Braun, G. Kubsch, E.E.B. Campbell, and I.V. Hertel, J. Phys. Chem. **99** (1995) 15335.

11.49 J. Macek, Phys. Rev. Lett. **23** (1969) 1.

11.50 J. Andrä, Phys. Rev. Lett. **25** (1970) 325.

11.51 D.J. Burns and W.H. Hancock, J. Opt. Soc. Am. **45** (1973) 553.

11.52 H.G. Berry, L.J. Curtis, D.G. Ellis, and R.M. Schectman, Phys. Rev. Lett. **32** (1974) 751.

11.53 H.J. Andrä, in *Fundamental Processes of Atomic Dynamics*, J.S. Briggs, H. Kleinpoppen, and H.O. Lutz (eds.), Plenum Press, New York and London 1988, p. 631.

11.54 H.J. Andrä, R. Fröhling, H.J. Plöhn, and J.D. Silver, Phys. Rev. Lett. **37** (1976) 1212.

11.55 H. Schröder and E. Kupfer, Z. Phys. A **279** (1976) 13.

11.56 H. Griem, *Spectral Line Broadening by Plasmas*, Academic Press, New York 1974.

11.57 D. Mihalas, *Stellar Atmospheres*, Freeman, San Francisco, 1970.

11.58 J. Cooper, in *Spectral Line Shapes, Vol. 2*, K. Burnett (ed.), de Gruyter, Berlin 1983, p. 737.

11.59 N. Allard and J. Kielkopf, Rev. Mod. Phys. **54** (1982) 1103.

11.60 A. Gallagher, in *Atomic, Molecular & Optical Physics Handbook*, G.W. Drake (ed.), AIP Press, Woodbury, New York 1996, p. 220.

11.61 G. Peach, in *Atomic, Molecular & Optical Physics Handbook*, G.W. Drake (ed.), AIP Press, Woodbury, New York 1996, p. 669.

11.62 K. Burnett, Phys. Rep. **118** (1985) 339.

11.63 E.L. Lewis, J.M. Salter, and M. Harris, J. Phys. B **14** (1981) L173.

11.64 E.L. Lewis, M. Harris, W.J. Alford, J. Cooper, and K. Burnett, J. Phys. B **16** (1983) 553.

11.65 G. Nienhuis, J. Phys. B **16** (1983) 1.

11.66 W.J. Alford, N. Andersen, K. Burnett, and J. Cooper, Phys. Rev. A **30** (1984) 2366.

11.67 J.L. Carlsten, A. Szöke, and M.G. Raymer, Phys. Rev. A **15** (1977) 1029.

11.68 P. Thomann, K. Burnett, and J. Cooper, Phys. Rev. Lett. **45** (1980) 1325.

11.69 W.J. Alford, K. Burnett, and J. Cooper, Phys. Rev. A **27** (1983) 1310.

11.70 W.J. Alford, N. Andersen, M. Belsley, J. Cooper, D.M. Warrington, and K. Burnett, Phys. Rev. A **31** (1985) 3012.

11.71 W. Behmenburg, Phys. Scr. **36** (1987) 300.

11.72 E. Paul-Kwiek and E. Czuchaj, Eur. Phys.J. D **3** (1998) 163.

11.73 A. Corney and J.V.M. McGinley, J. Phys. B **14** (1981) 3047.

11.74 H. Harima, K. Takechibana, and Y. Urano, J. Phys. B **15** (1982) 3679.

11.75 H. Harima, Y. Fukuzo, K. Takechibana, and Y. Urano, J. Phys. B **14** (1981) 3069.

11.76 S. Ni, Z. Phys. D **38** (1996) 303.

11.77 C. Bousquet, N. Bras, and Y. Majdi, J. Phys. B **17** (1984) 1831.

11.78 J. Grosser, J. Phys. B **14** (1981) 1449.

11.79 M.O. Hale, I.V. Hertel, and S.R. Leone, Phys. Rev. Lett. **53** (1984) 2296.

11.80 W. Bussert, D. Neuschäfer, and S.R. Leone, J. Chem. Phys. **87** (1987) 3833.

11.81 W. Bussert and S.R. Leone, Chem. Phys. Lett. **138** (1987) 276.

11.82 R.L. Robinson, L.J. Kovalenko, C.J. Smith, and S.R. Leone, J. Chem. Phys. **92** (1990) 5260.

11.83 J.P.J. Driessen, C.J. Smith, and S.R. Leone, J. Chem. Phys. **95** (1991) 8183.

11.84 A.Z. Devdariani and A.L. Zagrebin, Chem. Phys. Lett. **131** (1986) 197.

11.85 B. Pouilly, J.M. Robbe, and M.L. Alexander, J. Chem. Phys. **91** (1989) 1658.

11.86 G.C. Schatz, L.J. Kovalenko, and S.R. Leone, J. Chem. Phys. **91** (1989) 6961.

11.87 R.L. Dubbs, P.S. Julienne, and F.H. Mies, J. Chem. Phys. **93** (1990) 8784.

11.88 M.H. Alexander, J. Chem. Phys. **96** (1992) 6672.

11.89 J.P.J. Driessen and S.R. Leone, J. Chem. Phys. **96** (1992) 6136.

11.90 J.G. Kircz, R. Morgenstern, and G. Nienhuis, Phys. Rev. Lett. **48** (1982) 610.

11.91 J. Weiner, F. Masnou-Seeuws, and A. Giusti-Suzor, Adv. At. Mol. Opt. Phys. **26** (1989) 209.

11.92 B. Huynh, O. Dulieu, and F. Masnou-Seeuws, Phys. Rev. A **57** (1998) 2.

11.93 H.A.J. Meijer, T.J.C. Pelgrim, H.G.M. Heideman, R. Morgenstern, and N. Andersen, J. Chem. Phys. **90** (1989) 738.

11.94 J.J. Blangé, J.M. Zijlstra, A. Amelink, X. Urbain, H. Rudolph, P. van der Straten, H.C.W. Beijerinck, and H.G.M. Heideman, Phys. Rev. Lett. **78** (1997) 3089.

11.95 T. Fujimoto and S.A. Kazantsev, Plasma Phys. Control. Fusion **39** (1997) 1267.

11.96 W. Hanle, Z. Phys. **30** (1924) 93.

11.97 H. Mark and R. Wierl, Z. Phys. **55** (1929) 156.

11.98 S. Trajmar, I. Kanik, M.A. Khakoo, L.R. LeClair, I. Bray, D. Fursa, and G. Csanak, J. Phys. B **32** (1999) 2801.

11.99 H. Kleinpoppen and E. Kraiss, Phys. Rev. Lett. **20** (1968) 361

11.100 A. Werner and K-H. Schartner, J. Phys. B **29** (1996) 125.

11.101 W. Kedzierski, J.W. McConkey, K. Bartschat, D.V. Fursa, and I. Bray, Phys. Rev. A (2000) in preparation.

11.102 A. Götz, W. Mehlhorn, A. Raeker, and K. Bartschat, J. Phys. B **29** (1996) 4699.

11.103 H. Merabet, M. Bailey, R. Bruch, I. Bray, D.V. Fursa, J.W. McConkey, and P. Hammond, Phys. Rev. A **60** (1999) 1187.

11.104 E. Haug, So. Phys. **71** (1981) 77.

11.105 Y. Itikawa, R. Srivastava, and K. Sakimoto, Phys. Rev. A **44** (1991) 7195.

11.106 K. Bartschat, Comp. Phys. Commun. **75** (1993) 219.

11.107 A.S. Shlyaptseva, R.C. Mancini, P. Neill, P. Beiersdorfer, J.R. Crespo López-Urrutia, and K. Widmann, Phys. Rev. A **57** (1998) 888.

11.108 Y. Nojiri and B. Deutsch, Phys. Rev. Lett. **51** (1983) 180.

11.109 A. Zelenski, *et al.*, Nucl. Instr. and Meth. A **334** (1993) 285.

11.110 L.W. Anderson, Nucl. Instr. and Meth. **167** (1979) 363.

11.111 L.W. Anderson, Nucl. Instr. and Meth. **402** (1998) 179.

11.112 P.G. Sona, Energia Nucl. **14** (1970) 295.

11.113 *Helion 97 - from Quark to Life*, Proceedings of the 7th RCNP International Workshop on Polarized $^3$He Beams and Gas Targets and Their Application, Kobe, Japan, January 20-24, 1997, M. Tanaka (ed.), Nucl. Instr. and Meth. A **402** (1998) 179-499.

11.114 N. Andersen and S.E. Nielsen, Nucl. Instr. and Meth. A **402** (1998) 473.

11.115 S.E. Nielsen and T.H. Rod, J. Phys. B **30** (1997) 3833.

# SELECTION OF HISTORICAL PAPERS (1925–1976)

# Introductory Summaries

We have selected for Part III some of the early, pathbreaking publications in this field written during the period 1925–1976. Below we give a brief description of the papers, motivating our selection. Because of space limitations, some sections from the longer papers have been omitted, and sometimes short initial notes of the research results have been chosen instead of the longer follow-up publications that give more results and further important details. Consequently, the book is accompanied by a CD-ROM that contains all the papers of Part III in full, the follow-up papers of some of the Letters, and a few related publications. The cases for which more information may be found on the CD-ROM than in the book are marked with an asterisk.

*1.  W. Kossel and C. Gerthsen, *Prüfung von D-Leuchten, das von einem nahezu parallelen Elektronenbündel angeregt ist, auf Polarisation*, Ann. Phys. (Leipzig) **77** (1925) 273–286.

This pioneering paper is the first one in the literature on atomic collisions that investigates the possibility of polarization of electron-impact radiation, with the Na D-line as an example. The introduction of the paper outlines why it is meaningful to ask this question, and already this first paper addresses the value of the polarization that can be expected at the excitation threshold. The experiment failed, however, within the error bars, to detect any significant light polarization. Of course, the fine-structure and hyperfine-structure depolarization effects, which severely reduce the upper and lower limits within which the D-line polarization is allowed to vary, were not known at the time of this investigation. This problem was not elucidated in a fully satisfactory way until the work of Percival and Seaton 33 years later (see below).

*2.  H.W.B. Skinner, *On the Excitation of Polarised Light by Electron Impact*, Proc. Roy. Soc. A **112** (1926) 642–660.

*3.  H.W.B. Skinner and E.T.S. Appleyard, *On the Excitation of Polarised Light by Electron Impact. II.–Mercury*, Proc. Roy. Soc. A **117** (1927) 224–244.

In these two classic papers by Skinner and by Skinner and Appleyard, communicated to the Royal Society by Rutherford in 1926 and 1927, the light polarization of a series of transitions in mercury was investigated. The motivation for the study is summarized as: "Polarisation measurements ... give a powerful means of investigating the collision of an electron with an atom in the case where excitation of the atom takes place, and it is in this that their main interest lies". The qualitative observations of the first paper are discussed in terms of angular-momentum properties, with some success but also with some failures. The most spectacular example in the latter category is the 2537 Å intercombination line in mercury, which shows a sign of polarization opposite to the one predicted. A possible effect of the fact that "the impacting electron may possess angular momentum of spin" is speculated on. (Note that the first paper was published shortly after the Uhlenbeck-Goudschmidt paper on electron spin, also from 1926.)

The second paper presents the first systematic, quantitative measurements of light polarizations for a series of mercury lines. The experiment allowed for an external, varying magnetic field along or perpendicular to the direction of the electron beam, thereby enabling studies of the Hanle effect for selected transitions. The authors summarize their findings in this way: "The polarised lines show a characteristic variation of the polarisation with the electron speed; the most salient feature being a maximum of polarisation for an electron velocity corresponding to a few volts above the excitation point. A reversed polarisation for high electron velocities is found in most cases. It is shown that the various polarisation effects may be interpreted to give information about the dynamics of the collision process in which an atom is excited by electron impact." The latter conclusion is based partly on discussions with Oppenheimer, who published the following series of theoretical papers on this problem.

4.  J.R. Oppenheimer, *Zur Quantenmechanik der Richtungsentartung*, Z. Physik **43** (1927) 27–46.

5.  J.R. Oppenheimer, *On the Quantum Theory of the Polarisation of Impact Radiation*, Proc. Nat. Acad. Sci. **13** (1927) 800–805.

*6.  J.R. Oppenheimer, *On the Quantum Theory of Electronic Impacts*, Phys. Rev. **32** (1928) 361–376.

In these papers Oppenheimer develops various aspects of the quantum theory of electron-impact excitation. He discusses in particular the role of spin-flip for excitation of the mercury triplet states and shows that, at least qualitatively, the polarization of the 2537 Å intercombination line can be understood in this way. His concluding sentence in the first paper is almost prophetic, considering the central role of the electron–mercury system in the subsequent development (see Chapter 7). "Aber wenn auch diese Erklärung schon prinzipiell als befriedigend anzusehen sein sollte, so würde sie zu ihrer Bestätigung doch einer viel eingehenderen Untersuchung bedürfen (But although this explanation, in principle, might already be considered satisfactory, a much more comprehensive investigation would be necessary for its confirmation)."

*7.  K. Steiner, *Die Polarisation des Elektronenstoßleuchtens bei Edelgasen*,
     Z. Physik **52** (1928) 516–530.
The paper reports on the first polarization measurements for electron-impact
excitation of helium and neon. Steiner compares his results for the rare gases
with Skinner's data for mercury and makes an attempt to explain them theoretically.

8.  W.G. Penney, *Effect of Nuclear Spin on the Radiation Excited by
    Electron Impact*, Phys. Rev. **18** (1932) 231–237.
Penney's paper "is concerned with showing that no theory of polarisation
of radiation excited by electron impact can be complete without taking into
account nuclear spin ..... Since the nuclear spin of sodium is not known, we
consider chiefly the case of mercury." Calculations are made for the mercury 2537 Å intercombination line, using the first-order cross sections and
approximate wave functions. "The results do not agree at all with experimental data in the region where any measurements have been made." Penney's conclusion that "first-order cross-sections simply will not do for this
type of calculation ..." was amply confirmed by the later developments, as
seen in many examples of Chapter 7. He also states: "Perhaps the result that
the D lines of sodium are unpolarised when excited by electrons can be attributed to the large nuclear spin, the exact magnitude of which, however, is
not known."

*9.  H. Bethe, *Polarisation des Stoßleuchtens*, Handbuch der Physik **24**,
     508–515, Springer, Berlin 1933.
This early review chapter gives an overview of theoretical and experimental
results for electron-impact-induced light polarization as of 1933. Bethe tries
to explain various experimental results, pointing out the importance of the
momentum-transfer direction, in the spirit of the first Born approximation.
He also indicates the sign change in the light polarization when the projectile
energy is increased from threshold toward the high-energy limit. Also note
the footnote on page 509, pointing out that his result agrees with those of
Skinner, Appleyard, and Oppenheimer, while Steiner's attempt is discarded.
("Ein anderer theoretischer Versuch von Steiner ... ist verkehrt.")

   After Bethe's summary, not much happened in this area of atomic physics
for more than twenty years. Probably as a result of this lack of activity, this
chapter was omitted in the book by H.A. Bethe and E.E. Salpeter, *Quantum
Mechanics of One- And Two-Electron Atoms*, Springer, Berlin 1957.

    10. N.F. Ramsey, *Collision Alignment of Molecules, Atoms, and Nuclei*, Phys. Rev. **98** (1955) 1853–1854.

To the best of our knowledge, in this visionary Letter to the Editor, Norman F. Ramsey was the first to propose the idea of exploring alignment effects as a tool for the study of collision phenomena involving atoms, molecules, and nuclei.

   *11. I.C. Percival and M.J. Seaton, *The Polarization of Atomic Line Radiation Excited by Electron Impact*, Phil. Trans. Roy. Soc. **251** (1958) 113–138.

In this classic paper, Percival and Seaton develop the theory of electron-impact excitation beyond the early work of Oppenheimer and Penney and calculate expressions for integral cross sections and polarizations in terms of the cross sections for individual magnetic sublevels, taking spin effects into account. They also demonstrate how cases in which the fine-structure or hyperfine-structure separations are comparable to the linewidth can be evaluated. Based on their theory, Percival and Seaton discuss the polarization results of Kossel and Gerthsen for the Na D-line, and of Skinner and Appleyard for various transitions in Hg.

    Recently, K. Bartschat and G. Csanak (Comm. At. Mol. **32** (1996) 233), discussed the Percival–Seaton criticism of the "Oppenheimer–Penney" theory. They point out that Oppenheimer's formulation indeed contains all the necessary coherence terms to fulfill the "principle of spectroscopic stability", i.e., the necessary smooth transition between the extreme cases where the natural linewidth is very small or very large compared to the fine-structure or hyperfine-structure separations. This principle is ignored in Penney's formulation, which only deals with cases where the natural linewidth can be neglected, thereby yielding the maximum depolarization of the emitted radiation.

    12. K. Jost and J. Kessler, *Production of Highly Polarized Electron Beams by Low-Energy Scattering*, Phys. Rev. Lett. **15** (1965) 575–577.

   *13. K. Jost and J. Kessler, *Zur Polarisation langsamer Elektronen durch Streuung an Quecksilber zwischen 180 und 1700 eV*, Z. Physik **195** (1966) 1–12.

This is the first experimental observation of very high polarization for electrons elastically scattered from an atom, mercury, at specific scattering angles. The Letter reports polarizations ranging from −79% to +84% for electron-impact energies of several hundred eV. The electron-spin polarization was measured by Mott scattering, after the scattered electrons were accelerated to 120 keV. The follow-up paper gives details on the experimental setup and more examples of dramatic polarization curves, depending critically on electron energy and scattering angle.

14. B. Bederson, *The "Perfect" Scattering Experiment. I*, Comm. At. Mol. Phys. **1** (1969) 41–45.

15. B. Bederson, *The "Perfect" Scattering Experiment. II*, Comm. At. Mol. Phys. **1** (1969) 65–69.

*16. B. Bederson, *Electron-Atom Excitation with Spin Analysis*, Comm. At. Mol. Phys. **2** (1970) 160–164.

The three Comments by Bederson were instrumental in clarifying for the scientific community the concept of a "perfect" or "complete" scattering experiment as an experiment, or a group of experiments, in which the quantum-mechanical scattering amplitudes are fully determined. He demonstrates how such an experiment in general goes beyond the determination of total or differential cross sections and extracts information on the relative sizes and phases of the scattering amplitudes. Comment (I) discusses the case of elastic scattering of an electron on a hypothetical alkali atom (without hyperfine structure) for which determination of three parameters are needed for completeness. Comment (II) addresses S $\rightarrow$ P excitation, a case in which three parameters are needed if the projectile and the target are spinless. The information on the relative phase angle between the two scattering amplitudes can be obtained, for example, by analyzing the polarization of the light emitted in the subsequent P $\rightarrow$ S optical decay. If both the projectile and the target have spin $\frac{1}{2}$, as in electron–alkali-atom collisions, four amplitudes come into play. This results in a set of seven independent parameters that need to be determined for a complete experiment. The third Comment elaborates on this discussion and stresses the simplifications originating from the fact that the characteristic times for excitation, fine-structure interaction, and optical decay fall into three distinct domains. Interestingly, the three Comments were written before any experiments had been carried out involving low-energy spin-polarized electron beams. For a summary of the dramatic progress during the following 25 years in this area, see N. Andersen and K. Bartschat, Comm. At. Mol. Phys. **29** (1993) 157.

17. J. Kessler and J. Lorenz, *Experimental Verification of the Fano Effect*, Phys. Rev. Lett. **24** (1970) 87–88.

*18. U. Heinzmann, J. Kessler, and J. Lorenz, *Elektronen-Spinpolarisation unpolarisierter Cäsiumatome mit zirkularpolarisiertem Licht*, Z. Physik **240** (1970) 42–61.

In 1969 Fano predicted that highly polarized electrons could be produced by photoionization of unpolarized atomic beams. The papers report a photoelectron polarization of up to 100 % from cesium atoms exposed to circularly polarized light with a wavelength around the minimum of the photoabsorption cross section. Apart from the theoretical interest, the Fano effect is a very efficient way of producing highly polarized electron beams in a relatively simple way. The follow-up paper reports a dramatic variation in the electron polarization measured when the wavelength of the ionizing radiation is varied over the 2300 − 3200 Å range.

19. D.H. Jaecks, D.H. Crandall, and R.H. McKnight, *Measurement of Differential Charge-Transfer Cross Sections and Probabilities by Photon–Particle Coincidence Technique*, Phys. Rev. Lett. **25** (1970) 491–494.

This paper reports the first photon–particle coincidence experiment for low-energy heavy-particle collisions. Lyman-$\alpha$ photons emitted after electron transfer in proton–helium collisions were measured at a fixed impact parameter (or fixed value of the product $E\theta$ of impact energy and scattering angle) in the 4–20 keV impact energy range.

20. J. Macek and D.H. Jaecks, *Theory of Atomic Photon–Particle Coincidence Measurements*, Phys. Rev. A **4** (1971) 2288–2300.

The theory of atomic photon–particle coincidence measurements was put on a solid foundation by this much-cited and influential paper, in which the fundamental equations relating the measured coincidence rates to excitation amplitudes were derived. The equations incorporate the polarization of the radiation, the fine and hyperfine structure of the atomic levels, the coherence of the radiation, and the time variation of the radiation intensity. A simple, semiclassical model is used to interpret the radiation from a $^1$P $\rightarrow$ $^1$S transition emitted perpendicular to the scattering plane. Furthermore, the application of the theory to results for Lyman-$\alpha$ radiation and $^1$P $\rightarrow$ $^1$S transitions in helium is demonstrated. A few printing errors (J.H. Macek 1999, private communication) have been corrected in the copy reproduced here.

21. G. Rahmat, G. Vassilev, J. Baudon, and J. Barat, *Differential Measurement of the He $3\,^3P$ Excitation, in $He^+$–He Collisions, by Using an Ion–Photon Coincidence Method*, Phys. Rev. Lett. **26** (1971) 1411–1413.

The effect of channel interferences among quasi-molecular levels, causing oscillations in differential excitation cross sections for the $3^3$P level in He$^+$–He collisions at $200 - 350$ eV, were demonstrated in the first photon–particle coincidence study of this system.

*22. J. Wykes, *The Variation with Electron Scattering Angle of the Polarization of Atomic Line Radiation Excited by Electron Impact*, J. Phys. B **5** (1972) 1126–1137.

This early theoretical paper investigates how electron–photon angular correlations in a coincidence experiment, together with the differential cross section, give information on the relative phases of the various scattering amplitudes involved and thus may represent a complete observation of an excitation event. It is pointed out that such an experiment provides "a more sensitive test of theoretical predictions than the usual noncoincident polarization measurements whilst still retaining their essentially relative nature free from normalization difficulties."

23. M. Eminyan, K.B. MacAdam, J. Slevin, and H. Kleinpoppen,
    *Measurements of Complex Excitation Amplitudes in Electron–Helium
    Collisions by Angular Correlations Using a Coincidence Method*,
    Phys. Rev. Lett. **31** (1973) 576–579.

*24. M. Eminyan, K.B. MacAdam, J. Slevin, and H. Kleinpoppen,
    *Electron–Photon Angular Correlations in Electron–Helium Collisions:
    Measurements of Complex Excitation Amplitudes, Atomic Orientation
    and Alignment*, J. Phys. B **7** (1974) 1519–1542.

The Letter and the accompanying full publication present the first electron–
photon angular correlation study of excitation in electron–atom collisions,
here the $2^1$P and $3^1$P levels in helium. The Letter briefly points out the
new important aspects of this kind of investigation, while the paper carefully
discusses the implications of the measurements. These include the significant
deviation of the results from the predictions of the first Born approximation
and a discussion of how the parameter that escapes detection by this method,
namely the sign of the angular momentum transfer, may be obtained from a
circular-polarization analysis of the light emitted in a direction perpendicular
to the scattering plane.

*25. U. Fano and J. Macek, *Impact Excitation and Polarization of the
    Emitted Light*, Rev. Mod. Phys. **45** (1973) 553–573.

A comprehensive review of anisotropies in light emission is given. The elegant formulation disentangles geometrical and dynamical effects in a way
that stresses the extraction of data on the alignment and orientation of the
radiating atoms from the anisotropy and polarization of the radiation pattern. In addition, the time development of the alignment and orientation in
the interval between excitation and emission is accounted for. This seminal
review paper was published at a very timely stage in the development of the
field. Perhaps more than any other paper, it strongly influenced the way of
thinking in a large fraction of the subsequent literature on collision-induced
polarization effects. Some printing errors in Eqs. (18) were pointed out by
J.S. Briggs, J. Macek, and K. Taulbjerg (J. Phys. B **12** (1979) 1457). The
corrected part of the equation reads (corrections are underlined)

$$A_{2+}^{\mathrm{det}} = \ldots A_{1+}^{\mathrm{col}} \left( \sin\theta \, \underline{\sin}\phi \, \sin 2\psi \underline{-} \sin\theta \, \cos\theta \, \underline{\cos}\phi \, \cos 2\psi \right) \ldots$$

26. G.F. Hanne and J. Kessler, *Direct Observation of Exchange
    Scattering by Spin Flip of Polarized Electrons in Excitation of Mercury*,
    Phys. Rev. Lett. **33** (1974) 341–343.

*27. G.F. Hanne and J. Kessler, *Study of Exchange Excitation in Mercury
    by Means of Polarized Electrons I. Experiment*, J. Phys. B **9** (1976)
    791–804.

The electron-spin flip associated with exchange scattering is directly observed
for the first time. A beam of polarized electrons excites mercury atoms from
the $6^1$S$_0$ ground state to the $6^3$P$_{0,1,2}$ states (not resolved individually in

the experiment) and the polarization of the forwardly scattered electrons is determined. (The influence of spin-orbit coupling is only appreciable for scattering angles beyond 30°.) Exchange scattering dominates between threshold and 8 eV incident energy, thus resulting in a reversal of the direction of polarization after the scattering ($P'/P < 0$). In the full paper, the individual fine-structure states $6^3\mathrm{P}_1$ and $6^3\mathrm{P}_2$ are resolved, and the results show an indication of the "fine-structure effect".

28. J. Macek and I.V. Hertel, *Theory of Electron Scattering from Laser Excited Atoms*, J. Phys. B **7** (1974) 2173–2188.

This paper outlines the theory of electron scattering from laser-excited atoms in the language of state multipoles, as developed by Fano and Macek. It is shown that all state multipoles can be determined by suitable measurements with a variety of choices for the direction and polarization of the pumping light. The results are expressed in terms of the scattering amplitudes for the time-inverse scattering process. Most of the subsequent work on scattering from optically prepared atomic states rests on the framework outlined in this paper.

29. G. Vassilev, G. Rahmat, J. Slevin, and J. Baudon, *Measurements of Photon Polarization and Angular Correlations for He⁺–He Collisions Using an Ion–Photon Coincidence Technique*, Phys. Rev. Lett. **34** (1975) 444–447.

This Letter reports the first polarized-photon–scattered-ion coincidence measurement, as well as the first angular correlation measurement for direct excitation of the $3^3\mathrm{P}$ and $2^1\mathrm{P}$ states in helium, respectively, in He⁺–He collisions at an impact energy of 150 eV. Only the linear polarization was measured.

30. D.H. Jaecks, F.J. Eriksen, W. de Rijk, and J. Macek, *Polarized-Photon, Scattered-Atom Coincidence Measurements in He⁺–He Collisions*, Phys. Rev. Lett. **35** (1975) 723–725.

This Letter reports polarized-photon–scattered-atom coincidence measurement of the $3^3\mathrm{P}$ and $2^1\mathrm{P}$ states in helium, populated by electron transfer in He⁺–He collisions at an impact energy of 3.0 keV. Only the linear polarization was determined.

31. M.C. Standage and H. Kleinpoppen, *Photon Vector Polarization and Coherence Parameters in an Electron–Photon Coincidence Experiment on Helium*, Phys. Rev. Lett. **36** (1976) 577–580.

This Letter reports the first complete Stokes-parameter analysis using the polarized-photon–scattered-particle coincidence technique. The process studied was electron-impact excitation of the $3^1\mathrm{P}$ state in helium at an electron-impact energy of 80 eV and scattering angles ranging from $-15.0°$ to $+27.5°$. The data represent the first experimental determination of the degree of polarization.

# 3. *Prüfung von D-Leuchten, das von einem nahezu parallelen Elektronenbündel angeregt ist, auf Polarisation;*
## *von W. Kossel und C. Gerthsen.*

### A. Fragestellung.

Von der Lichterregung durch Elektronenstoß ist die energetische Seite, bei der man einfach mit den Termbeträgen zu tun hat, wohl bekannt. Es fehlt aber jede nähere Kenntnis über die Art und den Ablauf des Vorganges, der die Energie von dem fremden auf das Atomelektron überträgt. Die Ausschaltung aller Vorgänge, die nicht in einer nach der Quantentheorie zulässigen Situation beider Elektronen endigen, scheint klassische Behandlung auszuschließen und die Entdeckung Ramsauers scheint unmittelbar am Schicksal fremder Elektronen zu zeigen, daß ihre Wechselwirkung mit den Atomelektronen anders ist, als man klassisch erwarten sollte. Um nun etwas Weiteres zu erfahren, haben wir uns die Frage gestellt, ob in der Emission, die einem ganz einfachen Anregungsvorgange folgt, irgendeine „Erinnerung" an die Richtung vorhanden ist, aus der das stoßende Elektron kam, ob das von einem parallelen Elektronenbündel angeregte Licht eine merkliche Polarisation zeigt, deren Richtung von der Richtung des Bündels abhängt. Soweit uns bekannt ist, sind hierüber noch keine Versuche angestellt worden. Die bekannte Polarisation der Röntgenbremsstrahlung darf unmittelbar auf die Bewegung der fremden Elektronen selbst zurückgeführt werden; das Fehlen der Polarisation in der Eigenstrahlung der Antikathode betrifft einen verwickelten Prozeß, bei dem die Bewegung eines anderen Elektrons, als des unmittelbar „getroffenen" und entfernten die Energie für die Ausstrahlung liefert. Am ehesten ist solche Erinnerung zu vermuten, wenn man lediglich ein Elektron durch den Stoß um eine Quantenstufe hebt und unmittelbar zurückfallen läßt: also bei der Anregung von Resonanzlinien. Wir wählten zunächst die $D$-Linie des Natriums. Im Ver-

**274**                    *W. Kossel u. C. Gerthsen.*

gleich mit der Quecksilberlinie 2537 Å.-E. hat sie zwar die Nachteile verwickelterer Struktur und längerer Lebensdauer, aber die Vorzüge, daß man die Vorgänge stets sichtbar vor sich hat und in einem großen Spannungsbereich nur das Licht der einen gewollten Linie beobachtet. Aus der Tafel der Anregespannungen in folgender Tabelle ist zu entnehmen, daß der einfache $D$-Anregungsvorgang über einen Bereich von mehr als 1 Volt allein bleibt. Die dann neu auftretenden Linien[1]) sind sämtlich unsichtbar (eingeklammert), erst 2 Volt über der Anregespannung der $D$-Linie kommt anderes sichtbares Licht hinzu.

| Volt | Term | Termfolgen | Wellenlängen |
|---|---|---|---|
| 2,095 | $2p$ | $2p \rightarrow 1s$ | 5890 |
| 3,16 | $2s$ | $2s \rightarrow 2p \rightarrow 1s$ | (11404), 5890 |
| 3,59 | $3d$ | $3d \rightarrow 2p \rightarrow 1s$ | (8196); 5890 |
| 3,72 | $3p$ | $3p \rightarrow 1s$; $3p \rightarrow 2s \rightarrow 2p \rightarrow 1s$ | (3302); (22057), (11404), 5890 |
| 4,09 | $3s$ | $3s \rightarrow 2p \rightarrow 1s$; $3s \rightarrow 3p$ usw. | 6161 usw.; $(3,4\,\mu)$ usw. |
| 4,24 | $4d$ | $4d \rightarrow 3p \rightarrow$; $4d \rightarrow 2p \rightarrow 1s$ | (23361) usw.; 5688, 5890 |
| 4,25 | $4b$ | $4b \rightarrow 3d \rightarrow$; $4b \rightarrow 2p \rightarrow 1s$ | (18459) usw.; 5676 usw., 5890 |
| 4,31 | $4p$ | $4p \rightarrow 1s$; $4p \rightarrow 2p \rightarrow 1s$ | (2853); 5533, 5890 |
| — | — | $4p \rightarrow 3s \rightarrow$ | $(5,43\,\mu)$ usw. |
| 4,47 | $4s$ | $4s \rightarrow 2p \rightarrow 1s$ | 5154, 5890 |
| 5,56 | $5d$ | $5d \rightarrow 2p \rightarrow 1s$ | 4983, 5890 |
| — | — | — — — — — — — | |
| 5,10 | $\infty$ | Ablösung | |

Daß schon die einfachste Quantenvorschrift, nämlich die Plancksche Gleichung, die Energie und Frequenz verbindet, unter Umständen zu sehr bestimmten Polarisationsvorschriften führen kann, ist am Beispiel des quasielastisch isotrop gebundenen Elektrons leicht zu übersehen. Das zu einfacher harmonischer Schwingung fähige Elektron werde von einem fremden „angestoßen". Das dabei wirksame Kraftgesetz bleibe offen, es werde aber angenommen, daß die Dauer der Energieübertragung kurz sei gegen die Schwingungsdauer, daß die übertragene Energie also noch am Ende des Stoßvorganges bis auf einen verschwindenden Bruchteil rein kinetisch sei und daß Energie und Impuls erhalten bleibe. Es heiße:

    die Anfangsgeschwindigkeit des stoßenden $v$,

    die Endgeschwindigkeit des gestoßenen    $u$,

    der Winkel zwischen $u$ und $v$     $(u\,v) = \vartheta$.

---

1) Es sind der Sicherheit halber auch Übergänge notiert, die nur in fremden Feldern stattfinden ($np \rightarrow mp$); von Doppelniveaus ist nur ein Vertreter angeführt.

*Prüfung von D-Leuchten auf Polarisation.*      **275**

Dann gilt, da die Massen der beiden am Stoß teilnehmenden Massenpunkte gleich sind: $u = v \cos \vartheta$. Das getroffene Elektron (das nach den obigen Annahmen am Ende des Stoßvorgangs erst um einen verschwindend kleinen Bruchteil der Amplitude aus seiner Ruhelage gerückt ist) schwinge nun in der Richtung der ihm erteilten Geschwindigkeit $u$ aus.

Unter so bestimmten Annahmen ist also die Schwingungsrichtung des Oszillators fest mit dem Verhältnis der von ihm übernommenen Energie (die $T_2$ heiße) zu der lebendigen Kraft des heranfliegenden Teilchens ($T_1$) gegeben. Benutzen wir ein Elektronenbündel einheitlicher Geschwindigkeit, also einheitlichen $T_1$, um auf gebundene Elektronen zu schießen, die nur bestimmte einheitliche Energiebeträge $T_2$, nämlich das einer allein anregbaren Frequenz $\nu$ zukommende $h\nu$, aufnehmen, den Rest aber dem stoßenden Elektron lassen, so sind nur Schwingungen möglich, die mit der Flugrichtung $\mathfrak{v}$ einen ganz bestimmten Winkel $\vartheta$ einschließen, der durch

$$\cos^2 \vartheta = \frac{h\nu}{T_1} = \frac{1}{\beta}$$

gegeben ist. Für die charakteristische Variable des Versuchs, das Verhältnis der gegebenen zur Anregeenergie, werde die Abkürzung $\beta = \dfrac{T_1}{h\nu}$ benutzt.

Man würde für diesen primitiven Mechanismus also folgendes Bild erhalten: Erreicht $T_1$ gerade den Wert $h\nu$, ist also $\beta = 1$, so muß $T_1$ ganz an das gestoßene Elektron abgegeben werden. Das ist nur bei zentralem Stoß möglich: unmittelbar an der Anregegrenze sollte das Licht völlig so polarisiert sein, daß der elektrische Vektor der Flugrichtung des Elektronenstrahls parallel ist. Wächst $\beta$ über 1, so wird ein Kreiskegel von Schwingungsrichtungen um die Flugrichtung möglich, dessen Öffnungswinkel mit $T_1$ anwächst. Für den Grenzfall sehr hoher Geschwindigkeiten würden nur Stöße erlaubt sein, die das getroffene Teilchen quer zur Flugrichtung hinausschleudern. Indes kann dann die Emission der Resonanzlinie auf so viel verwickelten Wegen zustande kommen, daß dieser Grenzfall für die Beobachtung des einfachsten Vorganges nicht interessiert. Was aber in der Nähe der Anregungsgrenze beobachtet werden sollte, ist völlig bestimmt.

Ein senkrecht zur Flugrichtung blickender Beobachter wird bei Stößen, die in der zur Blickrichtung senkrechten Ebene verlaufen, ein Intensitätenverhältnis $\dfrac{I_{/\!/}}{I_\perp} = \dfrac{1}{\beta - 1}$ und einen Polarisationsgrad

$$\varkappa = \frac{I_{/\!/} - I_\perp}{I_{/\!/} + I_\perp} = \frac{2}{\beta} - 1$$

finden müssen. Für Stöße, deren Ebene gegen die Blickrichtung beliebig geneigt ist, kommen nur die senkrecht zur Blickrichtung stehenden Komponenten ins Spiel und die Mittelung ergibt:

$$\varkappa = \frac{3 - \beta}{1 + \beta}.$$

Beim Verhältnis $\beta = 1{,}5$, etwa einer Anregung der bei 2 Volt ansprechenden $D$-Linie mit einem Elektronenstrahl von 3 Volt, sollte dieser einfachste Stoßmechanismus also einen Polarisationsgrad von 60 Proz. ergeben.

Ob ein solcher einfacher Vorgang von vornherein als wahrscheinlich gelten kann, soll hier nicht ausführlich besprochen werden; wir erinnern nur daran, daß bei der Dispersion trotz aller Möglichkeit formaler Angleichung an die extreme Quantentheorie eine ganz nahe Berührung mit dem Verhalten harmonischer Eigenschwingungen der Elektronen mit den Sprungfrequenzen gegeben ist und nach der Ramsauerschen Entdeckung die Wechselwirkung zwischen einem fliegenden und einem gebundenen Elektron durchaus noch nicht völlig bekannt ist. Bei Polarisationsfragen bleibt ja selbst im Gebiet der Zeemaneffekte nichts anderes übrig, als mitten zwischen den Sprungmannigfaltigkeiten an das Verhalten harmonischer Schwingungen zu erinnern, um Aussagen über die Feldrichtungen zu erhalten. Die Härte, die darin liegt, daß das getroffene Elektron nur auf Stoßvorgänge eingehen darf, die ihm die von der Quantentheorie vorgeschriebene Energie überliefern, tritt natürlich besonders hervor,

---

1) *Anm. b. d. Korr.* Wir fügen hinzu, daß auch bei Annahme Coulombscher Kräfte zwischen stoßendem und quasielastischem Elektron die Bedingungen dafür, daß das letztere als frei behandelt werden darf, bei den in diesen Versuchen vorliegenden Geschwindigkeiten und Eigenschwingungsdauern mit ziemlich guter Annäherung erfüllt sind. Vgl. für diese Bedingungen die mit quasielastischen Elektronen arbeitende Elektronenverlangsamungstheorie von Bohr (Phil. Mag. 25. 1914; M. v. Laue, Handbuch der Radiologie VI).

*Prüfung von D-Leuchten auf Polarisation.*     **277**

wenn daneben die Stöße, die wirklich stattfinden, in so einfacher Weise wie Vorgänge zwischen freien Teilchen behandelt werden. Wir führen also die obige Überlegung vor allem aus dem Grunde an, weil auf einem gänzlich unbekannten Gebiet zunächst die Folgen der einfachsten Vorstellung bedacht werden müssen, die allenfalls in Frage kommt. Das Ergebnis der Versuche spricht nun durchaus *gegen* diese primitive Vorstellung vom momentanen „Anschlagen" einer quasielastischen Schwingung, bei dem die Quantenenergie übertragen wird: von der danach zu erwartenden recht beträchtlichen Polarisation ist nichts aufzufinden.

### B. Versuchsanordnung.

Um das Rohr rasch und bequem öffnen und doch den Rohrteil, in dem der Na-Dampf beobachtet wurde, beliebig lange erhitzen zu können, wurde dem Rohr eine lange Form gegeben, so daß weit vom erhitzten Ende eine Kittung lag, nach deren Lösung alle Innenteile mit einem Griff herausgenommen werden konnten. An dem von Kühlwasser durchflossenem Kopfteil $W$ (Fig. 1 a. f. S.), der in das obere Ende des 35 cm langen Glasrohres $R$ eingekittet werden kann, sitzt das lange Tragerohr $T$, das an seinem unteren Ende die ganze Elektrodenanordnung trägt. Die elektrischen Zuleitungen liegen im Inneren von $T$, abgedichtetdurch eine Siegellackfüllung, die durch eine in Höhe des Wassergefäßes liegende, von den Leitungen durchsetzte Siebplatte (in der Figur nur angedeutet sichtbar) gestützt ist. In der Fig. 1 geben a und b Vorder- und Seitenansicht der Elektrodenteile, für deren Anordnung die mechanische Festigkeit und Einstellbarkeit wichtig war. Die Elektronen aus der Glühkathode $K$ (Platin, Wehneltsche Mischung von Ca, Sr, Ba), die von den starken Backen $BB$ gehalten wird, werden durch eine Voranode $A_1$ weggezogen und mit Hilfe der zweiten Anode $A_2$ auf die gewünschte Geschwindigkeit gebracht. Sie durchlaufen damit das mit $A_2$ verbundene, mit zwei Innenblenden $b_1$, $b_2$ versehene engere Rohr $C$, den Beobachtungszylinder, in dem sie durch das längliche Fenster $F$ beobachtet werden. Zur Beurteilung der Dampfdichte des Na ist ein Thermoelement $Th$ eingeführt. Mit der hier gegebenen Anordnung wurden die eigentlichen Prü-

sichtlich so gewählt, um eine etwaige Polarisation durch Reflexion des im Gasvolum erzeugten Lichtes an den Wänden für sich hervortreten zu lassen; wir schließen, daß der hierdurch erzeugte Polarisationsgrad sich unter den Beträgen hält, die bei den Helligkeiten des reinen Strahls überhaupt nachgewiesen werden konnten.

Im folgenden sind einige typische Beispiele von Beobachtungen bei verschiedenen Temperaturen gekürzt in Tabellenform vereinigt. Die vorletzte Spalte „$\varDelta V$" gibt in Volt den Überschuß der Strahlgeschwindigkeit über die Spannung, bei der die $D$-Linie auftauchte, die letzte Spalte „Empfindlichkeit" den bei jedem Versuch mittels des Plattensatzes festgestellten Polarisationsgrad an, bei dem sich in dem verwendeten Licht noch Streifen wahrnehmen ließen, also die Grenze, unter der eine etwaige Polarisation des primären Lichtes geblieben ist.

| Temperatur | Aussehen | $\varDelta V$ | Empfindlichkeit |
|---|---|---|---|
| 240 °C | Zur Empfindlichkeitsprüfung ganz stark geheizt, sehr hell, diffuse aber noch deutliche Strahlgrenzen | > 3 | < 2 % |
| 220 ° | etwas weich erscheinende Strahlgrenze, daneben leichtes Resonanzlicht | 0,8 | 6 % |
| 205 ° | sehr scharf begrenzt, Resonanz sehr gering | 0,7 | 5 %, 6 % |
| 185 ° | gut definiert, nicht ganz frei von Resonanz | | 6 % |
| 170 ° | gute Grenzen | 0,5 | 6 %, 8 % |
| 165 ° | sehr dunkel; sehr scharf begrenzt; Messung über $b_2$ | 1,2 | 11 % |
| 150 ° | sehr lichtschwach, guter Strahl | $\leqq 1$ | 8 %, 11 % |

### Zusammenfassung.

An reinem $D$-Licht, das von einem nahe parallelen Elektronenbündel erregt ist, ist keine Polarisation nachweisbar; im Leuchten ist also keine „Erinnerung" an die Richtung zu finden, aus der das stoßende Elektron kam.

Kiel, Physikalisches Institut, April 1925.

(Eingegangen 1. Mai 1925.)

642

# On the Excitation of Polarised Light by Electron Impact.

By H. W. B. SKINNER, Ph.D., M.A., Exhibition of 1851, Senior Student.

(Communicated by Sir Ernest Rutherford, Pres.R.S.—Received July 1, 1926.)

(PLATE 21.)

### § 1. *Introduction.*

In the course of some experiments* on polarisation effects shown by mercury lines, emitted from a low-pressure electron-maintained arc, it was found that the yellow mercury lines λ 5770, 5791 are weakly polarised even in the absence of a magnetic field, the direction of the maximum electric vector being parallel to the direction of the discharge. From the general characteristics of the effect, it appeared likely that the polarisation is due to the partly directed character of the electron tracks in the arc ; and, in this way, one was led to the view that an electron is capable of exciting an atom to the emission of polarised light. The present paper describes an attempt to investigate this point more thoroughly.

While the work was in progress, two papers have appeared in which a search for signs of polarisation in the light excited by electron impact has been made. The first is by Kossel and Gerthsen[†] who examined the case of the D lines of Sodium with a negative result. This was confirmed by Ellett, Foote and Mohler,[‡] who also investigated the case of the mercury line λ 2537 with a positive result, which will be described subsequently. These experiments, however, only dealt with a few individual spectral lines. In the present work data have been obtained for all the more prominent lines of the mercury spectrum.[§]

### § 2. *The Source of Light.*

It is clear that the chief necessity for the experiments is an electron tube which is capable of producing an intense, perfectly directed stream of electrons.

After some trials, a tube was designed which has proved to work well. The design is shown in fig. 1 (approximately a quarter of actual size) and fig. 2 (half of actual size).

It is an electron tube which runs on mercury vapour at a pressure of one-

---

* An account of these is in the press.

† ' Ann. d. Physik,' vol. 77, p. 273 (1925).

‡ ' Phys. Rev.,' vol. 27, p. 31 (1926).

§ Preliminary results were published in ' Nature,' vol. 117, p. 418 (1926).

*Excitation of Polarised Light by Electron Impact.*    651

The lines whose series are given in brackets are considerably weaker and inseparable companions to the lines above them.

The calculated values of the fifth column will receive explanation later and are connected with the values (column 1) of the change of the quantum number $j$ (which represents the angular momentum of the atom) calculated for an absorption switch.

### § 5. *Discussion of Results.*

It can scarcely be doubted that the polarisation effects in question have their origin in the fact that the electron stream is unidirectional.   We must therefore assume that in the case of many spectral lines, at any rate, the impact of an electron on an atom has the result of exciting the atom to produce plane-polarised light.   The effect is analogous in some respects to the known result that the X-rays emitted from an ordinary target are polarised.   But the analogy seems rather superficial, because it has been shown* that this polarisation is only found in the case of the continuous X-ray spectrum, and not for the lines.

Assuming that an atom is in a position to emit plane-polarised light, one would expect that the effects of an external field on it will be independent of the mechanism by which it has reached this excited state.   This fact suggests an obvious analogy between the results we have described and the results found in experiments of the type initiated by Wood and Ellett.†   These experiments deal with observations of the polarisation of resonance radiation and the effects of a magnetic field on it.   In their case, the excitation of the atoms is by the absorption of radiation, and in our case by electron impact, but the effects of an external field on the polarisation may be expected to be the same.

This actually proves to be the case.   The most completely studied case of resonance is that of the Mercury Line λ 2537, for which observations have been made most completely by Hanle‡ and von Keussler,§ who have found the depolarisation and rotation effects exactly as we have described them.

The theory of the effect which Hanle has given is the following:—Suppose there is a linear vibrator in the atom which is emitting the light as on the classical theory.   If a magnetic field is applied in a direction at right angles to the direction of vibration, a Larmor precession will be superposed on the motion. Since, further, we must assume that the vibration is damped, the motion will be of the type shown in fig. 8.   It can be seen from this that the plane of polarisa-

* Bassler, ' Ann. d. Physik,' vol. 28, p. 808 (1909).
† ' Phys. Rev.,' vol. 24, p. 243 (1924).
‡ ' Z. f. Physik,' vol. 37, p. 93 (1925).
§ ' Phys. Zeit.,' vol. 27, p. 331 (1926).

2 x 2

tion of the total light emitted can be considered as rotated to a direction indicated by the arrow and the total light partially depolarised.  If a Babinet

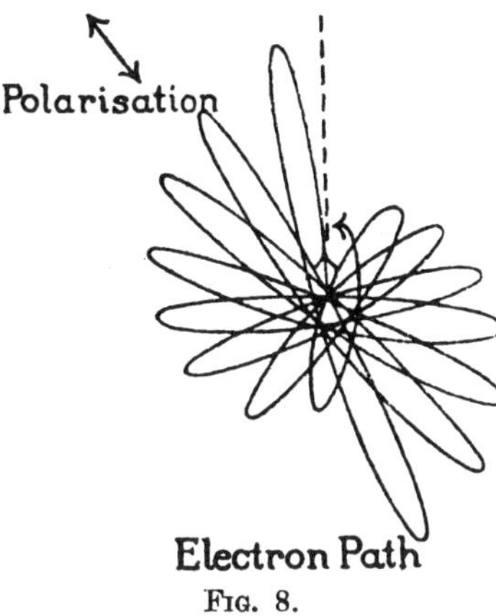



Fig. 8.

compensator is used for detection, it is easy to calculate that the apparent
rotation will be given by the relation

$$\tan 2\delta = \frac{\omega}{2\pi\beta},$$

where $\omega$ is the magnetic rotation frequency corresponding to the field strength
H, and $\beta$ is the damping constant of the oscillation, defined by the statement
that the amplitude of the vibration is given by

$$a = a_0 e^{-\beta t}.$$

The apparent depolarisation is also given by the formula

$$\left(\frac{\omega}{2\pi\beta}\right)^2 = \frac{\Pi_0^2}{\Pi^2} - 1,$$

where $\Pi_0$ is the percentage polarisation in a zero field and $\Pi$ the percentage
in the field H.

In the case of the singlet $2^1P_1$—$3^1D_2$ combination, $\lambda 5791$, there is no ambiguity
as to what we shall take for $\omega$.  The Zeeman splitting is normal and so we naturally take the Larmor frequency.  In the case of the line $\lambda\,5770$ ($2^1P_1$—$3^3\,D_2$),
the Zeeman effect is anomalous, since the $^3D_2$ term has a Landé splitting
factor ($g$) of 7/6, that for the $^1P_1$ term being of course 1.  But since the value
7/6 is not very different from 1, it would seem that in this case also we shall
not be much in error if we again take the Larmor frequency for $\omega$.

We are now in a position to interpret the curves of the rotation and

*Excitation of Polarised Light by Electron Impact.*    653

depolarisation which have been given in fig. 7.  The curves should, according to the theory, be straight lines, which within the limits of error they are.  The slope of these lines gives in each case the value of $\omega/2\pi\beta$ for a field of 1 gauss.  The values obtained also agree well.  From them one can calculate $\beta$, but it is of more significance to calculate $\tau = 1/\beta$.  $\tau$, on the classical theory, is the time taken for the vibrations to die down to $1/e$th of their initial value.  It corresponds on the quantum theory to the mean life of the atom in the excited state.

We obtain, in fact, the following values for the mean life of the mercury atom in the $3^1D_2$ and $3^3D_2$ states before the switch to the $2^1P_1$ state.

From rotation        $= 2\cdot85 \pm 0\cdot15 \ 10^{-8}$ sec.

From depolarisation $= 2\cdot88 \pm 0\cdot15 \ 10^{-8}$ sec.

We may therefore take $2\cdot9 \ 10^{-8}$ sec. as the value of $\tau$ for these states.

The value found by von Keussler for the $2^3P_1$ state of mercury is $1\cdot12 \ 10^{-7}$ sec.,* and that of the mean of the $2^2P_1$ and $2^2P_2$ states of sodium is considerably less.  These results are in general agreement with the results of Wien on the mean time of life of an atom in the excited state.

We may here also note that if the $\tau$ for the $3^1D_2$ and $3^3D_2$ states are not the same, we should not expect the curves of fig. 7 to be straight lines.  The fact that they are approximately so shows that the $\tau$'s must be nearly, at any rate, the same as has already been mentioned.

The $\tau$'s for the other lines have not so far been determined, but the method described evidently admits of a wide application.

These effects of a magnetic field on the polarisation have been interpreted on Hanle's classical theory.  We are here dealing with such weak magnetic fields that the atoms are not space-quantised.  For if they were, it has been pointed out by Hanle that the rotation effect could not exist.  For this is essentially an interference effect between the light of different frequencies emitted by an atom in a magnetic field, and he concludes that in the case of these weak fields all the Zeeman components of a line must be emitted by the same atom.  We have here, then, a case to which the rigid quantum theory seems quite inapplicable.

This remark is of some importance when we attempt to frame a quantum theory of the mechanism by which an electron can excite an atom to emit polarised light.  For it shows that we cannot hope to do so while the atom remains in a degenerate state.  On this ground the theory of Ellett, Foote and Mohler would seem to be open to criticism.  They predict theoretically that

---

* Their value is further confirmed as regards order of magnitude by the experiments of Ellett, Foote and Mohler which have been mentioned.

654    H. W. B. Skinner.

polarisation should only occur for the transitions in which $j$, the quantum number representing the angular momentum of the atom, changes by one unit ; and this proves to be by no means the case.

The simplest method of treatment would seem to be to suppose a magnetic field parallel to the electron beam, of sufficient strength to ensure orientation of the atoms, but not strong enough to split the lines appreciably. We have seen that experimentally such a field produces no effect on the polarisation, and Heisenberg* has shown, in the case of resonance radiation, from arguments based on a generalised correspondence principle, that this may be expected.

We now come to the fundamental basis of the theory. To simplify matters we shall at first suppose that the impacting electron has just that velocity necessary for excitation of the line in question, and no more. This means that after the collision the electron is reduced to rest. One may reasonably assume that the angular momentum which is given by the electron to the atom must be in a direction at right angles to the initial direction of motion of the electron.†

Since the angular momentum transferred during the collision is at right angles to the direction of the electron stream, it is at right angles to the magnetic field which we are supposing to be applied in the direction of the stream.

We represent by $j$ the quantum number expressing the total angular momentum of the atom, and by $m$ its component in the direction of the magnetic field H. It follows that in the transitions induced by the impact of an electron moving in the direction of H that the change in $m$, namely,

$$\Delta m = 0. \tag{1}$$

This equation therefore expresses the condition that in the excitation of light by electron impact the angular momentum communicated must be at right angles to the direction of motion of the impacting electron previous to collision.

Now the magnetic field has the property of splitting the levels of the atom. The way in which the splitting takes place is well known from the analysis of the Zeeman effect. Actually, the normal unexcited state of the mercury atom, the $1^1S_0$ state, is not split, the magnetic quantum number of this state being given by $m = 0$. In general the states will, however, be split, and $m$ will have for them, in the case of the mercury atom, the values $0, \pm 1, \pm 2 \ldots \pm (j - \tfrac{1}{2})$.

We may represent these states diagrammatically, for the sake of simplicity

---

* ' Z. f. Physik,' vol. **31**, p. 617 (1925).

† This point also forms the basis of the theory of Ellett, Foote and Mohler.

*Excitation of Polarised Light by Electron Impact.*    655

only three terms being taken—the $1^1S_0$ term, the $2^1P_1$ term and the $3^1D_2$ term (fig. 9).

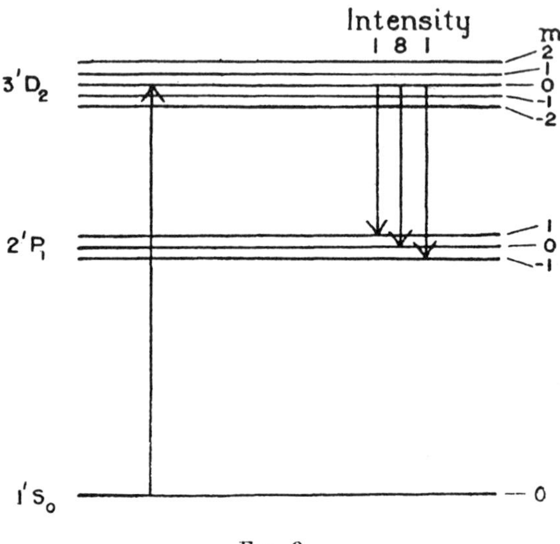


Fig. 9.

Now the atom is initially in the $1^1S_0$ state and has $m = 0$. Using equation (1) we see that by the impact of an electron it must be transferred (for example) into the $3^1D_2$ term ($m = 0$). This is represented by the upward arrow.

The atom is now free to emit light, and, according to the rules of the Zeeman effect, it is known that the following transitions can occur, for instance to the $2^1P_1$ state.

(*a*) $m = 0$ corresponding to light polarised parallel to H ($\parallel$).

(*b*) $m = \pm 1$ corresponding to light polarised perpendicular to H ($\perp$). Since there are no atoms in the $^1D_2$ state corresponding to $m = \pm 1, \pm 2$, the lines which will be emitted are represented by the downward arrows.

The intensities of these Zeeman components are calculable on the basis of the " Summation rules " of Ornstein, Burger and Dorgelo. A useful account is given by Kronig* of the application to the case of the Zeeman components.

We have in our case only three Zeeman components emitted, one polarised $\parallel$ and the other two polarised $\perp$. The net polarisation of the emitted light is thus calculable. In the case of all transitions involving $\Delta j = \pm 1$ (as in the example given) the intensity of the component polarised $\parallel$ is greater than the sum of the intensities of the components polarised $\perp$. The net polarisation is thus

* 'Z. f. Physik,' vol. 31, p. 885 (1925).

a ‖ polarisation, the value of which varies with the precise $j$ and $k$ values involved in the transition.

However, in the case of the transitions $\Delta j = 0$, the state of affairs is quite different. The essential distinction is that the analysis of the Zeeman effect has shown that the transitions

$$\Delta j = 0 \qquad \Delta m = 0$$

do not occur, and this also follows from the theory of Kronig.

If the atom is excited by electron impact, it will, as before, be in an $m = 0$ state. In the emission of a line corresponding to $\Delta j = 0$, therefore, only the transitions

$$\Delta m = \pm 1$$

can take place, and the light will be polarised perpendicular to the direction of motion of the electron.

These considerations are based on the assumption that the electron has just enough energy to excite the atom, and no more. However, observations at such velocities are not possible on account of the faintness of the light, and hence the experimental results apply to electrons of higher speed. If the electron has a finite velocity after the impact, it is clear that it will be possible for it to transfer angular momentum in a direction parallel to that of its initial motion to the atom.

Unfortunately, it is not possible to calculate precisely the ratio of the amount of angular momentum perpendicular to the initial direction of motion of the electron transferred in a number of collisions to the amount transferred in the perpendicular direction. The reason for this is, that in the general case under consideration, the relations of energy and momentum do not suffice. We require a more detailed knowledge of the forces acting on the electron during impact, and this at present is lacking.

However, it would seem reasonable to suppose that the probablity of a collision in which angular momentum in a direction parallel to the initial direction of motion of the electron is transferred will increase with the emerging velocity of the electron. The magnitude of the polarisation effects will, therefore, be expected to decrease with the increasing initial velocity of the electron. In the case of $\lambda$ 5770, 5791 a steady decrease with increasing velocity was observed (see fig. 5).

In the case of $\lambda$ 5770, 5791 we have already sufficient material for a rough numerical test between theory and experiments. The polarisation of these lines is parallel to the stream. We require the degree of polarisation when the

## *Excitation of Polarised Light by Electron Impact.*     657

electron speed corresponds to the excitation voltage of the line, and this can only be obtained by extrapolation of the results obtained with greater electron speeds. In fact, an extrapolation back to 9 volts by means of an exponential formula (see fig 5) gives a value of about 65 per cent. of polarisation. This corresponds very nearly to the calculated value of 60 per cent.

In the cases of the other lines, we have at the present moment to fall back on a qualitative test. The values of the polarisation calculated for electrons of speed corresponding to the excitation potential have been inserted in Table I and may be compared with the experimental results for an electron speed of 20 volts.

It is seen that, in spite of some exceptional cases to be mentioned later, there is a general agreement both as regards the direction of the plane of polarisation and the relative magnitudes of the polarisation.

We may here also notice that the case of the D lines of sodium also agrees with the theory. This case has been investigated by Kossel and Gerthsen,* who found no polarisation effect, and their result was confirmed by Ellett, Foote and Mohler.* The normal $1^2S_1$ state of sodium splits in a magnetic field into two states with $m = \pm\frac{1}{2}$, and one sees at once that on this account the effects will be smaller than in the case of mercury with its unsplit normal state. Actually, calculation shows 0 per cent. and 60 per cent. with a mean (bearing in mind the relative intensities of the D lines) of 20 per cent.; and remembering the rapid decrease of the percentage polarisation with increasing speed of electrons, an effect of this magnitude might well escape detection.

It will be seen, therefore, that the theory goes a considerable way towards explaining the facts. But returning now to the mercury lines, it appears from Table I that there are four lines for which the observed polarisation seems definitely not to agree with the calculated. These are :—

### Table II.

| $\lambda$ | | | Series. |
|---|---|---|---|
| 2537 | .. | .. | $1^1S_0 - 2^3P_1$ |
| 4047 | .. | .. | $2^3P_0 - 2^3S_1$ |
| 4358 | .. | .. | $2^3P_1 - 2^3S_1$ |
| 2967 | .. | .. | $2^3P_0 - 3^3D_1$ |

Of these the most flagrant case is $\lambda$ 2537, which shows a $\perp$ polarisation of 30 per cent. at 7 volts in place of the theoretical value of 100 per cent. $\parallel$ at

* *Loc. cit.*

658                    H. W. B. Skinner.

the excitation voltage.   In the other cases the plane of polarisation is correctly given, but the observed magnitude is smaller than expected.

It will be noticed that all these lines involve impact switches from the normal $1^1S_0$ state of mercury for which $\Delta j = 1$, while for the lines for which the theory seems to hold, in the impact transition, $\Delta j = 0$, 2 or 3.*

Let us take the case of $\lambda\,2537$ as the clearest example for which the theory fails. The observed $\perp$ polarisation, if our picture is correct at all, shows that the probability of the excitation of the states $2^3P_1$ ($m = \pm 1$) must be greater than the probability of the excitation of the state $2^3P_1$ ($m = 0$).

It is not definitely proved yet whether or not this $\perp$ polarisation occurs when the speed of the electrons corresponds exactly to the excitation voltage.   All that is proved is that there is a $\perp$ polarisation at an electron speed of 7 volts. This fact may provide a loophole for the applicability of the theory ;   for, as we saw, if the electron goes away from the excited atom with a finite velocity, the impact switches $\Delta m = \pm 1$ are possible, and we have no means of determining the relative probabilities of these switches and of the impact switch $\Delta m = 0$.

But although further work on this point is needed, it would seem fair to state that the experiments make it appear unlikely that, for an electron speed of $4 \cdot 9$ volts (the excitation potential), the line $\lambda\,2537$ will be found polarised 100 per cent. $\parallel$.   If this is not the case, there would be a definite disagreement with the theory suggested above.  We may perhaps put forward some tentative speculations on this point.

The difficulty might be avoided by making the assumption that the impacting electron may possess angular momentum of spin.  This is the model proposed by Uhlenbeck and Goudsmid† which has proved successful in resolving many of the difficulties connected with the multiplicity of spectra and the Zeeman effect.   On this hypothesis, when an electron in an atom is orientated by the magnetic field within the atom, it adds a half of a quantum of angular momentum in the direction of the field, or in the opposite direction.   We may perhaps extend this case by supposing that a free electron when orientated by an external field has half a quantum of angular momentum in the direction of the field or in the opposite direction.

In the present problem, if we suppose, as before, that there is a magnetic field, in the direction of the electron stream, it is clear that, with the hypothesis of the spinning electron, we have the possibility that an electron (by changing

---

* Except in the case of the mean line $\lambda\,5461$ ($2^3P_2 - 2^3S_1$), for which the calculated polarisation is too small for any decision to be possible.

† 'Nature,' vol. 117, p. 264 (1926).

### *Excitation of Polarised Light by Electron Impact.*     659

its direction of orientation) may on impact transfer to the atom *one* quantum of angular momentum in the direction of the field.   (We call this process " Process B.")

Now it is just in the case of the lines which involve impact switches $\Delta j = \pm 1$ (*i.e.* the atom, on impact, gains or loses one quantum of angular momentum) that the simple theory suggested above appears to break down, and this fact is suggestive.

Of course, we still have the possibility of the transfer of any number of quanta of angular momentum in a direction *perpendicular* to the field, as described previously (" Process A ").   We may suppose, therefore, that a line like $\lambda$ 2537 which involves an impact switch $\Delta j = \pm 1$ may be excited either by Process A or by Process B, and the degree of polarisation to be expected will be indeterminate.

In the case of the lines involving impact switches $\Delta j = 2, 3$, we have seen that, experimentally, the part played in excitation by Process B, if any, must be small.   Since, by Process B, exactly one quantum of angular momentum must be transferred, it is evident that Process A must be concerned in every excitation of this type.   If both processes were concerned in an excitation, the angular momentum transferred by process A would not be quantised.*   If, therefore, we were to make the hypothesis that the angular momenta transferred by Process A and Process B are quantised individually, the simple theory involving Process A only, which was described above, might be retained for these lines.

Experiments on the excitation of polarised light by electron impact are still in progress, and it is hoped that before long it will be possible to subject these speculations to a more rigid test.


### *Summary.*

1. An electron tube producing an intense unidirectional stream of electrons of slow speed is used for the excitation of the mercury spectrum, and polarisation measurements are made on the light emitted from the tube in a direction at right angles to the direction of the stream.   It is found that with an electron speed corresponding to 20 volts many of the mercury lines are partially plane-polarised, most with direction of the maximum electric vector parallel to the stream, but some in the perpendicular direction.   The experiments of


---

* For example, in the case of an impact transition in which $\Delta j = 2$, if one quantum of angular momentum parallel to the field is transferred by Process B, it would be necessary for $\sqrt{3}$ quanta perpendicular to the field to be transferred by Process A.

660    *Excitation of Polarised Light by Electron Impact.*

Ellett, Foote and Mohler, who found that $\lambda$ 2537 was polarised in the latter direction, are confirmed.   In the cases of $\lambda$ 5770, 5791 the polarisation parallel to the stream decreases rapidly as the speed of the electrons increases.

2. The application of a magnetic field in the direction of the stream has no influence on the polarisation ;   but a weak field of the order of 2 gauss causes two effects to appear which increase with the field strength, namely, (*a*) a depolarisation and (*b*) a rotation of the plane of polarisation.   These are investigated in the case of the lines $\lambda$ 5770, 5791.

3. The magnetic effects are interpreted satisfactorily on a theory of Hanle put forward in connection with work on resonance radiation.   Each effect (*a*) and (*b*) leads to determination of $\tau$, the mean life of the atom in the excited state, and these values are concordant.

4. The polarisation effects are due to the direct excitation of polarised light by electron impact.   An attempt is made to picture this process.   The theory is based on the fact that, in excitation, an electron may, if its speed corresponds nearly to the excitation voltage, be expected to transfer angular momentum to the atom in a direction perpendicular to the initial direction of motion of the electron.   The facts in the case of most of the lines agree well with the theoretical expectation, but in a few cases the theory seems to be inadequate.

In conclusion, the author wishes to express his thanks to Prof. Sir Ernest Rutherford, O.M., P.R.S., for his constant interest and helpful criticism, and to Mr. R. H. Fowler, F.R.S., for valuable discussion.

**224**

# On the Excitation of Polarised Light by Electron Impact. II.—Mercury.

By H. W. B. Skinner and E. T. S. Appleyard.

(Communicated by Sir Ernest Rutherford, P.R.S.—Received September 26, 1927.
—Revised October 26, 1927.)

## § 1. *Introduction.*

In a number of recent papers* descriptions have been given of how the light from atoms excited by a directed stream of electrons is polarised, in the absence of any external field of force. In particular we may refer to the paper by one of the present writers (Paper I), where some of the characteristics of the effect in mercury were discussed. It was found that the different lines of the spectrum were differently polarised, both in magnitude and direction. The behaviour of the polarisation when a magnetic field is applied to the mercury atoms was found to be identical with the effect produced by a similar field on the polarisation when the light is excited by the absorption of plane polarised light, the electric vector of which is parallel to the direction of the stream of electrons. In this way, the polarisation of the light excited by electron impact was shown to be closely related to the polarisation of resonance radiation, which had been investigated by Wood and Ellett† and others. In the paper referred to, the measurement of the polarisation of the spectral lines was, except in an individual instance, only qualitative. Since its appearance two papers have appeared in which polarisation measurements of the mercury lines excited by electron impact have appeared. Eldridge and Olsen‡ gave qualitative results, and these confirmed those which we had given. Quarder§ gives quantitative measurements which are also for the most part in general accord. The aim of the present paper is to describe quantitative results, and, as will be seen, these are not in all cases in detailed agreement with those given by Quarder. The second aim is to determine the variation of the polarisation with the velocity of the electron stream, a point which has not been previously attempted.

In Paper I it was shown that the magnitude of the polarisation is intimately

---

* Kossel and Gerthsen, ' Ann. d. Physik,' vol. 77, p. 273 (1925); Ellett, Foote and Mohler, ' Phys. Rev.,' vol. 27, p. 31 (1926); Skinner, ' Roy. Soc. Proc.,' A, vol. 112, p. 642 (1926).

† ' Phys. Rev.,' vol. 24, p. 243 (1924); Hanle, ' Z. f. Physik,' vol. 30, p. 93 (1924).

‡ ' Phys. Rev.,' vol. 28, p. 1150 (1926).

§ ' Z. f. Physik,' vol. 41, p. 674 (1927).

*Excitation of Polarised Light by Electron Impact.*    **225**

connected with the dynamics of the collision process.   By the electron impact, the atom is raised from its normal state into an upper quantum state, and by its return into the normal state, or into another state, light is emitted.   The magnitude of the polarisation depends on both of these processes, but the second process, the actual emission of the light, is well understood.   Polarisation measurements therefore give a powerful means of investigating the collision of an electron with an atom in the case when excitation of the atom takes place, and it is in this that their main interest lies.

### § 2. *The Electron Tube.*

The tube used to provide a powerful, directed stream of electrons was the same as that described in Paper I, to which we refer for details.   A stream of electrons from an oxide-coated filament is defined in the usual way by slits and passes vertically into a field-free box with a hole cut in the side for observation.   A light trap is provided so that no polarisation errors can come from the reflexion of the light inside the tube.   The only difference from the previous tube is that the window in the present experiments was of clear fused quartz, fixed on with hard enamel.   The ground-glass joint by which the filament could be removed was used dry, being sealed round the outside with wax. With these alterations it was possible to bake out the tube to some extent. Since the tube runs at a fairly high current, the charging up of surface films on the electrodes was found to be troublesome, as it prevented precise definition of the speed of the electron stream.   It was found, however, that these films are easily and rapidly removed completely by running the tube at a potential of about 2,000 volts.   The vanishing of the films is easily tested.   In mercury the green line starts to be excited at 7·7 volts, and the yellow at 8·8.   The visual effect of these lines together is a bluish white, while if the yellow lines are absent the colour is green.   It was possible to obtain streams of pure green colour passing throughout the length of the observation chamber without apparent diminution of intensity, thus showing that the velocity of the electrons was sufficiently constant along the length of the stream.

The definition of the stream can be controlled by choice of the potentials applied to the various electrodes of the tube (see Paper I).   It was possible to obtain a stream which traversed the field-free observation box practically without spreading.

For the exact determination of the velocity of the beam, it is necessary to take into account the existence of contact potentials.   The magnitude of these was found in several ways, *e.g.*, (1) determination of the ionisation potential

*Excitation of Polarised Light by Electron Impact.*    237

of polarisation, and (*b*) the fainter, less accurately measurable, for which it should be correct only to within about 8 per cent. The values for lines of the second class are indicated by brackets, *e.g.* (19). The column $V_0$ gives the corresponding critical voltages for the excitation of the lines. The polarisations $\Pi_{16}$ measured by Quarder (*loc. cit.*) for an exciting voltage of 16 volts are added for comparison. The two sets of values agree in a general way, but there are many cases of detailed disagreement. A possible cause of the discrepancy has been suggested. The remaining column P is theoretical and will be discussed later.

### § 5. *Theoretical Discussion.*

In Paper I, it was definitely proved that the polarisation effects we are considering are determined simply by the directed character of the electron stream. A further important property, established experimentally there, is that as far as the effects of an applied magnetic field are concerned,* the behaviour of the polarisation of the light excited by electron impact is identical with the behaviour of the polarisation of the light excited by resonance when the electric vector of the absorbed radiation is taken parallel to the direction of the electron stream. In particular, in both cases, the polarisation is unaffected by a field parallel to the electron stream (or electric vector). Now it was known that in the case of resonance, correct values for the amount of the polarisation were only obtained on the classical quantum theory in the case when there is a magnetic field applied. For the case when there is no field, one had merely to assume the effect independent of the presence of a field applied parallel to the electric vector.

It was therefore clear that to obtain a corresponding theory of the excitation of polarised light by electron impact, the case when the magnetic field is applied parallel to the electron stream must be treated, and that the independence of the effect on a field applied parallel to the electron stream must be assumed. This assumption may be called the assumption of spectroscopic stability.

The theory of Paper I is based on the principle of the conservation of angular momentum for the collision. Consider first the limiting case when the velocity of the electron after impact is zero—*i.e.*, the electron has just the minimum energy necessary for excitation. Following Ellett, Foote and Mohler (*loc. cit.*), we make the assumption that, in this case, the vector change of angular momentum of the atom is in a direction at right angles to the initial direction

---

* A slight mistake in Part I of this paper, which deals with the magnetic effects, may here be corrected. On pp. 652, 653, $\omega/2\pi\beta$ should be $2\pi\omega/\beta$. The values given for $\tau$ are correct.

of motion of the electron ; or, in other words, the electron cannot carry away from the collision any angular momentum. This assumption may be extended to the general case by supposing that if $v_0$ is the initial velocity of the electron and $\Delta v$ the change in velocity it suffers during the collision, then *the angular momentum change of the atom is at right angles to* $\Delta v$ (assumption A).

This assumption implies a restriction of the type of orbit which the electron can describe during the excitation process, as is easily seen by considering particular examples. But in the case of large electron velocities, when the energy of excitation is not sufficient to diminish appreciably the energy of the electron, assumption A would appear to follow necessarily from symmetry considerations. In the limiting case of small final velocities, it follows from the classical mechanics.*

Applied in the limiting case when the electron has only the minimum energy necessary for excitation, assumption A leads immediately to the following relation for the change in the magnetic quantum number $m$ of the atom, caused by the impact :—

$$\Delta m = 0. \tag{1}$$

Using this, one obtains the polarisation directly from a knowledge of the intensities of the Zeeman components of the line in question. For a detailed account of the theory, we refer to Paper I.

It is important to note that *the relation* (1) *also follows, for the " ideal " case when the velocity of the electron after collision is parallel to the velocity before the collision.* This result is independent of the magnitude of $v_0$, and even of the general assumption A. For it seems impossible to doubt, on any theory, that, in this particular case, the angular momentum transferred must be at right angles to $v_0$. We thus see that there is a *correlation between the magnitude of the observed polarisation and the degree to which the electrons are scattered when the excitation takes place.*

When the energy of the exciting electron is greater than the critical energy, a scattering of the electrons must be expected to take place. Thus, $\Delta v$ is not necessarily parallel to $v_0$, and the relation (1) is not necessarily true. One anticipated that the most probable direction for $\Delta v$ would change from a direction parallel to $v_0$ for small velocities, to one perpendicular to $v_0$ for large velocities. Thus one was led to expect that the percentage polarisation of a

---

* We are indebted to Mr. P. M. S. Blackett for pointing out that assumption A is only true classically if the law of force between the electron and the atom varies more rapidly than the inverse third power of the distance between them. But since the atom is neutral, we must expect this condition to be satisfied.

*Excitation of Polarised Light by Electron Impact.*     239

given spectrum line would decrease progressively with increasing electron velocities. A quantitative prediction was not possible, since we do not know how the electrons of a given velocity are scattered on collision.

In a recent paper Oppenheimer[*] has considered the polarisation effects from the standpoint of the new quantum theory. He has concluded :—

(a) That the polarisation of the light excited by a single collision is independent of a magnetic field parallel to $\Delta v$. When the polarisation is large the most probable direction for $\Delta v$ is nearly parallel to $v_0$, and thus the independence of the observed effect on a field parallel to $v_0$ is interpreted. In this way Oppenheimer has established the assumption of spectroscopic stability. The effect of a field in a direction inclined to the electron stream is also correctly calculated.

(b) That the method of calculation by means of the angular momentum principle, and in particular assumption A, are correct.

Thus Oppenheimer has shown that the theory of Paper I is right within the limits of his method of calculation.

Actually the experiments have shown that the polarisation *decreases to zero* from a maximum value as the velocity of the electrons is diminished to the critical value. Thus the theory fails definitely in the one case where it can be applied quantitatively. But Oppenheimer's method of approximation leaves open the possibility of the failure of his conclusions, *in the case of very small final electron velocities.* It is probable, therefore, that this effect may receive an interpretation on the quantum mechanics,[†] though it is inconsistent with the classical.

For excitation by electrons whose velocity is not too small, we are undoubtedly justified in applying the theory of Paper I. The large value of the polarisation at the maximum therefore shows definitely that, at the corresponding velocity, a large fraction of the electrons are scattered nearly straight forward. With this conclusion we may compare the observation of Dymond,[‡] who showed directly that in helium, with moderate electron velocities, a large fraction is scattered nearly straight forward when excitation takes place.

For excitation by electrons of small velocity, we are on uncertain ground, as the experiments have shown that the theory of Paper I is untenable in this region. But it seems reasonable to assume, even in this case, a correlation between the value of the polarisation and the degree of scattering of the

[*] ' Z. f. Physik,' vol. 43, p. 27 (1927).
[†] Dr. J. R. Oppenheimer has very kindly informed us of this fact.
[‡] ' Phys. Rev.,' vol. 29, p. 433 (1927).

240          H. W. B. Skinner and E. T. S. Appleyard.

electrons. If this is granted, we may conclude that, as the velocity of the electrons is diminished from a value corresponding to the maximum of the polarisation curves, the electrons are scattered more and more violently, until in the limiting case, corresponding to the zero value of the polarisation, the angular distribution of the electrons must be more or less uniform.

We are thus led to postulate a probable anomalous scattering of the electrons of velocity slightly in excess of the minimum excitation velocity.* It is interesting to note that the final velocity of these electrons is of the same order of magnitude as the velocity of the electrons for which the Ramsauer effect is observed. Further, that the method of approximation used by Oppenheimer does not suffice for a theory of the Ramsauer effect.

We now come to the consideration of the relative magnitudes of the polarisation for the various lines, and we shall be dealing with the case where the velocity of the exciting electron is sufficiently great for the theory of Paper I to be valid. If we take the " *ideal* " case when the electron is not deflected during the collision, we have stated that the polarisation may be calculated from equation 1. In this way, we obtain a standard value P for the polarisation. The actually observed polarisation will be less than P owing to the scattering of the electrons, and we cannot calculate how much less it will be. On the other hand, it seems reasonable to suppose that the scattering will not be very different in the case of the various spectrum lines. Thus we may compare with the *relative* values of P the *relative* values of the polarisation observed, for example, at the maxima of the polarisation curves.

In Paper I, a qualitative comparison of this sort (but using the values of the polarisation corresponding to a definite electron velocity instead of the maximum values) was made, and it was found that, up to a point, there was good agreement. The results are amplified by those of the present paper. In Table I the values of P are inserted, and it is seen that, in most cases, and especially for the lines for which the excited state is a singlet state, the maximum polarisation $\Pi_{max.}$ is roughly expressed by the relation

$$\Pi_{max.} = \tfrac{1}{2}\,P.$$

In particular we may note the absence of polarisation in the cases of the lines $\lambda\,4078$ $(2\,^3P_1 - 2\,^1S_0)$ and $\lambda\,4108$ $(2\,^1P_1 - 4\,^1S_0)$. Further, if we extrapolate the curves back to the excitation points in the manner of fig. 13, we obtain, in most cases, values not far removed from P.

But as was noted in Paper I, there are several flagrant exceptions to this rule.

* An experiment for the direct verification of this effect will be undertaken shortly.

*Excitation of Polarised Light by Electron Impact.*    241

The lines $\lambda$ 2537 (1 $^1S_0$ — 2 $^3P_1$), $\lambda$ 4047 (2 $^3P_0$ — 2 $^3S_1$), and $\lambda$ 4358 (2 $^3P_1$ — 2 $^3S_1$) and a higher member of the same series are seen to violate it completely·

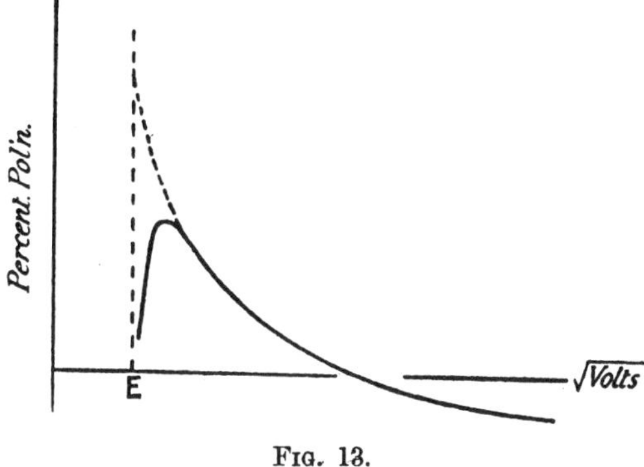


Fig. 13.

The first of these lines is polarised in the wrong direction, as was first observed by Ellett, Foote, and Mohler (*loc. cit.*).   The others are polarised in the right direction but much more weakly than would be expected.   The line $\lambda$ 2967 (2 $^3P_0$ — 3 $^3D_1$) and the next member of its series and the line $\lambda$ 2654 (2 $^3P_1$ — 4 $^3D_1$) are polarised distinctly more weakly than would be expected, and it seems probable that the lines of the series beginning with $\lambda$ 5461 (2 $^3P_2$ — 3 $^1S_1$) can also be included among the exceptional lines in spite of the low value of P.

In Paper I, it was suggested that the reason for this breakdown of the classical theory of the effect is that we have neglected to take into account the *spin* of the exciting electron.   If the direction of the spin axis of the electron is not the same before and after the collision, angular momentum must have been transferred to the atom by a process not considered in the theory. Thus the assumption allows the modification of equation 1 and so provides a possibility of the interpretation of the anomalous results.   Oppenheimer's calculations have strongly supported this point of view.   He has pointed out that the anomalous lines in question all correspond to *intercombination switches* from the normal singlet state of mercury on excitation (*i.e.*, the lines are those which correspond to the excitation of the atom into a triplet state).   He has suggested that the spin comes in as a "*resonance effect*" in Heisenberg's* sense.   This leads to the assumption that the orbital angular momentum of the exciting electron must be *separately* coupled with the orbital angular momentum of the atom, its spin angular momentum with the spin

* ' Z. f. Physik,' vol. 41, p. 239 (1927).

242        H. W. B. Skinner and E. T. S. Appleyard.

angular momentum of the atom.   In other words, in the " ideal " calculation, the change in the orbital angular momentum $\Delta l$ of the atom on excitation (instead of the vector $\Delta j$ in the calculation when spin is neglected) must be at right angles to the electron stream.   Since the angle between $l$ and $j$ is known, the direction of $j$ corresponding to the excited state is fixed.

In this way, it is possible to estimate the order of the correction which has to be applied to the values P for the intercombination lines.   And in fact the calculation shows that the correction will be greatest for $\lambda$ 2537, as is observed. The angle between $j$ and $l$ which determines the correction is only considerable in the case of the lines for which $j$ is equal to 1 in the excited state, and these in fact are the exceptional lines.   In the case of lines for which $j$ is 2 in the excited state the correction will be small ; but it is interesting to note that, in an especially favourable case, it is detectable.   The lines $\lambda$ 2652 (2 $^3$P$_1$ — 4 $^3$D$_2$) and $\lambda$ 2655 (2 $^3$P$_1$ — 4 $^1$D$_2$) are a close pair, for one of which the excited state is a triplet state and for the other a singlet state.   Any errors in measurement (but the purely photographic) must be the same for both these lines. Their polarisation curves are shown in fig. 14, and it is seen that $\lambda$ 2655 is

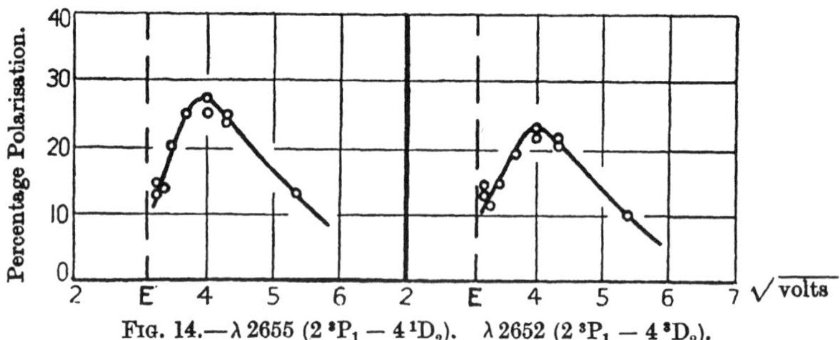



FIG. 14.—$\lambda$ 2655 (2 $^3$P$_1$ — 4 $^1$D$_2$).   $\lambda$ 2652 (2 $^3$P$_1$ — 4 $^3$D$_2$).

definitely about 4 per cent. more strongly polarised than $\lambda$ 2652.   Thus it seems that qualitatively the theory of the separate coupling of the " orbital " and " spin " angular momenta describes the facts very satisfactorily.   But it should be pointed out that there seems to be a difficulty in applying the theory to a non-degenerate system, and hence an exact calculation of the effect of the spin on the values P has not yet been made.   It would seem that in spite of this, the case for the necessity of the consideration of the spin of the exciting electron in the excitation process is established.

There remains for consideration the polarisation effects for very high velocities of excitation.   Quarder (*loc. cit.*), measuring the polarisation of the total light emitted from an electron tube, working with mercury, has shown that for small

*Excitation of Polarised Light by Electron Impact.*     243

velocities the light is polarised with the electric vector parallel to the stream (*i.e.*, positively). As the electron speed increases the polarisation sinks to zero and subsequently becomes more and more negative. No spectroscopic analysis was used. Our curves show the beginnings of this effect in the case of individual spectrum lines. In nearly all cases where there is a large positive polarisation for small electron speeds, for a large electron speed the polarisation is negative. If it is negative for small speeds, it is positive for large, and if the polarisation is zero for small speeds, it remains zero. An exception, however, occurs in the case of the lines $\lambda$ 3650 ($2\,{}^3P_2 - 3\,{}^3D_3$) and $\lambda$ 3021 ($2\,{}^3P_2 - 4\,{}^3D_3$) These lines show a positive polarisation for small speeds which decreases with the speed. For large velocities, it still remains positive.

The general lines of an explanation of this effect are clear and have been pointed out by Oppenheimer. For moderate velocities of the exciting electron, the most probable direction for $\Delta v$ is nearly parallel to the direction of the electron stream. For very large velocities, on the other hand, it is nearly perpendicular to the electron stream. Hence a reversal of the direction of the polarisation for very large velocities is to be anticipated. The exceptional effect in the case of $\lambda\lambda$ 3650, 3021 is not accounted for ; but it is possible that even in this case, the reversal takes place for higher electron velocities.

Oppenheimer has remarked in this connection that this effect would scarcely be anticipated, owing to the fact that one would expect most of the light emitted from atoms excited by fast electrons to come, not from direct excitation, but from a " cascade " effect. This view is supported by the work of Dymond (*loc. cit.*), who showed that the probabilities of the switches in helium, corresponding to the excitation of the stronger lines, are measurable for small electron velocities, but vanish for large velocities. If the " cascade " effect is large in mercury, one would not expect to find any appreciable polarisation for high velocities, since a " cascade " emission would appear to imply that the direction of the electron stream cannot appreciably affect the characteristics of the light emitted. Hence it seems that we must assume that, in mercury at any rate, a considerable part of the excitation is direct, even for high excitation velocities.

*Summary.*

The paper continues the study of the polarisation of the mercury lines excited by a directed stream of electrons, which was begun in a previous paper. Photometric determinations of the percentage polarisation of the individual lines are described, the velocity of the exciting electron stream being varied from 9 to 200 volts. The different lines show different polarisation characteristics,

R 2

244     *Excitation of Polarised Light by Electron Impact.*

some being unpolarised, some polarised with the electric vector in a direction parallel to the electron stream, and some polarised in the perpendicular direction. The polarised lines show a characteristic variation of the polarisation with the electron speed ; the most salient feature being a maximum of polarisation for an electron velocity corresponding to a few volts above the excitation point. A reversed polarisation for high electron velocities is found in most cases. It is shown that the various polarisation effects may be interpreted to give information about the dynamics of the collision process in which an atom is excited by electron impact.

It is hoped shortly to give an account of the excitation of polarised light in helium.

We had the advantage of discussing the interpretation of the results with Dr. J. R. Oppenheimer, and are much indebted to him for his contributions. In conclusion, we should like to thank Mr. R. H. Fowler, F.R.S., for valuable discussion, and Sir Ernest Rutherford, P.R.S., for providing the means for carrying out the work and for his constant interest. The work was performed during the tenure by one of us of a Senior 1851 Exhibition Studentship, and he is greatly indebted to the Commissioners for their assistance.

# Zur Quantenmechanik der Richtungsentartung.

Von **J. R. Oppenheimer** in Göttingen.

(Eingegangen am 8. März 1927.)

Es wird gezeigt, daß man in der Quantenmechanik die Störungen entarteter Systeme ohne „Hilfsfelder" untersuchen kann. Die spontane Ausstrahlung solcher gestörten Systeme wird nach der Diracschen Theorie berechnet. Es ergeben sich als Spezialfälle die Heisenbergschen Polarisationsregeln für das Resonanzleuchten, die Skinnerschen für das Stoßleuchten, die Hanleschen für den Einfluß äußerer Felder. Die Theorie wird auf verschiedene Probleme des Stoßleuchtens angewandt; die anomale Polarisation der Quecksilberresonanzlinie wird als Resonanzeffekt gedeutet.

Es hat sich als möglich erwiesen, die übliche Beschreibung eines Atoms durch stationäre Zustände, Übergänge, Einsteinkoeffizienten usw. in die Quantenmechanik zu übertragen. So kann man das Verhalten eines Atoms einer Störung gegenüber immer noch durch die Änderung der Energiewerte und die erzeugten Übergangswahrscheinlichkeiten angeben. Aber diese „Wahrscheinlichkeiten" unterscheiden sich auf charakteristische Weise von den klassischen, da sie nicht mehr unbedingt voneinander unabhängig sind. Dies kommt z. B. in den Störungen eines entarteten Systems vor: Die Störung erzeugt bestimmte Übergänge, und im Laufe der Zeit kehrt das Atom zum Normalzustand durch spontane Ausstrahlung zurück; aber die entsprechenden „Übergangswahrscheinlichkeiten" dieses zweiten Prozesses müssen als gegenseitig voneinander abhängig betrachtet werden. Wir können daher Abweichungen von der alten Quantentheorie erwarten, und wir werden zeigen, daß diese gerade die erforderlichen sind, um eine hinreichende Theorie der Polarisation des Resonanzleuchtens und des Stoßleuchtens zu erzielen.

Von diesem Gesichtspunkt aus betrachtet, unterscheiden sich die entarteten Systeme darin, daß die Eigenfunktionen nicht durch die Angabe der nicht entarteten Quantenzahlen $n_i$ allein bestimmt sind. Im allgemeinen werden wir zwar wissen, daß für ein gegebenes $n_i$ alle Werte der entarteten Quantenzahlen $\sigma$ gleichwahrscheinlich sind. Aber, wenn $\psi(n, \sigma)$ eine entsprechende Eigenfunktion ist, so ist dann

$$\sum_{\sigma'}^{N} a(\sigma\sigma')\, \psi(n, \sigma') \tag{1}$$

28                                J. R. Oppenheimer,

auch eine Eigenfunktion, wo die Summe über alle die $N$ Zustände des Grundterms erstreckt ist und wo die $N^2$ Konstanten $a_{m\,m'}$ den $^1/_2\,N(N+1)$-Bedingungen genügen:

$$\sum_{\sigma'}^{N} a_{\sigma\,\sigma'}\,a_{\sigma''\,\sigma'}^{*} = \delta_{\sigma\sigma''}, \qquad (2)$$

sonst aber unbestimmt sind. Für den Fall eines Atoms in der Abwesenheit äußerer Felder z. B. ist dies gleichbedeutend mit der klassischen Aussage, daß die Achse der Richtungsquantelung unbestimmt ist, da irgend eine Eigenfunktion eines entarteten Systems durch eine lineare Kombination der Eigenfunktionen für dieselbe Energie und für eine beliebige Stellung des Achsensystems dargestellt werden kann. Für ein Wasserstoffatom z. B. sind die Eigenfunktionen $c\,P_l^{(m)}\,(\cos\vartheta)\,\varepsilon^{i\,m\,\varphi}$, und man kann leicht zeigen, daß diese sich bei einer beliebigen Rotation in eine lineare Kombination der neuen $P_l^{(m)}\,(\cos\vartheta)\,\varepsilon^{i\,m\,\varphi}$ für dieselbe Haupt- und azimutale Quantenzahl umwandeln.

Wir können jetzt zwischen Störungen unterscheiden, je nachdem sie die Entartung aufheben oder nicht [z. B. Elektronenstoß, Licht][1]. Man kann zeigen[2], daß man im ersteren Falle die $a_{\sigma\,\sigma'}$ so wählen muß, daß die $^1/_2\,N(N-1)$ Bedingungen des Verschwindens derjenigen Matrixkomponenten $E_{\sigma\sigma'}$ der Störungsenergie erfüllt sind, die den Übergängen zwischen Zuständen, die ursprünglich zu demselben Energiewert gehörten, entsprechen:

$$(1-\delta_{\sigma\,\sigma'})\sum_{\sigma''}^{N}\sum_{\sigma'''}^{N} a_{\sigma\,\sigma''}\,E_{\sigma''\,\sigma'''}\,a_{\sigma'\,\sigma'''}^{*} = 0. \qquad (2\,\mathrm{a})$$

Falls einige dieser Bedingungen identisch erfüllt sind, so ist die Entartung nicht völlig aufgehoben. Als Beispiel nehmen wir das Stern-Gerlach-Experiment; hier dürfen wir die Polarachse für die Molekeln, auch bevor sie in das magnetische Feld eingetreten sind, parallel der Feldrichtung wählen; dann genügen $a_{\sigma\,\sigma'} = \delta_{\sigma\,\sigma'}$ den Gleichungen (2) und (2 a). Wir bemerken, daß die Wellengleichung für dieses Experiment in drei Gleichungen zerfällt: erstens eine für den zu der magnetischen Quantenzahl $m$ konjugierten Winkel, zweitens eine für die übrigen atomaren Koordinaten

---

[1] Die Energiewerte des gestörten Atoms sind in allen Näherungen die des ungestörten. M. Born, ZS. f. Phys. **38**, 803, 1926.

[2] Vgl. die Behandlung des Starkeffekts nach der Methode der säkularen Störungen. E. Schrödinger, Ann. d. Phys. **80**, 437, 1926. Ferner M. Born, W. Heisenberg, P. Jordan, Quantenmechanik II, Kap. 3, § 2; ZS. f. Phys. **35**, 585, 1926.

und drittens eine für die translatorischen Koordinaten der Molekel. Die
letzte lautet:

$$\varDelta\,\psi\,(x\,y\,z) + \frac{8\,\pi^2\,M}{h}\left[E_t - \frac{m\,e\,H}{4\,\pi\,\mu}\right]\psi = 0,$$

wo $E_t$ die ursprüngliche translatorische Energie, $M$ die Molekülmasse,
$\mu$ die Elektronenmasse und $H$ das Feld bedeuten; sie zeigt, daß die Ein-
stellungsenergie [1]) der Moleküle im Felde aus der translatorischen ent-
nommen wird: die dem Felde parallel eingestellten Moleküle werden
beschleunigt, die antiparallelen verlangsamt. Ferner, wenn ein zusätz-
liches, homogenes, magnetisches Feld vorhanden ist, so sind die zwei
Strahlen nicht von gleicher Geschwindigkeit; das Verhältnis ist

$$\left(\frac{E_t + \varDelta\,E_m}{E_t - \varDelta\,E_m}\right)^{1/2},$$

wo $E_t$ die translatorische Energie, $\varDelta\,E_m$ die Aufspaltungsenergie ist.

Wir betrachten jetzt Störungen, die die Entartungen nicht aufheben.
Seien die stationären Zustände des Atoms $(n_1, n_2 \ldots; \sigma_1, \sigma_2 \ldots) = (n_i, \sigma_k)$
$= (n, \sigma)$, die Energie $E = E(n_1, n_2 \ldots) = E(n)$ und die Anzahl der
Zustände mit dem Termwert $E = E(n)$ gleich $N$. Ursprünglich dürfen
wir annehmen, daß das Atom im Normalzustand $(n^0)$ ist, und es wird
mit gleicher Wahrscheinlichkeit in jedem der $N^{(0)}$ Zustände $(n^{(0)}, \sigma^{(0)})$
sich befinden. Die Eigenfunktion des ungestörten Atoms können wir,
wenn wir mit $\psi\,(n, \sigma)$ die Eigenfunktionen für die Quantenzahlen $n, \sigma$ in
einem beliebigen Achsensystem bezeichnen, folgendermaßen schreiben:

$$\left.\begin{aligned}
\psi^{(0)} &= \sum_{\sigma_i'}^{N^{(0)}} \varepsilon^{i\,\gamma_{\sigma'}\,n^{(0)}}\,\overline{\psi}\,(n^{(0)}, \sigma')\\[2mm]
\overline{\psi}\,(n^{(0)}, \sigma') &= \sum_{\sigma_i}^{N^{(0)}} a\,(\sigma', \sigma; n^{(0)})\,\psi\,(n^{(0)}, \sigma),
\end{aligned}\right\} \quad (3)$$

wo die $a\,(\sigma', \sigma)$ der Bedingung (2) genügen. Es ist wichtig, zu bemerken,
daß wir die Phasenkonstanten $\gamma_{\sigma\,n^{(0)}}$ nicht kennen und daß wir in unserem
Endergebnis über sie werden mitteln müssen. Das hat aber mit der Entartung des Systems nichts zu tun; es ist der formale Ausdruck der
Tatsache, daß die Geschichte (d. h. der zeitliche Verlauf der Hamilton-
schen Funktion) für die verschiedenen Atome verschieden ist, und daß
deshalb die Zeit für diese Atome nicht von einem gemeinsamen Nullpunkt

---

[1]) Für die klassische Besprechung dieses Experiments siehe A. Einstein
und P. Ehrenfest, ZS. f. Phys. **11**, 31, 1922. Nach dem Mechanismus der Ein-
stellung hat es aber hier keinen Sinn zu fragen, denn das System darf, auch bevor
es in das Feld eingetreten ist, als eingestellt betrachtet werden.

30                          J. R. Oppenheimer,

aus gerechnet werden darf. Nun ist die Eigenfunktion (3) wegen der Willkürlichkeit der $a\,(\sigma, \sigma')$ nicht völlig bestimmt: Wir müssen also zeigen, daß trotz dieser Unbestimmtheit das Verhalten des Systems eindeutig festgelegt werden kann.

Die Eigenfunktion des gestörten Atoms ist durch

$$\psi^{(1)} = \sum_{n_i} \sum_{\sigma_i'}^{N} \overline{\omega}\,(n^{(0)}; n, \sigma')\,\overline{\psi}\,(n, \sigma') \tag{4}$$

gegeben, wo die $\overline{\omega}\,(n^{(0)}; n, \sigma)$ die Übergangswahrscheinlichkeitsamplituden $\psi^{(0)} \to \overline{\psi}\,(n, \sigma)$ sind. Diese Amplituden hängen natürlich von der Art der Störung ab und können aus der Störungsenergie abgeleitet worden sein. Diese Energie wird im allgemeinen von den dynamischen Koordinaten des störenden Systems und von denen des Atoms gleichzeitig abhängen. Ist z. B. das störende System ein Elektron, so wird sie von der Lage des Elektrons abhängen; ist es Strahlung, so hängt sie von der Intensität und den Phasen dieser Strahlung ab. Aber für jeden Übergang des störenden Systems ist diese Störungsenergie als Funktion der atomaren Koordinaten allein gegeben: die Matrixkomponente der Energie für den betreffenden Übergang des störenden Systems. Wir werden später zeigen, daß diese Übergänge voneinander unabhängig sind, und können uns daher auf die Betrachtung einer einzigen Funktion — etwa $V$ — beschränken[1]). Dann wird

$$\overline{\omega}\,(n, \sigma) = \frac{-\,2\,\pi\,i}{h} \int d\tau\,\psi^{(0)}\,V\,\overline{\psi}^{*}\,(n, \sigma).$$

Nun können wir die Funktionen $\overline{\psi}\,(n, \sigma)$ einer Transformation (1) unterwerfen. Dabei transformieren sich auch die $\overline{\omega}\,(n, \sigma)$, aber sie transformieren sich kontragredient zu den $\overline{\psi}\,(n, \sigma)$, so daß $\psi^{(1)}$ bei der Transformation invariant bleibt:

$$\left.\begin{aligned}
&\sum_{\sigma_i'}^{N} \overline{\omega}\,(n^{(0)}; n, \sigma')\,\overline{\psi}\,(n, \sigma') \\
&\to \sum_{\sigma_i'}^{N} \sum_{\sigma_i}^{N} \sum_{\sigma_i''}^{N} \omega\,(n^{0}; n, \sigma)\,a^{*}\,(\sigma', \sigma; n)\,a\,(\sigma', \sigma''; n)\,\psi\,(n, \sigma'') \\
&= \sum_{\sigma_i}^{N} \omega\,(n^{(0)}; n, \sigma)\,\psi\,(n, \sigma)
\end{aligned}\right\} \tag{5}$$

nach (2). Dabei ist

$$\omega\,(n^{(0)}; n, \sigma) = \frac{-\,2\,\pi\,i}{h} \int d\tau\,\psi^{(0)}\,V\,\psi^{*}\,(n, \sigma) \tag{5a}$$

gesetzt.

---

[1]) Dabei ist $V$ im allgemeinen ein Operator und eine Funktion der Zeit.

Zur Quantenmechanik der Richtungsentartung.        31

Setzen wir jetzt (5a), (5) und (3) in (4) ein, so erhalten wir

$$\psi^{(1)} = \sum_{n_i^{(1)}} \psi^{(1)}\,(n^{(1)})$$

$$= \sum_{n_i^{(1)}} \sum_{\sigma_i'}^{N^{(0)}} \sum_{\sigma_i^{(0)}}^{N^{(0)}} \sum_{\sigma_i^{(1)}}^{N^{(1)}} a\,(\sigma',\sigma^{(0)};n^{(0)})\,\varepsilon^{\,i\,\gamma_\sigma n^{(0)}}\,\omega\,(n^{(0)},\sigma^{(0)};n^{(1)},\sigma^{(1)})\,\psi\,(n^{(1)},\sigma^{(1)}),\quad (6)$$

$\psi^{(1)}$ hängt also von den $a\,(\sigma,\sigma';n^0)$, nicht aber von den $a\,(\sigma,\sigma';n^{(1)})$ ab: Für gegebene Eigenfunktionen des Anfangssystems ist die gestörte Eigenfunktion eindeutig bestimmt: sie hängt nicht von der Wahl der Achse ab. Aber da auch die Phasen in $\psi^{(1)}$ durch die Angabe von $\psi^{(0)}$ völlig bestimmt sind, darf man nicht über sie unabhängig mitteln; es folgt daraus, daß die Übergänge (und daher auch die Strahlung) von den $N$ Zuständen $(n,\sigma)$ nicht als unabhängig betrachtet werden können.

Wir berechnen jetzt die Intensität der spontanen Strahlung von der Frequenz $v\,(n^{(1)},n^{(2)})$, die dem Übergang $n_i^{(1)} \rightarrow n_i^{(2)}$ entspricht und die parallel einem beliebigen Vektor $\mathfrak{s}$ polarisiert ist. Sei $\mathfrak{M}$ der Vektor des elektrischen Moments des Atoms und sei $(\mathfrak{M}.\mathfrak{s}) = M_s$, dann wird[1] diese Intensität

$$I\,(\mathfrak{s};n^{(1)},n^{(2)}) = \frac{16\,\pi^3\,v^4\,(n^{(1)},n^{(2)})}{c^3} \sum_{\sigma_i''}^{N^{(2)}} \left| \int d\tau\,\psi^{(1)}\,(n^{(1)})\,M_s\,\overline{\psi}^*\,(n^{(2)},\sigma'') \right|^2,$$

$$\overline{\psi}\,(n^{(2)},\sigma'') = \sum_{\sigma_i'''}^{N^{(2)}} a\,(\sigma'',\sigma''';n^{(2)})\,\psi\,(n^{(2)},\sigma''').$$

$$\left.\right\} (7)$$

In $I$ gehen jetzt $a\,(\sigma,\sigma';n^{(2)})$ und die $a\,(\sigma,\sigma';n^{(0)})$ ein. Mitteln wir aber über die unabhängigen Phasen $\gamma_{\sigma n^{(1)}}$, so wird mit (6) und (2)

$$I\,(\mathfrak{s};n^{(1)},n^{(2)}) = \frac{16\,\pi^3\,v^4\,(n^{(1)},n^{(2)})}{c^3} \sum_{\substack{\sigma^{(0)}\\i}}^{N^{(0)}} \sum_{\substack{\sigma^{(2)}\\i}}^{N^{(2)}} |\,M_s\,(n^{(0)},\sigma^{(0)};n^{(2)},\sigma^{(2)})\,|^2,$$

$$M_s\,(n^{(0)},\sigma^{(0)};n^{(2)},\sigma^{(2)}) = \sum_{\substack{\upsilon^{(1)}\\i}}^{N^{(1)}} \omega\,(n^0,\sigma^0;n^{(1)},\sigma^{(1)}) \int d\tau\,\psi\,(n^{(1)},\sigma^{(1)})\,M_s\,\psi^*\,(n^{(2)},\sigma^{(2)}).$$

$$\left.\right\} (8)$$

Das ist jetzt von den $a$ unabhängig. Daß man, um zu bestimmten Ergebnissen zu gelangen, die unbekannten Phasen einführen und dann wegmitteln muß, wird wohl daher rühren, daß in der heutigen Theorie die Zeit immer als Parameter und nicht als dynamische Koordinate behandelt

---

[1] P. Dirac, Proc. Roy. Soc. **114**, A, 243, 1927.

32                                    J. R Oppenheimer,

wird. In dem Falle, wo für eine bestimmte Wahl dieses Systems alle $\omega$ verschwinden, außer einem, kann man offenbar für d i e s e Wahl die Strahlung klassisch — mit unabhängigen Wahrscheinlichkeiten — berechnen.

Die Formel (8) gibt die Strahlung für einen einzigen Übergang des störenden Systems an, und die gesamte Strahlung wird durch Summieren bzw. Integrieren über alle solche Übergänge erhalten. Daß es berechtigt ist, diese Übergänge als voneinander unabhängig zu betrachten, zeigt man analytisch auf dieselbe Weise, wie wir die Unabhängigkeit der Übergänge von $(n^{(1)})$ nach den $N^{(2)}$ Zuständen $(n^{(2)}, \sigma^{(2)})$ bestätigt haben. Falls das störende System ein Elektronenstrahl ist, so bedeutet diese Unabhängigkeit physikalisch, daß die Übergänge, die durch die verschiedenen möglichen Ablenkungen des Elektrons verursacht sind, nicht untereinander gekoppelt sind. Das muß so sein, da man, indem man sich eines genügend schwachen Strahles bedient und genügend viele Experimente macht, prinzipiell eine Korrelation zwischen der Ablenkung des Elektrons und dem durch das Atom emittierten Licht finden kann.

In der Ableitung von (8) haben wir angenommen, daß die unregelmäßigen kleinen Felder, die immer vorhanden sind, zu vernachlässigen und daß die Übergänge zwischen den verschiedenen $n_i^{(1)}$-Niveaus unabhängig waren. Um dies zu rechtfertigen, müssen wir die spontane Ausstrahlung etwas näher betrachten; dies wird uns auch eine Erklärung der bekannten Effekte der äußeren Felder auf die Polarisation des Resonanz- und Stoßleuchtens liefern.

Wir können uns wieder auf einen einzigen Übergang des störenden Systems beschränken, und daher auch auf ein einziges Wertsystem der $n_i^{(1)}$. Der Einfachheit halber dürfen wir ferner annehmen, daß dieser Zustand der erste angeregte Zustand des Atoms ist; dann können wir alle spontanen Übergänge vernachlässigen, außer denen zum Normalzustand $n^{(0)}$. Nach den vorhergehenden Betrachtungen können wir auch den Normalzustand als einfach voraussetzen ($N^{(0)} = 1$); seine Energie sei $E^{(0)} = h\nu_0$. Wir nehmen dann an, daß ein kleines Feld vorhanden ist, das die Entartung des Systems aufhebt, indem es die $N^{(1)} = N$ Zustände $(n^{(1)}, \sigma^{(1)})$ aufspaltet; ihre Energien seien $E(n_i^{(1)}, \sigma_i^{(1)}) = h\nu_0$. Endlich dürfen wir die spontanen Übergänge zwischen den Zuständen $(n^{(1)}, \sigma^{(1)})$ vernachlässigen wegen des Faktors $(\Delta \nu)^3$ in den Einstein- $A$-Koeffizienten. Nun darf man $\psi^{(1)}(n, \sigma)$ als Produkt einer Zeitfunktion und einer Funktion der Koordinaten $x_j$ des Atoms schreiben:

$$\psi(n, \sigma) = b_\sigma(t)\, u(n, \sigma; x_j).$$

Zur Quantenmechanik der Richtungsentartung.    33

Dann ist $b_\sigma(t)$ die Wahrscheinlichkeitsamplitude dafür, daß zur Zeit $t$ das Atom im $(n,\sigma)$ Zustand ist, und daß es nicht gestrahlt hat. Nun wird im Laufe der Zeit das Atom zum Grundzustand zurückkehren; sei $b_{vs}(t)$ die Amplitude dafür, daß zur Zeit $t$ das Atom im Normalzustand ist und daß es ein Quant der Frequenz $v$ und Polarisation $s$ ausgestrahlt hat. Die Größen $b$ können als Funktionen der atomaren Quantenzahlen, der Zeit, und der Anzahlen $Z_{vs}$ der Quanten der Frequenz $v$ und Polarisation $s$ aufgefaßt werden:

$$b_\sigma(t) = b(n^{(1)}, \sigma; t; 0, 0 \ldots 0),$$
$$b_{vs}(t) = b(n^{(0)}; t; 0, 0 \ldots 0; Z_{vs} = 1 \ldots 0 \ldots 0).$$

Nach Dirac genügen die $b$ den Differentialgleichungen [l. c. Gleichung (18)]

$$\frac{2\pi i}{h} \dot{b}(n, \sigma; t; Z_{v_1 s_1}, Z_{v_2 s_2} \ldots) \tag{9}$$
$$= \sum F(n, \sigma, Z_{v_1 s_1}, Z_{v_2 s_2} \ldots; n', \sigma', Z'_{v_1 s_1}, Z'_{v_2 s_2} \ldots) \, b(n', \sigma'; t; Z'_{v_1 s_1}, Z'_{v_2 s_2} \ldots),$$

wo die Summe $\sum$ über alle stationären Zustände des Atoms und alle Zustände $v, s$ der Strahlung zu erstrecken ist, und wo $F$ die Hamiltonsche Funktion des Gesamtsystems ist. Nun verschwinden alle Komponenten von $F$ für $(n, n') \neq (n^{(1)}, n^{(0)})$ und ferner alle Diagonalglieder für $n = n^{(1)}$, $Z_{vs} \neq 0$ und endlich alle für $\sum Z > 1$. Die übrigen Komponenten sind:

$$F(n^{(1)}, \sigma; 0, 0 \ldots 0 \ldots 0; n^{(1)}, \sigma'; 0, 0 \ldots 0 \ldots 0) = \delta_{\sigma\sigma'} \, h v_\sigma,$$
$$F(n^{(0)}; 0, 0 \ldots Z_{vs} = 1 \ldots 0; n^{(0)}; 0, 0 \ldots Z_{vs} = 1 \ldots 0) = h(v_0 + v),$$
$$F(n^{(1)}, \sigma; 0, 0 \ldots 0 \ldots 0; n^{(0)}; 0, 0 \ldots Z_{vs} = 1 \ldots 0, 0)$$
$$= F^*(n^{(0)}; 0, 0 \ldots Z_{vs} = 1 \ldots 0, 0; n^{(1)}, \sigma; 0, 0 \ldots 0 \ldots 0) = l \dot{M}_s(n^{(1)} \sigma; n^{(0)})$$

wo

$$\dot{M}_s(n^{(1)}, \sigma; n^{(0)}) = \int d\tau \, u(n^{(1)}, \sigma) \, \dot{M}_s \, u^*(n^{(0)}),$$

und wo

$$l = \frac{1}{c}\left(\frac{hv}{\pi c \, q_{vs}}\right)^{1/2}$$

ist, wo $g_{vs}$ die Anzahl der Zustände der Strahlung im Bereich $ds\,dv$ ist. Das Fehlen des Faktors $\sqrt{2}$ gegenüber Dirac rührt daher, daß wir die Richtungen der Ausstrahlung nicht angeben, die Polarisation dagegen mit dem Vektor $s$ beschreiben. Wir führen jetzt, um die Gleichungen von $q$ zu befreien,

$$\xi_\sigma = b_\sigma \, e^{2\pi i v_\sigma t},$$
$$\xi_{vs} = b_{vs} \, q_{vs}^{1/2} \, e^{2\pi i (v_0 + v) t}$$

34                                    J. R. Oppenheimer,

ein. Dabei ist $|\xi_{v\delta}(t)|^2 \, dv \, d\delta$ die Wahrscheinlichkeit dafür, daß zur Zeit $t$ das Atom im Normalzustand ist, und daß ein Quant im Bereich $dv \, d\delta$ ausgestrahlt worden ist. Dann nimmt (9) die einfache Form an

$$\left.\begin{aligned}
i\,\dot{\xi}_\sigma &= \int d\delta \int dv \, g \, \dot{M}_\delta\,(n^{(1)},\sigma^{(1)};\, n^{(0)})\,\xi_{v\delta}\, e^{-2\pi i t(v+v_0-v_\sigma)}, \\
i\,\dot{\xi}_{v\delta} &= \sum_{\sigma_i^{(1)}}^{N} \dot{M}_\delta^*\,(n^{(1)},\sigma^{(1)};\, n^{(0)})\,g\,\xi_\sigma\, e^{2\pi i t(v+v_0-v_\sigma)}.
\end{aligned}\right\} \quad (10)$$

Setzt man für $g$ seinen berechneten Wert $\dfrac{2}{c}\left(\dfrac{\pi v}{hc}\right)^{1/2}$ ein, so sieht man, daß die Integrale nach $v$ nicht konvergieren. In den meisten Fällen sind sie aber halbkonvergent, und man kann $v$ durch seinen Wert an der Resonanzstelle $(v_\sigma - v_0)$ ersetzen. Da man immer $v$ durch einen von $(v_\sigma - v_0)$ nur wenig abweichenden Mittelwert ersetzen kann, und da der Punkt für die folgenden Betrachtungen nicht sehr wesentlich ist, werden wir $g$ durch $g_\sigma = \dfrac{2}{c}\left(\dfrac{\pi\,(v_\sigma - v_0)}{hc}\right)^{1/2}$ gleich in (10) ersetzen[1].

Diese Gleichungen lösen wir durch den Ansatz

$$\xi_\sigma = \sum_{\varkappa=0}^{N} p_{\sigma\varkappa}\,\varepsilon^{\alpha_{\sigma\varkappa}t} \tag{11}$$

für die Anfangsbedingungen

$$\xi_\sigma(0) = \sum_\varkappa p_{\sigma\varkappa} = \varepsilon^{i\gamma n^{(0)}}\,\omega\,(n^{(0)};\, n^{(1)},\sigma^{(1)}); \quad \sigma = 1,2\ldots N, \tag{12a}$$

$$\xi_{v\delta}(0) = 0. \tag{12b}$$

Die $N^2$ Größen $\alpha_{\sigma\varkappa}$ sind durch die $N(N-1)+N$ Gleichungen gegeben:

$$[\alpha_{\sigma\varkappa} - \alpha_{\sigma'\varkappa} + 2\pi i\,(v_{\sigma'}-v_\sigma)]\,\mu_{\sigma\sigma'} = 0; \quad \left.\begin{aligned}\sigma' &= 1,2\ldots N, \\ \sigma' &\neq \sigma, \\ \varkappa &= 1,2\ldots N,\end{aligned}\right\} \tag{13a}$$

$$\left.\begin{aligned}
|\,\mu_{\sigma\sigma'} + \delta_{\sigma\sigma'}\,\alpha_{\sigma\varkappa}\,| &= 0; \quad \varkappa = 1,2\ldots N, \\
\mu_{\sigma\sigma'} &= \frac{1}{2}\,g_\sigma g_{\sigma'}\int d\delta\,\dot{M}_\delta\,(n^{(0)};\, n^{(1)},\sigma)\,\dot{M}_\delta^*\,(n^{(0)};\, n^{(1)},\sigma').
\end{aligned}\right\} \tag{13b}$$

Die $p_{\sigma\varkappa}$ sind durch die $N$ Gleichungen (12a) und durch die $N(N-1)$ Gleichungen

$$\sum_{\sigma_i'}^{N} [\mu_{\sigma\sigma'} + \delta_{\sigma\sigma'}\,\alpha_{\sigma\varkappa}]\,p_{\sigma'\varkappa} = 0; \quad \left.\begin{aligned}\sigma &= 1,2\ldots N-1, \\ \varkappa &= 1,2\ldots N\end{aligned}\right\} \tag{14}$$

gegeben.

---

[1] Für eine kurze Besprechung der hier betonten Schwierigkeiten siehe P. Dirac, Proc. Roy. Soc. (im Erscheinen).

Zur Quantenmechanik der Richtungsentartung.      35

Wir bemerken, daß, falls für irgend ein $\sigma$ alle die $(1 - \delta_{\sigma\sigma'})\,\mu_{\sigma\sigma'}$ verschwinden, (11) die einfache Form

$$\xi_\sigma = \xi_\sigma(0)\,\varepsilon^{-^1/_2 A_\sigma t}; \quad A_\sigma = 2\,\mu_{\sigma\sigma} \tag{15}$$

annimmt, wo $A$ der Einsteinkoeffizient für $(n^{(1)}, \sigma) \to n^{(0)}$ ist. Ferner, wenn die Terme $\nu_\sigma$ weit auseinander rücken, nähern sich die $\xi_\sigma(t)$ der Form

$$\xi_\sigma(0)\,\varepsilon^{-^1/_2 A_\sigma t + i\beta_\sigma t}; \quad \Im(\beta_\sigma) = 0,$$

d. h. die Kopplung zwischen den Zuständen verschwindet. Falls endlich das Feld verschwindet, so wird $\alpha_{\sigma\varkappa} = \alpha_\varkappa$ eine Wurzel der säkularen Gleichung

$$|\mu_{\sigma\sigma'} + \delta_{\sigma\sigma'}\,\alpha_\varkappa| = 0. \tag{16}$$

Man kann leicht bestätigen, daß in diesem Falle die $\alpha_\varkappa$ reell und negativ für reelle $\mu_{\sigma\sigma'}$ sind.

Wir werden jetzt die Ausstrahlung für den Fall $\mu_{\sigma\sigma'} \neq 0$ und für ein nicht verschwindendes Feld untersuchen; zu diesem Zwecke wird es bequem sein, $N = 2$ zu setzen, so daß wir die Lösung (14) explizit hinschreiben können. Damit werden wir den Starkeffekt erster Ordnung für die $P$-Terme quantitativ behandeln können. Hier wird:

$$\left.\begin{aligned}
\alpha_{j\,2} &= \frac{1}{2}\left\{-\mu_{11} - \mu_{22} + 2\mu_j' + \Delta_j\right\}, \\[6pt]
\alpha_{j\,2} &= \frac{1}{2}\left\{-\mu_{11} - \mu_{22} - 2\mu_j' + \Delta_j\right\}, \\[6pt]
\Delta_1 &= -\Delta_2 = 2\pi i\,(\nu_1 - \nu_2), \\[6pt]
\mu_j' &= +\sqrt{(\mu_{22} - \mu_{11})^2 + 2\Delta_j(\mu_{22} - \mu_{11}) + \Delta_j^2 + 4\mu_{12}\mu_{21}}\,.
\end{aligned}\right\} \tag{17}$$

Ferner für $\mu_{12} = \mu_{21}$:

$$\left.\begin{aligned}
p_{11} &= -\frac{1}{2\mu_1'}\left\{(\mu_{11} + \alpha_{12})\,\xi_1(0) + \mu_{12}\,\xi_2(0)\right\}, \\[6pt]
p_{12} &= \frac{1}{2\mu_1'}\left\{(\mu_{11} + \alpha_{11})\,\xi_1(0) + \mu_{12}\,\xi_2(0)\right\}, \\[6pt]
p_{21} &= \frac{1}{2\mu_1'}\left\{(\mu_{11} + \alpha_{11})\,\xi_2(0) - \mu_{12}\,\xi_1(0)\right\}, \\[6pt]
p_{22} &= -\frac{1}{2\mu_1'}\left\{(\mu_{11} + \alpha_{12})\,\xi_2(0) - \mu_{12}\,\xi_1(0)\right\}.
\end{aligned}\right\} \tag{18}$$

36 $\qquad$ J. R. Oppenheimer,

Aus (10) und (12b) können wir $\mathit{\Xi}_s = \int d\nu\, \xi_{\nu s}^{(\infty)}\, \xi_{\nu s}^{*\,(\infty)}$ berechnen, d. h. die Gesamtwahrscheinlichkeit, daß Licht von der Polarisation $s$ während der Rückkehr des Systems in den Normalzustand ausgestrahlt worden ist[1]):

$$\left.\begin{aligned}
\mathit{\Xi}_s &= \sum_{i,j}^{N} \frac{-\,\tau_i\,\tau_j^{*}}{\alpha_i + \alpha_j^{*}}, \\
\alpha_j &= \alpha_{\sigma j} + 2\,\pi\,i\,(\nu_0 - \nu_\sigma)^2), \\
\tau_j &= \sum_{\sigma}^{N} g_\sigma\, p_{\sigma j}\, \dot{M}_s\,(n^{(0)};\, n^{(1)},\, \sigma).
\end{aligned}\right\} \tag{19}$$

Für den Fall, daß die $(1 - \delta_{\sigma\sigma'})\,\mu_{\sigma\sigma'}$ alle verschwinden, nimmt (19) die einfachere Form an:

$$\left.\begin{aligned}
\mathit{\Xi}_s &= \sum_{i,j}^{N} \frac{(\mu_{ii} + \mu_{jj})\,(\overline{\tau_i\,\tau_j^{*}} + \overline{\tau_i^{*}\,\tau_j}) + 2\,\varDelta_{ij}\,(\overline{\tau_i\,\tau_j^{*}} - \overline{\tau_i^{*}\,\tau_j})}{(\mu_{ii} + \mu_{jj})^2 + |\varDelta_{ij}|^2}, \\
\overline{\tau_j} &= g_j\, \dot{M}_s\,(n^{(0)};\, n^{(1)}, j), \\
\varDelta_{ij} &= 2\,\pi\,i\,(\nu_i - \nu_j).
\end{aligned}\right\} \tag{20}$$

Diese letzte Form zeigt in klarer Weise, wie das wachsende Feld die Kopplung der Emission von den verschiedenen Zuständen ($\sigma$) zerstört.

Die Formel (8) liefert eine Rechtfertigung der Theorie von Heisenberg[3]) über die Polarisation der Resonanzstrahlung. Denn, wie wir gesagt haben, falls wir das Achsensystem (und daher auch die Eigenfunktionen) so wählen, daß für gegebene $n_i^{(0)}$, $\sigma_i^{(0)}$ und $n_i^{(1)}$ nur ein $\omega$ von Null verschieden ist, so können die Übergänge von dem Zustand $(\sigma_i^{(1)})$ klassisch — d. h. als unabhängige Übergänge — berechnet werden. Für linear polarisiertes Licht z. B. können wir leicht ein solches System angeben, da, wenn die Polarachse des Systems parallel zu dem elektrischen Vektor des Lichtes liegt, nur die Übergänge, die die magnetische Quantenzahl unverändert lassen, stattfinden werden. Wenn daher keine weitere

---

[1]) Dieses betrifft einen einzigen Zeitpunkt der ursprünglichen Störung. Für die ganze Störung müßte man $\xi_{\nu s}\,(\infty)$ über die Zeit der Störung integrieren. Die Verteilung der Strahlung über $\nu$ und $\sigma$ bleibt aber dabei ungeändert.

[2]) Vgl. (13a).

[3]) W. Heisenberg, ZS. f. Phys. **31**, 617, 1925; ferner J. H. van Vleck, Proc. Nat. Acad. **11**, 612, 1925. Die Berechtigung der Heisenbergschen Regeln ist auf Grund der klassischen Quantentheorie von L. Nordheim versucht worden. Das Ergebnis war jedoch nicht in jeder Hinsicht befriedigend: L. Nordheim, ZS. f. Phys. **33**, 729, 1925. Herrn Nordheim bin ich für interessante Besprechungen zu bestem Dank verpflichtet.

Entartung vorhanden ist, so wird die Heisenbergsche Regel gelten. Aber wir bemerken, daß sie nur dann gilt: bestrahlte man atomaren Wasserstoff mit der ersten Balmerlinie, so würde die Ausstrahlung nicht durch diese Regel berechenbar sein. Ähnliche Betrachtungen können für zirkular polarisiertes Licht durchgeführt werden.

Ein Fall, der experimentell ziemlich gründlich untersucht worden ist[1]), ist der, wo die ursprüngliche Störung durch einen gerichteten Elektronenstrahl erfolgt. Wir haben gezeigt, daß das durch die verschiedenen möglichen Ablenkungen der Elektronen erzeugte Licht unabhängig behandelt werden kann. Dann können wir eine einfache, der Heisenbergschen analoge Regel angeben, die immer dann gilt, wenn das atomare System nicht in $k$ und $j$ entartet ist. Das Elektron möge die Anfangsgeschwindigkeit $\mathfrak{v}_1$ und die Endgeschwindigkeit $\mathfrak{v}_2$ haben; dann darf die Polarisation der emittierten Strahlung klassisch, d. h. mit unabhängigen Wahrscheinlichkeiten berechnet werden, indem man das System als durch ein magnetisches Feld parallel dem Vektor $\varDelta\mathfrak{v} = \mathfrak{v}_1 - \mathfrak{v}_2$ richtungsgequantelt betrachtet und das Elektron durch eine polarisierte Lichtwelle mit elektrischem Vektor in derselben Richtung ersetzt. Zunächst werden wir einen Beweis dieser Aussage skizzieren, und sie dann auf einige einfache Fälle anwenden.

Der Einfachheit halber vernachlässigen wir zunächst den Elektroneneigendrehimpuls und untersuchen nur nachträglich die dadurch erzeugte Abänderung unserer Ergebnisse. Seien $R$, $\varTheta$, $\varPhi$ die Koordinaten des freien Elektrons in dem oben definierten Koordinatensystem, d. h. $\varTheta = 0$ für den Radiusvektor parallel zu $\varDelta\mathfrak{v}$. Seien ferner $\varphi_1 \ldots \varphi_u$ die Azimute der atomaren Elektronen um $\varDelta v$, $p_1 \ldots p_u$ die konjugierten Impulse. In der potentiellen Energie des ungestörten Systems werden dann nur die Differenzen $\varphi_i - \varphi_j$ eingehen. Führen wir also

$$\varphi = \frac{1}{u}\sum\varphi_i, \; \varphi_j' = \frac{1}{2}(\varphi_j - \varphi_1), \; P = \sum p_i; \; p_j = p_j - p_1$$

als neue Variabeln ein, so können wir von der Schrödingerfunktion des Atoms eine Funktion $\varepsilon^{im\varphi}$ von $\varphi$ allein abspalten; der andere Faktor hängt nicht von $\varphi$ ab. Für gegebene andere Quantenzahlen $n_i$ stellt dann $\varphi$ das Azimut des Atoms um $\varDelta v$, und $m$ die Komponente des Dreh-

---

[1]) W. Kossel und Gerthsen, Ann. d. Phys. **77**, 273, 1925; A. Ellett, P. Foote und F. Mohler, Phys. Rev. **27**, 31, 1926; H. Skinner, Proc. Roy. Soc. (A) **112**, 642, 1926; B. Quarder, ZS. f. Phys., im Erscheinen.

38                                    J. R. Oppenheimer,

impulses in dieser Richtung dar. Dann ist die Amplitude $\omega$ ($n_i^{(0)}$, $m^{(0)}$; $n_i^{(1)}$, $m^{(1)}$) proportional dem Ausdruck[1])

$$\int_0^{2\pi} \int_0^{\pi} d\Theta \sin\Theta \, d\Phi \, V(n_i^{(0)}, m^{(0)}; n_i^{(1)}, m^{(1)}; : R, \Theta, \Phi) \, \varepsilon^{i\,K\,R\cos\Theta}, \qquad (21)$$

wo $K$ eine durch den Ablenkungswinkel bestimmte Konstante[2]) und $V$ die entsprechende Matrixkomponente der Störungsenergie des Elektrons und des Atoms ist. Diese Störungsenergie wird im allgemeinen eine sehr komplizierte Funktion der atomaren Koordinaten sein; aber nachdem wir sie über alle diese Koordinaten, außer $\varphi$, integriert haben, wird sie eine Funktion $F$ von $R$, $\Theta$ und $\cos\gamma$ allein, wo $\gamma = \varphi - \Phi$ ist. Sei nun

$$F(\gamma, R, \Theta) = \sum_{l=0}^{\infty} c_l(R, \Theta) \cos l\gamma,$$

dann wird

$$\left. \begin{aligned} & V(n_i^{(0)}, m^{(0)}; n_i^{(1)}, m^{(1)}; R, \Theta, \Phi) \\ & = \sum_l \frac{c_l}{\pi} \int d\varphi \, \{\cos l\Phi \cos l\varphi + \sin l\Phi \sin l\varphi\} \, e^{i\varphi(m^{(0)}-m^{(1)})} \\ & = c_{m^{(1)}\,m^{(0)}}(R, \Theta) \, e^{i\,\Phi\,(m^{(1)}-m^{(0)})}. \end{aligned} \right\} \qquad (22)$$

Aus (21) und (22) folgt das Verschwinden aller durch das Elektron erzeugten $\omega$, außer denen für $m^{(0)} = m^{(1)}$; und das rechtfertigt die oben angegebene Regel.

Dieses Ergebnis kann man unmittelbar auf Stoßionisation durch geladene Teilchen anwenden; im allgemeinen ist die Berechnung der $c_l(R)$ sehr verwickelt, aber für die härteren $\delta$-Teilchen, die durch schnelle $\alpha$-Teilchen aus leichten Atomen losgerissen werden, kann man zeigen, daß die Verteilung der $\delta$-Teilchen gleichmäßig über alle Winkel wird[3]).

---

[1]) M. Born, Göttinger Nachrichten 1926, S. 146. — Herrn Prof. Born bin ich für die Gelegenheit, diese Arbeit vor ihrem Erscheinen sehen zu dürfen, zu Dank verpflichtet.

[2]) $K^2 = \dfrac{4\pi^2 m}{h^2}\,(v_1^2 + v_2^2 - 2\,v_1 v_2 \cos\delta)$, wo $\delta$ den Ablenkungswinkel, $m$ die Elektronenmasse bedeutet.

[3]) Dies Ergebnis wurde unter Vernachlässigung der relativistischen Mechanik erhalten. Man findet für diesen Grenzfall, daß die Verteilung der $\delta$-Teilchen durch $(1 - 2\lambda^{-1}\,\varDelta\lambda \cos\vartheta \ldots)$ gegeben ist, wo der Polarabstand $\vartheta$ im oben definierten Koordinatensystem gemessen wird, und wo $\varDelta\lambda = \left(\dfrac{1}{\lambda_1} - \dfrac{1}{\lambda_2}\right)^{-1}$ ist, wo $\lambda$, $\lambda_1$, $\lambda_2$ die de Broglieschen Wellenlängen des $\delta$-Teilchens bzw. des ankommenden $\alpha$-Teilchens bzw. des auslaufenden $\alpha$-Teilchens sind. Für die analoge photoelektrische Rechnung und die Matrixkomponenten siehe G. Wentzel, ZS. f. Phys. **40**, 574, 1926, und R. Oppenheimer, ebenda **41**, 268, 1927.

Zur Quantenmechanik der Richtungsentartung.     39

Dieses Resultat könnte experimentell geprüft werden, indem man die Teilchen in einem Gas bei niedrigem Drucke zählt.

Wenn $v_2$ klein ist — d. h. das Elektron nur die nötige Anregungsenergie hat —, so liegt die Polarachse des oben eingeführten Systems für alle Ablenkungen zu dem Elektronenstrahl parallel. So erhalten wir in diesem Falle das Skinnersche Ergebnis, daß die Komponente des Drehimpulses des Atoms parallel dem Strahl sich nicht ändern kann. Die Polarisation für ein gegebenes $v_1$ hängt also von der Verteilung der Vektoren $\Delta \mathfrak{v}$ ab. In einfachen Fällen kann man diese Verteilung berechnen, aber einige Ergebnisse können wir geometrisch und ohne auf die Einzelheiten einzugehen einsehen: Wenn die Anfangsenergie des Elektrons nicht mehr als zweimal so groß wie die Anregungsenergie ($\Delta E$) ist, so muß die Polarisation in demselben Sinne wie bei der Anregungsspannung sein, was auch das Ablenkungsspektrum der Elektronen sein mag. Ferner, für sehr große Anfangsgeschwindigkeiten werden die $\Delta \mathfrak{v}$ fast alle zu dem Strahl senkrecht liegen. Denn wenn wir mit $\delta$ den Ablenkungswinkel des Elektrons bezeichnen, so ist die senkrechte Komponente $(\perp)$ von $\Delta \mathfrak{v}$ gleich $v_2 \sin \delta$, die parallele Komponente $(\parallel)$ dagegen [1] $|\mathfrak{v}_1 - \mathfrak{v}_2| \cos \delta \simeq \dfrac{2\, \Delta E}{m\,(v_1 + v_2)} \cos \delta$. Aber für große Geschwindigkeiten wird $\delta$ außer für einen sehr kleinen Bruchteil der Stöße sehr klein sein; deshalb

$$\frac{(\perp)}{(\parallel)} \sim \frac{m\, v_2^2}{\Delta E}\, \delta.$$

Es wird daher plausibel, daß die Polarisation aller Linien für hohe Geschwindigkeiten sich umkehrt [2]. Das können wir bestätigen, indem wir die Rechnung für einen einfachen Fall durchführen. Sei ein Wasserstoffatom ursprünglich im Normalzustand, und sei es durch Elektronenstoß zum ($n = 2$, $k = 1$)-Zustand angeregt. Die relative Wahrscheinlichkeit einer Ablenkung zwischen $\delta$ und $\delta + d\delta$ ist dann [3]

$$\omega^2\,(\delta)\, d\delta$$
$$= \sin \delta\, d\delta\, \{(v_1^2 + v_2^2 - 2\,v_1 v_2 \cos \delta)(v_0^2 + v_1^2 + v_2^2 - 2\,v_1 v_2 \cos \delta)^6\}^{-1}; \quad (23)$$

---

[1] Für kleine Ablenkungswinkel.

[2] B. Quarder findet für große $v_1$ eine Gesamtpolarisation senkrecht zu dem Elektronenstrahl. Das ist jedoch wegen der Kaskadensprünge schwer verständlich. Vgl. B. Quarder, l. c.

[3] M. Born, l. c. Formel (29).

40                                    J. R. Oppenheimer,

wo $v_0 = \dfrac{3\,\pi\,e^2}{h}$ und $e$ die Elektronenladung ist.   Der Mittelwert des
Quadrats des Sinus des Winkels zwischen $\varDelta\mathfrak{v}$ und der Richtung des
Elektronenstrahls ist dann

$$P_\perp = v_1^2 \left\{ \int \omega^2\,(\delta)\cdot d\,\delta \right\}^{-1} \left\{ \int \frac{\omega^2\,(\delta)\,\sin^2\delta\,d\,\delta}{v_1^2 + v_2^2 - 2\,v_1\,v_2\cos\delta} \right\}.$$

Man kann dann leicht beweisen, daß im Grenzfall $\dfrac{m\,v_2^2}{\varDelta E} \to \infty$, $P_\perp \to 1$.
Die senkrechte Komponente wächst nur sehr langsam mit der Ge-
schwindigkeit an; ihr Verhältnis zur parallelen Komponente wächst wie
$\left(\dfrac{8}{5}\ln\dfrac{v_1}{v_0}\right)$ für große $v_1$ an.   Wenn $\omega^2\,(\delta)$ wesentlich steiler von $\delta = 0$
abfiele wie in (23), so würde die Umkehrung der Polarisation nicht zu-
stande kommen.   Man kann aber einsehen, daß $\omega\,(\delta)$ im allgemeinen ungefähr die Form (23) hat.   Es ist nicht möglich, eine allgemein gültige
Formel für die Änderung der Polarisation mit der Geschwindigkeit der
Elektronen anzugeben.   Aber aus dem oben Gesagten geht hervor, daß die
Polarisation für eine Geschwindigkeit $v_1 > 2\sqrt{\dfrac{\varDelta E}{m}}$ verschwinden muß.

Bevor wir die Modifikationen untersuchen, die durch den Elektronenmagnete verursacht werden, dürfen wir vielleicht einige Beispiele der
Anwendung von (19) und (20) zur Berechnung der Wirkung äußerer
Felder auf die Polarisation angeben.   Es ist klar, daß, mit Ausnahme
von Wasserstoff, wo das Abschirmungsdublett zusammenfällt, diese
Effekte unabhängig davon sein werden, ob das Licht durch Elektronenstoß oder durch Strahlung erzeugt worden ist.

a) Ein einfaches Beispiel ist folgendes: Polarisiertes Licht regt ein
normales „klassisches“ [1]) Wasserstoffatom an, das sich in einem Magnet-
feld $\mathfrak{H}$ befindet, das in der Richtung $\gamma$ mit dem elektrischen Vektor $\mathfrak{E}$
des Lichtes liegt.   Zu berechnen ist die Intensitätsverteilung der spontanen
ersten Lymanlinie, die dem Vektor $\mathfrak{s}$ parallel polarisiert ist.   Sei $s$ der
Polarabstand von $\mathfrak{s}$, $\varrho$ der Winkel zwischen seiner Projektion auf eine Ebene
senkrecht zur ($\mathfrak{E}$, $\mathfrak{H}$)-Ebene und dieser letzten Ebene.   Die $\mu_{ij}$ bilden
eine Diagonalmatrix, und wir können daher (15) gebrauchen.   Setzen wir

$$A = 2\,\mu_{11} = 2\,\mu_{22} = 2\,\mu_{33}$$

---

[1]) D. h. ohne Magnetelektron.

Zur Quantenmechanik der Richtungsentartung.    41

gleich dem Einsteinkoeffizienten, so erhalten wir aus (20) für die relative Intensität des parallel zu dem Vektor $\mathfrak{s}$ polarisierten Lichtes

$$\left.\begin{aligned}\cos^2 \gamma \cos^2 s + \frac{1}{2}\sin^2 \gamma \sin^2 s + \frac{1}{2}\frac{\sin 2\gamma \sin 2s \left(A^2 \cos \varrho + \dfrac{eH}{2m}\sin \varrho\right)}{A^2 + \dfrac{e^2}{4m^2}\cdot H^2} \\ + \frac{1}{2}\frac{\sin^2 \gamma \sin^2 s \left(A^2 \cos 2\varrho + \dfrac{eHA}{m}\sin 2\varrho\right)}{A^2 + \dfrac{e^2}{m^2}\cdot H}\,.\end{aligned}\right\} \quad (24)$$

So bewirkt ein Feld parallel dem elektrischen Vektor des Lichtes $(\cos \gamma = 1)$ keine Störung der Polarisation. Ein senkrechtes Feld dagegen reduziert, falls man senkrecht zu $\mathfrak{H}$ und senkrecht zu $\mathfrak{E}$ $(\varrho = 0)$ beobachtet, die Intensität auf ihren Halbwert; falls man parallel $\mathfrak{H}$, senkrecht $\mathfrak{E}$ $(s = \pi/2)$ beobachtet, so bewirkt das Feld außer der Depolarisation auch noch eine Rotation der Polarisationsebene um einen Winkel $\lambda$, der durch

$$\operatorname{tang} 2\lambda = \frac{eH}{mA}$$

gegeben ist. Das ist aber gerade die Formel, die Hanle [2]) aus der Oszillatorentheorie abgeleitet hat; (24) gibt natürlich auch die Effekte im schrägen Felde und mit schräger Beobachtung an. Die Polarisation der betreffenden ersten Lymanlinie sollte durch ein Feld von 70 Gauß um $15^0$ gedreht werden.

b) Als zweites Beispiel betrachten wir den Einfluß eines elektrischen Feldes auf dieselbe, dieses Mal aber durch Elektronenstoß angeregte Lymanlinie. Wir nehmen an, daß die Elektronen gerichtet sind und daß sie ungefähr die Anregungsenergie ursprünglich besitzen. Wir wählen die Polarachse dem Felde $\mathfrak{F}$ parallel und messen $\varphi$ und $\varrho$ von der $(\mathfrak{F}, \mathfrak{v}_1)$-Ebene; dann haben wir die Eigenfunktionen nach der Integrabilitätsbedingung (2 a) zu wählen. Sie sind

$$\left.\begin{aligned}\psi_1 &= \frac{b}{\sqrt{2}}e^{-\frac{r}{2a}}\left[\frac{r}{2a}(1 - \cos \vartheta) - 1\right], \\ \psi_2 &= \frac{b}{\sqrt{2}}e^{-\frac{r}{2a}}\left[\frac{r}{2a}(1 + \cos \vartheta) - 1\right], \\ \psi_3 &= b\,e^{-\frac{r}{2a}}\left[\frac{r}{2a}\sin \vartheta \cos \varphi\right], \\ \psi_4 &= b\,e^{-\frac{r}{2a}}\left[\frac{r}{2a}\sin \vartheta \sin \varphi\right],\end{aligned}\right\} \quad (25)$$

42                                 J. R. Oppenheimer,

mit $b = (8\,\pi\,a^3)^{-1/2}$ und $a = \dfrac{h^2}{4\,\pi^2\,m\,e^2}$. Die entsprechenden $\omega$ sind [1]:

$$\left.\begin{aligned}
\omega_1 &= \omega_0\left[-\frac{i}{\sqrt{6}} - \frac{1}{\sqrt{2}}\cos\gamma\right], \\[2mm]
\omega_2 &= \omega_0\left[-\frac{i}{\sqrt{6}} + \frac{1}{\sqrt{2}}\cos\gamma\right], \\[2mm]
\omega_3 &= \omega_0\sin\gamma, \\[2mm]
\omega_4 &= 0.
\end{aligned}\right\} \qquad (26)$$

Die $p$-Matrix ist wegen

$$p_{13} = p_{31} = p_{32} = p_{23} = p_{4j} = 0$$

reduzierbar. Setzen wir

$$\mu_{11} = \mu_{22} = -\mu_{12} = -\mu_{21} = \tfrac{1}{2}\mu_{33} = \mu;$$

$$\delta = \frac{3\,\pi\,i\,e\,a\,F}{2\,h}; \qquad \mu' = +\sqrt{\mu^2 + \delta^2},$$

so wird mit (13a), (13b) und (14)

$$\left.\begin{aligned}
\alpha_{11} &= -\mu + \mu' - \delta; & p_{11} &= \frac{\omega_0}{2\,\mu'}\left\{-\frac{\alpha_{12}\,i}{\sqrt{6}} - \frac{\alpha_{21}}{\sqrt{2}}\cos\gamma\right\}, \\[2mm]
\alpha_{12} &= -\mu - \mu' - \delta; & p_{12} &= \frac{\omega_0}{2\,\mu'}\left\{\frac{\alpha_{11}\,i}{\sqrt{6}} + \frac{\alpha_{22}}{\sqrt{2}}\cos\gamma\right\}, \\[2mm]
\alpha_{21} &= -\mu + \mu' + \delta; & p_{21} &= \frac{\omega_0}{2\,\mu'}\left\{-\frac{\alpha_{22}\,i}{\sqrt{6}} + \frac{\alpha_{11}}{\sqrt{2}}\cos\gamma\right\}, \\[2mm]
\alpha_{22} &= -\mu - \mu' + \delta; & p_{22} &= \frac{\omega_0}{2\,\mu'}\left\{\frac{\alpha_{21}\,i}{\sqrt{6}} - \frac{\alpha_{12}}{\sqrt{2}}\cos\gamma\right\}, \\[2mm]
\alpha_{33} &= -2\,\mu; & p_{33} &= \omega_0\sin\gamma;
\end{aligned}\right\} \quad (27)$$

---

[1] Man setze in M. Born, l. c. Gleichungen (17), (25), 27)

$$k_1 \sim 0, \quad a\,k_0 = \frac{\sqrt{3}}{2}, \quad \alpha_1 = \gamma, \quad \beta_1 = 0.$$

Dann sieht man, daß die Anregungsamplituden $\pi_1$, $\pi_2$, $\pi_3$, $\pi_4$ für $u^{(1)}_{04}$, $u^{(1)}_{03}$, $u^{(1)}_{01}$ $u^{(1)}_{02}$ sich wie $\sin\gamma$, $\cos\gamma$, $\dfrac{-i}{\sqrt{3}}$, 0 verhalten. Andererseits ist $\omega_1 = \dfrac{1}{\sqrt{2}}(-\pi_2 + \pi_3)$, $\omega_2 = \dfrac{1}{\sqrt{2}}(\pi_2 + \pi_3)$, $\omega_3 = \pi_1$, $\omega_4 = \pi_4$. Daraus folgt (25).

**Zur Quantenmechanik der Richtungsentartung.**     43

ferner mit (19), (25) und (27)

$$\tau_1 = -\,\omega_0\,\sqrt{\frac{3\,\mu}{4\,\pi}}\left\{\left(1-\frac{\mu}{\mu'}\right)\cos s\cos\gamma + \frac{i\,\delta}{\mu'\sqrt{3}}\cos s\right\},$$

$$\tau_2 = \omega_0\,\sqrt{\frac{3\,\mu}{4\,\pi}}\left\{\left(1-\frac{\mu}{\mu'}\right)\cos s\cos\gamma + \frac{i\,\delta}{\mu'\sqrt{3}}\cos s\right\},$$

$$\tau_3 = \omega_0\,\sqrt{\frac{3\,\mu}{4\,\pi}}\,\sin\gamma\,\sin s\,\cos\varrho.$$

Für die Strahlung parallel zu $\mathfrak{s}$ polarisiert wird für $\delta \neq 0$

$$\Xi_s = \frac{3\,\omega_0^2}{4\,\pi}\left\{\cos^2\gamma\,\cos^2 s + \sin^2\gamma\,\sin^2 s\,\cos^2\varrho\right\}$$

$$+ \frac{\omega_0^2}{4\,\pi}\left\{\cos^2 s\right\} + \frac{3\,\omega_0^2}{\pi}\,\frac{1}{8-\delta^2}\left\{\sin 2\,\gamma\,\sin 2\,s\,\cos\varrho\right\} \quad (28)$$

und für $\delta = 0$

$$\Xi_s = \frac{3\,\omega_0^2}{4\,\pi}\,\cos^2(\mathfrak{s}\cdot\mathfrak{v}_1).$$

Vernachlässigt man das zweite Glied in (28) (d. h. betrachtet man Resonanzleuchten statt Stoßleuchten), so sieht man, daß in diesem Falle nur ein
schräges Feld eine Depolarisation oder Drehung der Polarisationsebene
verursachen kann, im Gegensatz zu den Beobachtungen von Hanle[1] an
dem Starkeffekt zweiter Ordnung bei Natrium. Man sieht ferner, daß
ein Drittel der Strahlung eine Polarisation hat, die von der Anregung
unabhängig ist. Wir haben hier anscheinend ein Versagen des Prinzips
der spektroskopischen Stabilität, das folgendermaßen zu erklären ist: Für
verschwindendes Feld fällt der metastabile Teil der angeregten Eigenfunktion nicht exponentiell ab — die Strahlung ist gerade die, die man
beobachten würde, falls nur der $k = 1$-Zustand angeregt wäre. Wenn
man jetzt ein kleines Feld anlegt, so kehrt der metastabile Zustand
langsam (mit einer Abklingungskonstante, die gleich der Frequenz-
aufspaltung ist) in den Normalzustand zurück: seine Metastabilität wird
zerstört. Aber die Lebensdauer wächst unbegrenzt an, wenn das Feld
verschwindet. Experimentell würde man bei „Abwesenheit" eines
äußeren Feldes den zweiten Term in (28) als unpolarisierte Strahlung

---

[1] W. Hanle, ZS. f. Phys. **35**, 346, 1926.

44                              J. R. Oppenheimer,

beobachten; ein Feld parallel dem Elektronenstrahl würde die Polarisation steigern, ein senkrechtes sie vermindern.  Hier haben wir einen
Fall, wo der Einfluß eines elektrischen Feldes anders ist, je nachdem die Linie durch polarisiertes Licht oder durch Elektronenstoß
angeregt ist, da im ersteren Falle der zweite Term in (28) nicht vorhanden sein würde.

In diesen Überlegungen haben wir den Drehimpuls der Elektronen
in dem atomaren System und des stoßenden Elektrons vernachlässigt.
Die atomaren Elektronenmagnete ändern natürlich die Emissionsmöglich-
keiten, aber sie ändern nichts Wesentliches in den früheren Betrachtungen:
die Strahlung, die durch einen Stoß erzeugt wird, ist im allgemeinen
dieselbe, welche durch eine Lichtwelle mit elektrischem Vektor parallel zu
dem antisymmetrischen Geschwindigkeitsvektor $\varDelta \mathfrak{v}$ erzeugt wird, und
kann klassisch — d. h. mit unabhängigen Wahrscheinlichkeiten — berechnet werden, indem man die Polarachse für das atomare System
parallel $\mathfrak{E}$ legt.  Es sollte betont werden, daß die Strahlung quantenmechanisch — d. h. mit interferierenden Wahrscheinlichkeiten — in allen
Fällen prinzipiell berechnet werden kann[1]).  Nehmen wir jetzt z. B. den
Fall der $D$-Linien, wo die Matrixkomponenten des elektrischen Moments
quantenmechanisch ausgerechnet worden sind[2]).  Wir haben oben gezeigt,
daß die Störungsenergie nur von der Komponente des elektrischen
Moments parallel dem antisymmetrischen Geschwindigkeitsvektor abhängt;
die Theorie sagte dann voraus, daß nur Übergänge stattfinden, wo
diese Komponente des Bahndrehimpulses sich nicht ändert.  Aus der
Arbeit von Heisenberg und Jordan entnehmen wir, daß in der
neuen Theorie diese Komponente des Moments und daher auch die
Störungsfunktion selbst nur nicht verschwindende Matrixkomponenten
hat für die Übergänge, wo diese Komponente des Gesamtdrehimpulses
konstant bleibt.  Dieses zeigt wieder die Berechtigung der Skinnerschen
Annahme.

Zunächst möchte es scheinen, daß der Drehimpuls des stoßenden
Elektrons keinen merklichen Einfluß haben könnte, da erstens die betreffenden Störungsglieder außerordentlich klein sind, und da zweitens
man erwarten könnte, daß der Drehimpuls mit dem Bahnimpuls des
stoßenden Elektrons selbst gekoppelt sein würde.  Skinner hat aber

---

[1]) Das gilt auch, natürlich, für allgemeine Störungen: z. B. elliptisch polarisiertes Licht.

[2]) W. Heisenberg und P. Jordan, ZS. f. Phys. **37**, 263, 1926; vgl. ins-
besondere Gleichung (30).

vermutet, daß dieser Drehimpuls die anomale Polarisation der Quecksilberresonanzlinie (und die geringe Polarisation einiger anderer Linien) erklären könnte. Es scheint festgestellt zu sein, daß diese Polarisation weder einem Kaskadensprung aus dem $2\,{}^3S_1$-Niveau[1]), noch zerstreuter Strahlung zuzuschreiben ist[2]); und die Polarisation ist für Spannungen gemessen worden[3]), wo sie nach den vorhergehenden Betrachtungen denselben Sinn haben muß, wie für die Anregungsspannung selbst. Wir müssen also mit Skinner schließen, daß die klassische Theorie diesen Effekt nicht zu erklären vermag. Um uns einer nicht ganz genauen Schreibweise zu bedienen, können wir sagen, daß das stoßende Elektron die Richtung seines Drehimpulses während des Stoßes ändern müßte. Nun entsprechen die Anregungen der anomalen Linien in allen Fällen Interkombinationsübergängen, und man wird vermuten, daß es daher nicht berechtigt ist, die Störung des stoßenden Elektrons gegenüber der Kopplung zwischen dem atomaren Elektronendrehimpulsvektor und dem atomaren Bahnimpulsvektor als klein zu betrachten. Wir können das etwas anders formulieren: die Wahrscheinlichkeit einer „Vertauschung" des stoßenden Elektrons mit einem atomaren Elektron ist nicht zu vernachlässigen gegenüber der Wahrscheinlichkeit eines Interkombinationsübergangs im Atom. Wir müßten also den Eigendrehimpuls des stoßenden Elektrons mit dem atomaren Eigendrehimpuls als gekoppelt betrachten[4]). Dann würden Übergänge, bei denen der Betrag des gesamten Eigendrehimpulsvektors sich änderte, nur selten vorkommen. Das bedeutet aber, daß für die betreffenden Anregungen der Drehimpuls des stoßenden Elektrons sich „umdrehen" muß. Es ist dann nicht schwer zu sehen, daß man auf diese Weise die kleine senkrechte Polarisation der Resonanzlinie erklären kann, da in erster Näherung die Bahnimpulse der atomaren und des stoßenden Elektrons bzw. der Eigenimpulse dieser Elektronen untereinander gekoppelt sind. Nur nachträglich ist der Gesamtimpuls mit dem Gesamteigendrehimpuls gekoppelt. Man hat also zunächst den Stoß ohne Berücksichtigung der Eigenimpulse auszurechnen, und findet, wie oben gezeigt, daß der Bahnimpuls senkrecht auf $\varDelta\mathfrak{v}$ liegen muß. Dieser muß dann mit dem Eigenimpulsvektor so gekoppelt sein, daß der Gesamtdrehimpuls der des betreffenden Terms ist. Nach dieser Theorie sollten die von ${}^3S$-Termen ausgehenden Linien fast unpolarisiert sein,

---

[1]) A. Ellett, P. Foote, F. Mohler, l. c.

[2]) H. Skinner, l. c.

[3]) B. Quarder, ZS. f. Phys., l. c.

[4]) Vgl. W. Heisenberg, ZS. f. Phys. **41**, 251, 1927.

46    J. R. Oppenheimer, Zur Quantenmechanik der Richtungsentartung.

die von $^3D$-Termen ausgehenden dagegen fast ihre „berechnete" Polarisation haben.  Das stimmt auch mit den Experimenten überein. Nach dieser Theorie würde ferner ein Wasserstoffion trotz eines Eigendrehimpulses eine normal polarisierte Resonanzlinie erregen.  Aber wenn auch diese Erklärung schon prinzipiell als befriedigend anzusehen sein sollte, so würde sie zu ihrer Bestätigung doch einer viel eingehenderen Untersuchung bedürfen.

Herrn Prof. M. Born bin ich für seine fördernde Besprechung herzlich dankbar.

Göttingen, Institut für theoretische Physik.

# ON THE QUANTUM THEORY OF THE POLARIZATION OF IMPACT RADIATION

### By J. R. Oppenheimer*

JEFFERSON PHYSICAL LABORATORY, HARVARD UNIVERSITY

Communicated October 26, 1927

The general theory of the polarization of light excited by electronic impact has been given in a previous paper.[1] In particular, it is shown that the polarization computed on the quantum mechanics is independent of the axis chosen for describing the directional degeneracy of the atom. It is further shown that, if one neglects the spin of the impacting electron and if one assumes that the perturbing interaction energy is small, the polarization of the emitted light may be computed "as if" only those states were excited, for which the component of angular momentum in a certain direction is the same as in the normal state; and the direction is that of the vector difference of the velocities of the incident and scattered electron.   If this result is applied to electrons with just the necessary energy to excite the atom, it yields the momentum rule formulated by Skinner; the excitation does not change the component of angular momentum in the direction of the electron beam.   Experimentally[2] one finds that these rules fail in two ways: (1) They do not apply to intercombination lines; in particular, they fail conspicuously for radiation from the $^3S_1$ and $^3P_1$ states of mercury; and (2) the observed polarization tends, as the energy of the exciting electron is diminished, first, towards a maximum, which in the case of normal lines is proportional to and of the same sign as that given by the rule, and, as the energy is further decreased, tends to zero.   In this paper we shall show that the first of these discrepancies may be explained when one takes into account the spin of the impacting electron and the Heisenberg resonance principle;[3] and that the second follows when one discards the assumption that the perturbing energy is small. The theory may thus be made to give a complete account of the experimental results.

We shall first show that, as the velocity of the exciting electron approaches that of the resonance potential in question, the polarization of the excited line tends to zero. If $\psi_n(X)$ be the wave functions of the atom, we may write[4] the wave function for the system atom + electron:

$$\Psi(X; r, \vartheta, \varphi) = \sum_{n=0} \psi_n(X)\chi_n(r, \vartheta, \varphi) = \sum_{n=0} \sum_{p=0} \sum_{s=-p}^{s=+p} \psi_n(X) Y_{ps}(\vartheta, \varphi) y_{nps}(r) \quad (1)$$

where $\qquad r^{-2}(r^2 y'_{nps})' + [\lambda_n^2 - p(p+1)r^{-2}]y_{nps} + \alpha_{nps}(r) = 0 \qquad (2)$

$$\alpha_{nps}(r) = \int_0^{2\pi} d\varphi \int_0^{\pi} \sin \vartheta d\vartheta Y_{ps}^*(\vartheta, \varphi) \sum_{n'=0} V_{nn'} \chi_{n'}. \qquad (3)$$

Here $\dfrac{h^2\lambda_n^2}{8\pi^2 m}$ is the excess of the energy of the exciting electron over the resonance energy; $r, \vartheta, \varphi$, are the polar coördinates of the electron with the polar axis chosen in the same arbitrary direction as for the atom; $Y_{ps}$ are the normalized spherical harmonics; and $V$ is the interaction energy of atom and electron. Whenever $\lambda_n^2$ is positive, $y_{nps}$ will be the amplitude of the probability of the excitation of the state $n$, which is associated with that electronic wave which behaves asymptotically like

$$\frac{e^{-i\lambda_n r}}{r} \cdot Y_{ps}(\vartheta, \varphi). \qquad (4)$$

Thus $y_{n00}$ is the amplitude for the excitation which is independent of the initial direction of the electron; if $y_{nps} = 0$ for $p \neq 0$, the light emitted from the state $n$ would be unpolarized. We have thus to show that

$$\lim_{\lambda_n \to 0} y_{nps}/y_{n00} \longrightarrow 0 \qquad \text{For } p \neq 0, n \neq 0. \qquad (5)$$

Now, uniformly in $\lambda_n \longrightarrow 0$, for $n \neq 0$,

$$\alpha_{nps} \leqslant A/r^{\alpha+\epsilon} \text{ for } r \geqslant r_0; \qquad \alpha_{nps} \leqslant A/r \text{ for } r \leqslant r_0; \qquad \epsilon > 0. \qquad (6)$$

For the non-diagonal terms this is shown in l. c. 2. For the diagonal term $V_{nn}\chi_n$ it follows from the fact that the dipole moment of the atom has no diagonal elements; and this excludes only the case of atomic hydrogen, which has a first order Stark effect.

If we now use, as in l. c. 2, for the solution of (2),

$$y_{nps} = c_+ \xi^{-1/2} J_{+p+1/2}(\xi) + c_- \xi^{-1/2} J_{-p-1/2}(\xi) + I(\xi); \quad \xi = \lambda_n r; \quad (7)$$

with
$$I(\xi) = \int_{+\infty}^{\xi} dx\, G(x,\,\xi)\, \alpha_{nps}(x/\lambda_n);\tag{8}$$

$$G(x,\,\xi) = \frac{J_{+p+1/2}(x)\, J_{-p-1/2}(\xi) \,-\, J_{-p-1/2}(x)\, J_{+p+1/2}(\xi)}{J_{+p+1/2}(x)\, J'_{-p-1/2}(x) \,-\, J'_{+p+1/2}(x)\, J_{-p-1/2}(x)}\tag{9}$$

we find, for the asymptotic outgoing wave associated with that solution $(F)$ which is everywhere finite, and which corresponds to no incident wave— since there are initially no atoms in the excited state $n$—

$$2c_{nps}/\xi.\,e^{-i(\xi+p/2\pi)}\tag{10}$$

where
$$c_{nps} = \lim_{\xi \to 0} \int_{+\infty}^{\xi} dx\, F(x,\,\xi)\alpha_{nps}(x/\lambda_n)\tag{11}$$

$$F(x,\,\xi) = \frac{-G(x,\,\xi).\,\xi^{1/2}}{J_{-p-1/2}(\xi)}.\tag{12}$$

With
$$\lim_{\xi \to 0} F(x,\,\xi) \leqslant Ax^{\alpha+p} \quad \text{for } x \leqslant x_0;\tag{13a}$$

$$\lim_{\xi \to 0} F(x,\,\xi) \leqslant Ax \quad \text{for } x \geqslant x_0;\tag{13b}$$

this gives
$$\lim_{\lambda_n \to 0} c_{nps}/c_{n00} \sim \lambda^p \longrightarrow 0 \quad \text{for } p \neq 0.\tag{14}$$

This establishes our first point.

As was explained in l. c. 1,[5] the spin of the exciting electron is important only because of Heisenberg's resonance: there is a chance that the colliding electron will change places with one of the valence electrons of the atom. For normal lines this effect is not important, since in the excitation of these lines the magnitude of the total spin vector of the system will not change; the probability of interchange of electrons will thus be of the order of the interaction energy of the spins compared with that of the atomic spin and atomic orbit, and this will be negligible. It is thus permissible, for these cases, to couple the atomic spin and atomic orbital angular momentum before applying the perturbation; and this leads at once to Skinner's rules for the *total* angular momentum of the atom. For intercombination excitations the situation is different: the total spin vector can only be conserved if the spin of the impacting electron changes during the excitation. This form of excitation $(A)$ will be more probable than that $(B)$ in which the atomic coupling remains undisturbed, in the same way in which normal transitions are more probable the intercombination transitions in an atom. We may thus expect that when a state can be excited by $(A)$, the momentum rules will fail, and that the correct polarization must be computed by coupling spin and angular momentum *after* considering the perturbation; when $(A)$ is impossible, the lines should be considerably weaker, and should be polarized in the same way, but not so completely, as predicted by the momentum rules.[6]

We shall now show how to compute the polarization of lines excited by scheme $(A)$. If we treat the collision problem with neglect of all spins, we find[7] that the component of atomic orbital angular momentum parallel to the vector difference of velocities of incident and scattered electron remains unchanged by the excitation. We thus know $k$, $s$ and the component in this direction, $k_z$, of $k$; what we want to know is the probability amplitude of $j$, $k$, $s$, $j_z = m$. For from these amplitudes, for given $j$ and all $m$, we can calculate, by the methods of l. c. 1, the polarization of the spontaneous radiation from the state $j$, $k$, $s$. Thus we need only the transformation functions from the quantities $k$, $s$, $k_z$, $s_z = m - k_z$, to $j$, $k$, $s$, $m$. This problem can be solved by the canonical methods of the quantum mechanics. For if we can write $j$ as a function of the quantities $k$, $s$, $k_z$, $s_z$, and other quantities whose laws of commutation with these are known, then the required amplitude will be given by

$$\{j(k', s', k_z', m' \ldots .) - j'\}(j', k', s', m'/k', s', k_z', m') = 0. \quad (15)$$

Now

$$j(j + 1) = k(k + 1) + s(s + 1) - 2(k.s); \quad (k.s) = k_x s_x + k_y s_y + k_z s_z; \quad (16)$$

so that

$$j(j + 1) = k(k + 1) + s(s + 1) - 2[k_x s_x + k_y s_y + k_z s_z]. \quad (17)$$

Instead of solving (15) directly as a matrix equation, we may, by repeated applications of the formula

$$(x'/z') = \int (x'/y') \, dy'(y'/z') \quad (18)$$

obtain    $\gamma(j', k_z', m') = (j', k', s', m'/k', s', k_z', m') =$

$$\sum_{k_x', k_y', s_x', s_y', q_x', q_y'} (s_x'/s_z')(s_y'/s_z')(k_x'/k_z')(k_y'/k_z')\delta_{m'-k_z'}^{s_z'} \, \delta_{s_x' k_x'}^{q_x'} \, \delta_{s_y' k_y'}^{q_y'} \cdots$$

$$\delta_{2q_x' + 2q_y' + 2k_z' s_z'}^{j'(j'+1) - s'(s'+1) - k'(k'+1)}. \quad (19)$$

The moduli of the functions

$$(s_x'/s_z') \ldots \ldots (k_y'/k_z')$$

are known, but their phases are indeterminate, and one must average over them in the final result. The phases of

$$(s_x'/s_z') \quad \text{and} \quad (k_x'/k_z')$$

are independent of each other, but those of

$$(s'_x/s'_z), \quad (s'_y/s'_z) \quad \text{and of} \quad (k'_x/k'_z), \quad (k'_y/k'_z)$$

are not.  One has thus to average only over the independent phases of $(s'_x/s'_z)$ and $(k'_x/k'_z)$.

We may apply these results to the case of mercury, which has been studied in detail by Skinner.[2]  Let us take first the simple case of the $^3S_1$ states.  Here

$$k' = 0, \quad j' = s' = 1; \quad m' = \pm 1, 0;$$

and one readily finds

$$\left|\, \gamma(1, 0, 0)\, \right|^2 = \left|\, \gamma(1, 0, \pm 1)\, \right|^2 = 1. \tag{20}$$

The radiation from this state is thus unpolarized on scheme $(A)$.  If it were excited by $(B)$, it would be completely polarized.  Actually Skinner finds that the polarization is either absent or extremely weak.

Another case of interest is that of $^3P_1$, which gives the resonance line.  Here $j' = k' = s' = 1; m' = 0, \pm 1;$

$$(0/0) = \int_0^{2\pi} d\varphi \int_0^{\pi} \sin \vartheta d\vartheta \, \sin \vartheta \, \cos(\varphi + \beta) \, \cos \vartheta = 0 \tag{21}$$

so that $\gamma(1, 0, 0) = 0$.  Further, (19) gives

$$\left|\, \gamma(1, 0, \pm 1)\, \right|^2 = 1/2. \tag{22}$$

The polarization of the resonance line is thus of opposite sign and similar magnitude to that given by the momentum rules.  That is what is found.[2]  By the same methods one finds for the $^3P_2$ terms:

$$\left.\begin{array}{l} \gamma(2, 0, 0) = \gamma(2, 0, \pm 2) = 0, \\ \left|\, \gamma(2, 0, \pm 1)\, \right|^2 = 1/2. \end{array}\right\} \tag{23}$$

For the $^3D_1$ and $^3D_2$ states the excitation $(A)$ is impossible.  Lines from these states should, therefore, be weaker than the others, and their polarization should fall rather below that of the momentum rules.  For $^3D_1$ this has been reported by Skinner.[2]  For the $^3D_3$ the excitation $(A)$ is possible; the states $m = 0, \pm 1$ are all excited, but the polarization is only slightly less than that of the momentum rules.  This is also confirmed.

It may be mentioned that the transformation function $\gamma(j', k'_z, m')$ enables one to answer the question: Given that an atom has a certain energy in a weak magnetic field; what is the probability that it will have a given energy (Paschen-Back level) in a strong field?

In conclusion we may summarize the theoretical predictions on the polarization of impact radiation: For very high electronic velocities the line has,[1] insofar as the excitation is direct (in principle this can be separated from cascade effects), a polarization $-p$. As the velocity of the exciting electron is diminished, this polarization also diminishes, and changes sign at a voltage of the order of 200. The polarization then increases with diminishing velocity, and the curve, if extrapolated to the resonance velocity, gives a polarization $+p$. When the velocity of the electron approaches that of the resonance potential, the polarization grows more slowly, and has a maximum within a volt or so of the resonance potential; thereafter it approaches zero. The polarization $p$ may, for normal lines, and for intercombination lines for which the excitation $(A)$ is impossible, be computed by the momentum rule; for other inter-combination lines it is more nearly given by the method (19) outlined above.

[*] NATIONAL RESEARCH FELLOW.

[1] Oppenheimer, J. R., *Z. Physik*, **43**, 27, 1927.  Referred to as l. c., 1.

[2] For the former discrepancy see Skinner, H. W. B., *Roy. Soc. Proc.*, **A112**, 642, 1926.  I am much indebted to Dr. Skinner for informing me of his recent results by letter.

[3] L. c., 1, p. 46.

[4] See a paper on the Ramsauer effect, referred to as l. c., 2, to appear shortly in the *Z. Physik*.

[5] P. 45.

[6] The polarization should be less marked in part because of the relative importance of cascade excitation.

[7] L. c., 1, p. 38.

# EFFECT OF NUCLEAR SPIN ON THE RADIATION EXCITED BY ELECTRON IMPACT

By W. G. Penney*

Department of Physics, University of Wisconsin

Communicated February 13, 1932

Although the polarisation of resonance radiation of components in nuclear spin multiplets has been worked out quite satisfactorily, the treatment is not so complete when the method of excitation is electron impact. It seemed worth while to investigate this point in greater detail. In the case of resonance radiation Ellett,[1] Ellett and MacNair,[2] and Larrick and Heydenburg[3] have worked out the particular case of the resonance line 2537 of mercury, the calculations allowing for the different nuclear spins of the different isotopes. The general method of attack is very simple in principle: the probability of an atom being raised from the ground state by the incident radiation to a certain Zeeman component of a higher level is found. The probability that the atom then falls to a Zeeman component of a lower level is obtained and from these two it is simple to find the polarisation of the radiation emitted. A similar method can be used if the atom is excited by electron impact. There is, however, a considerable complication introduced by the possibility of an atomic and an impinging electron changing over, with or without a change in the total spin of the atom.

The simplest case to work out would be that of an atom with one valence electron, whose nucleus had a spin $1/2$, for example $H$, or even with a general nuclear spin $i$, as would apply to the alkali metals. A further simplification would be to assume that the ground state of the atom had $J = 0$, as then $f = i$ for this level. For experiments on electron impact mercury, helium, zinc and cadmium are especially suitable and these are all two electron systems with a ground state $J = 0$. The actual experimental material is rather meager, sodium and mercury being the only two metals on which any data are available. That the sodium $D$ lines are nearly or completely unpolarised has been shown by several observers.[4] For mercury 2537 there is, however, a definite polarisation depending very much on the electron velocity. Since the nuclear spin of sodium is not known we consider chiefly the case of mercury.

The question of the polarisation of the radiation excited by electron impact has been considered by Oppenheimer[5] and it was claimed in his paper that the methods of the quantum theory give a good account of all the experimental data. However, as pointed out by the same author,[6] the proof that the polarisation tends to zero as the exciting energy is decreased to the excitation potential is not valid because electron interchange is neglected. The rest of the paper of Oppenheimer is mainly qualitative and on pursuing the calculations further the results are disappointing. The present paper is concerned with showing that no theory of polarisation of radiation excited by electron impact can be complete without taking into account nuclear spin and exact expressions are given in a few cases. An endeavor is made to make calculations for mercury 2537, using the first order cross-section and approximate wave functions. The results do not agree at all with experimental data in the region where any measurements have been made. The conclusion is that first order cross-sections simply will not do for this type of calculation, although possibly not such a bad approximation at energies of one or two hundred volts. This conclusion is not surprising but it is interesting to have a case where first order cross-sections fail so badly, since as far as the total cross-section is concerned the results with first order cross-sections are surprisingly good.

*The Atomic Wave Functions.*—The type of coupling we assume is that $L$ is coupled to $S$ to give a resultant $J$ which is then coupled to $i$ to give a resultant $f$. The values $f$ can assume are then $|J + i| \ldots |J - i|$. First of all we neglect the nuclear spin entirely and adopt for the atom the system of quantisation $(L, S, M_L, M_S)$, that is, a system appropriate to the case when a strong field is acting on the atom. By introducing the orbit-spin interaction $(\mathbf{L}.\mathbf{S})$ and making the field strength tend to zero, we get a system of quantisation $(L, J, M_J)$ and can express the wave functions as

$$\psi(L, J, M_J) = \sum_{M_S} S^{LS}_{JM_L M_S} \psi(L, S, M_L, M_S), \qquad (1)$$

where the coefficients $S$ are given by Wigner.[7] For two electron systems, when the singlet levels are close to the triplets, it is necessary to consider all four levels as forming one group and then the easiest way of finding the coefficients $S$ is that of Houston.[8] For Hg, Cd, Zn, it is necessary to make this refinement or else the intersystem lines are not possible and we are very much interested in these.

Introducing the nuclear spin $i$, we can use a system of quantisation $(L, J, M_J, i, M_i)$, which would be suitable if there were a weak magnetic field acting on the atom, strong enough to break down the coupling between $J$ and $i$, but too weak to affect that between $L$ and $S$. The wave functions for this system of quantisation will be

$$\psi(L, J, M_J, i, M_i) = \psi(L, J, M_J)e^{iM_i\phi} I(M_i), \qquad (2)$$

where $I(M_i)$ is the spin function for the nucleus in the state $M_i$. Introducing the coupling $(\mathbf{J}.\mathbf{i})$ and making the field strength tend to zero, we arrive at the system of quantisation needed in our problem $(L, J, f, M_f)$ and the wave functions are

$$\psi(L, J, f, M_f) \;=\; \sum_{M_i} S^{J_i}_{fM_JM_i}\, \psi(L, J, i, M_J, M_i), \qquad (3)$$

where $S^{J_i}_{fM_JM_i}$ is obtained from (2) by making the suitable changes in the symbols. Equation (3) is particularly simple for states with $J = 0$. Then

$$\psi(L, J, f, M_f) \;=\; \psi(L, 0, i, 0, M_i). \qquad (4)$$

*Excitation by Impact.*—It is necessary now to calculate the probability that an electron impinging on the atom should excite it from one of the Zeeman components $(f', M_f')$ of the ground state $1^1S_0$ to one of the Zeeman components $(f, M_f)$ of a higher energy level. In the case of resonance radiation it is necessary to consider the intensity distribution within the line itself. This is quite unnecessary with electron impact since here there is nothing corresponding to resonance, except just at the excitation potential, where the thermal spread of the electrons is quite enough to give uniform excitation over all the hyperfine components. By considering the incident and scattered electrons as plane waves (a good approximation for energies more than 100 volts and a fair approximation at lower energies, not too small) and not considering nuclear spin at all, the author[9] has shown that the chance of exciting the Zeeman component $M_J$ of any excited level J is given by

$$A_{M_J} = \int_0^{\pi} \sin\delta\, d\delta\; k'/k[\,|\,a_1(f_{M_J} - G_{M_J}) + a_3 g_{M_J}\,|^2 \qquad (5)$$
$$+\, |\, a_2 g_{M_J - 1}\,|^2 + |\, a_4 g_{M_J + 1}\,|^2],$$

where $kh/2\pi m$ and $k'h/2\pi m$ are the velocities of the incident and scattered electrons, $\delta$ being the angle of scattering. The coefficients $a$ are the transformation coefficients in the special case when singlets and triplets must be considered together, and the formulae for them are given by Houston.[8] $f$, $G$ and $g$ are complicated definite integrals involving the electron velocity, but whose explicit forms are not needed here. $G_0$ and $g_0$ are by far the most important of the $G_M$ and $g_M$ integrals, but even these are negligible at energies of 30 or 40 volts upward. At the excitation potential $f_{\pm 1}$ vanish but $f_0$ is quite large. As the energy increases $f_{\pm 1}$ increase and $f_0$ diminishes, until for very fast electrons (say 1000 volts) $f_0$ is inappreciable compared with $f_{\pm 1}$.

The effect of the introduction of a nuclear spin will be to modify this formula. From equations (3), (4) and (5) we see

$$A_{f'M_f'fM_f} \;=\; A_{iM_i'fM_f} \;=\; |\,S^{J_i}_{fM_JM_i'}\,|^2\, A_{M_J},$$

since $f = i$ in the ground state. Here $M_J = M_f - M_i'$. Summing over all the excitations to the level $(f, M_f)$ and remembering that $A_1 = A_{-1}$, we find that the total probability of this level being excited is

$$A_{fM_f} = A_0 \mid S^{Ji}_{f0M_f} \mid^2 + A_1[\mid S^{Ji}_{f1M_f-1} \mid^2 + \mid S^{Ji}_{f-1\,M_f+1} \mid^2].$$

*Polarisation of Radiation.*—We have now to calculate the probability that an atom in an excited state $(f, M_f)$ falls to a lower level $(f', M'_f)$. For the parallel components we have $M'_f = M_f$ and for the perpendicular $M'_f = M_f \pm 1$.   Knowing the number in the excited state and the relative intensities of the Zeeman components arising from this level, the product of the two gives the contribution to the parallel and perpendicular components from this $(f, M_f)$.   By summing over $M_f$ we can obtain the polarisation of any hyperfine line and by summing again over $f$ and over the different isotopes we get the polarisation of the unresolved line as it is usually observed.   The relative intensities of the Zeeman components follow immediately from the usual Hönl formulae.[10]   Using these and the populations in the excited states, as already explained, we get immediately the polarisation

$$P = (I_1 - I_2)/(I_1 + I_2),$$

where $I_1$ is the intensity parallel and $I_2$ is the intensity perpendicular to the electron stream, when the radiation is observed at right angles to the electron beam.   We find the following expressions for the polarisation of the radiation emitted in the jump $J = 1$ to $J = 0$ and $f$ to $f'$. The results are summarized in the table.   Here $\lambda$ and $\mu$ denote the probability of exciting the components $M_J = 0$ and $M_J = \pm 1$, respectively, in the case where there is no nuclear spin.   $A_1$ and $A_{\pm 1}$ of equation (5) are thus first approximations to $\lambda$ and $\mu$.

TABLE

| $i$ | $f \longrightarrow f'$ | | $P$ |
|---|---|---|---|
| 0 | 1 | 0 | $(\lambda - 2\mu)/(\lambda + 2\mu)$ |
| $3/2$ | $5/2$ | $3/2$ | $(\lambda - 26\mu)/25(\lambda + 2\mu)$ |
| | $3/2$ | $3/2$ | $(7\lambda - 82\mu)/75(\lambda + 2\mu)$ |
| | $1/2$ | $3/2$ | $-1/3$ |
| $1/2$ | $3/2$ | $1/2$ | $(\lambda - 4\mu)/3(\lambda + 2\mu)$ |
| | $1/2$ | $1/2$ | $-1/3$ |

On summing over $f$ for each value of $i$, we find the polarisation for $i = 3/2$.

$$P = -(\lambda + 224\mu)/225(\lambda + 2\mu)$$

and for $i = 1/2$

$$P = (\lambda - 10\mu)/9(\lambda + 2\mu).$$

*Discussion.*—For the singlet states, where the wave functions with very good approximation are antisymmetrical in the spins, $\mu$ is zero at the excitation potential. Hence the introduction of a nuclear spin·has actually lowered the polarisation (per cent) from 100 for $i = 0$ to 11 for $i = 1/2$ and to $-1/4$ for $i = 3/2$. As the excitation energy increases, $\mu$ increases and for each isotope the polarisation decreases and hence that of the unresolved line also. For very high velocities $\lambda$ vanishes compared with $\mu$ and again the polarisation, instead of being $-100$ is $-56$ and $-50$, respectively. Our results extend the rules given by Skinner.[11] The rule that at the excitation potential only those states are excited for which $\Delta M_J = 0$, and the polarisation is therefore 100 per cent, is seen to require severe modification. It applies only when there is no nuclear spin and also only to singlet states, where there is no possibility of an electron interchange, with a change of spin of the atom. Again, at high velocities the rule $\Delta M_J = \pm 1$, and the polarisation is $-100$, applies only to cases where $i = 0$ (not merely to singlet levels this time because interchange is unimportant at high velocities).

Explanation is needed in the case of intersystem lines and an actual example will make the situation clear. With mercury, the line $1{}^1S_0 - 2{}^3P_1$ ($\lambda$ 2537) is possible because the $2{}^3P_1$ wave function is built up of parts, one antisymmetrical in the spins and the rest symmetrical. In other words, the (**L.S**) coupling has introduced a term antisymmetrical in the spins into the wave function, and it is this part which enables combinations with the ground state to occur. When the atom is excited by electrons with energy less than about 30 volts, the terms which are antisymmetrical in the orbits also contribute on account of interchange, most often with the spin of the emerging electron opposite to that of the incident. Since the antisymmetrical terms are much the larger in the $2{}^3P_1$ wave function, interchange plays a very important part here. A small error in estimating the interchange term leads to a large error in the polarisation and this is probably the reason why our calculations (described later) are so much in error. It does seem remarkable that 2537 is unpolarised at the excitation potential because there are so many features involved which have appreciable effects. It is clear that the effect of the nuclear spin is to diminish the polarisation and the greater the nuclear spin the more complete is the depolarisation. Perhaps the result that the $D$ lines of sodium are unpolarised when excited by electron impact can be attributed to the large nuclear spin, the exact magnitude of which, however, is not known.

It may be mentioned that a theoretical interpretation of the polarisation of the mercury lines in the visible spectrum, which has been observed quite accurately,[11] is impossible at present on account of the cascade effect. One would expect that this objection is not very serious when applied to the 1850 line, since it is so intense that atoms passing through the $2{}^1P_1$ state on their way down to the ground state are relatively few in number compared with those excited directly. However, the difficulties of working in this region make measurements on this line almost impossible.

*Calculations on 2537.*—In the hope that at least a rough check might be obtained on the theory given here, the first order cross-sections in the case of mercury 2537 were estimated and the polarisation calculated as a function of the electron energy. The calculations have been described in a previous paper.[9] The results are most disappointing and are so far from reality that they are only briefly described. The first number is the polarisation per cent for $i = 0$, the second for $i = {}^1/_2$ and the third for $i = {}^3/_2$. The number in brackets is for the unresolved line correcting for the relative abundance and different nuclear spins of the isotopes as given by Schuler and Keyston.[12] At the excitation potential the polarisation is $-92$, $-53$, $-48$, $(-80)$; it rises rapidly to $-34$, $-33$, $-33$, $(-34)$ at 20 volts and then falls slowly to $-54$, $-40$, $-39$, $(-50)$ at 100 volts, becoming more and more negative with increasing energy. According to various observers,[11] there is no polarisation at the excitation potential but it falls steeply to $-30$ at 8 volts, rises to zero and is 10 at 20 volts (where the cascade effects may be considerable). At high velocities it probably becomes more and more negative. The disagreement is striking and shows that care is needed in using first order cross-sections. It seems quite certain that first order cross-sections will not do for calculations on polarisation, although it must be admitted that we have chosen about the worst possible case, where the exchange integrals have a much magnified importance.

In conclusion I wish to express my thanks to Prof. J. H. Van Vleck for many useful suggestions during the course of this work.

*Summary.*—It is shown that no theory of the polarisation of radiation excited by electron impact can be complete without taking into account nuclear spin. Exact expressions are given in a few special cases and applied to the particular case of mercury. The anomalous behavior of 2537 is shown to be due to the exaggerated importance which electron exchange has in exciting the $2^3P_1$ level. An attempt is made to calculate the polarisation using first order cross-sections but these are not accurate enough for this purpose.

* Commonwealth Fund Fellow.

[1] A. Ellett, *Phys. Rev.*, **35**, 588 (1930).

[2] A. Ellett and W. A. MacNair, *Phys. Rev.*, **31**, 180 (1928).

[3] L. Larrick and N. P. Heydenburg, *Phys. Rev.*, **39**, 289 (1932).

[4] W. Kossel and C. Gerthsen, *Ann. d. Phys.*, **77**, 273 (1925); A. Ellett, P. D. Foote and F. L. Mohler, *Phys. Rev.*, **27**, 31 (1926).

[5] J. R. Oppenheimer, *Proc. Nat. Acad. Sci.*, **13**, 800 (1927).

[6] J. R. Oppenheimer, *Phys. Rev.*, **32**, 361 (1928).

[7] E. Wigner, *Gruppentheorie*, p. 206.

[8] W. H. Houston, *Phys. Rev.*, **33**, 297 (1928).

[9] W. G. Penney, *Phys. Rev.*, **39**, 467 (1932).

[10] See for example, H. Weyl, *Gruppentheorie*, p. 159; H. Hönl, *Zeit. f. Phys.*, **31**, 340 (1925).

[11] H. W. B. Skinner, *Proc. Roy. Soc.*, **A112**, 642 (1926); H. W. B. Skinner and E. T. S. Appleyard, *Proc. Roy. Soc.*, **A117**, 224 (1927); B. Quarder, *Zeit. f. Phys.*, **41**, 674 (1927); A. Ellett, P. D. Foote and F. L. Mohler, *Phys. Rev.*, **27**, 31 (1926); J. A. Eldridge and H. F. Olson, *Phys. Rev.* **28**, 1151 (1926).

[12] H. Schüler and J. E. Keyston, *Zeit. f. Phys.*, **72**, 423 (1931).

Reprinted from THE PHYSICAL REVIEW, Vol. 98, No. 6, 1853-1854, June 15, 1955
Printed in U. S. A.

# Collision Alignment of Molecules, Atoms, and Nuclei

NORMAN F. RAMSEY[*]

*Clarendon Laboratory, Oxford, England, and Harvard University, Cambridge, Massachusetts*

(Received April 4, 1955)

RECENTLY, there have been a number of experiments on the polarization of scattered beams of particles. Such polarization is reasonable in scattering since the axial spin vector of the particle can be associated with the axial angular momentum vector that is related to the angle of scattering. However, in the transmitted beam no simple polarization is expected since there is no axial vector with which the particle spin can be associated. However, it is the purpose of this note to point out that if the particle spin is greater than $\frac{1}{2}$, the orientation properties frequently described as alignment[1,2] can be achieved in the transmitted beam, e.g., the spin orientation states corresponding to $M_I = \pm 1$ relative to the direction of motion may be more occupied than the state $M_I = 0$. This alignment of the transmitted beam is related to the more complex orientations achievable in the scattered beam when the spin is greater than $\frac{1}{2}$.[3] However, from an experimental point of view alignment of the transmitted beam instead of a scattered beam has the advantage that the transmitted beam intensity is ordinarily much greater. For particles of spin $I$, the order $k$ of the degree of orientation[2] of the transmitted beam may include any even number up to $2I$.

With beams of molecules attenuated by a scattering vapor, the geometrical shape of the molecule leads to a considerable dependence of the cross section upon $M_J$. For example, with beams of linear molecules attenuated by a monatomic scattering vapor whose kinetic velocities are considerably less than those of the molecular beam, factors of two or more in cross section may exist between the $M_J = 0$ and $M_J = \pm J$ states. For such a case amounts of molecular alignment can be achieved for an attenuation of the beam by only a factor of two or so. For higher atomic velocities of the attenuating gas the amount of alignment is reduced but is still appreciable. In any specific case the amount of alignment produced depends on the molecular shape, the mean velocity of the incident molecule (no alignment can result from pure $S$ scattering), the mean velocity of the scattering atom, and the mean rotational angular velocity of the molecules.

Atoms in electronic $P$ or higher states similarly should be aligned in atomic beams that have been attenuated by scattering.

In experiments with either atoms or molecules, the existence of collision alignment should be observable in many ways. For example, in molecular beam magnetic resonance experiments with molecular hydrogen,[4] separate nuclear resonances occur for different values of $M_J$ so the relative intensity of these resonances should be modified by attenuation with a scattering gas. Alternatively, observations of the apparent cross section in successive scattering attenuators should indicate the occurrence alignment by a diminution in the apparent cross section in the last attenuator in comparison with the first. In such an experiment, however, care must be taken to distinguish this effect from those corresponding to the changes in molecular velocity distribution by the attenuation.

In addition to its value as a tool for the study of collision phenomena and molecular shapes, collision alignment should be useful in some cases as a replacement for the inhomogeneous deflecting fields in molecular and atomic beam magnetic resonance experiments. For example, if an atomic beam passes through two successive attenuation regions, the total attenuation will be less if the atom retains its orientation than if it is reoriented between the two attenuation regions. Consequently, a diminution in transmitted beam intensity indicates the presence of a resonance reorientation of the atom. An atomic beam magnetic resonance experiment with collision alignment of the atoms possesses the advantage that much broader and more intense beams can be used than with the conventional molecular beam resonance method.

At least in principle, collision alignment of transmitted nuclear particles could be achieved for nuclei of spin greater than $\frac{1}{2}$. However, the significance of collision alignment in nuclear research is diminished by the fact that most high-energy nuclear particles have spins of 0 or $\frac{1}{2}$. Although the deuteron has a spin of 1, nuclear attenuation experiments are made more difficult by its electrical charge and consequent ionization energy losses. However, some alignment of the transmitted beam should occur in high-energy deuteron total cross-section experiments[5] and a slight departure from simple exponential attenuation should arise from alignment. Even with incident nucleons of spin 0 or $\frac{1}{2}$, alignment considerations affect nuclear reaction problems. For example, Bohr and Mottelson[6] find nuclear surface deformations as large as 30 percent for some nuclei while the transit time of an incident high-energy nucleon past a nucleus is small compared to the rotational period of the deformed surface. Consequently, even from purely geometrical considerations nuclear reactions will take place dominantly with nuclei in preferred states of alignment.

[*] Partially supported by the John Simon Guggenheim Memorial Foundation.
[1] B. Bleaney, Phil. Mag. **42**, 441 (1951) and O. J. Poppema, thesis, Rijksuniversiteit te Groningen, 1954 (unpublished).
[2] S. R. De Groot, Physica **18**, 1201 (1952).
[3] L. Wolfenstein, Phys. Rev. **92**, 123 (1953); W. Lakin and L. Wolfenstein, Phys. Rev. **90**, 365(A) (1953); and A. Simon, Phys. Rev. **90**, 326 (1955).
[4] Kolsky, Phipps, Ramsey, and Silsbee, Phys. Rev. **87**, 395 (1952).
[5] Millburn, Birnbaum, Crandall, and Schecter, Phys. Rev. **95**, 1268 (1954).
[6] A. Bohr and B. R. Mottelson, Kgl. Danske Videnskab. Selskab, Mat.-fys. Medd. **27**, No. 16 (1953).

[ 113 ]

# THE POLARIZATION OF ATOMIC LINE RADIATION EXCITED BY ELECTRON IMPACT

By I. C. PERCIVAL and M. J. SEATON

*Department of Physics, University College London*

(*Communicated by H. S. W. Massey, F.R.S.—Received 6 February* 1958)

## CONTENTS



The dipole radiation emitted by an atom excited by a unidirectional electron beam has a non-uniform angular distribution which is simply related to the percentage polarization $P$ of the radiation emitted perpendicular to the beam. $P$ was first calculated using the Oppenheimer-Penney (O.-P.) theory. In this theory the probability of excitation of an upper quantum state and the probability of subsequent emission of a polarized photon from such a state are considered independently.

$P$ is finally expressed in terms of the cross-sections $Q_{|M_L|}$ for excitation of states of definite component of angular momentum along the direction of the electron beam. In general, $P$ is dependent on *detailed numerical calculations of* $Q_{|M|}$, but the selection rule $\Delta M_L = 0$ removes this dependence at threshold.

In the O.-P. theory allowance may be made for fine structure and hyperfine structure, but the theory is ambiguous when the f.s. or h.f.s. separations are comparable with the line width. A theory is therefore developed which is based on the calculation of the probability of a polarized photon being emitted by the complete system of atom + electron. The ambiguity of the O.-P. theory is removed by integration over line profiles, but the expressions reduce to O.-P. expressions when the f.s. or h.f.s. separations are much smaller or much larger than the line width. The *Ly*α line of hydrogen is an intermediate case for which the line widths and the h.f.s. separations are comparable.

Assuming the validity of the Born approximation, a simple expression is obtained which allows the $Q_{|M_L|}$ to be calculated from the angular distribution of the scattered electrons.

Theoretical predictions are compared with experimental results. For the Na$D$ lines the predicted polarization is small enough to escape experimental detection. Polarizations observed by Skinner & Appleyard in 1927 for various Hg lines rise to maxima with decreasing electron energy, and then tend to values close to zero at threshold. These experimental results at low energies appear to be inexplicable in terms of the reactions considered, but if the polarization curve above the maximum is extrapolated to threshold, the theory and experiment are found to be in reasonable agreement. Further experimental work is thought to be desirable.

## 1. Introduction

Atomic line radiation excited by an electron beam will, in general, be polarized and will have an anisotropic angular distribution. If the beam is in the $Oz$-direction the radiation may be considered to be due to an electric dipole in the $Oz$-direction and two electric dipoles of equal strength in the $Ox$ and $Oy$-directions. Let $I(\theta)$ be the radiation intensity per unit solid angle in a direction making an angle $\theta$ with $Oz$ and let $I^{\shortparallel}$ and $I^{\perp}$ be the intensities, in a direction perpendicular to $Oz$, with electric vectors respectively parallel and perpendicular to $Oz$. The percentage polarization is defined by

$$P = \frac{100(I^{\shortparallel} - I^{\perp})}{I^{\shortparallel} + I^{\perp}}. \tag{1.1}$$

Using the fact that the intensity of dipole radiation, in a direction making an angle $\chi$ with the dipole axis, is proportional to $\sin^2 \chi$ one obtains

$$I(\theta) = \bar{I}\left[\frac{3(100 - P\cos^2\theta)}{300 - P}\right]. \tag{1.2}$$

The total intensity, integrated over all angles, is $4\pi\bar{I}$. The polarization $P$ may therefore be determined either by measuring $I^{\shortparallel}$ and $I^{\perp}$ or by obtaining the photon angular distribution $I(\theta)$ (Smit 1935). In order to determine absolute cross-sections by optical methods the quantity $\bar{I}$ must be determined. To do this one may combine measurements of $P$ with an absolute measurement of $I(\theta)$ for any one direction.

Experimental determinations of $P$ for a number of cases were made about 30 years ago.† Since that time little experimental work of this type has been attempted but a recent

---

† Kossel & Gerthsen (1925) (Na); Skinner (1926) (Hg); Ellett, Foote & Mohler (1926) (Hg and Na); Eldridge & Olson (1926) (Hg); Quarder (1927) (Hg); Skinner & Appleyard (1927) (Hg); Steiner (1928) (He and Ne); Hanle & Quarder (1929) (Ne); Smit (1935) (He).

## POLARIZATION OF ATOMIC IMPACT RADIATION     115

revival of interest in the subject has resulted from the possibility of using polarization measurements in conjunction with microwave studies (Dehmelt 1956; Lamb 1957; Lamb & Maiman 1957).

The theory of polarization of impact radiation was first developed by Oppenheimer (1927 $a, b$, 1928). The essential idea is to calculate the probabilities of exciting individual quantum states and the probabilities of emission of polarized photons in transitions from these states. The first attempt at detailed calculation was made by Penney (1932) who showed that it was necessary to allow not only for electron spin and fine structure (f.s.) but also for nuclear spin and hyperfine structure (h.f.s.). The theory developed by Oppenheimer and by Penney will be referred to as the O.-P. theory. (For previous discussion of this theory see Bethe (1933) and Lamb (1957).)

We distinguish three atomic energy levels: an *initial* level $a$ (usually the ground level), an *upper* level $b$ populated after collision and before photon emission, and the *final* level $c$ reached after photon emission. We assume the experimental conditions to be such that there is a completely isotropic distribution of spin directions in the incident beam and in the initial states of the atom.

To simplify the discussion we also assume that the initial level of the atom has zero orbital angular momentum. The anisotropy of the problem is then introduced entirely through that of the motion of the incident electron. One calculates cross-sections for exciting quantum states of the upper level with definite orbital angular momentum component $M_L$, the quantization axis being in the direction of the incident beam. If the upper level has well defined f.s. and possibly h.f.s. the upper states must be described in a representation in which spin and orbital angular momenta are coupled. One must therefore develop the algebraic theory required to express cross-sections for excitation of vector-coupled states in terms of those for excitation of $M_L$ states. The corresponding algebraic theory required for the radiative transition probabilities is well known (Condon & Shortley 1951); we shall find it convenient to use expressions in a form obtained by using tensor operator methods (Racah 1942, 1943).

The interaction producing the collisional transition is assumed to be such that total spin and total orbital angular momenta are separately conserved. Since the total component of orbital angular momentum is assumed to be zero the angular momentum of the scattered electron, after exciting the upper state $M_L$, must be equal to $-M_L \hbar$. This rule enables us to calculate the threshold polarization. At the excitation threshold the scattered electron has zero velocity and hence zero orbital angular momentum and therefore only states with $M_L = 0$ can be excited. (A formal proof may be obtained by the method of Wigner (1948).) Threshold polarizations can therefore be calculated without detailed calculation of cross-sections. At energies above the threshold comparison of theory and experiment should provide a sensitive test of the accuracy of calculated cross-sections. In the limit of high energies the relative cross-sections for excitation of various $M_L$ states may be obtained using Born's approximation (in cases for which no change of atom spin occurs during the collision). The high-energy results are of less interest, however, since appreciable population of the upper level by cascade transitions makes comparison with experiment more difficult.

The polarization and angular distribution of emitted photons is related to the angular distribution of scattered electrons. Thus, if the angular distribution of the electrons is isotropic the polarization must be equal to its threshold value. Assuming the validity of Born's approximation a simple relation between polarization and electron angular distribution may be obtained (§ 6·1).

Since we assume that there are no preferred spin directions the complete system becomes more isotropic when additional spin variables are included. Thus, the calculated polarization for an isotope with nuclear spin is generally smaller than that for an isotope of the same chemical element without nuclear spin. It is shown, however, that uncertainty arises in the application of the O.-P. theory if there is doubt about the manner of specifying quantum states. Such uncertainties arise whenever the f.s. or h.f.s. splittings are of a magnitude comparable with the natural line widths, and in the limit of vanishingly small f.s. or h.f.s. separations the O.-P. theory fails to give unambiguous predictions (§ 3·7). A re-examination of the fundamental ideas of the theory appeared desirable, first, because the h.f.s. separations for excited states of hydrogen are less than the line widths, and, secondly, because certain experimental results for Hg are not consistent with theoretical predictions (§ 7·3). Our approach (first suggested to us by Mr L. Castillejo) (§ 4) is to consider the probabilities of photons of definite polarization being emitted by the entire system of atom and colliding electron. Allowing for radiation damping we obtain an expression involving integrals over the line profiles. The O.-P. expressions with h.f.s. are then obtained when there is negligible overlap between the profiles of the h.f.s. components and the O.-P. expressions calculated with neglect of nuclear spin are obtained when the h.f.s. profiles overlap completely. The polarization may also be calculated for intermediate cases.

Although some features of the observed results for Hg are shown to be in satisfactory agreement with theory (§ 7·3), we are unable to offer a theoretical explanation of the fact that the observed polarizations of a number of lines appear to go to zero at excitation thresholds. It is considered that the experiments should be repeated using modern techniques.

## 2. The Oppenheimer–Penney theory

### 2·1. *Radiative transition probabilities*

Let $b$ represent either a simple energy level or a group of closely spaced energy levels and denote the quantum states of $b$ by $\psi_\beta$; the number of such states is equal to the statistical weight $\omega_b$ of $b$. Let $c$ represent a second level or group of levels with states $\psi_\gamma$. And let $\xi$ stand for $x$, $y$ or $z$. Photons having polarization and directional distribution characteristic of the emission by an electric dipole lying in the $\xi$-direction will be referred to as $\xi$-photons. The probability per unit time of an atom in state $\beta$ undergoing the transition $\beta \to \gamma$ with emission of a $\xi$-photon is

$$A_\xi(\beta \to \gamma) = C(\nu_{bc}) \,|\, (\gamma\,|\,\xi\,|\,\beta)\,|^2, \tag{2·1}$$

where

$$C(\nu) = \frac{64\pi^4 e^2 \nu^3}{3hc^3} \tag{2·2}$$

and where $h\nu_{bc} = E_b - E_c$. We here neglect possible energy differences within $b$ or $c$, assuming such differences to be much smaller than $h\nu_{bc}$. The total probability of emission of a $\xi$-photon of frequency $\nu_{bc}$ in transition from $\beta$ is

$$A_\xi(\beta) = \sum_\gamma A_\xi(\beta \to \gamma). \tag{2·3}$$

## POLARIZATION OF ATOMIC IMPACT RADIATION     117

The total probability of emission of a photon of frequency $\nu_{bc}$, obtained on summing (3) over $\xi$, is the same for all quantum states $\beta$ of $b$. We therefore put

$$A(b) = \sum_{\xi} A_{\xi}(\beta). \tag{2·4}$$

### 2·2. *The percentage polarization*

For light propagated in the $Ox$-direction let $I^{\perp}$ and $I'$ be the intensities with electric vector, respectively, perpendicular and parallel to $Oz$. Then

$$I^{\perp} = D \sum_{\beta} n(\beta) A_{y}(\beta) \tag{2·5}$$

and

$$I' = D \sum_{\beta} n(\beta) A_{z}(\beta), \tag{2·6}$$

where $n(\beta)$ is the number density of atoms in state $\beta$ and $D$ is a constant. The percentage polarization is

$$P = 100 \frac{I' - I^{\perp}}{I' + I^{\perp}}. \tag{2·7}$$

Assuming symmetry about $Oz$ we have $A_x = A_y = \frac{1}{2}(A - A_z)$ and $\bar{I} = I' + 2I^{\perp}$, where

$$\bar{I} = Dn(b) A(b) \tag{2·8}$$

and

$$n(b) = \sum_{\beta} n(\beta). \tag{2·9}$$

The expression for the polarization may therefore be written

$$P - 100 \frac{3I' - \bar{I}}{I' + \bar{I}}. \tag{2·10}$$

### 2·3. *Calculation of populations and of rate coefficients*

Suppose that a beam of electrons with electron number density $n(e)$ and velocity $v_a$ is incident on atoms in level $a$ and that the number density $n(\alpha)$ of atoms in quantum states $\alpha$ of $a$ is the same for all $\alpha$. Assume that the states $\beta$ are excited only by collisional transitions from $a$ and that they are depopulated only by radiative transitions to $c$. We denote the cross-sections for $\alpha \to \beta$ transitions by $Q(\alpha \to \beta)$ and put

$$Q(\beta) = \omega_a^{-1} \sum_{\alpha} Q(\alpha \to \beta). \tag{2·11}$$

The rate of transitions to $\beta$ is then $n(e) \, n(a) \, K(\beta)$, where the *rate coefficient* $K(\beta)$ is

$$K(\beta) = v_a \, Q(\beta). \tag{2·12}$$

Equating the number entering $\beta$ to the number leaving, which is $A(b) \, n(\beta)$, we obtain

$$n(\beta) = \frac{n(e) \, n(a) \, K(\beta)}{A(b)}. \tag{2·13}$$

The rate of emission of $\xi$-photons from $\beta$ is

$$n(\beta) \, A_{\xi}(\beta) = n(e) \, n(a) \, K_{\xi}(\beta), \tag{2·14}$$

where the rate coefficient is

$$K_{\xi}(\beta) = \frac{A_{\xi}(\beta)}{A(b)} \, v_a \, Q(\beta). \tag{2·15}$$

The total rate coefficient for emission of $\xi$-photons is

$$K_\xi = \sum_\beta K_\xi(\beta) = \frac{v_a}{A} \sum_\beta A_\xi(\beta)\, Q(\beta) \tag{2.16}$$

and the total rate coefficient for emission of all photons is

$$K = \sum_\xi K_\xi = v_a Q, \tag{2.17}$$

where

$$Q = \sum_\beta Q(\beta). \tag{2.18}$$

The radiation intensity being proportional to the rate coefficient, we have in place of (2.10)

$$P = 100 \frac{3K_z - K}{K_z + K}. \tag{2.19}$$

### 3. Expressions for the polarization

#### 3.1. *Angular momentum conservation*

In order to apply the O.-P. theory we require cross-sections for excitation of the various quantum states of the upper level. We consider the $LS$ coupling representations

$$\beta = \Delta S L M_S M_L \tag{3.1}$$

for the upper states, and

$$\alpha = \Delta' S' L' M_S' M_L' \tag{3.2}$$

for the initial states. We here use $\Delta$ and $\Delta'$ to denote all quantum numbers other than those specifying the angular momenta; for brevity the quantum numbers $\Delta$ and $\Delta'$ will be omitted in most of the following discussion.

To calculate the cross-section $Q(SLM_SM_L)$ we average over spins $m_s'$ of the incident electron and sum over spins $m_s$ of the scattered electron. We then have

$$Q(SLM_SM_L)$$

$$= \frac{v_b}{v_a} \frac{1}{2(2S'+1)(2L'+1)} \sum_{M_S'M_L'm_s m_s'} \int |f_{SLM_SM_Lm_s}(S'L'M_S'M_L'm_s' \mathbf{k}_a \,|\, \hat{\mathbf{k}}_b)|^2 \, \mathrm{d}\omega(\hat{\mathbf{k}}_b), \tag{3.3}$$

where $f_{\beta m_s}$ is the *scattering amplitude* (which is defined in terms of the asymptotic form of the wave function in §4.2) and where the integration is over all directions $\hat{\mathbf{k}}_b$ of the scattered electrons. The incident electron is in the direction of the vector $\hat{\mathbf{k}}_a$.

We now make two simplifying assumptions: (i) The interaction potential producing the transition is assumed not to involve spin co-ordinates and (ii) the initial state $\alpha$ is assumed to have zero orbital angular momentum (i.e. we put $L' = 0$). The first of these gives us the spin conservation condition

$$M_S + m_s = M_S' + m_s'.$$

Since orbital and spin angular momenta are separately conserved and since there is no preferred spin direction, $Q(SLM_SM_L)$ is independent of $M_S$. Furthermore, owing to symmetry about $Oz$, $Q(SLM_SM_L)$ does not depend on the sign of $M_L$. We may therefore put

$$Q(SLM_SM_L) = (2S+1)^{-1} Q_{|M_L|}, \tag{3.4}$$

giving

$$Q(SL) = \sum_{M_L} Q_{|M_L|}. \tag{3.5}$$

## POLARIZATION OF ATOMIC IMPACT RADIATION    119

To obtain the conservation condition for orbital angular momentum we use the expansion

$$f_{\beta m_s}(\alpha m_s' \mathbf{k}_a \mid \hat{\mathbf{k}}_b) = \sum_{lm_l} Y_{lm}(\hat{\mathbf{k}}_b)\, f(\beta l m_s, m_l, \alpha m_s' \mathbf{k}_a), \tag{3.6}$$

where $Y_{lm_l}$ is a normalized spherical harmonic. Taking $\mathbf{k}_a$ to be in the direction of the quantization axis $Oz$ the initial total orbital angular momentum in the $Oz$-direction is zero and the conservation condition is therefore

$$M_L + m_l = 0. \tag{3.7}$$

The two conservation conditions give us the relation

$$\frac{v_b}{v_a}\frac{1}{2(2S'+1)} \sum_{\alpha m_s, m_s'} \int f_{\beta m_s}^*(\alpha m_s' \mathbf{k}_a \mid \hat{\mathbf{k}}_b)\, f_{\beta m_s}^0(\alpha m_s' \mathbf{k}_a \mid \hat{\mathbf{k}}_b)\, d\omega(\hat{\mathbf{k}}_b)$$
$$= \delta(\beta, \beta^0)\, Q(\beta), \tag{3.8}$$

where $\beta = SLM_SM_L$ and $\beta^0 = SLM_S^0 M_L^0$.

### 3.2. Cross-sections for fine-structure levels

With a weak spin-orbit interaction we use, in place of (3.1), the representation

$$\beta = SLJM_J. \tag{3.9}$$

The transformation relation for the scattering amplitude is

$$f_{SLJM_Jm_s} = \sum_{M_S M_L} C_{M_S M_L M_J}^{S\ L\ J}\, f_{SLM_S M_L m_s}, \tag{3.10}$$

where $C$ is a vector coupling or Clebsch–Gordan coefficient. Using (3.8) we obtain the cross-section transformation relation

$$Q(SLJM_J) = \sum_{M_S M_L} (C_{M_S M_L M_J}^{S\ L\ J})^2\, Q(SLM_S M_L). \tag{3.11}$$

It should be emphasized that this relation is valid only so long as $LS$ coupling may be assumed and so long as the initial state of the atom has zero orbital angular momentum.

The cross-section for excitation of the level $SLJ$ is obtained in summing (3.11) over $M_J$. Using (3.4) and the relation

$$\sum_{m_1 m} C_{m_1 m_2 m}^{j_1\ j_2\ j}\, C_{m_1 m_2' m}^{j_1\ j_2'\ j} = \frac{2j+1}{2j_2+1} \delta_{j_2 j_2'} \delta_{m_2 m_2'} \tag{3.12}$$

(Racah 1942), we obtain

$$Q(SLJ) = \frac{2J+1}{(2S+1)(2L+1)}\, Q(SL). \tag{3.13}$$

It is seen that the cross-sections for excitation of the levels $SLJ$ are proportional to the statistical weights $(2J+1)$.

### 3.3. Cross-sections for hyperfine-structure levels

The representation for h.f.s. states is

$$\beta = SLJIFM_F, \tag{3.14}$$

where $I$ is the nuclear spin and $F$ the resultant of $J$ and $I$. For the cross-sections we could use the transformation

$$Q(SLJIFM_F) = (2I+1)^{-1} \sum_{M_J M_I} (C_{M_J M_I M_F}^{J\ I\ F})^2\, Q(SLJM_J). \tag{3.15}$$

120                    I. C. PERCIVAL AND M. J. SEATON ON THE

It is convenient to consider the representation

$$\beta = SITLM_T M_J, \tag{3.16}$$

in which $S$ and $I$ are coupled to give a total atomic spin angular momentum $T$. In this representation we have

$$Q(SITLM_T M_L) = \frac{1}{(2S+1)\,(2I+1)}\, Q_{|M_L|}. \tag{3.17}$$

The representations (3.14), (3.16) are related by the transformation

$$(SITLM_T M_L \,|\, SLJIFM_F) = C^{T\ L\ F}_{M_T M_L M_F}(SI(T)\,LF \,|\, SL(J)\,IF), \tag{3.18}$$

where    $(SI(T)\,LF \,|\, SL(J)\,IF) = [(2T+1)\,(2J+1)]^{\frac{1}{2}}(-1)^{F+S-T-J}\,W(SLIF;\,JT)$    (3.19)

(Racah 1943, equation (5)), $W$ being a Racah coefficient.†  We then have

$$Q(SLJIFM_F) = \frac{2J+1}{(2S+1)\,(2I+1)}\sum_{TM_T M_L}(2T+1)\,[W(SLIF;\,JT)\,C^{T\ L\ F}_{M_T M_L M_F}]^2\,Q_{|M_L|}. \tag{3.20}$$

3.4. *Transition probability transformation relations*

The transition relations are obtained in the most convenient form on expressing the transition probabilities in terms of matrix elements of the tensor operator $\mathbf{T}_1$ with components‡

$$T_{1\mu} = \sum_i r_i C_{1\mu}(\hat{\mathbf{r}}_i), \tag{3.21}$$

where the sum is over all atomic electrons and where

$$C_{\lambda\mu}(\hat{\mathbf{r}}) = \left(\frac{4\pi}{2\lambda+1}\right)^{\frac{1}{2}} Y_{\lambda\mu}(\hat{\mathbf{r}}), \tag{3.22}$$

$Y_{\lambda\mu}$ being a normalized spherical harmonic.  Putting

$$A^{(\mu)}(\beta \to \gamma) = C(\nu_{bc})\,|\,(\gamma\,|\,T_{1\mu}\,|\,\beta)\,|^2, \tag{3.23}$$

we have for electric dipole transitions

$$A_z(\beta \to \gamma) = A^{(0)}(\beta \to \gamma) \tag{3.24}$$

and

$$A(\beta \to \gamma) = \sum_\mu A^{(\mu)}(\beta \to \gamma). \tag{3.25}$$

Since $\mathbf{T}_1$ does not operate on spin variables we have

$$A_z(SLM_S M_L \to S''L'') = \delta(S, S'')\,\delta(M_L, O)\,A(SL \to SL''). \tag{3.26}$$

For transitions between two states $JM_J$, $J''M_J''$ we have, using the theory of tensor operators (Racah 1942, equation (29))

$$A^{(\mu)}(JM_J \to J''M_J'') = (C^{J''\ 1\ J}_{M_J'\ -\mu M_J})^2\,A(J \to J''), \tag{3.27}$$

where

$$A(J \to J'') = \sum_{M_J''\mu} A^{(\mu)}(JM_J \to J''M_J''). \tag{3.28}$$

It should be emphasized that the quantum numbers $SL$, $S''L''$ do not occur in these expressions. Equations (3.27), (3.28) may therefore be used for intercombination electric dipole transitions which can occur only if there are departures from $LS$ coupling.

---

† Tables of these coefficients are available (Simon, Vander Sluis & Biedenharn 1954; Obi *et al.* 1953).

‡ We consider, in effect, the matrix elements of $(x \pm iy)/\sqrt{2}$ and of $z$ instead of those of $x$, $y$ and $z$.

## POLARIZATION OF ATOMIC IMPACT RADIATION    121

In order to calculate the polarization of the total light in an optically allowed multiplet $SL \rightarrow SL''$, assuming the lines of the multiplet to be unresolved, it is necessary to know the relative values of the transition probabilities $A(SLJ \rightarrow SL''J'')$. Using tensor operator methods (Racah 1942, § 5) the required relation is obtained in the form

$$A(SLJ \rightarrow SL''J'') = (2L+1)\,(2J''+1)\,W^2(LJL''J''; S1)\,A(SL \rightarrow SL''), \tag{3.29}$$

where $W$ is a Racah coefficient.

For transitions between h.f.s. states we obtain in a similar way

$$\left.\begin{array}{l} A^{(\mu)}(JIFM_F \rightarrow J''IF''M_F'') = (C^{F'',\,1\,\,F}_{M_F'' \,-\mu\, M_F})^2\, A(JIF \rightarrow J''IF''), \\[2mm] A(JIF \rightarrow J''IF'') = (2J+1)\,(2F''+1)\,W^2(JFJ''F''; I1)\,A(J \rightarrow J''). \end{array}\right\} \tag{3.30}$$

### 3.5. *Expressions for the polarization without h.f.s.*

We continue to assume an initial state with $L' = 0$. With an upper state $SLJ$ and a final state $J''$ we have

$$K_z(SLJ \rightarrow J'') = \frac{v_a A(SLJ \rightarrow J'')}{(2S+1)\,A(SLJ)} \sum_{M_S M_L M_J} [C^{S\,\,L\,\,J}_{M_S M_L M_J} C^{J'',\,1\,J}_{M_J'\,0\,M_J}]^2\, Q_{|M_L|}, \tag{3.31}$$

$$K(SLJ \rightarrow J'') = \frac{v_a A(SLJ \rightarrow J'')}{(2S+1)\,A(SLJ)} \frac{2J+1}{2L+1} \sum_{M_L} Q_{|M_L|}. \tag{3.32}$$

For the total light in the multiplet $SL \rightarrow SL''$ we have

$$K_z(SL \rightarrow SL'')$$
$$= \frac{v_a A(SL \rightarrow SL'')}{(2S+1)\,A(SL)}\,(2L+1) \sum_{JJ''M_S M_L M_J} (2J''+1)\,[C^{S\,\,L\,\,J}_{M_S M_L M_J} C^{J'',\,1\,J}_{M_J'\,0\,M_J} W(LJL''J''; S1)]^2\, Q_{|M_L|}, \tag{3.33}$$

$$K(SL \rightarrow SL'') = \frac{v_a A(SL \rightarrow SL'')}{(2S+1)\,A(SL)} \sum_M Q_{|M|}. \tag{3.34}$$

These formulae are of interest for He, which has zero nuclear spin, and for Hg isotopes having zero nuclear spin. We therefore consider the atomic spin $S$ to be 0 or 1. With upper states for which $L = 0$ the polarization is always zero in $LS$ coupling. For upper $P$ states $(L = 1)$ the polarization formula may be written

$$P = 100 \frac{G(Q_0 - Q_1)}{h_0 Q_0 + h_1 Q_1} \tag{3.35}$$

and for upper $D$ states $(L = 2)$

$$P = 100 \frac{G(Q_0 + Q_1 - 2Q_2)}{h_0 Q_0 + h_1 Q_1 + h_2 Q_2}. \tag{3.36}$$

The coefficients $G$, $h_0$, $h_1$ and $h_2$ are given in tables 1 and 2.

### 3.6. *Expressions for the polarization of $^2P \rightarrow {}^2S$ transitions with h.f.s.*

Particular interest attaches to the $Ly\alpha$ line of hydrogen and to the resonance lines of the alkali metals. We therefore consider the case of an initial $^2S$ state, an upper $^2P$ state and a final $^2S$ state.

From the unitary character of the transformation (3.18) we have

$$\sum_{TM_T M_L} (2T+1)\,(2J+1)\,W^2(SLIF; JT)\,(C^{T,\,L\,\,F}_{M_T M_L M_F})^2 = 1. \tag{3.37}$$

122                    I. C. PERCIVAL AND M. J. SEATON ON THE

With $L = 1$, $S = \frac{1}{2}$ the cross-section expression (3·20) may therefore be written

$$Q(^2\mathrm{P}, JIFM_F) = \frac{1}{2(2I+1)} \{\tau(\tfrac{1}{2}1JIFM_F)\, Q_0 + [1 - \tau(\tfrac{1}{2}1JIFM_F)]\, Q_1\}, \qquad (3\cdot38)$$

with

$$\tau(\tfrac{1}{2}1JIFM_F) = \sum_T (2T+1)(2J+1)[W(\tfrac{1}{2}1IF; JT)C^{T\ IF}_{M_F 0 M_F}]^2. \qquad (3\cdot39)$$

TABLE 1. POLARIZATION FORMULAE FOR He LINES $SLJ \to J'$

$$P(S0J \to J') = 0, \quad P(S1J \to J') = 100\,\frac{G(Q_0 - Q_1)}{h_0 Q_0 + h_1 Q_1}, \quad P(S2J \to J') = 100\,\frac{G(Q_0 + Q_1 - 2Q_2)}{h_0 Q_0 + h_1 Q_1 + h_2 Q_2}$$

| $SLJ$ | $J'$ | $G$ | $h_0$ | $h_1$ | $h_2$ |
|---|---|---|---|---|---|
| 011 | 0 | 1 | 1 | 1 | — |
|  | 1 | −1 | 1 | 3 | — |
|  | 2 | 1 | 7 | 13 | — |
| 022 | 1 | 3 | 5 | 9 | 6 |
|  | 2 | −3 | 3 | 7 | 10 |
|  | 3 | 3 | 15 | 29 | 26 |
| 110 | 1 | 0 | — | — | — |
| 111 | 0 | −1 | 1 | 3 | — |
|  | 1 | 1 | 3 | 5 | — |
|  | 2 | −1 | 13 | 27 | — |
| 112 | 1 | 21 | 47 | 73 | — |
|  | 2 | −7 | 11 | 29 | — |
|  | 3 | 1 | 7 | 13 | — |
| 121 | 0 | 3 | 5 | 9 | 6 |
|  | 1 | −3 | 7 | 15 | 18 |
|  | 2 | 3 | 41 | 81 | 78 |
| 122 | 1 | 3 | 9 | 17 | 14 |
|  | 2 | −3 | 7 | 15 | 18 |
|  | 3 | 3 | 29 | 57 | 54 |
| 123 | 2 | 18 | 41 | 76 | 58 |
|  | 3 | −9 | 11 | 25 | 34 |
|  | 4 | 3 | 15 | 29 | 26 |

TABLE 2. POLARIZATION FORMULAE FOR He MULTIPLETS $SL \to SL'$

$$P(S0 \to S1) = 0, \quad P(S1 \to SL') = 100\,\frac{G(Q_0 - Q_1)}{h_0 Q_0 + h_1 Q_1}, \quad P(S2 \to SL') = 100G(Q_0 + Q_1 - 2Q_2)$$

| $SL$ | $L'$ | $G$ | $h_0$ | $h_1$ | $h_2$ |
|---|---|---|---|---|---|
| 01 | 0 | 1 | 1 | 1 | — |
|  | 2 | 1 | 7 | 13 | — |
| 02 | 1 | 3 | 5 | 9 | 6 |
|  | 3 | 3 | 15 | 29 | 26 |
| 11 | 0 | 15 | 41 | 67 | — |
|  | 2 | 3 | 73 | 143 | — |
| 12 | 1 | 213 | 671 | 1271 | 1058 |
|  | 3 | 213 | 2171 | 4271 | 4058 |

The transition probabilities are given by†

$$A_z(^2\mathrm{P}, JIFM_F \to {}^2\mathrm{S}) = \tau(\tfrac{1}{2}1JIFM_F)\, A, \qquad (3\cdot40)$$

where $A = A(^2\mathrm{P} \to {}^2\mathrm{S})$. We therefore have

$$K_z(^2\mathrm{P}_J \to {}^2\mathrm{S}) = \frac{v_a}{2(2I+1)} \sum_{FM_F} \tau(\tfrac{1}{2}1JIFM_F)\{Q_1 + \tau(\tfrac{1}{2}1JIFM_F)(Q_0 - Q_1)\}. \quad \blacksquare\ (3\cdot41)$$

† This may be obtained on putting $L'' = 0$, $S = \frac{1}{2}$ in (3·29) and (3·30), or may be obtained directly using the transformation (3·18).

POLARIZATION OF ATOMIC IMPACT RADIATION    123

Using (3·12) and the unitary relation for the transformation (3·19) we have

$$\sum_{FM_F} \tau(\tfrac{1}{2}1JIFM_F) = (2J+1)\,(2I+1) \tag{3·42}$$

and therefore

$$K_z(^2P_J \to {}^2S) = \frac{v_a}{2(2I+1)}\{(2J+1)\,(2I+1)\,Q_1 + \mu(\tfrac{1}{2}1JI)\,(Q_0 - Q_1)\}, \tag{3·43}$$

with

$$\mu(\tfrac{1}{2}1JI) = \sum_{FM_F} [\tau(\tfrac{1}{2}1JIFM_F)]^2. \tag{3·44}$$

Using

$$K(^2P_J \to {}^2S) = \frac{2J+1}{6}\{2Q_1 + Q_0\}, \tag{3·45}$$

the polarization formula is found to be

$$P(^2P_J \to {}^2S) = \frac{100[9\mu(\tfrac{1}{2}1JI) - (2I+1)\,(2J+1)]\,(Q_0 - Q_1)}{[3\mu(\tfrac{1}{2}1JI) + (2I+1)\,(2J+1)]\,Q_0 + 3[(2I+1)\,(2J+1) - \mu(\tfrac{1}{2}1JI)]\,Q_1}. \tag{3·46}$$

TABLE 3. POLARIZATION FORMULAE FOR $^2P_J \to {}^2S$ TRANSITIONS

$$P(^2P_{\frac{1}{2}} \to {}^2S) = 0, \quad P(^2P_{\frac{3}{2}} \to {}^2S) = 100\,\frac{G(Q_0 - Q_1)}{h_0 Q_0 + h_1 Q_1}$$

| $I$ | $G$ | $h_0$ | $h_1$ | $P_0(^2P_{\frac{3}{2}} \to {}^2S)$ | $P_\infty(^2P_{\frac{3}{2}} \to {}^2S)$ |
|---|---|---|---|---|---|
| 0 | 3 | 5 | 7 | 60·0 | −42·9 |
| $\frac{1}{2}$ | 15 | 37 | 59 | 40·5 | −25·4 |
| 1 | 33 | 161 | 289 | 20·5 | −11·4 |
| $\frac{3}{2}$ | 81 | 427 | 773 | 19·0 | −10·5 |

TABLE 4. POLARIZATION FORMULAE FOR $^2P \to {}^2S$ MULTIPLETS

$$P(^2P \to {}^2S) = 100\,\frac{G(Q_0 - Q_1)}{h_0 Q_0 + h_1 Q_1}$$

| $I$ | $G$ | $h_0$ | $h_1$ | $P_0(^2P \to {}^2S)$ | $P_\infty(^2P \to {}^2S)$ |
|---|---|---|---|---|---|
| 0 | 3 | 7 | 11 | 42·9 | −27·3 |
| $\frac{1}{2}$ | 15 | 53 | 91 | 28·3 | −16·5 |
| 1 | 33 | 236 | 439 | 14·0 | − 7·5 |
| $\frac{3}{2}$ | 27 | 209 | 391 | 12·9 | − 6·9 |

To obtain the polarization for the total light from the $^2P \to {}^2S$ multiplet we sum (3·43) and (3·45) over $J$. We obtain

$$P(^2P \to {}^2S) = \frac{100[9\mu(\tfrac{1}{2}1I) - 6(2I+1)]\,(Q_0 - Q_1)}{[3\mu(\tfrac{1}{2}1I) + 6(2I+1)]\,Q_0 + 3[6(2I+1) - \mu(\tfrac{1}{2}1I)]\,Q_1} \tag{3·47}$$

with

$$\mu(\tfrac{1}{2}1I) = \sum_J \mu(\tfrac{1}{2}1JI). \tag{3·48}$$

For purposes of tabulation the polarization formulae are conveniently written as in equation (3·35).

The coefficients have been evaluated for $I = 0, \tfrac{1}{2}, 1$ and $\tfrac{3}{2}$. For all cases it is found that $P(^2P_{\frac{1}{2}} \to {}^2S)$ is zero. The coefficients $G$, $h_0$ and $h_1$ for $P(^2P_{\frac{3}{2}} \to {}^2S)$ are given in table 3 and those for $P(^2P \to {}^2S)$ are given in table 4. We include values of $P_0$, the threshold polarization $((Q_1/Q_0) = 0)$, and of $P_\infty$, the polarization in the limit of high energy, neglecting cascade $((Q_1/Q_0) \to \infty)$.

### 3·7. *The O.-P. theory and the principle of spectroscopic stability*

The requirement of the principle of spectroscopic stability is that quantum mechanical expressions for observable quantities should be independent of the choice of representation. Thus, in the expression for the polarization the same results should be obtained using either one of two different representations so long as both diagonalize the energy matrix for the atom. It will be shown that the O.-P. theory does not always satisfy this requirement.

Consider first the case of $^2\mathrm{P} \to {}^2\mathrm{S}$ transitions with $I = 0$. If the spin-orbit and relativistic energy is sufficiently large to give two completely separate f.s. levels we may use the representation[†]

$$\beta = {}^2\mathrm{P}, JM_J. \tag{3.49}$$

We then have

$$K_z = \frac{v_a}{2} \sum_{M_S M_L J M_J} [C^{\frac{1}{2}\,1\,J}_{M_S M_L M_J} C^{\frac{1}{2}\,1\,J}_{M_J\,0\,M_J}]^2 Q_{|M_L|}, \tag{3.50}$$

which gives

$$P({}^2\mathrm{P} \to {}^2\mathrm{S}) = \frac{300(Q_0 - Q_1)}{7Q_0 + 11Q_1}. \tag{3.51}$$

But the magnitude of the f.s. energy does not appear in the O.-P. theory. In the limit of vanishingly small f.s. energy one should be able to use either the representation (3·49) or the representation

$$\beta = {}^2\mathrm{P}, M_S M_L. \tag{3.52}$$

The second of these gives, in consequence of (3·4) and (3·26)

$$K_z({}^2\mathrm{P} \to {}^2\mathrm{S}) = v_a Q_0 \tag{3.53}$$

and

$$P({}^2\mathrm{P} \to {}^2\mathrm{S}) = 100 \frac{Q_0 - Q_1}{Q_0 + Q_1}. \tag{3.54}$$

The O.-P. theory therefore fails to give unambiguous results in the limit of small f.s. energy. With $I \neq 0$ a similar failure occurs in the limit of small h.f.s. energy.

In the next section it will be shown that these ambiguities arise from assuming that one may calculate separately the probabilities of exciting individual quantum states and the probabilities of emission of polarized photons in transitions from these states.

### 4. Emission of radiation by the complete system

In order to obtain an improved theory we consider the probability of polarized photons being emitted by the complete system of atom plus colliding electron. We proceed by generalizing the usual Bremsstrahlung formula.

### 4·1. *Bremsstrahlung emission by an electron in a central field*

Consider that an electron with energy $E = \tfrac{1}{2}mv^2$ is incident in the direction $\hat{\mathbf{k}}$ on a scattering centre at the origin of the co-ordinate system. The initial wave function has asymptotic form

$$\psi(\mathbf{k} \mid \mathbf{r}) \sim \mathrm{e}^{\mathrm{i}\mathbf{k}\cdot\mathbf{r}} + f(\mathbf{k} \mid \hat{\mathbf{r}}) \frac{\mathrm{e}^{\mathrm{i}kr}}{r}, \tag{4.1}$$

where $k = mv/\hbar$ and where $\hat{\mathbf{r}}$ is the unit radial vector $\mathbf{r}/r$. Let a $\xi$-photon be emitted with energy $h\nu = (E - E')$ in the range $\mathrm{d}(h\nu)$, the electron being scattered with energy

$$E' = \hbar^2 k'^2/2m$$

---

[†] It is assumed at the same time that the f.s. energy is not large enough to cause significant departures from $LS$ coupling.

## POLARIZATION OF ATOMIC IMPACT RADIATION     129

### 4·6. *Final results for the rate coefficients*

An important feature of (4·32) is that the sum over $\beta$ occurs inside the modulus; had this sum occurred outside the modulus we would have been able to integrate (4·32) over the line profile to obtain equation (2·16) of the O.-P. theory.

The rate coefficient for all polarizations

$$K(\nu) = \sum_{\xi} K_{\xi}(\nu),\tag{4·33}$$

may be simplified using the relation

$$C(\nu_{bc}) \sum_{\gamma\xi} (\gamma\,|\,\xi\,|\,\beta)^{*}\,(\gamma\,|\,\xi\,|\,\beta^{0}) = \delta(\beta,\beta^{0})\,A(b)\,;\tag{4·34}$$

this may be proved by the methods of Condon & Shortley (1951, pp. 20 and 71) or by the methods of Racah (1942). Using the expression (4·1) for the cross-section and assuming $b$ and $c$ to be single energy levels we obtain

$$K(\nu)\,\mathrm{d}(h\nu) = \frac{v_a\,Q(b)\,A(b)\,\mathrm{d}\nu}{[2\pi(\nu-\nu_{bc})]^2 - [\tfrac{1}{2}A(b)]^2},\tag{4·35}$$

which is the correct expression for the line profile. It follows from (4·35) that the *width* of the line, defined as the frequency separation of the positions of the profile for which the intensity is half maximum intensity, is $(A/2\pi)$.

Integration of (4·32) over the line profiles may be carried out using the integral (for the sake of simplicity it is assumed that all states $\alpha$ have identical energies):

$$\int_{-\infty}^{\infty} \frac{\mathrm{d}\nu}{[2\pi i(\nu-\nu_{\beta^{0}\gamma}) - \tfrac{1}{2}A]^{*}\,[2\pi i(\nu-\nu_{\beta\gamma}) - \tfrac{1}{2}A]} = \frac{1}{2\pi i\nu_{\beta\beta^{0}} + A},\tag{4·36}$$

where

$$h\nu_{\beta\beta^{0}} = h(\nu_{\beta\gamma} - \nu_{\beta^{0}\gamma}) = E_{\beta} - E_{\beta^{0}}\tag{4·37}$$

and $\beta$ and $\beta^{0}$ are two states of $b$. A convenient expression is obtained on introducing unit complex vectors $\mathbf{F}(\beta)$ satisfying

$$\mathbf{F}^{*}(\beta^{0})\,.\,\mathbf{F}(\beta) = \frac{A}{2\pi i\nu_{\beta\beta^{0}} + A}.\tag{4·38}$$

The rate coefficient for emission of $\xi$-photons in $b\to c$ transitions is then

$$K_{\xi}(b\to c) = \frac{v_b\,C(\nu_{bc})}{A\omega_a} \sum_{\alpha\gamma} \int \Big|\, \sum_{\beta} (\gamma\,|\,\xi\,|\,\beta)\,\mathbf{F}(\beta)\,f_{\beta}(\alpha\mathbf{k}_a\,|\,\hat{\mathbf{k}})\,\Big|^2\,\mathrm{d}\omega(\hat{\mathbf{k}}).\tag{4·39}$$

The rate coefficient for all polarizations is obtained on summing (4·39) over $\xi$. Using (4·10) and (4·34) we obtain $K(b\to c) = v_a\,Q(b)$ in agreement with the O.-P. theory. Further discussion of (4·39) may therefore be restricted to the case $\xi = z$.

### 5. Reconsideration of $^2P \to {}^2S$ transitions

In § 3·7 we showed, taking $^2P \to {}^2S$ transitions as an example, that the O.-P. theory did not always satisfy the principle of spectroscopic stability. The case of $^2P \to {}^2S$ transitions will be reconsidered in the present section using the expression (4·39).

### 5·1. *Zero nuclear spin*

We consider initial states $\alpha = \Delta'\,{}^2S, M_J'$, upper states $\beta = \Delta\,{}^2P, JM_J$ and final states $\gamma = \Delta''\,{}^2S, M_J''$. The f.s. energy separation of the upper states will be $h\delta\nu = E({}^2P_{\frac{3}{2}}) - E({}^2P_{\frac{1}{2}})$. The polarization depends on the ratio

$$\epsilon = 2\pi\delta\nu/A \tag{5·1}$$

of f.s. separation to line width. From (4·39) we have

$$K_z^{(\epsilon)}({}^2P \to {}^2S) = \frac{v_b\, C(\nu_{bc})}{4A}$$

$$\times \sum_{M_J M_J'' m_s' m_s} \int \Big| \sum_{JM_J} ({}^2S, M_J'' \,|\, z \,|\, {}^2P, JM_J)\, \mathbf{F}(J)\, f_{{}^2P, JM_J m_s}({}^2S, M_J' m_s' \mathbf{k}_a \,|\, \hat{\mathbf{k}}) \Big|^2 \, d\omega(\hat{\mathbf{k}}) \tag{5·2}$$

with

$$\mathbf{F}^*(J) \cdot \mathbf{F}(J) = 1 \tag{5·3}$$

and

$$\mathbf{F}^*(\tfrac{1}{2}) \cdot \mathbf{F}(\tfrac{3}{2}) = [\mathbf{F}^*(\tfrac{3}{2}) \cdot \mathbf{F}(\tfrac{1}{2})]^* = \frac{1}{1+\mathrm{i}\epsilon}. \tag{5·4}$$

We now use the relations

$$({}^2S, M_J'' \,|\, z \,|\, {}^2P, JM_J) = C^{\frac{1}{2}\,1\,J}_{M_J\,0\,M_J}\, \delta(M_J'', M_J)\, ({}^2S, M_S \,|\, z \,|\, {}^2P, M_S M_L = 0), \tag{5·5}$$

$$A = C(\nu_{bc}) \,|\, ({}^2S, M_S \,|\, z \,|\, {}^2P, M_S M_L = 0) \,|^2, \tag{5·6}$$

$$f_{{}^2P, JM_J m_s} = \sum_{M_S M_L} C^{\frac{1}{2}\,1\,J}_{M_S\,M_L\,M_J}\, f_{{}^2P, M_S M_L m_s} \tag{5·7}$$

and

$$\delta(M_S M_L, M_S^0 M_L^0)\, \tfrac{1}{2} Q_{|M_L|} = \frac{v_b}{4v_a} \sum_{M_J' m_s m_s'} \int f^*_{M_S M_L m_s}(M_J' m_s' \mathbf{k}_a \,|\, \hat{\mathbf{k}})\, f_{M_S^0 M_L^0 m_s}(M_J' m_s' \mathbf{k}_a \,|\, \hat{\mathbf{k}})\, d\omega(\hat{\mathbf{k}}), \tag{5·8}$$

equation (5·8) following from (3·4) and (3·8). Substitution in (5·2) gives

$$K_z^{(\epsilon)} = \frac{v_a}{2} \sum_{M_S M_L M_J} \Big| \sum_J C^{\frac{1}{2}\,1\,J}_{M_S\,M_L\,M_J}\, C^{\frac{1}{2}\,1\,J}_{M_J\,0\,M_J}\, \mathbf{F}(J) \Big|^2 Q_{|M_L|}. \tag{5·9}$$

We consider the two limiting cases of $\epsilon = 0$ and of $\epsilon \to \infty$. For $\epsilon \ll 1$ the line profiles overlap completely and all vectors $\mathbf{F}(J)$, $\mathbf{F}^*(J^0)$ are parallel. We have therefore

$$K_z^{(0)} = \frac{v_a}{2} \sum_{M_S M_L M_J} \Big| \sum_J C^{\frac{1}{2}\,1\,J}_{M_S\,M_L\,M_J}\, C^{\frac{1}{2}\,1\,J}_{M_J\,0\,M_J} \Big|^2 Q_{|M_L|}. \tag{5·10}$$

From the orthonormality relations for the $C$-coefficients the sum over $J$ is equal to $\delta(M_S, M_J)\, \delta(M_L, 0)$. Therefore

$$K_z^{(0)} = v_a Q_0, \tag{5·11}$$

which agrees with the result obtained in the O.-P. theory if we use the representation $\beta = {}^2P, M_S M_L$. For $\epsilon \gg 1$, on the other hand, there is negligible overlap between the line profiles, and the vectors $\mathbf{F}^*(\tfrac{1}{2})$, $\mathbf{F}(\tfrac{3}{2})$ are orthogonal. We have therefore

$$K_z^{(\infty)} = \frac{v_a}{2} \sum_{M_S M_L JM_J} [C^{\frac{1}{2}\,1\,J}_{M_S\,M_L\,M_J}\, C^{\frac{1}{2}\,1\,J}_{M_J\,0\,M_J}]^2 Q_{|M_L|}, \tag{5·12}$$

which agrees with the result obtained in the O.-P. theory using the representation $\beta = {}^2P, JM_J$.

On expanding (5·9) we readily obtain, for any $\epsilon$,

$$K_z^{(\epsilon)} = \frac{K_z^{(0)} + \epsilon^2 K_z^{(\infty)}}{1+\epsilon^2}. \tag{5·13}$$

## POLARIZATION OF ATOMIC IMPACT RADIATION    131

### 5·2. *Nuclear spin and hyperfine structure*

We assume the f.s. separation to be large compared with the line width. We then obtain the O.-P. theory expression for $\beta = {}^2P, JIFM_F$ if the h.f.s. separations are also much greater than the line width and the O.-P. expression for $\beta = {}^2P, JM_J$ if the h.f.s. separations are much smaller than the line width.

For $I = \frac{1}{2}$ each f.s. level is split into two h.f.s. levels with separation $\delta v_J$. For each value of $J$ we obtain as in the previous section

$$K_z^{(\epsilon,J)}(J) = \frac{K_z^{(0)}(J) + \epsilon_J^2 K_z^{(\infty)}(J)}{1 + \epsilon_J^2}, \tag{5·14}$$

where $\epsilon_J^2 = (2\pi \delta v_J / A)^2$.

### 5·3. *Numerical results for hydrogen*

The value of $A$ for $2p \to 1s$ transitions is $6·25 \times 10^8 \, \mathrm{sec}^{-1}$ giving a line width

$$(A/2\pi c) = 3·3 \times 10^{-3} \, \mathrm{cm}^{-1}.$$

The $2p$ f.s. separation is $0·36 \, \mathrm{cm}^{-1}$.

The h.f.s. separations are $(\delta v_{\frac{3}{2}}/c) = 1·97 \times 10^{-3} \, \mathrm{cm}^{-1}$ and $(\delta v_{\frac{1}{2}}/c) = 0·79 \times 10^{-3} \, \mathrm{cm}^{-1}$. The h.f.s. separations are therefore smaller than the line width but of comparable order of magnitude. We must therefore use (5·13) in order to make an exact calculation of the polarization. The values of $\epsilon_J^2$ are $\epsilon_{\frac{3}{2}}^2 = 0·356$ and $\epsilon_{\frac{1}{2}}^2 = 0·057$. For transitions from $2p_{\frac{1}{2}}$ the polarization is always zero. For $2p_{\frac{3}{2}}$ we obtain

$$P(2p_{\frac{3}{2}}) = 100(Q_0 - Q_1)/(1·694Q_0 + 2·388Q_1), \tag{5·15}$$

compared with $\quad P(2p_{\frac{3}{2}}) = 100(Q_0 - Q_1)/(1·667Q_0 + 2·333Q_1)$

in the O.-P. theory neglecting h.f.s. and

$$P(2p_{\frac{3}{2}}) = 100(Q_0 - Q_1)/(2·467Q_0 + 3·933Q_1)$$

in the O.-P. theory including h.f.s. For the total radiation from the $2p$ level we obtain

$$P(2p) = 100(Q_0 - Q_1)/(2·375Q_0 + 3·749Q_1), \tag{5·16}$$

compared with $100(Q_0 - Q_1)/(2·333Q_0 + 3·667Q_1)$ in the O.-P. theory without h.f.s. and $100(Q_0 - Q_1)/(3·533Q_0 + 6·067Q_1)$ in the O.-P. theory with h.f.s.

It is seen that the exact expressions (5·15), (5·16) differ little from those obtained neglecting h.f.s. This result has been used by Khashaba & Massey (1958) who give the polarization as a function of energy, the cross-sections being calculated using a distorted wave approximation.

For atoms other than hydrogen the h.f.s. separations will generally be large and the theory with inclusion of h.f.s. should be used for all but highly excited states.

For hydrogen we may check that the collision time is small compared with the radiative lifetime; putting $R = v_{2p}/A(2p)$ the condition is that $R$ should be much larger than the Bohr radius $a_0$. We obtain $R = 1·8 \times 10^7 \mathscr{E}^{\frac{1}{2}} a_0$ where $\mathscr{E}$ is the energy of the scattered electron measured in electron volts. The required condition is therefore satisfied for all cases of practical importance.

### 6. The Born approximation

We consider the use of the Born approximation for the calculation of cross-sections for excitation of $M_L$ states. We start by considering transitions in hydrogen.

#### 6·1. *The Born approximation for hydrogen*

For transitions $1s \to nlm$ the Born differential cross-section per unit solid angle is

$$I_{nlm}(K) = \frac{4k_n}{K^4 k_1} \, | \, (nlm \, | \, e^{i\mathbf{K}\cdot\mathbf{r}} \, | \, 1s) \, |^2, \tag{6·1}$$

where
$$\mathbf{K} = \mathbf{k}_n - \mathbf{k}_1 \tag{6·2}$$

is the change of momentum vector. It is supposed that $\mathbf{k}_1$ is in the direction of the quantization axis; $I_{nlm}$ then depends only on the energy and on the angle between $\mathbf{k}_1$ and $\mathbf{k}_n$.

The total cross-section is

$$Q(nlm) = \frac{2\pi}{k_1 k_n} \int_{K_1}^{K_2} I_{nlm}(K) \, K \, dK, \tag{6·3}$$

where $K_1 = (k_1 - k_n)$ and $K_2 = (k_1 + k_n)$ (Mott & Massey 1949, p. 226). Using (4·17) and (4·20) we obtain

$$(nlm \, | \, e^{i\mathbf{K}\cdot\mathbf{r}} \, | \, 1s) = 4\pi \sum_{\lambda\mu} Y^*_{\lambda\mu}(\hat{\mathbf{K}}) \, i^\lambda (nlm \, | \, Y_{\lambda\mu}(\hat{\mathbf{r}}) f_\lambda(Kr) \, | \, 1s) \tag{6·4}$$

and, on carrying out the angular integrations,

$$(nlm \, | \, e^{i\mathbf{K}\cdot\mathbf{r}} \, | \, 1s) = (2l+1)^{\frac{1}{2}} C^*_{lm}(\hat{\mathbf{K}}) \, i^l (nl \, | \, f_l(Kr) \, | \, 1s), \tag{6·5}$$

where $C_{lm}$ is the spherical harmonic operator defined by (3·22) and where

$$(nl \, | \, f_l(Kr) \, | \, 1s) = \int_0^\infty P_{nl}(r) \, f_l(Kr) \, P_{1s}(r) \, dr. \tag{6·6}$$

Hence
$$I_{nlm} = \frac{4k_n}{K^4 k_1} (2l+1) \, | \, C_{lm}(\hat{\mathbf{K}}) \, |^2 \, | \, (nl \, | \, f_l(Kr) \, | \, 1s) \, |^2. \tag{6·7}$$

The differential cross-section $I_{nl}$ is obtained on summing $I_{nlm}$ over $m$. Since $P_l(1) = 1$ we obtain from (3·22) and (4·20)

$$\sum_m | \, C_{lm}(\hat{\mathbf{K}}) \, |^2 = 1 \tag{6·8}$$

and therefore
$$I_{nl}(K) = \frac{4k_n}{K^4 k_1} (2l+1) \, | \, (nl \, | \, f_l \, | \, 1s) \, |^2. \tag{6·9}$$

Hence
$$I_{nlm}(K) = | \, C_{lm}(\hat{\mathbf{K}}) \, |^2 I_{nl}(K) \tag{6·10}$$

and
$$Q(nlm) = \frac{2\pi}{k_1 k_n} \int_{K_1}^{K_2} | \, C_{lm}(\hat{\mathbf{K}}) \, |^2 I_{nl}(K) \, K \, dK. \tag{6·11}$$

We may put
$$| \, C_{lm}(\hat{\mathbf{K}}) \, | = \left(\frac{2}{2l+1}\right)^{\frac{1}{2}} | \, \mathscr{P}_l^{|m|}(\cos\omega) \, |, \tag{6·12}$$

where $\mathscr{P}_l^{|m|}$ is a normalized associated Legendre polynomial, and where $\omega$ is the angle between $\mathbf{K}$ and $\mathbf{k}_1$. Using (6·2) we have $\cos\omega = (\hat{\mathbf{K}}\cdot\hat{\mathbf{k}}_1) = (K^2 + k_1^2 - k_n^2)/2Kk_1$. We thus obtain

$$Q(nlm) = \frac{4\pi}{(2l+1) \, k_1 k_n} \int_{K_1}^{K_2} \left| \mathscr{P}_l^{|m|}\left(\frac{K^2 + 2E_{n1}}{2Kk_1}\right) \right|^2 I_{nl}(K) \, K \, dK, \tag{6·13}$$

## POLARIZATION OF ATOMIC IMPACT RADIATION     133

where $E_{n1} = \frac{1}{2}(k_1^2 - k_n^2)$. This relation between $Q(nlm)$ and $I_{nl}$ may be applied to systems other than hydrogen. It may be noted that, assuming the validity of the Born approximation, (6·13) may be used to determine $Q(nlm)$ from measured angular distributions.

### 6·2. *The high-energy limit*

For any finite value of $K$ we obtain $[(K^2 + 2E_{n1})/2Kk_1] \to 0$ as $k_1 \to \infty$. Since

$$Q(nl) = \frac{2\pi}{k_1 k_n} \int_{K_1}^{K_2} I_{nl}(K)\, K\, \mathrm{d}K, \tag{6·14}$$

we obtain from (6·13)

$$\lim_{k_1 \to \infty}\left[\frac{Q(nlm)}{Q(nl)}\right] = \left(\frac{2}{2l+1}\right)\,|\,\mathscr{P}_l^{|m|}(0)\,|^2. \tag{6·15}$$

Numerical values of this limit are given in table 5.

TABLE 5

| $L$ | $\|M_L\|$ | $\left(\dfrac{2}{2L+1}\right)\|\mathscr{P}_L^{\|M_L\|}(0)\|^2$ | $L$ | $\|M_L\|$ | $\left(\dfrac{2}{2L+1}\right)\|\mathscr{P}_L^{\|M_L\|}(0)\|^2$ |
|---|---|---|---|---|---|
| 1 | 1 | $\frac{1}{2}$ | 3 | 3 | $\frac{5}{16}$ |
|   | 0 | 0 |   | 2 | 0 |
| 2 | 2 | $\frac{3}{8}$ |   | 1 | $\frac{3}{16}$ |
|   | 1 | 0 |   | 0 | 0 |
|   | 0 | $\frac{1}{4}$ |   |   |   |

The physical significance of this result is that at high energies most of the scattering occurs at small angles between $\mathbf{k}_n$ and $\mathbf{k}_1$; as $k_1 \to \infty$ the change-of-momentum vector $\mathbf{K}$ is therefore at right angles to $\mathbf{k}_1$. The polarization observed at high energies will generally be less than that predicted, due to the upper state being populated by cascade. For the alkali metals, however, the cross-sections for excitation of resonance lines will be much larger than all other inelastic cross-sections and the observed polarization at high energies may not be far different from that calculated neglecting cascade.

### 6·3. *Born calculations for* $3\,^1\mathrm{D} \to {}^1\mathrm{P}$ *transitions in* He

With separable He functions

$$\left.\begin{aligned}
\psi(1s^2) &= \psi_{1s}(\mathbf{r}_1)\,\psi_{1s}(\mathbf{r}_2), \quad \psi_{1s}(\mathbf{r}) = (4\pi)^{-\frac{1}{2}}r^{-1}P_{1s}(r),\\
\psi(1snl\,^1L) &= 2^{-\frac{1}{2}}[\psi_{1s'}(\mathbf{r}_1)\,\psi_{nlm}(\mathbf{r}_2) + \psi_{1s'}(\mathbf{r}_2)\,\psi_{nlm}(\mathbf{r}_1)],\\
\psi_{1s'}(\mathbf{r}) &= (4\pi)^{-\frac{1}{2}}r^{-1}P_{1s'}(r), \quad \psi_{nlm}(\mathbf{r}) = Y_{lm}(\hat{\mathbf{r}})\,r^{-1}P_{nl}(r),
\end{aligned}\right\} \tag{6·16}$$

we obtain

$$I_{nl} = \frac{8k_n(2l+1)}{K^4 k_1}\,|\,(1s'\,|\,1s)\,|^2\,|\,(nl\,|\,f_i\,|\,1s)\,|^2. \tag{6·17}$$

We use the ground-state function

$$P_{1s} = 2Z^{\frac{3}{2}}r\,\mathrm{e}^{-Zr} \quad (Z = 27/16),$$

a He$^+$ function for $P_{1s'}$ and a hydrogen $3d$ radial function to obtain

$$(3d\,|\,f_2\,|\,1s) = \frac{128Z^{\frac{3}{2}}K^2\alpha}{27(30)^{\frac{1}{2}}(\alpha^2 + K^2)^4} \quad (\alpha = Z + \tfrac{1}{3}) \tag{6·18}$$

and

$$|\,(1s'\,|\,1s)\,| = \frac{64(2Z)^3}{(2+Z)^6} = 0.9785. \tag{6·19}$$

134                 I. C. PERCIVAL AND M. J. SEATON ON THE

The total cross-section is shown in figure 1. Our results are in agreement with those of Massey & Mohr (1931). The shape of the curve is very similar to that of the experimental curves for $4\,^1\mathrm{D}$ and $5\,^1\mathrm{D}$ (see Bates, Fundaminsky, Leech & Massey 1950) but at $100\,\mathrm{eV}$ the calculated $3\,^1\mathrm{D}$ cross-section is $0\cdot5\times10^{-3}\pi a_0^2$ compared with the much larger value of $2\cdot5\times10^{-3}\pi a_0^2$ measured by Dr D. W. O. Heddle (private communication). The discrepancy

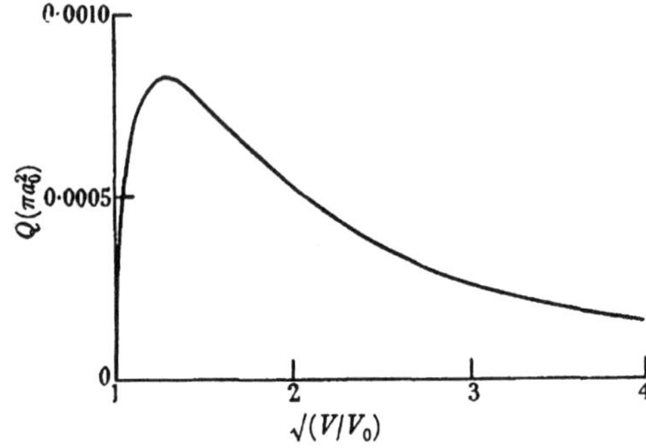


FIGURE 1. Cross-section $Q$ for excitation of He $3\,^1\mathrm{D}$ as a function of $\sqrt{(V/V_0)}$ ($V=$ energy of incident electron, $V_0=$ threshold energy). Born approximation.

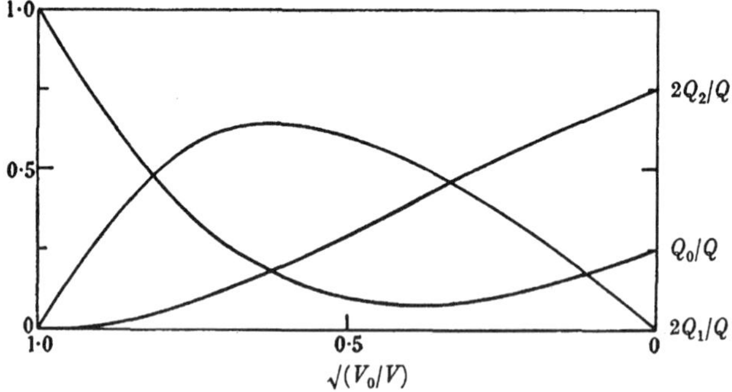


FIGURE 2. Cross-sections $Q_{|M_L|}$ for excitation of He $3\,^1\mathrm{D}$, $M_L$ as functions of $\sqrt{(V_0/V)}$ ($V_0=$ threshold energy, $V=$ energy of incident electron). Curves give $Q_0/Q$, $2Q_1/Q$ and $2Q_2/Q$, where $Q=Q_0+2Q_1+2Q_2$. Born approximation.

may be due to the approximate nature of the wave functions used rather than to the Born approximation being unreliable. The ratios of the cross-sections $Q_{|M_L|}$ may be more accurate than the absolute value of the total cross-section. The integrals in (6·13) have been evaluated numerically; the ratios $Q_0/Q$, $2Q_1/Q$ and $2Q_2/Q$ are shown in figure 2. For the polarization we obtain by the method of §§ 2 and 3,

$$P(^1\mathrm{D}\to{}^1\mathrm{P}) = \frac{300(Q_0+Q_1-2Q_2)}{5Q_0+9Q_1+6Q_2}, \tag{6·20}$$

the nuclear spin being zero for He. The calculated polarization curve is shown in figure 3.

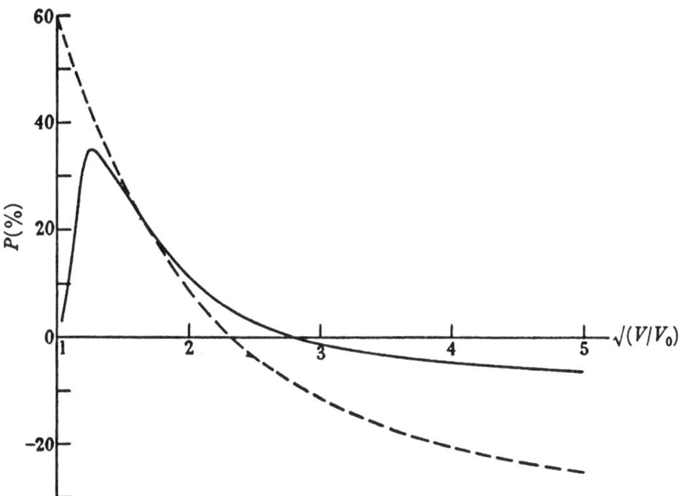


FIGURE 3. Percentage polarization for $^1D \rightarrow {}^1P$ transitions as functions of $\sqrt{(V/V_0)}$. Dashed curve gives calculated result for He $3\,^1D \rightarrow {}^1P$ (Born approximation). Full line curve gives experimental result for Hg $7\,^1D \rightarrow 6\,^1P$ (Skinner & Appleyard 1927).

### 7. COMPARISON WITH EXPERIMENT

#### 7·1. *The sodium D lines*

Polarization of the $D$ lines excited by electron impact has not been detected (Kossel & Gerthsen 1925; Ellett, Foote & Mohler 1926); in the energy range studied the polarization may be less than 2 %. It was first suggested by Penney (1932) that the small polarization is due to h.f.s. but detailed calculations could not be made at that time since the value of $I$ for sodium was unknown. It is now known that for sodium $I = \frac{3}{2}$. We see from table 4 that the predicted polarization is $+13$ % at threshold and $-7$ % at high energies (neglecting cascade).

Just beyond the threshold the polarization may decrease rapidly, since even at low energies many partial waves are important for the calculation of the cross-section (Seaton 1955). The chances of detecting polarization should be better if the $3p_{\frac{1}{2}} \rightarrow 3s$ line were isolated, since for this the values are $+19$ % at threshold and $-10·5$ % at high energies. Further experimental work would be worthwhile.

#### 7·2. *Transitions in* He

The only published results giving the polarization as a function of electron energy are those of Lamb & Maiman (1957) for $3\,^3P \rightarrow 2\,^3S$. These are shown in figure 4. The calculated threshold polarization is $36·6$ %. Lamb (1957) has calculated the threshold polarization as a function of the strength of an applied magnetic field. In the limit of zero field strength his results agree with ours. Further measurements at lower energies are required in order to check whether the polarization tends to the calculated threshold value.

### 7·3. *Transitions in* Hg

Mercury consists of a mixture of isotopes with normal abundances of 70 % for $I = 0$, 17 % for $I = \frac{1}{2}$ and 13 % for $I = \frac{3}{2}$. For a first approximation we may neglect the isotopes with $I \neq 0$ and neglect departures from $LS$ coupling in calculating collision cross-sections; this is the approximation of table 1. An idea of the error introduced by this approximation is given by the work of Penney (1932) for the transitions $6s\,6p\,{}^3\mathrm{P}_1 \to 6s^2\,{}^1\mathrm{S}_0$. From table 2 we obtain a threshold† polarization $P = -100\,\%$ while with allowance for departures from $LS$ coupling Penney obtains $P = -92\,\%$ for $I = 0$, $-53\,\%$ for $I = \frac{1}{2}$ and $-48\,\%$ for $I = \frac{3}{2}$; with the normal isotope mixture Penney's calculations give $P = -80\,\%$.

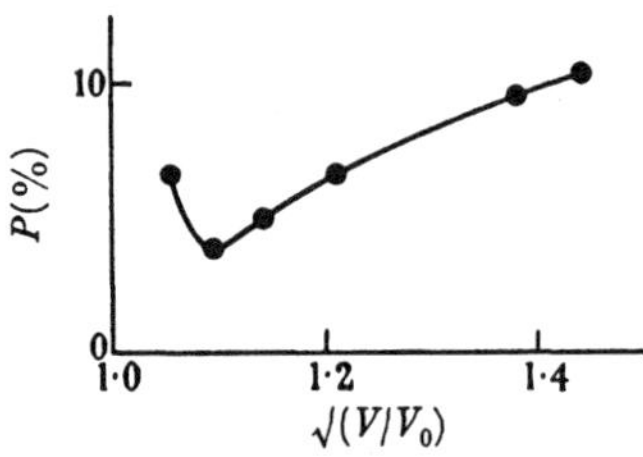


FIGURE 4. Polarization curve for He $3\,{}^3\mathrm{P} - 2\,{}^3\mathrm{S}$ as measured by Lamb & Maiman (1957).

Typical observed polarizations, taken from the work of Skinner & Appleyard (1927), are plotted as functions of $\sqrt{(V/V_0)}$ in figure 5, $V$ being the energy of the incident electrons, $V_0$ the threshold energy. Very similar polarizations are observed for lines belonging to the same spectral series. It might be expected that the polarization curves for Hg, with ground configuration $6s^2$, would resemble those of similar lines in He, which has a ground $1s^2$ configuration. In figure 3 we have therefore plotted the observed polarization for Hg $7\,{}^1\mathrm{D}_2 \to 6\,{}^1\mathrm{P}_1$ and the Born approximation for the polarization of He $3\,{}^1\mathrm{D}_2 \to {}^1\mathrm{P}_1$, both being given as functions of $\sqrt{(V/V_0)}$.

The most puzzling feature of the experimental results is that with decreasing energy the observed polarization rises to a maximum and then decreases, tending to values close to zero at threshold. There is no evident theoretical explanation for this anomalous behaviour at energies less than those at which maximum polarization occurs (an explanation advanced by Oppenheimer (1927 b) has been shown by Lamb (1957) to be fallacious). If we consider the observed results only for energies greater than those of maximum polarization one may make plausible extrapolations (dashed curves of figure 5) which are in reasonable agreement with the calculated threshold polarizations of table 1 (marked with crosses in figure 5). Referring to figure 3 we see that, apart from the anomalous low-energy experimental results, the agreement between the two curves is quite good, the discrepancy at high energies being very probably due mainly to cascade. A further prediction of the theory is that the polarization should be zero for upper S states. The observations for Hg give zero polarization for upper ${}^1\mathrm{S}$ states and zero for ${}^3\mathrm{S}_1 \to {}^3\mathrm{P}_2$ transitions. For ${}^3\mathrm{S}_1 \to {}^3\mathrm{P}_1$ and ${}^3\mathrm{S}_1 \to {}^3\mathrm{P}_0$

† We do not discuss further the numerical results of Penney for energies above threshold, since the Born–Oppenheimer approximation used has since been shown to be unreliable in this type of calculation (Bates *et al.* 1950).

## POLARIZATION OF ATOMIC IMPACT RADIATION     137

the observed polarization is zero at moderate and high energies but rises to maxima at low energies of $-12\%$ for $7\,^3S_1 \to 6\,^3P_1$ and of $+8\%$ for $7\,^3S_1 \to 6\,^3P_0$. These small non-zero values for upper S states may be due to departures from $LS$ coupling.

We may conclude that, apart from the anomalous low-energy results, theory and experiment for Hg are in reasonable accord. At energies less than those at which maximum polarization occurs the total light intensity will be small (see figures 1 and 3) and any background unpolarized light will become increasingly important. The problem of collimating the electron beam also becomes more difficult at low energies. Since there still

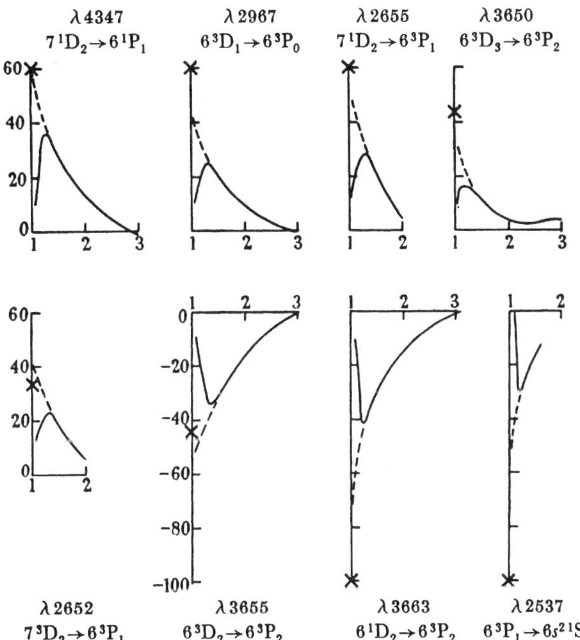



FIGURE 5. Polarization of Hg lines as functions of $\sqrt{(V/V_0)}$. Full line curves give experimental results of Skinner & Appleyard (1927). Crosses give calculated threshold polarizations (for isotopes with $I = 0$ and assuming $LS$ coupling for the collision process). Dashed curves are extrapolations of full line curves neglecting experimental results for energies less than those at which maximum polarization was measured.

appears to be no explanation of the disagreement with theory at low energies in terms of the reaction $Hg + e \to Hg + e + h\nu$ alone, more complicated pressure-dependent reactions may be involved, and further experimental work using modern techniques would be desirable.

*Note added in proof*, 20 August 1958: Baranger & Gerjuoy (1958) have discussed the polarization of impact radiation from the standpoint of a compound ion model. Fite & Brackmann (1958) have measured the polarization of $Ly\alpha$ radiation excited by electron impact.

138                    I. C. PERCIVAL AND M. J. SEATON

We wish to thank Mr L. Castillejo and Professor H. S. W. Massey, F.R.S., for a number of helpful discussions about the contents of the present paper. We are greatly indebted to Dr A. Fonda for informing us of some work he had done on these problems and for several stimulating discussions. Our work was supported in part by the Atomic Energy Research Establishment, to whom we are indebted for permission to publish.


REFERENCES

Baranger, E. & Gerjuoy, E. 1958 *Proc. Phys. Soc.* **72**, 326.
Bates, D. R., Fundaminsky, A., Leech, J. W. & Massey, H. S. W. 1950 *Phil. Trans.* A, **243**, 93.
Bethe, H. A. 1933 *Hand. Physik.* **24**, part 1, pp. 508–515, 2nd ed. Berlin: Springer.
Condon, E. U. & Shortley, G. H. 1951 *The theory of atomic spectra.* (Reprint of 1st ed. 1935.) Cambridge University Press.
Dehmelt, H. G. 1956 *Phys. Rev.* **103**, 1125.
Eldridge, J. A. & Olson, H. F. 1926 *Phys. Rev.* **28**, 1151.
Ellett, A., Foote, P. D. & Mohler, F. L. 1926 *Phys. Rev.* **27**, 31.
Fite, W. L. & Brackmann, R. T. 1958 *Phys. Rev.* (in course of publication).
Hanle, W. & Quarder, B. 1929 *Z. Phys.* **54**, 819.
Heitler, W. 1954 *The quantum theory of radiation* (3rd ed., chap. 5). Oxford University Press.
Khashaba, S. & Massey, H. S. W. 1958 *Proc. Phys. Soc.* **71**, 574.
Kossel, W. & Gerthsen, C. 1925 *Ann. Phys., Lpz.,* **77**, 273.
Lamb, W. E. 1957 *Phys. Rev.* **105**, 559.
Lamb, W. E. & Maiman, T. H. 1957 *Phys. Rev.* **105**, 573.
Massey, H. S. W. & Mohr, C. B. O. 1931 *Proc. Roy. Soc.* A, **132**, 605.
Mott, N. F. & Massey, H. S. W. 1949 *Theory of atomic collisions*, 2nd ed. Oxford: Clarendon Press.
Obi, S., Ishidzu, T., Horie, H., Yanagawa, S., Tanabe, Y. & Sato, M. 1953 *Ann. Obs. astr. Tokyo,* **3**, no. 3 and succeeding issues.
Oppenheimer, J. R. 1927*a* *Z. Phys.* **43**, 27.
Oppenheimer, J. R. 1927*b* *Proc. Nat. Acad. Sci.* **13**, 800.
Oppenheimer, J. R. 1928 *Phys. Rev.* **32**, 361.
Penney, W. G. 1932 *Proc. Nat. Acad. Sci.* **18**, 231.
Quarder, B. 1927 *Z. Phys.* **41**, 674.
Racah, G. 1942 *Phys. Rev.* **62**, 438.
Racah, G. 1943 *Phys. Rev.* **63**, 367.
Seaton, M. J. 1955 *Proc. Phys. Soc.* A, **67**, 457.
Simon, A., Vander Sluis, J. H. & Biedenharn, L. C. 1954 *U.S. Atomic Energy Commission Report* ORNL-1679 (Oak Ridge, Tenn.).
Skinner, H. W. B. 1926 *Proc. Roy. Soc.* A, **112**, 642.
Skinner, H. W. B. & Appleyard, E. T. S. 1927 *Proc. Roy. Soc.* A, **117**, 224.
Smit, J. A. 1935 *Physica*, **2**, 104.
Steiner, K. 1928 *Z. Phys.* **52**, 516.
Wigner, E. P. 1948 *Phys. Rev.* **73**, 1002.

# PHYSICAL REVIEW
# LETTERS

VOLUME 15       4 OCTOBER 1965       NUMBER 14

## PRODUCTION OF HIGHLY POLARIZED ELECTRON BEAMS BY LOW-ENERGY SCATTERING

K. Jost and J. Kessler

Physikalisches Institut der Technischen Hochschule, Karlsruhe, Germany
(Received 29 July 1965)

The problem of producing polarized electron beams has been fascinating many physicists during the last years. The aim of such experiments consists in producing electron beams with a polarization as high as possible and an intensity as high as possible. Most of the attempts at producing highly polarized electron beams failed, however. For example, all the experiments based on ejection of oriented electrons from magnetized materials by the photoelectric effect, field emission, etc., have been of little success.[1] Except for the method discussed here, there has been only one successful method: Polarized atom beams have been produced by the Stern-Gerlach experiment and oriented electrons have been obtained therefrom by photoionization.[2]

Therefore, the recent theoretical results of Holzwarth and Meister[3] become especially interesting. They predicted "the possibility of the production of almost totally polarized beams of electrons" by scattering electrons of low energy (within a large range about 1 keV) by mercury atoms. Similar results were obtained by Bunyan and Schonfelder.[4] The polarization attainable in this way is predicted to be a rapidly oscillating function of scattering angle[5] (Fig. 1 gives an example), in contrast to the results at higher energies: Here the Mott theory of scattering by the screened and by the unscreened Coulomb field yields rather smooth curves. For checking theory we have performed an experiment in which the angular resolution was suf-

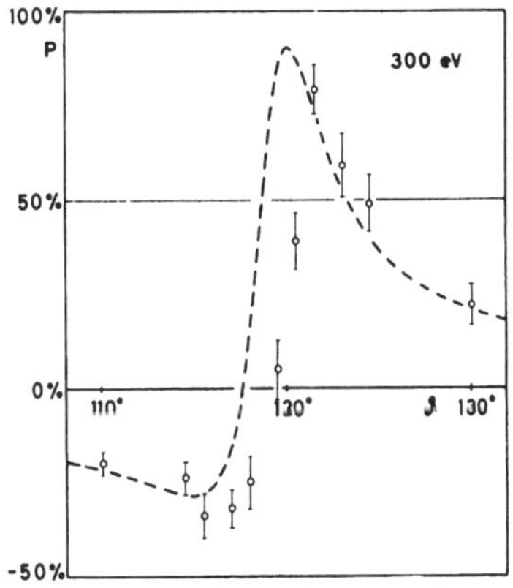



FIG. 1. Example of the production of a highly polarized electron beam by scattering from screened Hg atoms. Dashed line, theoretical results. Points, measured values with statistical error and angular resolution of ±1°.

ficient to demonstrate the high polarization values concentrated in the peaks of the kind plotted in Fig. 1. In this way electron beams of about 80% polarization were produced and essential features of the theory were confirmed.

The general layout of the apparatus can be described as follows (Fig. 2): A monoenergetic electron beam of some 100 μA crosses a mer-

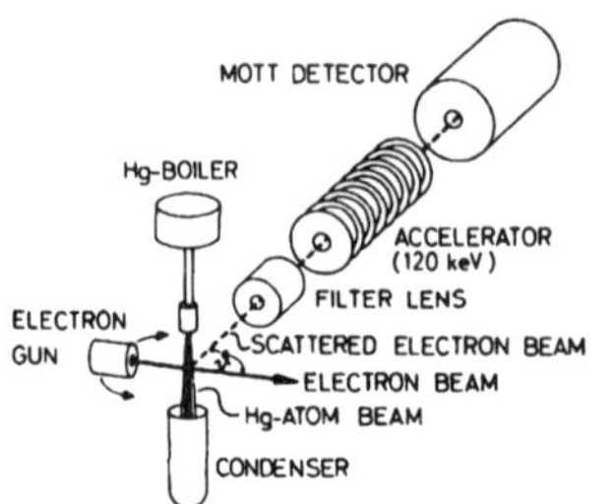


FIG. 2. Schematic diagram of the apparatus (perspective, not to scale).

cury-atom beam which scatters part of the electron beam. The electron energy can be varied between 100 and 1000 eV. By rotating the electron gun the scattering angles can be varied continuously between 0° and 155°. The angular resolution is about ±1° (half-amplitude). The scattered electron beam is expected to be polarized, the direction of the polarization vector being perpendicular to the plane of scattering. The scattered electrons are accelerated to 120 keV, in order that the analysis of the electron polarization can be made by the Mott scattering method.[6]

As the theoretical calculations of the polarization effects only refer to elastically scattered electrons, an energy analyzer was put behind the first scattering chamber to reject all those electrons which had lost energy. For this purpose an intermediate-image filter lens of the kind described by Simpson and Marton was used.[7] This compares favorably with the conventional type of filter lens because the filtered beam leaves it with a small angular divergence so that it can be postaccelerated conveniently. Instrumental asymmetries could be easily eliminated in several ways; e.g. measurements could be made symmetrically about the angle 0°.

Figure 1 gives an example of a measurement in a range in which theory predicts a sudden rise from negative values of the polarization to high positive values. The theoretical results are confirmed on the whole. It is true that the peak values of our measured polarizations are here and in other cases somewhat lower than the theoretical ones. Furthermore, there is a slight shift of the measured values towards higher angles, which has been found in our earlier measurements[8] of the cross section,

Table I. Experimental examples of further combinations of electron energy $E$ and scattering angle $\theta$ resulting in peak values of the polarization $P$.

| $E$ (eV) | $\theta$ (deg) | $P$ (%) |
|---|---|---|
| 180 | 90 | +68 |
| 230 | 84 | −73 |
| 900 | 97.5 | +84 |
| 900 | 102 | −79 |

too. By further experiments we hope to find out whether the small discrepancies still existing are due to theory or to experiment.

In Table I the peak values of the polarization from our further measurements are compiled. Of course we have not yet found all the peaks existing.

It was the aim of the present work to check the theory and not to produce a high-intensity source of polarized electrons. So no special attempts to maximize the intensity have been made and the currents at the peak values of the polarization were in the range of $10^{-13}$ A.

Taking the results presented here together with our earlier results[8] on the cross sections and the careful work done at Mainz University[9] where polarizations averaged over an angular range of 6° have been measured, we can say that the theoretical results are well confirmed. The occasional deviations from theory found by the Mainz group[10] (e.g. at 300 eV and 120°-140°) are not in contradiction to this statement, as they probably may be explained by plural scattering: At the low electron energies, we had to choose a rather low density of the atom beam in order to avoid just the same deviations caused by plural scattering.

So we can say that the theoretical results are confirmed and that scattering of low-energy electrons represents a convenient method of producing electron beams of high polarization.

The authors are indebted to the Deutsche Forschungsgemeinschaft for support.

[1]R. L. Long, V. W. Hughes, J. S. Greenberg, I. Ames, and R. L. Christensen, Phys. Rev. 138, A1630 (1965). Further references are to be found here.

[2]R. L. Long, Jr., W. Raith, and V. W. Hughes, Phys. Rev. Letters 15, 1 (1965).

[3]G. Holzwarth and H. J. Meister, "Tables of Asymmetry, Cross Section, and Related Functions for Mott Scattering of Electrons by Screened Gold and Mercury

VOLUME 15, NUMBER 14          PHYSICAL REVIEW LETTERS          4 OCTOBER 1965

Nuclei," München, Germany, 1964 (unpublished);
G. Holzwarth and H. J. Meister, Nucl. Phys. **59**, 56 (1964).

[4]P. J. Bunyan and J. L. Schonfelder, Proc. Phys. Soc. (London) **85**, 455 (1965).

[5]There is also a rapid oscillation of the polarization with regard to energy.

[6]A. R. Brosi, A. I. Galonsky, B. H. Ketelle, and H. B. Willard, Nucl. Phys. **33**, 353 (1962).

[7]J. A. Simpson and L. Marton, Rev. Sci. Instr. **32**, 802 (1961); J. A. Simpson, Rev. Sci. Instr. **32**, 1283 (1961); J. Kessler and H. Lindner, Z. Physik angew. **18**, 7 (1964).

[8]J. Kessler and H. Lindner, Z. Physik **183**, 1 (1965).

[9]H. Deichsel and E. Reichert, Z. Physik **185**, 169 (1965), and earlier work.

[10]H. Steidl, E. Reichert, and H. Deichsel, Phys. Letters **17**, 31 (1965).

# The "Perfect" Scattering Experiment. I

The present level of sophistication of experimental atomic collision physics is such that it is now possible to re-examine the collision experiment to see precisely how much information can actually be extracted from it. The usual goals of determinations of differential and total cross sections, while of course of great practical value, do not always supply information which can be directly and uniquely compared with collision theory. Specifically this relates to the fact that spatial and spin orientation of the collision partners generally affect the interaction, and therefore without suitable preparation of collision partners, followed by suitable analysis after the collision, some sort of averaging process over several different interactions occurs.

To illustrate this, consider the conceptually simplest collision experiment, that of the elastic scattering of an electron by an atom possessing a single valence electron in an $s$-state, no nuclear spin, no core angular momentum, and no spin–orbit interaction during the collision[1,2] (i.e., an "ideal" alkali metal atom). In time-independent collision theory, the interaction is describable in terms of a pair of velocity-independent potentials or pseudopotentials which can indeed possess velocity dependence. These potentials are characterized as singlet and triplet, representing the effective electron–atom interaction seen when the two participating electrons possess an effective spin 0 or 1 during the interaction, respectively. Each of these potentials has a scattering amplitude associated with it, i.e., a scattered wave function possessing phase and magnitude which is determined, relative to the unperturbed wave, by the interaction potential. Of the four quantities involved, three are observable, in principle at least (the fourth representing an arbitrary phase factor). One can consider the three independent quantities to be the two magnitudes and the relative phase between the singlet and triplet scattering amplitudes, which we call $S$ and $T$, respectively.

Alternatively one can refer to the direct and exchange scattering amplitudes $f$, $g$ which are related to $T$, $S$ by

$$T = f - g; \quad S = f + g.$$

In the usual elastic cross-section measurement, made with unpolarized electrons and atoms, the "average" differential cross section

$$\sigma(\theta) = \tfrac{3}{4}|f - g|^2 + \tfrac{1}{4}|f + g|^2$$

is measured. This determination establishes only a single relation for $f, g$ at each energy and angle. Clearly, two additional independent observations relating to scattering at a given energy and angle are required in order to experimentally determine $f, g$ (or $S, T$). In fact, one can write down a number of independent (and a least theoretically feasible) experiments, which can be performed in order to fully determine $f, g$ and, further, to achieve a much-to-be-desired redundancy in this determination. The redundancy allows for an internal consistency check on the basic validity of time-independent collision theory; more practically, it serves to check the reliability of the several independent, and quite difficult experiments.

A list of such experiments, and the quantities they measure, is presented in Table I. In the table $e$ and $A$ refer to the electron and the atom respectively, $(\uparrow\downarrow)$ represents an unpolarized beam, and $(\uparrow)$ and $(\downarrow)$ represent polarized beams. The table reveals some interesting choices we have in designing a sequence of experiments of varying degrees of difficulty.

TABLE I. Tabulation of possible elastic collision experiments, electron–atom, involving various combinations of polarized and analyzed beams.

| Expt. No. | Type of experiment | Quantity observed |
|---|---|---|
| I | $e(\uparrow\downarrow) + A(\uparrow\downarrow) \rightarrow e(\uparrow\downarrow) + A(\uparrow\downarrow)$ | $\tfrac{3}{4}|f - g|^2 + \tfrac{1}{4}|f + g|^2$ |
| IIa | $e(\uparrow\downarrow) + A(\uparrow) \rightarrow e(\uparrow) + A(\uparrow)$ | $\tfrac{1}{2}|f - g|^2$ |
| b | $\rightarrow e(\downarrow) + A(\uparrow)$ | $\tfrac{1}{2}|f|^2$ |
| c | $\rightarrow e(\uparrow) + A(\downarrow)$ | $\tfrac{1}{2}|g|^2$ |
| IIIa | $e(\uparrow) + A(\uparrow\downarrow) \rightarrow e(\uparrow) + A(\uparrow)$ | $\tfrac{1}{2}|f - g|^2$ |
| b | $\rightarrow e(\uparrow) + A(\downarrow)$ | $\tfrac{1}{2}|f|^2$ |
| c | $\rightarrow e(\downarrow) + A(\uparrow)$ | $\tfrac{1}{2}|g|^2$ |
| IV | $e(\uparrow) + A(\uparrow) \rightarrow e(\uparrow) + A(\uparrow)$ | $|f - g|^2$ |
| V | $e(\uparrow) + A(\downarrow) \rightarrow e(\uparrow) + A(\downarrow)$ | $|f|^2$ |
| VI | $e(\uparrow) + A(\downarrow) \rightarrow e(\downarrow) + A(\uparrow)$ | $|g|^2$ |

Experiment I is, of course, the full differential measurement with unpolarized beams. Experiments II, III use either unpolarized electrons and polarized atoms, or vice versa. The quantity measured depends upon whether observation is made on the scattered atom or electron, or both. The former type of experiment, involving spin analysis of the scattered atom, has been performed at New York University[3] for electrons elastically scattered by potassium at several energies between 0.5 eV and the first inelastic threshold. It is not surprising that this should be the first spin-analyzed scattering experiment with electrons, since spin-state selection and analysis of the neutral atom using an inhomogeneous magnetic field is far simpler to accomplish than spin selection and analysis of the free electron.

The recoil experiment measures IIc ($\frac{1}{2}|g|^2$) and IIa + IIb. Note that the latter experiment is equivalent to I − IIc, and so recoil combined with a full differential measurement determines two out of the three parameters characterizing $f$ and $g$.

To go further one must somehow employ polarized electrons. The simplest such routes are via experiments II, III, i.e., scattering unpolarized electrons by polarized atoms with spin analysis of the scattered electrons, or vice versa. Experiment II is complementary to the recoil experiment. Using a Mott analyzer one measures the polarization of the scattered electrons, i.e., IIa + IIc − Ib, which yields

$$\tfrac{1}{2}|f - g|^2 + \tfrac{1}{2}|g|^2 - \tfrac{1}{2}|f|^2 .$$

This quantity, when subtracted from I, gives $|f|^2$. Thus it is seen that experiment II with electron spin analysis, combined with a full differential measurement, also determines two out of the three parameters characterizing $f$ and $g$. A combination of recoil plus electron spin analysis represents a complete determination of $f$ and $g$. S. J. Smith at JILA is currently conducting such an experiment with lithium.

The alternate route, via III, is seen to yield the same information as II. Thus, using polarized electrons and unpolarized atoms with spin analysis of the scattered electrons one obtains $|g|^2$. This experiment effectively requires triple scattering, if one assumes that some type of scattering must be used, in the present state of the art, to either produce or analyze polarized electrons. Spin analysis of the atom, which yields $|f|^2$, is a variant of the recoil technique, but with the immense disadvantage of the limitation on electron current imposed by its polarization. At present, maximum polarized electron currents obtainable are of the order of $10^{-9}$ A, a current which is too small to be practical for a recoil-type experiment. From practical considerations it seems then that experiments II are a more promising set than III.

Finally there are the three experiments IV, V, VI which involve preparing *both* electron and atom beams. Of these, IV is a special case since it does not require spin analysis and therefore constitutes only "double" scattering (the elimination of the need for spin analysis is a consequence of our assumption that spin flips can only occur by exchange. If spin–orbit effects are possible, e.g., near a resonance, this assumption fails). Experiment IV does constitute a third independent measurement involving $f$ and $g$. Such an experiment on potassium is in progress at NYU. The polarized electrons are to be produced by low-energy Mott scattering from mercury.[4]

Experiments V, VI involve polarization and analysis of both beams and clearly these, plus say I, also represent a full determination of $f$ and $g$. Such experiments on atomic hydrogen, using polarized electrons from photoionized aligned alkalis,[5] are in progress at Yale.

Analogous experiments involving heavy-particle collisions yield similar information. Thus, elastic alkali–alkali experiments with spin analysis, analogous to experiment II, are being performed by Kleppner at MIT,[6] from which information concerning the singlet–triplet interaction difference at large internuclear separation can be obtained.

All such elastic experiments involving some form of spin analysis are clearly in their infancy. One can scarcely doubt that our ability to deal with low-energy polarized beams, both charged and neutral, will improve rapidly in the next few years. The next portion of this discussion will deal with analogous experiments involving *inelastic* collisions.

BENJAMIN BEDERSON

### References


1. Many of the ideas in this discussion were originally presented by H. Kleinpoppen (1967, unpublished).
2. The still simpler positron–atom case is yet beyond our experimental capabilities, and will not be discussed here.
3. R. E. Collins, B. Bederson, M. Goldstein, and K. Rubin, Phys. Letters **27A**, 440 (1968); also Phys. Rev. Letters **19**, 1366 (1967).
4. J. Kessler, Rev. Mod. Phys. **41**, 3 (1969).
5. V. W. Hughes, M. S. Lubell, M. Posner, and W. Raith, in *Fifth International Conference on the Physics of Electronic and Atomic Collisions: Abstracts of Papers*, edited by I. P. Flaks (Publishing House Nauka, Leningrad, 1967), p. 544.
6. D. E. Pritchard, D. C. Burnham, and D. Kleppner, Phys. Rev. Letters, **19**, 1363 (1967).

# The "Perfect" Scattering Experiment. II

In part I of this Comment I discussed the use of spin-polarized and analyzed beams in the study of electron–atom elastic collisions. The relation of the observables to the scattering amplitudes was discussed. It was shown that in the simplest of all electron–atom scattering experiments, i.e., those involving a hypothetical alkali atom with no hyperfine structure (hfs), three independent experiments are required in order to fully determine the parameters obtained from potential scattering theory. These parameters can be the magnitudes of $f,g$ (the direct and exchange scattering amplitudes), and their phase difference. The observables depend upon the manner in which the spin orientations of the incoming electrons and atoms are prepared. For example, one complete determination would consist of (a) a differential measurement involving unpolarized beams, (b) an exchange measurement, in which a polarized electron or atom changes its spin state in the collision, (c) a pure triplet interaction, in which the spins of the incoming electron and atom are parallel before, and consequently after, the collision.

I would now like to extend this discussion to inelastic (and ionizing) collisions. Let us ask the same question, namely, how many independent differential scattering experiments are required in order to fully determine all the parameters which are calculable from potential scattering theory? Rubin and Walker[1] have analyzed this problem in detail. Again, consider the same hypothetical alkali atom, with no hfs. Consider also the simplest of excitations, namely, $S \rightarrow P$.

For the moment let us neglect all spin effects, and consider an even more hypothetical system involving spinless electron and spinless atom. The atom undergoes a transition from a state of zero angular momentum to angular momentum of unity, in units of $\hbar$. If the incoming electron is moving in the $z$-direction, the magnetic quantum number of the excited atom is either $\pm 1$ or $0$ referred to the $z$-axis. Accordingly, there are two scattering amplitudes, $f^1$ and $f^0$ ($f^{+1}$ equals $f^{-1}$ from symmetry considerations), which describe the probability that the excited atom is left with its angular momentum aligned in the $z$-direction or at right angles to it.

The full differential cross section is

$$\sigma = \sigma^0 + 2\sigma^1 = |f^0|^2 + 2|f^1|^2,$$

and a complete determination would require measurements of $\sigma$ as well as of either $|f^0|^2$ or $|f^1|^2$. This latter determination would be achievable in principle using coincidence between the scattered electron and the photon resulting from subsequent $P \rightarrow S$ decay, with polarization analysis of the coincident photon. Were the lifetime of the excited atom long enough, a spin analysis of the excited atom would also yield the same information.

Giving the electrons spin introduces the same situation as in the elastic case, that is, direct ($f$) and exchange ($g$) scattering amplitudes for each final angular-momentum projection state. There then become altogether four scattering amplitudes and three observable phases, involving differences between $f^1, g^1, f^0, g^0$. It should be pointed out that the excitation process involves only orbital angular momentum, as before. The spin–orbit coupling in the excited states only takes hold after the excitation is complete.

Rubin and Walker[1] have shown how the various magnitudes and phases (seven in all) can be obtained by suitable combinations of electron and atom spin alignments combined with electron, atom, and photon polarization analyses.

In a sense, the first experiments to observe more than the gross inelastic differential cross sections were the observations of the polarization of resonance radiation, starting with Skinner and Appleyard, but achieving quantitative status only with the recent work of McFarland, Kleinpoppen, and others.[2] The fractional polarization of resonance radiation, when observed at right angles to the incident electron direction, is[3]

$$P = 3\left[\frac{\sigma^0 - \sigma^1}{7\sigma^0 + 11\sigma^1}\right] \tag{1}$$

for alkalis without hfs. In Eq. (1), $\sigma^0$ and $\sigma^1$ are equal to the *total* cross sections for excitation with $M_L = 0, \pm 1$. For example,

$$\sigma^0 = \tfrac{3}{4}|F^0 - G^0|^2 + \tfrac{1}{4}|F^0 + G^0|^2,$$

where $F^0$, $G^0$ are the *total* direct and exchange scattering amplitudes, i.e.

$$|F^0|^2 = \int_{\text{all angles}} |f^0|^2 \, d\Omega.$$

The principal concern in such experiments in the past has been with observation of $P$ at or near threshold, where simple momentum conservation arguments require that $F^1$, $G^1 \rightarrow 0$, so that $P(\text{threshold}) = 3/7$. The high-energy limit has been less well studied. A conservation argument also applies here, and requires that $F^0$, $G^0 \rightarrow 0$, so that the high-energy limit for $P$ for the alkalis without hfs should approach $-3/11$. The recent experiments of Kleinpoppen and coworkers[4] are in excellent agreement with the predicted *threshold* values, taking hfs into account, for $\text{Li}^6$, $\text{Li}^7$, and $\text{Na}^{23}$.

It should be noted that such experiments, despite their importance and elegance, yield little information, beyond the threshold $P$-values. They represent integral cross-section determinations, and as such do not yield direct information concerning $f,g$. In addition they cannot distinguish exchange from direct scattering.

The second step towards the ultimate goal of obtaining complete information from experiment has been the work of Rubin *et al.*[5] at New York University. Here, the recoil technique is employed to prepare the spin state of the atom before scattering, and to analyze the final spin state of the scattered atom after being excited, and after having decayed back to the ground state with its spin possibly disoriented. The spin-flip is due to a combination of processes. First, exchange contributes to the spin-flip. Second, the distribution of orbital angular momentum orientations in the excited state, combined with the spin–orbit interaction of the excited electron causes spin-flips upon subsequent decay. The probability for spin-flip depends upon the relative magnitudes and phases of $f^1, f^0, g^1, g^0$, and, of course, upon the (known) relative transition probabilities among the various fine-structure levels of $^2P_{1/2,3/2}$ and $^2S_{1/2}$. Rubin *et al.*[5] have shown that the relative differential spin-flip cross section of scattered atoms is

$$R(\theta) \equiv \frac{\sigma_{\text{spin-flip}}}{\sigma} = \frac{4}{9} + \frac{1}{18}\left[\frac{|g^0|^2 + 10|g^1|^2 - 8\sigma^1}{\sigma^0 + 2\sigma^1}\right]. \qquad (2)$$

This is the recoil experiment analog to Eq. (1). Note that $f,g$ do not appear symmetrically in Eq. (2), so that under certain conditions exchange excitation can be observed. Further, $R$ represents a differential determination, i.e., is a function of electron polar scattering angle, so that considerably more information is retrievable from the data than in the optical case.

The recoil analog to the optical threshold behavior is

$$R_{\text{threshold}} = \frac{4}{9} + \frac{1}{18}\frac{|g^0|^2}{\sigma^0},$$

while the high-energy limit is

$$R_{\text{high energy}} \to \frac{2}{9},$$

where it is assumed that $g^1$, $g^0$ both vanish at high energies. Goldstein *et al.*[6] have recently measured $R$ for angles between $0°$ and $20°$ in potassium, from threshold to about 15 eV. Such measurements will be extended to other alkalis as well.

Beyond the recoil experiment, it is necessary to use either polarized electrons, or coincidence between the polarization and the angular distributions of the emitted photons, or both. For example, the coincidence experiment would enable observation of interference, i.e., phase factors, between $f^1$, $g^1$ and $f^0$, $g^0$.

Such experiments without polarized electrons are feasible today; with polarized electrons, not yet. [It should be emphatically pointed out that, despite the generally optimistic tone of these articles, there has not yet been *any* scattering experiments in low-energy atomic physics employing polarized electrons, as opposed to *producing* polarized electrons at low energies, which of course has been accomplished.]

Analogous experiments involving ionization, rather than excitation, are also in progress. Here the recent trend has been towards the use of coincidence [an inheritance of atomic physicists from our nuclear brethren] between angular and energy distributions of the two electrons. Glassgold[7] has discussed a simple case involving high-energy electrons incident upon atomic hydrogen. Ehrhart has already succeeded in performing energy–angle correlations in helium.[8] Amaldi *et al.*[9] are currently conducting coincidence experiments using two monochromators, each capable of angular variation.

BENJAMIN BEDERSON

**References**

1. K. Rubin and J. Walker (1967, unpublished; a summary of their results is contained in Appendix I of Ref. 5).
2. H. B. W. Skinner and E. T. S. Appleyard, Proc. Roy. Soc. (London) **A117**, 224 (1927).
  H. Hafner, H. Kleinpoppen, and H. Krüger, Phys. Letters **18**, 270 (1965); also *IVth International Conference on the Physics of Electronic and Atomic Collisions, Quebec, 1965*, Book of Abstracts (Science Bookcrafters, Inc., Hastings-on-Hudson, N.Y.), p. 386.
  D. Haidt, H. Kleipoppen, and H. Krüger, *ibid.*, p. 390.
  R. H. McFarland, *Vth International Conference on the Physics of Electronic and Atomic Collisions, Leningrad, 1967*, Book of Abstracts (Four Continents Book Corp., New York), p. 498.
3. M. J. Seaton, in *Excitation Electronique d'une Vapeur Atomique* (Centre National de la Recherche Scientifique, Paris, 1967), pp. 21–53.
4. H. Kleinpoppen, *Vth International Conference on the Physics of Electronic and Atomic Collisions, Leningrad, 1967*, Book of Abstracts (Four Continents Book Corp., New York), p. 538.
5. K. Rubin, B. Bederson, M. Goldstein, and R. E. Collins, Phys. Rev. **182**, 201 (1969); also, International Conference on Atomic Physics, New York University, 1968; Abstracts of Contributed Papers, edited by V. W. Cohen, and G. zu Putlitz (unpublished), p. 131.
  B. Bederson, R. E. Collins, M. Goldstein, and K. Rubin, in *Physics of One- and Two-Electron Atoms*, edited by Bopp and Kleinpoppen (North-Holland Publishing Co., Amsterdam, 1969).
6. M. Goldstein *et al.* (to be published).
7. A. E. Glassgold, *Vth International Conference on the Physics of Electronic and Atomic Collisions, Leningrad, 1967*, Book of Abstracts (Four Continents Book Corp., New York), p. 646.
8. H. Ehrhardt, T. Tekaat, and K. Willmann, International Conference on Atomic Physics, New York University, 1968; Abstracts of Contributed Papers, edited by V. W. Cohen and G. zu Putlitz (unpublished), p. 125.
9. U. Amaldi, Jr., A. Egidi, R. Marconero, and G. Pizzola, "Use of a Two-Channeltron Coincidence in a New Line of Research in Atomic Physics", Technical Report, Istituto Superiore di Sanità, Rome [ISS 68/40], 1968.

VOLUME 24, NUMBER 3     **PHYSICAL REVIEW LETTERS**     19 JANUARY 1970

## EXPERIMENTAL VERIFICATION OF THE FANO EFFECT

J. Kessler and J. Lorenz

Physikalisches Institut der Universität Karlsruhe, Karlsruhe, Germany

(Received 8 December 1969)

An absolute measurement of the electron polarization obtained by photoionization of cesium by circularly polarized light is reported. The measured polarization of $65 \pm 15\%$ confirms the prediction made by Fano and the assumptions of Seaton on spin-orbit interaction in photoionization of alkali atoms. They also show that it is possible to construct a photoionization source of polarized electrons without using a polarized atom beam.

Recently Fano[1] predicted an effect which is very interesting from a theoretical viewpoint as well as for practical purposes: Photoelectrons emitted by alkali atoms exposed to circularly polarized light should be highly polarized if the wavelength of the light falls within a broad spectral band around the minimum of the photoabsorption cross section. Fano made a numerical analysis for cesium resulting in a polarization $\geq 85\%$ for $2750 \lesssim \lambda \lesssim 3150$ Å.

The basic idea of the Fano effect can be illustrated as follows: The unpolarized cesium-atom beam is considered as a mixture of equal numbers of ground-state atoms with spins parallel ("spin up") and antiparallel ("spin down") to the direction of the incident light. In Fig. 1 the transitions from the ground state to the continuum $P$ states[2] due to $\sigma^+$ light are indicated, showing that there is a certain probability for spin flip $m_s = -\frac{1}{2} \to m_s = +\frac{1}{2}$ in the transitions 2 and 3. The cross section for producing spin-up and spin-down electrons by these transitions can be shown to be proportional to $\frac{2}{9}(R_3 - R_1)^2$ and $\frac{1}{9}(R_3 + 2R_1)^2$, respectively. $R_1$ and $R_3$ are the radial parts of the matrix elements for transitions to the $j = \frac{1}{2}$ and $\frac{3}{2}$ states; they differ slightly due to spin-orbit interaction. The cross section for transition 1 which results in spin-up electrons is proportional to $R_3^2$.

The polarization of the photoelectrons is therefore given by

$$P = \frac{\sigma^\uparrow - \sigma^\downarrow}{\sigma^\uparrow + \sigma^\downarrow} = \frac{R_3^2 + \frac{2}{9}(R_3 - R_1)^2 - \frac{1}{9}(R_3 + 2R_1)^2}{R_3^2 + \frac{2}{9}(R_3 - R_1)^2 + \frac{1}{9}(R_3 + 2R_1)^2}. \quad (1)$$

Since spin-orbit coupling is a small interaction, one finds $R_1 \approx R_3$; thus the cross sections for producing spin-up and spin-down electrons are not very different from each other. Significant polarization of the photoelectrons can therefore only be expected near the minimum of the photoionization cross section $\sigma^\uparrow + \sigma^\downarrow$, since only there is the relative difference of spin-up and spin-down electrons appreciable, analogous to the circumstances in electron scattering.[3]

An indirect experimental verification of Fano's ideas was given by Lubell and Raith.[4] These authors did not observe the electron polarization, but they measured the ion intensities $I^+$ and $I^-$ obtained by ionizing polarized cesium atoms by circularly polarized light with positive and negative photon helicities. Transitions 1' and 2 + 3 in Fig. 1 were observed. With the results for the cross sections given above we therefore obtain (for completely polarized Cs atoms)

$$\delta = \frac{I^+ - I^-}{I^+ + I^-} = \frac{R_3^2 - \frac{2}{9}(R_3 - R_1)^2 - \frac{1}{9}(R_3 + 2R_1)^2}{R_3^2 + \frac{2}{9}(R_3 - R_1)^2 + \frac{1}{9}(R_3 + 2R_1)^2}, \quad (2)$$

which differs from Eq. (1) by a minus sign. The authors made a relative measurement of $\delta$ after determining their atomic-beam polarization from the difference of the ion intensities. Eq. (2) was confirmed.

In the present paper an absolute measurement of the electron polarization, Eq. (1), is reported. Figure 2 gives a schematic diagram of the apparatus. A cesium-atom beam is crossed by a beam of circularly polarized light. The photoelectrons are extracted by a weak electric field of about 10 V/cm perpendicular to the plane defined by the two beams and pass through an electron-optical focusing system which also removes

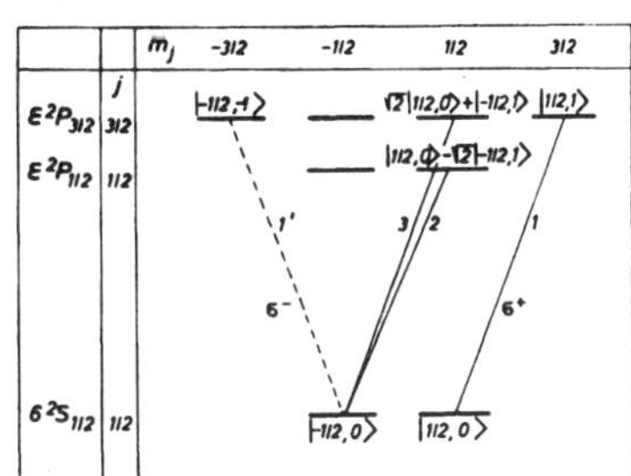


FIG. 1. States, wave functions $|m_s m_j\rangle$ [cf. L. Schiff, Quantum Mechanics (McGraw-Hill Book Company, Inc., New York, 1955), p. 291], and transitions connected with photoabsorption in cesium.

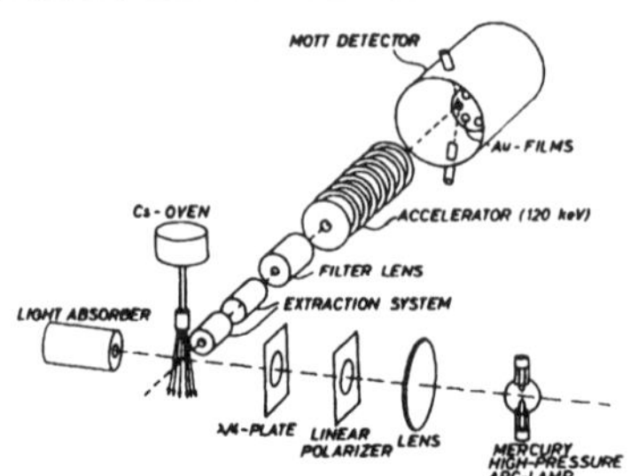


FIG. 2. Schematic diagram of the apparatus.

spurious electrons. After postacceleration to 120 keV they enter a Mott detector where their polarization is analyzed.

For the production of the cesium atom beam the oven described by Eitel, Jost, and Kessler[5] was used after making a few modifications. As light source a mercury high-pressure arc lamp was used together with a polarizing foil and a quarter-wave plate, but without a filter. The main intensity of the lamp is concentrated in the spectral range where Fano's theoretical polarization is between 90 and 100%. Correcting the the oretical polarization curve for the intensity distribution and circular polarization of the incident light yields an average value of 80% for the spin polarization to be expected.

Before the results proved to be reliable, several difficulties had to be overcome. The measured electron polarization turned out to depend on the temperature of the cesium oven, i.e. the density of the beam, unless this temperature was chosen lower than about 140°C. According to the results of Popescu et al.,[6] this is due to spurious electrons resulting from photoionization of $Cs_2$ and from collisions of ground-state atoms with those in excited states which are produced by irradiation with our broad spectral band. These processes depend strongly on the beam density. Another serious problem was the electrons originating from inner walls of the apparatus. They

gave rise to polarization values which were unreproducible and too low. Finally a design of the electron optical system was found which satisfactorily suppressed these electrons.

Due to the low oven temperature only about $10^4$ photoelectrons were produced per second. The average value of the polarization was $(65 \pm 15)\%$. The measured values (10 runs) varied between these limits of error because of the spurious effects described in the last paragraph. (The statistical error is much smaller, since $4.2 \times 10^4$ electrons were counted). The experiment confirms Fano's prediction on the possibility of producing highly polarized electrons by photoionization of unpolarized atomic beams. All the sources of polarized electrons constructed so far on the basis of photoionization[7] start from polarized atomic beams.

Measurements of the Fano effect along the spectrum with good resolution and with an improved apparatus are now being made. The observed electron polarization will provide a detailed check of the theoretical assumptions made in describing the influence of spin-orbit coupling on photoabsorption by cesium vapor.[8,9]

We gratefully acknowledge helpful discussions with Dr. A. Müllensiefen, the help of Cand. Phys. U. Heinzmann in evaluating the measurements, and support by the Deutsche Forschungsgemeinschaft.

[1]U. Fano, Phys. Rev. 178, 131 (1969).
[2]The continuum P states have the same energy!
[3]J. Kessler, Rev. Mod. Phys. 41, 3 (1969).
[4]M. S. Lubell and W. Raith, Phys. Rev. Letters 23, 211 (1969).
[5]W. Eitel, K. Jost and J. Kessler, Z. Physik 209, 348 (1968).
[6]I. Popescu, C. Ghita, A. Popescu, and G. Musa, Ann. Physik 18, 103 (1966).
[7]R. L. Long, Jr., W. Raith, and V. W. Hughes, Phys. Rev. Letters 15, 1 (1965); P. Coiffet, Compt. Rend. 264B, 160, 454 (1967); G. Baum and U. Koch, Nucl. Instr. Methods 71, 189 (1969).
[8]E. Fermi, Z. Physik 59, 680 (1930).
[9]M. J. Seaton, Proc. Roy. Soc. (London), Ser. A 208, 418 (1951).

# PHYSICAL REVIEW
# LETTERS

VOLUME 25          24 AUGUST 1970          NUMBER 8

## MEASUREMENT OF DIFFERENTIAL CHARGE-TRANSFER CROSS SECTIONS AND PROBABILITIES BY PHOTON-PARTICLE COINCIDENCE TECHNIQUE*

D. H. Jaecks, D. H. Crandall, and R. H. McKnight
Behlen Laboratory of Physics, University of Nebraska, Lincoln, Nebraska 68508
(Received 25 June 1970)

The probabilities and differential cross sections for charge transfer to the $2p$ state of hydrogen in 4- to 20- keV proton-helium collisions have been measured for specific impact parameters by detecting in delayed coincidence the scattered neutral hydrogen atom and the resulting $2p$-$1s$ Lyman-$\alpha$ photon. These $2p$ probabilities are compared with the total charge transfer and $2s$ state probabilities, and with the coupled atomic state calculations of Sin Fai Lam.

In recent years, the study of electron charge-transfer processes in ion-atom collisions has been undertaken by measuring differential cross sections and charge-transfer probabilities.[1] In these experiments the total number of neutral particles scattered into specific angles is determined, with the probability defined as the number of neutral atoms scattered into a given angle divided by the total number of particles scattered into the same angle. In these experiments no distinction has been made among various possible excited states that can be formed in the process. To interpret these experimental results in terms of a collision model it is essential that the specific excited state of the neutral particle and ion formed in the charge-transfer process is known.

We wish to report new measurements of electron charge-transfer probabilities in proton-helium collisions, where the excited state of the resulting fast neutral atom is determined by a photon-particle coincidence technique. These probability measurements, $P_{2p}$ for H($2p$) formation, have been made for definite impact parameters of the $H^+$-He system. Specific impact parameters are related to a particular angle of scattering using a screened Coulomb potential as calculated by Dose.[2]

Unlike $He^+$-rare-gas systems that are under investigation,[1] the $H^+$-He collision system is not dominated by curve crossings; so one might expect a coupled state theory, based on an atomic eigenfunction expansion, to describe the collision process sufficiently well. Sin Fai Lam[3] has calculated $P_{2p}$ for specific impact parameters using an expansion of products of atomic eigenfunctions centered about both the target and projectile. Included in the expansion are the $1s$, $2s$, and $2p$ states centered about the projectile and the ground-state functions of He and $He^+$. One of the motivations of this work was to check this calculation in some detail. By measuring probabilities rather than total cross sections one can make a more direct comparison between theory and ex-

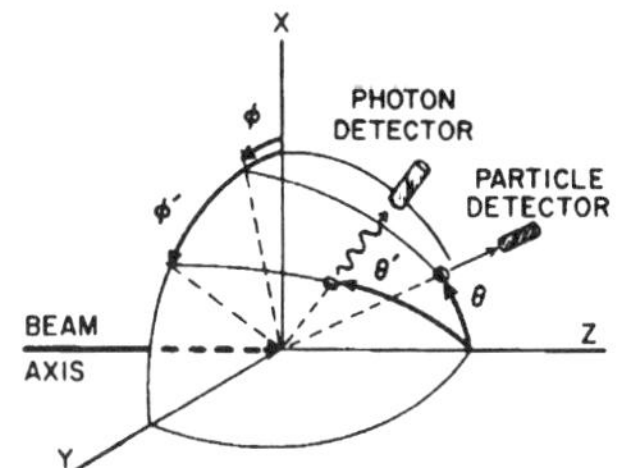


FIG. 1. Basic collision geometry. Primed and unprimed coordinates indicate the photon–detector and particle–detector positions.

VOLUME 25, NUMBER 8          PHYSICAL REVIEW LETTERS          24 AUGUST 1970

periment. The measured excited-state probabilities should provide a more sensitive test of the theory than the total charge transfer probabilities, which includes all states.

The basic scheme of the measurement technique is illustrated in Fig. 1. The photon detector is a solar blind photomultiplier whose position is defined by coordinates $(\theta', \varphi')$. The Cu-Be particle detector position has coordinates $(\theta,$

$\varphi)$. When an H($2p$) atom is formed and scattered into a direction $\theta, \varphi$ in a charge-transfer collision, it decays essentially at the collision center emitting a Lyman-$\alpha$ photon into a direction $\theta'$, $\varphi'$. The delayed coincidence rate, $dN_c$, between the Lyman-$\alpha$ photon and the ground-state H atom after collision with an He target, has been given by Macek[4] in terms of the collision parameters,

$$dN_c = n_A n_0 v (3/8\pi)\tfrac{1}{8}[7\sigma_0 + 11\sigma_1 - 3(\sigma_0 - \sigma_1)\cos^2\theta' - 3\sigma_1 \sin^2\theta' \cos2(\varphi - \varphi')$$
$$- 3\sqrt{2}\ \mathrm{Re}\langle a_0 a_1{}^* \rangle \sin2\theta' \cos(\varphi - \varphi')]d\Omega d\Omega', \qquad (1)$$

where $n_A$ is the number of target atoms in the observed scattering volume, $v$ is the velocity of the incident proton beam, and $n_0$ the number of particles in a unit volume of the beam. The scattering amplitudes of the H($2p$) magnetic substates are $a_0$ and $a_1$, where $\sigma_0 = |a_0|^2$ and $\sigma_1 = |a_1|^2$.

In the present experiment the scattering plane is defined as the $y$-$z$ plane and the detector is situated on the $x$ axis. If the detectors are small so that we can write $\theta' \simeq 90°$ and $\varphi - \varphi' \simeq 90°$, the coincidence rate becomes

$$dN_c \simeq n_A n_0 v (3/8\pi)(7/9)[\sigma_0 + 2\sigma_1]d\Omega d\Omega'. \qquad (2)$$

We see that the coincidence rate is proportional to the differential cross section, $\sigma = \sigma_0 + 2\sigma_1$. It is to be noted that other positions of the photon detector relative to the neutral detector will not give coincidence rates proportional to the differential cross section and in general will depend upon the matrix elements $\langle a_0 a_1{}^* \rangle$. If the detectors are not small, Eq. (1) must be integrated over $d\Omega$ and $d\Omega'$.

The measured charge-transfer probabilities are shown in Figs. 2 and 3. In this experiment $P_{2p}$ is defined as the number of hydrogen atoms formed in the $2p$ state and scattered into a specific angle, divided by the total number of particles scattered into that angle. Only the probabilities are shown; however, differential cross sections were determined at the same time.

Measurements were made for constant values of $\theta T$, where $\theta$ is the laboratory scattering angle and $T$ the laboratory kinetic energy. For small-angle scattering, a constant $\theta T$ corresponds to constant values of the impact parameter, which for the data shown, is 0.07 Å. The present measured total charge-transfer probabilities $P_0$, shown in Fig. 3, are in excellent agreement with the measurements of Helbig and Everhart.[5] The values of $P_{2s}$ are the same as those previously reported by this group except that corrections

have been made for the finite lifetime of the $2s$ state in the electric quench field.[6] Both phototube efficiencies were determined by normalizing to previously measured total charge-transfer cross sections for $2p$ and $2s$ formation.[7,8]

The first thing to note is that $P_{2s} + P_{2p}$ is only about 1% of $P_0$ indicating that most of the charge transfer is to the ground state of hydrogen. The magnitude of the $P_{2p}$ theoretical results of Sin Fai Lam are ten times as large as the present measurements, although the two agree in phase. Comparing the theoretical and experimental values of $P_{2s}$, one finds a difference in phase as well as magnitude that is ten times as large.

This magnitude and phase difference may be due to the fact that Sin Fai Lam did not include in his expansion the direct excitation channels

$$H^+ + He \rightarrow H^+ + He(2^1s, 2^1p)$$

which are nearly degenerate to the charge-trans-

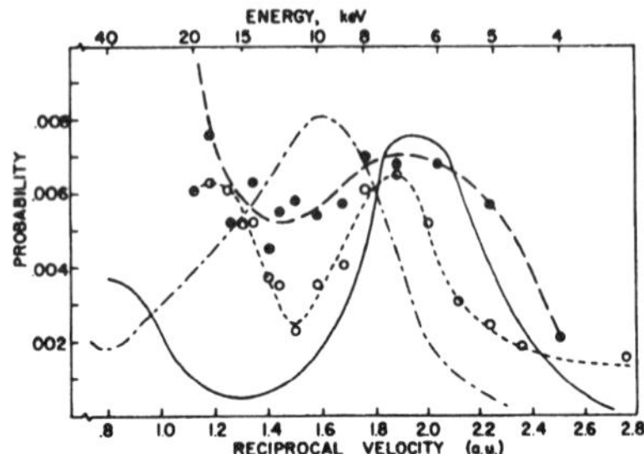



FIG. 2. Electron capture probabilities versus reciprocal proton velocity in atomic units. Present $P_{2p}$ measurements, closed circles; previous $P_{2s}$ measurements, open circles; Sin Fai Lam's theory: (i) $P_{2p}/10$, solid line, and (ii) $P_{2s}/10$, long and short dashed line.

VOLUME 25, NUMBER 8     PHYSICAL REVIEW LETTERS     24 AUGUST 1970

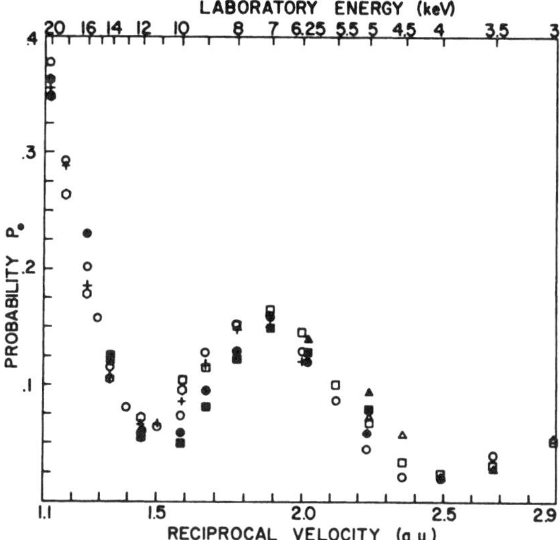


FIG. 3. Total charge-transfer probabilities, $P_0$. Present results: $\theta T = 50$ keV deg, open hexagons; $\theta T = 30$ keV deg, crosses; $\theta T = 20$ keV deg, open circles; $\theta T = 10$ keV deg, open squares; and $\theta T = 5$ keV deg, open triangles. Everhart *et al.*: Same notation as above only closed symbols.

fer channels[9]

$$H^+ \mid Ho - H(2\sigma, 2p) \mid He^+.$$

Putting in these nearly degenerate excitation channels should decrease the theoretical results and bring them more in line with the present experiment.[10]

As pointed out by Cheshire,[11] another important consideration is the fact that the atomic-state expansions give the incorrect nuclear charge in the united-atom limit. Cheshire tried, with some success, to take this into account for $H^+ + H$ collisions by varying the nuclear charge of the system so as to give the correct united-atom limit. Since the present probabilities are measured essentially at the united-atom limit (0.07 Å = impact parameter) it is important that the wave function has the proper behavior in this region. It would be interesting to see this approach used for the $H^+$-He system.

The present results also strongly indicate that to properly describe close collisions resulting in excited-state formation, one must include all the nearly degenerate channels as well as the nondegenerate channels whose amplitudes are large.

This is also suggested by the recent theoretical work of Flannery[10] who has calculated the excitation probabilities of helium by proton impact, but has ignored the charge-transfer channels measured in the present experiment. His theoretical results for the excitation cross sections are too large and inclusion of the nearly degenerate charge-transfer channels should bring his results into better agreement with experiment.

---

*Work supported by the National Science Foundation.

[1]*Proceedings of the Sixth International Conference on the Physics of Electronic and Atomic Collisions,* Boston, July–August 1969 (Massachusetts Institute of Technology, Cambridge, Mass., 1969), Sessions B-5 and B-14.

[2]V. Dose, Helv. Phys. Acta **41**, 261 (1968).

[3]L. T. Sin Fai Lam, Proc. Phys. Soc., London **92**, 67 (1967).

[4]J. Macek, Bull. Amer. Phys. Soc. **15**, 73 (1970).

[5]H. F. Helbig and E. Everhart, Phys. Rev. **136**, A674 (1964).

[6]D. H. Jaecks, R. H. McKnight, and D. H. Crandall, in *Proceedings of the Sixth International Conference on the Physics of Electronic and Atomic Collisions,*

VOLUME 25, NUMBER 8          PHYSICAL REVIEW LETTERS          24 AUGUST 1970

*Boston, July-August 1969* (Massachusetts Institute of Technology, Cambridge, Mass., 1969), p. 862

[7]D. H. Jaecks, B. Van Zyl, and R. Geballe, Phys. Rev. **137**, A340 (1965).

[8]E. P. Andreev, A. Ankudinov, and S. V. Bobashev, Zh. Eksp. Teor. Fiz. **50**, 565 (1966) [Sov. Phys.-JETP **23**, 375 (1966)].

[9]H. H. Michels, J. Chem. Phys. **44**, 3834 (1966).

[10]M. R. Flannery, J. Phys. B: Proc. Phys. Soc., London **3**, 306 (1970).

[11]I. M. Cheshire, J. Phys. B: Proc. Phys. Soc., London **1**, 428 (1968).

PHYSICAL REVIEW A          VOLUME 4, NUMBER 6          DECEMBER 1971

# Theory of Atomic Photon-Particle Coincidence Measurements*

Joseph Macek and D. H. Jaecks

*Behlen Laboratory of Physics, The University of Nebraska, Lincoln, Nebraska* 68508

(Received 18 March 1971)

The theory of measurements in which photons are detected in delayed coincidence with a scattered particle is developed in a form specifically applicable to atomic collisions. Equations are obtained which relate the measured coincidence rate to excitation amplitudes. These equations incorporate the polarization of the radiation, the fine and hyperfine structure of the atomic levels, the coherence of the radiation, and the time dependence of the radiation intensity. A semiclassical model for certain transitions is introduced to illustrate new features of coincidence measurements. The Ly-$\alpha$ transitions in hydrogen and $^1P - {}^1S$ transitions in He are treated in detail to illustrate the general theory.

## I. INTRODUCTION

Recently, coincidence techniques have been used to measure the amplitudes that describe atomic scattering processes. Three different types of measurements, distinguished by the types of particles detected, have been employed. Erhardt[1] has detected scattered electrons in coincidence with secondary electrons ejected in an ionizing collision of electrons with atomic helium. This represents a particle-particle coincidence measurement, and the data from such experiments can be directly interpreted in terms of differential ionization cross sections. Sheridan[2] has proposed measurements of photons emitted in the $D - P$ transition of He in coincidence with a photon emitted in the subsequent $P \rightarrow S$ decay. This constitutes a photon-photon coincidence measurement, and the data can be interpreted in terms of the probabilities for these radiative transitions.[3] Jaecks *et al.*[4] have measured Ly-$\alpha$ photons emitted in the decay of the $2p$ states of hydrogen atoms formed by the charge exchange reaction

$$H^+ + A \rightarrow H^* + A^*$$

where $A$ represents any target atom. This represents a photon-particle coincidence measurement, and, although similar experiments have been employed in nuclear physics, a theory relating experimental data to scattering amplitudes has not been given in a form directly applicable to atomic collisions. The theory[5] developed for nuclear studies is not directly applicable to atomic studies since nuclear studies have concentrated on determinations of the multipolarity of the electromagnetic transitions. In experiments of interest in atomic collision studies, the multipolarity of the radiative transitions is known, and experiments are conducted to determine the parameters of the population of states produced by the collision.

Recently, the theory of photon-particle coincidences has been developed in a form directly applicable to elementary particle reactions.[6] This theory emphasizes the determination of scattering amplitudes (or density matrix elements) from experimental data, and is more directly applicable to atomic studies than is the theory developed for nuclear studies. Even this theory is not directly applicable, however, since the "beating" of radiation from different fine- and hyperfine-structure levels introduces oscillatory terms[7] in the radiative decay of excited atoms, whereas the theory of Ref. 5 assumes a purely exponential decay. In this paper we develop the theory of photon-particle coincidence in a form directly applicable to atomic collisions, taking into account the fine and hyperfine structure of atomic levels.

The first consistent treatment relating the number of photons emitted after an atomic collision to excitation cross sections was given by Percival and Seaton.[8] They did not consider the detection of scattered particles and photons in coincidence and therefore used a time independent theory with the further condition of incoherent excitation of magnetic substates. Since we are interested in coincidence experiments, and since modern electronics with resolution times of the order of nanoseconds are used in such experiments, it is necessary to consider the time dependence of the radiation. Furthermore the incoming particle beam and the axis of the particle detector define a plane, which we shall call the scattering plane, so that the geometry possesses only reflection symmetry in this plane rather than rotational symmetry about the beam axis. As a consequence of this lower symmetry the magnetic substates are coherently excited and interference between amplitudes referring to different magnetic sublevels occurs. The consequences of this interference is the subject of the study reported in Ref. 6. Although we develop the present theory from first principles without drawing on the theory previously developed to treat elementary particle decays, our final results essentially combine the existing theories of angular correlations and time resolved spectroscopy to obtain expressions for the number of coincidences in a form directly applicable to atomic collisions. This theory is presented in Sec. II. The theory relates the number of coincidences measured for a given orientation of photon and particle detectors to the amplitudes describing the formation of the decaying states. The amplitudes are treated as parameters to be fitted to the experimental data. The atomic level structure and the radiative decay widths are assumed known, although they may also be treated as fitting parameters.

Section IV treats the time dependence in more detail than is given in Sec. II. Earlier treatments[9] of the time dependence of the radiation in cross section measurements considered the time development of the number of atoms in excited states by means of rate equations, and then supposed that the number of photons observed was proportional to the number of excited atoms at each instant of time. This supposition is rigorously correct only when the number of photons integrated over all angles of emission and summed over all polarizations is measured. Most experiments detect light emitted in a definite direction and polarization corrections are needed to relate the measured intensity to the number of excited atoms. It has been shown that the polarization is time dependent,[7] and this time dependence has not been considered in earlier theories. Section IV treats the time dependence in detail and defines conditions under which a rate equation approach applies.

The discussion in Sec. II is quite general and the theory presented there applies to a wide variety of collisional processes. The development is necessarily mathematical and somewhat abstract; consequently it is difficult to relate the final equations to a simple physical picture of the production of line radiation excited by atomic collisions. Under certain conditions the angular distribution of light can be described in terms of a simple classical source and this source related directly to the collisional process. This construction provides a ready picture of the otherwise obtuse results of Sec. II and is discussed in Sec. III.

Although our theory has as its main objective the development of expressions for photon-particle coincidences, it also applies to noncoincidence experiments which measure cross sections by optical techniques. In particular, the theory extends the treatment of Percival and Seaton[8] to include expressions for the polarization of radiation emitted in charge-exchange collisions. We find that as long as $LS$ coupling holds and as long as the emitting atom remains in view of the photon detector for a time long compared to the mean life, Percival and Seaton's expressions apply. The appropriate modifications when one or the other of these assumptions fails are discussed in Secs. II and IV.

### II. GENERAL THEORY OF PHOTON-PARTICLE COINCIDENCES

#### A. $LS$ Coupling

The general theory of photon-particle coincidences proceeds most rigorously from the quantum theory of radiation.[3] The alternative approach of Franken,[10] however, provides a more simple but entirely adequate point of departure, which we will follow here. This theory assumes that cascade contributions are negligible, and that the light detection system detects all radiation with a frequency near the central frequency of a fine-structure or hyperfine-structure multiplet. The sensitive frequency range is assumed to be sufficiently broad that little intensity is contained in the undetected wings of the emission lines. These same assump-

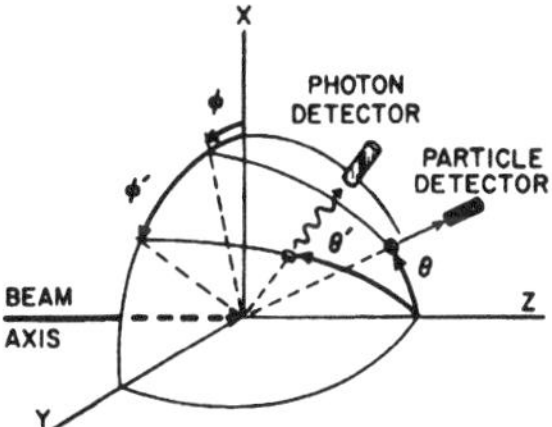


FIG. 1. Schematic collision geometry. Primed and unprimed coordinates indicate the photon detector and particle detector positions.

tions are involved in Ref. 7 ; indeed the derivation in this section follows the earlier discussion very closely except that cylindrical symmetry is not assumed. However, the formulas of Ref. 7 can be recovered from those derived here by integrating Eq. (21) of this paper over the azimuthal directions of the detected particle. This integration implies cylindrical symmetry, and therefore some of the results we obtain here, for example, expressions for the sum of Clebsch-Gordan coefficients, can be applied to the analysis of atomic lifetime measurements using beam-foil techniques.

We consider a collision in which a molecule, ion, or electron strikes an atom or molecule. In the collision some excited atoms are produced which subsequently decay by photon emission. The experiment measures the number of these photons in coincidence with some scattered particle, which may be an electron, atom, or molecule. We seek an expression relating the coincidence rate $dN_c$ to the amplitudes describing the excitation process. We stress here that the collision may involve molecules of any complexity, but the photons must come from an atom. We suppose that the atomic levels are adequately described by $LS$ coupling.

The coincidence rate depends upon the position of the photon and particle detectors. A schematic experimental arrangement with an associated coordinate system is shown in Fig. 1. The incoming beam axis is taken to be the $z$ axis, and the $x$-$z$ plane is located arbitrarily. The angular coordinates of the particle detector relative to this coordinate system are denoted by $\theta$ and $\phi$, and the coordinates of the photon detector by $\theta'$ and $\phi'$. We seek an expression for $dN_c$ in terms of these four angles. In all practical cases the collision takes place in a time that is short compared to the radiative lifetime. Thus the wave function for the excited atom at $t = 0$ is

$$|\psi) = \sum_{JFM_F} a(JFM_F)\,|\,JFM_F)\,, \qquad (1)$$

where $|\,JFM_F)$ is the state vector describing a particular atomic state with total angular momentum $F$ and electron angular momentum $J$, and $a(JFM_F)$ is the amplitude for producing the excited state. Because $a(JFM_F)$ is a scattering amplitude, it depends upon $\theta$ and $\phi$, the coordinates of the particle detector. The state vector (1) is not an eigenstate of the Hamiltonian. Consequently at a later time the atom is in the state

$$|\psi) = \sum_{JFM_F} a(JFM_F)\,|\,JFM_F)e^{-(\gamma/2+iE_{JF}/\hbar)t}\,, \qquad (2)$$

where the factor $e^{-\gamma t/2}$ is included to account for the decay of the upper level population. Here $1/\gamma$ is the mean life of the excited atom. Note that the time is measured from the instant of collision.

The coincidence rate is proportional to the square of the dipole matrix element describing the decay integrated over the resolution time of the circuits

$$dN_C = vn_0\,n_A\,(e^2\,\omega^3/2\pi hc^3)\sum_0\int_0^{\Delta t} dt\,|\,(0\,|\,\hat{\epsilon}\cdot\vec{X}\,|\,\psi)\,|^2\,, \qquad (3)$$

where $n_0$ is the density of particles in the incoming beam, $v$ is the velocity of the beam, $n_A$ is the number of target atoms in the scattering volume, $\omega$ is the frequency of the emitted light, $\Delta t$ is the resolution time of the circuits, $\hat{\epsilon}$ is the polarization vector of the detected light, and $(0\,|$ is the state vector for the lower levels reached in the decay. The quantum numbers of the lower levels are summed over after squaring the matrix element. In Eq. (3) we have supposed that all photons emitted in a definite direction in the time interval $0 - \Delta t$ are detected. In Read's[11] application of coincidence techniques to lifetime measurements, photons emitted at a much later time are detected. Then the limits of the time integration are $t_i$ and $t_i + \Delta t_i$, where $t_i$ is the delay time beyond that required for the transit of the scattered particle to the detector. In Sec. IV we will consider specific experimental geometries with $t_i$ unequal to zero. The polarization vector $\hat{\epsilon}$ is determined by the geometry of the photon detection system, and depends upon the coordinates $\theta'$ and $\phi'$ of the photon detector. Since the amplitudes $a(JFM_F)$ have dimensions of length the quantity $dN_c/d\Omega\,d\Omega'$ is seen to be a rate equal to the number of counts per unit of time.

Writing $\hat{\epsilon}$ in terms of its spherical components, substituting Eq. (2) into Eq. (3) and denoting the angular momentum quantum numbers of the lower levels explicitly, we find for $dN_c/d\Omega d\Omega'$

$$\frac{dN_c}{d\Omega d\Omega'} = vn_0 n_A(e^2\omega^3/2\pi hc^3)$$

$$\times \sum_{qq'JFM_FJ'F'M'_FM L_0 M_S 0 M_{I0}} a^*(J'F'M'_F)\,a(JFM_F)$$

$$\times(-)^{q+q'}\,\epsilon^{*}_{-q'}\,\epsilon_{-q}$$

$$\times(L_0 M_{L_0} S_0 M_{S_0} I_0 M_{I_0}\,|\,X'_q\,|\,LS(J')IF'M'_F)^*$$

$$\times(L_0 M_{L_0} S_0 M_{S_0} I_0 M_{I_0}\,|\,X_q\,|\,LS(J)IFM_F)$$

$$\times\int_0^{\Delta t} dt\, e^{-(\gamma+i\omega_{JFJ'F'})t}\,, \qquad (4)$$

where

$$\omega_{JFJF'}=(E_{JF}-E_{J'F'})/\hbar.$$

Equation (4) is just Eq. (1.2) of Ref. 10 integrated over the resolution time.

Let us consider Eq. (4) in some detail. Basically it contains three different types of terms: (a) the scattering amplitudes describing the excitation process, (b) the dipole matrix elements describing the decay, and (c) the time-dependent factor. The scattering amplitudes are here considered as unknown parameters to be fitted to the experimental data. Because the amplitudes describe the scattering, they are functions of $\theta$ and $\phi$, the coordinates of the particle detector. Later we will use the transformation properties of the amplitudes under rotations about the incoming beam axis to consider unknown parameters which are functions of $\theta$ only.

The dipole matrix elements and the polarization vectors constitute the second type of term. They describe the radiative decay of the upper levels. The polarization vectors are determined by the geometrical arrangement of the photon detection system, and are therefore known completely. The value of the dipole transition matrix element is assumed known from theory or from other experimental data. It can also be treated as a fitting parameter, however.

The third term describes the time dependence of the emitted light. Note that in addition to the usual exponential factor $e^{-\gamma t}$, it contains an oscillatory factor $e^{i\Omega t}$, where we have used $\Omega$ to denote any one of the frequencies $\omega_{JFJ'F'}$. This oscillatory factor describes the modulation of the light intensity due to interference of radiation from the coherently excited levels of the fine-structure or hyperfine-structure multiplet. Effects of this term have been observed in radiation from atoms excited by beam-foil collisions.[11,12] Percival and Seaton's[6] theory also contains this particular term, but with $\Delta t$ taken equal to infinity. This is appropriate for experiments in which all light emitted in the particular transition under study is collected regardless of the time of emission. It is not appropriate for a coincidence experiment, or for a charge-exchange experiment where the emitting atom may be viewed for a time short compared to the decay time.

Equation (4) is the basic equation of our theory.

However, it is not yet in a form suitable for application. In line with our supposition that $LS$ coupling holds, we may reduce the number of independent parameters by expressing the amplitude and the state vectors in the $LM_L$ representation. The state vectors $|\,LS(J)IFM_F)$ are written in terms of the state vectors $|\,LM_L SM_S IM_I)$ according to the usual vector coupling rules

$$|\,LS(J)IFM_F) = \sum_{M_L M_S M_I M_J} |\,LM_L SM_S I M_I)$$

$$\times(LM_L SM_S\,|\,LSJM_J)(JM_J IM_I\,|\,JIFM_F)\,. \qquad (5)$$

Similarly, we have

$$a(JFM_F) = \sum_{\bar{M}_L \bar{M}_S M_I M_J} a(L\bar{M}_L S\bar{M}_S IM_I)$$

$$\times(L\bar{M}_L S\bar{M}_S\,|\,LSJM_J)(JM_J IM_I\,|\,JIFM_F)\,. \qquad (6)$$

The amplitudes $a(LM_L SM_S IM_I)$ occur in Eq. (4) as products

$$a^*(LM'_L SM'_S IM'_I)\, a(LM_L SM_S IM_I)\,. \qquad (7)$$

To obtain the number of photons counted in a unit time interval, Eq. (4) and hence also Eq. (7) is averaged over initial states and summed over final states. We denote this sum and average by $S$. To evaluate this sum we now express the amplitudes explicitly as the matrix elements of transition operators,

$$a(LM_L SM_S IM_I) = (i\vec{k}_i\,|\,T\,|\,\vec{k}_f \vec{K}_1 \vec{K}_2 \cdots f,\, LM_L SM_S IM_I)\,,$$

where $i$ is the set of quantum numbers describing the internal state of both particles before collision, $\vec{k}_i$ is the momentum vector of the incoming particle, $\vec{k}_f$ is the momentum vector of the outgoing particle detected by the particle detector, the $\vec{K}_i$'s are the momentum vectors of all particles not detected by the particle detector, $f$ represents the set of all quantum numbers describing the internal state of all particles except the radiating atom, and $LM_L SM_S IM_I$ refer to the excited states of the atom emitting the detected radiation. The sum denoted by $S$ then corresponds to an average over $i$, a sum over $f$, and an integration over all momentum vectors $\vec{K}_i$ of the undetected particles

$$S\, a^*(LM'_L SM'_S IM'_I) a(LM_L SM_S IM_I)$$

$$=\frac{1}{W_i}\sum_i \sum_f \int \prod_r d\hat{K}_r$$

$$\times(i\vec{k}_i\,|\,T\,|\,\vec{k}_f \vec{K}_1 \vec{K}_2 \cdots f,\, LM'_L SM'_S IM'_I)$$

$$\times(i,\vec{k}_i\,|\,T\,|\,\vec{k}_f \vec{K}_1 \vec{K}_2 \cdots f,\, LM_L SM_S IM_I)\,, \qquad (8)$$

where $W_i$ is the statistical weight of the initial state.

Several details in this expression should be

noted.  First, notice that both $T$-matrix elements refer to the same final state of all particles except the emitting atom.  The initial states are the same since the wave function, Eq. (1), refers to the state formed in a particular collision, i. e., a collision in which the initial state is specified by a definite set of quantum numbers $i$.  The final states of the nonemitting particles, whose quantum numbers are denoted by $f$, could in general be different in the two matrix elements; however, the condition that the collision time is much shorter than the lifetime of the decaying states implies that any change of state of the nonemitting particles cannot affect the radiative decay of the atom whose radiation is detected.  Thus the $f$ quantum numbers are also the same.  Further, since the $z$ projection of the total electronic spin is conserved in the collision, and since both matrix elements contain the same $i$ and $f$ quantum numbers, we must have

$$M_S = M_{S'} \ . \tag{9}$$

Similarly we have

$$M_I = M_{I'} \ . \tag{10}$$

To factor out the $\phi$ dependence of the product of amplitudes, we consider the transformation of this product under two different rotations: (a) any rotation of the arbitrary coordinate variables $\vec{r}_s$ integrated over in forming the $T$-matrix elements, and (b) a rotation of the directions of all outgoing particles about the beam axis.  The first transformation cannot affect the value of the matrix element, but the second rotation does since it describes a new physical arrangement.  The effect of the second rotation is the same as the effect of the combined rotations (a) and (b).  Under these simultaneous rotations the plane-wave factors $|\vec{k}_f)$ and $|\vec{K}_i)$ do not change since they are functions of the dot product $\vec{K}_s \cdot \vec{r}_s$ only.  The states labeled by the internal quantum numbers $i$, $f$, $M_L$, $M_S$, and $M_I$ transform as

$$|i) \rightarrow e^{iM_I\phi}|i) \ ,$$
$$|f, M_LM_SM_I) \rightarrow e^{i(M_f+M_L+M_S+M_I)\phi}|f, M_LM_SM_I) \ . \tag{11}$$

These transformation equations and Eqs. (9) and (10) imply that

$$S\, a^*(M_L'\cdot M_S'M_I')\, a(M_LM_SM_I)$$

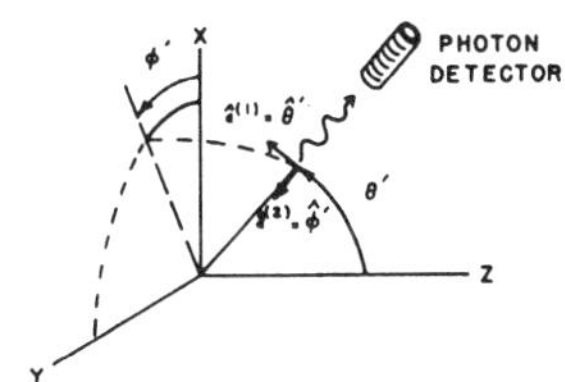



FIG. 2.  Standard polarization vectors $\hat{\epsilon}^{(1)}$ and $\hat{\epsilon}^{(2)}$.

$$= \delta_{M_S'M_S}\, \delta_{M_I'M_I}\langle a_{M_L'}\, a_{M_L}\rangle e^{i(M_L-M_L')\phi}/(2S+1)(2I+1) \ , \tag{12}$$

where we have used brackets to denote the average previously denoted by $S$, and where we have introduced amplitudes $a_{M_L}$.  These amplitudes differ from the amplitudes $a(LM_LSM_SIM_I)$ in that $a_{M_L}$ is normalized so that $|a_{M_L}|^2$ is equal to the partial cross section for exciting the $M_L$th magnetic substate of the decaying states, and $a_{M_L}$ is a function of $\theta$ only.

A similar consideration of the transformation properties of the amplitudes under reflections in the scattering plane shows that

$$\langle a_{M_L'}\, a_{M_L}\rangle = (-1)^{M_L-M_L'}\langle a_{-M_L'}a_{-M_L}\rangle \ . \tag{13}$$

Here and throughout this paper we are assuming the Condon-Shortley phase convention.  Equations (12) and (13) enable the coincidence rate to be expressed in terms of a minimum number of parameters $\langle a_{M_L}\, a_{M_L}\rangle$ which depend only upon the angular coordinate $\theta$.  The parameters are often denoted as density matrix elements $\rho_{M_L'M_L}$.  However the present notation emphasizes their connection with the scattering amplitudes more commonly used in excitation studies.

Now consider the polarization vector $\hat{\epsilon}$.  For every direction of the outgoing photon, only two polarization directions are linearly independent.  It is convenient to express an arbitrary polarization vector $\hat{\epsilon}$ in terms of two independent unit vectors perpendicular to the detector axis.  We choose (Fig. 2) one vector $\hat{\epsilon}^{(1)}$ lying in plane formed incoming beam and the photon detector axis and pointing in the direction of increasing $\theta'$.  The other vector $\hat{\epsilon}^{(2)}$ is perpendicular to both $\hat{\epsilon}^{(1)}$ and the photon detector axis, and points in direction of increasing $\phi'$.  The spherical components of these two vectors are given in Table I.  An arbitrary polarization vector $\hat{\epsilon}$ oriented at an angle $\beta$ with respect to $\hat{\epsilon}^{(1)}$ is then given by

$$\hat{\epsilon} = \hat{\epsilon}^{(1)}\cos\beta + \hat{\epsilon}^{(2)}\sin\beta \ . \tag{14}$$

Substituting Eqs. (5), (6), (12), (14), and the expressions for $\hat{\epsilon}^{(1)}$ from Table I into Eq. (4), and using the Wigner-Eckart theorem to evaluate the dipole matrix element, we obtain

TABLE I.  Spherical-tensor components of the two polarization vectors $\hat{\epsilon}^{(1)}$ and $\hat{\epsilon}^{(2)}$.

| $q =$ | 1 | 0 | $-1$ |
|---|---|---|---|
| $\hat{\epsilon}^{(1)}$ | $-\cos\theta'\, e^{i\phi'}/\sqrt{2}$ | $-\sin\theta'$ | $\cos\theta'\, e^{-i\phi'}/\sqrt{2}$ |
| $\hat{\epsilon}^{(2)}$ | $-i\, e^{i\phi'}/\sqrt{2}$ | $0$ | $-i\, e^{-i\phi'}/\sqrt{2}$ |

$$dN_c = v n_0 n_A (3\gamma'/8\pi)\{A_{00}\cos^2\beta + A_{11}\sin^2\beta + (A_{11}-A_{00})\cos^2\beta\cos^2\theta'$$

$$+ \sqrt{2}\,\mathrm{Re}\,A_{01}[\sin2\theta'\cos^2\beta\cos(\phi-\phi') + \sin2\beta\sin\theta'\sin(\phi-\phi')] - \mathrm{Re}\,A_{1-1}[(\cos^2\beta\cos^2\theta' - \sin^2\beta)$$

$$\times\ \cos2(\phi-\phi') + \sin2\beta\cos\theta'\sin2(\phi-\phi')]\}\,d\Omega\,d\Omega' , \qquad (15)$$

where

$$A_{q\,q'} = \sum_{JFJ'F'M_LM_L'} U(qq'M_LM_L'JFJ'F'LL_0)\langle a_{M_L'}a_{M_L}\rangle \int_0^{\Delta t} dt\,\exp[-(\gamma + i\,\omega_{JFJ'F'})t] , \qquad (16)$$

with

$$U(qq'M_LM_L'JFJF'LL_0) = \sum_{M_F M_J M_F' M_J' M_I M_J' M_S M_I M_L M_{I_0} M_{S_0} M_L M_L'} (L\bar{M}_LSM_{S_0}\mid LSJM_J)(JM_JIM_{I_0}\mid JIFM_F)$$

$$\times(L_0M_{L_0}1q\mid L_01L\bar{M}_L)(L\bar{M}_L'SM_{S_0}\mid LSJ'M_J')(J'M_J'IM_{I_0}\mid J'IF'M_F')(L_0M_{L_0}1q'\mid L_01L\bar{M}_L')$$

$$\times(LM_LSM_S\mid LSJ\bar{M}_J)(J\bar{M}_JIM_I\mid JIFM_F)(LM_L'SM_S\mid LSJ'\bar{M}_J')(J'\bar{M}_J'IM_I\mid J'IF'M_F')[1/(2S+1)(2I+1)] , \qquad (17)$$

and where

$$\gamma' = (8\pi/3)\,|\,(L\,\|\,X_q\,\|\,L_0)\,|^2/(2L+1) \qquad (18)$$

is the partial decay width for producing the detected photon. In general $\gamma'$ is not equal to the decay width $\gamma$, whose reciprocal is related to the mean life of the excited atom. The quantity $\gamma'$ is given by the branching ratio $B$ for the observed transition times $\gamma$,

$$\gamma' = B\gamma . \qquad (19)$$

The sum which represents $U$ in Eq. (17) can be simplified by standard techniques. The reduction of $U$ to a simple form has been carried out in general in the Appendix. To evaluate the sum representing $U$ we need only identify the quantum num-bers in Eqs. (A1) and (A12) of the Appendix. Comparing Eq. (17) and Eq. (A1) we find that $U$ is given by Eq. (A12) if we identify $j_i$, $k_i$, and $l_i$ according to

$$j_6 = k_6 = j_2 = k_2 = L , \qquad k_4 = F' ,$$
$$l_2 = l_5 = S , \qquad j_4 = F ,$$
$$l_3 = l_4 = I , \qquad m_1 = q ,$$
$$k_3 = k_5 = J' , \qquad m_2 = M_L ,$$
$$k_1 = j_1 = 1 , \qquad r_1 = q' ,$$
$$l_6 = L_0 , \qquad r_2 = M_{L'} ,$$
$$j_3 = j_5 = J .$$

Then we find for $U$

$$U(qq'M_LM_L'JFJ'F'LL_0) = [(2J+1)(2J'+1)(2F+1)(2F'+1)(2L+1)/(2S+1)(2I+1)]\,(-)^{L_0+q-M_L}$$

$$\times\sum_{\chi\nu}(2\chi+1)(-)^{-\chi}\begin{Bmatrix}L & L & \chi\\ J' & J & S\end{Bmatrix}^2\begin{Bmatrix}J & J' & \chi\\ F' & F & I\end{Bmatrix}^2\begin{Bmatrix}L & L & \chi\\ 1 & 1 & L_0\end{Bmatrix}\begin{pmatrix}L & L & \chi\\ -M_L' & M_L & \nu\end{pmatrix}\begin{pmatrix}1 & 1 & \chi\\ -q & q' & -\nu\end{pmatrix} . \qquad (20)$$

Equation (20) is convenient for both the numerical evaluation of $U$ and for the study of the symmetry properties of $U$ under interchanges of the indices. The sum on the right-hand side can be easily evaluated numerically since $\chi$ takes on only the values 0, 1, and 2. Tables of the 3- and 6-$j$ symbols when one of the arguments is 0, 1, or 2 are given by Edmunds.[13]

The symmetry properties of the $U$ coefficients can be used to reduce the number of different terms. From the symmetry of the 6-$j$ symbols it follows that $U$ is invariant to the simultaneous interchange of $q$ and $q'$ and $M_L$ and $M_L'$.

Further, $U$ is invariant to a simultaneous change of sign of $q$, $q'$, $M_L$, and $M_L'$. We have already used these symmetry relations and Eq. (13) to obtain the expression in Eq. (15).

When no polarization measurement is performed the number of coincidences is proportional to $dN_c$ of Eq. (15) summed over two independent polarizations. Summing Eq. (15) over two polarization directions, say $\beta$ and $\beta + \frac{1}{2}\pi$, gives

$$dN_c = v n_0 n_A \gamma'\,(3/8\pi)[A_{00} + A_{11} + (A_{11}-A_{00})\cos^2\theta'$$

$$+ \sqrt{2}\,\mathrm{Re}\,A_{01}\sin2\theta'\cos(\phi-\phi')$$

$$+A_{1\text{-}1}\sin^2\theta'\cos2(\phi-\phi')\,]\,d\Omega d\Omega' \ . \qquad (21)$$

Equation (21) has been used previously by one of the authors to interpret data on the charge exchange of protons on helium.[4] When Eq. (21) is integrated over $d\phi$ or $d\phi'$ the last two terms average to zero, and Eq. (21) reduces to the results of Percival and Seaton provided $\Delta t$ is taken to be infinite in Eq. (16).

Equations (15) and (21) apply to a wide variety of atomic processes. For all such processes, the information obtainable in a coincidence experiment may be expressed in terms of the four quantities $A_{00}$, $A_{11}$, ${\rm Re}A_{01}$, and $A_{1\text{-}1}$. The theory given above then relates these quantities to the density matrix elements $\langle a_{M'_L} a_{M_L}\rangle$ which describe the excitation process. Alternatively, the density matrix elements may be calculated theoretically and the quantities $A_{q'q}$ then calculated from Eq. (16). Theoretical calculations have usually been directed toward obtaining cross sections, or equivalently, the quantities $A_{00}$ and $A_{11}$. In order to more thoroughly relate theory and experiment it is desirable to present the quantities $A_{01}$ and $A_{1\text{-}1}$ as well.

To give some indication of the significance of the quantities $A_{01}$ and $A_{1\text{-}1}$ we consider in this section, and in Sec. IV, two special cases which illustrate the general features of the angular distribution. We consider radiation from $P\to S$ transitions excited by an atomic collision where the initial states of both incoming particles are $S$ states, and where there are only two particles in the final state with the nonradiating particle also in an $S$ state. Then reflection symmetry in the plane of scattering implies that $a_1=-a_{\text{-}1}$. Thus we have

$$\langle a_{\text{-}1}a_1\rangle = -|a_1|^2 = -\sigma_1 \qquad (22)$$

and the quantity $A_{1\text{-}1}$ can be written in terms of the partial cross section for exciting the $M_L=\pm1$ states.

### 1. $^1P\to{}^1S$ Transitions in He

We now consider the $^1P\to{}^1S$ transitions in He, excited by electron or proton impact, for example. This case is interesting because the $^1P$ states of He[4] have no fine or hyperfine structure, therefore the time-dependent exponential is just $e^{-\gamma t}$ and is common to all terms in Eq. (16). Also the lifetime of these excited states is short compared to experimentally realizable resolution times so we may take $\Delta t$ to be infinite. Then using Eq. (22) in Eq. (16) we find

$$A_{00}=\sigma_0/\gamma \ ,$$

$$A_{11}=-A_{1\text{-}1}=\sigma_1/\gamma \ , \qquad (23)$$

$$A_{01}={\rm Re}\langle a_0 a_1\rangle/\gamma \ .$$

When the photon detector is placed so that it views photons emitted perpendicular to the scattering plane, then both $\theta'$ and $\phi-\phi'$ equal $90°$ and Eq. (21) becomes

$$dN_c=vn_0n_A(\gamma'/\gamma)(3/8\pi)[\sigma_0+2\sigma_1]\,d\Omega d\Omega' \ . \qquad (24)$$

We see that the number of coincidences for this particular geometry is proportional to the cross section summed over magnetic substates. This result will be further discussed in Sec. IV.

### 2. Ly-$\alpha$ Transition in H

Because of the simplicity of the hydrogen atom, the study of atomic excitation processes involving it have always occupied an important position in the theoretical and experimental literature. Coincidence measurements of $A_{01}$ and $A_{1\text{-}1}$ will further our understanding of these excitation processes since the relative phase of $a_0$ and $a_1$ can be extracted if $A_{00}$, $A_{11}$, and $A_{01}$ are known.

The Ly-$\alpha$ radiation is emitted from excited atoms in the $2p$ states. These states have both fine and hyperfine structure, consequently the time-dependent factor in Eq. (16) does not factor out as is the case for $^1P\to{}^1S$ transitions in He. Since the hyperfine structure of both the $^2P_{1/2}$ and the $^2P_{3/2}$ levels is small compared to their decay widths, we may set $\omega_{JFJF'}$ equal to zero to a good approximation. Further since the precession period $1/\omega_{JJ'}$ and the mean life $1/\gamma$ are much shorter than commonly employed resolution times, the time integral in Eq. (16) becomes approximately

$$\int_0^\infty dt\,\exp[-(\gamma+i\omega_{JJ'})t]=\frac{1}{\gamma+i\omega_{JJ'}}$$

$$\approx\begin{cases}0\ , & J\ne J'\\[4pt]1/\gamma\ , & J=J'\ .\end{cases} \qquad (25)$$

With these approximations we find

$$A_{00}-(5\sigma_0+4\sigma_1)/9\gamma \ ,$$

$$A_{11}=(2\sigma_0+7\sigma_1)/9\gamma \ ,$$

$$A_{01}={\rm Re}\langle a_0 a_1\rangle/3\gamma \ , \qquad (26)$$

$$A_{1\text{-}1}=-\sigma_1/3\gamma \ .$$

When the photon detector axis is perpendicular to the scattering plane, Eq. (21) becomes

$$dN_c=vn_0n_A(3/8\pi)(7/9)[\,\sigma_0+2\sigma_1]\,d\Omega d\Omega' \ . \qquad (27)$$

This result is identical to Eq. (24) except for the factor of $\tfrac{7}{9}$. This factor represents a reduction in the number of photons emitted perpendicular to the scattering plane because the fine structure causes the angular distribution to smear out. This observation will be further developed in Sec. IV.

### B. *LS Coupling Violated in Collision*

With a redefinition of $A_{qq'}$, Eqs. (15) and (21) also apply when *LS* coupling does not hold. Two somewhat different situations may occur here: (a) The radiating atom may violate *LS* coupling, or (b) some nonradiating atom taking part in the collision may violate *LS* coupling. We consider case (a) first.

#### 1. *Radiating Atom Violating LS Coupling*

In this case the scattering amplitudes referring to states with different $J$ values are no longer related by the transformation of Eq. (6), but are treated as independent fitting parameters. Similarly the level width is also a function of $J$. Now a particular experiment may resolve the light from levels of the same multiplet with different $J$ quantum numbers. Therefore, we are interested in expressions analogous to Eq. (16), but with $J$ no longer summed over. Even when the light from different $J$ levels is not resolved, the breakdown

of *LS* coupling implies that the splittings are large compared to the level widths. Then $\omega_{JJ'}$ in Eq. (16) is large compared to $\gamma$ and the time-dependent factors for $J \neq J'$ are small compared to those with $J = J'$ and terms with $J \neq J'$ can be neglected, i.e., the radiation from different $J$ levels adds incoherently. Then the contribution from all levels with different $J$ values is just the sum of the individual contributions. Only interference of radiation from different hyperfine levels need be considered. The required expression for $A_{qq'}$ is then obtained by replacing $L$ by $J$, $J$ by $F$, $J'$ by $F'$, $S$ by $I$, $I$ by 0, $L_0$ by $J_0$, $M_L$ by $M_J$, and $M_{L'}$ by $M_{J'}$ in Eqs. (16) and (20), with a corresponding identification for the primed quantum numbers. We then have for $A_{q'q}$

$$A_{qq'} = \sum_{FF'M_JM'_J} U(qq'\,M_J M'_J\,FF'JJ_0)\langle a_{M'_J}a_{M_J}\rangle$$

$$\times \int_0^{\Delta t} dt\,\exp[-(\gamma + i\omega_{FF'})t]\ ,\qquad (23)$$

where

$$U(qq'\,M_J M'_J\,FF'JJ_0) = [(2F+1)(2F'+1)(2J+1)/(2I+1)]\,(-)^{J_0 + q - M_J}$$

$$\times \sum_{\chi\nu}(2\chi+1)(-1)^{-\chi}\begin{Bmatrix} J & J & \chi \\ F' & F & I \end{Bmatrix}^2 \begin{Bmatrix} J & J & \chi \\ 1 & 1 & J_0 \end{Bmatrix}\begin{pmatrix} J & J & \chi \\ -M'_J & M_J & \nu \end{pmatrix}\begin{pmatrix} 1 & 1 & \chi \\ -q & q' & \nu \end{pmatrix}\ .\qquad (29)$$

#### 2. *Nonradiating Atom Violating LS Coupling*

Our assumptions leading to Eqs. (15) and (21) do not hold when the target atom or any atom formed in the collision violates the *LS* coupling rules, even though the radiating atom obeys them. For example, the Ly-$\alpha$ radiation formed by charge transfer of protons on Xe originates from the $2p$ state of the hydrogen atom, which obeys *LS* coupling, but the states of Xe and Xe$^+$ do not. In this case total electronic spin is not a good quantum number of the system and Eq. (12) no longer holds.

Three different situations may occur which are not described by Eq. (12): (a) The $T$ operator in Eq. (8) may depend explicitly upon spin; (b) the states $i$ and $f$ may not obey *LS* coupling rules; or (c) the state vectors for the states $i$ and $f$ may approximately obey *LS* coupling, but some of the substates of their multiplets may not be energetically accessible. In all three cases the amplitudes

$a(JM_J)$, where $JM_J$ refer to the state of the radiating atom, can still be related to amplitudes $a(LM_LSM_S)$ by a Clebsch-Gordan transformation. Now, however, Eq. (9) no longer holds and amplitudes referring to different $M_S$ states can interfere. Because of this interference, no reduction in the number of unknown parameters ensues from the transformation to the $LM_LSM_S$ representation. Thus we use the $JM_J$ representation for amplitudes describing the formation of the radiating atom as well. Because the radiative decay does not change the spin of the excited atom, however, we will use the *LS* representation to evaluate the dipole matrix element. Our expression for $A_{qq'}$ in this case differs from Eq. (29) only in that account is taken of interference between amplitudes referring to different $J$ levels, since the different $J$ levels may only be separated by a spacing of the same order as their widths. All manipulations made in deducing Eq. (17) can be repeated for this case and one finds

$$A_{qq'} = \sum_{JJ'FF'M_JM'_J} U(qq'M_JM'_JJJ'FF'LL_0)\langle a_{J'M'_J}a_{JM_J}\rangle\int_0^{\Delta t} dt\,\exp[-(\gamma + i\omega_{JFJ'F'})t]\ ,\qquad (30)$$

where

$$U(qq'M_JM'_J\,JJ'FF'LL_0) = \{[(2J+1)(2J'+1)]^{1/2}(2F+1)(2F'+1)(2L+1)[1/(2I+1)(2S+1)]\}\,(-1)^{S+L_0+q-M_J}$$

$$\times \sum_{\chi\nu} (2\chi + 1) \begin{Bmatrix} J & J' & \chi \\ F' & F & I \end{Bmatrix}^2 \begin{Bmatrix} J & J' & \chi \\ L & L & S \end{Bmatrix} \begin{Bmatrix} L & L & \chi \\ 1 & 1 & L_0 \end{Bmatrix} \begin{pmatrix} J' & J & \chi \\ -M_{J'} & M_J & \nu \end{pmatrix} \begin{pmatrix} 1 & 1 & \chi \\ -q & q' & -\nu \end{pmatrix} \ . \tag{31}$$

To apply this equation to a noncoincidence measurement of Ly-$\alpha$ radiation produced in charge exchange, $d\Omega$ in Eq. (15) is integrated over, and only the quantities $A_{00}$ and $A_{11}$ enter into the expressions for the number of photons. These expressions, even though they apply to Ly-$\alpha$ radiation, now differ from Percival and Seaton's.[8] They differ because electronic spin is no longer a good quantum number of the target-projectile system, even though it is a good quantum number for the states of hydrogen atom alone.

### III. SEMICLASSICAL MODEL OF ANGULAR DISTRIBUTION

The theory of the preceding section relates the angular distribution of the coincidences to the parameters of the initial-state wave function. The connection between the experimentally observed quantity, the angular distribution, and the initial-state wave function is somewhat remote, involving many mathematical manipulations which are difficult to interpret physically. To aid our understanding of the theory, we consider an alternative approach, which proceeds by constructing a classical source with the angular distribution given by Eq. (15). This approach was also used by Percival and Seaton.[8,14] Their source consisted of one oscillator oriented along the beam axis and two equal strength oscillators oriented perpendicular to each other and to the beam axis. The three oscillators emitted incoherently thus maintaining cylindrical symmetry about the beam axis. In our case the source is more complicated and will in general be made up of oscillators radiating coherently [to represent the sums in Eqs. (3) and (12) performed before squaring], and oscillators radiating incoherently (to represent the sums performed after squaring). It is clear that the equivalent source for the most general case is rather complicated, and little can be gained by constructing it. However, the presence of oscillators radiating coherently represents the new feature of the source with the angular distribution of Eq. (15). This feature can be illustrated, and its significance for future investigations discussed, for the special cases of Sec. II.

Consider first the $^1P - {}^1S$ transitions in He excited by electron or proton impact, for example. The collision prepares the atom in the state

$$\psi = a_0 \psi_0 + a_1 \psi_1 + a_{-1} \psi_{-1} \ . \tag{32}$$

$\psi$ is invariant to reflections in the plane of scattering, and therefore

$$a_1 = -a_{-1} \ , \tag{33}$$

with the Condon-Shortley phase convention.

Thus

$$\psi = a_0 \psi_0 + a_1(\psi_1 - \psi_{-1}) = a_0 \psi_z - a_1 \sqrt{2}\, \psi_x \ , \tag{34}$$

where the $z$ and $x$ subscripts denote the angular part of the wave functions. Now we can write $\psi$ in terms of wave functions referred to a new $z$ axis, called here the $z'$ axis, oriented at an angle $\alpha$ to the old $z$ axis and lying in the plane of scattering. Then $\psi$ becomes

$$\psi = [a_0 \cos\alpha + a_1 \sqrt{2}\, \sin\alpha\,]\psi_{z'}$$
$$+ [a_0 \sin\alpha - a_1 \sqrt{2}\, \cos\alpha\,]\psi_{x'}$$
$$= a_{z'} \psi_{z'} + a_{x'} \psi_{x'} \ . \tag{35}$$

Now $\alpha$ can always be chosen so that $a_{z'}$ and $a_{x'}$ differ in phase by 90°:

$$\tan 2\alpha = \frac{2\sqrt{2}\ \mathrm{Re}(a_0 a_1^*)}{\sigma_0 - 2\sigma_1} \ . \tag{36}$$

With this choice of $\alpha$ the angular distribution of the radiation is proportional to

$$|\,(0\,|\,\hat{\epsilon}\cdot\vec{x}\,|\,\psi)\,|^2 = (\hat{\epsilon}\cdot\vec{d}_1)^2 + (\hat{\epsilon}\cdot\vec{d}_2)^2 \ ,$$

where

$$\vec{d}_1 = (0\,|\,x\,|\,\psi_x)\,|\,a_{x'}\,|\,\hat{x}' \ ,$$

$$\vec{d}_2 = (0\,|\,z\,|\,\psi_z)\,|\,a_{z'}\,|\,\hat{z}' \ .$$

This distribution is identical to the average over one period of the radiation from two classical oscillators $\vec{d}_1$ and $\vec{d}_2$, oscillating 90° out of phase, oriented perpendicular to each and lying in the scattering plane as shown in Fig. 3(a).

The length of the two oscillators is proportional to the magnitudes of $a_{z'}$ and $a_{x'}$. As the dipoles oscillate they trace out an ellipse. The radiation emitted by the oscillators may be viewed perpendicular to the scattering plane with a polarizer oriented at an angle $\beta'$ with respect to the major axis of the ellipse (or at an angle $\beta = \beta' + \alpha$ to the beam axis). The intensity observed is proportional to

$$I \propto |\,a_{z'}\,|^2 \cos^2\beta' + |\,a_x\,|^2 \sin^2\beta' \ . \tag{37}$$

A schematic polar plot illustrating the variation of $I$ with $\beta'$ appears in Fig. 3(b). From this figure we easily read off $|\,a_{z'}\,|$ and $|\,a_{x'}\,|$. Since we know that $a_{x'}$ and $a_{z'}$ differ in phase by 90°, we have determined $\psi$ up to an over-all phase factor (actually one does not know whether the dipole vector rotates clockwise or counterclockwise; this requires a measurement of the circular polarization).

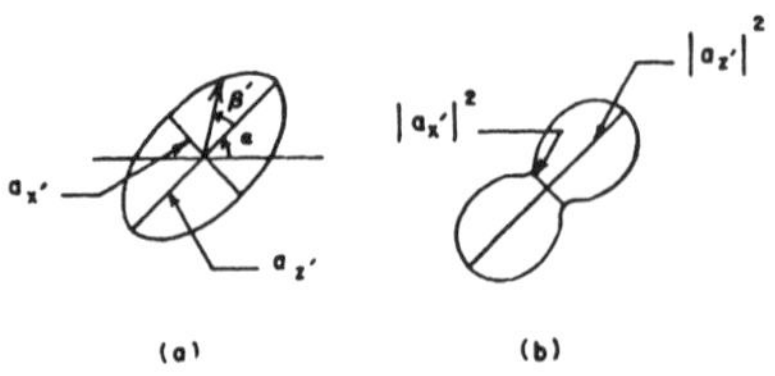

FIG. 3. Polarization of light emitted perpendicular to the scattering plane for $^1P - {}^1S$ transitions in He. (a) gives the measured distribution and (b) shows the equivalent classical source.

This construction also shows why the radiation from $^1P \rightarrow {}^1S$ transitions viewed perpendicular to the scattering plane is proportional to the total (for a given scattering angle) cross section. Effectively one sees the full length of both oscillators. At any other angle we see a projection of the oscillators which distorts the length of one compared to the other.

When the levels are split by fine or hyperfine structure, Eq. (35) still represents the wave function of the excited atom at $t = 0$. However, since $\psi_{x'}$ and $\psi_{z'}$ are not eigenstates of the target, at later times they evolve into spin dependent linear combinations of $\psi_{x'}$, $\psi_{y'}$, and $\psi_{z'}$. This evolution tends to smear out the angular distribution, however the dipole vectors at $t = 0$ still serve to characterize the collisional excitation process.

Our classical construction suggests that the experimental data for $P \rightarrow S$ transitions may be graphically presented in terms of the two dipole vectors just discussed. The vectors are directly measured only for $P \rightarrow S$ transitions with no fine- or hyperfine-structure depolarization, as for the $^1P - {}^1S$ transitions in He[4].

When fine structure or hyperfine structure is present the amplitude $a_0$ and $a_1$ can be extracted from Eq. (15) provided Eq. (22) holds. Then the dipole vectors may be constructed using Eqs. (35) and (36). A determination of these vectors over a wide range of energies will provide substantially new information on atomic collision processes, in particular, it will provide information on how excitation processes depart from predictions of the Born theory. In the Born approximation, the ellipse traced out by the dipole vectors degenerates into a straight line parallel to the momentum-transfer vector. Departures are manifested by (a) a rotation of the line from an orientation along the momentum transfer axis, (b) a broadening of the ellipse, (c) a contraction or expansion of the degenerate ellipse away from the magnitude predicted by the Born theory. Experimental determination of the relative importance of these three types of departures will provide substantially new information on the excitation process.

As a second example of the usefulness of the data provided by coincidence experiments, consider excitation or charge transfer in heavy-ion collisions in the adiabatic region. Here the change of the projection of the electronic angular momentum along the internuclear axis serves to classify the excitation mechanism. Direct excitations, which occur because of the translational kinetic energy perturbation to the temporarily formed molecular system, obey the $\Delta\lambda = 0$ rule, while those which occur because of the rotational coupling obey the $\Delta\lambda = \pm 1$ rule. Here $\lambda$ is the projection of the electronic angular momentum on the internuclear axis. As long as the relation (33) holds, the $\Delta\lambda = 0$ excitations will correspond to a dipole vector oriented along the direction of the outgoing particle while $\Delta\lambda = \pm 1$ excitations correspond to the dipole vector oriented perpendicular to direction of the outgoing particle. Coincidence techniques thus make possible the experimental differentiation of these two mechanisms.

### IV. TIME INTEGRATION

As we have suggested in the Introduction, the theory of Sec. II may be applied to noncoincidence experiments as well as coincidence experiments. One only need integrate Eq. (16) over the solid angle of the particle detector. The resulting equations, however, do not reduce to the Percival and Seaton theory unless one further assumes that the emitting atom remains in view of the photon detector for a time long compared to the mean life of the excited atom. This condition is not always met in charge exchange measurements and corrections for the finite lifetime of the atom must be made. Corrections are usually made by solving the classical rate equations[9] for the number of excited atoms. This approach is appropriate as long as the experiment actually measures a quantity proportional to the number of excited atoms. Such measurements may be accomplished by measuring the light intensity integrated over all angles and summed over both polarizations; however, one often wishes to measure the polarization of the radiation as well as its integrated intensity. This requires a measurement of a light intensity which is not proportional to the number of atoms in the excited state. The instantaneous intensity could be obtained here using rate equations for the density matrix describing the initial-state population.[15] Alternatively, one may simply integrate Eq. (16) over time in a manner appropriate to the particular experimental geometry. Since this latter approach is more direct, and since the classical rate equation approach has often been used without any discussion of the applicability of the procedure, we

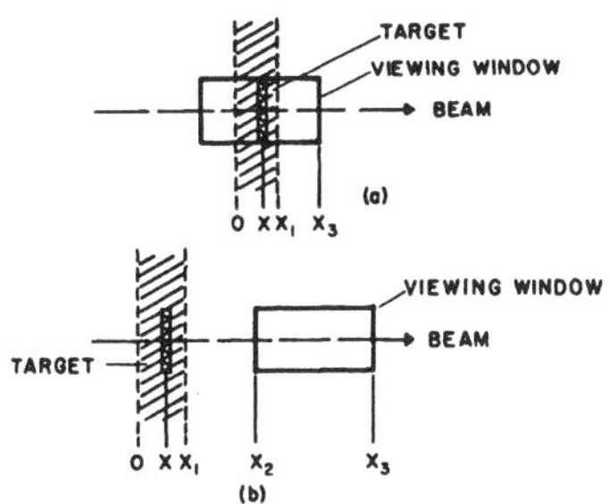

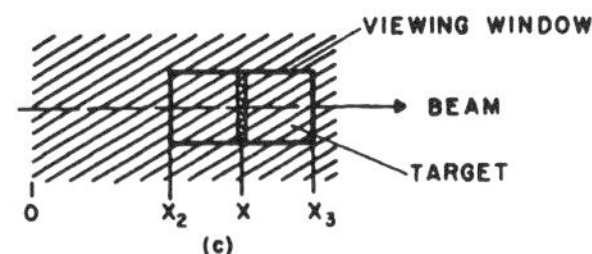



FIG. 4. Schematic collision geometries requiring different time integrations. (a) Crossed beam, (b) decay curve measurement, and (c) flooded collision chamber.

will discuss the time integrations for the three commonly encountered geometries shown in Fig. 4.

The schematic diagrams in Fig. 4 contain three parts: (a) the target gas configuration, (b) the proton beam, and (c) the schematic viewing window. The length of the viewing window is the parameter of most importance here. This length is represented geometrically, but its effective length is not always the geometrical length. If $v$ is the velocity of the incoming proton and if $t_d$ is the delay time between the detection of a photon and the detection of a particle, then only photons emitted when the atom is at a distance greater than $(t_d - \Delta t/2)v$ but less than $(t_d + \Delta t/2)$ from the particle detector are counted as real coincidences. Here $\Delta t$ is the resolution time. Thus $\Delta d = v\Delta t$ essentially measures the length of the viewing window as long as $\Delta d$ is less than the geometrical length viewed by the photon detector. With this interpretation of the schematic diagrams, we consider the three experimental configurations.

(a) Crossed beam. Here the proton crosses a beam of neutral particles and picks up an electron in an excited state at some point $x$ in the gas beam. The number of atoms so formed is proportional to $n_A(x)$ the number of atoms in a thin volume element of thickness $dx$ whose sides are perpendicular to the proton beam. Taking the zero of time when the atom is at the leftmost edge of the gas target, the probability that the excited atom formed at $x$ will

decay in a time $dt$ is proportional to

$$e^{-\gamma(t-x/v)} dt, \quad t > x/v. \tag{38}$$

A typical time-dependent factor in Eq. (16) also contains $e^{i\Omega t}$, where $\Omega$ is short for $\omega_{JFJ'F'}$. To obtain the number of photons detected we must integrate the time-dependent factor over $x$ and $t$, weighted by $n_A(x)$. The factor

$$n_A \int_0^{\Delta t} e^{-\Gamma t}$$

in Eq. (16) is then replaced by

$$\int_0^{x_1} dx\, n_A(x) \int_{x/v}^{x_3/v} dt\, e^{-\Gamma(t-x/v)}, \tag{39}$$

where $\Gamma = \gamma + i\Omega$.

(b) Decay-curve measurements. In this case all excited atoms are formed outside of the viewing window. The integrals are identical to Eq. (39) except that the lower limit $x/v$ is replaced by $x_2/v$.

(c) Flooded collision chamber. Here some atoms are formed prior to entering the viewing region, as in case (b), and some are formed in the viewing region, as in case (a). The number of photons counted is obtained by adding the two contributions. The combined integral is

$$\int_{x_2}^{x_3} dx\, n_A(x) \int_{x/v}^{x_3/v} dt\, e^{-\Gamma(t-x/v)}$$
$$+ \int_0^{x_2} dx\, n_A(x) \int_{x_2/v}^{x_3/v} dt\, e^{-\Gamma(t-x/v)}. \tag{40}$$

To connect these expressions with those obtained by a rate equation approach, we will evaluate the integrals for the cases considered by Hughes, et al.[9] [case (b)] and by Jaecks[9] [case (c)]. In case (b) considered by Hughes et al.,[9] the condition $(x_3 - x_2)/v \ll 1/\gamma$ holds, so the integral over $t$ was not performed. Then setting $n_A(x) = \text{const}$ and replacing $t$ by $x_2/v$ we get for the integral over $x$

$$-e^{-\Gamma x_2/v}(1 - e^{\Gamma x_1/v}) = e^{-\Gamma(x_2-x_1)/v}(1 - e^{-\Gamma x_1/v}). \tag{41}$$

When $\Gamma = \gamma$, this is precisely the result obtained by Hughes et al.[9] However, since Eq. (16) contains terms with $\Omega \neq 0$, the result obtained from a rate equation approach can only be applied when the light intensity being measured is proportional to the intensity integrated over all directions and summed over polarizations, or when corrections introduced by the nonvanishing of $\Omega$ are negligible. Such corrections are small when the ratio $\gamma/\Omega$ is small compared to unity, since then the coefficients of the oscillatory terms are small compared to those of the nonoscillatory terms. The corrections are also small if $\gamma/\Omega$ is much larger than unity, since then the atom decays before the oscillations produce a noticeable effect. Between these two limits the corrections must be evaluated for each geometrical arrangement. Use of a rate equation approach by Hughes et al. is quite correct, however, since they measured the radiation

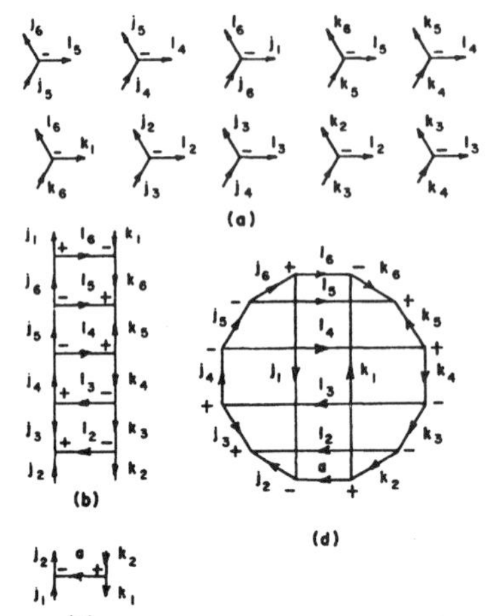

FIG. 5. Evaluation of a sum of Clebsch–Gordan co-efficients by diagrams. The ten diagrams in Fig. 5(a) represent the products of ten Clebsch–Gordan coefficients. The diagrams are joined to form Fig. 5(b), which is contracted with Fig. 5(c) to obtain Fig. 5(d).

from an $S$ state, which is unpolarized and isotropic. For such radiation, terms containing $\Omega \neq 0$ do not enter into the expression for the angular distribution.

In the expression considered by Jaecks[9] the condition $(x_3 - x_2)/v \ll 1/\gamma$ also applies and the integration over $t$ in Eq. (40) simply replaces $t$ by $x_2/v$ and multiplies the integrand by the factor $(x_3 - x_2)/v$. The first term of Eq. (40) is second order in $(x_3 - x_2)/v$ and is negligible compared to the second term, which becomes, upon taking $n_A(x) = \text{const}$,

$$(1 - e^{-\Gamma x_2/v})/\Gamma . \tag{42}$$

When $\Gamma = \gamma$ this equals the result obtained from the rate equation theory. When $x \to \infty$, we get just $1/\Gamma$ which is the result obtained by Percival and Seaton.[8] Note that $x \to \infty$, or more precisely, $x_2/v \gg \gamma$, is realized by allowing the proton beam to pass through a long gas-filled collision chamber. Thus Percival and Seaton's theory applies here even though a particular atom remains in view of the photon detector for only a short time. Since Gaily et al.[9] actually ensured that $x_2/v \gg \gamma$, their use of Percival and Seaton's theory is fully justified.

### APPENDIX: EVALUATION OF SUM OVER CLEBSCH-GORDAN COEFFICIENTS

We evaluate the sum

$$C = \sum_B (j_6 m_6 l_5 n_5 \mid j_6 l_5 j_5 m_5)(j_5 m_5 l_4 n_4 \mid j_5 l_4 j_4 m_4)$$

$$\times (l_6 n_6 j_1 m_1 \mid l_6 j_1 j_6 m_6)(k_6 r_6 l_5 n_5 \mid k_6 l_5 k_5 r_5)$$

$$\times (k_5 r_5 l_4 n_4 \mid k_5 l_4 k_4 r_4)(l_6 n_6 k_1 r_1 \mid l_6 k_1 k_6 r_6)$$

$$\times (j_2 m_2 l_2 n_2 \mid j_2 l_2 j_3 m_3)(j_3 m_3 l_3 n_3 \mid j_3 l_3 j_4 m_4)$$

$$\times (k_2 r_2 l_2 n_2 \mid k_2 l_2 k_3 r_3)(k_3 r_3 l_3 n_3 \mid k_3 l_3 k_4 r_4) , \tag{A1}$$

where $B$ stands for the set $m_3 m_4 m_5 m_6\, n_2 n_3 n_4 n_5 n_6\, r_3 r_4 r_5 r_6$, using the diagrammatic technique of Yutsis.[16] We first write the Clebsch-Gordan coefficients in terms of the Wigner $3-j$ symbols[13]

$$(j_1 m_1 j_2 m_2 \mid j_1 j_2 j_3 m_3) = (-)^{j_1 - j_2 + j_3}(j_3)^{1/2}(-)^{j_3 - m_3}$$

$$\times \begin{pmatrix} j_1 & j_2 & j_3 \\ m_1 & m_2 & -m_3 \end{pmatrix} , \tag{A2}$$

where $(j_3)$ equals $(2j_3 + 1)$. This converts the sum over Clebsch-Gordan coefficients into a sum over $3-j$ symbols:

$$C = A(-)^{R_1} \sum_B{}' \begin{pmatrix} j_6 & l_5 & j_5 \\ m_6 & n_5 & -m_5 \end{pmatrix}\begin{pmatrix} j_5 & l_4 & j_4 \\ m_5 & n_4 & -m_4 \end{pmatrix}$$

$$\times \begin{pmatrix} l_6 & j_1 & j_6 \\ n_6 & m_1 & -m_6 \end{pmatrix}\begin{pmatrix} k_6 & l_5 & k_5 \\ r_6 & n_5 & -r_5 \end{pmatrix}\begin{pmatrix} k_5 & l_4 & k_4 \\ r_5 & n_4 & -r_4 \end{pmatrix}\begin{pmatrix} l_6 & k_1 & k_6 \\ n_6 & r_1 & -r_6 \end{pmatrix}$$

$$\times \begin{pmatrix} j_2 & l_2 & j_3 \\ m_2 & n_2 & -m_3 \end{pmatrix}\begin{pmatrix} j_3 & l_3 & j_4 \\ m_3 & n_3 & -m_4 \end{pmatrix}\begin{pmatrix} k_2 & l_2 & k_3 \\ r_2 & n_2 & -r_3 \end{pmatrix}\begin{pmatrix} k_3 & l_3 & k_4 \\ r_3 & n_3 & -r_4 \end{pmatrix},$$

$$\tag{A3}$$

where

$$A = [(j_3)(j_4)^2(j_5)(j_6)(k_3)(k_4)^2(k_5)(k_6)]^{1/2} \tag{A4}$$

and

$$R_1 = -j_1 + j_2 + 2j_3 - j_4 + 2j_5 + 2j_6 - k_1 + k_2 + 2k_3 - k_4$$

$$+ 2k_5 + 2k_6 + l_2 + l_3 + l_4 + l_5 + l_6 - r_2 - m_1 . \tag{A5}$$

The prime on the summation sign indicates that for each magnetic quantum number $m$ summed over, there corresponds the factor $(-)^{j-m}$.

The diagrams corresponding to the sum over $3-j$ symbols in Eq. (A3) are shown in Fig. 5(a). These diagrams are joined according to Yutsis's[16] rules to form Fig. 5(b). Interchanges performed to relate Figs. 5(a) and 5(b) show that the coefficient represented by Fig. 5(a) equals the coefficient represented by Fig. 5(b) multiplied by the factor

$$(-)^{j_2 + j_4 + k_1 + k_4 + 2k_6 + l_2 + l_3 + l_4 + l_5 + l_6} . \tag{A6}$$

The diagram in Fig. 5(b) is contracted with the diagram in Fig. 5(c) to form the diagram in Fig. 5(d), which is recognized as an $18-j$ symbol of the second kind multiplied by the factor

$$(-)^{j_1 + j_4 + 2(j_3 + j_5 + j_6) + k_1 - k_4 + 2(k_2 + k_3 + k_5)} . \tag{A7}$$

According to Yutsis's rules, the sum represented by Fig. 5(a) is then given, to within a sign factor, by the $18-j$ coefficient multiplied by the two Clebsch-Gordan coefficients represented in Fig.

5(c) and summed over $a$. Combining this result with the factors Eqs. (A4), (A3), (A6), and (A7), we have

$$C = A(-)^{R_2} \sum_{a\mu} (a)(-)^{a-\mu} \begin{bmatrix} j_1 & j_2 & j_3 & j_4 & j_5 & j_6 \\ a & l_2 & l_3 & l_4 & l_5 & l_6 \\ k_1 & k_2 & k_3 & k_4 & k_5 & k_6 \end{bmatrix}$$

$$\times \begin{pmatrix} j_1 & j_2 & a \\ -m_1 & m_2 & -\mu \end{pmatrix} \begin{pmatrix} a & k_1 & k_2 \\ \mu & r_1 & -r_2 \end{pmatrix} , \quad \text{(A8)}$$

where $[\ \ ]$ represents an $18-j$ symbol of the second kind and $R_2$ is given by

$$R_2 = 2j_2 + j_4 - k_1 + k_2 - k_4 - r_2 - m_1 . \quad \text{(A9)}$$

In Eq. (A9) we have used relations such as

$$2(j_5 + j_6 + l_5) = \text{even}$$

to equate some sign factors to unity.

The $18-j$ coefficient can be written directly in terms of Racah coefficients[16]

$$\begin{bmatrix} j_1 & j_2 & j_3 & j_4 & j_5 & j_6 \\ a & l_2 & l_3 & l_4 & l_5 & l_6 \\ k_1 & k_2 & k_3 & k_4 & k_5 & k_6 \end{bmatrix} = (-)^{R_3} \sum_{x} (x)(-)^{2x} \begin{Bmatrix} j_1 k_1 x \\ k_2 j_2 a \end{Bmatrix}$$

$$\times \begin{Bmatrix} j_2 k_2 x \\ k_3 j_3 l_2 \end{Bmatrix} \begin{Bmatrix} j_3 k_3 x \\ k_4 j_4 l_3 \end{Bmatrix} \begin{Bmatrix} j_4 k_4 x \\ k_5 j_5 l_4 \end{Bmatrix} \begin{Bmatrix} j_5 k_5 x \\ k_6 j_6 l_5 \end{Bmatrix}$$

$$\times \begin{Bmatrix} j_6 k_6 x \\ k_1 j_1 l_6 \end{Bmatrix} , \quad \text{(A10)}$$

where $R_3 = \sum_i (j_i + k_i + 1_i)$, with $l_1$ set equal to $a$.

Substituting Eq. (A10) in Eq. (A8), we see that the sum over $a$ and $\mu$ can be carried out since[13]

$$\sum_{a\mu} (-)^{2a-\mu} \begin{Bmatrix} j_1 k_1 x \\ k_2 j_2 a \end{Bmatrix} \begin{pmatrix} j_1 & j_2 & a \\ -m_1 & m_2 & -\mu \end{pmatrix} \begin{pmatrix} a & k_1 & k_2 \\ \mu & r_1 & -r_2 \end{pmatrix}$$

$$= (-)^{k_2 + k_1 - 3x - m_2 + r_2 + 2m_1} \sum_{\nu} \begin{pmatrix} k_2 & j_2 & x \\ -r_2 & m_2 & \nu \end{pmatrix} \begin{pmatrix} j_1 & k_1 & x \\ -m_1 & r_1 & -\nu \end{pmatrix} . \quad \text{(A11)}$$

Substituting Eqs. (A10) and (A11) into Eq. (A8) then gives our final result

$$C = A(-)^R \sum_{x\nu} (x)(-)^{-x} \begin{Bmatrix} j_2 k_2 x \\ k_3 j_3 l_2 \end{Bmatrix} \begin{Bmatrix} j_3 k_3 x \\ k_4 j_4 l_3 \end{Bmatrix} \begin{Bmatrix} j_4 k_4 x \\ k_5 j_5 l_4 \end{Bmatrix}$$

$$\times \begin{Bmatrix} j_5 k_5 x \\ k_6 j_6 l_5 \end{Bmatrix} \begin{Bmatrix} j_6 k_6 x \\ k_1 j_1 l_6 \end{Bmatrix} \begin{pmatrix} k_2 & j_2 & x \\ -r_2 & m_2 & \nu \end{pmatrix} \begin{pmatrix} j_1 & k_1 & x \\ -m_1 & r_1 & -\nu \end{pmatrix} , \quad \text{(A12)}$$

where

$$R = R_3 - a + 2j_2 + j_4 + 2k_2 - k_4 + m_1 - m_2 .$$

---

*Supported in part by a grant from the National Science Foundation.

[1]H. Ehrhardt, M. Schulz, T. Tekaat, and K. Willmann, Phys. Rev. Letters 22, 89 (1969).

[2]J. R. Sheridan, Rev. Sci. Instr. 40, 060 (1909).

[3]V. Weisskopf and E. Wigner, Z. Physik 65, 18 (1930).

[4]D. H. Jaecks, D. H. Crandall, and R. H. McKnight, Phys. Rev. Letters 25, 491 (1970).

[5]L. C. Biedenharn, in Nuclear Spectroscopy, Part B, edited by Fay Ajzenberg-Selove (Academic, New York, 1960).

[6]K. Schilling, P. Seyboth, and G. Wolf, Nucl. Phys. B15, 397 (1970).

[7]J. Macek, Phys. Rev. Letters 23, 1 (1969); Phys. Rev. A 1, 618 (1970).

[8]I. Percival and M. Seaton, Phil. Trans. Roy. Soc. London A251, 113 (1958).

[9]R. H. Hughes, H. R. Dawson, B. M. Doughty, D. B. Kay, and C. A. Stigers, Phys. Rev. 146, 53 (1966), D. Jaecks, Ph.D. thesis (University of Washington, 1964) (unpublished); T. D. Gaily, D. H. Jaecks, and R. Geballe, Phys. Rev. 167, 81 (1968).

[10]P. Franken, Phys. Rev. 121, 508 (1961).

[11]F. H. Read and R. E. Imhof, Chem. Phys. Letters 13, 652 (1969).

[12]H. J. Andra, Phys. Rev. Letters 25, 325 (1970).

[13]A. Edmunds, Angular Momentum in Quantum Mechanics (Princeton U.P., Princeton, New Jersey, 1957).

[14]J. A. Smit, Physica 2, 104 (1935).

[15]T. Hadishi, Phys. Rev. 102, 16 (1967).

[16]A. P. Yutsis, I. B. Levinson, and V. V. Vanagas, Theory of Angular Momentum (National Science Foundation, Washington, D. C., 1960).

# PHYSICAL REVIEW
# LETTERS

| Volume 26 | 7 JUNE 1971 | Number 23 |
| --- | --- | --- |

## Differential Measurement of the He $3^3P$ Excitation, in He$^+$-He Collisions, by Using an Ion-Photon Coincidence Method

G. Rahmat, G. Vassilev, J. Baudon, and M. Barat

*Institut d'Electronique Fondamentale, Laboratoire associé au Centre National de la Recherche Scientifique, Université Paris XI, 91 Orsay, France*

(Received 2 April 1971)

In order to determine the differential cross section for excitation of a given radiative level, in ion-atom collisions, the scattered ions are detected at a given angle, in coincidence with the photons emitted by the excited target atoms. The method has been applied to the He$^+$-He collision at laboratory energies of 200, 300, and 350 eV in the angular range 5–18 deg, for the $3^3P$ level of He. At 200 and 300 eV, the differential cross sections exhibit an interference pattern which is characteristic of a curve crossing. The behavior is more complicated at 350 eV because, we believe, more than two molecular states are involved.

Excitation processes occuring in the mean energy range (few hundred eV) of ion-atom collisions are generally explained by transitions which occur in the neighborhood of a crossing point of nonadiabatic-potential curves of the molecular-ion–atom system. The usual experimental study of such processes consists of analyzing the energy of the scattered ions at given angles and impact energies. However this method provides, in most cases, somewhat ambiguous data because of the lack in energy resolution, which makes it impossible to isolate a given excited level, especially if a high–quantum-number excited level is considered. In the present experiment this difficulty has been overcome by detecting the photons emitted by the excited atom in coincidence with the ions scattered at a given angle $\theta$.[1] Then the overall energy resolution is that provided by the optical analysis and is much better than the resolution of an electrostatic analyzer.

A rather similar technique has been recently used by Jaecks, Crandall, and McKnight[2] to investigate differential cross sections of capture into excited states in H$^+$-He collisions. This method has been applied to the excitation of the

$3^3P$ level in He in He$^+$-He collisions at impact energies of 200, 300, and 350 eV. The apparatus has been described elsewhere[3] and only its main features are reported here. A He$^+$ ion beam is crossed at right angles by a He atomic beam, and the collision volume is located at the focal point of a parabolic mirror which gathers 70% of the emitted photons. The wavelength of the investigated line ($3^3P$-$2^3S$, $\lambda = 3889$ Å) is selected by an interference filter, and the photons are counted by a low-noise photomultiplier. Two circular movable slits select a given scattering angle and, in order to reduce the accidental coincidence rate, an ion-energy analysis is performed by a cylindrical-mirror–type electrostatic analyzer; the emerging ions are then counted by a tubular particle multiplier. The delay time $\Delta t$ between a photon-channel pulse and the subsequent ion-channel pulse is converted into pulse height by a time-to-amplitude converter; the heights of the output pulses are analyzed in a multichannel analyzer. The final result is a "coincidence spectrum" (Fig. 1) on which the count numbers are plotted as a function of $\Delta t$. A maximum count number can be observed for the delay time $\Delta t_0$ corresponding to an ion

1411

VOLUME 26, NUMBER 23     PHYSICAL REVIEW LETTERS     7 JUNE 1971

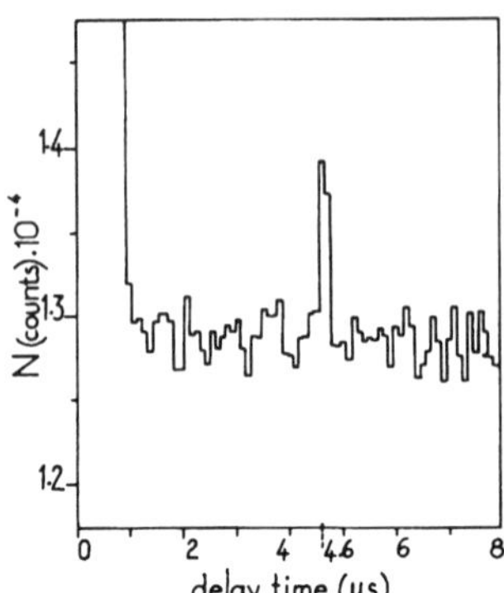

FIG. 1. Coincidence spectrum obtained at $E = 350$ eV, $\theta = 7$ deg. A maximum count number (real coincidences) is observed at the delay time $\Delta t_0 = 4.6$ $\mu$sec. (The increasing count rate from accidental coincidences toward zero time delay is due to electronics.) The running time is about 12 h and the signal-to-noise ratio is 17.

and a photon produced by the same collision, while accidental coincidences provide a base line surmounted by a statistical noise. In our operating conditions, the time resolution was good enough so that the width of the coincidence peak was about equal to the lifetime of the excited state (0.1 $\mu$sec). The running times necessary to obtain a satisfactory signal-to-noise ratio (4 or 5) are several hours (4-10 h). For given energy $E$ and angle $\theta$, the true coincidence number $N_c$ and the total photon number $N(h\nu)$ are measured; the differential cross section $\sigma$ is then given by

$$\sigma(E, \theta) = n(E, \theta) f(\theta),$$

where $n(E, \theta) = N_c/N(h\nu)$ and $f(\theta)$ is an instrumental factor. This function $f$ has not yet been determined, but it can be expected to vary slowly and monotonically with $\theta$, and then the structures which are observed on the curves $n(\theta)$ can be considered to occur also on $\sigma(\theta)$.

Keeping in mind the double-crossing model proposed by Rosenthal and Foley[4] in order to explain the oscillation in the total excitation cross section $Q$,[5,3] we have chosen values of the impact energy corresponding either to minima of $Q$ (200 and 300 eV) or to a maximum (350 eV).

Figure 2 shows $n$ as a function of the reduced collisional parameter $\tau = E\theta$. At the energies $E = 200$ eV and $E = 300$ eV, the differential cross

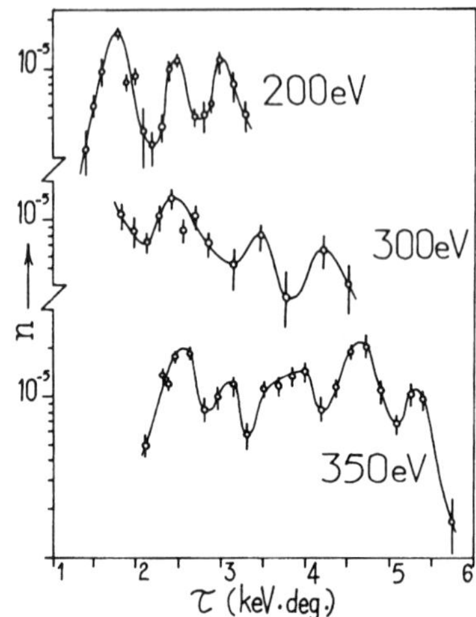


FIG. 2. Ratio $n = N_c/N(h\nu)$ as a function of $\tau = E\theta$, for laboratory impact energies of 200, 300, and 350 eV. As explained in text, $n(\tau)$ has about the same behavior as the excitation differential cross section. The threshold of the excitation can be located (from the 200-eV curve) at about $\tau_0 = 1600$ eV deg.

sections exhibit a regular oscillation which is characteristic of a crossing point of the diabatic-potential curves corresponding to the ground state $(A\,^2\Sigma_g)$ and to the excited state $(^2\Sigma_g)$ of the $He_2^+$ system, which leads, at infinite internuclear distances, to $He^+ + He(3\,^3P)$. The results obtained at 200 eV allow the determination of the apparent threshold of the process at about $\tau_0 = 1.6$ keV deg. The internuclear distance $R_x$ of the crossing point has been evaluated at about 1.4 a.u. from the ground-state–potential curve.[7] This value is qualitatively in good agreement with the predicted $\Sigma$-$\Sigma$ crossing given by the energy diagram of the $He_2^+$ system.[7] If the maxima of $n$ are labeled by the index numbers $N = 0, 1, 2, 2, \cdots$ and the minima by $N = \frac{1}{2}, \frac{3}{2}, \cdots$, then the phase shift between the two interfering waves is $\varphi = 2\pi N$. As can be observed in Fig. 3, the quantity $NE^{1/2}$ is approximately a linear function of $\tau$, and the phase shift can be written $\varphi \simeq CE^{1/2} \times (\tau - \tau_0)$, where $C$ is a constant. From the semiclassical approximation it can be shown[8] that $\varphi$ is related to the impact parameter by

$$\Delta b = b_1 - b_2 = \frac{\partial}{\partial \tau}\left[\left(\frac{E}{2\mu}\right)^{1/2}\hbar\varphi\right],$$

VOLUME 26, NUMBER 23     **PHYSICAL REVIEW LETTERS**     7 JUNE 1971

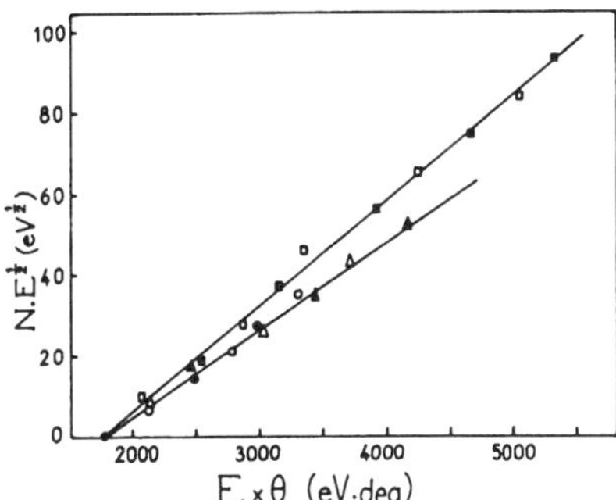

FIG. 3. The quantity $NE^{1/2}$ is plotted as a function of $\tau = E\theta$. $N$ is the index number for the extrema of $n(\tau)$ ($N = 0, 1, 2, \cdots$ for maxima; $\frac{1}{2}, \frac{3}{2}, \cdots$ for minima). Open symbols are maxima, closed are minima: circles, 200 eV; triangles, 300 eV; squares, 350 eV.

where $\mu$ is the reduced mass and $b_1$ and $b_2$ are the values of the impact parameter $b$ corresponding to the two scattering potentials for a given scattering angle $\theta$. The linear dependence of $NE^{1/2}$ on $\tau$ shows that $\Delta b$ is approximately independent of $\tau$: $\Delta b \approx 0.62$ a.u. At 350 eV, the behavior of the differential cross section is more complicated: Oscillations appear also in this case, but they seem to be distorted by another phenomenon. In this case the quantity $NE^{1/2}$ is still a linear function of $\tau - \tau_0$ (see Fig. 3), but the slope $\partial(NE^{1/2})/\partial\tau$ and the values of

$\Delta b$ are slightly larger than for the previous energies. With regard to the diabatic correlation diagram of the $He_2^+$ system,[9] the mechanisms causing the two curve crossings can be considered as possible mechanisms for the He $(1s3p,$ $^3P)$ excitation, corresponding to two electrons being elevated from the $(1s\sigma_g)(2p\sigma_u)^2\,{}^2\Sigma_g$ antibonding input channel to $(1s\sigma_g)^2(4d\sigma_g)\,{}^2\Sigma_g$ and $(1s\sigma_g)(4d\pi_g)^2\,{}^2\Pi_g$.

In an attempt to clarify the parts played by short-range and long-range crossings, we are currently investigating the "companion" level $(3p, {}^1P)$.

[1] R. L. Smick and W. W. Smith, in *Proceedings of the Sixth International Conference on the Physics of Electronic and Atomic Collisions*, Cambridge, Massachusetts, 27 July–2 August 1969 (Massachusetts Institute of Technology, Cambridge, Mass., 1969), p. 306.

[2] D. H. Jaecks, D. H. Crandall, and R. H. McKnight, Phys. Rev. Lett. **25**, 491 (1970).

[3] G. Vassilev, J. Baudon, G. Rahmat, and M. Barat, C. R. Acad. Sci. **271**, 1170 (1970).

[4] H. Rosenthal and H. M. Foley, Phys. Rev. Lett. **23**, 1480 (1969).

[5] S. Dworetsky, R. Novick, W. W. Smith, and N. Tolk, Phys. Rev. Lett. **18**, 939 (1967).

[6] E. Everhart, G. Stone, and J. Carbone, Phys. Rev. **99**, 1287 (1955).

[7] W. Lichten Phys. Rev. **131**, 229 (1963).

[8] D. Coffey, Jr., D. C. Lorents, and F. T. Smith, Phys. Rev. **107**, 201 (1969), and A **2**, 549(E) (1970).

[9] G. Vassilev, J. Baudon, G. Rahmat, and M. Barat, to be published.

J. Phys. B: Atom. Molec. Phys., Vol. 5, June 1972.  Printed in Great Britain.  © 1972

# The variation with electron scattering angle of the polarization of atomic line radiation excited by electron impact

J WYKES

Department of Physics, University of Edinburgh, Edinburgh

MS received 28 September 1971

**Abstract.** Information on both the relative magnitudes and phases of the various scattering amplitudes can be obtained by measuring the angular correlation of inelastically scattered electrons with the emitted photons in the electron impact excitation of atomic line radiation. As an example, excitation of sodium resonance radiation by electron impact at energies close to threshold is examined using close coupling $T$ matrices of Karule and Peterkop.

## 1. Introduction

The theory of the polarization of atomic line radiation excited by electron impact was first developed by Oppenheimer and Penney (see, for example, Penney 1932) who treated the excitation and the emission of radiation as completely separate processes. By treating the emission as coming from the whole system of electron plus atom, Percival and Seaton (1958) were able to clear up some unsatisfactory aspects of the former theory. In both these theories the emphasis is on the 'total' polarization that is that obtained on averaging over all directions of the scattered electron. In the present paper the detailed coordinates of photon polarization with scattered electron direction is examined. In an unpolarized situation, that is one where we are not concerned with measuring the spin correlation of either electron or atom, knowledge of this correlation together with the differential scattering cross section represents an almost complete observation of the excitation event. It is shown that a measurement of this correlation in a coincidence experiment gives information on the relative phases of the various scattering amplitudes involved in the excitation. Electron–photon coincidence experiments are now practical propositions (Imhof and Read 1971) and a measurement of the type described above would provide a more sensitive test of theoretical predications than the usual non-coincident polarization measurements whilst still retaining their essentially relative nature free from normalization difficulties.

In § 2 the basic theory is reviewed briefly, the basic result of the Percival and Seaton theory (1958) being obtained simply by considering the coherent decay of the excited atomic states, the role of the free electron being reduced to producing the coherent excitation of the atom. In § 3 the photon polarization is considered as a function of the direction of emission and direction of electron scattering for an atom with nuclear spin and weak spin–orbit coupling. In § 4 the particular case of excitation of the sodium resonance lines is treated in detail, the various electron scattering amplitudes being obtained from the $T$ matrices of Karule and Peterkop (1965) calculated in a 3s–3p close coupling approximation  General relationships are presented in the appendix.

the components of the total orbital and spin angular momenta along the quantization axis, might not be separately conserved in which case (11) is no longer valid. For resonance states lasting for times of the order of the lifetime of the excited atomic states that is $\Delta t \sim 10^{-8}$ s, $\Delta E \sim 10^{-7}$ eV the whole description of the radiative decay would be invalid and we would then have to consider the evolution and decay of the atom plus electron complex.

Resonances of energy widths $\Delta E \sim 10^{-3}$ eV are on the limit of observability with present day experimental techniques whereas those with $\Delta E \sim 10^{-7}$ eV would not be observable in any forseeable experiment so we can say that in most experiments the effects of long lived resonant states would be averaged out.

## 5. Conclusions

We have considered the excitation of unpolarized sodium atoms by unpolarized electrons at incident energies close to the excitation threshold and correlated the direction of the scattered electron with the polarization of the photons. It was shown that information on the relative magnitudes and phases of the various scattering amplitudes involved can be obtained by observing such a correlation between the polarization of photons emitted at an angle $\Theta$ to the incident electron direction, provided $\Theta \neq \frac{1}{2}n\pi$ ($n = 0, 1, 2, \ldots$). Effects that could be observed using external magnetic fields, as in analogous optical excitation experiments, have not been considered in the present paper owing to the difficulty of observing the scattered electron direction experimentally in such situations.

## Acknowledgments

I would like to thank Professor P S Farago and Dr F H Read for preliminary discussions and for stimulating my interest in this work as an experimentally feasible topic, and also Professor P G Burke for suggesting part of this work. I am indebted to Professor Farago for constant advice and guidance.

## Appendix

The general relationship (10) for electron excitation $S'L' \rightarrow SL$ followed by electric dipole radiation $SL \rightarrow S''L''$ may be written out explicitly as follows

$$w(\theta\phi ; \Theta\Phi, \hat{e}) = K\Sigma A(M_L M_S, JFM_F, q)A(\overline{M}_L\overline{M}_S, \overline{J}\overline{F}\overline{M}_F, \bar{q})$$

$$\times \operatorname{Re}\left(1 + \epsilon^2(JF, \overline{J}\overline{F})\right)^{-1}(D(\Theta\Phi\phi\hat{e})f/\Gamma', \Gamma ; \theta)f^*(\Gamma', \Gamma ; \theta)) \tag{A1}$$

where the $A$ contain the properties of the coupling scheme and are given by

$$A(M_L M_S, JFM_F, q) = (-1)^{-2J - F}\{(2J'' + 1)(2F + 1)\}^{1/2}W(LJ, L''J'' ; S1)$$

$$C^{F'', M''_F}_{F1, M_F q}C^{F, M_F}_{J1, M_J M_I}C^{J, M_J}_{SL, M_S M_L} \tag{A2}$$

VOLUME 31, NUMBER 9      PHYSICAL REVIEW LETTERS      27 AUGUST 1973

# Measurements of Complex Excitation Amplitudes in Electron-Helium Collisions by Angular Correlations Using a Coincidence Method*

M. Eminyan,† K. B. MacAdam,‡ J. Slevin, and H. Kleinpoppen
*Physics Department, University of Stirling, Stirling, Scotland*
(Received 25 June 1973)

Angular correlations were measured by delayed coincidences between electrons of incident energy about 80 eV scattered inelastically from helium and photons from excited $3\,^1P$ and $2\,^1P$ states. From the angular correlation in each case we deduce the ratio of the differential cross sections for exciting the magnetic sublevels of the excited state and the phase between the corresponding excitation amplitudes. We compare the atomic radiation patterns with those predicted by the Born approximation.

Valuable information about inelastic electron-atom scattering processes can be obtained from electron-photon angular correlations measured by coincidence techniques.[1,2] This paper reports preliminary measurements of electron-photon coincidences in a crossed electron-helium beam experiment.

The results are significant for several reasons.

Volume 31, Number 9    PHYSICAL REVIEW LETTERS    27 August 1973

Previous experimental work in electron-atom collisions has usually involved averages over significant parameters in the scattering. Differential-cross-section measurements do not distinguish the excitations to degenerate magnetic sublevels. Measurements of line polarization in excitation-function experiments separate the contributions of magnetic sublevels but obtain only integral cross sections.[3,4] Furthermore, neither differential cross sections nor the polarization of impact radiation allow the relative phases of excitation amplitudes for degenerate magnetic sublevels to be deduced. Yet the phases depend in a nontrivial way on the dynamics of the scattering and are related to the transfer of angular momentum to the atom.[5] Their measurement, therefore, is expected to provide a new test of electron-atom scattering theories.

An $n\,^1P$ state of He[4] excited from the ground state $(1\,^1S)$ by electron bombardment, in which the electron is scattered in a particular direction $(\theta_e, \varphi_e)$ relative to the incident beam,[6] can be described as a coherent superposition of sublevels as follows:

$$\psi = a_1|11\rangle + a_0|10\rangle + a_{-1}|1-1\rangle. \qquad (1)$$

Amplitudes $a_M(\theta_e, \varphi_e, E)$ describe the excitation to particular magnetic sublevels $|LM\rangle$ of the $n\,^1P$ state $(L=1)$ by electrons of kinetic energy $E$. In the absence of spin effects, atoms excited to the $n\,^1P$ state are prepared in an identical way whenever electrons are scattered into a small solid angle around $(\theta_e, \varphi_e)$. The photons emitted in the decay of these states can be separated from those emitted by $n\,^1P$ states produced by electrons scattered in other directions, and hence having different excitation amplitudes, by detecting them in delayed coincidence[7,8] with the inelastically scattered electrons. The angular distribution of time-correlated photons can be used to extract the complex excitation amplitudes $a_M$. Since the scattering process has mirror symmetry in the plane of scattering, parity invariance requires that $-a_{-1}=a_1$. Of the remaining two independent complex quantities $a_1$ and $a_0$, the overall phase is not observable, and, in the absence of absolute cross-section measurements, only the ratio of magnitudes of the amplitudes and their relative phase can be measured.

The angular correlation given by Macek and Jaecks[1] for the case in which electrons are scattered at an angle $(\theta_e, 0)$ and photons from the $n\,^1P - n'\,^1S$ decay are counted, without regard to

polarization, at angles $(\theta_\gamma, \pi)$ can be written

$$N \propto \lambda \sin^2\theta_\gamma + (1-\lambda)\cos^2\theta_\gamma$$
$$- 2[\lambda(1-\lambda)]^{1/2}\cos\chi \sin\theta_\gamma \cos\theta_\gamma. \qquad (2)$$

Here $\lambda = |a_0|^2/(2|a_1|^2 + |a_0|^2) = \sigma_0/\sigma$, where $\sigma$ is the differential cross section for excitation of the $n\,^1P$ state. $\sigma_0$ is the differential cross section for excitation of the $M=0$ sublevel. $\chi$ is the relative phase between $a_1$ and $a_0$. Equation (2) illustrates the dependence of the radiation intensity on the real quantities $\lambda$ and $\chi$ that can be measured. The first Born approximation,[3] or indeed any theory which results in a $\Delta M_L = 0$ selection rule along the momentum transfer direction in the excitation, implies that $\lambda = \cos^2\theta_K$ and $\chi = 0$. $\theta_K$ is the angle of the vector $\vec{K}$ describing the momentum transferred to the atom in the collision.

We have studied the excitation of the $3\,^1P$ and $2\,^1P$ states of He[4]. The apparatus is shown in Fig. 1. We measured the angular distribution of 501.6-nm photons $(3\,^1P\text{-}2\,^1S)$ from the $3\,^1P$ state excited by an energy-selected incident electron beam of kinetic energy 77.7 eV. Electrons scattered into a small solid angle at $\theta_e = 30°$, $\varphi_e = 0°$ were energy selected to reject all except those corresponding to $n=3$ excitations. Photons were detected by a photomultiplier with a narrow-band interference filter which isolated the 501.6-nm line. The photomultiplier could be rotated through the angular range $\theta_\gamma = 30°$ to $130°$ in the scattering plane on the opposite side of the electron beam from the scattered electrons, i.e., $\varphi_\gamma = \pi$. The helium target was a beam emitted from a multichannel capillary source.

The angular distribution of 58.4-nm photons

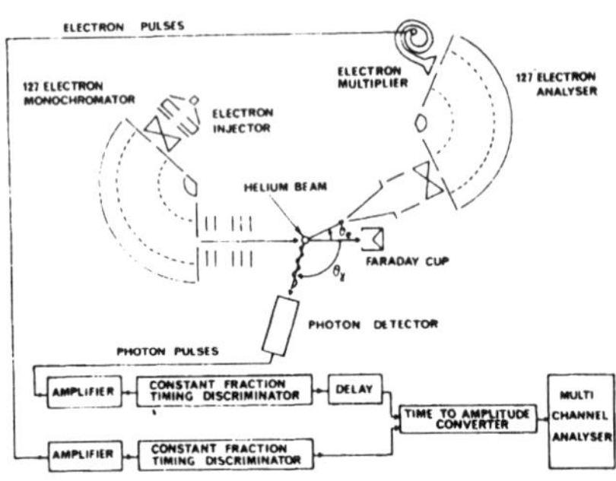


FIG. 1. Diagram of the apparatus (not to scale). The helium beam emerges perpendicular to the plane of the diagram, which represents the plane of scattering.

($2\,^1P$-$1\,^1S$) from the $2\,^1P$ state excited at 80.0 eV by electrons scattered at $\theta_e = 16°$ was also measured. The uv photons were detected by a channel electron multiplier which could be rotated through the same angular range. No wavelength selection was necessary in this case because the coincidence method allowed the 58.4-nm photons to be identified uniquely by their correlation in time with scattered electrons having the correct energy loss to have excited the $2\,^1P$ state. Signals were much stronger for the $2\,^1P$ state because the decay to the ground state was observed.

The coincidence method renders the results insensitive to cascading from higher excited states, imperfect energy resolution of the $n = 3$ terms, and background counts. The total number of real coincidences was scaled in such a way as to make the resulting values of $N$ in Eq. (1) insensitive to small variations of electron beam current, target density, and efficiency of the electron detector. It was verified directly that the target density was sufficiently low to eliminate 501.6-nm photons resulting from the trapping of 53.7-nm resonance radiation from the $3\,^1P$-$1\,^1S$ decay. The data for $2\,^1P$ excitations were taken at the same target density. Three further possible sources of systematic error were considered: (1) Photons of the correct wavelength leaving the interaction region in various directions and scattered from nearby surfaces could enter the photon detector and give a false contribution to the true coincidence rate. To minimize this, surfaces within view of the photon detector were blackened. Reflection coefficients are very small

for 58.4-nm light in any case. (2) Spurious electronic noise could affect coincidence rates. The coincidence electronics recorded the delay-time spectrum over a range much longer than the time spanned by the real coincidences. The uniformity of the background of random coincidences during the experiment was a strong indication that spurious events were absent. (3) The results could be affected by changes in the sensitivity of the photon detector. However, the reproducibility of the data demonstrated the absence of any significant drift.

Angular-correlation curves are shown in Figs. 2 and 3. For the $3\,^1P$ data of Fig. 2 each point represents 12 to 48 h integration, and for the $2\,^1P$ data of Fig. 3 each point represents 1 to 12 integration. The full curves are the result of computer least-squares fits of the data with photon Eq. (1), allowing for the angular width of the ton aperture. The Born-approximation predictions, averaged over the angular width of the photon detector, are given in the dashed curves. It should be emphasized that these predictions are based not on any approximate wave functions but only on the fundamental selection rule $\Delta M_L = 0$ along the momentum-transfer axis. The results are given numerically in Table I.

An angular-correlation curve represents a radiation pattern which, for $\chi = 0$, is that of a classical linearly polarized electric dipole lying in the plane of scattering. The radiation intensity should vanish in the direction of the dipole moment. The Born approximation predicts a zero of the angular-correlation curve at the momentum-transfer angle $\theta_K$. The momentum-transfer angles are compared in Table I with the angles $\theta_{\min}$ of the minima of the observed radiation patterns. The data for both $3\,^1P$ and $2\,^1P$ excitations

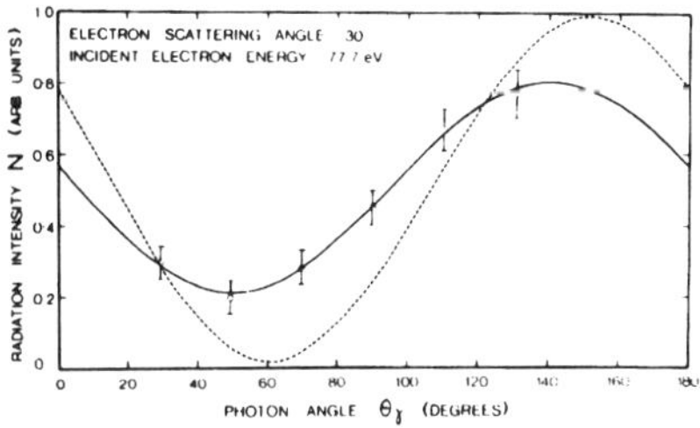



FIG. 2. Electron-photon angular correlation for the $3\,^1P$ state expressed as the angular distribution of 501.6-nm photons radiated by atoms excited at a particular scattering angle [cf. Eq. (1)]. Error bars represent 1 standard deviation in the counting statistics. Full curve, least-squares fit to the data; dashed curve, prediction of the Born approximation.

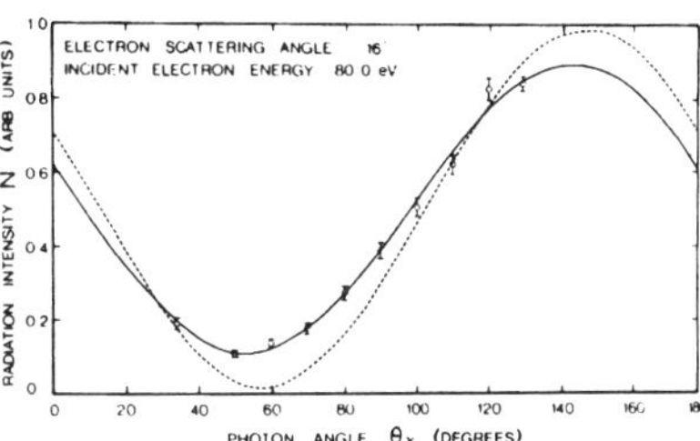



FIG. 3. Electron-photon angular correlation for the $2\,^1P$ state expressed as the angular distribution of 58.4-nm photons. Notation as in Fig. 2.

VOLUME 31, NUMBER 9          PHYSICAL REVIEW LETTERS          27 AUGUST 1973

TABLE I. Experimental results and comparison with the Born approximation (BA) (see text). The angles given for $\theta_{\min}$ in the Born approximation are the angles $\theta_K$ of the momentum transfer.

| Excited state | $E$ (eV) | $\theta_e$ (deg) | $\lambda$ Expt [a] | BA | $\|\chi\|$ (deg) Expt [a] | BA | $\theta_{\min}$ (deg) Expt [a] | BA |
|---|---|---|---|---|---|---|---|---|
| $3\,^1P$ | 77.7 | 30 | $0.45 \pm 0.01$ | 0.231 | $53 \pm 2$ | 0 | $50 \pm 2$ | 61.3 |
| $2\,^1P$ | 80.0 | 16 | $0.39 \pm 0.02$ | 0.290 | $38 \pm 2$ | 0 | $53 \pm 2$ | 57.4 |

[a] Uncertainties quoted for $\lambda$ and $\chi$ represent 1 standard deviation.

indicate significant deviations from the predictions of the Born approximation.

Work is continuing to measure the dependence of $\lambda$ and $\chi$ on $E$ and $\theta_e$ for both excited states. The results will map out in greater detail than heretofore possible the deviations of various electron-atom collision theories from experiment.

The authors are grateful to Professor R. Novick and Dr. M. Levitt for their assistance in the early stages of the experiment and to Professor P. G. Burke for helpful discussions.

*Work supported by the U. S. Air Force Office of Scientific Research under Grant No. AFOSR-68-1454, the British Science Research Council, and the University Grants Committee.

†European Science Exchange Fellow. Permanent address: U. E. R. de Physique, Université de Paris, Paris 7, France.

‡Frank Knox Memorial Traveling Fellow of Harvard University, 1971–1972.

[1] J. Macek and D. H. Jaecks, Phys. Rev. A **4**, 2288 (1971).

[2] J. Wykes, J. Phys. B: Proc. Phys. Soc., London **5**, 1126 (1972).

[3] B. L. Moiseiwitsch and S. J. Smith, Rev. Mod. Phys. **40**, 238 (1968).

[4] H. Kleinpoppen, in *Physics of One- and Two-Electron Atoms*, edited by F. Bopp and H. Kleinpoppen (North-Holland, Amsterdam, 1969), pp. 612–631.

[5] It can be shown by elementary angular-momentum algebra that for the wave function $\psi$ of Eq. (1) the expectation value of the component of atomic orbital angular momentum perpendicular to the plane of scattering is $-2[\lambda(1-\lambda)^{1/2}\sin\chi$.

[6] The spherical polar coordinate system is centered on the intersection of the helium and electron beams with the direction of the latter defining the polar axis and the axis of quantization. Zenith and azimuth angles are measured in the usual way. The scattering plane, defined by the incident electron beam and the axis of the electron analyzer, can be described by $\varphi = 0$.

[7] R. E. Bell, in *Alpha, Beta, and Gamma Ray Spectroscopy*, edited by K. Siegbahn (North-Holland, Amsterdam, 1966), Vol. 2, pp. 905–929.

[8] R. E. Imhof and F. H. Read, J. Phys. B: Proc. Phys. Soc., London **4**, 450 (1971).

# REVIEWS OF
# MODERN PHYSICS

VOLUME 45, NUMBER 4                                    OCTOBER 1973

## Impact Excitation and Polarization of the Emitted Light*

**U. Fano**

*Department of Physics, The University of Chicago, Chicago, Illinois 60637*

**Joseph H. Macek**

*Department of Physics, University of Nebraska, Lincoln, Nebraska 68508*

A new formulation is presented for the angular distribution and the polarization of light excited by atomic and electronic collisions and modulated in time by the action of internal and external fields. The formulation disentangles geometrical and dynamical effects and stresses the extraction of data on the alignment and orientation of radiating atoms from observations of the emitted light. The treatment is set in the context of recent experimental and theoretical literature and points to new avenues of research.

## CONTENTS



## I. INTRODUCTION

Excitation of an atom or molecule by collision in a gas, or by passage through a foil, leaves it generally in an anisotropic state. Light emitted in the subsequent decay manifests this anisotropy through its angular distribution and polarization. Measurements of the angular distribution and/or polarization of emitted light have been used since the '20s (S26, SA27) to determine the anisotropy of the source atom or molecule. The recent emphasis on this type of analysis (PS58, K69, MJ71), and the introduction of techniques for observing variations of the anisotropy during the light emission (HN65, A70, BH71, BS71, LDAF71) prompt us to review the theoretical connection between the radiated pattern and the relevant source parameters.

These parameters consist normally of an *orientation* vector proportional to the average angular momentum of the source atom, $\langle \mathbf{J} \rangle$, and of an *alignment* tensor whose components are proportional to the mean values of quadratic expressions in $J_x, J_y, J_z$, such as $\langle J_z^2 - J_y^2 \rangle$.

The theory will be articulated from an operational point of view to show, in the first place, just what characteristics of the radiating atoms are determined by observing the intensity and the polarization of the light emitted in various directions. In a second step the dynamical aspects of the emission process will be disentangled from the geometrical aspects. That is, we will show how the emission averaged over all directions depends on dynamical factors such as the line strength, while the anisotropy and polarization depend only on the alignment and orientation of the excited atoms. Then it will be shown how the number of independent nonzero alignment and orientation parameters is often restricted by the geometry of the exciting collision. Finally, we shall develop the theory of the time dependence of alignment and orientation in the interval between excitation and emission. It will emerge that the action of external fields during this interval may perturb the alignment and orientation of excited atoms to the extent of interlinking them with other parameters such as octupole or hexadecapole moments induced by the collision. Internal fields, on the other hand, yield reversible exchanges of orientation and alignment among orbits and electronic or nuclear spins. We shall thus endeavor to present a unified, geometrical picture of the phenomena, illustrated by experimental results and pointing to the limits of current advances and to opportunities for developing new theories and experimental designs.

554    REVIEWS OF MODERN PHYSICS · OCTOBER 1973

In our problem, as in many others involving the angular distribution of reaction products, one faces considerable difficulty in directing attention to the main geometrical observable features of the phenomenon without resorting to unfamiliar formalisms. Most current treatments still proceed by summing or averaging explicitly over all unobserved magnetic quantum numbers; this method is elementary but laborious and furthermore fails to interpret the simplicity of the results. Compact formulations of the theory have existed since the early 1950's but they rely on somewhat abstract formulations using density matrices and extensive applications of Racah algebra (see, e.g., Chap. 19 of FR59, Sec. 3 of FS68). In this article we strive for a transparent formulation without explicit use of density matrices; to this end we shall use a minimum of Wigner–Racah algebra, introducing it by qualitative discussion where it becomes essential.

The connection between the anisotropy of the radiation source and the collision process from which it arises will be illustrated by outlining the theory of alignment by electron collision. Particular mention will be made of the striking loss of alignment which often occurs over a narrow energy band above threshold.

In the interest of brevity we are excluding from our subject several of its interesting aspects. Among these are the important influence of successive photon emissions by the same atom and the effects of magnetic resonance upon light emission. We also deal only with light emitted in electric dipole processes even though extension to other processes would be straightforward. This paper refers to "atoms" as the sources of light but is meant to apply, in essence, also to radiation by molecules and, in fact, by nuclei as well.

The first part of this article deals with light emitted in transitions between two sharply defined energy levels, that is, with light observed by high resolving power spectroscopy. Sharp definition of the energy levels requires, of course, that the time of emission be determined with correspondingly low resolution. The second part will consider the phenomena that occur, e.g., under the conditions of beam foil spectroscopy (B67) or when photons are detected in delayed coincidence with a colliding particle (M69, A70, IR71, McJ71, MJ71). Sharper definition of the time interval between a collision and the following light emission implies here a lower spectral resolution in determining the initial energy of the radiating atom. The observed light originates then typically from a nonstationary state and is modulated in time as represented by the coherent superposition of excited stationary states with different, unresolved energy eigenvalues. Our formulation will yield directly the modulation of the several alignment and orientation parameters.

Historically, the development of this subject has proceeded along two converging lines. On the one hand, the emission of light from nonstationary states has been extensively studied in connection with the Hanle effect (H23, Br33), optical pumping (H72), and level crossing experiments. Since the states are usually excited by optical radiation, the early theoretical developments (Br33, F61) treated the entire process of absorption and emission from the point of view of the quantum theory of radiation without reference to orientation or alignment of the atoms. On the other hand, measurements of static properties of radiation fields, such as their polarization and angular distribution have been used mainly to detect or verify simple source properties of collision excited atoms (S26, SA27, PS58).

Later experiments were aimed at demonstrating modulated decays following collision excitation. A direct observation of the modulation following excitation by a pulsed electron beam was made by Hadeishi and Nierenberg (HN65), but equivalent results had been obtained previously by Pebay-Peyroula and by Aleksandrov through modulation of the electron beam (PP63, A64) and by Series and co-workers through optical excitation (CS64, DKW64). Determination of source parameters was not emphasized in these experiments. Correspondingly, the theory of Franken (F61) and Kelly (K66) dealt mainly with the modulation aspects of impulsively excited line radiation.

The theory of light modulation was concerned initially, in this and other work (e.g., DS61, S67), with atoms in magnetic fields. With the advent of beam foil spectroscopy and coincidence techniques (IR71, McJ71) interest focused on emission from atoms in field-free regions and on the determination of source parameters (M69, A70, BH71, IR71, LDAF71, McJ71, MJ71). The theory related to such measurements essentially combines the treatment of modulated decays with the theory of angular correlations, and closely resembles the theory of perturbed angular correlations of nuclear physics (KMS64, FS68). Owing to the known dipole nature of most atomic transitions and to the simple vector coupling nature of fine and hyperfine interactions, the extraction of source parameters from data becomes an exercise in angular momentum recoupling, and can be carried out in a rather complete fashion. In addition, atoms respond to external magnetic fields in a way that provides additional information on the excited state populations.

A time dependence of the anisotropy and polarization of the emitted light occurs also when the source atoms interact appreciably with one another or with other atoms during the emission process. Typically the polarization of light emitted by a sufficiently dense medium following a pulsed excitation relaxes exponentially, whether or not it is also modulated by the mechanics of single atoms. The relaxation effects have been reviewed in (H72) and are not included in the present article which deals only with emission by isolated atoms, i.e., by gases in the low-pressure limit. We note, however, that the treatment of relaxation centers, like ours, on the orientation and alignment of the source atoms—or on the equivalent dipole and quadrupole moments—because such tensorial quantities relax independently of one another in isotropic media. In particular, Barrat (B59) showed that the linear polarization of light emitted by an optically pumped dense gas decays faster than the total light intensity; the intensity depends, of course, on scalar properties of the source atoms while the polarization depends on their tensor alignment. The difference of relaxation rates of different tensorial parameters has been emphasized in more recent studies (DP65, O65, HS67) which represent the excited state of the source atoms by a density matrix expanded into irreducible tensor components (F57).

Our review concentrates on recent developments and emphasizes their potential for measuring the source parameters of collision excited atoms. Accordingly, we do not review the quantum theory of radiation from nonstationary states, but use its results. We stress instead the extraction of source parameters from data and their interpretation in a more direct manner than has been done previously.

## II. LIGHT EMISSION IN THE DECAY OF A STATIONARY STATE

The intensity $I$ measured by an ideal detector sensitive to light with the polarization vector $\boldsymbol{\epsilon}$ is proportional to $\sum_{m_f} \langle |(f | \boldsymbol{\epsilon}^* \cdot \mathbf{r} | i)|^2 \rangle$, where $\mathbf{r}$ is the transition dipole operator of the atom, $\sum_{m_f}$ indicates summation over all values of the final state magnetic quantum number $m_f$, and $\langle \ \rangle$ indicates averaging over the initial $m_i$. The average is weighted by excitation amplitudes whose values are regarded here as unknown parameters, to be determined by measuring the light intensity $I$. Circular and elliptical polarizations are represented by complex vectors $\boldsymbol{\epsilon}$. We express $I$ more precisely in the form

$$I = C \sum_{m_f} \langle (i' | \boldsymbol{\epsilon} \cdot \mathbf{r}' | f)(f | \boldsymbol{\epsilon}^* \cdot \mathbf{r} | i) \rangle, \qquad (1)$$

where one of the $\mathbf{r}$ and one of the $i$ have been primed to distinguish the integration variables and quantum numbers of the two matrix elements. The factor

$$C = e^2 \omega_{fi}^4 / 2\pi c^3 R^2 \qquad (2)$$

incorporates the light frequency $\omega_{fi}$, the detector's distance from the source atom $R$, and the other factors required to express $I$ as a power flux.

Equation (1) warrants some discussion. Its main element is the average value of a physical quantity, that is, the mean or quantum mechanical expectation value of that quantity. The average pertains to the excited state generated by the collision. The notation allows this state to be represented by the coherent superposition of states with different magnetic quantum numbers, $m_i$, $m_i'$. Circumstances will be pointed out later in which this superposition becomes essential, but actually we need not concern ourselves here with it or, more generally, with the calculation of the average starting from collision theory. On the contrary, this paper concentrates on the averages themselves. That is, we deal with their determination from observational data, with the restrictions imposed on their values by experimental symmetries, and with theoretical relations between the averages of different physical variables.

As an incidental remark, we also note that the exponential decay of light intensity following the collision has been omitted from Eq. (1) for simplicity. This decay should be integrated over if the time resolution of the experiment were much longer than the mean life of the excitation. Similarly, Eq. (1) pertains to emission by a single excited atom and should be multiplied by an appropriate factor, including, e.g., a collision cross section, whenever it is to represent the light received from a macroscopic source.

The expression (1) is to be transformed so as to separate its dependence on the direction and polarization response of the detector from the anisotropy of the excited state expressed by the averaging $\langle (i' | \cdots | i) \rangle$. The transformation consists of three steps of increasing complexity: (a) taking explicit advantage of the isotropy that results from the summation over the unobserved final states, (b) disentangling the dependence of (1) on the direction of $(\boldsymbol{\epsilon}, \boldsymbol{\epsilon}^*)$ from its dependence on $(\mathbf{r}, \mathbf{r}')$ by recoupling these vectors, and (c) applying the Wigner–Eckart theorem to replace the dependence on $(\mathbf{r}, \mathbf{r}')$ by the more familiar expressions of alignment and orientation as mean values of angular momentum operators.

(a) The sum over projections onto the several degenerate states of the final energy level, indicated by $\sum_{m_f} |f)(f|$, constitutes a single projection operator $P_f(\mathbf{r}', \mathbf{r})$ which is a *scalar*, i.e., is invariant under joint rotation of $\mathbf{r}'$ and $\mathbf{r}$. With this notation, Eq. (1) takes the form

$$I = C \langle (i' | \boldsymbol{\epsilon} \cdot \mathbf{r}' P_f(\mathbf{r}', \mathbf{r}) \boldsymbol{\epsilon}^* \cdot \mathbf{r} | i) \rangle. \qquad (3)$$

(b) The product of scalar products $\boldsymbol{\epsilon} \cdot \mathbf{r}' \boldsymbol{\epsilon}^* \cdot \mathbf{r}$ can be recoupled into the form $\sum_k Q^{(k)}(\boldsymbol{\epsilon}, \boldsymbol{\epsilon}^*) \cdot R^{(k)}(\mathbf{r}', \mathbf{r})$, such that each factor $Q^{(k)}$ can be pulled out of the matrix element (3). Here $k$ represents the degree of an irreducible tensor, that is, $k=0$ indicates a scalar, $k=1$ a vector, and $k=2$ a quadrupole moment. The recoupling can be performed in our problem by elementary vector algebra, without resorting to more general procedures, since the polarization vector $\boldsymbol{\epsilon}$ is restricted to the plane perpendicular to the direction of observation. We take this direction as the $\hat{\zeta}$ axis of a "detector frame" of coordinates $(\xi, \eta, \zeta)$. We also take the $\hat{\xi}$ axis of this frame along the major axis of

556    Reviews of Modern Physics · October 1973

the general elliptical polarization selected by the detector and represented by $\boldsymbol{\ell}$. In this frame we can then set $\boldsymbol{\ell} = (\cos\beta,\, i\sin\beta,\, 0)$, whereby $\beta = 0$ represents selection of linear polarization and $\beta = \tfrac{1}{4}\pi$ selection of right circular polarization. [When the polarization analyzer consists of a Nicol prism, or an analogous linear polarization filter, and of a $\tfrac{1}{4}$-wavelength plate, $\beta$ is the angle between optical axes of Nicol and plate, see, e.g., (F49).] These definitions yield

$$\boldsymbol{\ell}\cdot\mathbf{r}'\boldsymbol{\ell}^{*}\cdot\mathbf{r} = \xi'\xi\cos^2\beta + \eta'\eta\sin^2\beta - i(\xi'\eta - \eta'\xi)\sin\beta\cos\beta.$$
$$(4)$$

The last term of Eq. (4), which peaks at $\beta = \pm\tfrac{1}{4}\pi$ and therefore represents the dependence of $I$ on circular polarization, is proportional to the $\zeta$ component of the vector product $\mathbf{r}' \times \mathbf{r}$. The other terms can be rearranged so as to separate a contribution that does not depend on the polarization parameter $\beta$ as well as contributions that depend separately on scalar and quadrupole moment combination of $\mathbf{r}'$ and $\mathbf{r}$. This rearrangement transforms Eq. (4) into

$$\boldsymbol{\ell}\cdot\mathbf{r}'\boldsymbol{\ell}^{*}\cdot\mathbf{r} = \tfrac{1}{2}\big[(\xi'\xi + \eta'\eta) + (\xi'\xi - \eta'\eta)\cos 2\beta$$
$$-i(\xi'\eta - \eta'\xi)\sin 2\beta\big]$$
$$= \tfrac{1}{3}\mathbf{r}'\cdot\mathbf{r} - \tfrac{1}{6}(3\zeta'\zeta - \mathbf{r}'\cdot\mathbf{r}) + \tfrac{1}{2}(\xi'\xi - \eta'\eta)\cos 2\beta$$
$$+\tfrac{1}{2}i^{-1}\mathbf{r}' \times \mathbf{r}\cdot\boldsymbol{\zeta}\sin 2\beta. \quad (5)$$

Substituting this expression into (3) we obtain

$$I = C\tfrac{1}{3}\{\langle(i' \mid \mathbf{r}'\cdot\mathbf{r}\, P_f(\mathbf{r}',\mathbf{r})\mid i)\rangle$$
$$-\tfrac{1}{2}\langle(i' \mid (3\zeta'\zeta - \mathbf{r}'\cdot\mathbf{r})P_f(\mathbf{r}',\mathbf{r})\mid i)\rangle$$
$$+\tfrac{3}{2}\langle(i' \mid (\xi'\xi - \eta'\eta)P_f(\mathbf{r}',\mathbf{r})\mid i)\rangle\cos 2\beta$$
$$+\tfrac{3}{2}\langle(i' \mid i^{-1}(\mathbf{r}' \times \mathbf{r})\cdot\boldsymbol{\zeta}\, P_f(\mathbf{r}',\mathbf{r})\mid i)\rangle\sin 2\beta\}. \quad (6)$$

The first term in the braces of Eq. (6) depends on the matrix elements of a scalar operator and therefore has the same value for all degenerate states $\mid i)$; in fact it coincides with the Condon–Shortley "line strength" parameter to within a factor $e^2(2j_i+1)$. This term represents the average emission over all directions and polarizations and will be called $S$. The second term depends on the direction $\boldsymbol{\zeta}$ of the detector and represents the anisotropy of emission, still averaged over the polarization. The third and fourth terms represent the anisotropy with linear and with circular polarization, respectively.

Each term of Eq. (6) is proportional to the mean value of an operator of the irreducible tensor type which may be indicated by a two-index symbol $S^{[k]}{}_q$. Thus we shall indicate by $S^{[0]}{}_0$ the scalar operator $\mathbf{r}'\cdot\mathbf{r}\, P_f$ with the mean value $\langle S^{[0]}{}_0\rangle = S$. Similarly, $i^{-1}\mathbf{r}' \times \mathbf{r}\cdot\boldsymbol{\zeta}\, P_f$ is the $\zeta$ component $S^{[1]}{}_0$ of a vector operator, whose index $k=1$ indicates "vector" while $q=0$ indicates "invariance under rotations about the $\zeta$ axis." Further, $(3\zeta'\zeta - \mathbf{r}'\cdot\mathbf{r})P_f = S^{[2]}{}_0$ is the axially

symmetric component of a quadrupole moment tensor and $(\xi'\xi - \eta'\eta)P_f = S^{[2]}{}_{2+}$ is another component of the same tensor that transforms like $\cos 2\varphi$ under rotations about $\zeta$.

Real tensor components $S^{[k]}{}_{q\pm} = N_{kq\pm}[S^{[k]}{}_q \pm S^{[k]}{}_{-q}]$, with $q>0$, are used in this article in preference to the complex components $S^{[k]}{}_q$ commonly used in analytical work. Here $N$ is a normalization factor which we adjust to simplify the final expressions even though this entails nonunitary transformation matrices for coordinate rotations. The definitions of $S^{[2]}{}_0$ and $S^{[2]}{}_{2+}$ given above imply $N_{20}/N_{22+} = 6^{1/2}$. In the following we shall express the matrix elements and the mean values of $S^{[1]}{}_0$, $S^{[2]}{}_0$, and $S^{[2]}{}_{2+}$ in terms of the matrices and mean values of corresponding tensor components that are functions of the angular momentum operator $\mathbf{J}$.

(c) The mean values of the squared-dipole operators in expression (6) depend on the one hand on the alignment and orientation of the excited state generated by the collision and on the other hand on the dynamics of the dipole transition and, through it, on characteristics of the final state. These two dependences will be sorted out, using the fact that the mean values of any two tensorial operators $S^{[k]}{}_q$ and $T^{[k]}{}_q$ with the same indices $k$ and $q$ depend equally on the alignment and orientation of the excited atom. Specifically the Wigner–Eckart theorem states that the ratio of matrix elements $(i' \mid S^{[k]}{}_q \mid i)$ and $(i' \mid T^{[k]}{}_q \mid i)$ is independent of the magnetic quantum numbers $m_i'$ and $m_i$; indeed this ratio equals the ratio of "reduced matrix elements" $(i \parallel S^{[k]} \parallel i)$ and $(i \parallel T^{[k]} \parallel i)$, each of which is wholly independent of $m_i'$, $m_i$, and of $q$. the (unknown) alignment and orientation of the state $i$ determine what averaging should be taken over $m_i$ and $m_i'$; since this averaging is the same for the matrices of any two $S^{[k]}{}_q$ and $T^{[k]}{}_q$ we can write

$$\langle(i' \mid S^{[k]}{}_q \mid i)\rangle = \langle(i' \mid T^{[k]}{}_q \mid i)\rangle$$
$$\times (i \parallel S^{[k]} \parallel i)/(i \parallel T^{[k]} \parallel i). \quad (7)$$

The structure of Eq. (7) is designed to sort out the anisotropy of the excited atom from the dynamics of light emission. On its left-hand side the operators $S^{[k]}{}_q$ will be the functions of $\mathbf{r}$, $\mathbf{r}'$ and $P_f$ whose averages must be entered in Eq. (6); these averages depend both on the state of the excited atom and on the emission process. On the right-hand side we shall enter operators $T^{[k]}{}_q$ constructed as products of the angular momentum components $J_\xi$, $J_\eta$, $J_\zeta$, because each of these components is a constant of motion of an isolated atom, such that their averages depend only on the state of the excited atom and not on the dynamics of light emission. The averages $\langle(i' \mid T^{[k]}{}_q \mid i)\rangle$ in Eq. (7) will then depend on alignment and orientation only, while the ratios of reduced matrix elements will depend on the dynamics of dipole transitions irrespective of alignment or orientation. Equation (7) is particularly convenient because it allows us considerable freedom in defining the operators $S^{[k]}_q$ and $T^{[k]}_q$ insofar as normalization factors cancel out in this equation; normalization factors matter only in the application of coordinate transformations.

Notice now that the dynamics of light emission expresses itself through the dipole transition matrix elements $(f\,|\,\mathbf{r}\,|\,i)$, and that the matrices of *all tensors* $S^{[k]}_q$ in our problem are expressed in terms of *the same* $(f\,|\,\mathbf{r}\,|\,i)$. The tensors $S^{[k]}_q$ with different $k$ and $q$ differ only in the coupling of the vectors $\mathbf{r}'$ and $\mathbf{r}$ to form a scalar, a vector, or a tensor. Hence the dependence of reduced matrix elements on $k$ involves only the coupling of vector operators, that is, geometrical considerations, and can be worked out by Racah algebra in terms of $6j$ coefficients.

The relevant tensorial considerations can be sketched as follows. The matrix element $(f\,|\,\mathbf{r}\,|\,i)$ of the operator $\mathbf{r}$ (which is a vector; i.e., an irreducible tensor with $k=1$) is constructed with wave functions of angular momenta $j_f$ and $j_i$; we indicate this tensorial structure by the 3-digit symbol $(j_f j_i)1$ which may be read "$j_f$ and $j_i$ coupled to $k=1$." The construction of the matrix of a tensor $S^{[k]}$ of degree $k$, as the product of $\mathbf{r}'$ and $\mathbf{r}$ with the matrices $(i'\,|\,\mathbf{r}'\,|\,f)$ and $(f\,|\,\mathbf{r}\,|\,i)$, is then indicated by the composite symbol $[(j_i j_f)1(j_f j_i)1]^{(k)}$. On the other hand, we want to obtain the matrix of $S^{[k]}$ between states $(i'\,|$ and $|\,i)$ after coupling the final states into the scalar $P_f$ (with $k=0$); this second construct is indicated by the composite symbol $[(j_i j_i)k(j_f j_f)0]^{(k)}$. The transformation from the first to the second of these alternative tensorial constructs is indicated in Racah algebra by a "recoupling coefficient" which is an element of an orthogonal transformation matrix and is indicated by

$$((j_i j_i)k(j_f j_f)0\,|\,(j_i j_f)1(j_f j_i)1)^{(k)}.$$

The reduced matrix element $(i\,||\,S^{[k]}\,||\,i)$ depends on $k$ exclusively through this recoupling coefficient, whose value is usually expressed in terms of the standard $6j$-coefficient

$$\begin{Bmatrix} j_i & j_i & k \\ 1 & 1 & j_f \end{Bmatrix}.$$

[The relevant expression is given by Eq. (15.15) of (FR59) or Eq. (7.1.1) of (E57).] The same considerations apply to the reduced matrix elements $(i\,||\,T^{[k]}\,||\,i)$, except that the tensors $T^{[k]}$ are constructed with the vector operators $\mathbf{J}$ whose matrices $(i''\,|\,\mathbf{J}\,|\,i)$ are diagonal in the quantum number $j_i$. Therefore $j_i$ replaces $j_f$ in the recoupling coefficient applicable to $T^{[k]}$. Thus one obtains the dependence on $k$ of the ratio of matrices of $S^{[k]}$ and $T^{[k]}$ in the

form

$$\frac{(i\,||\,S^{[k]}\,||\,i)}{(i\,||\,T^{[k]}\,||\,i)}$$

$$= (-1)^{j_i-j_f}\left(\begin{Bmatrix} j_i & j_i & k \\ 1 & 1 & j_f \end{Bmatrix} \Big/ \begin{Bmatrix} j_i & j_i & k \\ 1 & 1 & j_i \end{Bmatrix}\right)$$

$$\times \frac{(i\,||\,S^{[0]}\,||\,i)}{(i\,||\,T^{[0]}\,||\,i)},$$

$$= h^{(k)}(j_i,j_f)\,(i\,||\,S^{[0]}\,||\,i)/(i\,||\,T^{[0]}\,||\,i). \tag{8}$$

The dependence of this ratio on the quantum number $j_f$ of the final state reflects the loss of angular and polarization dependence which results from summing over $m_f$. The values of the $6j$ coefficients are tabulated in (RBMW59). The phase normalization factor $(-1)^{j_i-j_f}$ is required for (8) to hold identically for $k=0$.

In the particular case of $k=0$, that is, for scalar operators, inspection of Eq. (6) identifies the scalar $S^{[0]}_0$ as the operator $\mathbf{r}'\cdot\mathbf{r}\,P_f$, whose nonzero matrix elements have been called $S$ and are independent of $m_i$. The corresponding scalar $T^{[0]}_0$ constructed with $\mathbf{J}$ is $\mathbf{J}^2$, whose nonzero matrix elements equal $j_i(j_i+1)$ for all states $|\,i)$. Hence we can write

$$(i\,||\,S^{[0]}\,||\,i)/(i\,||\,T^{[0]}\,||\,i)=S/j_i(j_i+1) \tag{9}$$

and substitute this ratio in (8).

In the final application to Eq. (6) we replace each component of the vectors $\mathbf{r}'$ or $\mathbf{r}$ by the corresponding component of $\mathbf{J}$ and $P_f$ by 1, compensating for this substitution by the factor $h^{(k)}S/j_i(j_i+1)$. Thus we obtain

$$I=\tfrac{1}{2}CS\{1-\tfrac{1}{2}h^{(2)}(j_i,j_f)[\langle(i'\,|\,3J_\zeta^2-\mathbf{J}^2\,|\,i)\rangle/j_i(j_i+1)]$$

$$+\tfrac{3}{2}h^{(2)}(j_i,j_f)[\langle(i'\,|\,J_\xi^2-J_\eta^2\,|\,i)\rangle/j_i(j_i+1)]\cos 2\beta$$

$$+\tfrac{3}{2}h^{(1)}(j_i,j_f)[\langle(i'\,|\,J_\zeta\,|\,i)\rangle/j_i(j_i+1)]\sin 2\beta\}, \tag{10}$$

where the formula $i^{-1}\mathbf{J}\times\mathbf{J}=\mathbf{J}$ has been used. The second term in the braces shows that the anisotropy of emission is proportional to the alignment of the emitter along the direction of emission. The third and fourth terms show the linear and circular polarization to be proportional to the alignment within the plane of polarization $(\xi,\eta)$ and to the component of the orientation $\langle\mathbf{J}\rangle$ in the direction of emission, respectively.

Experimental analysis of the *variation of light intensity* as a function of the detector's direction $\hat{f}$ determines the dependence of the alignment parameter $\langle(i'\,|\,3J_\zeta^2-\mathbf{J}^2\,|\,i)\rangle$ on $\hat{f}$ and thus reconstructs the direction and magnitude of the axes of *the alignment tensor*. The same result is achieved, often with higher sensitivity, by determining $\langle(i'\,|\,J_\xi^2-J_\eta^2\,|\,i)\rangle$ through measurements of the intensity variation that accompanies the rotation of a linear polarization analyzer,

558    Reviews of Modern Physics · October 1973

that is, the rotation of the $\hat{\xi}$ axis, for two fixed directions $\hat{\zeta}$. *Measurement of the circular polarization* observed as a function of the direction $\hat{\zeta}$ determines instead the various components of *the orientation vector* $\langle (i' \mid \mathbf{J} \mid i) \rangle$.

To simplify the notation we introduce in the detector frame two alignment parameters

$$A_0{}^{\text{det}} = \langle (i' \mid 3J_{\zeta}{}^2 - \mathbf{J}^2 \mid i) \rangle / j_i(j_i+1), \qquad (11)$$

$$A_{2+}{}^{\text{det}} = \langle (i' \mid J_{\xi}{}^2 - J_{\eta}{}^2 \mid i) \rangle / j_i(j_i+1), \qquad (12)$$

and one orientation parameter

$$O_0{}^{\text{det}} = \langle (i' \mid J_{\zeta} \mid i) \rangle / j_i(j_i+1), \qquad (13)$$

whereby Eq. (10) takes the form

$$I = \tfrac{1}{2}CS\{1 - \tfrac{1}{2}h^{(2)}(j_i, j_f)A_0{}^{\text{det}} + \tfrac{3}{2}h^{(2)}(j_i, j_f)A_{2+}{}^{\text{det}}\cos 2\beta$$
$$+ \tfrac{3}{2}h^{(1)}(j_i, j_f)O_0{}^{\text{det}}\sin 2\beta\}. \quad (14)$$

The most familiar type of polarization experiment deals with light excited by collisions that produce only alignment along a single axis $\hat{z}$, as described in the following paragraphs. Light is observed at 90°, that is, in a direction $\hat{\zeta}$ perpendicular to $\hat{z}$, with a linear polarization filter, that is, setting $\beta = 0$. One sets the $\hat{\xi}$ axis alternately parallel and perpendicular to $\hat{z}$ thus obtaining two intensity measurements $I_{||}$ and $I_{\perp}$ and defines the degree of polarization as $P = (I_{||} - I_{\perp})/(I_{||} + I_{\perp})$. In the notation of Eq. (14), $I_{||}$ differs from $I_{\perp}$ by sign reversal of $A_{2+}{}^{\text{det}}$. This gives

$$P = \frac{I_{||} - I_{\perp}}{I_{||} + I_{\perp}} = \frac{3h^{(2)}(j_i, j_f)A_{2+}{}^{\text{det}}}{2 - h^{(2)}(j_i, j_f)A_0{}^{\text{det}}}, \qquad (15)$$

with reference to $\hat{\xi}$ parallel to $\hat{z}$.

### II.1. Alignment and Orientation by Collision

The radiator's anisotropy is represented in Eq. (14) by components of the alignment tensor and of the orientation vector in the detector frame $(\xi, \eta, \zeta)$. The collision process determines, of course, the components of this tensor and vector in a "collision frame" $(x, y, z)$, whose $\hat{z}$ axis usually coincides with the direction of an incident particle beam. Components of the alignment tensor in this frame will be denoted by $A^{\text{col}}$.

In the simplest cases (e.g., an unpolarized particle beam incident on gas molecules without detection of scattered or recoil particles, or an ion beam emerging perpendicularly from a foil surface) the experimental arrangement identifies only this $\hat{z}$ axis and has cylindrical symmetry about it. Under these circumstances, the alignment tensor has a single nonzero component, namely $A_0{}^{\text{col}}$. Only in this familiar, but simple, case—or under equivalent circumstances—does symmetry require the excited state to be representable as the incoherent superposition of pure states

with different magnetic quantum numbers $m_i$. Here we have

$$A_0{}^{\text{col}} = \sum_{m_i}[3m_i{}^2 - j_i(j_i+1)]\sigma(m_i)/j_i(j_i+1)\sum_i\sigma(m_i),$$
$$(16)$$

where $\sigma(m_i)$ indicates the partial cross section for excitation of the state $\mid j_i m_i \rangle$.

From the value of $A_0{}^{\text{col}}$ we obtain the parameters $A_0{}^{\text{det}}$ and $A_{2+}{}^{\text{det}}$ in the detector frame by a coordinate transformation. [One must allow here for our non-standard normalization of tensor components. The appropriate transformation formulas can be obtained by transforming separately each Cartesian component of $\mathbf{J}$ in the expression of $\mathbf{A}$.] The result is

$$A_0{}^{\text{det}} = A_0{}^{\text{col}}\tfrac{1}{2}(3\cos^2\theta - 1),$$
$$A_{2+}{}^{\text{det}} = A_0{}^{\text{col}}\tfrac{3}{2}\sin^2\theta\cos 2\psi, \qquad (\cdot 17)$$

where $\cos\theta = \hat{\zeta}\cdot\hat{z}$ and $\psi$ is the angle between the detector's axis $\hat{\xi}$ and the plane $(\hat{\zeta}\hat{z})$. In this same case the *orientation* vector $\langle (i' \mid \mathbf{J} \mid i) \rangle$ *vanishes* altogether, because it is an axial vector (pseudovector) and no such quantity can be identified in a frame characterized by a single incidence vector $\hat{z}$, unless the incident particles have nonzero helicity.

Equation (17) contains two well-known results. The source anisotropy has no effect upon the light intensity emitted at the "magic angle" $\theta = \arccos(\tfrac{1}{3})^{1/2} = 54°44'$, where $A_0{}^{\text{det}}$ vanishes. That is, intensity measurements in this direction are independent of alignment. On the other hand, the polarization vanishes in the forward or backward directions, $\theta = 0°$ or 180°, under the condition of symmetry about the collision axis, irrespective of the source alignment.

The collision frame loses its axial symmetry when the scattering direction (or a recoil direction) $\hat{z}'$ is observed in coincidence with the light emission. In beam foil excitation, the state of an atom emerging from the foil depends presumably on its interaction with the surface layers of the foil and hence on the orientation of this surface; therefore one can spoil the axial symmetry by tilting the normal to the foil, $\hat{z}'$, away from the beam axis $\hat{z}$.

When an axis $\hat{z}' \neq \hat{z}$ is thus singled out we may lay the $\hat{x}$ axis in the plane $(\hat{z}\hat{z}')$ and the collision identifies the axial vector $\hat{z} \times \hat{z}'$ parallel to $\hat{y}$. The *orientation* parameter $O_{1-}{}^{\text{col}} \propto \langle (i' \mid J_y \mid i) \rangle$ is now generally *nonzero* and there occur two additional nonzero alignment components, namely, $A_{2+}{}^{\text{col}} \propto \langle (i' \mid J_x{}^2 - J_z{}^2 \mid i) \rangle$ and $A_{1+}{}^{\text{col}} = \langle (i' \mid J_zJ_x + J_xJ_z \mid i) \rangle / j_i(j_i+1)$. (It is just the occurrence of such nonzero components which implies *coherent superposition* of states with different $m_i$ values.) The residual symmetry still causes all other components of orientation and alignment to vanish. A frame transformation yields now

$$O_0{}^{\text{det}} = O_{1-}{}^{\text{col}} \sin\theta \sin\phi,$$

$$A_0{}^{\text{det}} = A_0{}^{\text{col}}\tfrac{1}{2}(3\cos^2\theta-1) + A_{1+}{}^{\text{col}}\tfrac{3}{2}\sin 2\theta \cos\phi$$
$$+ A_{2+}{}^{\text{col}}\tfrac{3}{2}\sin^2\theta \cos 2\phi,$$

$$A_{2+}{}^{\text{det}} = A_0{}^{\text{col}}\tfrac{1}{2}\sin^2\theta \cos 2\psi + A_{1+}{}^{\text{col}}\{\sin\theta \cos\phi \sin 2\psi$$
$$+ \sin\theta \cos\theta \sin\phi \cos 2\psi\}$$
$$+ A_{2+}{}^{\text{col}}\{\tfrac{1}{2}(1+\cos^2\theta)\cos 2\phi \cos 2\psi$$
$$- \cos\theta \sin 2\phi \sin 2\psi\}, \quad (18)$$

where $\phi$ is the angle between the planes $(\hat{z}\hat{x})$ [i.e., $(\hat{z}\hat{z}')$] and $(\hat{z}\hat{\zeta})$. That is, $\theta$ and $\phi$ are the polar coordinates of the light detector while $\psi$ is the third Euler angle, required to identify the orientation of a linear polarization analyzer. Other components of $\mathbf{O}^{\text{det}}$ and $\mathbf{A}^{\text{det}}$ are not given here, being irrelevant to light observations.

The several components of the orientation vector and alignment tensor are independent parameters in the sense that the value of any one of them may be changed by a change of the excitation process while the others remain fixed. There are, however, upper limits to the sum of their squares. Such an upper limit is reached for example for the pure state identified in a given frame by $m_i = j_i$. In this event we have $(i|J_z|i) = j_i$ and $(i|3J_z{}^2 - J^2|i) = j_i(2j_i-1)$ but all other components of $\mathbf{O}^{\text{col}}$ and of $\mathbf{A}^{\text{col}}$ vanish. One can also see directly that full circular polarization would imply total absence of any linear polarization, and vice versa.

For purposes of illustration we consider some simple cases which have been studied experimentally and treated successfully by earlier procedures, reformulating their treatment in the language of this article. In one example, the scattered particle is detected after collision in coincidence with the emitted photon, but just in the direction of incidence, i.e., at $0°$ deflection. This arrangement was suggested by Imhoff and Read (IR71) to test detailed theoretical predictions of (PS58) and was then successfully applied (KAR72). The arrangement implies not only that the $\hat{z}'$ axis coincides with $\hat{z}$, thus restoring axial symmetry, but also that the collision imparts to the atom no angular momentum component parallel to the $\hat{z}$ axis (except for possible spin exchange). If the target atom was initially in a $^1S$ state, without any angular momentum, the collision will then leave it in a state with $m_i = 0$ and, therefore, with $\langle 3J_z{}^2 - J^2\rangle = -j_i(j_i+1)$ and $A_0{}^{\text{col}} = -1$. Light emission by decay to any final state with $j_f = 0$, which occurs by a dipole process when $j_i = 1$, has then full linear polarization according to (PS58). In the present formulation the same result is retrieved by setting $A_0{}^{\text{col}} = -1$ in Eq. (17), $j_i = 1$ and $j_f = 0$ in (8) which yields $h^{(2)}(1,0) = -2$, and $\beta = 0$ for detection of linear polarization. The intensity

formula (14) gives then

$$I = CS \sin^2\theta[\tfrac{1}{2}(1+\cos 2\psi)], \quad (19)$$

which is the result expected for a linear polarization detector placed along $\hat{\zeta}$, i.e., at an angle $\theta$ with respect to a radiating dipole parallel to $\hat{z}$, and with its $\hat{\xi}$ axis at an angle $\psi$ with respect to the $(\zeta z)$ plane. The degree of polarization $P$ given by Eq. (15) reduces now to $\cos 2\psi$, and more specifically to 1 when $\hat{\xi}$ lies in the $(\zeta z)$ plane, i.e., when $\psi = 0$, as assumed by the definition of $P$.

Another application of angular momentum and other symmetry considerations is afforded by measurements of Lyman-$\alpha$ excitation by charge transfer collision of protons with helium. The experiment by McKnight and Jaecks (JCMc70, McJ71) was designed to measure a coincidence rate between photons and scattered H atoms, which would be proportional to the excitation cross section and independent of unknown parameters such as the polarization of the emitted light. The design utilized the invariance of the collision process under reflection in the scattering plane, $(zz') \equiv (xz)$. Excitation of the $p_y$ state, odd under this reflection, is then forbidden provided the He$^+$ ion is left in its even-parity ground state. Since the excluded $p_y$ state has $J_y = 0$, the $p$ state which is actually excited must have $J_y{}^2 = 1$. This circumstance was exploited by placing the photon counter, i.e., the $\zeta$ axis of the detector frame, along the $y$ axis of the collision frame. Accordingly we substitute in Eq. (11) $\langle J_\zeta{}^2\rangle = J_y{}^2 = 1$ and $J^2 = j_i(j_i+1) = 2$ and thus obtain $A_0{}^{\text{det}} = \tfrac{1}{2}$. Recalling that $h^{(2)} = -2$ for a $p \to s$ optical transition and averaging Eq. (14) over $\beta$, as appropriate to a counter insensitive to polarization, we have finally

$$I = CS\tfrac{1}{6}, \quad (20)$$

which is indeed independent of unknown parameters.

### II.2. Excitation by Electron Impact

Basic experiments on light excited by electron impact (S26, SA27) and their theoretical framework (O27) date from the mid-1920's. Theory provides clear predictions both for high collision energies and for energies very close to threshold. Interpolation between the results valid in the two limits seems plausible but the resulting prediction was soon found erroneous. We outline this development to illustrate both the power and the limitations of simple geometrical and dynamical considerations.

At high collision energies the Born approximation, which treats a collision as an impulsive transfer of momentum $\mathbf{q}$ by the incident electron, prevails; this transfer provides for target excitation. The vector $\mathbf{q}$, which can be determined for each collision by observing the direction and energy of the scattered electron, is

560    Reviews of Modern Physics · October 1973

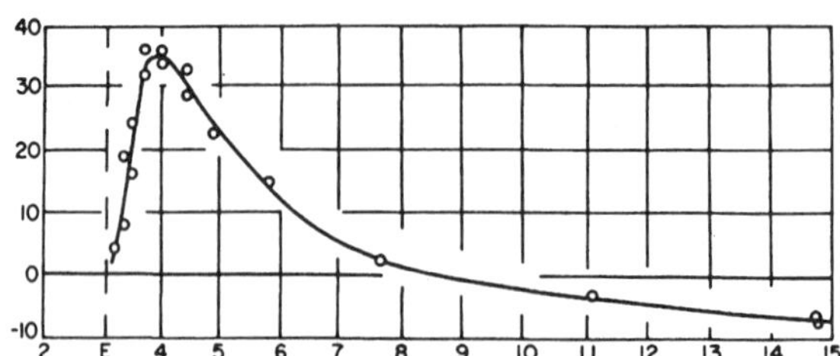

Fig. 1. Percent linear polarization of mercury light observed at 90° from the direction of incidence (SA27). Note the sign reversal of the polarization at $\sim$80 eV and its disappearance near threshold. The energy scale [in $(eV)^{1/2}$] is nonlinear and the energy resolution not very high.

the only collision parameter on which alignment and orientation can depend. Since $\mathbf{q}$ is a polar vector, this implies that *no orientation* can occur in this approximation. Further, symmetry about the direction of $\mathbf{q}$ implies that the only nonzero component of the alignment tensor, in a frame with axis $\mathbf{q}$, is $A_0^{\hat{q}} = \langle\langle i' \mid 3(\mathbf{J}\cdot\hat{q})^2 - \mathbf{J}^2 \mid i\rangle\rangle / j_i(j_i+1)$. Finally, the transfer of momentum $\mathbf{q}$ cannot impart to the target any nonzero value of $\mathbf{J}\cdot\hat{q}$, from which follows $A_0^{\hat{q}} = -1$. Transformation to the collision frame, with $\ell$ in the direction of impact and $\hat{x}$ in the plane of scattering, then yields

$$A_0^{\text{ool}} = -\tfrac{1}{2}[3(\hat{q}\cdot\ell)^2 - 1],$$

$$A_{1+}^{\text{ool}} = -\tfrac{1}{2}\hat{q}\cdot\ell[1 - (\hat{q}\cdot\ell)^2]^{1/2},$$

$$A_{2+}^{\text{ool}} = -\tfrac{1}{2}[1 - (\hat{q}\cdot\ell)^2]. \qquad (21)$$

Normally, the scattering direction remains undetected and an average must be taken over the direction $\hat{q}$. Recalling that $A_{1+}^{\text{ool}}$ and $A_{2+}^{\text{ool}}$ are proportional to $\langle J_x J_z + J_z J_x\rangle$ and $\langle J_x^2 - J_y^2\rangle$, respectively, and that the orientation of the $\hat{x}$ and $\hat{y}$ axes rotates about $\ell$ in the course of averaging over $\hat{q}$, we see that $A_{1+}^{\text{ool}}$ and $A_{2+}^{\text{ool}}$ average out. We are thus left, as expected, with a single nonzero parameter, namely,

$$A_0^{\text{ool}} = -\tfrac{1}{2}[3\langle(\hat{q}\cdot\ell)^2\rangle - 1]. \qquad (22)$$

A final step of the theory utilizes energy-momentum considerations in the collision (I71) to predict that $\hat{q}$ is predominantly transverse to $\ell$ at very high impact energies and parallel to $\ell$ at lower energies. The resulting *sign reversal* of $A_0^{\text{ool}}$ occurs as the energy decreases through a range of the order of ten times the threshold energy (Be33). In the high-energy limit we have $\langle(\hat{q}\cdot\ell)^2\rangle \rightarrow 0$ and $A_0^{\text{ool}} \rightarrow \tfrac{1}{2}$.

In the opposite, low-energy limit, i.e., just above threshold, the electron must be scattered inelastically into an $s$ state; otherwise it would have to escape by tunneling through a centrifugal potential barrier. Hence the departing electron does not weigh on the angular momentum balance, apart from spin exchange which is often disregarded. Any nonzero alignment of the excited atom reflects then only the contribution of the incident electron, whose orbital momentum component vanishes in the direction of incidence. Setting thus $\langle J_z^2\rangle = 0$, i.e., *disregarding spin exchange*,

we have

$$(A_0^{\text{ool}})_{\text{thresh}} = -1. \qquad (23)$$

Remarkably, this value coincides with the low-energy limit of Eq. (22), since $\hat{q}$ must be parallel to $\ell$ at threshold. This coincidence may lead to the surmise that $A_0^{\text{ool}}$ should interpolate smoothly, and that the Born approximation result might hold uninterruptedly—if only approximately—down to threshold. Early measurements of the dependence of light polarization upon the velocity of incident electrons (SA27) disproved this surmise. As shown in Fig. 1, the polarization seems to disappear when the incident energy drops to within a few electron volts of threshold.

This remarkable drop of polarization near threshold, in apparent disagreement with a clearcut theoretical prediction, was studied in increasing detail between 1955 and 1970 through a sequence of experiments, mostly on helium (LM57, McF64, HK67, McF67, SFG67). Actually, none of these experiments could measure the degree of light polarization at energies sufficiently close to threshold to allow unequivocal application of the argument leading to Eq. (23). Note that since the scattered electron leaves the atom in an excited state with a radius of $\sim n^2$ atomic units ($n$ is the principal quantum number of the state less its quantum defect and $n^2$ its reciprocal binding energy in units of 13.6 eV), the centrifugal barrier which hinders its escape with orbital quantum number $l>0$ lies at radial distances larger than $n^2$ and peaks at

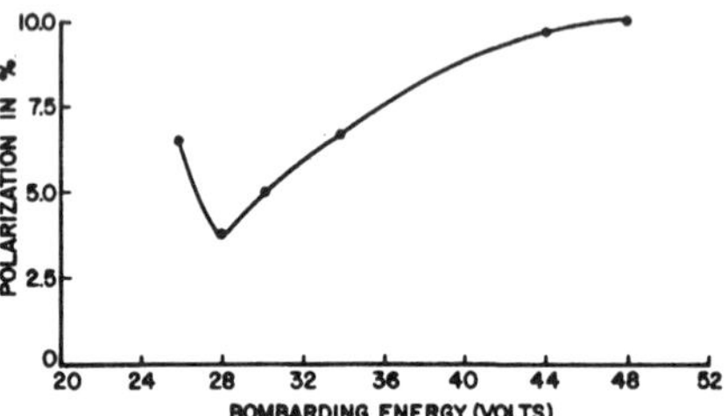

Fig. 2. Same as Fig. 1 for the 3889-Å line of helium (LM57). The scale is linear and greatly expanded with respect to Fig. 1. Note the low minimum and the incipient rise toward the threshold which lies at 23 eV.

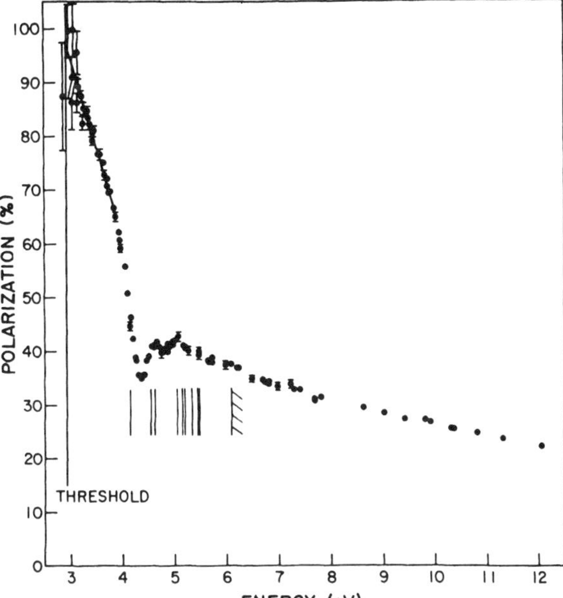

FIG. 3. Polarization of the calcium 4227-Å resonance line ($^1P_1^\circ \to {}^1S_0$) as a function of the energy of exciting electrons (EG73). Note the rapid drop from the maximum at threshold and the unsmoothness in the range of expected, but unresolved, resonances.

energies smaller than $l(l+1)/n^4$ a.u. Equation (23) may then apply only within a fraction of 1 eV above threshold. The experimental evidence provided, e.g., by the left-hand portion of the curve in Fig. 2, is thus not inconsistent with Eq. (23). Similar evidence is found in the recent data of (OHK72) on mercury. In other instances, particularly in the excitation of the resonance lines of the alkalis and alkaline earths (HKK65, EG72, EG73), the observed polarization tends rather clearly to the appropriate threshold value, as shown in Fig. 3.

Assuming then that Eq. (23) does hold at threshold, the experimental data indicate that $A_0^{00l}$ drops in many cases to a very low value shortly above threshold, returning then to its Born approximation value only at higher energies. This drop has not been explained in any detail but is probably related to the occurrence of conspicuous resonances in inelastic electron–atom cross sections near threshold. At resonance the colliding electron and the one being excited in the atom remain strongly correlated for a time interval sufficient to allow extensive exchange of angular momentum between them, with a resulting decrease of $A_0^{00l}$. The necessary time interval is of the order of the reciprocal optical frequency for transitions between excited states, i.e.,

$\lesssim 10^{14}$ sec. Various, albeit fragmentary, evidence on hand points to the conclusion that such extensive correlations occur normally in the process of excitation by electron impact and persist in the spectrum over a range $\gtrsim 10$ eV above threshold.

Indeed careful extensive studies of optical excitation by highly monochromatic electrons might prove very effective in establishing and refining our knowledge of these correlations. The occurrence of resonances near threshold, superimposed on a smooth dependence of polarization on impact energy, has been demonstrated clearly in a very recent experiment on the excitation of the resonance line of Ba$^+$ (CTD73). As seen from these results, in Fig. 4, the resonances fail in this case to reduce the average polarization to a near zero value. Figure 4 also demonstrates a smooth variation of polarization as a function of collision energy (on a logarithmic scale) from $\sim 7$ to 500 eV, with sign reversal at an energy $\sim 10$ times threshold in agreement with the crude rule indicated above. The Ba$^+$ data are quite analogous in this respect to those of Ca (EG73) which extrapolate at high energy to the 100% negative polarization limit predicted by Eq. (15) with $h^{(2)} = -2$ and with the limiting value $A_0^{00l} \to \frac{1}{2}$. By contrast, the Ba$^+$ and Ca data differ in

562    REVIEWS OF MODERN PHYSICS · OCTOBER 1973

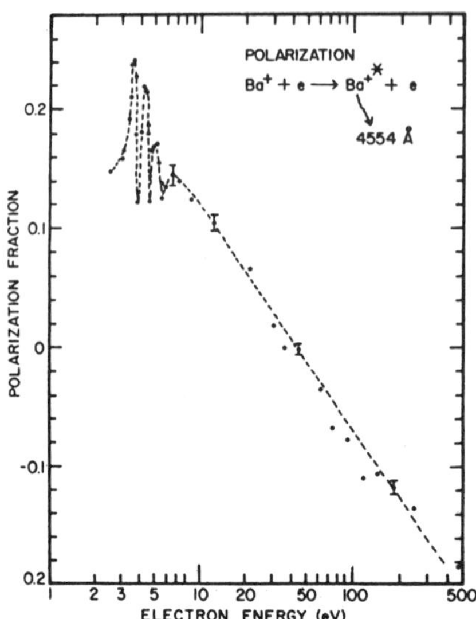



FIG. 4. Polarization of the 4554-Å Ba⁺ resonance line ($^2P_{3/2}{}^\circ\rightarrow$ $^2S_{1/2}$) as a function of the energy of exciting electrons (CTD73). Note the clearly resolved resonances and the smooth variation and sign reversal of the polarization on the logarithmic scale.

that the polarization shown in Fig. 3 drops rapidly from threshold to a region where barely resolved resonances appear to exist.

*Note added in proof:* Private communications have been received regarding current experiments that extend and complement the material of this section. Each of these experiments pertains to the polarization of helium lines excited by electron collisions.

Heddle and collaborators (to be published in *J. Phys. B*) have increased the resolution of incidence energy to the point of detecting changes of polarization resulting from 0.01 eV step-ups in energy. The polarization of the lines studied in this work drops sharply from threshold over a few energy steps and then exhibits a number of resonances within a few eV.

Resonances have also been detected in the helium light polarization at much higher bombarding energy by A. Defrance (University of Rennes, France). These resonances occur near 58 eV and correspond to the formation of triply excited He⁻ levels, whose presence had been detected in several other processes since 1965.

Coincidence detection of photon and scattered electron has been achieved by Kleinpoppen and collaborators [*Phys. Rev. Lett.* (to be published)]. The collision energy was ~80 eV, that is, sufficiently low to yield departures from the predictions of Born approximation theory. As explained above, Born approximation yields

an alignment tensor with axial symmetry about the momentum transfer q and with a single nonzero component $A_0{}^{\hat q}=-1$. Departures from this prediction may be described in terms of changes of direction and magnitude of the principal axes of the alignment tensor (MJ71). (A principal-axes coordinate frame is characterized by $A_2{}_-{}^{\text{prime}}=A_1{}_+{}^{\text{prime}}=A_1{}_-{}^{\text{prime}}=0$.) The coincidence experiment found a principal axis to depart from q by about 8° in the plane of electron scattering. The value of $A_0{}^{\text{prime}}$ also departs from $-1$. These initial observations are confined to photon emission in the scattering plane.

### III. EMISSION BY ATOMS IN NONSTATIONARY STATES

Light emission by excited atoms has been studied widely in recent years under conditions of high time resolution, which determine the time interval between excitation and emission to within $10^{-10}$ or even $10^{-11}$ sec (H67). When the atoms travel in a beam of known velocity, this time interval is determined by the distance between the point of collision and the field of view of the light detector. Alternatively one may time the collision by detecting a scattered particle and the emitted light by photon counting in delayed coincidence.

Sharp definition of the time of excitation requires the collision to be of short duration and hence capable of populating coherently a range of energy levels. Similarly, a sharp definition of the time of emission within an interval $\Delta\tau$ prevents the spectral analysis of the light from resolving levels spaced much closer than $\hbar/\Delta\tau$. The observed light originates then from a nonstationary state represented by the coherent superposition of stationary states with different, unresolved, energy eigenvalues. We discuss in this second part how the intensity, anisotropy, and polarization of the emitted light are modulated in accord with the coherent superposition of radiation with different, unresolved frequencies.

As in Sec. II we will express the anisotropy and polarization of light in terms of the alignment and orientation of the excited states. The time modulation of these parameters provides information on both the energy and eigenfunctions of the unresolved levels as well as on their superposition generated initially by the collision. The energy splittings of the unresolved levels are due in general to hyperfine interactions, spin–orbit coupling, or the action of external fields. The time modulation of atomic orientation and alignment may be interpreted as due to oscillations or precession of electron orbitals under influence of the interactions.

A typical collision excites the orbital motion of electrons leaving the electronic and nuclear spins unaffected. (Collisions with electron exchange or rearrangement are exceptional in this respect, since they may introduce an initial spin polarization.) We deal here, then, with the alignment and/or orientation of the orbital motion which is responsible for the subsequent light emission by electric dipole transitions. By contrast, spin–orbit interactions and external fields have the effect of disturbing the initial alignment and orientation. The observations of light with high spectral resolution, considered in Sec. I, display the mean orbital orientation and alignment in individual hyperfine or fine structure levels. The observations with high time resolution to be considered in this part display instead the time dependence of the orbital anisotropy.

It should be noted that spin–orbit interactions and external fields affect only the anisotropy of the orbital motion, but do not affect the rate of light emission. Therefore the variations of alignment and orientation to be considered here merely modulate the anisotropy of the exponential decay. This decay is normally characterized by a single constant $\Gamma$, pertaining to the unresolved multiplet. Exceptional in this respect are the experiments where levels with different orbital classification also remain unresolved, because these levels radiate at intrinsically different rates. The decay constant $\Gamma$ then becomes a matrix. The hydrogenic systems provide an outstanding example of this effect because their orbitals with equal $n$ and different $l$ quantum numbers are nearly degenerate. Normally, however, the interactions that change the orbital classification are so strong as to yield easily resolved term splittings. The resulting modulations of light emission are then too fast for resolution by current techniques.

The theory of the modulation phenomena is still fragmentary. We present here a general formulation and develop it analytically for a few simple examples. We shall also indicate where complications arise in other problems and how they may be treated. The analytical approach to be followed here utilizes the shortcuts to Racah algebra introduced in Sec. II but its physical substance and its main results are common to the theory of perturbed angular correlations of nuclear radiations, as presented, e.g., in Sec. 9 of (FS68).

### III.1. General Approach

The anisotropy and polarization of electric dipole radiation are related by Eq. (6) to the mean values of irreducible components $S^{[k]}_q$ of the tensor

$$\mathbf{r}' P_f(\mathbf{r}', \mathbf{r}) \mathbf{r} = \mathbf{r}' \sum_f |f\rangle \langle f| \mathbf{r}, \qquad (24)$$

where $f$ represents the quantum numbers of the final state of the radiating atom. We have seen in Sec. II how the Wigner–Eckart theorem relates the mean values of $S^{[k]}_q$ to those of the irreducible tensor components $T^{[k]}_q$ constructed as products of angular momentum operators. The mean values $\langle T^{[k]}_q \rangle$ constitute alignment and orientation parameters of the excited atom. The following treatment of nonstationary

states deals explicitly with the time dependence of the mean values of $S^{[k]}_q$ but the relationship to the mean values of $T^{[k]}_q$ remains unchanged. This relationship insures that the modulation of the emitted light depends only on $\langle T^{[k]}_q \rangle$, and hence on the state of the excited atoms, and not on the emission process itself.

To study the time dependence of the anisotropy and polarization of the emitted light, we evaluate the mean values of $S^{[k]}_q$ at the time $t$ of emission. Recall, in this connection, that the averaging process indicated in Eq. (6) by $\langle\langle (i' \cdots i) \rangle\rangle$ pertains to the state of excitation generated by the collision process at the initial time $t = 0$. Under these circumstances it appears convenient to introduce the time dependence using the Heisenberg representation. This is done by replacing each of the parameters of Eq. (6) according to the prescription

$$\langle\langle (i' | S^{[k]}_q | i) \rangle\rangle \rightarrow \langle\langle (i' | \exp(iHt/\hbar) S^{[k]}_q$$
$$\times \exp(-iHt/\hbar) | i) \rangle\rangle, \qquad (25)$$

where $H$ is the Hamiltonian operator of the atom. [Note that one arrives at the same prescription in the Schrödinger representation by regarding the states $(i' |$ and $| i)$ as time-dependent, the dependence being represented by $\exp(\pm iHt/\hbar)$, respectively.] From this point of view, the original formula (6) pertains to the case where $(i' |$ and $| i)$ are stationary states with same energy eigenvalue $E_i$. In that event the operator $H$ may be replaced by $E_i$; the two factors $\exp(\pm iE_it/\hbar)$ cancel then in Eq. (25) and $\langle S^{[k]}_q \rangle$ becomes time independent. Our task consists of evaluating the time dependence introduced in Eq. (25) when $(i' |$ and $| i)$ are not stationary. More specifically, we shall express the right-hand side of Eq. (25) in terms of its value for $t = 0$ which has been discussed in Sec. II.

The modulations which we consider result from interactions that are not only weak but also of little relevance to the initial excitation by collision. This is why we relate the initial excitation to states $| i \rangle$ that would be stationary only in the absence of weak interactions. In fact, these states differ from the eigenstates $| n \rangle$ of the complete Hamiltonian with energy levels $E_n$. The evaluation of the time-dependent parameters in Eq. (25) requires us then to take into account not only the level splittings but also the connection between the stationary states $| n \rangle$ and the nonstationary states $| i \rangle$.

We consider particularly atomic excitations in which the states of both sets, $| i \rangle$ and $| n \rangle$, are fully identified by specific—although different—sets of angular momentum quantum numbers. Stationary states of hyperfine structure are thus identified as $| F M_F \rangle$, where $\mathbf{F}$ is the constant resultant of two coupled momenta, namely, $\mathbf{I}$ of the nucleus and $\mathbf{J}$ of the electrons. On the other hand, $\mathbf{I}$ and $\mathbf{J}$ are uncoupled in the $| i \rangle$ states which may thus be labeled by the

564    REVIEWS OF MODERN PHYSICS · OCTOBER 1973

magnetic quantum numbers of the separate nuclear and electronic states, $|M_I M_J)$. Similarly, the stationary states of an atom in an external field are classified by their magnetic quantum number in a frame with its $\tilde{z}$ axis parallel to the field, whereas the $|i)$ states are naturally identified in the collision frame introduced in Sec. II. In this class of problems the connection between the sets of stationary and nonstationary states is fully determined by angular momentum theory or by other general considerations independent of dynamical details. We single out this class because it is amenable to analytical treatment, thus providing a framework for the introduction of more complex phenomena.

The interactions which modulate the light emission fall into two broad types, namely, interactions among internal orbits and spins and interactions of the atom with external fields. As noted above, the light emission depends on the orientation and alignment of electron orbits. Its modulation by internal couplings reflects then reversible exchanges of angular momentum between orbits and spins, and the accompanying exchange of alignment and orientation. Usually the exciting collision leaves the spins unpolarized. In this event the spins can only draw from the initial alignment and orientation of the orbits, thus reducing it albeit reversibly without contributing any input. Moreover, the scalar character of the initial spin distribution and of the interaction Hamiltonian leaves unchanged the tensorial character of each $S^{[k]}_q$ operator of the orbits. Therefore, in this case the interaction merely modulates the magnitude of each separate parameter $\langle S^{[k]}_q \rangle$.

In contrast, the presence of an external field introduces a new reference frame and thus tends to interlink the modulations of the different parameters $\langle S^{[k]}_q \rangle$. We shall deal first with the simple case of a weak field Zeeman effect, where the modulation may be represented as a precession of the collision frame about the field direction $\tilde{z}$.

We have emphasized that the modulation of light emission reflects the nonstationary character of the excited state. However, the evaluation of the parameters $\langle S^{[k]}_q \rangle$ must also take into account that observations with low spectral resolution generally detect simultaneously light emitted in transitions to different final states. These different emissions are superposed incoherently, barring separate observations of the final state. Their existence is included in the theory by specifying that the summation in the definition of the operator $P_f$ in Eq. (24) extends to include all unresolved final states.

By way of illustration consider transitions leading to unresolved hyperfine levels $|IJ_f F_f M_{F_f})$. The sum over final states runs here over both $F_f$ and $M_{F_f}$, at fixed $J_f$. This set of states spans the same Hilbert space as the set of uncoupled states $|IM_I J_f M_{J_f})$. Therefore the operator $P_f$ factors into two scalars,

one operating on electrons and the other on the nuclear spin,

$$P_f = \sum_{F_f M_{F_f}} |F_f M_{F_f})(F_f M_{F_f}|$$
$$= \left[\sum_{M_I} |IM_I)(IM_I|\right]\left[\sum_{M_{J_f}} |J_f M_{J_f})(J_f M_{J_f}|\right]. \tag{26}$$

The last factor of Eq. (26) represents a projection on the electronic final states with $J = J_f$ and the preceding one is simply the unit operator in the space of nuclear states with the fixed spin $I$. The operator $P_f$ defined by Eq. (26) should be entered in Eq. (24) together with the electron dipole operators $\mathbf{r}$ and $\mathbf{r}'$. Each of the resulting tensorial operators $S^{[k]}_q$ factors then into an irreducible tensor component operating only on the electron variables multiplied by the unit operator of nuclear spin.

In this example we would then construct the operator $T^{[k]}_q$ from components of $\mathbf{J}$ rather than from components of $\mathbf{F} = \mathbf{I} + \mathbf{J}$. Similarly, when fine structure is also unresolved one should further factor the scalar $P_J$ of electronic variables into its orbital and spin parts. Each tensor component $S^{[k]}_q$ would then factor into a tensorial part operating on orbital variables only, multiplied by the unit operator of electron spin. The operators $T^{[k]}_q$ would be constructed from the components of $\mathbf{L}$ rather than of $\mathbf{J} = \mathbf{L} + \mathbf{S}$.

### III.2. Weak Field Effect on Emission from a Single Level

As an initial example, we consider here an excited state which would be stationary, as in Sec. II, with angular momentum quantum number $j$, except for the action of a magnetic field $\mathbf{B}$. The strength of $\mathbf{B}$ is sufficiently small for us to neglect any field-induced coupling to other excited states; that is, the mean value of the magnetic interaction is far smaller than the separation of zero-field levels. The excited state has then a definite gyromagnetic ratio $\gamma_i$ and the Hamiltonian has the eigenvalues

$$E_i - \hbar \gamma_i B \bar{m}_i, \tag{27}$$

where $\bar{m}_i$ is the magnetic quantum number in a frame with $\tilde{z}$ axis parallel to $\mathbf{B}$. The set of eigenstates $|n)$ of the Hamiltonian is identified here as $|j_i \bar{m}_i)$, with $j_i$ fixed.

In this case one may consider the tensor components $S^{[k]}_q$ in the frame with axis $\tilde{z}$. The matrix elements of the time-dependent operator are then easily evaluated as follows:

$$(j_i \bar{m}_i' | \exp(iHt/\hbar) S^{[k]}_{\bar{q}} \exp(-iHt/\hbar) | j_i \bar{m}_i)$$
$$= \exp[i(E_i/\hbar - \gamma_i B \bar{m}_i')t](j_i \bar{m}_i' | S^{[k]}_{\bar{q}} | j_i \bar{m}_i)$$
$$\times \exp[-i(E_i/\hbar - \gamma_i B \bar{m}_i)t]$$
$$= \exp[-i\gamma_i B(\bar{m}_i' - \bar{m}_i)t](j_i \bar{m}_i' | S^{[k]}_{\bar{q}} | j_i m_i). \tag{28}$$

Since a selection rule on the matrices of tensorial

operators requires the matrix element to vanish unless $\bar{m}_i{}' - \bar{m}_i = \bar{q}$, the time-dependent factors of Eq. (28) depend only on $\bar{q}$, rather than on the separate $\bar{m}_i{}'$ and $\bar{m}_i$, and can be factored out of the mean value, yielding

$$\langle (i' \mid \exp(iHt/\hbar)\, S^{[k]}{}_{\bar{q}}\, \exp(-iHt/\hbar) \mid i) \rangle$$
$$= \exp(-i\bar{q}\gamma_i Bt)\,\langle (i' \mid S^{[k]}{}_{\bar{q}} \mid i) \rangle. \quad (29)$$

This equation represents the familiar result that the field cannot alter the mean value of any irreducible tensor component in the $\bar{z}$ frame but merely causes its complex phase to change in the course of time. The phase change is the same as that which results from a rotation of $(\bar{x}, \bar{y})$ axes about $\bar{z}$ at the rate of $-\gamma_i B$ rad/sec.

Once this result is established, it becomes unnecessary to consider the tensorial components $S^{[k]}{}_{\bar{q}}$ in the $\bar{z}$ frame explicitly. It is sufficient to consider the parameters $\langle S^{[k]}{}_q \rangle$ calculated at $t = 0$ in the collision frame, as in Sec. II, and then to state that each of these parameters remains constant in a rotating frame which coincides with the collision frame at $t = 0$ and precesses about the field direction at the uniform rate of $-\gamma_i B$ rad/sec. The effect of this precession can be combined with the eventual transformation to the detector frame represented by Eq. (18).

A direct demonstration of this Larmor precession was provided by Hadeishi and Nierenberg's (HN65) observation of the 3261-Å line of Cd excited by a pulsed electron beam in a magnetic field **B** of 0.88 G. The electron pulse, lasting $\sim 2$ nsec, excites the atoms and starts the timing of the photon counting. The time-resolved counts in Fig. 5 clearly show that the light intensity reaching a counter placed at 90° from $\hat{B}$ precesses at twice the Larmor frequency; this is the result to be expected since the intensity depends linearly on $A_0{}^{\text{det}}$ according to Eq. (14) and $A_0{}^{\text{det}}$ depends in turn on $A_{\bar{q}}{}^{\hat{B}}$ with $\bar{q} = \pm 2$ in the geometry

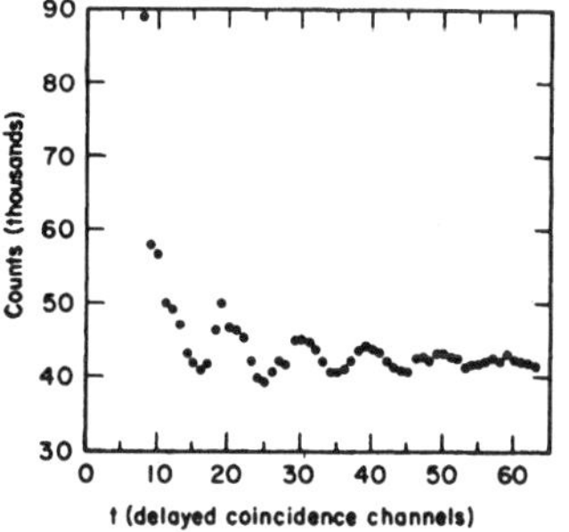


FIG. 5. Larmor precession in the decay of the 3261-Å transition in Cd excited by pulsed electrons, as observed by Hadeishi and Nierenberg (HN65).

of this experiment. Similar modulations were obtained by Dodd, Kaul, and Warrington (DKW64) using pulsed optical excitation. Equivalent results have also been obtained by periodic modulation of the primary source of excitation, a procedure that is experimentally easier although less direct than excitation by a single pulse (BB61, K61, PP63, A64, CS64).

The result (29) has been derived by Nedelec (N65) for atoms excited by electron impact, using a density matrix representation of the state of the excited atom equivalent to the treatment of the present paper. In fact, density matrix treatments serve to obtain equivalent results in various forms for any particle of interest (see, e.g., F57) and have also permitted the generalization of the Larmor precession Eq. (29) to describe the effects of nonuniform external fields of arbitrary multipolarity (F64).

### III.3. Modulation Due to Unresolved Hyperfine Structure

The next example concerns the time dependence of the alignment or orientation parameters

$$\langle (i' \mid \exp(iHt/\hbar)\, S^{[k]}{}_q\, \exp(-iHt/\hbar) \mid i) \rangle \quad (30)$$

which results from the coherent superposition of the states of a hyperfine multiplet. In this case, both the operators $S^{[k]}{}_q$ and the sets of states $(i' \mid$ and $\mid i)$ to be entered in Eq. (30) are defined with reference to uncoupled states $\mid IM_I JM_J )$ whereas the Hamiltonian $H$ is diagonal in the basis of coupled states $\mid IJFM_F )$. The evaluation of Eq. (30) will then proceed through a sequence of transformations involving two reciprocal changes of coupling. Starting from the matrix of $S^{[k]}{}_q$ in the uncoupled representation we shall transform it to the coupled representation $\mid IJFM_F )$. Next we shall multiply this matrix by the matrices of $\exp(\pm iHt/\hbar)$ which are now diagonal. The product will be transformed back to the uncoupled representation and will then finally be averaged according to $\langle (i' \mid \cdots \mid i) \rangle$.

Our task is made easier by the invariance of the interaction Hamiltonian under joint rotation of electronic and nuclear coordinates. Owing to this invariance, the product of each operator $S^{[k]}{}_q$ by $\exp(\pm iHt/\hbar)$ is a new tensorial operator $\tilde{S}^{[k]}{}_q$ with the same indices $k$ and $q$. Moreover the calculation of the matrix transformations and products can be carried out independently of the magnetic quantum numbers which pertain to a specific coordinate system; therefore it is sufficient to calculate in terms of reduced matrix elements.

Recalling that the operator $S^{[k]}{}_q$ actually consists of an electronic factor identical to that considered in Sec. II multiplied by a unit operator of the nuclear spin, we write its single nonzero reduced matrix element in the uncoupled basis

$$(I \parallel 1^{[0]} \parallel I)(J \parallel S^{[k]} \parallel J). \quad (31)$$

(45) no longer belong to the matrix of a single irreducible tensor component $S^{[k]}{}_q$. This result is not surprising because the Hamiltonian no longer has spherical invariance in the presence of the field **B**.

Formally one verifies this result through a calculation designed to invert the Wigner–Eckart theorem. If Eq. (45) represented the matrix element of a single irreducible tensor component, the reduced matrix element of this tensor would be obtained by multiplying Eq. (45) by the $3j$ coefficient

$$(-1)^{F'-\bar{M}_{F'}}\begin{pmatrix} F' & k' & F \\ -\bar{M}_{F'} & \bar{q} & \bar{M}_F \end{pmatrix}(2k'+1),$$

summing over $\bar{M}_{F'}$ and $\bar{M}_F$ and applying a closure property of the $3j$ coefficients. The essential step of this procedure consists of calculating the double summation

$$\sum_{M_{F'}M_F}\begin{pmatrix} F' & k & F \\ -\bar{M}_{F'} & \bar{q} & \bar{M}_F \end{pmatrix}$$
$$\times \exp\left[i(E_{F'}-\gamma_{F'}B\bar{M}_{F'}\hbar-E_F+\gamma_F B\bar{M}_F\hbar)t/\hbar\right]$$
$$\times \begin{pmatrix} F' & k' & F \\ -\bar{M}_{F'} & \bar{q} & \bar{M}_F \end{pmatrix}(2k'+1). \quad (46)$$

If we had $\gamma_{F'}=\gamma_F$, the time-dependent factor would reduce to $\exp\left[i(E_{F'}-E_F-\gamma\bar{q}B\hbar)t/\hbar\right]$ and factor out, after which the summation would reduce to $\delta_{k'k}$, thus yielding the desired reduced matrix element with $k'=k$. In fact, we have $\gamma_{F'}\neq\gamma_F$, the exponential does not factor, and Eq. (46) has a nonzero value, in general, for several values of $k'$ other than $k$.

Physically, this means that the time-dependent operator $\exp(iHt/\hbar)S^{[k]}{}_{\bar{q}}\exp(-iHt/\hbar)$ splits into a number of tensor components $S^{[k']}{}_{\bar{q}}$ with the same value of $\bar{q}$ but with different $k'$. As we know, light emission depends directly only on the alignment and orientation parameters $\langle S^{[k]}{}_q\rangle$ with $k=2$ or 1; we find here that the value of these parameters observed at $t\neq 0$ may depend on the value at $t=0$ of different parameters $S^{[k']}{}_q$ with $k'>2$. This remark extends the range of data on excitation by collision which can be obtained, in principle, from studies of light emission in the presence of external fields. The opportunity for this extension has been mentioned previously (W72) but remains to be exploited experimentally and theoretically.

More specifically, an excited atom with angular momentum $j_i$ can display $2^{k'}$-pole moments with all values of $k'\leq 2j_i$. For $j_i=1$ this means just a dipole and a quadrupole moment, represented by the orientation **O** and alignment **A** which are determined by analysis of the emitted light. For $j_i>1$, the atom may have additional nonzero moments (octupole, hexadecapole, etc.) which are not normally observed

through optical emission. However, the presence of an external field causes the time-modulation of each multipole moment to depend on the multipoles of different orders. The determination of **O** and **A** at time $t$ may thus provide evidence on the $2^{k'}$-pole moments with $k'>2$ generated by collision at $t=0$.

An illustration of this principle has been treated theoretically and verified experimentally by Lombardi (L69), though under different circumstances. Atoms were excited and aligned—but *not* oriented—by electron impact and subjected to an inhomogeneous electric field acting, in effect, crosswise upon their quadrupole moment. The emitted light was found to be circularly polarized, thus showing a nonzero orientation **O** to have arisen from the coupling of an initial alignment **A** with a quadrupolar field.

### III.7. Effect of a Strong Field

An external field, magnetic or electric, is said to be "strong" when its interaction with an atom is comparable to the energy separation of zero-field levels with different angular momenta. In this event, the energy levels in the presence of the field are no longer classified by a total angular momentum quantum number. (Angular momentum quantum numbers may become relevant again at very strong fields, sufficient to overcome internal couplings completely.) It then becomes necessary to solve a Schrödinger equation to establish the connection between the actual stationary states and the basis of uncoupled states which serve to construct the operators $S^{[k]}{}_q$ and to determine their averages. The problem of calculating the alignment and orientation resembles in this respect that which arises when fine and hyperfine structure interactions have comparable strength and which is treated by solving the Schrödinger equation (41). In addition, however, the external field spoils the invariance of the Hamiltonian. Hence the alignment and orientation parameters which relate to the light emission at time $t$ may depend on the initial values of parameters $\langle S^{[k]}{}_q\rangle$ with $k>2$. This more complicated problem does not appear to have been studied.

* Work supported in part by the U.S. Atomic Energy Commission, Contract No. COO-1674-73, and in part by the National Science Foundation.

## REFERENCES

A69    I. D. Abella, in *Progress in Optics*, edited by E. Wolf (North-Holland, Amsterdam, 1969), Vol. VII, p. 158.

A52    K. Alder, Helv. Phys. Acta **25**, 235 (1952).

A64    E. B. Aleksandrov, Opt. Spektrosk. **16**, 377 (1964); **14**, 436 (1963) [transl.: Opt. Spectrosc. **16**, 209 (1964); **14**, 233 (1963)].

A70    H. J. Andrä, Phys. Rev. Lett. **25**, 325 (1970); see also Nucl. Instr. Meth. **90**, 343 (1971).

A71    L. Armstrong, *Theory of Hyperfine Structure of*

|  | *Free Atoms* (Wiley–Interscience, New York, (1971). |
| B59 | J. P. Barrat, J. Phys. Radium **20**, 541 and 633 (1959). |
| B67 | *Beam Foil Spectroscopy*, edited by S. Bashkin (Gordon & Breach, New York, 1967); see also Nucl. Instr. Meth. **90**, 1–368 (1971). |
| BB61 | W. E. Bell and A. L. Bloom, Phys. Rev. Lett. **6**, 280 (1961). |
| Be33 | H. A. Bethe, *Handbuch der Physik*, edited by H. Geiger and K. Scheel (Springer, Berlin, 1933), Vol. XXIV/1, p. 508ff. |
| BH71 | D. J. Burns and W. H. Hancock, Phys. Rev. Lett. **27**, 370 (1971). |
| BH73 | D. J. Burns and W. H. Hancock, J. Opt. Soc. Am. **63**, 241 (1973). |
| Br33 | G. Breit, Rev. Mod. Phys. **5**, 91 (1933). |
| BS71 | H. G. Berry and J. L. Subtil, Phys. Rev. Lett. **27**, 1103 (1971). |
| BSC72 | H. G. Berry, J. L. Subtil, and M. Carré, J. Phys. Radium **33**, 950 (1972). |
| BSPAWG73 | H. G. Berry, J. L. Subtil, E. A. Pinnington, H. J. Andrä, W. Wittman, and A. Gaupp, Phys. Rev. A **7**, 1609 (1973). |
| CS64 | A. Corney and G. W. Series, Proc. Phys. Soc. (London) **83**, 207 and 213 (1964). |
| CTD73 | D. H. Crandall, P. O. Taylor, and G. H. Dunn (private communication). |
| DKW64 | J. N. Dodd, R. D. Kaul, and R. Warrington, Proc. Phys. Soc. London **84**, 176 (1964). |
| DP65 | M. I. D'yakonov and V. I. Perel', Zh. Eksp. Teor. Fiz. **48**, 345 (1965) [Sov. Phys. JETP **21**, 227 (1965)]. |
| DS61 | J. N. Dodd and G. W. Series, Proc. R. Soc. Lond. A**263**, 353 (1961). |
| E57 | A. Edmonds, *Angular Momentum in Quantum Mechanics* (Princeton U. P., Princeton, N.J., 1957). |
| EG72 | E. A. Enemark and A. C. Gallagher, Phys. Rev. A **6**, 192 (1972). |
| EG73 | V. J. Ehlers and A. C. Gallagher, Phys. Rev. A **7**, 1573 (1973). |
| F49 | U. Fano, J. Opt. Soc. Am. **39**, 859 (1949). |
| F57 | U. Fano, Rev. Mod. Phys. **29**, 74 (1957). |
| F61 | P. A. Franken, Phys. Rev. **121**, 508 (1961). |
| F64 | U. Fano, Phys. Rev. **133**, B828 (1964). |
| FR59 | U. Fano and G. Racah, *Irreducible Tensorial Sets* (Academic, New York, 1959). |
| FS67 | D. R. Flower and M. J. Seaton, Proc. Phys. Soc. Lond. **91**, 59 (1967). |
| FS68 | H. Frauenfelder and R. M. Steffen, in *Alpha-, Beta-, and Gamma-Ray Spectroscopy*, edited by K. Siegbahn (North-Holland, Amsterdam, 1968), 2nd ed., Vol. 2, Chap. 19A. |
| H23 | W. Hanle, Naturwissenschaften **11**, 691 (1923). |
| H67 | L. Heroux, Phys. Rev. **153**, 156 (1967). |
| H72 | W. Happer, Rev. Mod. Phys. **44**, 169 (1972). |
| HK67 | D. W. O. Heddle and R. G. W. Keesing, Proc. R. Soc. Lond. A**299**, 212 (1967). |
| HKK65 | H. Hafner, H. Kleinpoppen, and H. Krüger, Phys. Lett. **18**, 270 (1965). |
| HK167 | H. Hafner and H. Kleinpoppen, Z. Physik. **198**, 315 (1967). |
| HN65 | T. Hadeishi and W. Nierenberg, Phys. Rev. Lett. **14**, 891 (1965). |
| HPS73 | S. Haroche, J. A. Paisner, and A. L. Schawlow, Phys. Rev. Lett. **30**, 948 (1973). |
| HS67 | W. Happer and E. B. Saloman, Phys. Rev. **160**, 23 (1967). |
| I71 | M. Inokuti, Rev. Mod. Phys. **43**, 297 (1971). |
| IR71 | R. E. Imhof and F. H. Read, J. Phys. B **4**, 450 (1971). |
| J72 | V. Jacobs, J. Phys. B **5**, 2257 (1972). |
| JCMc70 | D. H. Jaecks, D. H. Crandall, and R. H. McKnight, Phys. Rev. Lett. **25**, 491 (1970). |
| K61 | A. Kastler, C. R. Acad. Sci. (Paris) **252**, 2396 (1961). |
| K66 | R. L. Kelly, Phys. Rev. **147**, 376 (1966). |
| K69 | H. Kleinpoppen, in *Physics of One- and Two-Electron Atoms* (North-Holland, Amsterdam, 1969), p. 613. |
| KAR72 | G. C. King, A. Adams, and F. H. Read, J. Phys. B **5**, L254 (1972). |
| KMS64 | B. Karlsson, E. Matthias, and K. Siegbahn, *Perturbed Angular Correlations* (North-Holland, Amsterdam, 1964). |
| KN67 | H. Kleinpoppen and R. Neugart, Z. Phys. **198**, 321 (1967). |
| L69 | M. Lombardi, J. Phys. Radium **30**, 631 (1969). |
| L73 | L. Q. Lambert, Phys. Rev. B **7**, 1834 (1973). |
| LCA69 | L. Q. Lambert, A. Compaan, and I. D. Abella, Phys. Lett. A **30**, 153 (1969). |
| LCA71 | L. Q. Lambert, A. Compaan, and I. D. Abella, Phys. Rev. A **4**, 2022 (1971). |
| LDAF71 | D. J. Lynch, C. W. Drake, M. J. Alguard, and C. E. Fairchild, Phys. Rev. Lett. **26**, 1211 (1971). |
| LM57 | W. E. Lamb and T. H. Maiman, Phys. Rev. **105**, 573 (1957). |
| M69 | J. Macek, Phys. Rev. Lett. **23**, 1 (1969). |
| M70 | J. Macek, Phys. Rev. A **1**, 618 (1970). |
| McF64 | R. H. McFarland, Phys. Rev. **133**A, 986 (1964). |
| McF67 | R. H. McFarland, Phys. Rev. **156**, 55 (1967). |
| McJ71 | R. McKnight and D. H. Jaecks, Phys. Rev. A **4**, 2281 (1971). |
| MJ71 | J. Macek and D. H. Jaecks, Phys. Rev. A **4**, 2288 (1971). |
| N66 | O. Nedelec, J. Phys. Radium **27**, 660 (1966). |
| O27 | J. R. Oppenheimer, Z. Phys. **43**, 27 (1927). |
| O65 | A. Omont, J. Phys. Radium **26**, 26 (1965). |
| OHK72 | T. W. Ottley, D. R. Denne, and H. Kleinpoppen, Phys. Rev. Lett. **29**, 1646 (1972). |
| PP63 | J. J. Pebay-Peyroula, C. R. Acad. Sci. (Paris) **257**, 3130 (1963). |
| PS58 | I. C. Percival and M. J. Seaton, Philos. Trans. R. Soc. London. A**251**, 113 (1958) |
| RBMW59 | M. Rotenberg, R. Bivins, N. Metropolis, and J. K. Wooten, Jr., *The 3-j and 6-j Symbols* (Technology Press, MIT, Cambridge, Mass., 1959). |
| S26 | H. W. B. Skinner, Proc. R. Soc. Lond. A**112**, 642 (1926). |
| S67 | G. W. Series, Physica **33**, 138 (1967). |
| SA27 | H. W. B. Skinner and E. T. S. Appleyard, Proc. R. Soc. Lond. A**117**, 224 (1927). |
| SFG67 | E. A. Soltysik, A. Y. Fournier, and R. L. Gray, Phys. Rev. **153**, 152 (1967). |
| TAW73 | K. Tillmann, H. J. Andrä and W. Wittmann, Phys. Rev. Lett. **30**, 155 (1973). |
| W72 | J. W. Wooten, University of Nebraska thesis (unpublished). |

VOLUME 33, NUMBER 6    **PHYSICAL REVIEW LETTERS**    5 AUGUST 1974

# Direct Observation of Exchange Scattering by Spin Flip of Polarized Electrons in Excitation of Mercury

G. F. Hanne and J. Kessler

*Physikalisches Institut der Universität Münster, Münster, Germany*
(Received 13 May 1974)

Exchange scattering has been directly observed in the excitation of the $6^3P$ state of mercury by measuring the spin flip of polarized electrons. Between threshold and 8 eV exchange scattering dominates, resulting in a reversal of the direction of polarization of the scattered electrons. Above 8 eV the influence of exchange decreases rapidly for scattering in the forward direction.

In a conventional electron-atom scattering experiment it is impossible to decide whether the observed electron has suffered a "direct" or an "exchange" collision. Thus the direct and exchange cross sections $|f(\Theta)|^2$ and $|g(\Theta)|^2$ cannot be measured separately. This is different, if the colliding electrons are distinguishable, i.e., if their spins are antiparallel and each electron retains its spin direction during the scattering event. In this case, by determining its spin direction, the observed electron can definitely be stated to have, or to have not, suffered an exchange collision.

This is one of the reasons why scattering experiments with polarized particles are so appealing. Such experiments are feasible today, as the

pioneer work of Bederson's group[1] shows. The authors used spin-state-selected alkali atoms as the target and observed the angular distribution and the change of spin state of the recoil atoms. A group at the Joint Institute for Laboratory Astrophysics,[2] which measured cross sections for direct elastic scattering at a few angles, also used state-selected alkali atoms and observed the spins of the scattered electrons. For atoms other than alkalis no such experiments have been made. They can, of course, only be performed with atoms which have unsaturated spins.

Another approach to the separate measurement of direct and exchange scattering can be made by bombarding unpolarized atoms by polarized electrons, but "there has not yet been any scattering

VOLUME 33, NUMBER 6     PHYSICAL REVIEW LETTERS     5 AUGUST 1974

experiment in low-energy atomic physics employing polarized electrons."[3] This paper reports an experimental study of electron exchange in the excitation of mercury atoms by slow polarized electrons. Inelastic exchange scattering leading to the $6^3P$ state of this "two-electron atom" was directly observed by measuring the depolarization of the scattered electrons.

Numerical results for the size of the effects to be expected were not available. We were, however, encouraged by the following considerations to perform the experiment: Theoretical calculations[4] have shown that the cross section for the $6^3P$ excitation in the range up to about 5 eV above threshold should be dominated by exchange. At higher energies, direct scattering becomes increasingly important. (Direct scattering is caused by the fact that there is a certain $^1P_1$ admixture to the $^3P_1$ state.) If this is correct, the polarization of the scattered electrons should be affected below 10 eV as a result of the spin flips occurring by exchange excitation of the $^3P$ state, whereas above about 10 eV the scattered electrons should still have the initial polarization.

The ratio of the final and initial polarization should be $P'/P = -\frac{1}{3}$, if exchange scattering were the only way to excite the $^3P$ state. This is because for electrons with spin $+\frac{1}{2}$ the probability for excitation of the atomic substates $M_S = +1$ (opposite spin directions of observed and incident electron) is twice as large as that for excitation of $M_S = 0$ (equal spin directions of these electrons), as can be shown by a simple calculation. $M_S = -1$ cannot be excited by electrons with spin $+\frac{1}{2}$ if spin conservation is valid. Since there is, however, a certain percentage of direct scattering due to the $^1P_1$ admixture, the probability of non–spin-flip processes is somewhat increased,

so that the ratio $P'/P$ is shifted to more positive values. The exact value of $P'/P$ yields the extent to which direct and exchange processes contribute to the excitation [cf. Eq. (2) at the end].

In the present experiment the depolarization of the electrons scattered in the forward direction has been determined. For these electrons exchange scattering is the only possible spin-flip mechanism; no other spin flips due to spin-orbit coupling occur. This can be concluded from the measurements of spin polarization resulting from elastic and inelastic electron scattering on mercury.[5,6] They showed that the influence of spin-orbit coupling is appreciable only for scattering angles larger than 30°.

Figure 1 gives a schematic diagram of the apparatus. Transversely polarized electrons resulting from elastic scattering by a mercury-vapor beam[7] ($E = 80$ eV, $\Theta = 80°$, $P = 0.22$) are decelerated to 5–20 eV and focused on a second mercury-vapor beam (target density $10^{-2}$–$10^{-3}$ Torr). The intensity of the incident polarized electron beam measured after deceleration to 5 eV and selection by the spectrometer was approximately $5 \times 10^{-11}$ A. The counting rate of the inelastically scattered electrons finally was 60–200 counts/min.

Figure 2 shows the experimental results. The error bars include the statistical error and the reproducibility of the measured polarizations $P$ and $P'$, as well as the reproducibility of the energy. At energies below 8 eV strong depolariza-

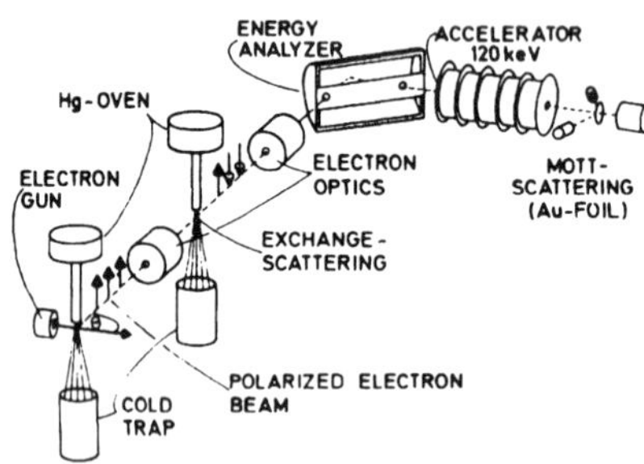


FIG. 1. Schematic diagram of the apparatus.

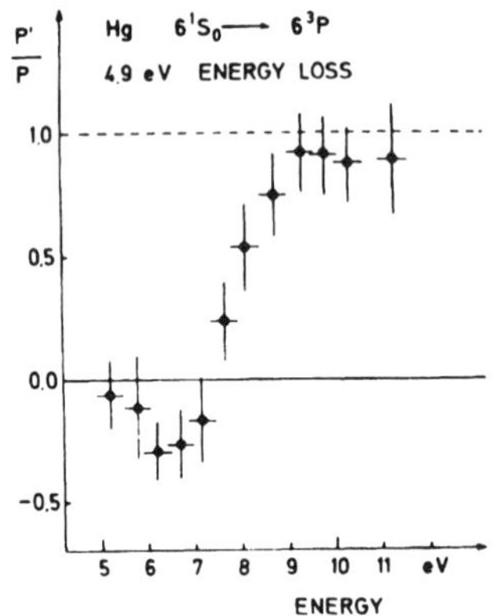


FIG. 2. Measured values of the depolarization ratios versus incident energy.

VOLUME 33, NUMBER 6    PHYSICAL REVIEW LETTERS    5 AUGUST 1974

tion or even reversal of the direction of polarization has been observed. This agrees with the theoretical prediction that exchange scattering dominates at these energies. Above 8 eV, however, the number of spin-flip processes decreases more rapidly than expected from Ref. 4: Above 9 eV significant depolarization could no longer be observed.

We want to point out that the theoretical calculation[4] refers to the total excitation cross section only. Theoretical predictions of exchange cross sections at small angles are controversial and depend strongly on the approximation made.[8,9] The present measurements show that the influence of exchange scattering is very strong even at small angles.

Apart from improving the accuracy we hope to resolve the $^3P_0$, $^3P_1$, and $^3P_2$ transitions in the forthcoming measurements. Making the assump-

tion that the spin-orbit relaxation time is long compared with the excitation time (which is not well-established for Hg), we found for transversely polarized electrons the depolarization ratios $P'/P$ to be

$$-\frac{|g_0|^2}{|g_0|^2+2|g_1|^2}, \quad -\frac{|g_1|^2}{|g_0|^2+2|g_1|^2},$$

$$-\frac{3|g_1|^2+2|g_0|^2}{5|g_0|^2+10|g_1|^2},$$

(1)

for $^3P_0$, $^3P_1$, and $^3P_2$, respectively. Here $g_0$ and $g_1$ are the amplitudes for inelastic exchange scattering with excitation of the $M_L = 0$ and $M_L = \pm 1$ substates of $^3P$. If the polarization is parallel to the quantization axis the rotation of the coordinates transforms $|g_1|^2$ and $|g_0|^2$ into $|g_0|^2$ and $2|g_1|^2 - |g_0|^2$, respectively. When the admixture of the $^1P_1$ state to the $^3P_1$ state is taken into account, the second of the above expressions is replaced by

(2)

$$\frac{P'}{P} = \frac{\beta^2(|f_0{}^s - g_0{}^s|^2 + 2|f_1{}^s - g_1{}^s|^2) - \alpha^2|g_1|^2}{\beta^2(|f_0{}^s - g_0{}^s|^2 + 2|f_1{}^s - g_1{}^s|^2) + \alpha^2(|g_0|^2 + 2|g_1|^2)},$$

where $f^s$ and $g^s$ are the direct and exchange scattering amplitudes for excitation of the $^1P$ state, and $\alpha$ and $\beta$ are the coupling coefficients.[4] Since at the 4.9-eV energy loss studied here the transitions to the $^3P_1$ state are dominant, it was mainly expression (2) which was measured in the present experiment. The contribution of the other channels ($^3P_0$ and $^3P_2$) is roughly estimated to be less than 40% (the overall energy resolution was 0.8 eV).

At threshold, according to angular momentum conservation, only the amplitudes with $M_L = 0$ can be excited. Thus the limiting value for the $^3P_1$ state should be $P'/P = 0$, which is obviously confirmed by the experiment. Resolution of the $^3P_0$ and $^3P_2$ transitions would yield more information about $g_1$ and $g_0$.

We gratefully acknowledge support by the Deutsche Forschungsgemeinschaft.

[1]B. Bederson, in *Atomic Physics 3*, edited by S. J. Smith and G. K. Walters (Plenum, New York, 1973), Vol. 3, p. 401, and further references quoted therein.

[2]D. Hils, M. V. McCusker, H. Kleinpoppen, and S. J. Smith, Phys. Rev. Lett. **29**, 398 (1972).

[3]B. Bederson, Comments At. Mol. Phys. **1**, 65 (1969).

[4]J. C. McConnell and B. L. Moiseiwitsch, J. Phys. B: Proc. Phys. Soc., London **1**, 406 (1968).

[5]W. Eitel and J. Kessler, Z. Phys. **241**, 355 (1971).

[6]J. Kessler, Rev. Mod. Phys. **41**, 3 (1969).

[7]K. Jost and J. Kessler, Z. Phys. **195**, 1 (1966).

[8]J. C. Steelhammer and S. Lipsky, J. Chem. Phys. **53**, 1445 (1970).

[9]R. A. Bonham, J. Chem. Phys. **57**, 1604 (1972).

J. Phys. B: Atom. Molec. Phys., Vol. 7, No. 16, 1974. Printed in Great Britain. © 1974

# Theory of electron scattering from laser excited atoms

Joseph Macek†§ and I V Hertel‡

† Fachbereich Physik der Freien Universität Berlin, 1 Berlin 33, Boltzmannstr 20, Germany
‡ Fachbereich Physik der Universität Trier-Kaiserslautern, 675 Kaiserslautern, Pfaffenbergstr, Germany

Received 13 May 1974

**Abstract.** An expression is derived for the intensity of electrons inelastically scattered from an atomic state prepared by optical pumping. The expression is a sum of state multipoles, the coefficient of each multipole being a spherical harmonic in the polar coordinates of the light direction for circularly polarized light, or the direction of the electric vector for linearly polarized light times a weight function depending on the population of the initial state. These weight functions are given in closed form for both linearly and circularly polarized light.

## 1. Introduction

Recent experiments by Hertel and Stoll (1974a) have demonstrated the feasibility of measuring elastic, inelastic and superelastic electron scattering from excited atomic states prepared by optical pumping with laser light. Their corresponding theoretical treatment (Hertel and Stoll 1974b) showed that the scattered electron intensity depended not only upon excitation cross sections, but also upon non-diagonal density matrix elements. Such experiments thus provide much more detailed information on scattering processes than do conventional cross section measurements. The earlier theoretical treatment (Hertel and Stoll 1974b) which used non-reduced density matrix elements, does not bring out clearly all of the possibilities provided by optical pumping. A more transparent picture emerges if one uses the language of state multipoles‖ (or equivalently, irreducible density matrix components). This paper discusses the experiment of Hertel and Stoll (1974a) in the language of state multipoles as developed by Fano and Macek (1973).

We show that all state multipoles can be determined by suitable measurements with a variety of choices for the direction and polarization of the pumping light. Section 2 describes the general theory and § 3 considers specific polarizations of the pumping light. There we give a simple expression for the population of the excited state prepared by optical pumping with linearly polarized light. Our expression turns out to be the square of a Clebsh–Gordan coefficient and this enables us to perform the average over magnetic substates in terms of a recoupling coefficient. Finally we discuss the possibilities for performing suitable normalizations to compare the scattering intensities at different geometries and polarizations.

§ Permanent address: Behlen Laboratory of Physics, The University of Nebraska, Lincoln, Neb. 68506. Supported in part by National Science Foundation Grant GP 39310.
‖ The book by Brink and Satchler (1968) gives a concise introduction to state multipoles.

2174        *J Macek and I V Hertel*

## 2. Basic Theory

### 2.1. General expression for the scattered electron intensity

Hertel and Stoll (1974b) considered the superelastic scattering from an optically pumped excited level $n_i$ to another level $n_j$ in terms of the time inverse scattering process $n_j \to n_i$. Since the $j$th level is not state selected except by measuring the energy of the superelastically scattered electrons, the initial level $n_j$ of the time inverse process is taken to have an isotropic distribution. The inverse scattering process is initially described in terms of scattering amplitudes $f_{ij}$ for transitions between eigenstates $i$ and $j$ of the $n_i$ and $n_j$ levels. One can view such scattering as preparing a particular linear combination of substates of $n_i$, since the asymptotic wavefunction has the form

$$\exp(i\mathbf{k}_j \cdot \mathbf{r})\psi_{m_j} + \frac{1}{r}\exp(ik_i r) \sum_{m_i} \psi_{m_i} f_{im_i,jm_j} = \exp(i\mathbf{k}_j \cdot \mathbf{r})\psi_{m_j} + \frac{1}{r}\exp(ik_i r)\Psi_i \tag{1}$$

where $\Psi_i$ is the wavefunction

$$\Psi_i = \sum_{m_i} \psi_{m_i} f_{im_i,jm_j} \tag{2}$$

and $\psi_{im_i}$ are the eigenfunctions corresponding to the magnetic substates of level $n_i$.

In the actual scattering process the level $n_i$ is prepared in a particular state by optical pumping. With right or left circularly polarized light this state is an eigenstate with magnetic quantum number $M_F = \pm F$, where $F$ is the total angular momentum quantum number of level $n_i$, and the axis of quantization is along the axis of the laser light. Then the scattered electron intensity is proportional to the component of the eigenstate $|FM_F\rangle$ contained in $\Psi_i$. Thus $I \propto |(FM_F|\Psi_i)|^2$. Since level $n_j$ is not state selected, the squared matrix element is to be averaged over all unresolved sublevels and substates of $n_j$. Furthermore, the electrons are not spin polarized nor is their final spin measured. Accordingly the squared matrix element is averaged over initial and final spins. For normalization purposes we also divide the squared matrix element by the similarly averaged $|(\Psi_i|\Psi_i)|^2$, that is, we normalize to an isotropic distribution.

We indicate all of these operations by enclosing the squared matrix element in brackets $\langle \ \rangle$. Thus we have

$$I = C\langle |(FM_F|\Psi_i)|^2 \rangle \tag{3}$$

where $C$ is a contant consisting of multiplicative factors such as solid angle, detection efficiencies, the number of atoms in the optically pumped level $n_i$, the flux of the incident electrons, and the total (at fixed scattering angle) differential superelastic scattering cross section $\sigma_{n_j n_i}$ for the transition $n_i \to n_j$. The last factor $\sigma_{n_j n_i}$ incorporates the sum over final states and average over initial states appropriate to excitation by scattering from an initial isotropic distribution. All of the effects of the anisotropy of the pumped level are contained in $\langle |(FM_F|\Psi_i)|^2 \rangle$.

Our objective is to express $I$ in terms of state multipoles defined as expectation values of irreducible tensors constructed from components of angular momentum vectors. To this end we (a) write equation (3) as the expectation value of the projection operator $\tau = |FM_F\rangle(FM_F|$, (b) write $\tau$ as a sum of its irreducible components $\tau_q^{[k]}$, (c) apply the Wigner–Eckart theorem to relate $\tau_q^{[k]}$ to operators $T_q^{[k]}$ constructed from components of angular momentum operators, and (d) transform these operators from the reference frame defined by the pumping light to the frame defined by the collision.

Equation (3) may be written in the form

$$I = C\langle(\Psi_i|\tau|\Psi_i)\rangle \tag{4}$$

where

$$\tau = |FM_F)(FM_F| \tag{5}$$

with $M_F = \pm F$ for circularly polarized light. For linearly polarized light the analysis of Hertel and Stoll (1974b) shows that $\tau$ is given by

$$\tau = \sum_{M_F} |FM_F)W(M_F)(FM_F| \tag{6}$$

where the axis of quantization is along the direction of linear polarization, and $W(M_F)$ is the relative population of the various $M_F$ eigenstates. The weight function $W(M_F)$ is calculated in § 2.

Equations (4), (5) and (6) have a simple interpretation. The time inverse scattering prepares a state of atom in the level $n_i$. Alternatively, optical pumping prepares a different state of the same level described by the population $W(M_F)$. The electron scattering from the optically pumped state is proportional to the degree of overlap of these two states as represented mathematically by equation (4).

The average denoted by brackets in equation (4) may be variously interpreted. The mean value is independent of the representation of $\Psi_i$, in particular $\Psi_i$ may be represented in eigenstates and the sum over amplitudes in its definition equation (2) incorporated into the average. Then $\langle(\Psi_i|\tau|\Psi_i)\rangle$ is written $\langle(i'|\tau|i)\rangle$ where $i'$ and $i$ refer to eigenstates and the average is now interpreted conventionally in terms of the density matrix $\rho$ for the population produced by the time inverse scattering process $\langle\tau\rangle = \mathrm{tr}\,\rho\tau$. For conciseness we employ the notation $\langle\tau\rangle$ for averaged quantities unless the expanded notation is needed for clarity. The matrix elements of $\rho$ can be given explicitly in terms of scattering amplitudes

$$\rho_{m_im_i} = \sum_{\substack{\text{unobserved}}} f_{im'_i,jm_j} f^*_{im_i,jm_j} \bigg/ \sum_{\substack{mi,\\ \text{unobserved}}} |f_{im_i,jm_j}|^2.$$

The irreducible components $\tau_q^{[k]}$ are related to $\tau$ by formally rewriting equation (6):

$$\tau = \sum_{kq} \tau_q^{[k]} \sum_{M_F} (kq|F-M_FFM_F)W(M_F)(-)^{k-F-M_F} \tag{7}$$

$$\tau_q^{[k]} = \sum_{M'_F,M''_F} |FM'_F)(FM''_F|(F-M''_FFM'_F|kq)(-)^{k-F-M''_F}. \tag{8}$$

Equation (7) is verified by direct substitution of equation (8) and the use of orthogonality relations for Clebsch–Gordan coefficients.

$\tau^{[k]}$ is seen to be an irreducible tensor operator, since it is the reduced form of the product of $|FM'_F)$ and $(-)^{M_F}(FM''_F|$ which transform under rotation like irreducible tensor operators of rank $F$. It should be noted that only $q = 0$ terms occur in equation (7).

Substituting equation (7) into equation (4) gives an expression for $I$ in terms of mean values of $\tau_q^{[k]}$ in the photon frame. Equation (4) becomes

$$I = C\sum_k \mathcal{W}(k)\langle\tau_0^{[k]}(\mathrm{ph})\rangle \tag{9}$$

2176        *J Macek and I V Hertel*

where

$$\mathscr{W}(k) = \sum_{M_F} (-)^{k-F-M_F} W(M_F)(F-M_F F M_F | kq) \tag{10}$$

and $\tau(\mathrm{ph})$ denotes the photon frame.

Equation (9) expresses the scattered electron intensity in terms of parameters $\langle \tau_0^{[k]}(\mathrm{ph}) \rangle$. The parameters are mean values of formally defined projection operators which may be calculated theoretically if $\Psi_i$ is known, but they do not bring out clearly the physical interpreation of the multipole moments. Furthermore, the definition is in terms of eigenstates to the quantum theory, consequently the multipoles appear to have no classical analogue. On the other hand the mean values of operators constructed from components of angular momentum operators relate to more readily interpreted quantities such as the mean angular momentum transferred to the atom during the collision (Fano and Macek 1973, Eminyan *et al* 1973). In addition, the mean values have a classical analogue; indeed in the classical calculations which represent atomic states as an ensemble of orbits (Burgess *et al* 1968) one could average the irreducible tensors constructed from classical angular momentum components over an ensemble of orbits. Accordingly we define real tensors $T_{q\pm}^{[k]}$ constructed from the components of angular momentum operators.

The tensors are defined by 'polarizing' (Falkoff and Uhlenbeck 1959, Brink and Satchler 1958) the real (Fano 1960) solid harmonics

$$r^k C_{kq\pm} = r^k (4\pi/(2k+1))^{1/2} Y_{q\pm}^{[k]}(\theta, \phi)$$
$$T_{q\pm}^{[k]}(F, F, \ldots) = \sum_{ij\ldots} (F_i F_j \ldots) \nabla_i \nabla_j \ldots r^k C_{kq\pm} \tag{11}$$

where $i$ and $j$ denote real rectangular coordinates. These operators are real polynomials in the components of $F$. The real quantities $Y_{q\pm}^{[k]}$ and $T_{q\pm}^{[k]}$ are used in preference to the more conventional complex quantities $Y_q^{(k)}$ and $T_q^{[k]}$, since the real set exhibits two important symmetry properties, hermiticity and reflection symmetry, explicitly.

Letting $p$ denote $\pm 1$ the real tensor operators $T_{qp}^{[k]}$ are connected to the usual complex ones by

$$T_{qp}^{[k]} = 1/\sqrt{2}(\mathrm{i})^{(p-1)/2}[(-)^q T_q^{[k]} + p T_{\bar q}^{[k]}] \qquad q > 0$$
$$T_{0+}^{[k]} = T_0^{[k]} \qquad T_{0-}^{[k]} = 0 \qquad q = 0. \tag{12}$$

The real spherical harmonics become with these relations:

$$\left. \begin{array}{l} Y_{q+}^{(k)}(\theta, \phi) = \pi^{-1/2}\mathscr{P}_{kq}(\cos\theta)\cos q\phi \\ Y_{q-}^{(k)}(\theta, \phi) = \pi^{-1/2}\mathscr{P}_{kq}(\cos\theta)\sin q\phi \end{array} \right\} \qquad q > 0 \tag{12a}$$

$$Y_{k+}^{(k)}(\theta, \phi) = (2\pi)^{-1/2}\mathscr{P}_{kq}(\cos\theta).$$

Here the $\mathscr{P}_{kq}$ are the normalized Legendre polynomials of Bethe and Salpeter (1957) which relate to the usual complex spherical harmonics by

$$Y_q^{(k)}(\theta, \phi) = (2\pi)^{-1/2}(-)^q \mathscr{P}_{kq}(\cos\theta)\exp(\mathrm{i}q\phi) \qquad q \geqslant 0.$$

One sees directly that the $Y_{qp}^{(k)}$ are real, and that they transform as $Y_{qp}^{(k)} \to p Y_{qp}^{(k)}$ under reflections in the $zx$ plane. The tensors $T_{q\pm}^{[k]}$ transform as $T_{qp}^{[k]} \to p(-)^k T_{qp}^{[k]}$ because each pair of $F_i$ and $\nabla_i$ in equation (11) changes sign under reflection.

In the third step (c) we apply the Wigner–Eckart theorem to replace the tensors $\tau_{q\pm}^{[k]}(\mathrm{ph})$ by the tensors $T_{q\pm}^{[k]}(\mathrm{ph})$. Specifically

$$\langle (i|\tau_{qp}^{[k]}|i)\rangle = \frac{(F\|\tau^{[k]}\|F)}{(F\|T^{[k]}\|F)}\langle(i'|T_{qp}^{[k]}|i)\rangle. \tag{13}$$

Equation (13) allows considerable flexibility in choosing $T_{qp}^{[k]}$. We use this freedom to choose alternate tensors $T_{qp}^{[k]}$ constructed from orbital, electronic or total angular momentum operators, or even the operator $\tau_{qp}^{[k]}$. This latter choice is acceptable (Happer 1972) since $\tau_{qp}^{[k]}$ is a projection operator onto combinations of eigenstates of the level $n_i$, but it is otherwise independent of atomic dynamics.

The calculation of the ratio of reduced matrix elements is most easily carried out in the complex representation. The operator $T_k^{[k]}$ is proportional to the $k$th power of the stepping operator $J_+ = J_x + iJ_y$. The operator $T_k^{[k]} = 2^{-k}\sqrt{[(2k)!]}J_+^k$ raises the magnetic quantum number of $|F - F)$ $k$ times, ie $T_k^{[k]}|F, -F) = C|F, -F+k)$, where

$$C = 2^{-k}[(2k)!]^{1/2}(-)^k\left[\frac{(2F)!k!}{(2F-k)!}\right]^{1/2}. \tag{14}$$

The reduced matrix element of $T^{[k]}$ is then given by (Fano and Racah p 79, 1968)

$$(F\|T^{[k]}\|F) = \frac{(F, -F+k|T_k^{[k]}|F, -F).(2F+1)^{1/2}}{(F, -F+k|F-F, kk)}$$

$$= 2^{-k}(-)^{2F}k!\left[\frac{(2F+k+1)!}{(2F-k)!}\right]^{1/2}. \tag{15}$$

The reduced matrix element of $\tau_q^{[k]}$ is easily seen to be

$$(-1)^{k-F-M_F}(k0|FM_F F-M_F)(2F+1)^{1/2}/(FM_F|FM_F k0) = (2k+1)^{1/2}.$$

The ratio of reduced matrix elements is then

$$\mathscr{V}^{(k)}(F) = \frac{(F\|\tau^{[k]}\|F)}{(F\|T^{[k]}\|F)} = \frac{2^k(2k+1)^{1/2}}{k!}\left[\frac{(2F-k)!}{(2F+k+1)!}\right]^{1/2}. \tag{16}$$

Substituting equations (16) and (13) into equation (9) gives the scattered electron intensity in terms of parameters $\langle T_0^{[k]}(\mathrm{Ph})\rangle$ defined in the photon frame.

$$I = C\sum_k \mathscr{W}(k)\mathscr{V}^{(k)}(F)\langle T_0^{[k]}(\mathrm{Ph})\rangle. \tag{17}$$

The final step consists of transforming to the collision frame. Since only $q = 0$ terms occur in equation (17) transformation to the collision frame introduces spherical harmonics rather than the more general rotation functions. Then

$$\langle T_0^{[k]}(\mathrm{ph})\rangle = \sqrt{(4\pi)}(2k+1)^{-1/2}\sum_{pq=0}^{k}\langle T_{qp}^{[k]}(\mathrm{col})\rangle Y_{qp}^{[k]}(\theta, \phi) \tag{18}$$

where $\theta, \phi$ are the spherical coordinates of the laser light axis for circularly polarized light and of the electric vector for linearly polarized light.

Since the collision system is invariant to reflections in the scattering plane we have

$$\langle T_q^{[k]}(\mathrm{col})\rangle = 0 \qquad \text{for } k = \text{even}$$
$$\langle T_q^{[k]}(\mathrm{col})\rangle - 0 \qquad \text{for } k = \text{odd}. \tag{19}$$

2178        *J Macek and I V Hertel*

Equation (17) together with the transformation relation (equation (18)) relates the scattered electron intensity to the $2(F+1)^2-1$, $F =$ integer or $2(F+3/2)(F+1/2)-1$, $F =$ half integer, state multipoles $\langle T_{qp}^{[k]}\rangle$. The scaler parameter $\langle T_0^{[0]}\rangle$ is used for normalization purposes and is not an independent parameter. With our normalization it just equals unity.

Equations (17) and (18) incorporate several numerical factors such as the normalization constants of the irreducible tensors and spherical harmonics as well as the explicit factors $\mathscr{V}(k)$ and $\mathscr{V}^{(k)}(F)$ which complicate the expression. For fluorescence excited by collisions, where only $k = 0$, 1 and 2 enter, the corresponding equation takes a simpler form when the fundamental parameters, the alignment and orientation, are defined as mean values of irreducible tensors with non-standard normalization (Fano and Macek 1973). We have not found a similarly convenient set for the general multipole moments, but for reference we record the relation of the $k = 0$, 1 and 2 parameters in equation (18) to the orientation and alignment parameters. They are

$$O_{1-}^{\text{col}} = \langle T_{1-}^{[1]}(\text{col})\rangle/F(F+1)$$

$$A_0^{\text{col}} = 2\langle T_0^{[2]}(\text{col})\rangle/F(F+1)$$

$$A_{1+}^{\text{col}} = 2(3)^{-1/2}\langle T_{1+}^{[2]}(\text{col})\rangle/F(F+1)$$

$$A_{2+}^{\text{col}} = 2(3)^{-1/2}\langle T_{2+}^{[2]}(\text{col})\rangle/F(F+1).$$

$$(20)$$

### 2.2. The hypothesis of Percival and Seaton

In their theory of the polarization of impact radiation, Percival and Seaton (1958) supposed that the hyperfine interaction plays no essential role during the collision. Then the state parameters do not depend in any significant way upon the nuclear spin, but only upon the electronic angular momentum. Accordingly the nuclear spin can be factored from the mean values describing the state. *Then the upper limit on k is either 2J or 2F whichever is less.*

To factor the nuclear part from the parameters we note that the average of $\langle \tau_q^{[k]}\rangle_{\text{nuc}}$ over the nuclear variables is itself an irreducible tensor of rank $k$ acting on the electronic coordinates only. Secondly the wavefunction $\Psi_i$ can be written in any representation, in particular the $JM_JIM_I$ representation. Thirdly the flexibility of equation (13) allows the tensors $T_q^{[k]}$ to be constructed from the components of $J$ rather than $F$. Then, since the tensors are nuclear spin scaler the dependence of $\Psi$ on $I$ in mean values $\langle T_q^{[k]}\rangle$ integrates out. We need only evaluate the reduced matrix element $(J\|\langle\tau^{[k]}\rangle_{\text{nuc}}\|J)$.

The operator $\tau^{[k]}$ is formed with the coupling scheme $[(IJ)F(IJ)F]^{(k)}$, but the reduced matrix element has the scheme $[(JJ)k(II)0]^{(k)}$. The recoupling coefficient

$$((IJ)F(IJ)F|(II)0(JJ)k)^{(k)}$$

effects the transformation between the two coupling schemes. Thus

$$(J\|\langle\tau^{[k]}\rangle_{\text{nuc}}\|J)$$

$$= ((IJ)F(IJ)F|(II)0(JJ)k)^{(k)}(J\|\tau^{[k]}\|J)(I\|1^{[0]}\|I)$$

$$= (-)^{I+J+F}(2F+1)\begin{Bmatrix} F & F & k \\ J & J & I \end{Bmatrix}(J\|\tau^{[k]}(J)\|J)$$

$$(21)$$

where $\tau^{[k]}(J)$ is constructed as in equation (8) from eigenkets of $J$ rather than $F$ and we

*Theory of electron scattering*      2179

have used the definition† $(I\|1^0\|I) = (2I+1)^{1/2}$. Similarly, the eigenfunction $\psi_i$ and amplitudes $f_{ij}$ in the definition of $\Psi_i$ refer to eigenstates of electronic angular momentum while the cross section $\sigma_{ij}$ in the multiplicative constant $c$ is the total superelastic cross section for excitation of a particular $J'$ level from an isotropic initial $J$ level. In essence one calculates ignoring hyperfine structure except for the recoupling coefficient in equation (20) and $\mathscr{W}(k)$ in equation (10). Since equation (13) with $F$ replaced by $J$ relates $\tau_q^{[k]}(J)$ to the new $T_q^{[k]}$ the expression for $I$ is exactly like equation (17) with $\mathscr{V}^{(k)}(F)$ replaced by

$$(-)^{I+J+F}(2F+1)\begin{Bmatrix} F & F & k \\ J & J & I \end{Bmatrix}\mathscr{V}^{(k)}(J) \tag{22}$$

and with $T_{qp}^{[k]}$ constructed from the elements of $J$ rather than $F = J + S$.

Similarly if the spin–orbit interaction is sufficiently weak that it also plays no essential role, $(J\|\tau^{[k]}(J)\|J)$ may be further factored to obtain

$$(J\|\langle\tau^{[k]}(J)\rangle_{\text{spin}}\|J) = (-)^{S+L+J}(2J+1)\begin{Bmatrix} J & J & k \\ L & L & S \end{Bmatrix}(L\|\tau^{[k]}\|L) \tag{23}$$

where $L$ is the orbital angular momentum. The corresponding operators $T_q^{[k]}$ are constructed from elements of $L$ rather than $J = L + S$, while $\sigma_{ij}$ is the total (at a specific scattering angle) superelastic cross section for excitation from an isotropic $L$ level. The factor $\mathscr{V}^{(k)}(F)$ in equation (17) is replaced by

$$(-)^{I+L+S+2J+F}(2F+1)(2J+1)\begin{Bmatrix} F & F & k \\ J & J & I \end{Bmatrix}\begin{Bmatrix} J & J & k \\ L & L & S \end{Bmatrix}\mathscr{V}^{(k)}(L). \tag{24}$$

*The largest value $k_{\max}$ of $k$ is now the smallest of $2L$, $2J$ or $2F$.*

The fine and hyperfine interactions are frequently comparable in strength. Under these circumstances $J$ is no longer a good quantum number, and one must determine the eigenstates by a dynamical calculation. The uncoupled states relate to the eigenstates by a unitary matrix $(IM_I SM_S IM_J|\alpha FM_F)$ diagonalizing the combined fine and hyperfine Hamiltonian. Here $\mathscr{V}^{(k)}(F)$ is replaced by (Fano and Macek 1973)

$$[(2I+1)(2S+1)]^{-1/2} \sum_{\substack{M_I M_S M_L \\ M'_L M_F M'_F}} (-)^{I-M_I+S-M_S}(FFkq|F-M_F FM_F)(\alpha FM_F|IM_I SM_S LM_L)$$

$$\times (\alpha F - M'_F|I - M_I S - M_S L - M'_L)(L - M'_L LM_L|LLkq)\mathscr{V}^{(k)}(L). \tag{25}$$

## 2.3. Connexion with the theory of Hertel and Stoll

To connect with the theory previously given, we rewrite equations 3.10 and 3.15 of Hertel and Stoll (1974b)

$$I = \frac{k_0}{k_i}\left(\sum_\mu q_{\mu\mu}(\theta)\right) \sum_{M\bar{M}} q_{M\bar{M}}(\theta)\bigg/\left(\sum_\mu q_{\mu\mu}(\theta)\right) K_{M\bar{M}} \tag{26}$$

with

$$K_{M\bar{M}} = \sum_{M_S M_I} (SM_S LM|JM_S + M)(SM_S L\bar{M}|JM_S + \bar{M})$$

$$\times (IM_I JM_S + M|FM_I + M_S + M)(IM_I JM_S + \bar{M}|FM_I + M_S + \bar{M})$$

$$\times \sum W(M_F)\mathscr{D}^F_{M_S + \bar{M} + M_I, M_F}\mathscr{D}^{F*}_{M_S + M + M_I, M_F}. \tag{27}$$

† We use the reduced matrix elements as defined by Fano and Racah (1959).

2180          *J Macek and I V Hertel*

The product of the rotation matrices can be recoupled to give a sum of single rotation functions. The subsequent summation over $M_F$ leads to

$$\sum_{M_F} \ldots = (-)^{M+M_S+M_I-F} \sum_k \mathscr{D}^k_{\overline{M}-M,0}\mathscr{W}(k)(FM_S+\overline{M}+M_IF-M_S-M-M_I|k\overline{M}-M)$$

where $\mathscr{W}(k)$ is given by equation (10).

Next we carry out the sum over $M_I$ and then over $M_S$ of equation (27) which each time gives a recoupling coefficient. Thus we obtain

$$K_{M\overline{M}} = \sum_{k=0}^{2L} \mathscr{W}(k)(-)^M \mathscr{D}^k_{\overline{M}-M,0}(L\overline{M}L-M|k\overline{M}-M)$$

$$\times (-)^{F+2J+2k+I+S}(2J+1)(2F+1)\begin{Bmatrix} J & J & k \\ L & L & S \end{Bmatrix}\begin{Bmatrix} F & F & k \\ J & J & I \end{Bmatrix}. \tag{28}$$

We insert this in equation (26) and note that the normalized density matrix element $q_{M\overline{M}}(\theta)/\Sigma_\mu q_{\mu\mu}(\theta)$ defined in equation 3.12 by Hertel and Stoll (1974b) just equals the mean value of the projection operator $|LM)(LM|$ representing one reducible component of $\tau(L)$.

We then may write equation (26) as

$$I = \frac{k_0}{k_i}\left(\sum_\mu q_{\mu\mu}(\theta)\right) \cdot \sum_{k=0}^{2L} (-)^{F+2J+2k+I+S+L}(2J+1)(2F+1)$$

$$\times \begin{Bmatrix} J & J & k \\ L & L & S \end{Bmatrix}\begin{Bmatrix} F & F & k \\ J & J & I \end{Bmatrix}\mathscr{W}(k)\sum_q \langle\tau_q^{[k]}(L,\text{col})\rangle\mathscr{D}^k_{q,0}. \tag{29}$$

Here the state multipoles in the collision frame are

$$\langle\tau_q^{[k]}(L,\text{col})\rangle = \sum_M (-)^{k-L-M}(L-MLM+q|kq)\frac{q_{MM+q}}{\Sigma_\mu q_{\mu\mu}}. \tag{30}$$

They may be transformed to the photon frame which leads to

$$I = C(2J+1)(2F+1)(-1)^{F+2J+I+S+L}$$

$$\times \sum_{k=0}^{2L} \begin{Bmatrix} F & F & k \\ J & J & I \end{Bmatrix}\begin{Bmatrix} J & J & k \\ L & L & S \end{Bmatrix}\mathscr{W}(k)\langle\tau_0^{[k]}(L,\text{Ph})\rangle \tag{31}$$

Equation (31) may be identified with equation (8). The factorized form of $\langle\tau_0^{[k]}\rangle$ in equation (31) corresponds to equation (24) and reflects the fact that Hertel and Stoll explicitly used the Percival–Seaton hypothesis.

For the case of an S $\rightarrow$ P transition (P being the laser-excited state) we give explicitly the multipole components $\langle\tau_q^{[k]}\rangle$ in table 1.


### 3. Optical pumping populations

*3.1. Circularly polarized light*

Optical pumping with circularly polarized light prepares an excited state population with $W(M_F)$ given by

$$W(M_F) = \delta_{M_F, \pm F} \tag{32}$$

*Theory of electron scattering*    2181

**Table 1.** Non zero multipole moment components in the collision frame:

$$C = 1/\sum_{\mu} q_{\mu\mu} \qquad q_{M\bar{M}}(\theta) = \sum_{\mathscr{S}} (2\mathscr{S} + 1)/4 f_M^{\mathscr{S}} f_{\bar{M}}^{\mathscr{S}*}.$$

$$\langle \tau_0^{[0]} \rangle = \frac{1}{\sqrt{3}} C \cdot \sum_M q_{MM} = 1/\sqrt{3}$$

$$\langle \tau_{1-}^{[1]} \rangle = 2 \cdot \sqrt{2} C(1011|11)\, \mathrm{Im}\, q_{01} = -2C\, \mathrm{Im}\, q_{01} = \sqrt{2}\, 0_{1-}^{\mathrm{col}}$$

$$\langle \tau_{0+}^{[2]} \rangle = -C[(1010|20)q_{00} - 2(111-1|20)q_{11}] = -\sqrt{\tfrac{2}{3}} C(q_{00} - q_{11}) = \sqrt{\tfrac{2}{3}} A_0^{\mathrm{col}}$$

$$\langle \tau_{1+}^{[2]} \rangle = -2\sqrt{2} C(1011|21)\, \mathrm{Re}\, q_{01} = -2C\, \mathrm{Re}\, q_{01} = \sqrt{2} A_{1+}^{\mathrm{col}}$$

$$\langle \tau_{2+}^{[2]} \rangle = -\sqrt{2} C(1111|22)q_{11} = -\sqrt{2} C q_{11} = \sqrt{2} A_{2+}^{\mathrm{col}}$$

where the sign of $F$ is $+$ for right and $-$ for left circularly polarized light with $Z$ axis along the direction of light propagation.

From equation (10) $\mathscr{W}(k)$ is simply given by

$$\mathscr{W}\pm(k) = (-)^{k-F\pm F}(F \mp F\, F \pm F|k0). \tag{33}$$

Here again $\pm$ refers to right and left circular polarization.

Since no $\mathscr{W}(k)$ vanish, all parameters $\langle T_{qp}^{[k]} \rangle$ are obtained from measurements at a variety of angles $\theta, \phi$ to determine the coefficients $Y_{qp}^{(k)}(\theta, \phi)$ in equation (18).

However, the spherical harmonics $Y_{q-}^{[k]}$ vanish for $\phi = 0$ (see equation (12a)) while the mean values of $T_{q+}^k$ vanish for $k = $ odd (see equation (19)). Thus only $k = $ even terms participate for $\phi = 0$ in the scattering intensity given by equation (18) and it is necessary to measure with the light axis out of the scattering plane to obtain odd multipoles.

We note that

$$\mathscr{W}^+(k) = (-)^k \mathscr{W}^-(k). \tag{34}$$

Thus, unless $\phi = 0$, the scattered electron intensity $I^+$ for right and $I^-$ for left circular polarization differ. The sum and difference of these intensities is given by

$$I^+ \pm I^- = \sum_{\substack{k = \text{even for } + \\ k = \text{odd for } -}} \mathscr{W}(F, k) \sum_{q=0}^{k} \langle T_{q\pm}^{[k]}(\text{col}) \rangle Y_{q\pm}^{[k]}(\theta, \phi) \tag{35}$$

with $\mathscr{W}(F, k) = 4\sqrt{\pi} \cdot C \cdot (2k+1)^{1/2} \mathscr{V}^k(F) \mathscr{W}^+(k)$. The sum depends only on even, the difference only on odd multipoles.

An experimentally favourable angle is $\phi = \pi/2$ (see figure 1(b)) where the scattering plane is kept perpendicular to the incident light direction, while $\theta$ is varied. From equations (35), (12a) and using $\langle T_{0-}^{[k]} \rangle = 0$ we obtain for $\phi = \pi/2$

$$I^+ - I^- = \pi^{-1/2} \sum_{k = \text{odd}} \mathscr{W}(F, k) \sum_{q=1}^{k} \langle T_{q-}^{[k]}(\text{col}) \rangle \mathscr{P}_{kq}(\cos\theta) \tag{36}$$

$$I^+ + I^- = (2\pi)^{-1/2} \sum_{k = \text{even}} \mathscr{W}(F, k) \sum_{q=0}^{k} \langle T_{q+}^{[k]}(\text{col}) \rangle \mathscr{P}_{kq}(\cos\theta). \tag{37}$$

The situation becomes particularly simple if we specialize to a $^2P_{3/2}$ state as in the case of the alkali metals. If the Percival–Seaton hypothesis holds only $k \leqslant 2$ multipoles occur in equations (36) and (37).

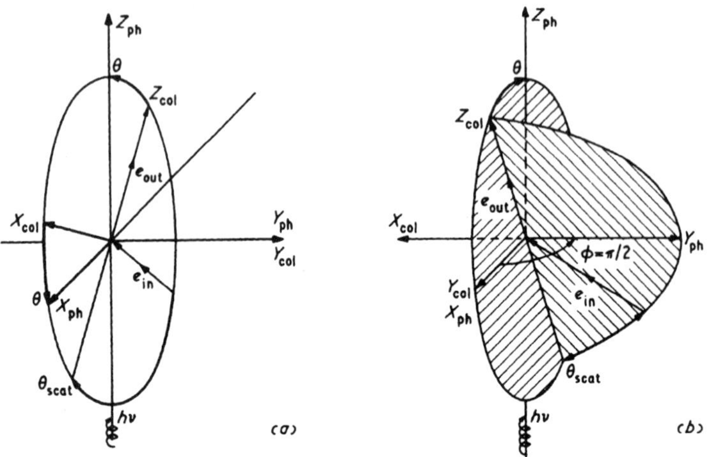

**Figure 1.** Collision and photon coordinate frames for circularly polarized light for the case
(*a*) when the light is incident in the scattering plane and (*b*) when the light is incident perpendicular to the scattering plane.

Thus for $\phi = \pi/2$ and a $^2$P state we have

$$I^+ - I^- = C_1 \langle T_1^{\{1\}}(\text{col}) \rangle \sin \theta \tag{38}$$

and

$$I^+ + I^- = C_2 + C_3 \langle T_0^{\{2\}}(\text{col}) \rangle (3 \cos^2 \theta - 1). \tag{39}$$

Any deviation from these angular distributions indicates a departure from the
Percival–Seaton hypothesis. If spin–orbit interaction plays a role multipoles up to
$k = 3$ would be measured in the $I^+ - I^-$ distribution. According to the discussion by
Burke and Mitchell (1974) this should be particularly noticeable in the region of narrow
resonances.

We want to extend the discussion one step further: Experimentally it is more easy
to keep the scattering plane fixed and vary the position of the electron gun and detector
within the scattering plane through an angle $\epsilon$ thereby varying $\theta$ as well as $\phi$ (figure 2).
In order to display the dependence of the intensities on $\epsilon$ we measure $\epsilon$ with respect to
a standard frame st whose $Z_{\text{st}} - x_{\text{st}}$ plane coincides with the collision plane. The light
beam is incident under $\phi_s = \pi/2$ on this standard plane and is inclined by an angle $\theta_s$.

Thus we may use equation (36) through (39) if we replace $\langle T_{q\pm}^{\{k\}}(\text{col}) \rangle$ by $\langle T_{q\pm}^{\{k\}}(\text{st}) \rangle$
and $\theta$ by $\theta_s$. The transformation to the collision frame is given by rotation of the
coordinate frame through the Euler angles $\beta = \epsilon$, $\alpha = \gamma = 0$.

From the standard formulas for the transformation of complex operators $T_q^{\{k\}}$ and the
definitions equation (12) we have

$$\langle T_{q'p'}^{\{k\}}(\text{st}) \rangle = \sum_{pq=0}^{k} \langle T_{qp}^{\{k\}}(\text{col}) \rangle \mathscr{D}_{qp,q'p'}^{(k)}(0, \epsilon, 0) \tag{40}$$

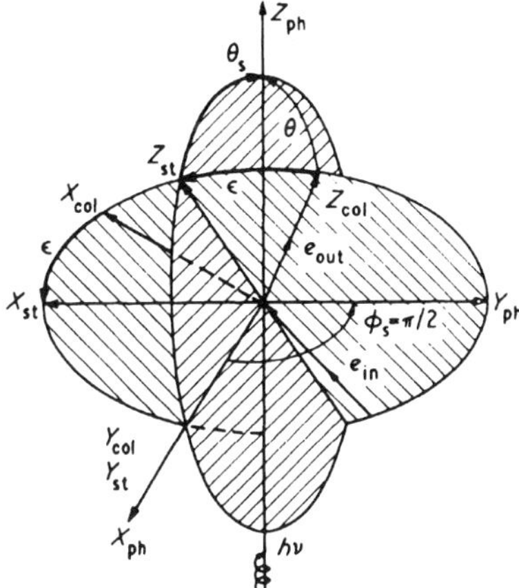

**Figure 2.** Collision and photon coordinate frames for cirularly polarized light. The $z_{st}x_{st}$ plane of the standard frame coincides with the $z_{col}x_{col}$ plane of the collision frame. The angle $\epsilon$ measures the direction of the outgoing electron with respect to the $z_{st}$ axis.

where the real rotation functions $\mathscr{D}^{(k)}_{qp,q'p'}$ are given by, for $q, q' > 0$

$$\mathscr{D}^{(k)}_{qp,q'p'} = (-)^{-q}[(-)^{q'}\cos(q\alpha + q'\gamma)d^{(k)}_{qq'}(\beta)$$

$$+ p'\cos(q\alpha - q'\gamma)d^{(k)}_{q-q'}(\beta)] \qquad \text{for } p = p'$$

$$\mathscr{D}^{(k)}_{qp,q'p'} = (-)^{-q}[p'(-)^{q'}\sin(q\alpha + q'\gamma)d^{(k)}_{qq'}(\beta)$$

$$+ \sin(q\alpha - q'\gamma)d^{(k)}_{q-q'}(\beta)] \qquad \text{for } p \neq p'$$

and for $q$ or $q' = 0$

$$\mathscr{D}^{k}_{qp,q'p'} = (-)^{q+q'}\sqrt{2}\,d^{(k)}_{qq'}\begin{cases} \cos(q\alpha + q'\gamma) & \text{for } p = p' \\ p'\sin(q\alpha + q'\gamma) & \text{for } p \neq p' \end{cases}$$

$$\mathscr{D}^{k}_{0^+,0^+} = d^{(k)}_{00}(\beta).$$

We now specialize again to the alkali $^2P_{3/2}$ case and obtain from equation (38)

$$I^+ - I^- = C_1 \sin\theta_s(d'_{11}(\epsilon) + d'_{-11}(\epsilon)\langle T^{[1]}_{1-}(\text{col})\rangle$$

$$= C_1 \sin\theta_s\langle T^{[1]}_{1-}(\text{col})\rangle \tag{41}$$

independent of $\epsilon$. Thus, any dependence on $\epsilon$ will indicate a departure from the Percival–Seaton hypothesis.

A normalization to $I^+ + I^-$ is possible if we choose $\theta_s$ to be the 'magic angle' $\cos\theta_S = 1/\sqrt{3}$. Then from equation (39) also $I^+ + I^-$ becomes independent of $\theta_S$ and

$$\frac{I^+ - I^-}{I^+ + I^-} = \frac{C_1\sqrt{2}\langle T^{[1]}_{1-}(\text{col})\rangle}{C_2\sqrt{3}} \tag{42}$$

which allows a sensitive test to the influence of spin–orbit interaction.

We conclude the discussion on circular polarized light by a short remark on Born's approximation: Since (in the direction of momentum transfer) the selection rule

2184        *J Macek and I V Hertel*

$\Delta M = 0$ holds, odd multipoles are zero using Born's amplitudes. Thus any observable difference $I^+ - I^-$ indicates that Born approximation cannot be applied.

### 3.2. Linearly polarized light

The natural coordinate system to use in this case has the axis of quantization along the electric vector, ie perpendicular to the light beam and in the plane of polarization. The equilibrium population of magnetic substates then satisfies a system of equations derived by Hertel and Stoll (1974b). They are

$$W(M_F)(A_{M_F,M_F+1} + A_{M_F,M_F-1})$$

$$= W(M_F-1)A_{M_F-1,M_F} + W(M_F+1)A_{M_F+1,M_F} \qquad \text{for } M_F \leqslant F-2 \quad (43)$$

and

$$W(F-1)A_{F-1,F-2} = W(F-2)A_{F-2,F-1}$$

where $A_{M_F,M'_F}$ is the probability for the optical transition from an upper state $FM_F$ to a lower state $F'M_{F'}$). These equations were solved for the special case $F = 3$. Here we derive the general solution.

We first show that if equation (43) holds then so does the equation

$$W(M_F+1)A_{M_F+1,M_F} = W(M_F)A_{M_F,M_F+1}. \qquad (44)$$

Equation (44) holds for $M_F = F-1$. It is only necessary to show that if equation (44) holds for $M_F = M_F$ it also holds for $M_F = M_F-1$. Substituting $W(M_F+1)$ from equation (44) into equation (43) yields

$$W(M_F)A_{M_F,M_F-1} = W(M_F-1)A_{M_F-1,M_F} \qquad (45)$$

establishing equation (44) generally.

The ratio $A_{M_F,M_F-1}/A_{M_F-1,M_F}$ simply equals the ratio of $3j$ symbols,

$$\frac{A_{M_F,M_F-1}}{A_{M_F-1,M_F}} = \begin{pmatrix} F & F' & 1 \\ -M_F & M_F-1 & 1 \end{pmatrix}^2 \bigg/ \begin{pmatrix} F' & F & 1 \\ -M_F & M_F-1 & 1 \end{pmatrix}^2. \qquad (46)$$

For the case of $F = F'+1$ by direct substitution one verifies that equation (44) is solved by

$$W(M_F) = (4F-1)\begin{pmatrix} F-1 & F & 2F-1 \\ M_F & -M_F & 0 \end{pmatrix}^2 = (F-1M_FF-M_F|2F-10)^2 \qquad (47)$$

where we have used the closed form expressions for the $3-j$ symbols (Brink and Satchler 1968) to evaluate the product $W(M_F)A_{M_F,M_F+1}/A_{M_F+1,M_F}$ for comparison with $W(M_F+1)$. Equation (47) agrees with the distribution calculated directly from equation (43) for the case of $F = 3$ (Hertel and Stoll 1974b).

With $W(M_F)$ from equation (30) the 'recoupled' weight function $\mathcal{W}(k)$ is expressed as a recoupling coefficient. We have

$$\mathcal{W}(k) = \sum_{M_F} (-)^{k-F-M_F}(F-1M_FF-M_F|2F-10)^2(F-M_FFM_F|k0)$$

$$= (-)^{2F+k+1}(2k+1)^{1/2}(4F-1)\begin{Bmatrix} 2F-1 & k & 2F-1 \\ F & F-1 & F \end{Bmatrix}$$

$$\times \begin{pmatrix} 2F-1 & k & 2F-1 \\ 0 & 0 & 0 \end{pmatrix}. \qquad (48)$$

*Theory of electron scattering*     2185

It is immediately apparent from the symmetry of the $3-j$ symbol in equation (48) that $\mathcal{W}(k) = 0$ unless $k =$ even. This is to be expected since linearly polarized light can prepare an aligned state, $k =$ even, but not an oriented one, $k =$ odd.

The case $F = F'$ leads by (44) and (46) to an equal population of all $|FM_F\rangle$ states and thus $w(k) = 0$ unless $k = 0$ and is of no interest in this paper. Finally $F < F'$ leads to no upper state population due to optical pumping.

The arguments of the spherical harmonics in equation (18) are the polar coordinates of the electric vector $E$ with respect to the collision plane. By varying the direction of this vector one may determine $I(\theta, \phi)$ at a sufficient number of points to invert equation (18) and determine all of the even multipoles. The plane of polarization of the incident linear polarized laser light may be rotated easily in the experiment, thus varying the direction of $E$.

In general both angles $\theta$ and $\phi$ vary, but the relative orientation of $E$ is best described by a third angle $\psi$ measuring the rotation of the plane of polarization from a standard position. It is convenient to exhibit this angle explicitly in equation (9).

Figure (3) exhibits the axes of the collision frame, the $X_{\mathrm{ph}}$ and $Z_{\mathrm{ph}}$ axes of the photon frame, the polar coordinates $\theta_{\hat{n}}\phi_{\hat{n}}$ of the laser light direction and the polar coordinates $\theta, \phi$ of the $E$ vector in the collision frame. The vector $E$ parallels the $Z$ axis of the old

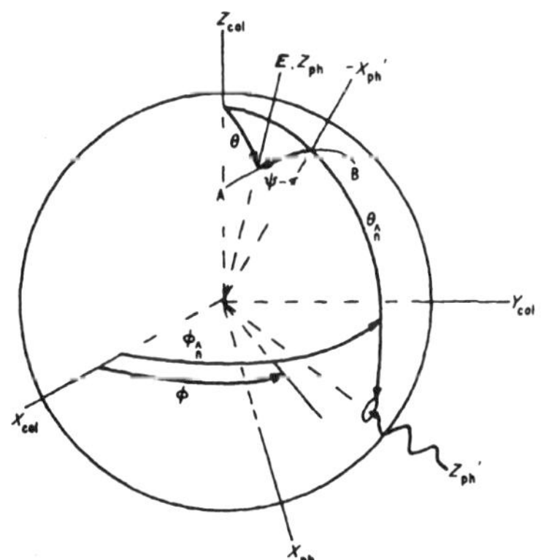


**Figure 3.** Collision and coordinate frames for linearly polarized light. Two photon frames are shown; the unprimed frame has its $z$ axis along the electric vector and the primed frame has its $z'$ axis along the incident light direction. The angle $\psi$ measures the orientation of a linear polarizer with respect to the plane of the outgoing electron and the incident light.

photon frame while the incident light direction parallels the $Z_{\mathrm{ph}}'$ axes. The corresponding $X_{\mathrm{ph}}$ and $X_{\mathrm{ph}}'$ axes lie in the $Z_{\mathrm{ph}} - Z$ and $Z_{\mathrm{ph}}' - Z$ planes. The arc AB is a portion of the circle representing the locus of $E$ vector directions obtained by rotating the linear polarizer about the light direction. Let $\psi$ be the angle between the $E$ vector and the $X_{\mathrm{ph}}'$ axis. The spherical coordinates of $E$ in the primed frame are $\pi/2, \psi$. From equation (18) we have

$$\langle T_0^{[k]}(\mathrm{ph})\rangle = 4\pi(2k+1)^{-1/2} \sum_{p'q=0}^{k} \langle T_{q'p}^{[k]}(\mathrm{ph}')\rangle\, Y_{q'p}^{[k]}(\pi/2, \psi). \tag{49}$$

2186        *J Macek and I V Hertel*

From equation (12a) $Y_{qp}^{(k)}(\pi/2, \psi) = 0$ for $k + q =$ odd. Since $\mathscr{W}(k) = 0$ for $k =$ odd $q$ has to be even $q = 2Q$. Then

$$Y_{q'p'}^{(k)}(\pi/2, \psi) \propto \begin{cases} \sin Q . 2\psi & p' = -1 \\ \cos Q . 2\psi & p' = +1 \end{cases} \tag{50}$$

$$Q = 0, 1, \ldots \frac{k}{2}.$$

We see that equation (17) with (49) represents a Fourier expansion of the electron intensity in a series of sines and cosines of $2\psi$. The maximum value of $q$ is $2[F]$, where $[F]$ is the integer part of $F$, thus a total of $2[F]$ Fourier coefficients can be extracted from measurements with fixed light direction but variable polarizer direction. Not all coefficients are independent since reflection symmetry implies that the intensity with light axis at $\phi_{\hat{n}}$ must equal that at $-\phi_{\hat{n}}$. This requirement takes a simple form when $\phi_{\hat{n}} = 0$. Then only $p =$ even terms or $\cos q\psi$ terms are non-zero.

The number of different non-zero Fourier coefficients provide a possible test of the hypothesis of Percival and Seaton (1958). If nuclear spin plays no essential role during the collision then $q_{max} = 2[J]$ rather than $2[F]$ when $F > J$. Similarly, if the electronic spin plays no role then $q_{max} = 2[L]$ for $J > L$.

Thus in the alkali $^2P$ case we have for $^2P_{1/2}$ no dependence on $\psi$ while for $^2P_{3/2}$ and $\phi_s = 0$

$$I = C_1' + C_2'\langle T_0^{[2]}(\text{ph}')\rangle + C_3'\langle T_2^{[2]}(\text{ph}')\rangle \cos 2\psi \tag{51}$$

is to be expected, as shown by the results of Hertel and Stoll (1974a) within the limits of error. We conclude the discussion of linearly polarized light by writing $\langle T_{qp}^{(k)}(\text{ph}')\rangle$ in the collision frame, thereby exhibiting the three angles $\theta_{\hat{n}}$, $\phi_{\hat{n}}$ and $\psi$ explicitly. This involves a rotation of the coordinate frame through Euler angles $\beta = \theta_{\hat{n}}$ and $\alpha = \phi_{\hat{n}}$.

Then in (49) we have to insert

$$\langle T_{q'p'}^{(k)}(\text{ph}')\rangle = \sum_{q=0}^{k} \mathscr{D}_{q}^{(k)}{}_{,q'p'}(\phi_{\hat{n}}, \theta_{\hat{n}}, 0)\langle T_q^{(k)}(\text{col})\rangle \tag{52}$$

where $\mathscr{D}_{q}^{(k)}{}_{,q'p'}$ is given by equation (40).

It should be noted that the product of the two rotations $(\psi, \pi/2, 0)$ and $(\phi_{\hat{n}}, \theta_{\hat{n}}, 0)$ does not equal the single rotation $(\phi, \theta, 0)$ of equation (18). There is no contradiction since one may rotate the coordinate system about $E$ without changing $T^{(k)}(\text{ph})$. The product of this latter rotation, appropriately chosen, and the two rotations $(\psi, \pi/2, 0)$, $(\phi_{\hat{n}}, \theta_{\hat{n}}, 0)$ equals the single rotation $(\phi, \theta, 0)$. Specifically one rotates the unprimed frame about $E$ so that $Y_{\text{ph}}$ lies along $Z_{\text{ph}}'$, then rotates through $(0, -\pi/2, -\psi)$ to bring the unprimed frame into correspondence with the primed frame. The subsequent rotation through $(0, -\theta_{\hat{n}}, -\phi_{\hat{n}})$ brings the unprimed frame into correspondence with the collision frame. The product of the three rotations is equivalent to the rotation $(0, -\theta, -\phi)$ bringing the unprimed frame into correspondence with the collision frame directly.

### 3.3. Elliptical polarization

In this case it is not possible to find a coordinate system in which the density matrix of the optically pumped level is diagonal. The weight functions $W$ are now a matrix and the projection operator $\tau$ is

$$\tau = \sum_{M_F M_F'} |FM_F) W(M_F, M_F') CFM_F'|$$

where $W(M_F, M_F')$ is the density matrix of the optically pumped excited level $n_i$. It depends upon the degree of elliptical polarization and the orientation of the polarization ellipse. The simultaneous variability of the ellipticity and the orientation of the ellipse offers additional possibilities for input data to a Fourier analysis. We have not found a general expression for $W$ in this case so further discussion will not be given here.


## 4. Experimental normalization

Since absolute measurements would be extremely difficult, electron scattering intensities for different angles of light incidence and status of polarization have to be normalized to the intensity in a suitable standard position. Although in principle a normalization to the cross section for inelastic scattering off the ground state is possible as described by Hertel and Stoll (1974a), their method relies highly on the assumption that the scattering volume does not change while changing the energy loss detected and/or the primary electron energy. This may be an adequate method for high incident electron energies. However, for low energies, a different normalization has to be applied.

Linear polarization may be used to give a standard intensity:

(1) If the incident light direction is in the scattering plane as in figure 1(*a*) and the $E$ vector rectangular to it (in the $Y_{col}$ direction) obviously the scattered electron intensity does not depend on the polar angle $\theta_{\hat{n}}$ of the collision frame. Thus this position may be used to normalize the scattered intensities for varying polarization angle $\psi$ with respect to the scattering plane and varying $\theta_{\hat{n}}$. Thus obtaining $I_{norm}(\theta_{\hat{n}}\psi)$ with $I_{norm}(\theta_{\hat{n}}, \pi/2) = 1$.

(2) For any direction of the incident light as in figure 1*b* and 2 we may choose the direction of the $E$ vector in the scattering plane ($\psi = 0$) and identify the scattered intensity thus with $I_{norm}(0, 0)$ and $I_{norm}(\epsilon, 0)$ for the situation of figures 1(*b*) and 2 respectively.

(3) Since the probability $P_c$ to excite atoms by circular polarized light is in general different from that by linear polarized light $P_L$, the ratio $P_c/P_L$ has to be determined in order to compare also the circular polarized light. Since that ratio should not depend too drastically on slight changes in the scattering volume, it may be measured by comparing the decrease of inelastic scattered electrons of the ground state.

(4) An independent method for normalizing electron scattering intensities for circular polarized light has been described in §2.1.


## 5. Application to the range of validity of Born approximation

Studies of the region of validity of the Born approximation are one application of the techniques discussed here. We have already remarked on the vanishing of $\langle T_1^{\{1\}} \rangle$ in the Born limit. Here we evaluate all of the parameters $\langle T_{qp}^{\{k\}}(col) \rangle$ in the Born approximation.

The Born theory treats the excitation as an impulsive transfer of momentum $K$ by the incident electron. Symmetry about the $K$ direction implies that only $q = 0$ multipoles are non-zero in a frame, the momentum transfer frame, with $Z$ axis along $K$. Furthermore, owing to reflection symmetry only $k =$ even terms are non-zero. To evaluate $\langle T^{[k]}_{qp}(\text{col})\rangle$ we need only evaluate $\langle T^{[k]}_{0+}(K)\rangle$ and rotate to the collision frame. The operator $T^{[k]}_{0+}(K)$ is a polynomial in $\mathbf{K}\cdot\mathbf{J}$ and $J^2$. In Born approximation, the collision transfers no net angular momentum along $\mathbf{K}$, thus $\langle(\mathbf{K}\cdot\mathbf{J})^n\rangle = 0$, $n = $ integer. It follows that the only term contributing to $\langle T^{[k]}_0(K)\rangle$ is $\mathbf{J}^k = [J^2]^{k/2}$.

The solid harmonic $r^k C_{kq}$ is given by

$$r^k C_{kq} = r^k P_k = (-1)^{k/2}\frac{(k-1)!!}{k!!}r^k + \ldots(\text{terms with } Z).$$

Inserting this expression into the definition of $T^{[k]}q$, gives

$$T^{[k]}_0 = (-)^{k/2}\frac{(k-1)!!}{k!!}k![J^2]^{k/2} + \ldots \text{terms with } J_z. \tag{53}$$

The mean value of $T^k_0$ in the momentum transfer frame is thus

$$\langle T^{[k]}_0(K)\rangle = (-)^{k/2}\frac{(k-1)!!}{k!!}k![j_i(j_i+1)]^{k/2} \tag{54}$$

where $j_i$ is the angular momentum of the initial state. In the collision frame we have

$$\langle T^{[k]}_{qp}(\text{col})\rangle = (-)^{k/2}\frac{(2k-1)!!}{2^k}[j_i(j_i+1)]^{k/2}\frac{\sqrt{4\pi}}{(2k+1)^{1/2}}Y^{(k)}_{qp}(\theta_K) \text{ for } p = +1, k = \text{even}$$

$$= 0 \qquad \text{for } p = -1, k = \text{odd} \tag{55}$$

where $\theta_K$ is the angle between the momentum transfer vector and the collision frame.

### Acknowledgment

We wish to thank Professor Fano for his interest in the subject.

One of us (JM) wishes to thank the Freie Universität, Berlin for support.

### References

Bethe H and Salpeter E 1957 *Encyclopedia of Physics* Vol 351 (Berlin: Springer-Verlag) p 431
Brink D M and Satchler G R 1968 *Angular Momentum* (Oxford: Clarendon Press) 2nd ed p 108
Burgess A and Percival I C 1968 *Adv. Atom. Molec. Phys.* 4 109–41 eds D R Bates and I Esterman (New York: Academic Press)
Burke P G and Mitchell J F B 1974 *J. Phys.* B: *Atom. Molec. Phys.* 7 214
Eminyan M, MacAdam K B, Slevin J and Kleinpoppen H 1972 *Phys. Rev. Lett.* 31 576–9
Falkoff D L and Uhlenbeck G E 1950 *Phys. Rev.* 79 323
Fano U 1960 *J. Math. Phys.* 1 417–423
Fano U and Macek J 1973 *Rev. Mod. Phys.* 45 553–73
Happer W 1972 *Rev. Mod. Phys.* 44 169–249
Hertel I V and Stoll W 1974a *J. Phys.* B: *Atom. Molec. Phys.* 7 570–82
—— 1974b *J. Phys.* B: *Atom. Molec. Phys.* 7 583–92
Percival I C and Seaton M J 1957 *Proc. Phys. Soc.* 53 644–62

VOLUME 34, NUMBER 8     **PHYSICAL REVIEW LETTERS**     24 FEBRUARY 1975

# Measurements of Photon Polarization and Angular Correlations for He$^+$ - He Collisions Coincidence Technique Using an Ion-Photon

G. Vassilev, G. Rahmat, J. Slevin,* and J. Baudon

*Laboratoire de Collisions Atomiques,† Université Paris-Sud, 91405 Orsay, France*

(Received 26 December 1974)

The polarization of photons from the transition $3^3P \rightarrow 2^3S$ and the angular distribution of radiation from the decay $2^1P \rightarrow 1^1S$ emitted in He$^+$ +He collisions at an incident energy of 150 eV and an ion scattering angle of 13.5° were measured by detecting the scattered ions and photons in delayed coincidence. From these measurements we conclude that the $3^3P$ excitation takes place through a $\Sigma$-$\Sigma$ radial coupling, and the $2^1P$ excitation predominantly through a $\Sigma$-$\Pi$ rotational coupling, probably at large internuclear separation.

The interpretation of excitation processes in ion-atom collisions at moderate energies (a few hundred eV) is based on a molecular description of the colliding-particle system. In most cases, the primary excitation mechanism consists in an interaction between ground and excited molecular states, which occurs at a rather small internuclear distance (a few bohrs). In general, many coherent scattering amplitudes, corresponding to several output channels, are produced at the same time by the collision and these amplitudes can interfere at large or infinite internuclear distances. Thus, in spite of the fact that the primary excitation mechanisms are now generally well understood, transition probabilities—or inelastic differential cross sections—for excitation to *one* specific atomic level are often difficult to determine because of these "secondary" interactions occurring at large distances. For the particular case of the excitation of a $^1P$ state of helium, in a He$^+$-He collision, for a specific scattering angle (the ion is assumed to remain in the $^1S$ state), different and coherent amplitudes $a_m(\theta, \Phi)$ are obtained for the different magnetic sublevels $m = -1, 0, +1$, with the symmetry relation $a_1 = -a_{-1}$ (see Macek and Jaecks[1]). As the collision time at moderate energies ($10^{-15}$ sec) is much shorter than the lifetime of the excited

VOLUME 34, NUMBER 8     PHYSICAL REVIEW LETTERS     24 FEBRUARY 1975

states, the two colliding particles can be considered to be far apart when the radiative decay occurs. Since each particular magnetic sublevel corresponds to a definite polarization of the light emitted in the transition $^1P \to {}^1S$, an analysis of this light provides information on magnetic-sublevel excitations. Measurements of the polarization have already been made for high-energy collisions and at low and moderate energies,[2-5] by measuring the intensity at 90° from the incident beam direction through a polarizer placed either parallel or perpendicular to that direction. However, since these polarization measurements were made without regard to the ions, the results are necessarily averaged over all ion scattering angles and information is lost in this averaging process. It is then clear that when the (polarized) photons are detected in coincidence with ions scattered in a *particular* direction, new information on the elementary collision process is obtained. Thus, if the final direction of the molecular axis is taken as the quantization axis for atomic states, each magnetic sublevel corresponds to a molecular state of definite $\Lambda$ symmetry ($\Lambda = 1$ for $m = \pm 1$ and $\Lambda = 0$ for $m = 0$). A measurement of the differential polarization of the coincident light or alternatively its angular distribution yields values of the relative probabilities for the excitation of a given $P$ state via $\Pi$ or $\Sigma$ molecular states, which are proportional to $|a_\Pi|^2$ and $|a_\Sigma|^2$ or $|a_{\pm 1}|^2$ and $|a_0|^2$, and in addition the relative phase $\varphi$ between $a_\Pi$ and $a_\Sigma$. For $^1P$ states the situation is more complicated because of the fine structure. While for such a light atom the spin is not expected to play any role in the collision itself, the spin-orbit interaction acts on the excited state causing $\vec{L}$ and $\vec{S}$ to precess around their resultant $\vec{J}$, and leads to a reduction in the polarization. However, this effect is quite independent of the collision process itself and can be evaluated *a priori*. Macek and Jaecks[1] have treated this problem in detail. It is the purpose of the present paper to report preliminary measurements of photon polarizations and angular correlations for $2^1P$ and $3^3P$ excitations in He$^+$-He collisions using an ion-photon coincidence technique.

Two experimental methods, already used for electron-atom collisions,[6,7] have been applied to analyze the coincidence light. In the first, photons are detected in a direction perpendicular to the collision plane, through a filter and a linear polarizer, and the number of ion-photon coincidences is measured as a function of the polarizer

orientation. In the second method, the spatial distribution of emitted photons (in coincidence with ions scattered at a given angle) is measured by rotating the photon detector around the collision volume. The ion-photon angular-correlation curve gives the same information as a measurement of the polarization.

In order to use the first method, the coincidence apparatus described previously[8] has been slightly modified (see Fig. 1): (i) A UV-HN PB polaroid has been inserted between the interference filter and the photomultiplier, and (ii) the scattered ions have been detected only within an azimuthal angular range of $\pm 20°$. The parabolic mirror, that was used to gather photons from a large solid angle (about $\pi$ sr), has not been removed for the present measurement: Its effect on the polarization has been calculated, and moreover it can be shown to be rather small when the optical device has a symmetry axis perpendicular to the reference collision plane. This experimental method is convenient for visible or near-uv light; it has been used for the $3^3P$-$2^3S$ line ($\lambda = 3889$ Å). The results obtained at an energy of 150 eV and a scattering angle of 13.5° (in the laboratory frame) are shown in Fig. 2. The coincidence number, calibrated with respect to the total number of scattered ions counted during the experimental run, is plotted as a function of

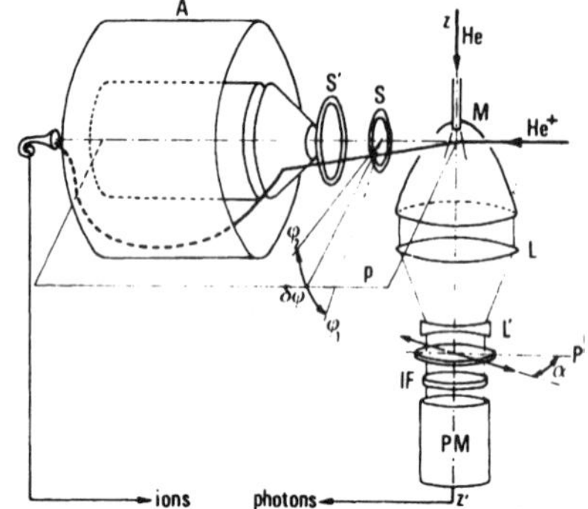


FIG. 1. Experimental setup for the determination of the linear polarization of the coincidence light. $M$ is the parabolic mirror, $LL'$ the quartz lenses, $P$ the polarizer, $IF$ the interference filter; the ion scattering direction is defined by slits $SS'$; $A$ is the electrostatic analyzer; $p$ is the collision plane, perpendicular to $z'z$.

VOLUME 34, NUMBER 8      PHYSICAL REVIEW LETTERS      24 FEBRUARY 1975

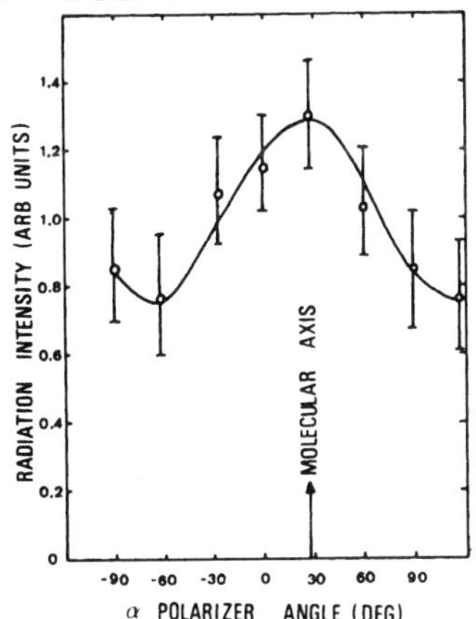

FIG. 2. Ion-photon coincidence rate as a function of the angle of the polarizer. $\alpha = 0$ corresponds to the incident beam direction. The error bars represent 1 standard deviation in the counting statistics. Full curve is a fit of a sinusoid of period $\pi$.

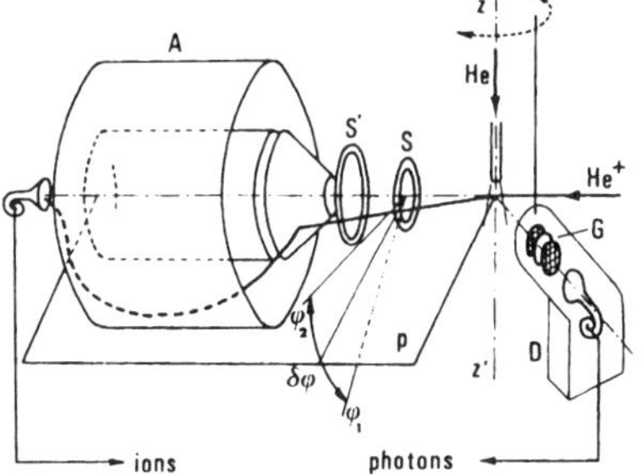

FIG. 3. Experimental arrangement for angular-correlation measurements on the $2^1P$-$1^1S$ resonance line ($\lambda = 584$ Å). The optics of Fig. 1 is replaced by the uv detector $D$ (electron multiplier); grids $G$ prevent the detection of charged particles. The rotation axis of $D$ is $z'z$; it is perpendicular to collision plane $p$.

the polarizer direction $\alpha$. Each datum point represents typically 24 h of integration and a real coincidence rate of about 2 counts/h. It is seen that a maximum of intensity is observed when the polarizer is parallel to the final molecular axis ($\alpha = 27°$). This can occur either when the scattering amplitude $a_\Pi$ vanishes, or when the relative phase $\varphi$ of $a_\Pi$ with respect to $a_\Sigma$ is $\pm\frac{1}{2}\pi$ rad. If it is assumed that $a_\Pi$ is negligible, and depolarization effects due to the spin-orbit coupling, to the mirror, and to Earth's magnetic field are taken into account, then the predicted ratio of intensities at $\alpha = -63°$ and $\alpha = +27°$ is 0.57. The observed ratio is $0.6 \pm 0.2$. This means that, in the present case, the excitation of the $3^3P$ state presumably takes place only via a $\Sigma$-$\Sigma$ radial coupling. This conclusion is consistent with qualitative theoretical predictions[9] but it must be remembered that those predictions concern the primary excitation mechanism at small internuclear distance, and that secondary couplings at large distance could change the $\Sigma$ and $\Pi$ contributions. Preliminary measurements on the $3^1P$ level actually show that the polarization is very different from the previous one, in spite of the fact that singlet and triplet $3P$ states are considered as identical in a simple correlation

diagram.

The second method, based on angular-correlation measurements, has been used to study the $2^1P$-$1^1S$ resonance line of helium ($\lambda = 584$ Å). For this experiment (Fig. 3) the parabolic mirror has been removed and the uv photons are detected by a Channeltron electron multiplier (Mullard 419 BL) which could be rotated around the collision volume. No wavelength selection was necessary in this case since the energy resolution of the analyzer was sufficient to separate $n = 2$ from $n \geq 3$ excitations. Thus ion-photon coincidences necessarily correspond to $2^1P$ excitation. Two grids biased at negative and positive voltages prevent charged particles from reaching the Channeltron aperture. However, fast or metastable atoms are detected within the angular range 0°–90° with respect to the incident beam direction. This makes the measurement more difficult, because of the increase in the accidental-coincidence rate, but it does not affect the true-coincidence number. It was verified directly that the target density was sufficiently low that pressure-dependent effects, such as resonance trapping of the 584-Å radiation, did not affect the experimental data. Figure 4 shows angular-correlation curves for an impact energy of 150 eV and a scattering angle of 13.5°. Integration times for each datum point are typically 24 h and the real coincidence rate is of the order of 2 to 10 counts/h. The experimental points have been obtained with two different azimuthal angu-

VOLUME 34, NUMBER 8     PHYSICAL REVIEW LETTERS     24 FEBRUARY 1975

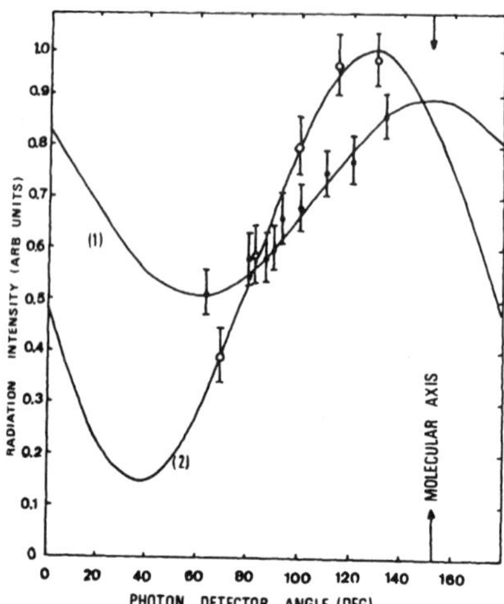



FIG. 4. Ion-photon angular correlation for an incident energy of $E = 150$ eV and scattering angle 13.5°. Error bars represent 1 standard deviation in the counting statistics. Filled circles, $\delta\varphi_i = 90°$; open circles, $\delta\varphi_i = 30°$. Curves (1) and (2) are least-squares fits to the data. The final molecular axis is at 153°, and the incident beam direction is at 0°.

lar spreads $\delta\varphi_i$ of the scattered ions. Each of the two curves, (1) and (2), represents a computer least-squares fit of the angular-correlation function,[10] integrated over the photon-detector aperture and the ion azimuthal angle $\delta\varphi_i$, to the experimental data points. Notice that, as expected, the angular distribution corresponding to the larger range of $\delta\varphi_i$ [curve (1)] is flatter (i.e., the polarization is reduced) than for curve (2). Within the experimental error, the maxima of both curves lie along the final molecular axis, and it is estimated that $|a_\Sigma|^2 \lesssim 10^{-1}|a_\Pi|^2$; i.e., the final molecular state is mainly a $\Pi$ state. This conclusion may seem rather surprising, because it is known that, at such a small energy,

the primary excitation essentially takes place *via* a $\Sigma_g$-$\Sigma_g$ coupling. However this $\Sigma_g$-excited-state potential curve lies very close to the $\Pi_g$ curve leading to He($2^1P$), along a rather large range of internuclear distance. Thus a secondary $\Sigma_g$-$\Pi_g$ coupling could explain the observed predominance of $|a_\Pi|$.

For $2^1P$ excitations, the amplitude of $|a_\Sigma|$ was not sufficiently great to permit us to determine a statistically significant value for the relative phase $\varphi$ between $a_\Sigma$ and $a_\Pi$. For the case of the $3^3P$ excitation, the reverse was true, i.e., $|a_\Sigma| \gg |a_\Pi|$, and again a reliable value of $\varphi$ could not be estimated. Work is continuing on both $2^1P$ and $3^3P$ excitations at different energies and scattering angles.

The authors are grateful to Dr. M. Barat for stimulating suggestions and comments. They also wish to thank Dr. B. Stern for helpful discussions.

*Permanent address: Department of Physics, University of Stirling, Stirling, Scotland. Collaboration made possible by a Royal Society Fellowship.

†Equipe de recherche du Centre National de la Recherche Scientifique.

[1]J. Macek and D. H. Jaecks, Phys. Rev. A 4, 2288 (1971).

[2]F. J. de Heer and J. Van den Bos, Physica (Utrecht) 31, 365 (1965).

[3]T. Andersen, A. K. Nielsen, and K. J. Olsen, Phys. Rev. A 10, 2174 (1974).

[4]N. H. Joek, C. W. White, S. H. Neff, and K. J. Olsen, Phys. Rev. Lett. 31, 671 (1973).

[5]T. D. Gaily, D. H. Jaecks, and R. Geballe, Phys. Rev. 167, 81 (1968).

[6]M. Eminyan, K. B. Mac Adam, J. Slevin, and H. Kleinpoppen, J. Phys. B: Proc. Phys. Soc., London 7, 1519 (1974).

[7]G. C. M. King, A. Adams, and F. H. Read, J. Phys. B: Proc. Phys. Soc., London 5, 254 (1972).

[8]G. Vassilev, J. Baudon, G. Rahmat, and M. Barat, Rev. Sci. Instrum. 42, 1222 (1971).

[9]M. Barat, D. Dhuicq, R. Francois, R. McCarroll, R. D. Piacentini, and A. Salin, J. Phys. B: Proc. Phys. Soc., London 5, 1343 (1972).

[10]The angular-correlation function is obtained by evaluating Eq. (15) in Ref. 1 for $^1P$ excited states. It is also given by Eq. (6) in Ref. 5.

VOLUME 35, NUMBER 11      PHYSICAL REVIEW LETTERS      15 SEPTEMBER 1975

# Polarized-Photon, Scattered-Atom Coincidence Measurements in He$^+$-He Collisions*

D. H. Jaecks, F. J. Eriksen, W. de Rijk, and J. Macek

*Behlen Laboratory of Physics, University of Nebraska-Lincoln, Lincoln, Nebraska 68508*

(Received 27 May 1975)

Photon–scattered-atom coincidence measurements with polarization analysis of the photons have been made for 3.0-keV He$^+$-He collisions leading to He ($3^3P$) excitation in a charge-transfer process. The phase difference between the scattering amplitudes for $m_l = 0$ and $m_l = 1$ magnetic sublevel formation is an approximately constant function of scattering angle. We propose a theory which predicts the constancy of this phase difference.

Only recently[1,2] have measurements of magnetic sublevel populations been made in ion-atom collisions where the scattering angle is a well-defined experimental parameter. Such measurements are of immediate interest since they provide the information necessary to decide between rotational or radial coupling interactions within the electron-promotion model.

Current measurements have employed a scattered-particle, polarized-photon coincidence technique; for example de Rijk, Eriksen, and Jaecks[1] have measured the polarization of 3889-Å radiation emitted in 3-keV He$^+$-He charge-exchange collisions, and Vassilev *et al.*[2] have reported similar polarization measurements, but of direct excitation, in 150-eV He$^+$-He collisions. Vassilev *et al.* have concluded on the basis of their data that $\sigma_1$ is small compared to $\sigma_0$, where $\sigma_1$ and $\sigma_0$ are the partial cross sections for exciting the $m_l = 1$ and $m_l = 0$ sublevels, respectively.

The purpose of this Letter is threefold: (1) to present data which show, contrary to the results at lower energy, that $\sigma_1 \approx \sigma_0$ for many scattering angles, (2) to show that the phase difference $\Delta\varphi$ between the $m_l = 0$ and $m_l = 1$ amplitudes is a nearly constant function of impact parameter (an extremely surprising result), and, finally, (3) to present a theory which predicts the constant dependence of $\Delta\varphi$ upon impact parameter.

The apparatus used for the measurement is shown in Fig. 1. A He$^+$ ion beam is crossed with a thermal beam of He atoms formed by a capillary array. Scattered atoms and ions are separated by a parallel-plate analyzer and the neutrals are detected by an electron multiplier (box-grid type). Perpendicular to the scattering plane, 3889-Å photons are collected and analyzed according to their linear polarization. The signal from the photomultiplier is then counted in delayed coincidence with the signal from the neutral-particle detector.

For a given "polarizer" angle $\beta$, the delayed coincidence rate $I(\beta)$ is interpreted using the results of Macek and Jaecks,[3]

$$I(\beta) = C\{28\sigma_0 + 26\sigma_1 + 15(2\sigma_0\sigma_1)^{1/2}\cos\Delta\varphi\,\sin2\beta + (30\sigma_1 - 15\sigma_0)\sin^2\beta\}, \quad (1)$$

where $C$ is a constant. Equation (1) in general describes an "hourglass"-shaped figure on a polar plot of $I(\beta)$ versus $\beta$.

Measured values of the number of coincidences per number of scattered neutrals are shown in Fig. 2 for six scattering angles. In all of the polar plots in Fig. 2, the He$^+$ beam is incident from the left along the $x$ axis and the scattered neutral particles are detected between 1.00° and 2.00° above the positive $x$ axis, i.e. in the first quadrant. Additional data at 1.00° were taken for polarizer angles of 22.5° and 112.5°, and these data confirm the shape of the curves. One datum point at a given scattering angle and polarizer angle represents an average of at least two 24-h determinations of the coincidence rate. The error bars in Fig. 2 are estimates of the standard deviation of the random-counting error.[4] An important feature of these curves is their rotation

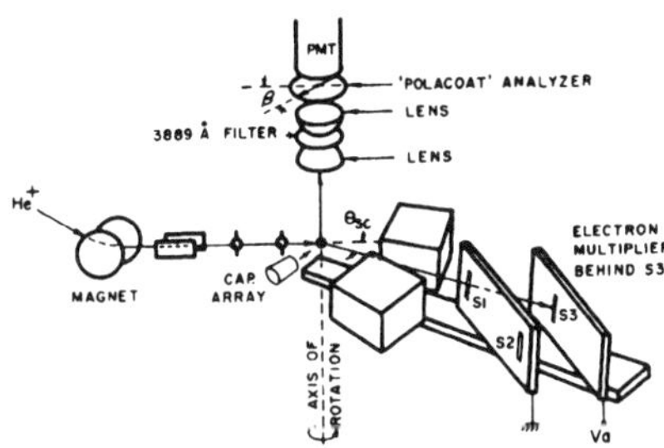


FIG. 1. Schematic diagram of the apparatus.

VOLUME 35, NUMBER 11     PHYSICAL REVIEW LETTERS     15 SEPTEMBER 1975

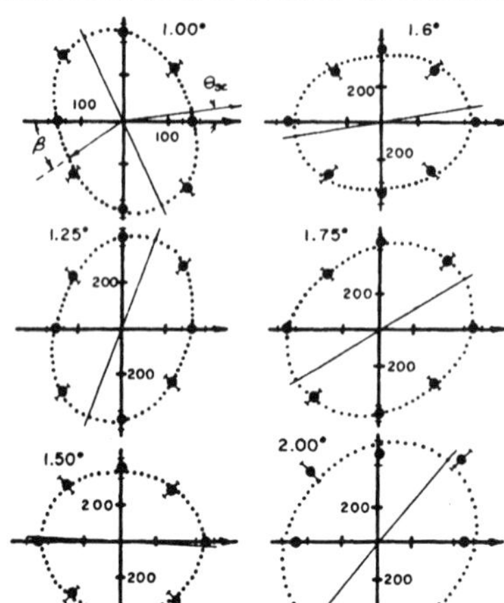

FIG. 2. Plots of $I(\beta)$ versus "polarizer" angle $\beta$ for six scattering angles. In all cases the incident beam direction is horizontal to the right and the neutral particle is detected above the positive $x$ axis. Numbers on the axes represent the scale of the figure in coincidences per $10^9$ scattered neutrals.

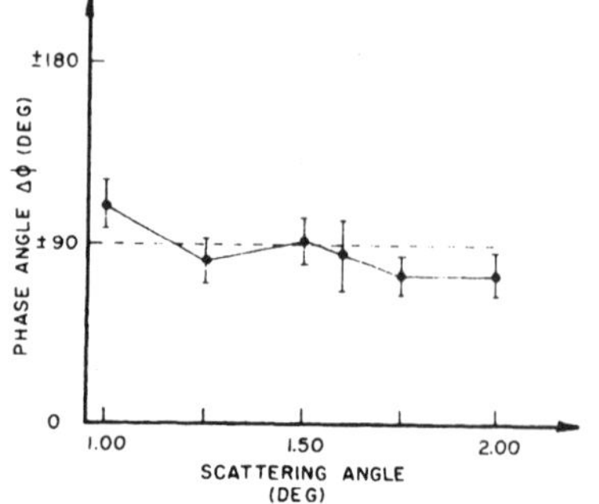

FIG. 3. Phase angle $\Delta\varphi$ versus laboratory scattering angle. Since only linear polarization analysis was made, the sign of the phase is undetermined.

as a function of scattering angle.

The dotted lines in Fig. 2 are computer least-squares best fits of Eq. (1) to the data. From these calculations one can determine $\Delta\varphi$ to within a sign. Figure 3 shows $\Delta\varphi$ versus laboratory scattering angle. One can also determine the quantities $C\sigma_0$ and $C\sigma_1$, the number of measured coincidences per scattered neutral particle for $m_l = 0$ and $m_l = 1$ excitation, respectively. The unknown constant $C$ is from Eq. (1) and depends, for example, on the efficiencies of the detectors. At a scattering angle of 1.50°, $C\sigma_0$ is about $4 \times 10^{-7}$ coincidences per scattered neutral particle, and there are typically $2 \times 10^9$ scattered neutrals per hour. For comparison of theory and experiment, more relevant parameters are $P_0 = C\sigma_0 P_{ex}$ and $P_1 = C\sigma_1 P_{ex}$, where $P_{ex}$ is the charge-exchange probability. The parameters $P_0$ and $P_1$ represent the number of measured coincidences per scattered particle, neutral or charged. Using our own measured values of $P_{ex}$, we have calculated $P_0$ and $P_1$ and they are shown in Fig. 4 versus laboratory scattering angle.

Clearly, $\Delta\varphi$ has the approximately constant value of 90°. This implies that the rotation of the polarization patterns is due as much to changes in the relative size of $\sigma_0$ and $\sigma_1$ as it is to changes in $\Delta\varphi$. Also, despite the fact that $\sigma_1$ is small relative to $\sigma_0$ in the vicinity of 1.60°, $\sigma_1$ is nearly equal to $\sigma_0$ at 1.00°, 1.25°, and 2.00°. This is to be contrasted to the conclusion of Vassilev et al. that $\sigma_1 \simeq 0$.

We interpret these results using the electron-promotion model, assuming a Landau-Zener transition at a crossing between the $1s\sigma_g(2p\sigma_u)^2$ and $(1s\sigma_g)^2 4d\sigma_g$ $He_2^+$ electronic energy curves at some internuclear distance $R_0$. The exact value of $R_0$ is unimportant for our analysis except to note that it is of the order of 1.0–1.5 a.u.

Because the mean radius of the $4d$ electron wave function is much larger than $R_0$, we suppose that the corresponding molecular energies are degenerate. On the incoming passage of the crossing, the $4d\sigma_g$ level is populated when the internuclear axis is oriented at an angle $\sin\alpha = b/R_0$, where $b$ is the impact parameter with respect to the incoming beam. Subsequent motion of the nuclei is accompanied only by a phase change of the wave function, since, in the approximation that the $4d\pi_g$ and $4d\sigma_g$ levels are degenerate, the charge distribution maintains the orientation it had at the time of the level crossing. A similar transition takes place on the outgoing passage of the crossing. With the assumption that the Landau-Zener transition amplitude $p$ is small compared to unity due to the two-electron nature of the transition, we have that the wave function

VOLUME 35, NUMBER 11          PHYSICAL REVIEW LETTERS          15 SEPTEMBER 1975

after the collision is

$$\psi \approx p\{\psi_2^{(i)} \exp(i/\hbar)[\int_{-\infty}^{t_i}E_1(t')dt' + \int_{t_i}^{\infty}E_2(t')dt'] + \psi_2^{(o)} \exp(i/\hbar)[\int_{-\infty}^{t_o}E_1(t')dt' + \int_{t_o}^{\infty}E_2(t')dt']\}$$
$$+ \psi_1' \exp(i/\hbar)\int_{-\infty}^{\infty}E_1(t')dt', \qquad (2)$$

where subscript 1 denotes the initial $1s\sigma_g(2p\sigma_u)^2$ level and subscript 2 denotes the $(1s\sigma_g)^24d\sigma_g$ level. The superscripts $i$ and $o$ indicate that $4d\sigma_g$ is evaluated with the symmetry axis along the internuclear axis at the in ($i$) or out ($o$) orientation.

A transformation of the $4d\sigma$ functions to a frame with the $z$ axis along the final-beam direction, taking into account that $4d\sigma_g$ and $4d\pi_g$ correlate with the $3p(m_l=0)$ and $3p(m_l=1)$ atomic orbitals respectively, gives the amplitude $a_0$ or $a_1$ that the final level is in the $m_l=0$ or $m_l=1$ magnetic substate. We find

$$a_0 = p2^{-1/2}(3\cos^2\alpha - 1)\cos A,$$
$$a_1 = ip3^{1/2}\sin\alpha\cos\alpha\sin A, \qquad (3)$$

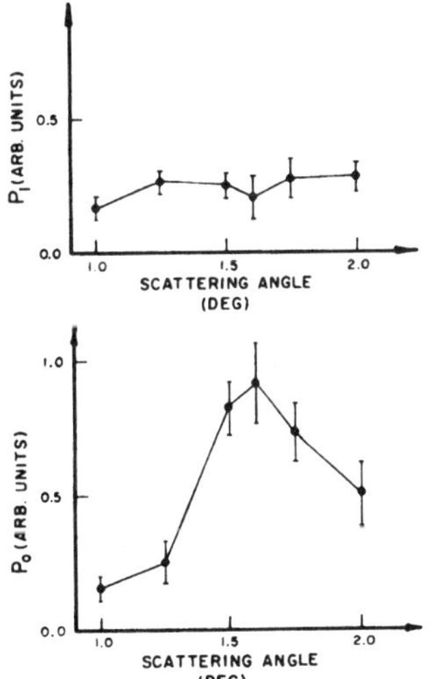

FIG. 4.  Probabilities $P_1$ and $P_0$ for exciting the $m_l$ =1 and $m_l=0$ substates as a function of laboratory scattering angle. Here $P_0$ and $P_1$ are proportional to the number of coincidences per billion total scattered particles. The number of total scattered particles equals scattered neutrals plus scattered ions.

aside from an overall phase factor. Here $A = \int_0^{t_o}\{E_2-E_1\}dt'/\hbar$.

From Eq. (3) we immediately see that $\Delta\varphi (= 90°)$ is a constant in agreement with the experimental results of Fig. 2. Slight variations around 90° are expected if other mechanisms contribute to the excitation. The model is in excellent accord with the measured phase difference, suggesting that $3p$ excitation proceeds via the assumed coupling. Note that the phase difference is independent of any assumed relation between impact parameter and scattering angle and any theoretical input concerning the detail of the potential energy curves including the value of $R_0$.

In contrast $|a_0|^2$ and $|a_1|^2$ are sensitive to all of those parameters. A detailed comparison of this model with the present data has not been made since the requisite potential curves are not available. Furthermore, there is some doubt as to whether a simple relation between scattering angle $\theta$ and impact parameter $b$ can be found. If we take $R_0\approx 1$ a.u., then since $a_0(b)$ cuts off sharply at $b=R_0$, one expects diffraction maxima and minima[5] separated by $\Delta\theta\approx\hbar/mvb = 1.57$ mrad $=0.09°$. Since this uncertainty in $\theta$ is comparable to the angular resolution of the experiment, a detailed calculation incorporating a partial-wave analysis of the scattering is required to fully compare theory and experiment.

*Work supported by the National Science Foundation.

[1] W. de Rijk, F. J. Eriksen, and D. H. Jaecks, Bull. Am. Phys. Soc. **19**, 1230 (1974).

[2] G. Vassilev, G. Rahmat, J. Slevin, and J. Baudon, Phys. Rev. Lett. **34**, 447 (1975).

[3] J. Macek and D. H. Jaecks, Phys. Rev. A **4**, 2288 (1971).

[4] H. D. Young, *Statistical Treatment of Experimental Data* (McGraw-Hill, New York, 1962), pp. 98 and 111.

[5] R. Shakeshaft and J. Macek, Phys. Rev. A **6**, 1876 (1972).

VOLUME 36, NUMBER 11          PHYSICAL REVIEW LETTERS          15 MARCH 1976

# Photon Vector Polarization and Coherence Parameters in an Electron-Photon Coincidence Experiment on Helium

M. C. Standage and H. Kleinpoppen

*Institute of Atomic Physics, University of Stirling, Stirling, FK9 4LA, Scotland*
(Received 3 December 1975)

We present linear and circular polarization measurements of $3^1P$-$2^1S$ (501.6 nm) photons detected in delayed coincidence with electrons inelastically scattered from the $3^1P$ level of helium. Vector polarization and coherence parameters of the coincident radiation provide direct experimental evidence of the coherent nature of the excitation process. Atomic orientation is obtained directly from the circular polarization measurements. Analysis of the data also yields $\lambda$ and $\chi$ parameters for $3^1P$ excitations, which are compared with previous data and theory.

The recent introduction of coincidence techniques, in which photons are detected in coincidence with electrons (ions) inelastically scattered from atoms, has yielded more detailed information on collision parameters and on coherence effects occurring in collisional excitation of atoms. Two methods have been used to analyze the photon radiation in these coincidence experiments. One method is to measure the polarization of coincident photons detected in a fixed direction with respect to a symmetry axis or plane for the experiment.[1-3] The other method used is to measure the angular distribution of coincident photons by rotating the photon detector in the scattering plane.[2, 4-7, 8]

In this Letter we report on linear and circular polarization measurements of photons emitted from the electron-impact–excited $3^1P$-$2^1S$ (501.6 nm) transition of helium which are detected in delayed coincidence with electrons inelastically scattered from the $3^1P$ level. Such measurements provide a further test of the theory of coincidence measurements[9-12] including, as we shall demonstrate, a direct experimental test of the coherent nature of the electron-impact excitation of atomic states. The apparatus (see Fig. 1), described in detail in Eminyan *et al.*,[5] has been modified so that the photons are detected perpendicular to the scattering plane. A photon detector has been mounted outside the vacuum chamber, the interaction region being viewed through a quartz window. A lens with a large collection solid angle (0.45 sr) focuses radiation from the interaction region onto the photon detector. The effect of the collection solid angle on the polarization is small (< 1%) for the experimental geometry used here. The photon detector consisted of a rotatable polarizer followed by a 501.6-nm interference filter (10 nm bandwidth) and a photomultiplier tube.

A Polacoat (type PL 40) Polaroid was used for linear analysis and a $\lambda/4$ ($\lambda = 501.6$ nm) low-order quartz plate was inserted to provide circular analysis. The handedness of the circular analyzer was determined with a Babinet-Soleil compensator. To minimize instrumental polarization, a Hanle-type depolarizer[13] was placed immediately after the polarizer. The overall instrumental polarization was small (< 1%). Since the real coincidence rate is normalized to the total scattered-electron count, the effect of variations in the target gas pressure and the incident electron beam current are eliminated. The absence of pressure-dependent effects such as radiation trapping was confirmed by carrying out runs at different target pressures.

In order to determine completely the polarization state of the coincident photons it is neces-

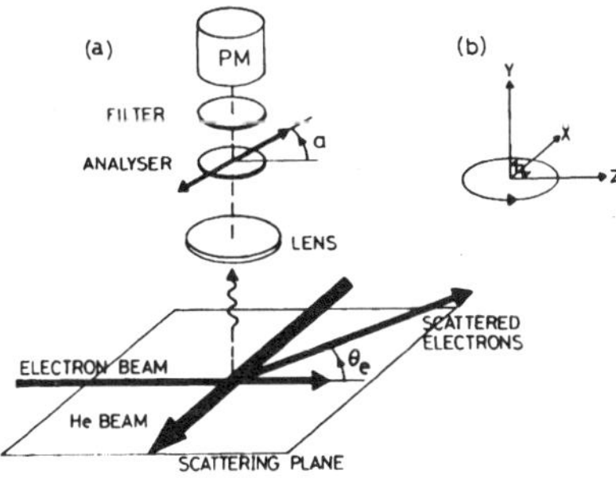



FIG. 1. Schematic diagram of experiment. The $x$-$z$ plane is the scattering plane; the photons are detected along the $y$ axis. Scattering angle $\theta_e$ and linear polarizer angle $\alpha$ are measured in the $x$-$z$ plane. Positive scattering angle is shown.

sary to introduce quantities such as the elements of the photon density matrix, or equivalently, the Stokes parameters which completely characterize their quantum mechanical state. These parameters $(I, P_1, P_2, P_3)$ have been extensively discussed in the literature.[14,15] Normalized to the total intensity $I$, the remaining three Stokes parameters are associated with linear and circular polarization measurements as follows: $P_1 = I(0°) - I(90°)$, $P_2 = I(45°) - I(135°)$, and $P_3 = I(RHC) - I(LHC)$, where RHC and LHC denote right- and left-hand circular polarizations, respectively. For photons observed perpendicular to the scattering plane, $I(\alpha)$ is the linearly polarized intensity component measured at an angle $\alpha$ to the incident electron beam direction. $P_1$ and $P_2$ are, respectively, linear polarizations measured with reference to the incident electron beam direction and at 45° against this direction, whereas $P_3$ is the circular polarization. The polarizations $P_1$, $P_2$, and $P_3$ can be regarded as components of a three-dimensional vector polarization $\vec{P} = (P_1, P_2, P_3)$ which has magnitude $|\vec{P}| = (P_1^2 + P_2^2 + P_3^2)^{1/2}$, where $0 \leqslant |\vec{P}| \leqslant 1$.[14] $|\vec{P}|$ is often referred to as the degree of polarization. Following Born and Wolf,[14] a further quantity which characterizes the polarization state of the coincidence radiation is obtained by defining a correlation factor

$$\mu = |\mu| e^{i\beta} = (P_2 - iP_3)/(1 - P_1^2)^{1/2}. \tag{1}$$

$\mu$ is a measure of the correlation between the two orthogonal linearly polarized components of the radiation, parallel and perpendicular to the incident electron beam. $|\mu|$ is the "degree of coherence" between the orthogonal components, and $\beta$ is their effective phase. It can be readily shown that $|\mu| \leqslant |\vec{P}| \leqslant 1$.[14] The significance of a degree of coherence and a degree of polarization equal to unity is that this can occur if and only if every detected photon is in the same polarization state.

For the electron-impact excitation of the $3^1P$ state the connection between the atomic target parameters and the photon polarization is particularly straightforward. Based upon the assumption of coherent excitation of the degenerate magnetic sublevels, the excited state is assumed to be a linear superposition of magnetic sublevels such that

$$|\psi(t=0)\rangle = a_1|11\rangle + a_0|10\rangle + a_{-1}|1-1\rangle, \tag{2}$$

where the amplitude $a_M$ describes the excitation to the $|M\rangle$ sublevel.[16] These amplitudes are functions of the incident electron energy $E_0$ and scattering angle $\theta_e$. The components of the vector po-

larization $\vec{P}$ can be expressed in terms of the parameters $\lambda$ and $\chi$,[12] where $\lambda = \sigma_0/\sigma$, the ratio of the $M = 0$ partial differential cross section to the total differential cross section, and $\chi$ is the quantum-mechanical phase difference between the excitation amplitudes $a_1$ and $a_0$. When photons emitted from excited helium atoms are detected in a direction perpendicular to the scattering plane, $P_1$, $P_2$, and $P_3$ reduce to

$$P_1 = 2\lambda - 1, \quad P_2 = -2[\lambda(1-\lambda)]^{1/2}\cos\chi,$$
$$P_3 = 2[\lambda(1-\lambda)]^{1/2}\sin\chi. \tag{3}$$

Several conclusions can be drawn from Eq. (3). In the coherent-excitation model the complex correlation factor becomes a pure phase factor with $|\mu| = 1$ and $\beta = \chi$. Thus the quantum-mechanical phase difference between the excited-state sublevels is obtained directly from the effective phase difference associated with the coincident photon radiation. In addition, the degree of polarization and degree of coherence are predicted to be unity independent of electron impact energy and scattering angle. These results only hold true provided the assumption of coherent excitation is valid. In the absence of coherent excitation the phase difference $\chi$ can be regarded as random, $P_2$ and $P_3$ are zero, $\beta$ is undetermined, $|\mu|$ is zero (for $P_1 \neq 0$), while $|P|$ is no longer independent of electron impact energy and scattering angle.

The connection between the vector polarization and the alignment and orientation parameters of Fano and Macek[11] is as follows: $A_0^{col} = -(1 + 3P_1)/4$, $A_{1+}^{col} = -P_2/2$, $A_{2+}^{col} = (P_1 - 1)/4$, and $O_{1-}^{col} = -P_3/2 = \frac{1}{2}\langle L_y \rangle$. Accordingly, the linear polarization measurements determine the components of the alignment tensor, whereas the circular polarization measurements determine only the orientation. We note here that $O_{1-}^{col} = \frac{1}{2}\langle L_y \rangle$, where $\langle L_y \rangle$ is the only nonzero component of orbital angular momentum transferred to the atom during the collision. Figure 2 shows linear- and circular-polarization data for the helium $3^1P$-$2^1S$ (501.6 nm) transition at an incident electron energy of 80 eV. Note the striking behavior of $P_3$; unlike the linear polarizations $P_1$ and $P_2$, the circular polarization changes sign as the electron scattering angle is changed from positive to negative [Fig. 1(a) shows positive scattering angle]. This result is a direct consequence of the parity invariance of the scattering process.[18] Also shown are the predictions of the Born approximation and a recent multichannel eikonal calcula-

VOLUME 36, NUMBER 11          PHYSICAL REVIEW LETTERS          15 MARCH 1976

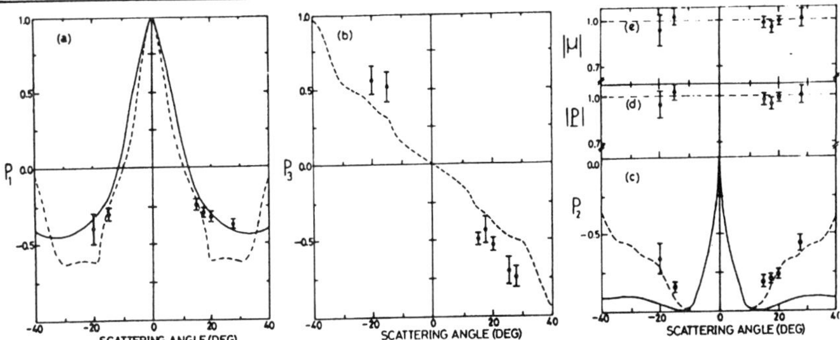

FIG. 2. (a)–(c) Experimental data for the vector polarization components $P_1$, $P_3$, and $P_2$, respectively, of He $3^1P$–$2^1S$ (501.6 nm) coincident photons at 80 eV incident electron energy versus electron scattering angle. Solid line, first Born approximation; dashed line, multichannel eikonal approximation (Ref. 19). (d) Degree of polarization. (e) Degree of coherence.

tion.[19]

The complex correlation factor and the degree of polarization have been determined from the experimental polarization data (see Fig. 2 and Table I). Within experimental uncertainty the degree of coherence and the degree of polarization are unity independent of scattering angle, confirming the predictions of the coherent-excitation model. Comparison with previous $3^1P$ ($E = 80$ eV) angular correlation measurements of Eminyan et al.[6] (see Table I) shows that over all there is good agreement with the values for $\lambda$ and $|\chi|$ obtained from the linear polarization data. Furthermore, there is good agreement between the $|\beta|$ and $|\chi|$ data. It should be emphasized that the extraction of $\lambda$ and $|\chi|$ parameters from angular-correlation or linear-polarization data is based upon the assumption of coherent excitation, whereas data for the correlation factor and the vector polarization are obtained from the linear- and circular-polarization measurements independently of any model for the atomic excitation process. Also included in Table I are atomic orientation data. Since the circular polarization observed in a direction perpendicular to the scattering plane is directly proportional to the atomic orientation, the observation of nonzero circular polarization provides direct evidence of the transfer of a net $y$ component of angular momentum to the atom during the collision. Note that there is good agreement between values for the atomic orientation computed from $\lambda$ and $|\chi|$ data and directly

TABLE I. Data for He $3^1P$ excitations at an incident electron energy of 80 eV. Data in parentheses are from $3^1P$–$1^1S$ (537 nm) angular correlation measurements (Ref. 6). Quoted experimental uncertainties are one standard deviation. $|O_{1-}{}^{col}|$ is computed from $\lambda$ and $|\chi|$ data.

| $\theta_e$ (deg) | $\lambda$ | $\|\chi\|$ (rad) | $\beta$ (rad) | $\|O_{1-}{}^{col}\|$ | $P_3/2(=-O_{1-}{}^{co})$ |
|---|---|---|---|---|---|
| −20 | 0.30 ± 0.05 | 0.76 ± 0.14 | 0.69 ± 0.13 | 0.31 ± 0.05 | 0.28 ± 0.05 |
| −15 | 0.35 ± 0.02 | 0.49 ± 0.07 | 0.55 ± 0.09 | 0.23 ± 0.03 | 0.26 ± 0.05 |
| 15 | 0.38 ± 0.02 | 0.58 ± 0.07 | −0.55 ± 0.04 | 0.27 ± 0.03 | −0.25 ± 0.04 |
|  | (0.38 ± 0.01) | (0.55 ± 0.02) |  | (0.25 ± 0.01) |  |
| 17.5 | 0.36 ± 0.02 | 0.62 ± 0.06 | −0.51 ± 0.09 | 0.28 ± 0.02 | −0.22 ± 0.05 |
| 20 | 0.34 ± 0.02 | 0.63 ± 0.05 | −0.62 ± 0.04 | 0.28 ± 0.02 | −0.27 ± 0.02 |
|  | (0.36 ± 0.02) | (0.64 ± 0.05) |  | (0.29 ± 0.02) |  |
| 27.5 | 0.36 ± 0.02 | 0.98 ± 0.07 | −0.93 ± 0.06 | 0.40 ± 0.03 | −0.38 + 0.04 |

VOLUME 36, NUMBER 11          PHYSICAL REVIEW LETTERS          15 MARCH 1976

measured values obtained from circular polarization measurements. Since $P_3$ is negative over the measured range of positive scattering angles, $\chi$ is also negative in agreement with the prediction of a multichannel eikonal approximation.[19]

In conclusion, we wish to point out that vector-polarization and coherence-parameter measurements, besides providing a more complete analysis of coincidence experiments, allow consistency checks to be made for those experiments where coherent excitation is expected ($|\vec{P}|$ and $|\mu|$ should be constant independent of electron energy or angle) and offer a means of investigating the nature of the excitation mechanism where coherent excitation may not apply (e.g., resonances and near-threshold excitations).

We acknowledge the support for this work provided by the Science Research Council (Great Britain). One of us (M.C.S.) acknowledges the support of the New Zealand University Grants Committee for a post-doctoral fellowship.

[1]G. C. M. King, A. Adams, and F. H. Read, J. Phys. B **5**, L254 (1972).

[2]G. Vassilev, G. Rahmat, J. Slevin, and J. Baudon, Phys. Rev. Lett. **34**, 447 (1975).

[3]D. H. Jaecks, F. J. Erikson, W. de Rijk, and J. Macek, Phys. Rev. Lett. **35**, 723 (1975).

[4]M. Eminyan, K. B. MacAdam, J. Slevin, and H. Kleinpoppen, Phys. Rev. Lett. **31**, 576 (1973).

[5]M. Eminyan, K. B. MacAdam, J. Slevin, and H. Kleinpoppen, J. Phys. B **7**, 1519 (1974).

[6]M. Eminyan, K. B. MacAdam, J. Slevin, M. C. Standage, and H. Kleinpoppen, J. Phys. B **8**, 2058 (1975).

[7]J. W. McConkey, K. H. Tan, P. S. Farago, and P. J. O. Teubner, in *Proceedings of the Ninth International Conference on the Physics of Electronic and Atomic Collisions, Seattle, Washington, 1975* (Univ. of Washington Press, Seattle, Wash., 1975).

[8]H. Arriola, P. J. O. Teubner, A. Ugbabe, and E. Weigold, J. Phys. B **8**, 1275 (1975).

[9]J. Wykes, J. Phys. B **5**, 1126 (1972).

[10]J. H. Macek and D. H. Jaecks, Phys. Rev. A **4**, 2288 (1971).

[11]U. Fano and J. H. Macek, Rev. Mod. Phys. **45**, 553 (1973).

[12]K. Blum and H. Kleinpoppen, J. Phys. B **8**, 922 (1975).

[13]W. Hanle, Z. Instrumentenkd. **51**, 488 (1931).

[14]M. Born and E. Wolf, *Principles of Optics* (Pergamon, New York, 1975), 5th ed.

[15]W. H. McMaster, Rev. Mod. Phys. **33**, 8 (1961).

[16]Since the scattering process has mirror symmetry in the scattering plane, $a_1 = -a_{-1}$.

[17]D. G. Ellis, J. Opt. Soc. Am. **63**, 1232 (1973).

[18]H. Kleinpoppen, K. Blum, and M. C. Standage, in *Proceedings of the Ninth International Conference on the Physics of Electronic and Atomic Collisions, Seattle, Washington, 1975* (Univ. of Washington Press, Seattle, Wash., 1975).

[19]M. R. Flannery and K. J. McCann, J. Phys. B **8**, 1716 (1975).

# Further Readings


1. M. Born and E. Wolf, *Principles of Optics* (4th edition), Pergamon Press, New York 1970.
2. W.A. Shurcliff, *Polarized Light: Production and Use*, Harvard University Press, Harvard 1962.
3. J. Kessler, *Polarized Electrons* (2nd edition), Springer, Heidelberg 1985.
4. I.E. McCarthy and E. Weigold, *Electron–Atom Collisions*, University Press, Cambridge 1995.
5. K. Blum, *Density Matrix Theory and Applications* (2nd edition), Plenum, New York 1996.
6. G.W.F. Drake (Ed.), *Atomic, Molecular, and Optical Physics Handbook*, AIP Press, Woodbury, New York 1996.
7. K. Bartschat (Ed.), *Computational Atomic Physics — Electron and Positron Scattering from Atoms and Ions*, Springer, Heidelberg 1996.
8. V.V. Balashov, A.N. Grum-Grzhimailo, and N.M Kabachnik, *Polarization and Correlation Phenomena in Atomic Collisions* Plenum, New York 2000.
9. N. Andersen, J. W. Gallagher and I.V. Hertel, *Collisional Alignment and Orientation of Atomic Outer Shells. I. Direct Excitation by Electron and Atom Impact*, Phys. Rep. **165** (1988) 1.
10. N. Andersen, J.T. Broad, E.E.B. Campbell, J.W. Gallagher and I.V. Hertel, *Collisional Alignment and Orientation of Atomic Outer Shells. II. Quasi-Molecular Excitation, and Beyond*, Phys. Rep. **278** (1997) 107.
11. N. Andersen, K. Bartschat, J.T. Broad and I.V. Hertel, *Collisional Alignment and Orientation of Atomic Outer Shells. III. Spin-Resolved Excitation*, Phys. Rep. **279** (1997) 251.
12. N. Andersen and K. Bartschat, *Complete Experiments in Atomic Collisions*, Adv. At. Mol. Opt. Phys. **36** (1996) 1.
13. N. Andersen and K. Bartschat, *Collisional Excitation of Atomic D States*, J. Phys. B **30** (1997) 5071.

# Subject Index

Springer Series on

# Atoms+Plasmas

Editors: G. Ecker   P. Lambropoulos   I.I. Sobel'man   H. Walther
Founding Editor: H.K.V. Lotsch